T0344881

Ecology of Invertebrate Diseases

Ecology of Invertebrate Diseases

Edited by

Ann E. Hajek
Cornell University, Ithaca, New York, US

David I. Shapiro-Ilan
USDA-ARS, Byron, Georgia, US

Registered Office(s)
John Wiley & Sons, Inc., 111 River Street, Hoboken, NJ 07030, USA
John Wiley & Sons Ltd, The Atrium, Southern Gate, Chichester, West Sussex, PO19 8SQ, UK

Editorial Office
9600 Garsington Road, Oxford, OX4 2DQ, UK

For details of our global editorial offices, customer services, and more information about Wiley products visit us at www.wiley.com.

Wiley also publishes its books in a variety of electronic formats and by print-on-demand. Some content that appears in standard print versions of this book may not be available in other formats.

Library of Congress Cataloging-in-Publication Data
Names: Hajek, Ann E., editor. | Shapiro-Ilan, David I., editor.
Title: Ecology of invertebrate diseases / Edited by Ann E. Hajek, David
  I. Shapiro-Ilan.
Description: Hoboken, NJ : John Wiley & Sons, 2017. | Includes
  bibliographical references and index. |
Identifiers: LCCN 2017023661 (print) | LCCN 2017035570 (ebook) | ISBN
  9781119256014 (pdf) | ISBN 9781119256069 (epub) | ISBN 9781119256076
  (cloth)
Subjects: LCSH: Invertebrates--Ecology. | Invertebrates--Diseases.
Classification: LCC QL364.4 (ebook) | LCC QL364.4 .E26 2017 (print) | DDC
  592.17/82--dc23
LC record available at https://lccn.loc.gov/2017023661

Cover Design: Wiley
Cover Image: Top photo by Ann E. Hajek; bottom photo by Ivan Hiltpold

Set in 10/12pt WarnockPro by SPi Global, Chennai, India
Printed in Singapore by C.O.S. Printers Pte Ltd

10 9 8 7 6 5 4 3 2 1

Contents

Contents

List of Contributors

James J. Becnel
USDA ARS CMAVE
Gainesville, FL, USA

Colin Berry
Cardiff School of Biosciences
Cardiff University
Cardiff, UK

Colleen A. Burge
Institute of Marine and Environmental
Technology
University of Maryland Baltimore
County
Baltimore, MD, USA

Raquel Campos-Herrera
MeditBio
University of Algarve
Faro, Portugal

Louela A. Castrillo
Department of Entomology
Cornell University
Ithaca, NY, USA

Jenny S. Cory
Department of Biological Sciences
Simon Fraser University
Burnaby, BC, Canada

Surendra K. Dara
University of California Cooperative
Extension
Division of Agriculture and Natural
Resources
San Luis Obispo, CA, USA

Pauline S. Deschodt
Department of Biological Sciences
Simon Fraser University
Burnaby, BC, Canada

Jørgen Eilenberg
Department of Plant and Environmental
Sciences
University of Copenhagen
Frederiksberg, Denmark

Bret D. Elderd
Department of Biological Sciences
Louisiana State University
Baton Rouge, LA, USA

James R. Fuxa
Louisiana State University (Retired)
Cypress, TX, USA

Itamar Glazer
Department of Entomology
ARO, Volcani Centre
Rishon LeZion, Israel

Tarryn A. Goble
Department of Entomology
Cornell University
Ithaca, NY, USA

Ann E. Hajek
Department of Entomology
Cornell University
Ithaca, NY, USA

Ivan Hiltpold
Department of Entomology and Wildlife
Ecology
University of Delaware
Newark, DE, USA

Gernot Hoch
Department of Forest Protection
BFW Austrian Research Centre for
Forests
Vienna, Austria

Trevor A. Jackson
AgResearch Ltd
Lincoln Research Centre
Christchurch, New Zealand

Annette Bruun Jensen
Department of Plant and Environmental
Sciences
University of Copenhagen
Frederiksberg, Denmark

Lawrence A. Lacey
IP Consulting International
Yakima, WA, USA

Edwin E. Lewis
Department of Entomology and
Nematology
University of California – Davis
Davis, CA, USA

Dana Ment
Department of Entomology
ARO, Volcani Centre
Rishon LeZion, Israel

Nicolai V. Meyling
Department of Plant and Environmental
Sciences
University of Copenhagen
Frederiksberg, Denmark

Maureen O'Callaghan
AgResearch Ltd
Lincoln Research Centre
Christchurch, New Zealand

Natalie D. Rivlin
Institute of Marine and Environmental
Technology
University of Maryland Baltimore
County
Baltimore, MD, USA

David I. Shapiro-Ilan
USDA-ARS, SEA
SE Fruit and Tree Nut Research Unit
Byron, GA, USA

Jeffrey D. Shields
Department of Aquatic Health Sciences
Virginia Institute of Marine Science
The College of William & Mary
Gloucester Point, VA, USA

Ikkei Shikano
Department of Entomology and Center
for Chemical Ecology
Pennsylvania State University
University Park, PA, USA

Amanda Shore-Maggio
Institute of Marine and Environmental
Technology
University of Maryland Baltimore
County
Baltimore, MD, USA

Leellen F. Solter
Illinois Natural History Survey
Prairie Research Institute
University of Illinois
Champaign, IL, USA

Trevor Williams
Instituto de Ecologia AC (INECOL)
Xalapa, Veracruz, Mexico

Preface

All have their worth and each contributes to the worth of the others.
J.R.R. Tolkien, *The Silmarillion*

When you try to study something in isolation, you find it hooked to everything else.
John Muir

It becomes necessary in any active scientific discipline to sit back every few years and take stock of the "state of the art." The time arrives to review recent progress, inspire new ideas, and propose critical and novel lines of research.

In 1963, Yoshinori ("Joe") Tanada laid the foundation for the current book with his chapter "Epizootiology of Infectious Diseases" in E.A. Steinhaus' *Insect Pathology: An Advanced Treatise*, a two-volume reference that defined the scope of invertebrate pathology. By 1987, after almost a quarter-century, the time had come for more than just a review of the "state of the art." Fuxa and Tanada formalized the new scientific discipline in their edited monograph, *Epizootiology of Insect Diseases*, organizing an emerging field of study by establishing its components, definitions, types of studies, and research methods.

Another 30 years have passed, and that time has come again. Much has happened since 1987 – science never stands still. New methods have opened doors unheard of in the 1980s, most notably in molecular biology. Detection and characterization of strands of DNA, RNA, and transposable genetic elements create almost unlimited research opportunities in ecology. New diseases have emerged, such as the mysterious colony collapse disorder of the honeybee, which has raised concern throughout much of the world. Previously characterized diseases have erupted again in devastating epizootics, notably MSX disease (*Haplosporidium nelsoni*) and dermo disease (*Perkinsus marinus*) in populations of oysters. New relationships have opened eyes – who would have thought that microsporidia are highly evolved fungi, not primitive protozoans that evolved before the advent of mitochondria? New concepts have arisen for invertebrate pathogens, contributing to theory in general ecology and host–pathogen coevolution.

Thus, we arrive at this new book. The reader, however, might ask, why *Ecology of Invertebrate Diseases* rather than *Epizootiology of Insect Diseases*? A definition of epizootiology borrowed from the 1987 volume, "the science of causes and forms of the mass phenomena of disease at all levels of intensity in a host population," allows for study of the total environment, including the host and pathogen populations, even a pathogen's environment inside its host. Epizootiology in turn is a subset of ecology, which was defined by H.G. Andrewartha as "the scientific study of the distribution and abundance of organisms," a definition that has evolved to embrace concepts of "population" and "ecosystem."

Simpler may be better. Epizootiology is the study of animal disease at the population level. It fits well into a broader mold of ecology, as outlined by Tolkien and Muir, if they may be paraphrased, that everything is "connected to everything else" and "contributes to the worth of the others." Perhaps these two authors were not trying to define ecology, but they might just as well have been.

The editors of this book realized that the "state of the science" has exceeded the scope of the 1987 monograph, thereby creating a multitude of opportunities to discuss new concepts and types of studies, not to mention the myriad of non-insect hosts of infectious disease. Even in a new volume, however, old questions arise, especially, why study ecology of pathogens and their hosts?

First and foremost, parasites are not just dirty little things living a disgusting lifestyle. They are highly evolved organisms – or near-organisms – as intricate and unique as any creature on the planet. They contribute to all, whether by culling the weak or by transporting bits of DNA, in relationships with their hosts ranging from near-benign to something out of a horror film. The reader who delves into this book will see the "worth" in these fascinating little creatures, whether prokaryotic, eukaryotic, or viral.

And, of course, invertebrate pathogens certainly are "hooked to everything" abiotic and biotic, even to humans, a web of life and environment and planet earth. Many such interactions almost defy belief, for that is life.

Science also is called upon to provide tangible benefit. The historical advantages of studying invertebrate diseases remain as important as ever – enhancing disease in pestiferous organisms and preventing disease in invertebrates useful to humans. Pathogens, even viruses, function as parasites with population-level and ecological characteristics, parasites that must be suppressed in populations of beneficial organisms or conserved or enhanced if they are to succeed in pest management. Critics, however, might say that such "germ warfare" against pests is passé, that we now have genetically modified crops, recombinant mosquitoes, and on and on. Perhaps. However, the biopesticide market, which is based on mass production of invertebrate pathogens, continues to grow. Moreover, nothing works in isolation. For example, understanding the dynamics of insect population resistance to disease facilitated management of resistance in widespread use of crops incorporating toxin genes and environmental risk assessment contributed to a safe, first release of a recombinant baculovirus.

Diseases of invertebrate hosts, many of them easy to culture, with their tiny sizes and short generation times, also serve as model systems giving insight into disease ecology in higher organisms – for example, contributions of pathogen reproductive rate, transmission, and virulence to epizootics.

So much for the "What?" and the "Why?" of this book – how about the "Who?" Many decisions were made in compiling this monograph, not the least of which is, where does one stop if everything is "hooked to everything else"? Researchers will find themselves fortunate in this volume's scientific writers, fortunate because they will recognize the names of the editors and authors, all outstanding pathologists, ecologists, or epizootiologists. Readers will appreciate as they peruse this book that, much like Tolkien's world, every author has indeed "contributed to the worth of the others."

James R. Fuxa
Louisiana State University (Retired)

Section I

Introduction

1

General Concepts in the Ecology of Invertebrate Diseases

Ann E. Hajek[1] and David I. Shapiro-Ilan[2]

[1] Department of Entomology, Cornell University, Ithaca, NY, USA
[2] USDA-ARS, SEA, SE Fruit and Tree Nut Research Unit, Byron, GA, USA

1.1 Introduction

With the advent of molecular methods, new species of pathogens and parasites are constantly being described, and as these new species are found, we are learning more about the ecology of new invertebrate diseases, as well as diseases known for many years. Parasitism is a specific and common life-history strategy, and understanding the activity of parasites is central to community and population ecology (Bonsall, 2004). Parasitism of invertebrate hosts also has practical sides, because diseases can help to control insects in an environmentally friendly manner, but we also need to understand the ecology of diseases killing beneficial invertebrates, ranging from pollinators to clams and shrimp, in order to protect managed populations.

The ecology of invertebrate diseases is often referred to as the epizootiology of invertebrate diseases; the word **epizootiology** is similar to the term epidemiology but refers specifically to "the science of causes and forms of the mass phenomenon of disease at all levels of intensity in an animal population" (Fuxa and Tanada, 1987). The ecology of animal diseases, with emphasis on vertebrates, has been treated in an edited book on disease ecology (Hudson et al., 2002), followed by books emphasizing community and ecosystem ecology (Collinge and Ray, 2006; Hatcher and Dunn, 2011; Ostfeld et al., 2014).

Disease ecology with an emphasis on invertebrates was first addressed by Steinhaus (1949), specifically in relation to insects, and the treatment of this subject developed depth and breadth with the publication of an edited volume by Fuxa and Tanada (1987). Around this time, Anderson and May (1981, 1982) created models to investigate factors driving the development of disease epizootics, with at least one system involving epizootics caused by a granulovirus in a forest-defoliating lepidopteran (Anderson and May, 1980). Today, studies of the ecology of invertebrate diseases are commonly conducted, often to understand the ecology underpinning control of invertebrate pests by pathogens, or to understand protection from pathogens for invertebrates valued by humans. In addition, ecological studies of invertebrate diseases are used to build theoretical insights into the causes and dynamics of all diseases. With the wealth of knowledge that

Ecology of Invertebrate Diseases, First Edition. Edited by Ann E. Hajek and David I. Shapiro-Ilan.
© 2018 John Wiley & Sons Ltd. Published 2018 by John Wiley & Sons Ltd.

has accumulated since the last synthesis on the ecology of invertebrate diseases in 1987, it is high time to pull together information on this subject. We are also broadening the focus of this book to include the ecology of diseases of all invertebrates and not only insects. Therefore, the hosts included in this book range from pest insects like grasshoppers and caterpillars to valued insects like bees, along with marine and soil invertebrates that are important to humans or ecosystems.

In this chapter, we will present and define the basic concepts on which this field of study is built. Concepts that will be defined will be consistent with definitions in the online glossary published by the Society for Invertebrate Pathology (Onstad et al., 2006).

1.1.1   What Is Disease?

There are numerous definitions for disease, but we consider **disease** to be a departure from the state of health or normality. Of course, this creates a very broad characterization, including multitudes of causes. However, this book will focus on **infectious** diseases, which are those diseases caused by living organisms. Invertebrates are also hosts to many **noninfectious diseases**, of which physical and chemical injuries, genetic diseases, and cancers are a few examples. An example of noninfectious disease impacting insects can occur due to exposure to pesticides. Noninfectious diseases are, however, outside of the material being covered in this book. Descriptive treatments of noninfectious diseases of a diversity of invertebrates can be found in Lewbart (2012) and Sparks (1972).

Returning to disease being a departure from health, this can be much more difficult to determine for invertebrates than for higher vertebrates. Diseases of invertebrates that cause subacute effects and which do not kill hosts could very well be regularly occurring at low prevalence but going undetected. Perhaps recent studies demonstrating the diversity of previously undetected and unrecognized pathogens occurring in honey bee (*Apis mellifera*) colonies that do not die (see Chapter 14) indicate that departures from health being caused by a diversity of parasites acting together can be quite common. The most frequent way that invertebrate diseases are recognized is due to the death of hosts, so emphasis in this field has been on acute diseases. However, in recent years, investigations have included the impact of disease on host fitness, both for pathogens causing chronic diseases and for acute pathogens, after infection and before host death.

The living organisms causing infectious diseases are **parasites**, which are organisms that live at a host's expense. This is a very successful life-history strategy as it has been estimated that the majority of species on earth have parasitic lifestyles (Price, 1980; Zimmer, 2000). **Pathogens** are defined as microorganisms capable of producing disease under normal conditions of host resistance and rarely living in close association with a host without producing some level of disease. Simply put, pathogens can be thought of as microscopic parasites. In this book, the main groups of pathogens that will be covered are viruses, bacteria, fungi, nematodes, and protists. Although many nematodes are not microscopic, the genera that constitute a special group – entomopathogenic nematodes – kill their hosts with the aid of symbiotic bacteria; these nematodes have traditionally been included and studied within the discipline of invertebrate pathology, and thus we include them as well. With the great diversity of pathogens and hosts being covered, of course examples will be missed. In addition, this is presently an expanding field of study, as new species of pathogens are constantly being found. One good example of this is the number of different pathogens that infect honey bees, with numerous examples discovered in recent years. We will focus on representatives from the diversity of pathogens and hosts for which we know the most about disease ecology.

Diseases may be **chronic** or **acute** (or somewhere in between). Chronic diseases are of a long duration, and thus the host is expected to survive a relatively long time before expiring, or to die of other causes before the disease can become fatal. Acute diseases of invertebrates are often of a short duration; host mortality or maximum severity is expected to occur within a relatively short time after infection. Certain pathogen groups, such as entomopathogenic nematodes (Heterorhabditidae and Steinernematidae) and their symbiotic bacteria tend to cause acute disease, whereas others, such as most Microsporidia, tend to cause chronic diseases. Within other pathogen groups, chronic versus acute diseases vary among combinations of pathogens and hosts.

1.1.2 Terminology and Measurements

Accurate use of terminology is critical to effective communication in science. In pathology, a number of terms have been used with variable meanings in the literature (Shapiro-Ilan et al., 2005). The terminology used in this book is supported by a widely accepted glossary by Onstad et al. (2006), which is based on an earlier glossary by Steinhaus and Martignoni (1970). Thus, we refer the reader to Onstad et al. (2006) as a reference for definitions that may not be spelled out in this chapter or other chapters within this book. Nonetheless, some of the more common terms in invertebrate pathology are defined and discussed in this section.

1.1.2.1 Prevalence/Incidence

Prevalence and incidence are examples of terms that have been variably defined in the literature. **Prevalence** refers to the total number or proportion of disease cases in a population at a given time. For example, if a survey of 10 000 pecan weevil (*Curculio caryae*) larvae in a population indicates that 4000 of the individuals are infected with the fungus *Beauveria bassiana*, then the prevalence rate is 40%. In contrast, **incidence** is the number or proportion of new cases of a disease within a population during a specific period of time. For example, in a given week, if 200 *C. caryae* larvae were found to be infected with *B. bassiana* within a population of 10 000 larvae, then the incidence rate would 2% for that week. Both terms are important for quantifying infection and disease levels over space and time. The difference lies in incidence emphasizing only new cases versus prevalence including both new and old cases. Therefore, incidence may be more useful in predicting risk or spread of disease within a certain timeframe, while prevalence provides an assessment of the full impact of a disease on a population at a given time.

1.1.2.2 Pathogenicity/Virulence

The terminology of pathogenicity and virulence has also been variably defined in the literature. However, definitions of pathogenicity and virulence in the field of invertebrate pathology have been largely consistent over time, and are in agreement with definitions found in the fields of human pathology and microbiology (Shapiro-Ilan et al., 2005). **Pathogenicity** is the quality or state of being pathogenic: the potential ability to produce disease. **Virulence** is the disease-producing power of an organism: the degree of pathogenicity within a population or species. Generally, the term pathogenicity is applied to entire populations or species, whereas virulence is usually intended for within-group or within-species comparisons. Yet, both terms can conceivably be applied across any taxonomic level (strain, species, genus, family, etc.), provided that pathogenicity is a qualitative and virulence a comparative measure.

Indeed, for a given pathogen and host, pathogenicity is absolute whereas virulence is variable (e.g., due to strain or environmental effects). Pathogenicity is an all-or-none phenomenon. An organism is either pathogenic to a host or it is not. Pathogenicity can be established based on Koch's postulates (Shapiro-Ilan et al., 2005; Lacey and Solter, 2012). Virulence can be measured using various approaches, such as LD_{50} or LC_{50} (dosage or concentration required to cause 50% mortality in the test organisms), LT_{50} (time required to cause 50% mortality in the test organisms), or comparative mortality or infection rates at a given dosage or concentration (Shapiro-Ilan et al., 2005).

1.1.2.3 Infection/Infectivity

An **infection** is the entry of a microorganism into a susceptible host. The presence of the organism may or may not cause overt disease (obvious pathological effects) in the host. If the infection is not immediately followed by overt disease then it is termed an attenuated infection. There are three types of attenuated infections: latent, carrier state, and microbial persistence. A latent infection is an asymptomatic infection that is in a dormant or stationary phase and is capable of being activated at a later time. In the case of viruses, a latent phase is also called an occult phase (and the virus is an occult virus at that stage). In a carrier state, the infection remains as an inapparent infection (no overt sign of its presence) in the current host but is capable of being transferred to other hosts. Microbial persistence is characterized by the continued presence of a pathogenic microorganism within the host in the absence of overt disease but following a period of overt disease.

Infectivity is the ability to produce infection. Thus, infectivity is a measurable characteristic. Infectivity can be measured by assessing the number of pathogenic units (e.g., spores, virus particles, infective juvenile nematodes, or other propagules) that have entered a host following exposure to a known quantity. Infectivity can be positively affected by the number of infection routes; infection routes include per os (through the mouth), through the cuticle or natural openings in the host, and vertical transmission (see later).

1.1.2.4 Immunity

Immunity is innate or induced resistance to a disease agent. It tends to be species-specific but can also vary within species. Invertebrates have immune systems, although predominantly without memories; they are not known to recover from infections and become resistant to further infections of the same pathogens, as vertebrates often can. Immune responses can be cellular (via hemocytes) or humoral (e.g., via antimicrobial peptides), or they may involve both humoral and cellular reactions (e.g., melanization) (Chapter 4). The effectiveness of immune responses may depend on strain or population variation, age and fitness, and stress factors such as environmental conditions or diet. For example, feeding an improper diet resulted in higher percentages of *Drosophila* being killed by a bacterium in their hemolymph (Galac and Lazzaro, 2011; Howick and Lazzaro, 2014). See Chapter 4 for a more in-depth discussion of immunity.

1.1.2.5 Transmission

Transmission is the transfer of a pathogen from a source to a new host (Chapter 3) (Andreadis, 1987; Onstad et al., 2006). The source of transmission may be an infected or uninfected organism or the environment. **Direct transmission** refers to transfer from an infected host to a susceptible host without intervention from other organisms. **Indirect transmission**, which is rare among invertebrate pathogens, relies on one or more intermediate hosts or vectors.

Transmission can be further classified into vertical and horizontal. **Vertical transmission** is direct transmission from parent to progeny. Transovum transmission occurs when the pathogen is transferred from parent to progeny on the surface of the egg (Andreadis, 1987). Transovarian transmission occurs when passage is known to occur within the ovary. Note that some authors consider transovarian transmission (transfer within the egg) to be a special case of transovum transmission (in or on the egg) (Onstad et al., 2006), but in more recent definitions the two terms are separated (on the egg versus in the egg, respectively) (see also discussion in Chapter 3). **Horizontal transmission** includes all forms of transmission except direct transfer from parent to offspring. Horizontal transmission can occur through a variety of mechanisms, including ingestion of food or fecal material, transfer from another organism (contact with another host including mating and cannibalism or from a non-host), and contact with the environment (e.g., air, water, soil, other organic matter).

Pathogen groups vary in the modes of transmission they employ. All five pathogen groups (viruses, bacteria, protists, fungi, and nematodes) can be transferred horizontally (Andreadis, 1987). However, the relative importance of horizontal versus vertical transmission varies by group, and vertical transmission has been reported in all groups except in entomopathogenic nematodes (Andreadis, 1987; Tanada and Kaya, 1993) (see Section III).

1.1.2.6 Epizootic and Enzootic Diseases

Some invertebrate diseases have the ability to increase in prevalence under specific conditions, and the term **epizootic** is used to refer to unusually large numbers of cases of disease in animal hosts (Fuxa and Tanada, 1987). Therefore, the term epizootic refers to a population-level phenomenon, and prior knowledge about the host–pathogen system is necessary in order to document that an epizootic is occurring (i.e., one must know what level of infection by a pathogen is considered "normal" in that system). So, even for a disease that is observed only rarely, low levels of infection can be called an epizootic. However, this term is usually used to refer to high levels of infection prevalence that have become very obvious when examining the host population. An enzootic disease is the opposite: a disease that is low in prevalence (Fuxa and Tanada, 1987). These phenomena also differ in time, as an epizootic occurs during a specific time, over which infection levels increase and then decrease, while enzootic disease is constantly present.

While epizootics can occur for diseases that are chronic as well as acute, we provide an example from an acute disease. An epizootic begins with an increase in the prevalence of infections – often a rapid increase (Fig. 1.1). Eventually, the infection prevalence peaks, after which few susceptible hosts remain to become infected and the infected hosts begin dying, so the infection prevalence decreases. Epizootics can have impacts on host populations due to both host mortality and reduced fecundity before death. When an epizootic ends, this does not mean that there are no more hosts

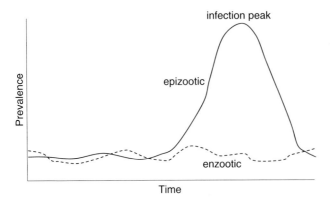

Fig. 1.1 Basic conceptual model of disease prevalence for an enzootic versus an epizootic disease. For the epizootic disease, there is an increase in infection to an infection peak, followed by a decrease. For acute invertebrate diseases, the decline after the peak is often caused by host mortality and decreased recruitment.

present, but only that the wave of increased and then decreased infection is over and, usually, the disease has dropped below a perception threshold. After an epizootic, there can be large amounts of pathogen inoculum in the environment (often in the soil), especially if the pathogen's lifecycle includes persistent stages.

It can be especially difficult to document the initiation and progress of an epizootic in invertebrates. The increase in infection can occur rapidly, and the peak in infection prevalence is therefore reached quickly. Often, by the time an epizootic becomes evident to humans, the majority of the infections have already occurred, and an abundance of cadavers or lack of hosts provides evidence that an epizootic has occurred. In such cases, tracing back to identify the conditions that resulted in epizootic levels of infection can be similar to environmental detective work. After epizootics have occurred, one possible approach to evaluating conditions leading to high prevalence of infection includes undertaking detailed field studies prior to future epizootics (hopefully, under conditions that will result in repeats of epizootics) or developing experiments to replicate epizootics, meanwhile answering specific questions regarding the epizootiology of the system.

1.1.2.7 Cycles of Infection

Disease cycles are composed of chains of events in the development of infectious disease (called "chains of infection" by the biomedical community) (Fig. 1.2.). These events consist of the following general stages. Details from individual types of pathogens and invertebrate hosts will be provided within chapters about different pathogen groups (especially in Section III).

1) The pathogen occurs in a **reservoir**, during periods when hosts are at low density or are in a stage that is not susceptible to infection, which could be caused by seasonality of the host. The repository for the pathogen during a period when infection is not possible can range from a chronically infected host (e.g., Eilenberg et al., 2013) to the occurrence of the pathogen in the cadaver of a host that has been killed or in an environmental habitat such as soil where the pathogen can persist across time (e.g., *Bacillus thuringiensis* spores are thought to survive indefinitely in the soil).

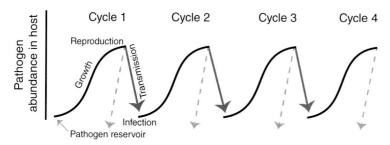

Fig. 1.2 Basic model indicating primary and secondary disease cycling. An individual host is infected from the pathogen reservoir (primary cycling). The pathogen develops within the host and then reproduces. The propagules are released to arrive at another host, where they either infect it or are deposited in the reservoir. This progression continues for four cycles in this figure, ending with the deposition of inocula in the reservoir. The second through fourth cycles are referred to as secondary cycling. The figure does not indicate the impact of the infections on either pathogen or host population density.

Decomposition rates of pathogens in environmental habitats affect the length of time in which horizontal transmission from environmental reservoirs can occur. Another type of reservoir for a pathogen occurs with vertically transmitted pathogens (see Section 1.1.2.6), which can persist within adults or in or on eggs.

2) The pathogen survives and comes into contact with the host (= **transmission**). Some pathogens do not actively move to reach a host (e.g., viruses, some bacteria, microsporidia) and are dispersed by environmental forces like wind, water, and rain. Some fungi actively initiate aerial dispersal (e.g., Entomophthoromycotina), but reaching a host is a random process, although fungi and Oomycota with zoospore stages actively move in water. Nematodes, however, can move in a directed manner to locate a host. It is also possible that through host movements, the pathogen contacts the host or is vectored to the host via another organism.

3) The pathogen must then infect the host (= **infection**), which in some cases requires that the pathogen gain access to the host gut (i.e., bacteria, viruses, microsporidia) or just to the host's cuticle (i.e., many fungi and nematodes). At this point, the pathogen usually differentiates and penetrates into the host, most often breaching the cuticle or gut and increasing within the body of the host.

4) The pathogen reproduces, often first producing stages for dispersing within the host (although pathogens differ in the tissues that they utilize (see Chapter 3)) and then stages that will survive outside of the host. The period during which a pathogen is present within a living infected host can vary considerably (see definition of acute/chronic), and the next cycle of infection can only begin once transmissive stages of the pathogen are released from the host.

Infections that start a period of disease by a pathogen, prior to which transmission was not occurring, are called **primary infections** (or primary transmission). For invertebrates, it is common that periods when pathogens are not actively infecting are determined by seasons, which often also determine the periods when hosts are present in appropriate stages for infection. For terrestrial invertebrates, primary infections can occur due to pathogen inocula that have overwintered in soilborne reservoirs, where the pathogen spends the period during which susceptible stages of host populations are not present and active. When a host is infected by inoculum produced in a host that season, this is called **secondary infection** (or secondary transmission); that is, infection of a healthy individual by propagules produced from an infected individual within the same season.

From infection to pathogen reproduction can be thought of as one cycle of infection (Fig. 1.2.). Then, when pathogen inocula produced from the first set of hosts are transmitted to infect a second set of hosts will be a second cycle of infection, frequently called recycling or secondary cycling. During epizootics, abiotic conditions, pathogen density, and host density all strongly influence how many disease cycles occur before the level of disease decreases to zero or some enzootic level. Models for the gypsy moth fungal pathogen *Entomophaga maimaiga* show that the number of cycles (i.e., from one healthy host becoming infected to a second healthy host becoming infected) can range from four to nine cycles per season (Hajek et al., 1993); the host in this system has only one generation per year, and the pathogen infects only larval stages that are active for approximately 2 months/year; temperature and moisture conditions were varied to arrive at this estimate.

1.1.2.8 R_0 and the Host Density Threshold

One concept that arose from the development of host pathogen models by Anderson and May (1981) was the idea of looking at infection levels across time to evaluate whether a disease is increasing or decreasing. R_0 (= R nought, the **basic reproduction number**) is a parameter calculated to estimate the basic reproductive success of a parasite. It represents the number of new (secondary) infections arising from one infection within a population that is wholly susceptible to the parasite (Roberts, 2007). This number helps to determine whether or not an **infectious disease** will increase in a host population. If $R_0 > 1$, then the number of infected hosts will increase, while if $R_0 < 1$, then the number of infected hosts will decrease over time. If the number of infections is too low, the pathogen may be eliminated from the system (ignoring whether the pathogen has persistent stages that are dormant in the environment). Therefore, $R_0 = 1$ can be thought of as a transmission threshold, above which a disease will increase in a host population and below which the disease will decrease.

R_0 can also be thought of as providing a measure of the relative fitness of a pathogen (Antolin, 2008); parasites with a higher R_0 can increase more rapidly. This parameter can be an especially appropriate measure for comparing different diseases or for comparing diseases in host populations over space or time. In fact, R_0s for individual host–pathogen systems are not static and are thought to vary over space and time (Altizer et al., 2006).

To work with R_0s, it is important to know the numbers of infected individuals both before and after transmission of a parasite. Perhaps of more applicability to work on invertebrate diseases is the level of infection in relation to host density. The **host density threshold** (also called the critical population size), referred to as N_T, is the minimum concentration of hosts necessary to sustain a given disease in a host population (Hatcher and Dunn, 2011). Below this threshold density, the disease will die out, while above it, infection will be maintained in the host population. Calculating N_T for a host requires knowing a bit about the host and the transmission process. Specifically, estimates of the disease transmission rate, the recovery rate from the disease (if at all; this is usually not assumed to occur with invertebrates), and the mortality rate of the host for noninfected individuals are needed in order for N_T to be readily determined. N_T is host species-specific, and this concept only applies to diseases that are density-dependent.

1.1.3 Factors Influencing the Ecology of Invertebrate Diseases

This book will concentrate on ecological interactions between hosts and pathogens and the environment. This relation has been displayed as a **disease triangle** (Fig. 1.3), in which all angles are critical to whether disease will occur and to the extent that it will occur.

The contributions of hosts, pathogens, and the environment (both abiotic and biotic) to disease ecology are covered in detail in individual chapters in this book (Section II). Here, we add a few comments about the impact of seasonality when invertebrates are hosts. The importance of seasonality to disease ecology has been discussed with an emphasis on vertebrate diseases (Altizer et al., 2006). However, as invertebrates are predominantly heterothermic organisms, their activity is highly linked with seasons (e.g., invertebrates frequently are active only during periods of time when their food is present and environmental conditions are favorable for growth of the host, as during spring and summer in temperate climates). Therefore, some of the environmental drivers impacting pathogen activity are often seasonal. Invertebrate animals also frequently include several very different growth forms, with changes from one form to another during partial or complete metamorphosis. Such conditions render invertebrates more complex as hosts, especially because pathogens frequently are only able to infect and grow in specific susceptible host stages (e.g., baculoviruses infect when being eaten by susceptible caterpillars, but once the caterpillars become pupae and then adults, the baculovirus cannot infect to initiate new infections). Thus, seasonal changes in the abiotic environment, accompanied by changes in host growth forms, have a huge impact on invertebrates and therefore on diseases of invertebrates.

Pathogens are also directly influenced by the environment when outside of hosts during dispersal, during persistence of inactive stages as reservoirs, or during transmission of infective stages. When dispersal is in part facilitated by a vector, the pathogen is directly or indirectly also influenced by that vector. Pathogens experience a specialized habitat within individual hosts where they interact with the host as a growth environment, including exposure to an immune response (Chapter 4). They can also be impacted indirectly by the abiotic environment being experienced by the host.

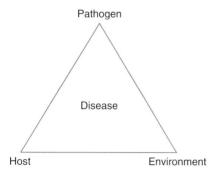

Fig. 1.3 The basic conceptual disease triangle model, indicating that the pathogen, host, and environment all must interact in order for disease to occur.

The density of hosts can influence whether infections are abundant for pathogens that are **density-dependent**. As for the gypsy moth nucleopolyhedrovirus, at higher host densities there can be an increased chance that successful transmission will occur, and in some cases, hosts can be more susceptible at higher densities (Reilly and Hajek, 2008). Density dependence is thought to be critical to sustained control of outbreak populations of pests, because infection levels will then increase after host populations increase. In contrast, populations of parasites can act **density-independently**, which means that equivalent levels of infection can occur regardless of the host density; a classic density-independent relationship for an infectious disease would be a situation where the levels of infection were influenced by weather conditions, regardless of host density. For example, with a fungal pathogen present in environmental reservoirs at high densities and under optimal temperature and moisture levels, high levels of infection can occur, in both high- and low-density populations.

1.1.3.1 Host Range

Host range is usually defined as the list of species used as hosts. Invertebrate pathogens span a continuum from highly host-specific species to species with broad host ranges, often with general trends for pathogen groups (see Section III), although the breadth of host ranges can also vary within groups. Perhaps the focus frequently has been on those species with broader host ranges because these can have a greater impact on host populations and when developed for pest control. Narrow host ranges are also characteristic of many pathogen species, including those with significant impact on important invertebrate hosts, although in many cases the breadth of natural host range simply has not been studied.

In general, the host range includes those species that are utilized in nature: this is called the **ecological host range** (also, natural, field, or realized host range). Peters (1996) used the term field host range to refer to the host range after inundative application of an entomopathogen, which would then be associated with high densities of pathogen inoculum, as would be present during the peak of an epizootic. This is important information because in the past, those species that could be infected in the laboratory under conditions optimal for the pathogen were often considered to be included in the host range. However, we now know that the range of hosts that can be infected in the laboratory is not the same as those that are naturally infected under field conditions. Therefore, this laboratory artifact is now called the laboratory or **physiological host range** and is known to provide only information about basic susceptibility (i.e., which host species can be infected when exposed under optimal conditions) and not the actual host species that are found infected in the field. Why can a host be infected in the laboratory and not the field? Besides the fact that conditions are optimized in the laboratory, this can often be due to the "laboratory" host not being present in nature at the same time and place as infective pathogen propagules and not displaying the correct behaviors to allow exposure to sufficiently high titers of the pathogen to become infected.

Host ranges of pathogens and parasites are important for numerous ecological reasons. Pathogen inocula that developed in one host species and were shed can spill over to infect other susceptible host species in the environment; this can be of environmental concern when using pathogens and parasites for biological control or if a pathogen develops in an abundant invasive species population and then spills over into susceptible native species. Host range can also impact the development of epizootics if inocula produced in one host species then infect a diversity of hosts, diluting the disease prevalence in the host species that was the initial source.

1.2 Types of Studies

In invertebrate pathology, as in all other disciplines, studies can be divided into two categories: observational and experimental. These may also be referred to as mensurative and manipulative, respectively. **Observational/mensurative studies** involve the measurement of a parameter at one or multiple points in time without disturbing the system (Hurlbert et al., 1984; Fuxa and Tanada, 1987; Campbell and Wraight, 2007). For example, observational studies most often measure pathogen prevalence or incidence over time. Observational studies are often analyzed for goodness of fit to a model, the only variable being space or time. One type of data generated from observational study is the life table, which follows specific cohorts over time in order to predict life expectancy and the relative influence of factors contributing to mortality. There are two types of life tables: age-specific (horizontal), which follow survivorship over time of a single cohort tied to a single generation, and time-specific, which is based on a multigenerational population at a given time (Southwood, 1978).

Experimental/manipulative studies involve disturbance of the system with two or more treatments (Hurlbert et al., 1984; Fuxa and Tanada, 1987; Campbell and Wraight, 2007). Manipulative experiments are often undertaken because confounding variables can be controlled. Experimental approaches tend to be more suited to hypothesis testing compared with observational studies, due to the ability to control variables during experiments. Of course, it is not possible to completely control all variables, especially in field experiments. Hence, in order to pin down ecological factors affecting invertebrate pathogens or invertebrate pathogen–host relationships, it is advantageous to make comparisons or test hypotheses in the laboratory first and then confirm results in the field.

One form of study in invertebrate pathology that may use observational or experimental data is focused on model development. A model is an idealized portrayal of reality that is used to understand a specific defined system. Given the intricacy of factors that contribute to epizootiology, mathematical models are required in order to understand and predict insect disease dynamics in natural populations. Furthermore, models may be highly useful in understanding and predicting the parameters required for success in microbial control applications. The basis for modeling the epizootiology of insect diseases and the underlying ecological theory are derived primarily from Anderson and May's work in the early 1980s (1980, 1981, 1982). Examples of models, types of models, and analysis in invertebrate pathology have been reviewed by Brown (1987), Onstad and Carruthers (1990), Hesketh et al., (2010), and Shapiro-Ilan et al. (2012); see also Chapter 12 for in-depth coverage.

1.3 Why Study the Ecology of Invertebrate Diseases?

One of the primary reasons to study the ecology of invertebrate pathogens is to enhance their use as microbial biological control agents. Invertebrate pathogens can be used in microbial control applications to control terrestrial and aquatic invertebrate pests. The pathogens can be applied via four approaches: classical/introduction biological control (the pathogen is introduced to an area for permanent establishment), inoculative (the pathogen is applied with the expectation of some recycling in host populations, but not permanent establishment), inundative (an acute "pesticidal approach," where recycling is not expected and seasonal applications are generally required), and conservation/environmental manipulation (the pathogens are not applied directly, but rather the system is manipulated to enhance endemic pathogen populations) (Eilenberg et al., 2001). Obtaining knowledge on invertebrate pathogen ecology and the factors that induce epizootics can greatly increase the efficacy of microbial control programs. See Chapter 11 for an expanded discussion of this subject. Approaches to the use of invertebrate pathogens for pest control have been recently reviewed by Lacey et al. (2015) and Lacey (2017), as well as other sources.

In contrast to the use of invertebrate pathogens to kill organisms, studies on invertebrate pathogen ecology are used to assist in preventing disease in beneficial organisms, such as by protecting terrestrial and aquatic species being used for food and feed, those species important to ecosystem services such as pollination, or those species that are simply of public interest, such as monarch butterflies (*Danaus plexippus*). Indeed, as pointed out in Chapter 9, the earliest studies of invertebrate diseases were based on the protection of beneficial invertebrates (e.g., fungal diseases in silkworms, *Bombyx mori*).

Another important impact of studies on invertebrate pathogen ecology is their usefulness as model systems for understanding ecological theory. Generally, invertebrate systems are easier to study than some other systems (e.g., due to the abundance and ease of collection of some invertebrates), and these systems are therefore attractive as models (Southwood, 1978). Invertebrate pathogen studies can address broad theoretical aspects of epizootiology/epidemiology. Some issues that are addressed include the regulation of wild or artificial (e.g., domesticated) host populations by pathogens and the mechanisms of regulation under different conditions (e.g., through reduced survival, fecundity, or both). Additionally, models pertaining to fundamental aspects of host–pathogen relationships can be elucidated; in some cases, such as when a symbiotic pathogen relationship is involved (e.g., with entomopathogenic nematodes and their bacterial symbionts), tripartite systems can be explored. Investigations of multitrophic levels and food web dynamics can also be advanced, such as recent studies in signaling ecology among three trophic levels (pathogen, herbivore invertebrate–host, and plant). Finally, both as model systems and in terms of practical application, the study of invertebrate pathogen ecology is important to understanding and controlling emerging diseases (see Chapter 16 for coverage), whether pathogens are thought to have evolved in situ or to have been introduced from somewhere else, often as a byproduct of human globalization.

1.4   What this Book Covers

This book begins by presenting definitions as well as some of the basic concepts covered today in the ecology of invertebrate diseases (Chapter 1; Ann Hajek and David Shapiro-Ilan). In fact, much of the general literature on disease ecology focuses on vertebrate diseases (not only in the context of human diseases). While general concepts are usually the same or similar, the various aspects have differing importance for vertebrates compared with invertebrates. For example, while diseases of wildebeest on the African savannas are impacted by rainy and dry seasons but the animals are always present, diseases of many invertebrates are impacted by which stages of the host are present and active at any given time of the year, in addition to the sensitivity to abiotic conditions. This first chapter identifies that this book will focus on infectious rather than noninfectious diseases and that its focus will be on pathogens, which we define as microscopic parasites. Following is a chapter discussing the major methods used for studies in the field of invertebrate ecology (Chapter 2; Raquel Campos and Lerry Lacey).

The chapters in Section II discuss the ecology of invertebrate diseases in the context of pathogens (Chapter 3; Lee Solter and James Becnel), hosts (Chapter 4; Louela Castrillo), abiotic environment (Chapter 5; Dana Ment, Ikkei Shikano, and Itamar Glazer), and biotic environment (Chapter 6; Jenny Cory and Pauline Deschodt). These are all components of the disease triangle (Fig. 1.3), and all must coordinate for a parasite to infect a host. Chapter 3 will compare and contrast the disease ecology of the diversity of groups of pathogens in association with pathogen life strategies, while Chapter 4 principally emphasizes the ecology of pathogens in the microhabitat within hosts, where they must contend with immune systems. Chapter 5 covers the influence of the abiotic environment. As a departure from Fuxa and Tanada's (1987) book on the epizootiology of insect diseases, the biotic environment is included separately in Chapter 6, as this subject is the focus of extensive study in recent years.

In Section III, chapters on the ecology of major pathogen groups infecting invertebrates cover viruses (Chapter 7; Trevor Williams), bacteria (Chapter 8; Trevor Jackson, Colin Berry, and Maureen O'Callaghan), fungi (Chapter 9; Ann Hajek and Nicolai Meyling), microsporidia (Chapter 10; Gernot Hoch and Lee Solter), and nematodes (Chapter 11; David Shapiro-Ilan, Ivan Hiltpold, and Ed Lewis). Of course, not all taxa that are pathogens of invertebrates are included, but the major groups have been presented and discussed. Throughout these chapters, the emphasis will be on diseases of insects as hosts, as this comprises a large portion of the research on invertebrate diseases.

Section IV includes chapters with some practical applications, beginning with one on modeling disease ecology (Chapter 12; Bret Elderd). Mathematical models are used to investigate disease dynamics at the population level, in order to understand what drives disease prevalence under different conditions. Models provide an opportunity to explore basic theories in disease ecology, but also, at times, to work toward predicting levels of infection, which can influence subjects covered in the following three chapters.

Pathogens have been used for the control of insect pests since the late 1800s, and there are currently many such pathogenic pest-control products on the market worldwide. Chapter 13, by Surendra K. Dara, Tarryn A. Goble, and David I. Shapiro-Ilan, discusses how insect diseases are used for pest control (i.e., microbial control). In particular, the authors discuss how pathogens can be mass-produced and applied to favor transmission, resulting in successful pest control. The next two chapters discuss the ecology of invertebrate pathogens that we seek to control. Chapter 14, by Jørgen Eilenberg and Annette Jensen, discusses diseases of terrestrial invertebrates (e.g., bees, silkworms, insects reared for food and feed) in light of their ecology and describes difficulties in controlling these diseases and how understanding disease ecology provides important insights. Next, Chapter 15, by Jeff Shields, describes the ecology of diseases of aquatic invertebrates, placing emphasis on marine systems in which diseases have wreaked havoc in invertebrates used to harvest human foods (e.g., oysters), often at high densities.

In Chapter 16, Colleen Burge, Amanda Shore-Maggio, and Natalie Rivlin discuss emerging infectious invertebrate diseases. These diseases have been of concern for the past several decades, including new plant diseases and diseases of vertebrates. Recognition of these diseases (e.g., coral and sea star diseases, and diseases of invertebrates accidentally moved from one country to another, to name only a few) is relatively new, and identification of causes has not always been possible. This review of the ecology of emerging invertebrate diseases is timely and much-needed.

We conclude with Chapter 17 (David Shapiro-Ilan and Ann E. Hajek), which summarizes the growing importance of invertebrate disease ecology. Pathogens are increasingly useful in controlling insect pests, but they themselves need to be controlled in populations of invertebrates of direct importance to humans (e.g., invertebrates managed for the production of food or feed or for ecosystem services). Thankfully, methods are available today that provide us better understandings of the ecology of invertebrate pathogens, allowing us to continue making advances in this field.

Acknowledgments

We thank S. van Nouhuys for helping to create Fig. 1.2 and B. Elderd for his input on the discussion of R_0.

References

Altizer, S., Dobson, A., Hosseini, P., Hudson, P., Pascual, M., Rohani, P., 2006. Seasonality and the dynamics of infectious diseases. Ecol. Lett. 9, 467–484.

Anderson, R.M., May, R.M., 1980. Infectious diseases and population cycles of forest insects. Science 210, 658–661.

Anderson, R.M., May, R.M., 1981. The population dynamics of microparasites and their invertebrate hosts. Phil. Trans. R. Soc. B Biol. Sci. 291, 451–524.

Anderson, R. M., May, R. M., 1982. Theoretical basis for the use of pathogens as biological control agents of pest species. Parasitology 84, 3–33.

Anderson, R.M., May, R.M., 1991. Infectious Diseases of Humans: Dynamics and Control. Oxford University Press, Oxford.

Andreadis, T.G., 1987. Transmission, in: Fuxa, J.R., Tanada, Y. (eds.), Epizootiology of Insect Diseases. John Wiley & Sons, New York, pp. 159–176.

Antolin, M.F., 2008. Unpacking β: within-host dynamics and the evolutionary ecology of pathogen transmission. Annu. Rev. Ecol. Evol. Syst. 39, 415–437.

Bonsall, M.B., 2004. The impact of diseases and pathogens on insect population dynamics. Physiol. Entomol. 29, 223–236.

Brown, G. C., 1987. Modeling, in: Fuxa, J.R., Tanada, Y. (eds.), Epizootiology of Insect Diseases. John Wiley & Sons, New York, pp. 43–68.

Campbell, J.F, Wraight, S.P, 2007. Experimental design: statistical considerations and analysis, in: Lacey, L.A., Kaya, H.K. (eds.), Field Manual of Techniques in Invertebrate Pathology, 2nd edn. Springer, Dordrecht, pp. 37–69.

Collinge, S.K., Ray, C., 2006. Disease Ecology: Community Structure and Pathogen Dynamics. Oxford University Press, Oxford.

Eilenberg, J., Hajek, A., Lomer, C., 2001. Suggestions for unifying the terminology in biological control. BioControl 46, 387–400.

Eilenberg, J., Thomsen, L., Jensen, A.B., 2013. A third way for entomophthoralean fungi to survive the winter: slow disease transmission between Individuals of the hibernating host. Insects 4, 392–403.

Fuxa, J.R., Tanada, Y., 1987 Epidemiological concepts applied to insect epizootiology, in: Fuxa, J.R., Tanada, Y. (eds.), Epizootiology of Insect Diseases. John Wiley & Sons, New York, pp. 3–21.

Galac, M.R., Lazzaro, B.P., 2011. Comparative pathology of bacteria in the genus *Providencia* to a natural host, *Drosophila melanogaster*. Microbes Infect. 13, 673–683.

Hajek, A.E., Larkin, T.S., Carruthers, R.I., Soper, R.S., 1993. Modeling the dynamics of *Entomophaga maimaiga* (Zygomycetes: Entomophthorales) epizootics in gypsy moth (Lepidoptera: Lymantriidae) populations. Environ. Entomol. 22, 1172–1187.

Hatcher, M.J., Dunn, A.M., 2011. Parasites in Ecological Communities: From Interactions to Ecosystems. Cambridge University Press, Cambridge.

Hesketh, H., Roy, H.E., Eilenberg, J., Pell, J.K., Hails, R.S, 2010. Challenges in modelling complexity of fungal entomopathogens in semi-natural populations of insects. BioControl 55, 55–73.

Howick, V.M., Lazzaro, B.P., 2014. Genotype and diet shape resistance and tolerance across distinct phases of bacterial infection. BMC Evol. Biol. 14, 56.

Hudson, P.J., Rizzoli, A., Grenfell, B.T., Heesterbeek, H., Dobson, A.P. (eds.), 2002. The Ecology of Wildlife Diseases. Oxford University Press, Oxford.

Hurlbert, S.H., 1984. Pseudoreplication and the design of ecological field experiments. Ecol. Monogr. 54, 187–211.

Lacey, L.A. (ed.), 2017. Microbial Agents for Control of Insect Pests: From Discovery to Commercial Development and Use. Academic Press, Amsterdam.

Lacey, L.A., Solter, L., 2012. Initial handling and diagnosis of diseased invertebrates, in: Lacey, L.A. (ed.), Manual of Techniques in Invertebrate Pathology, 2nd edn. Academic Press, San Diego, CA, pp. 1–14.

Lacey, L. A., Grzywacz, D., Shapiro-Ilan, D.I., Frutos, R., Brownbridge, M., Goettel, M.S., 2015. Insect pathogens as biological control agents: back to the future. J. Invertebr. Pathol. 132, 1–41.

Lewbart, G.A. (ed.), 2012. Invertebrate Medicine, 2nd edn. Wiley-Blackwell, Ames, IA.

Onstad, D.W., Carruthers, R.I., 1990. Epizootiological models of insect diseases. Annu. Rev. Entomol. 35, 399–419.

Onstad, D.W., Fuxa, J.R., Humber, R.A., Oestergaard, J., Shapiro-Ilan, D.I., Gouli, V.V., et al., 2006. Abridged Glossary of Terms Used in Invertebrate Pathology, 3rd edn. Society for Invertebrate Pathology. Available from: http://www.sipweb.org/resources/glossary.html (accessed May 8, 2017).

Ostfeld, R.S., Keesing, F., Eviner, V.T. (eds.), 2014. Infectious Disease Ecology: Effects of Ecosystems on Disease and of Disease on Ecosystems. Princeton University Press, Princeton, NJ.

Peters, A., 1996. The natural host range of *Steinernema* and *Heterorhabditis* spp. and their impact on insect populations. Biocontr. Sci. Technol. 6, 389–402.

Price, P., 1980. Evolutionary Biology of Parasites. Princeton University Press, Princeton, NJ.

Reilly, J.R., Hajek, A.E., 2008. Density-dependent resistance of the gypsy moth, *Lymantria dispar* to its nucleopolyhedrovirus, and the consequences for population dynamics. Oecologia 154, 691–701.

Roberts, M.G., 2007. The pluses and minuses of R_0. J. R. Soc. Interface 4, 949–961.

Shapiro-Ilan, D.I., Fuxa. J.R., Lacey, L.A., Onstad, D.W., Kaya, H.K., 2005. Definitions of pathogenicity and virulence in invertebrate pathology. J. Invertebr. Pathol. 88, 1–7.

Shapiro-Ilan, D.I., Bruck, D.J., Lacey, L.A., 2012. Principles of epizootiology and microbial control, in: Vega, F.E., Kaya, H.K. (eds.), Insect Pathology, 2nd edn. Elsevier, Amsterdam, pp. 29–72.

Southwood, T.R.E., 1978. Ecological Methods: With Particular Reference to the Study of Insect Populations. Chapman and Hall, London.

Sparks, A.K., 1972. Invertebrate Pathology: Noncommunicable Diseases. Academic Press, New York.

Steinhaus, E.A., 1949. Principles of Insect Pathology. McGraw-Hill, New York.

Steinhaus, E.A., Martignoni, M.E., 1970. An Abridged Glossary of Terms Used in Invertebrate Pathology, 2nd edn. USDA Forest Service, PNW Forest and Range Experiment Station.

Tanada, Y., Kaya, H.K., 1993. Insect Pathology. Academic Press, San Diego, CA.

Zimmer, C., 2000. Parasite Rex: Inside the Bizarre World of Nature's Most Dangerous Creatures. Free Press, New York.

2

Methods for Studying the Ecology of Invertebrate Diseases and Pathogens

Raquel Campos-Herrera[1] and Lawrence A. Lacey[2]

[1] *MeditBio, University of Algarve, Faro, Portugal*
[2] *IP Consulting International, Yakima, WA, USA*

2.1 Introduction

Observational and experimental methods commonly used for ecological studies of insects and other invertebrates in terrestrial habitats have been routinely employed for the estimation of the effects of entomopathogens on the natality, mortality, fertility, and dispersion of insect and mite pests (Hynes, 1970; Fuxa, 1987; Merritt and Cummins, 1996; Southwood and Henderson, 2000; Skovmand et al., 2007). In addition to quantitative sampling, determination of these effects is based on biotic (pathogen pathogenicity and virulence for specified insects, age and population densities of hosts) and abiotic (temperature, humidity, rainfall, habitat, ultraviolet (UV) radiation) factors for both hosts and pathogens. Advances in molecular biology and its implementation in the study of invertebrate pathology in a broad sense (identification and quantification, molecular recognition, regulation, etc.) (Gelernter and Federici, 1990; Shapiro et al., 1991; Hominick et al., 1997; Wong et al., 2007; Sunagawa et al., 2009) provide new opportunities to understand the biology and ecology of complex pathogenic interactions, which can have practical applications in areas as distinct as conservation biology and biological control. This chapter presents a broad array of techniques and methods employed in the study of the ecology and biology of invertebrate diseases and pathogens, illustrating their application in various systems. A number of other sources have addressed methodology in insect pathology, such as Stock et al. (2009), Lacey and Kaya (2007), and Lacey (2012), yet without a particular focus on ecology (as is the case for this chapter).

2.2 Traditional Methods for Studying Diseases

2.2.1 Sampling Goals

The goals of sampling invertebrate diseases include (i) determining the incidence (the number of new cases of a particular disease within a given period of time, in a population

Ecology of Invertebrate Diseases, First Edition. Edited by Ann E. Hajek and David I. Shapiro-Ilan.
© 2018 John Wiley & Sons Ltd. Published 2018 by John Wiley & Sons Ltd.

being studied; Onstad et al., 2006) and prevalence (the total number of cases of a particular disease at a given moment of time, in a given population; Onstad et al., 2006), (ii) determination of factors that govern outbreaks of disease and how this information can be used to forecast the onset of epizootics and their role in conservation biological control and integrated pest management, and (iii) the collection and isolation of causal agents of insect diseases for further development as microbial control agents (MCAs) or in preventing disease in beneficial invertebrates.

2.2.2 Sampling Regimes

To determine the prevalence of a given disease, sampling of the infected and uninfected host population should be established on a regular basis. The frequency and quantitative detail of sampling will depend on a variety of factors, including specific goals and available resources. Concomitantly, monitoring of the biotic and abiotic factors that govern or at least influence the activity of naturally occurring or applied entomopathogens can provide useful information on the organism's ecology. Pathogen biology, host–insect interactions, crop factors (cultivar, phylloplane factors, planting and harvest dates, and preceding crops), and parameters in aquatic habitats (depth, oxygen level, water quality and hardness, salinity, etc.) need to be determined, in addition to climatic factors (temperature, humidity, rainfall, and solar activity). Time of year and latitude will also influence the prevalence of entomopathogens. For example, apple orchards in Nova Scotia (Canada) require significantly fewer applications of the codling moth granulovirus (CpGV) to protect fruit from damage by a single generation of codling moth (*Cydia pomonella*) compared to apple orchards in Washington state (USA), where there may be two or more generations (Lacey et al., 2008).

2.2.3 Methodologies

The simplest method for detecting diseased invertebrates involves visual search. Often, the sampler who is well acquainted with a particular species or group of insects will be the first to spot unusual looking specimens. For example, the discovery of CpGV was made while sampling for arthropod natural enemies of *C. pomonella* in Chihuahua, Mexico (Tanada, 1963; Lacey et al., 2008). Infected larvae are most easily spotted during epizootics. Under optimal conditions of weather, dense host arthropod populations, and the presence of highly infectious entomopathogens, phenomenal epizootics can occur (Fuxa and Tanada, 1987; Shapiro-Ilan et al., 2012a). In addition to death and degradation of the host, infected insects may show other signs of infection such as differential coloration, emission of an odor unlike that of a noninfected host, an appearance of melting, and exhibition of altered behavior, such as climbing to higher positions on host plants (Lacey and Solter, 2012).

Some examples of readily detected epizootics include nucleopolyhedrovirus (NPV) in Lepidoptera and Hymenoptera and fungal pathogens in the Entomophthorales (e.g., spectacular outbreaks of the fungal pathogen of gypsy moth, *Lymantria dispar* and the NPV of the European pine sawfly, *Neodiprion sertifer*; Hajek et al., 2015; Hajek and van Frankenhuyzen, 2016). Another group of baculoviruses, the granuloviruses (GVs), are found only in Lepidoptera (Fig. 2.1a,b). Epizootics due to GVs are reported from diamondback moth (*Plutella xylostella*), cabbage butterfly (*Pieris rapae*), cabbage looper (*Trichoplusia ni*), and the bean shoot borer (*Epinotia aporema*). Additional species of

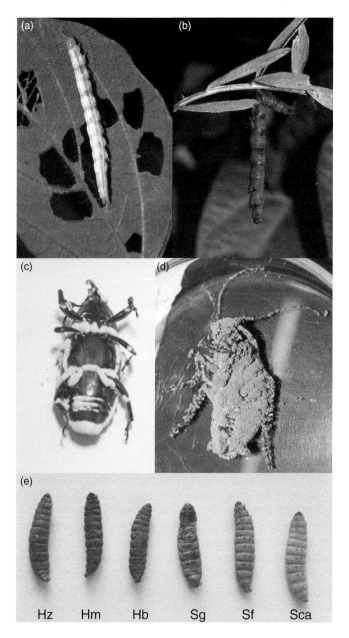

Fig. 2.1 Larva of the green clover worm moth, *Hypenas cabra*, infected with granulovirus (GV), with change of color to become increasingly whiter (a), before dying (b). Infection caused by the fungus *Beauveria bassiana* in an adult red weevil (*Rhynchophorus ferrugineus*) (c) and by *Metarhizium anisopliae* in an adult locust (d), with the distinct colorations of each genus. Larvae of the wax worm *Galleria mellonella* killed by entomopathogenic nematodes display different colors depending on the species: dark green/gray for *Heterorhabditis zealandica* (Hz), dark red for *H. megidis* (Hm) and *H. bacteriophora* (Hb), dark gray for *Steinernema glaseri* (Sg), brown for *Steinernema feltiae* (Sf), and creamy for *Steinernema carpocapsae* (Sca). *Source:* Courtesy of Gerry Carner, Surendra Dara, Stefan Jaronski, and Rubén Blanco-Pérez. (*See color plate section for the color representation of this figure.*)

entomopathogenic viruses that clearly show symptoms of infection are presented by Cory and Evans (2007), Eberle et al. (2012), and Grzywacz (2016). The progression of disease in baculovirus-infected populations includes cessation of feeding, fat body becoming increasingly whiter in moribund larvae, and, ultimately, death and disintegration of larvae.

The bacteria *Paenibacillus japonicum* and *P. lentimorbus* have been observed causing epizootics of milky spore disease in Japanese beetle (*Popillia japonica*) and other scarab larvae (Klein, 1992; Koppenhöfer and Wu, 2016). Infected larvae are easily distinguished by the milky appearance of the hemolymph. Although *Bacillus thuringiensis* (Bt) subspecies are not often observed causing natural epizootics, two of the most commercially successful subspecies, *B. thuringiensis kurstaki* (Btk) and *B. thuringiensis israelensis* (Bti), were isolated from outbreaks of disease caused by these bacteria in populations of *Ephestia kuhniella* and *Culex pipiens*, respectively (Kurstak, 1962; Goldberg and Margalit, 1977). Btk-killed larvae appear brown to black and may appear to disintegrate or become dry and brittle.

Fungal epizootics have been reported from a wide variety of insects and mites. These include such notable examples in the Entomophthorales as *Neozygites gossypii* in cotton aphid (*Aphis gossypii*) (Steinkraus, 2007), *Neozygites tanajoae* in green cassava mite (Delalibera, 2009), and *Entomophaga maimaiga* in gypsy moth (Hajek, 1999; Hajek et al., 2015; Hajek and von Frankenhuyzen, 2016). Disease outbreaks caused by hypocrealean fungi (*Metarhizium* spp., *Beauveria* spp., *Isaria* spp., *Lecanicillium* spp., *Nomuraea rileyi*, etc.) in host insects are mostly reported from species that have soil-inhabiting stages (Vittum et al., 1999; Klingen and Haukeland, 2006; Meyling and Eilenberg, 2007; Ormond et al., 2010; Pell et al., 2010). Infected insects can stand out due to the distinctive colors of each genus of fungus (i.e., green, gray, white, lavender) (Fig. 2.1c,d).

Outbreaks of disease caused by microsporidia are especially apparent in insects with transparent epidermis. What may appear to be a fungus-caused epizootic might actually be disproportionate signs of fungus-caused disease expressed late in the season after the majority of the uninfected population has emerged.

Epizootics caused by entomopathogenic nematodes (EPNs) include those produced by *Steinernema scapterisci* in mole crickets (*Scapteriscus* spp.). In addition to color changes, the surfaces of patently infected crickets swarm with infective juveniles (IJs). The difference in color due to EPN infection of other insect species depends on the genus of EPN. The color of larvae infected by *Heterorhabditis* spp. ranges from orange to deep red, and in some cases dark green-gray (i.e., *H. zealandica*), while those infected with *Steinernema* spp. appear cream, light tan, brown, and even dark gray (Koppenhöfer, 2007; Fig. 2.1e).

Outbreaks of parasitic nematodes in the family Mermithidae in mosquito larvae (*Romanomermis iyengari*, *Romanomermis culicivorax*, *Strelkovimermis spiculatus*) are distinguished by the presence of straight or rolled subadult nematodes in the hemocoel (Campos et al. 1993; Micieli and García, 1999; Platzer, 2007). Mosquito larvae usually live until the fourth instar before the nematodes emerge. As seen in microsporidian-infected mosquito larvae, the presence of larval mosquitoes parasitized by mermithids (*Romanomermis* spp.) in high percentages of the population late in the season may actually be due to the earlier emergence (and subsequent absence) of uninfected larvae.

2.2.3.1 Searching for Infected Insects Using General Entomological Sampling Methods

A broad array of equipment and strategies is available for the sampling of invertebrate populations (Merritt and Cummins, 1996; Southwood and Henderson, 2000). The use of equipment such as sweep nets, D-vac samplers, pitfall traps, pheromone- or kairomone-baited attractant traps, passive attractant traps, and beating trays will often sample predominantly uninfected insects. However, infected insects may also be obtained in this manner. Search of aquatic habitats can also be performed using general sampling techniques such as drift nets (to capture disturbed arthropods) and dredging of benthic substrata; these methods will be useful tools for obtaining both infected and uninfected specimens. For example, dredging of lake substrata and sieving of organisms from sediments have revealed chironomid larvae infected with viruses, *Ricketsiella*, and microsporidia (Stoltz et al., 1968; Federici et al., 1974, 1976; Majori et al., 1986). Infected larvae usually stand out markedly from uninfected conspecifics due to iridescence or whiteness in infected fat bodies.

Collection from the field and subsequent incubation may reveal infections in individuals that did not show signs of disease in the field (Southwood and Henderson, 2000; Cory and Evans, 2007). There can be several reasons for rearing field-collected material. If the goal is to identify the pathogen or detect rare pathogens, stressing by crowding, starving, or incubation under other suboptimal conditions may accelerate the development of entomopathogens that are present at low levels, occult, or in an eclipse period at the time of collection (Lacey and Solter, 2012). If the goal is instead to quantify levels of infection in field populations, it is important not to promote stress but instead to rear under conditions similar to those experienced by field populations; in particular, this is important if samples are collected from the field across time, with a goal of documenting levels of disease over time. In these cases, rearing must be carefully considered and any chances for transmission from infected to healthy after collection (or from collection and rearing supplies and equipment) should be prevented (e.g., rearing of insects in groups should be avoided if possible, because as some in the group die, infectious units can be produced and individuals that were healthy when collected can become infected, leading to incorrect quantification of infection levels). While, often, incubation occurs in the laboratory, some studies have gone so far as to create field insectaries in order to conduct culturing under field conditions (Woods et al., 1991; Hajek, 1997).

2.2.3.2 Selective Media

Soil, regardless of the apparent absence of insects, can be a rich source of entomopathogenic bacteria, entomopathogenic fungi (EPF), and EPNs (Klingen and Haukeland, 2006). In the cases of bacteria and fungi, selective media can be employed to suppress certain unwanted species. For example, nutrient bacteriological medium supplemented with penicillin and streptomycin or mycostatin suppresses a variety of non-spore-forming bacteria and fungi (Yousten et al., 1985). Similarly, mycological media to which the fungicide dodine (N-dodecylguanidinemonoacetate) has been added will suppress most saprophytic fungi, although *Metarhizium* spp., *Beauveria* spp., and some other species of Hypocreales will survive (Beilharz et al., 1982; Chase et al., 1986). Fernandes et al. (2010) obtained similar results with dodine-free CTC medium (potato dextrose agar, chloramphenicol, thiabendazole and cycloheximide) in the isolation of *Metarhizium acridum*. Plating serially diluted soil on to selective media and subsequent quantification of colony-forming units (CFUs) can facilitate estimates of viable propagules and evaluations of environmental impacts on entomopathogen populations (Inglis et al., 2012; Shapiro-Ilan et al., 2012b).

2.2.3.3 Extraction Methods

Extraction of nematodes from soil, regardless of species, can be done using the Baermann funnel and centrifugation using a sucrose gradient. Although these methods can be quantitative in terms of IJs per gram of soil, taxonomic expertise will be required to separate EPNs from free-living and plant-parasitic nematodes. Other methods for the extraction of nematodes from soil include sieving and insect baiting (see later) (Stock and Goodrich–Blair, 2012). Aside from nematodes, Undeen and Avery (1987) reported on the use of continuous centrifugation to extract microsporidia and other entomopathogens from large volumes of ditchwater. Similarly, Hajek et al. (2012) quantified resting spores of *E. maimaiga* in soil samples using a modified method of discontinuous Percoll density-gradient centrifugation.

2.2.3.4 Airborne Spore Sampling

During epizootics, some EPF produce prodigious amounts of conidia, some of which becomes airborne. Hollingsworth et al. (1995) and Steinkraus et al. (1996) described the use of Rotorod and Burkhard spore samplers to monitor EPF conidia in the air within cotton crops to predict epizootics and determine when to avoid insecticide application for control of cotton aphid (*A. gossypii*). This enabled epizootics caused by *Neozygites fresenii* to proceed, bringing the aphid under control while allowing other natural enemies of cotton pests to survive. Hajek et al. (1999) also used Burkhard volumetric spore traps to determine *E. maimaiga* spore numbers over time, as well as Rotorod spore samplers (Sampling Technologies) at different vertical locations (0.5, 1.5, and 3.0 m) in the forest canopy. From third instar to pupation of *L. dispar*, the presence of conidia in the air was episodic, and infection in the resident gypsy moth population increased only after the first peak of abundance in airborne conidia.

2.2.3.5 Insect Baiting

One of the most useful tools for the isolation of EPNs and EPF from soil is baiting with insects (Bedding and Akhurst, 1975; Zimmermann, 1986). Protocols for the use of bait insects for EPN and EPF detection are presented by Stock and Goodrich-Blair (2012) and Inglis et al. (2012), respectively. Larvae of the wax moth (*Galleria mellonella*) and mealworm (*Tenebrio molitor*) are the most commonly used bait species for general sampling of the microbiome. Typically, larvae are exposed to field-collected soil in closed containers, such as 0.5 L delicatessen cups with lids, for 3 or more days, after which dead larvae are removed. Larvae that are presumed to be EPN-killed are rinsed and placed on White traps until IJs, if present, emerge 7–10 days later (Stock and Goodrich–Blair, 2012). Larvae with signs of fungal infections are rinsed, surface-sterilized, and placed on water agar (Lacey and Solter, 2012). Isolation of conidia emerging from the cadaver can be accomplished using methods described by Inglis et al. (2012) and Lacey and Solter (2012). Additional approaches using host baits have investigated levels of infection when hosts are exposed to different environmental locations or conditions. Hajek et al. (2000) caged third- and fourth-instar *L. dispar* larvae on the soil and in understory foliage in forests where *L. dispar* and *E. maimaiga* had previously been detected. Infection with *E. maimaiga* was detected predominantly in larvae from ground cages.

Entomophaga maimaiga persists in forests as a reservoir of spores in soil at the bases of trees (Hajek, 1999). Late-instar *L. dispar* larvae descend from the canopy of host trees to wander and rest under leaf litter during daylight hours (Hajek, 2010). This behavior increases their risk of coming into contact with fungal spores. To detect non-target infections, Hajek et al. (2000) collected and reared larvae in leaf litter, understory vegetation, and on tree boles within a 200 cm radius around trunks of red oak (*Quercus rubra*). Among the 358 lepidopteran larvae that were collected, 67 were *L. dispar*; 37% of these died due to *E. maimaiga* infections, while infections in other species collected were nonexistent or very low.

2.2.3.6 Dispersal of Entomopathogens: Mark–Release–Recapture Method

In addition to sampling, for prevalence or incidence (as described before), dispersal of pathogens or infected hosts may be of interest; the ability to disperse is a key factor enabling entomopathogens to cause epizootics (Tanada, 1963). One of the main means of determining dispersal is the mark–release–recapture method (Fuxa, 1987; Southwood and Henderson, 2000). An example of this method is presented by Lacey et al. (1995) using field-collected *P. japonica*, the fungus *Metarhizium anisopliae* isolated from *P. japonica* larvae in the Azores (Martins, 1988; Martins et al., 1988), and florescent dye powders. Approximately 2×10^4 adult *P. japonica* were collected on the Azorean island of Terceira and divided into two groups of 10^4 beetles, one topically treated with *M. anisopliae* conidia (1 g conidia/10^4 beetles, ca. 10^6 conidia/beetle) and green florescent powder, the other with orange florescent powder alone or with powder and inert conidia. The beetles were released in the centers of concentric circles of eight kairomone-baited traps placed at 50, 100, and 500 m. Sixteen traps were located 1 km around the release point. Beetles in all traps were collected after 24, 48, and 72 hours, chilled, and counted in the laboratory under a UV lamp. Data were compiled according to color of dye, distance and direction from release point, and day of collection. A total of five such releases were made in two successive summers (1990 and 1991). The research demonstrated a significant negative effect of fungus treatment on numbers of beetles recovered and distance of dispersal relative to that of untreated beetles.

2.3 Molecular Tools to Assist in the Detection and Quantification of Pathogens and their Impact on the Host

In past decades, various molecular techniques have been incorporated in the ecological study of invertebrate pathogens and the diseases they cause. Mainly, the focus has been on identifying and quantifying the target organism in the host or on elucidating problems caused by the modification of host physiology by the pathogen. With advances in methodology and the availability of new equipment, these techniques have been improved and refined. In general, molecular tools applied to the detection and quantification of pathogens have in common the search for the best molecular marker to address the specific basic or applied questions. A **molecular marker** is based on the selection of a section of the genome or its derivative (i.e., rRNA, proteins) that is expected to be representative of the whole genome. It might contain (or not) a functional gene that can be transcribed and further translated into a specific protein. Hence, if we consider the possibilities that constitute the whole genome and its various levels of expression, we can isolate and characterize a wide variety of molecular markers, ranging from those that are highly polymorphic (i.e., mini- and microsatellites as fingerprints) and thus allow differentiation among individuals of the same species for behavioral studies, to those that are less variable (e.g., allozymes) and so allow the study of population dynamics.

In general, two types of molecules are used as molecular markers: proteins and nucleic acids. Despite the broad use of protein studies, in particular in the 1980s and 90s, some factors have generally decreased their use. For example, studies involving proteins require relatively large amounts of material (i.e., tissues, feces, etc.), and in the case of small organisms, sacrifice of the organism is required. Also, the presence and quantity of the target protein often depends on factors such as the stage of development of the host or pathogen or the exact extraction location (tissue). Additionally, from a technical point of view, once extracted, proteins are usually less stable than DNA, and their maintenance in a nondenatured state will be critical for reproducibility of the results (Beebee and Rowe, 2008). However, no one molecular marker is ideal for all applications, and selection of the most appropriate technique will depend on the research question.

2.3.1 Employment of Proteins: The Beginning of the Molecular Era in Invertebrate Pathology

Historically, proteins were the first target for molecular studies. Smithies (1955a, 1955b) developed a method to separate protein variants according to their mobility in an electrical field using starch gel as the supporting medium. This marked the beginning of the study of **allozymes** to measure protein polymorphism. This method was widely employed in the 1980s and 90s (May, 1992) in studies of invertebrate pathogens. For example, St. Leger et al. (1992) employed allozymes to determine the variability among germination and production of infection structures of 120 isolates of the entomopathogenic fungus *Metarhizium* spp. More recently, Mitchell et al. (2004) employed allozymes to investigate the impact of the bacterium *Pasteuria ramosa* on a natural population of the crustacean, *Daphnia magna*. For *P. ramosa*, differences in *D. magna* allozyme genotypes and infection of the bacterium were consistent with parasite-mediated selection. However, despite its successful legacy, the low variability detected by the allozyme approach is limiting; for example, studies involving conspecifics or paternity approaches that require the ability to detect individual differences are not suitable for this approach.

Another molecular approach involving proteins is the use of **immunoassays**. Antibodies are highly specific to molecules called antigens, which are usually proteins. Use of antibodies always requires the creation of an antibody in a vertebrate. Antibody–antigen reaction has multitudes of uses. For example, the antibodies can be "labeled" with heavy metals or fluorescent markers, allowing visualization of the reaction in a complex matrix such as soil. Broadly used in invertebrate pathology, antibodies can be employed in an **enzyme-linked immunosorbent assay (ELISA)** technique. The essence is that a sample to be screened for the target antigen is bound to the bottom of a surface, such as a plastic tray well, and the specific antibody is applied. The antibody will react with the antigen in a density-dependent manner. Thereafter, a second and generalized antibody is applied in the mixture. This second antibody is labeled with a reporter enzyme that usually produces a change in substrate coloration. Examples of ELISA protocols for insect pathogens include those developed for two pathogens of the gypsy moth (*L. dispar*): LdMNPV (Ma et al., 1984) and *E. maimaiga* (Hajek et al., 1991), and *M. anisopliae* (Guy and Rath, 1990). Other ELISA protocols have been developed for certain microsporidian pathogens in immune-suppressed humans (i.e., Bouladoux et al., 2003).

In some cases, use of antibodies in the **Western blot technique** has been successfully employed. The fundamental aspect of this method is that a protein mixture to be screened for the target antigen is first separated by polyacrylamide gel electrophoresis. After migration, the proteins are bound in a nylon membrane, which is exposed to the specific antibodies to reveal the presence of the target antigen by enzymatic reaction. For example, Segers et al. (1999) employed Western blot analysis in combination with other molecular biology techniques to investigate the extracellular subtilisin-like protease on fungal pathogens of various organisms, including insects and nematodes. Similarly, Zhang et al. (2008) employed Western blotting to investigate the fibrinogen-related proteins (FREPs) (i.e., hemolymph proteins) present in the freshwater snail (*Biomphalaria glabrata*), the intermediate host for the human blood fluke (*Schistosoma mansoni*). They discovered that the FREPs can recognize a wide range of pathogens (i.e., non-self-recognition), ranging from prokaryotes to eukaryotes, and that the diversity of the FREPs may thus represent differential functional specializations with respect to the target pathogen.

2.3.2 Techniques Based on the Nucleic Acids: the "Pre-Omics" Era

Nowadays, almost all DNA-based identification procedures are based on polymerase chain reaction (PCR) protocols. However, there are techniques that can be performed with or without a preliminary PCR step. For example, the method called **restriction fragment-length polymorphism (RFLP)** can use either total DNA or a fragment derived from a PCR amplification. In both cases, the DNA is digested with one (or more) restriction endonucleases, also called "restriction enzymes," to produce fragments of various sizes (Aquato et al., 1992). The variability of the fragment sizes will be determined by the location of the site of recognition for the enzyme. Mutation can alter the specific areas of enzyme recognition, and new areas can be generated or lost. Once the fragments are separated by agarose gel electrophoresis, the variability in the sizes of these sequences will determine a pattern that allows differentiation among populations or individuals. PCR–RFLP assays provide one of the most useful molecular tools for distinguishing between species, such as parasitic protists in the phylum Haplosporidia (i.e., *Bonamia ostreae, B. exitiosa*) (Hine et al., 2001; Cochennec–Laureau et al., 2003). RFLP has also been widely used in other groups, such as baculoviruses (Gelernter and Federici, 1990; Shapiro et al., 1991; Laitinen et al. 1996; Cooper et al., 2003) and EPNs (see revision by Hominik et al., 1997; Adams et al., 2006). In some cases, total DNA or RNA is employed. Procedures allowing identification of selected DNA products are known as "**Southern blotting**," whereas for RNA, they are called "**Northern blotting**."

In these cases, it is necessary to denature and transfer the material from the gel on to a nylon membrane. Thereafter, the material is incubated with the target probe of interest, which is labeled with radioactivity or exposed to enzymes. The presence of the target sequence is subsequently revealed by autoradiography or resulting enzymatic reaction, in the case of Southern and Northern blotting, respectively.

Another approach that may use total DNA is **minisatellite DNA fingerprinting**, which can be considered a specific application of the RFLP technique (Bruford et al., 1992). Since minisatellites are highly polymorphic, they are especially useful for evaluating differences among individuals of the same species. As the basis for minisatellite studies, short DNA sequences (46–60 bp) are repeated in blocks, and these areas are characterized by high mutation rates. The blocks are distributed at various nuclear loci in organisms, including invertebrates (Carvalho et al., 1991). Polymorphisms are derived from the change in the numbers of repeated sequences in a block, which provides fragment sequences of various sizes. These differences will be revealed after restriction digestion and migration in an electrophoresis gel. A modification of this technique is the **microsatellite fingerprint**. In this case, the microsatellite is used as a probe, consisting of a synthetic oligonucleotide and a sequence that is the product of a tandemly repeated simple sequence (TTCs, or CCAs). Despite the myriad possibilities that these techniques raise, the complexity of the methods and the difficulties of separating the bands in a reproducible and reliable way have limited their use.

The development of PCR-based methods has expanded the utility of molecular techniques. The **PCR** technique revolutionized the molecular assays of the last 3 decades. The development of the PCR technique was possible due to the discovery of thermo-stable DNA polymerase (i.e., *Taq* isolated from the extremophile bacterium *Thermus aquaticus*). PCR makes it possible to produce a considerably high amount of target DNA from a small amount of material (Hoelzel and Green, 1992). The reaction only requires a few elements to work: the deoxyribonucleotides, the primers that flank the target region of interest, the buffer that enables the reaction, and the enzyme that will synthetize the new sequences. There are three main steps to the reaction: (i) denaturation to separate the double-stranded DNA, (ii) annealing to bind the primers to the target area of interest, and (iii) the synthesis itself, when the actual production of complementary copies starts from the end of the primers (Mullis and Faloona, 1987; Mullis, 1990). Once protocols were established and equipment became widely available at various institutions, the simplicity of the reaction promoted its incorporation as a routine method in many laboratories. In most applications, the key step is the design and/or selection of the most appropriate set of primers, although some approaches, such as **randomly amplified polymorphic DNA (RAPD)** or **amplified fragment-length polymorphism (AFLP)**, do not require any previous information about the target area of study. Briefly, both use total DNA and amplification with short DNA sequences or primers. RAPDs represent the simplest version, where randomly selected 10 mer oligonucleotides are used as primers and directly generate a pattern that is visualized by electrophoresis in an agarose gel. RAPDs have been employed for species identification of invertebrate pathogens, such as EPNs (Gardner et al., 1994; Hashmi et al., 1996; Liu and Berry, 1996). The AFLP approach is considered more reliable, although it involves more steps (Vos et al., 1995). It requires first the digestion of the total DNA with two restriction enzymes, which generate overlapping ends. Then, primers are synthetized *in vitro* to be compatible with those ends, along with a few additional, randomly chosen nucleotides at the 3′ end that confer specificity and more reproducible results (Mueller and Wolfenbarger, 1999). An example of the use of AFLP comes in the estimation of an inbreeding coefficient, as shown by Kaunisto et al. (2013). These authors studied the relationship between genetic heterozygosity in the damselfly (*Calopteryx splendens*) population and the presence of a gregarine endoparasite. They found that the more homozygous an individual was, the more parasites were present, highlighting the relevance of heterozygosity in invertebrates for individuals' pathogen resistance. However, one of the limitations of both random-amplification techniques is the lack of separation between homozygotes and heterozygotes (Beebee and Rowe, 2008). Other limitations are the strong influence of minimal alterations of the protocol and difficulties with cross-laboratory standardization.

PCR-based microsatellites, also called in some cases **simple sequence repeats** (SSRs) (Tautz, 1989), are known to be highly polymorphic. Microsatellites are short runs of di-, tri-, or tetranucleotide repeats that are widespread in the genomes of most of organisms. However, they are without any apparent function, and hence, are considered neutral markers (Goldstein and Schlotterer, 1999). This PCR-based approach requires primers complementary to the two flanking areas of the selected locus (the position of a gene on a specific chromosome), which will be used during a conventional PCR to generate fragments not bigger than 300 bp. The resulting PCR products can be visualized by simple gel staining, labeled with radioactivity and visualized by autoradiography, or treated with fluorescent nucleotides and evaluated with automated DNA sequencers (Beebee and Rowe, 2008). Microsatellites have been explored in invertebrates, in relation with their pathogens, such as in a study by Alcivar-Warren et al. (2007), which produced a microsatellite-based linkage map for the Pacific whiteleg shrimp (*Litopenaeus vannamei*), in which the presence of pathogens (Taura syndrome virus (TSV) and white spot syndrome virus (WSSV)) was also found. Alcivar-Warren et al. (2007) produced the first low-density linkage map (ShrimpMap) for specific pathogen-free (SPF) shrimp, which can now be employed to advance conservation and evolutionary genomics. From an applied point of view, this tool will allow the selection of genes associated with fitness traits in both wild and cultured shrimp, with the goal of enhancing production. The limitation of this technique is the need to develop the primers and the unsuitability of already published protocols for new target organisms. In many cases, a study may require *de novo* development of the system. Because the process will imply the construction of the genomic library enriched in microsatellites from the target organisms, the time and cost are significantly important.

DNA sequencing, a pioneering method developed by Sanger et al. (1977), is linked to PCR-based approaches and arose as a milestone in the identification of organisms, including invertebrate pathogens. This technology, especially once automated, enabled advancement in the detection of minor changes in the genome, even the characterization of **single nucleotide polymorphisms (SNPs)**, which are considered the highest possible resolution of genetic differences (Morin et al., 1999). **DNA barcoding** is a PCR-based method that enables full sequences of the target area of the genome to be obtained, allowing the presence of SNPs to be explored. The selection of the primer set provides the specificity for the genome area to be studied in the selected organism. The first approach for DNA barcoding was **clone-sequencing**, which implies the insertion of selected PCR products into a plasmid. Transformation of bacteria with the plasmid + insert construct enables multiplication of the bacteria to generate thousands of copies, resulting in final plasmid DNA extraction and sequence analysis. An advanced version for screening of genetic identity is the PhyloChip (e.g., PhyloChip G2). This technology, developed for 16S rRNA gene microarray screening, allows the evaluation of 300 000 oligonucleotide probes, with a resolution of >8000 operational taxonomic units (Wilson et al., 2002; DeSantis et al., 2007). This technique was successfully employed in the assessment of the bacterial diversity associated with the white plague disease in the Caribbean coral (*Montastraea faveolata*) (Sunagawa et al., 2009). By comparing traditional cloning methods and the PhyloChip datasets, Sunagawa et al. (2009) observed an increase in diversity and a significant shift in community structure in *M. faveolata* under the presence of the disease, and these results served as a basis for further studies unraveling the mechanism of the disease. More recently, Kimes et al. (2013) employed ribosomal RNA gene sequencing to identify the microbiome (zooxanthellae, bacteria, and archaea) associated with healthy and yellow-band diseased *M. faveolata*, evaluating the temporal patterns in populations. This study confirmed that temporal differences between the associated microbiome in March and September were more important than differences related to healthy versus unhealthy populations, highlighting the critical role that seasons plays in the disease–host complex.

The contributions of the DNA sequencing approach and other early molecular methods established the basis for the next generation of sequencing (NGS) technology and its derivative (proteins). The public databases, such as GenBank (created in 1988 and currently maintained by the National Center for Biotechnology Information (NCBI)), are key resources for freely available molecular information that have allowed significant advances in molecular ecology tools and their applications to be made. Currently, new sequences derived in research must be submitted to public services such as GenBank to be accepted for publication.

Subsequent techniques employed the identification of SNPs for screening variability, such as searching for single pair mismatches (**denaturing gradient gel electrophoresis (DGGE)** and **thermal gradient gel electrophoresis (TGGE)**) or the variable conformation that differs in nucleotide positions (**single strand conformation polymorphism (SSCP)**). These techniques have less power to resolve differences than direct sequencing, but they can process multiples of samples in a simpler manner. In all these cases, PCR employs selected primers that flank the desired region, and the resulting fragment will migrate in a polyacrylamide gel in a buffer with an increasing gradient of the specific compound (i.e., formamide and urea) (DGGE) or in a gradient of increasing temperature (TGGE). Because each sequence differs in its double-strand stability, the denaturation point will be characteristic and the migration will stop at different points in the gel. Thereafter, simple staining will allow identification of the pattern and comparison between samples. If required, the bands can be rescued and sequenced for further confirmation. In the case of SSCP, the denatured PCR products migrate in nondenaturing gel electrophoresis, allowing separation by different three-dimensional conformations. In addition to these three techniques, **terminal restriction fragment-length polymorphism (PCR–T–RFLP)** can also provide information on the variability of multiples samples without the necessity of whole sequencing. The main advantage of T–RFLP with respect to the DGGE/TGGE technique is its ability to compare data from different runs (Nunan et al., 2005; Pompanon et al., 2012). An example of the use of T–RFLP in combination with cloning and sequencing of the 16S rRNA genes has revealed the high diversity of ctenophore bacterial communities, including the detection of various sequences similar to a sea anemone pathogen (Daniels and Breitbart, 2012). This study provided the first report of distinct and dynamic bacterial communities associated with the phylum Ctenophora, advancing the understanding of bacteria–marine invertebrate ecology.

Another widespread PCR-based protocol that allows species-specific identification is the use of **conventional PCR with species-specific primers** for a target host. This technique has been employed successfully in various invertebrate–pathogen systems, such as the detection of the virus ostreid herpes virus 1 (OsHV-1) from the oyster *Crassostrea gigas* (Renault et al. 2000; Arzul et al., 2002; Barbosa-Solomieu et al., 2004) and of bacteria (i.e., associated with juvenile oyster disease) from the oyster *Roseovarius crassostreae* (Maloy et al., 2005). This system can also use multiplex reactions (i.e., targeting two or more species in the same reaction by combining various specific sets of primers). However, new techniques such as quantitative real-time PCR (qPCR) and NGS are now displacing the use of conventional PCR-based methods. In addition to this, the so called "-omics" (i.e., genomics, proteomics, metabolomics) disciplines are rapidly evolving. These new fields aim to describe and characterize the genome (structure), proteome (functions), or metabolome (dynamics and associated chemicals) of an organism or group of organisms. Application of the various -omics in insect pathogen ecology will provide new insights into the complex interactions and new systems.

2.3.3 Advanced Techniques: qPCR, NGS, and the Arrival of the -Omics Era

qPCR can be employed for the species-specific identification and quantification of target pathogen species, either in the host or in the environment. This method provided an important route for studying diseases of invertebrates; previously, molecular methods could identify and compare pathogens, but with this technique it became possible to quantify pathogens from different sources. The elements comprising qPCR reactions are the same as for conventional PCR, but with the addition of selected linking fluorescent products (e.g., SYBR Green for double-stranded DNA or TaqMan probes designed to be specific for the primer-amplified sequence). During the repeated cycles, in both biochemistries, the amplification of the target area produces a fluorescence signal, which is proportional to the initial amount of the target. The signal will be exponentially amplified after every next cycle, and the critical point and advanced aspect of this technique is that the increment of fluorescence can be monitored over time with the progression of the reaction in the subsequent cycles of the PCR. Hence, it is possible to establish the moment when positive detection is recorded. To consider an amplification as positive, the fluorescence should pass a threshold, which is determined to be a signal significantly higher than the background signal. The cycle where the signal overpasses the threshold (Ct) will be an indicator of the initial amount, with high initial DNA quantities translated into low Ct values and vice versa. Running in parallel samples of known concentrations can facilitate absolute quantification of the target organisms. Bustin et al. (2009) suggested following the MIQE (Minimum Information for publication of Quantitative real-time PCR Experiments) guidelines for good practices and accurate presentation of methods and results obtained using qPCR techniques. Many of the invertebrate pathogen groups have had qPCR techniques developed for their identification and quantification. As an example, this method has been used for the evaluation of EPNs in the soil food web employing TaqMan probes. To date, >25 qPCR probes allow the study of EPN assemblages along with bacterial and fungal antagonists and r-selected, free-living nematode competitors in the field (Campos-Herrera et al., 2013a, 2015a). This powerful tool has also recently shown how a new citrus management regime alters the soil food web and severity of a pest–disease complex (Campos-Herrera et al., 2013b, 2014). The method has also been used to link the assemblage of EPN soil food webs in different eco-regions with different patterns of herbivory (Campos-Herrera et al., 2013c, 2016a), and has revealed interguild associations among different soil organisms (Campos-Herrera et al., 2012, 2015b). Besides the identification/quantification of species, qPCR can also be employed for the identification and quantification of the RNA expression of target genes, a process that requires the creation of complementary DNA (cDNA) transcripts from the target RNA. For example, Evans (2006) developed a qPCR array to evaluate the honey bee immunity response to target pathogens (bacteria, fungi, and protists). The specificity and small amount of material required in this method opens the possibilities of exploring the tissue-specificity of the pathogens and can therefore provide a better understanding of insect–pathogen interactions. Multiplexing (i.e., using two or more sets of primers/primers + probe in the same qPCR reaction) can be desirable if various species need to be screened in the same sample (e.g., to identify which species is the pathogen in the infection). This can decrease the cost in terms of reagents and time to provide the same final information. However, the multiplexing system may reduce the amplification of target organisms under certain circumstances, such as low initial quantities (Zijlstra and van Hoof, 2006). Hence, attention should be paid to the sensitivity of the system and, in particular, to situations where the detection of the pathogen could pose a quarantine risk for a food product.

Within the last few years, more sophisticated methodologies have arisen. For example, **digital qPCR** (dPCR) (Hindson et al., 2011) and **microfluidic systems** (also microfluidic droplet PCR) (Zhu et al., 2012; Chang et al., 2013) are novel methods that allow DNA amplification in a compartmentalized reaction in a minimal volume. These systems allow a substantial reduction in PCR processing time, reaction volume, and hence final cost (Hindson et al., 2011; Zhu et al., 2012; Chang et al., 2013). One step forward from these novel qPCR approaches is the combination of methods developed by Eastburn et al. (2015), termed microfluidic droplet enrichment for sequence analysis (MESA). This new system combines the isolates' genomic DNA fragments in microfluidic droplets and subsequent TaqMan PCR reactions, which allows the identifications of droplets containing a selected target sequence. Thereafter, these TaqMan-positive droplets are recovered via dielectrophoretic sorting, and are removed enzymatically prior to sequencing. Although these advanced techniques have not yet been applied to invertebrate pathogen studies, the rapid assimilation of other techniques by the field of invertebrate pathology suggests that new advances and techniques will employ these evolving methods in the near future.

Another advanced technique currently often used in invertebrate pathogen ecology is **NGS**, also called **high-throughput sequencing**. This technique is a step forward from traditional clone sequencing, because the NGS approach generates a full sequence derived from the areas flanked by selected primers without the necessity of pre-cloning and transforming steps. In general, NGS protocols are currently less time-consuming and cheaper than the traditional clone-sequencing approach (Pompanon et al., 2012; van der Heijden and Wagg, 2013). The available NGS platforms vary by approach and specifications; for example, 454 GS FLX (Life Sciences, Roche) employs pyrosequencing, HiSeq 2000 by Illumina (Solexa Technology) employs *de novo* synthesis, and AB SOLiDv4 (AgencourtTechnology) uses ligation in two-base coding (Liu et al., 2012). NGS is a technique widely used for **metagenomics** studies, when the focus is the detection of the maximum biodiversity of organisms within the same sample. Hence, it allows, for example, a comparison of community structure in different hosts or habitats. The fact that several companies provide NGS services allows for a significant reduction in reagent and equipment costs, which may contribute to the reduction in the final price per sample. Simultaneously, the size of the sequences generated (or "amplicons") is increasing to >400 bp, providing greater ability to screen for new species. In the NGS era, bioinformatic skills are required to deal with the thousands of sequences generated, and standardization of the units to molecular operational taxonomic units (MOTUs) is required. The NGS technique requires multiple alignments and classification, involving hierarchic associations within samples, to establish similar sequences belonging to the same species or target group (Blaxter et al., 2005), but in ecological scenarios, the taxa defined by MOTU methods need to be linked to the organisms and attributed to ecological characteristics up to a certain point.

Transcriptomics is the discipline that studies the transcriptome, which comprises all messenger RNA (mRNA) molecules in a single cell or a particular population of cells. The DNA microarray (biochip) allows the evaluation of expression level for various genes or the characterization of multiple regions of the genome at the same time, by using small quantities of DNA attached to a solid surface (Nature Biotechnology Editorial, 2006). There are various organism-specific transcriptome databases available that support the identification of particular genes differentially expressed in those cell populations. For example, with the use of whole-genome microarrays, comprising >20 000 genes, Wong et al. (2007) investigated the basis of the response of *Caenorhabditis elegans* to four pathogens, establishing how the responses to different infections are triggered. They highlighted both species-specific and common responses, and identified those immune defense genes involved, establishing the fundamental basis for unraveling the genetic invertebrate–pathogen interactions. There are also DNA microarrays available for the study of the diseases and immune responses of some mollusks produced as food (Gestal et al., 2008), and specifically of the interaction between viruses and invertebrates, such as WSSV infection in shrimp (He and Zhang, 2012). However, the emerging RNA-seq allows for the measurement of transcriptomes even for new model organisms with genomes that have not yet been sequenced (Wang et al., 2009), which greatly expands the possible uses for this technique. Xue et al. (2015) presented a recent example of advances in immune molecule detection and characterization for sea cucumbers, opening the avenue for these studies in the near future.

2.4 Traditional Versus Molecular Methods: Advantages and Limitations

Which are the best methods to investigate the ecology of invertebrate pathogens? As described previously, none of the traditional or molecular methods allow for the best solution for answering all questions when studying the ecology of invertebrate pathogens. Rather, a combination of traditional and molecular tools will allow a more comprehensive understanding of target systems, because in most cases, each method will provide different sides of the same story. From one side, traditional methods are usually very time-consuming and taxonomically specialist-dependent. From the other, molecular tools, once established and validated, allow anyone to identify diverse organisms (e.g., virus, bacteria, nematode, fungi, etc.) as long as they are proficient in the technique; this versatility could expand the scope of research questions being asked. However, molecular techniques are in general very expensive, as they require a variety of expensive reagents (i.e., buffers for PCR, DNA polymerase, TaqMan probes, etc.) and equipment of medium-high technology levels. Of greater importance to choosing which methods are appropriate is asking what are the ecological questions? By answering this, we can then search for the type of data we need to obtain, and immediately find the array of methods to be used. For example, if we are searching for new isolates in order to develop biological control agents, such as EPNs or EPF, the isolation of these new populations is critical, and traditional methods will be the first step. If the question involves multitrophic assemblages of pathogens in a system, molecular methods can be reliable.

We can illustrate the advantages and limitations in this dichotomy of traditional versus molecular methods by using EPNs as a model system in invertebrate pathogen ecology. EPN studies have largely been focused on the use of insect baits to detect the presence and abundance of EPNs in an area or experiment (Hominick, 2002). We can identify various biases in the detection and quantification of EPN presence by using this technique. The first obvious bias of this technique is the fact that not all EPN species can be isolated with the use of just one insect bait species. The most common insect employed is *G. mellonella*, but some nematode species are very host-specific, such as *S. scapterisci* and *Steinernema kushidai* (Adams and Nguyen, 2002). Interestingly, the screening for EPNs can be performed by using target insect pests as baits. This constitutes the first selection of the EPN that can be considered a candidate biocontrol agent, as well as establishing the *in situ* potential for control by native EPNs against this pest, as was illustrated for the pecan weevil (*Curculio caryae*; Shapiro-Ilan et al., 2003) and citrus root weevils (*Diaprepes abbreviatus*; Duncan et al., 2003). Additionally, attempts using two model sentinel insect species or baiting at two different temperatures, for example, have been employed to enhance the recovery of EPNs (Shapiro-Ilan et al., 2003; Mráček et al., 2005; Campos–Herrera and Gutiérrez, 2008). A drawback of the use of insects as baits is that the EPNs should be active against the sentinel host at the time that the baiting occurs and the EPNs should be able to outcompete a variety of other organisms also present in the same sample (Campos-Herrera et al., 2015b; Griffin, 2015; Lewis et al., 2015).

The inference of the numbers of EPNs in the soil sample can be established indirectly by recording the total insect cadavers produced per sample (Koppenhöfer et al., 1998). However, this method will still depend on the activity of EPNs in the soil (Griffin, 2015), with some limitations from the soil properties themselves. In this case, other techniques can be employed to extract the whole nematode community and identify the IJs: the only stage of the EPN cycle that is free-living in the soil (Adams and Nguyen, 2002). However, very few studies select this method to infer population density in the soil, since the required level of taxonomical expertise for identifying EPNs is very high (Hominick, 2002).

Recently, the development of species-specific primers and probes has established a system to quantify the number of IJs present in a soil sample. A pioneering study by Torr et al. (2007) illustrated the good relationship between the numbers of EPNs counted in soil samples extracted by traditional methods and the numbers retrieved when using qPCR methods: this study established the basis for the study of EPN soil food webs using qPCR. Several years later, Campos-Herrera et al. (2011a) provided a new set of primers and probes for studying EPN ecology in Florida citrus groves. The use of qPCR systems was more efficient in revealing sympatric distributions of EPN species than the traditional insect-baiting method. Interestingly, qPCR methods allowed the detection of low numbers of EPN in samples where the traditional insect bait failed (Campos-Herrera et al., 2016b). Moreover, this toolkit was coupled with the qPCR systems developed for other soil-inhabiting organisms, expanding our ability to investigate the EPN soil food web in various ecological settings (Campos-Herrera et al., 2013c, 2014, 2016a). For example, the soil can be simultaneously screened for the presence of nematophagous fungi (Pathak et al., 2012), ectoparasitic bacteria in the genus *Paenibacillus* (Campos-Herrera et al., 2011b), and other free-living nematodes that can compete with EPNs for insect cadavers (Campos-Herrera et al., 2012, 2015b).

However, despite the advantages offered by the qPCR tools, their use also biases the data in various ways. For example, qPCR will identify and quantify only those organisms for which the molecular toolkits are employed. Hence, it does not reveal the presence of those species not screened for, or species for which the qPCR was not developed (e.g., a new, undescribed species or a species that has not previously been evaluated using molecular tools). Therefore, in areas where EPN diversity is not well known, it is recommended to perform the traditional insect-bait method in order to have the opportunity to isolate new species or species that are not expected to occur in the area (Campos-Herrera et al., 2015a). In addition, the primers/probe combination is designed to be specific for a single species, but discovery of closely related species in the sampling area might increase the likelihood of cross-amplification, so the optimization of the method in a new system is strongly recommended (Campos-Herrera et al., 2015a). Finally, qPCR and insect-baiting may or may not be in agreement. For example, Jaffuel et al. (2016) evaluated the natural occurrence of EPNs in three crops under rotation (winter wheat, maize, and grass-clover ley) in a field site located in Therwil (northern Switzerland). In this study, EPN abundance measured with qPCR and insect baiting was the highest in winter wheat, with some discrepancies in its presence recorded for the other two crops (maize and grass-clover ley) using both techniques (Fig. 2.2). However, in a subsequent field experiment in southern Switzerland, the observations were completely opposite. In plots where *H. bacteriophora* was augmented, the qPCR method revealed high numbers of IJs, but no *H. bacteriophora* cadavers were produced when the soil was baited. Contrarily, when *S. feltiae* was augmented, qPCR detected lower numbers in the soil than for *H. bacteriophora*, but this species did produce host mortality upon soil baiting (Jaffuel et al. (2017)). Hence, more studies are needed comparing both techniques (qPCR and insect-bait) to establish the benefits and limitations of each. In summary, qPCR methods allow the quantification of EPN presence, detection of sympatric EPN occurrence, and identification of the presence of other members of the soil food web. However, the insect-bait technique can reveal the presence of new species and provide abundance and activity (ability to kill) data on the species that are present.

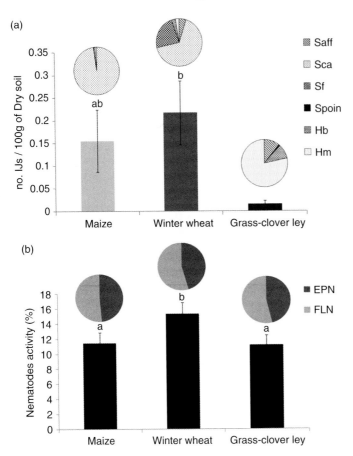

Fig. 2.2 Influence of crop type (maize, winter wheat, grass-clover ley) on the natural occurrence of entomopathogenic nematodes (EPNs). (a) qPCR quantification of the total number of EPN infective juveniles (IJs) of the species *Steinernema affine* (Saff), *S. carpocapsae* (Sca), *S. feltiae* (Sf), *S. poinari* (Spoin), *Heterorhabditis bacteriophora* (Hb), and *H. megidis* (Hm). (b) Activity recorded from *Galleria mellonella* baiting, as expressed by the average percentage of larval mortality producing nematode emergence (EPN + FLN). Data (means ± SEM) show the average value for the combined species-specific quantification of each of the organisms (EPN and free-living nematodes, FLNs) in the field trial, combining two different sampling periods in 2013. Lowercase letters indicate statistical differences. *Source:* Jaffuel et al. (2016). Reproduced with permission of Elsevier.

2.5 Advancing the Frontiers of Ecology using Pathogens and Diseases

Studies in invertebrate pathogen ecology can benefit from integration of other disciplines, such as geostatistics, multivariate statistics, and bioinformatics, which will expand the limits of the subject area. For example, through the integration of geostatistical approaches, we can illustrate the abundance of pathogens and related organisms on a regional scale (Fig. 2.3.). We will then be able to establish patterns of environmental variables or geomorphological limits for the distribution of the pathogens that can arise from this perspective. Spatial distribution analysis via SADIE (free software, SADIESshell 1.22,

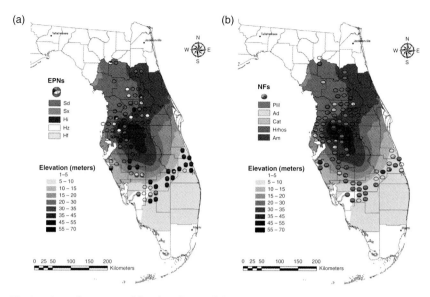

Fig. 2.3 Spatial patterns of the abundance of the most commonly encountered species of entomopathogenic nematodes (EPNs) and nematophagous fungi (NF) in natural areas across the Florida peninsula (*n* = 91 georeferenced localities), measured by species-specific primers and probes in qPCR assays. (a) EPNs: *Heterorhabditis floridensis* (Hf), *H. indica* (Hi), *H.zealandica* (Hz), *Steinernema diaprepesi* (Sd), and *S.* sp. *glaseri* group (Sx). (b) NF: *Purpureocillium lilacinus* (Plil), *Arthrobotrys dactyloides* (Ad), *A. musiformis* (Am), *Catenaria* sp. (Cat), and *Hirsutella rhossiliensis* (Hrhos). *Source:* Campos-Herrera et al. (2016a). Reproduced with permission of Elsevier. (*See color plate section for the color representation of this figure.*)

Kelvin F. Conrad and IACR–Rothamsted, 2001) can also add great potential to soil community studies (see review by Campos-Herrera et al., 2013a). SADIE facilitates the evaluation of spatial analysis based on a single index. To employ this tool, a correspondence between the space (with spatially referenced values: *x, y*) and the data of interest (e.g., abundance, incidence, presence, damage, etc.) must be provided, with the advantage that it is not necessary to have them in a systematic grid, with equidistant points conforming to the sites, and, hence, flexible designs can fit in the system (Perry, 1995, 1998). Further development of this technique allows the integration of maps and the evaluation of the association/dissociation of spatial patterns from two sets of data stemming from different organisms in the same area (Perry et al., 1999; Perry and Dixon, 2002). For example, analysis with SADIE was used to examine the spatial distribution of EPNs and members of the soil food web using qPCR tools in two habitats (citrus and natural areas) in the Florida peninsula (Campos-Herrera et al., 2012, 2013c, 2016a).

Bioinformatics facilitates advances in invertebrate pathogen ecology in many areas. For example, analyses of the whole-genome microarrays in response to pathogens have established the main genes involved in the infection process in the model nematode *C. elegans* (Wong et al., 2007). In reference to biodiversity studies, Sunagawa et al. (2009) employed advanced molecular tools to unravel the bacterial diversity associated with the white plague disease in the Caribbean coral *M. faveolata*, establishing the basis for further evaluation of causal agents and the mechanisms of the disease. Whole-genome sequencing of various invertebrate pathogens, such as EPNs (Bai et al., 2013) and EPF (Gao et al., 2011; Xiao et al., 2012; Agrawal et al., 2015), expands the potential of both basic and applied investigations using these resources. For example, advanced information on the genetic basis of virulence can be used in the screening and selection of pathogen populations that would be more tolerant to selected stresses or more active against certain pests. In addition, these tools can be employed to search for evolutionary links between the host and the pathogens using a molecular or functional approach.

2.6 Conclusion

To achieve a comprehensive understanding of invertebrate pathogen–host ecology, a combination of traditional methods from each discipline (virology, bacteriology, mycology, nematology, etc.) and other tools (molecular biology, geostatistics, modeling, etc.) is critical. By integrating various methodologies, we can unravel the molecular mechanisms behind diseases caused by invertebrate pathogens. In addition, we can place invertebrate–pathogen interactions into a regional context to determinate patterns of distribution linked to environmental and physical properties, such as those from the perspective of geology, geomorphology, and climate. As a result, we can model how changes in ecosystems can affect the advance of certain diseases in target invertebrates. Subsequently, this basic knowledge can be translated into applied studies toward improved microbial control, and can be used to solve critical problems related to the conservation of beneficial insects (e.g., honey bees decline) or other invertebrates directly related to food production (e.g., Crustacea and Mollusca diseases in fisheries). By exploring new systems and applying new tools and approaches, we envision a revolution in our comprehension of the interactions among invertebrates, their pathogens, and the environment in the near future.

Acknowledgments

The authors would like to thank Gerry Carner, Surendra Dara, Rubén Blanco-Pérez, and Stefan Jaronski for allowing us to include their pictures in this chapter. R. Campos-Herrera was financed as FCT Investigator Starting Grant IF/00552/2014/CP1234/CT0007 (Government of Portugal), hosted in Centro para os Recursos Biológicos e Alimentos Mediterrânicos (MeditBio), University of Algarve (Faro, Portugal).

References

Adams, B.J., Nguyen, K.B., 2002. Taxonomy and systematics, in: Gaugler, R. (ed.), Entomopathogenic Nematology. CABI, Wallingford, pp. 1–33.

Adams, B.J., Fodor, A., Koppenhöfer, H.S., Stackenbrandt, E., Stock, S.P., Klein, M.G., 2006. Biodiversity and systematic of nematode–bacterium entomopathogens. Biol. Control 38, 4–21.

Agrawal, Y., Khatri, I., Subramanian, S., Shenoy, B.D., 2015. Genome sequence, comparative analysis, and evolutionary insights into chitinases of entomopathogenic fungus *Hirsutella thompsonii*. Genome Biol. Evol. 7, 916–930.

Alcivar-Warren, A., Meehan-Meola, D., Park, S.W., Xu, Z., Delaney, M., Zuniga, G., 2007. ShrimpMap: a low-density, microsatellite-based linkage map of the pacific white leg shrimp, *Litopenaeus vannamei*: identification of sex-linked markers in linkage group 4. J. Shellfish Res. 26, 1259–1277.

Arzul, I., Renault, T., Thébault, A., Gérard, G., 2002. Detection of oyster herpes virus DNA and proteins in asymptomatic *Crassostrea gigas* adults. Virus Res. 84, 151–160.

Aquato, C.F., Noon, W.A., Begun, D.J., 1992. RFLP analysis using heterologous probes, in: Hoelzel, A.R. (ed.), Molecular Genetic Analysis of Populations: A Practical Approach. Oxford University Press, Oxford, pp. 115–158.

Bai, X., Adams, B.J., Ciche, T.A., Clifton, S., Gaugler, R., Kim, K., et al., 2013. A lover and a fighter: the genome sequence of an entomopathogenic nematode *Heterorhabditis bacteriophora*. PLoS ONE 8(7), e69618.

Barbosa-Solomieu, V., Miossec, L., Vazquez-Juarez, R., Ascencio-Valle, F., Renault, T., 2004. Diagnosis of Ostreid herpes virus 1 in fixed paraffin-embedded archival samples using PCR and *in situ* hybridisation. J. Virol. Meth. 119, 65–72.

Bedding, R.A., Akhurst, R.J., 1975. A simple technique for the detection of insect parasitic rhabditid nematodes in soil. Nematologica 21, 109–110.

Beebee, T.J.C., Rowe, G., 2008. An Introduction to Molecular Ecology, 2nd edn. Oxford University Press, New York.

Beilharz, V.C., Beilharz, D.G., Parbery, D.G., Swart, H.J., 1982. Dodine: a selective agent for certain soil fungi. Trans. Brit. Mycol. Soc. 79, 507–511.

Blaxter, M., Mann, J., Chapman, T., Thomas, F., Whitton, C., Floyd, R., Abebe, E., 2005. Defining operational taxonomic units using DNA barcode data. Philos. Trans. R. Soc. Lond. B Biol. Sci. 360, 1935–1943.

Bouladoux, N., Biligui, S., Desportes-Livage, I., 2003. A new monoclonal antibody enzyme-linked immunosorbent assay to measure in vitro multiplication of the microsporidium *Encephalitozoon intestinalis*. J. Microbiol. Methods 53, 377–385.

Bruford, M.W., Hanotte, O., Brookfield, J.F.Y., Burke, T., 1992. Single-locus and multilocus DNA fingerprinting, in: Hoelzel, A.R. (ed.), Molecular Genetic Analysis of Populations: A Practical Approach. Oxford University Press, Oxford, pp. 225–270.

Bustin, S.A., Benes, V., Garson, J.A., Hellemans, J., Hugget, J., Kubista, M., et al., 2009. The MIQE guidelines: minimum information for publication of quantitative real-time PCR experiments. Clin. Chem. 55, 611–622.

Campos, R.E., Maciá, A., García, J.J., 1993. Fluctuaciones estacionales de culícidos (Diptera) y sus enemigos naturales en zonas urbanas de los alrededores de La Plata, Provincia de Buenos Aires. Geotrópica 39, 55–66.

Campos–Herrera, R., Gutiérrez, C., 2008. Comparative study of entomopathogenic nematode isolation using *Galleria mellonella* (Pyralidae) and *Spodoptera littoralis* (Noctuidae) as baits. Biocontrol Sci. Techn. 18, 629–634.

Campos-Herrera, R., Johnson, E.G., Stuart, R.J., Graham, J.H., Duncan, L.W., 2011a. Long-term stability of entomopathogenic nematode spatial patterns in soil as measured by sentinel insects and real-time PCR assays. Ann. Appl. Biol. 158, 55–68.

Campos-Herrera, R., El-Borai, F.E., Stuart, R.J., Graham, J.H., Duncan, L.W., 2011b. Entomopathogenic nematodes, phoretic *Paenibacillus* spp., and the use of real time quantitative PCR to explore soil food webs in Florida citrus groves. J. Invertebr. Pathol. 108, 30–39.

Campos-Herrera, R., El–Borai, F.E., Duncan, L.W., 2012. Wide interguild relationships among entomopathogenic and free–living nematodes in soil as measured by real time qPCR. J. Invertebr. Pathol. 111, 126–135.

Campos-Herrera, R., Ali, J.G., Díaz, B.M., Duncan, L.W., 2013a. Analyzing spatial patterns linked to the ecology of herbivores and their natural enemies in the soil. Front. Plant Sci. 4, 1–18.

Campos-Herrera, R., Pathak, E., El-Borai, F.E., Schumann, A., Abd-Elgawad, M.M.M., Duncan, L.W. 2013b. New citriculture system suppresses native and augmented entomopathogenic nematodes. Biol. Control 66, 183–194.

Campos-Herrera, R., Pathak, E., El-Borai, F.E., Stuart, R.J., Gutiérrez, C., Rodríguez-Martín, J.A., Graham, et al., 2013c. Geospatial patterns of soil properties and the biological control potential of entomopathogenic nematodes in Florida citrus groves. Soil Biol. Biochem. 66, 163–174.

Campos-Herrera, R., El-Borai, F.E., Ebert, T.A., Schumann, A., Duncan, L.W., 2014. Management to control citrus greening alters the soil food web and severity of a pest-disease complex. Biol. Control 76, 41–51.

Campos-Herrera, R., Jaffuel, G., Chiriboga, X., Blanco-Pérez, R., Fesselet, M., Půža, V., et al., 2015a. Traditional and molecular detection methods reveal intense interguild competition and other multitrophic interactions associated with native entomopathogenic nematodes in Swiss tillage soils. Plant Soil 389, 237–255.

Campos-Herrera, R., Půža, V., Jaffuel, G., Blanco-Pérez, R., Čepulytė-Rakauskienė, R., Turlings, T.C.J., 2015b. Unraveling the intraguild competition between *Oscheius* spp. nematodes and entomopathogenic nematodes: implications for their natural distribution in Swiss agricultural soils. J. Invertebr. Pathol. 132, 216–227.

Campos-Herrera, R., El-Borai, F.E., Rodríguez Martín, J.A., Duncan, L.W., 2016a. Entomopathogenic nematode food web assemblages in Florida natural areas. Soil Biol. Biochem. 93, 105–114.

Campos-Herrera, R., Rodríguez Martín, J.A., Escuer, M., García-González, M.T., Duncan, L.W., Gutiérrez, C., 2016b. Entomopathogenic nematode food webs in an ancient, mining pollution gradient in Spain. Sci. Total Environ. 572, 312–323.

Carvalho, R.G., Maclean, N., Meyer, C.P., Guinez, R., Lindberg, D.R., 1991. Differentiation of aphid clones using DNA fingerprints from individual aphids. Proc. R. Soc. Lond. B 243, 109–114.

Chang, C.-M., Chang, W.-H., Wang, C.-H., Wang, J.-H., Maid, J.D., Lee, G.-B., 2013. Nucleic acid amplification using microfluidic systems. Lab. Chip. 13, 1225–1242.

Chase, A.R., Osborne, L.S., Ferguson, V.M., 1986. Selective isolation of the entomopathogenic fungi *Beauveria bassiana* and *Metarhizium anisopliae* from an artificial potting medium. Fla. Entomol. 69, 285–292.

Cochennec-Laureau, N., Reece, K.S., Berthe, F.C.J., Hine, P.M., 2003. *Mikrocytos roughleyi* taxonomic affiliation leads to the genus *Bonamia* (*Haplosporidia*). Dis. Aquat. Org. 54, 209–217.

Cooper, D., Cory, J.S., Myers, J.H., 2003. Hierarchical spatial structure of genetically variable nucleopolyhedroviruses infecting cyclic populations of western tent caterpillars. Mol. Ecol. 12, 881–890.

Cory, J.S., Evans, H.F., 2007. Viruses, in: Lacey, L.A., Kaya, H.K. (eds.), Field Manual of Techniques in Invertebrate Pathology: Application and Evaluation of Pathogens for Control of Insects and other Invertebrates, 2nd edn. Springer, Dordrecht, pp. 149–174.

Daniels, C., Breitbart, M., 2012. Bacterial communities associated with the ctenophores *Mnemiopsis leidyi* and *Beroe ovata*. FEMS Microbiol. Ecol. 82, 90–101.

Delalibera, I. Jr., 2009. Biological control of the cassava green mite in Africa with Brazilian isolates of the fungal pathogen *Neozygites tanajoae*, in: Hajek, A.E., Glare, T.R., O'Callaghan, M., (eds.), Use of Microbes for Control and Eradication of Invasive Arthropods. Springer, Dordrecht, pp. 259–269.

DeSantis, T.Z., Brodie, E.L., Moberg, J.P., Zubieta, I.X., Piceno, Y.M., Andersen, G.L., 2007. High–density universal 16S rRNA microarray analysis reveals broader diversity than typical clone library when sampling the environment. Microb. Ecol. 53, 371–383.

Duncan, L.W., Graham, J.H., Dunn, D.C., Zellers, J., McCoy, C.W., Nguyen, K., 2003. Incidence of endemic entomopathogenic nematodes following application of *Steinernema riobrave* for control of *Diaprepes abbreviatus*. J. Nematol. 35, 178–186.

Eastburn, D.J., Huang, Y., Pellegrino, M., Sciambi, A., Ptáček, L.J., Abate, A.R., 2015. Microfluidic droplet enrichment for targeted sequencing. Nucleic Acid Res. 1–8.

Eberle, K.E., Wennmann, J.T., Kleespies, R.G., Jehle, J.A., 2012. Basic techniques in insect virology, in: Lacey, L.A. (ed.), Manual of Techniques in Invertebrate Pathology, 2nd edn. Academic Press/Elsevier, San Diego, CA, pp. 15–74.

Evans, J.D., 2006. Beepath: an ordered quantitative-PCR array for exploring honeybee immunity and disease. J. Invertebr. Pathol. 93, 135–139.

Federici, B.A., Granados, R.R., Anthony, D.W., Hazard, E.I., 1974. An entomopoxvirus and nonoccluded virus-like particles in larvae of the chironomid. Goeldichironomus holoprasinus. J. Invertebr. Pathol. 23, 117–120.

Federici, B.A., Kramer, W.L., Mulla, M.S., 1976. A disease in *Chironomus frommeri* caused by a *Richettsiella*-like organism. Proc. Pap. Calif. Mosq. Control Assoc. 44, 123.

Fernandes, E.K.K., Keyser, C.A, Rangel, D.E.N., Foster, R.N., Roberts, D.E., 2010. CTC medium: a novel dodine-free selective medium for isolating entomopathogenic fungi, especially *Metarhizium acridum*, from soil. Biol. Control 54, 197–205.

Fuxa, J.R. 1987. Ecological methods, in: Fuxa, J.R., Tanada, Y. (eds.), Epizootiology of Insect Diseases. John Wiley & Sons, New York, pp. 23–41.

Fuxa, J.R, Tanada, Y., 1987. Epizootiology of Insect Diseases. John Wiley & Sons, New York.

Gao, Q., Jin, K., Ying, S.-H., Zhang, Y., Xiao, G., Shang, Y., et al., 2011. Genome sequencing and comparative transcriptomics of the model entomopathogenic fungi *Metarhizium anisopliae* and *M. acridum*. PLoS Genet. 7(1), e1001264.

Gardner, S.L., Stock, S.P., Kaya, H.K., 1994. A new species of *Heterorhabditis* from the Hawaiian Islands. J. Parasitol. 80, 100–106.

Gelernter, W.D., Federici, B.A., 1990. Virus epizootics in Californian populations of *Spodoptera exigua*: dominance of a single viral genotype. Biochem. System. Ecol. 18, 461–466.

Gestal, C., Roch, P., Renault, T., Pallavicini, A., Paillard, C., Novoa, B., et al., 2008. Study of diseases and the immune system of bivalves using molecular biology and genomics. Rev. Fish. Sci. 16(S1), 133–156.

Goldberg, L.H., Margalit, J., 1977. A bacterial spore demonstrating rapid larvicidal activity against *Anopheles sergenti*, *Uranotaenia unguiculata*, *Culex univittatus*, *Aedes aegypti* and *Culex pipiens*. Mosq. News 37, 355–358.

Goldstein, D.B., Schlotterer, C., 1999. Microsatellites: Evolution and Applications. Oxford University Press, Oxford.

Griffin, C.T., 2015. Behaviour and population dynamics of entomopathogenic nematodes following application, in: Campos-Herrera (ed.), Nematode Pathogenesis of Insects and Other Pests. Springer, Cham, pp. 139–163.

Grzywacz, D., 2016. Basic and applied research: baculovirus, in: Lacey, L.A. (ed.), Microbial Control of Insect and Mite Pests: From Theory to Practice. Academic Press, San Diego, CA, pp. 27–46.

Guy, P.L., Rath, A.C., 1990. Enzyme-linked immunosorbent assays (ELISA) to detect spore surface antigens of *Metarhizium anisopliae*. J. Invertebr. Pathol. 55, 435–436.

Hajek, A.E., 1997. Fungal and viral epizootics in gypsy moth (Lepidoptera: Lymantriidae) populations in central New York. Biol. Control 10, 58–68.

Hajek, A.E., 1999. Pathology and epizootiology of the Lepidoptera-specific mycopathogen *Entomophaga maimaiga*. Microbiol. Molecul. Biol. Rev. 63, 814–835.

Hajek, A.E., 2010. Larval behavior in *Lymantria dispar* increases the risk of fungal infection. Oecologia 126, 285–291.

Hajek, A.E., van Frankenhuyzen, K., 2016. Use of entomopathogens against forest pests, in: Lacey, L.A. (ed.), Microbial Control of Insect and Mite Pests: From Theory to Practice. Academic Press, San Diego, CA, pp. 313–330.

Hajek, A.E., Butt, T.M., Strelow, L.I., Gray, S.M., 1991. Detection of *Entomophaga maimaiga* (Zygomycetes: Entomophthorales) using enzyme-linked immunosorbent assay. J. Invertebr. Pathol. 58, 1–9.

Hajek, A.E., Olsen, C.H., Elkinton, J.S., 1999. Dynamics of airborne conidia of the gypsy moth (Lepidoptera: Lymantriidae) fungal pathogen *Entomophaga maimaiga* (Zygomycetes: Entomophthorales). Biol. Control 16, 111–117.

Hajek, A.E., Butler, L., Liebherr, J.K., Wheeler, M.M., 2000. Risk of infection by the fungal pathogen *Entomophaga maimaiga* among Lepidoptera on the forest floor. Environ. Entomol. 29, 645–650.

Hajek, A.E., Plymale, R.C., Reilly, J.R., 2012. Comparing two methods for quantifying soil-borne *Entomophaga maimaiga* resting spores. J. Invertebr. Pathol. 111, 193–195.

Hajek, A.E., Tobin, P.C., Haynes, K.J., 2015. Replacement of a dominant viral pathogen by a fungal pathogen does not alter the collapse of a regional forest insect outbreak. Oecologia 177, 785–797.

Hashmi, G., Glazer, I., Gaugler, R., 1996. Molecular comparison of entomopathogenic nematodes using randomly amplified polymorphic DNA (RAPD) markers. Fundam. App. Nematol. 19, 399–406.

He, Y., Zhang, X., 2012. Comprehensive characterization of viral miRNAs involved in white spot syndrome virus (WSSV) infection. RNA Biol. 9, 1019–1029.

Hindson, B., Ness, K.D., Masquelier, D.A., Belgrader, P., Heredia, N.J., Makarewiccz, A.J., et al., 2011 High-throughput droplet digital PCR system for absolute quantification of DNA copy number. Anal. Chem. 83, 8604–8610.

Hine, P.M., Cochennec-Laureau, N., Berthe, F.C.J., 2001. *Bonamia exitiosus* n. sp. (Haplosporidia) infecting flat oysters *Ostrea chilensis* (Philippi) in New Zealand. Dis. Aquat. Org. 47, 63–72.

Hoelzel, A.R., Green, A., 1992. Analysis of population-level variation by sequencing amplified DNA, in: Hoelzel, A.R. (ed.), Molecular Genetic Analysis of Populations: A Practical Approach. Oxford University Press, Oxford, pp. 159–187.

Hollingsworth, R.G., Steinkraus, D.G., McNewz, R.W., 1995. Sampling to predict fungal epizootics in cotton aphids (Homoptera: Aphididae). Environ. Entomol. 24, 1414–1421.

Hominick, W.M., 2002. Biogeography, in: Gaugler, R. (ed.), Entomopathogenic Nematology. CABI, Wallingford, pp. 115–143.

Hominick, W.M., Briscoe, B.R., del Pino, F.G., Heng, J., Hunt, D.J., Kozodoy, E., et al., 1997. Biosystematics of entomopathogenic nematodes: current status, protocols and definitions. J. Helminthol. 71, 271–298.

Hynes, H.B.N., 1970. The Ecology of Running Waters. University of Toronto Press, Toronto, ON.

Inglis, G.D., Enkerli, J., Goettel, M.S., 2012. Laboratory techniques used for entomopathogenic fungi: Hypocreales, in: Lacey, L.A. (ed.), Manual of Techniques in Invertebrate Pathology, 2nd edn. Academic Press/Elsevier, San Diego, CA, pp. 285–253.

Jaffuel, G., Mäder, P., Blanco-Pérez, R., Chiriboga, X., Fliessbach, A., Turlings, T.C.J., Campos-Herrera, R., 2016. Prevalence and activity of entomopathogenic nematodes and their antagonists in soils that are subject to different agricultural practices. Agri. Ecosys. Environ. 230, 329–340.

Jaffuel, G., Blanco-Pérez, R., Büchi, L., Mäder, P., Fliessbach, A., Charles, R, Degen, T., Turlings, T.C.J., Campos-Herrera, R., 2017. Effects of cover crops on the overwintering success of entomopathogenic nematodes and their antagonists. Applied Soil Ecology, 114, 62–73.

Kaunisto, K.M., Viitaniemi, H.M., Leder, E.H., Suhonen, J., 2013. Association between host's genetic diversity and parasite burden in damselflies. J. Evol. Biol. 26, 1784–1789.

Kimes, N.E., Johnson, W.R., Torralba, M., Nelson, K.E., Weil, E., Morris, P.J., 2013. The *Montastraea faveolata* microbiome: ecological and temporal influences on a Caribbean reef-building coral in decline. Environ. Microbiol. 15, 2082–2084.

Klein, M.G., 1992. Use of *Bacillus popilliae* in Japanese beetle control, in: Jackson, T.A., Glare, T.R. (eds.), Use of Pathogens in Scarab Pest Management. Intercept, Andover, pp. 179–189.

Klingen, I., Haukeland, S., 2006. The soil as a reservoir for natural enemies of pest insects and mites with emphasis on fungi and nematodes, in: Eilenberg, J., Hokkanen, H.M.T. (eds.), An Ecological and Societal Approach to Biological Control. Springer, Dordrecht, pp. 145–211.

Koppenhöfer, A.M., 2007. Nematodes, in: Lacey, L.A., Kaya, H.K. (eds.), Field Manual of Techniques in Invertebrate Pathology: Application and Evaluation of Pathogens for Control of Insects and other Invertebrates, 2nd edn. Springer, Dordrecht, pp. 249–264.

Koppenhöfer, A.M., Wu, S., 2016. Microbial control of insect pests of turfgrass, in: Lacey, L.A. (ed.), Microbial Control of Insect and Mite Pests: From Theory to Practice. Academic Press, San Diego, CA, pp. 331–341.

Koppenhöfer, A.M., Campbell, J.F., Kaya, H.K., Gaugler, R., 1998. Estimation of entomopathogenic nematodes population density in soil by correlation between bait inset mortality and nematode penetration. Fundam. App. Nematol. 21, 95–102.

Kurstak, E., 1962. Données sur l'épizootie bacterienne naturelle provoquée par un *Bacillus* du type *Bacillus thuringiensis* var. *alesti* sur *Ephestia kuhniella* Zeller. Entomophaga Mem. Hors Ser. 2, 245–247.

Lacey, L.A., Kaya, H.K., (eds.), 2007. Field Manual of Techniques in Invertebrate Pathology: Application and Evaluation of Pathogens for Control of Insects and other Invertebrates, 2nd edn. Springer, Dordrecht.

Lacey, L.A. (ed.), 2012. Manual of Techniques in Invertebrate Pathology, 2nd edition. Academic Press, San Diego, CA.

Lacey, L.A., Solter, L., 2012. Initial handling and diagnosis of diseased invertebrates, in: Lacey, L.A. (ed.), Manual of Techniques in Invertebrate Pathology, 2nd edn. Academic Press, San Diego, CA, pp. 1–14.

Lacey, L.A., Amaral, J.J., Coupland, J., Klein, M.G., Simões, A.M., 1995. Flight activity of *Popillia japonica* (Coleoptera: Scarabaeidae) after treatment with *Metarhizium anisopliae*. Biol. Control 5, 167–172.

Lacey, L.A., Thomson, D., Vincent, C., Arthurs, S.P., 2008. Codling moth granulovirus: a comprehensive review. Biocontrol Sci. Technol. 18, 639–663.

Laitinen, A.M., Otvos, I.S., Levin, D.B., 1996. Genotypic variation among wild isolates of Douglas-fir tussock moth (Lepidoptera: Lymantriidae) nuclear polyhedrosis virus. J. Econ. Entomol. 89, 640–647.

Lewis, E.E., Hazir, S., Hodson, A., Gulcu, B., 2015. Trophic relationships of entomopathogenic nematodes in agricultural habitats, in: Campos-Herrera, R. (ed.). Nematode Pathogenesis of Insects and Other Pests, Springer, Cham, pp. 139–163.

Liu, J., Berry, R.E., 1996. Phylogenetic analysis of the genus *Steinernema* by morphological characters and randomly amplified polymorphic DNA fragments. Fundam. App. Nematol. 19, 463–469.

Liu, L., Li, Y., Li, S., Hu, N., He, Y., Pong, R., et al., 2012. Comparison of next-generation sequencing systems. J. Biomed. Biotechnol. 251364.

Ma, M., Burkholder, J.K., Webb, R.E., Hsu, H.T., 1984. Plastic-bead ELISA: an inexpensive epidemiological tool for detecting gypsy moth (Lepidoptera: Lymantriidae) nuclear polyhedrosis virus. J. Econ. Entomol. 77, 537–540.

Majori, G., Ali, A., Donelli, G., Tangucci, F., Harkrider, R. J., 1986. The occurrence of a virus of the pox group in a field population of *Chironomus salinarius* Kieffer (Diptera: Chironomidae) in Italy. Fla. Entomol. 69, 418–421.

Maloy, A.P., Barber, B.J., Boettcher. K.J., 2005. A PCR-based diagnostic assay for the detection of *Roseovarius crassostreae* in *Crassostrea virginica* affected by juvenile oyster disease (JOD). Dis. Aquat. Org. 67, 155–162.

Martins, A.S.P., 1988. Fungos Entomopathogénicos como Agentes de Controlo Biológico e Perspectivas de Aplicação nos Açores. Master's Thesis, Universidade dos Açores, Ponta Delgada, Portugal.

Martins, A.,S.P., Paiva, M.R., Simões, N., 1988. Japanese beetle: monitoring in the Azores with semiochemicals. Ecol. Bull. 39, 101–103.

May, B., 1992. Starch gel electrophoresis of allozymes, in: Hoelzel, A.R. (ed.), Molecular Genetic Analysis of Populations: A Practical Approach. Oxford University Press, Oxford, pp. 1–27.

Merritt, R.W., Cummins, C.W., 1996. An Introduction to the Aquatic Insects of North America, 3rd edn. Kendall/Hunt, Dubuque, IA.

Meyling, N.V., Eilenberg, J., 2007. Ecology of the entomopathogenic fungi *Beauveria bassiana* and *Metarhizium anisopliae* in temperate agroecosystems: potential for conservation biological control. Biol. Control 43, 145–155.

Micieli, M.V., García, J.J., 1999. Estudios epizootiológicos de *Strelkovimermis spiculatus* Poinar y Camino, 1986 (Nematoda, Mermithidae) en una población natural de *Aedes albifasciatus* Macquart (Diptera, Culicidae) en la Argentina. Misc. Zool. 22.2, 31–37.

Mitchell, S.E., Read, A.F., Little, T.J., 2004. The effect of a pathogen epidemic on the genetic structure and reproductive strategy of the crustacean *Daphnia magna*. Ecol. Lett. 7, 848–858.

Morin, P.A., Saiz, R., Monjazeb, A., 1999. High-throughput single nucleotide polymorphism genotyping by fluorescent 5′ exonuclease assay. Biotechniques 27, 538–541.

Mráček, Z., Bečvář, S., Kindlmann, P., Jersákova, J., 2005. Habitat preference for entomopathogenic nematodes, their insect hosts and new faunistic records for the Czech Republic. Biol. Control 34, 27–37.

Mueller, U.G., Wolfenbarger, L.L., 1999. AFLP genotyping and fingerprinting. Trends Ecol. Evol. 14, 389–394.

Mullis, K.B., 1990. The unusual origin of the polymerase chain reaction. Sci. Am. 262, 36–43.

Mullis, K.B., Faloona, F.A., 1987. Specific synthesis of DNA *in vitro* via polymerase catalyzed chain reaction. Methods Enzymol. 155, 335–350.

Nature Biotechnology Editorial, 2006. Making the most of microarrays. Nat. Biotechnol. 24, 1039.

Nunan, N., Daniell, T.J., Singh, B.K., Papert, A., Mcnicol, J.W., Prosser, J.I., 2005. Links between plant and rhizoplane bacterial communities in grassland soils, characterized using molecular techniques. Appl. Environ. Microbiol. 71, 6784–6792.

Onstad, D.W., Fuxa, J.R., Humber, R.A., Oestergaard, J., Shapiro-Ilan, D.I., Gouli, V.V., et al., 2006. Abridged Glossary of Terms Used in Invertebrate Pathology, 3rd edn. Society for Invertebrate Pathology. Available from: http://www.sipweb.org/resources/glossary.html (accessed May 8, 2017).

Ormond, E.L., Thomas, A.P.M., Pugh, P.J.A., Pell, J.K., Roy, H.E., 2010. Fungal pathogen in time and space: the population dynamics of *Beauveria bassiana* in a conifer forest. FEMS Microbiol. Ecol. 74, 146–154.

Pathak, E., El-Borai, F.E., Campos-Herrera, R., Johnson, E.G., Stuart, R.J., Graham, J.H., Duncan, L.W., 2012. Use of real-time PCR to discriminate parasitic and saprophagous behaviour by nematophagous fungi. Fungal Biol. 116, 563–573.

Pell, J.K., Hannam, J.J., Steinkraus, D.C., 2010. Conservation biological control using fungal entomopathogens. BioControl 55, 187–198.

Perry, J.N., 1995. Spatial analysis by distance indices. J. Anim. Ecol. 64, 303–314.

Perry, J.N., 1998. Measures of spatial pattern for counts. Ecology 79, 1008–1017.

Perry, J.N., Dixon, P.M., 2002. A new method to measure spatial association for ecological count data. Ecoscience 9, 133–141.

Perry, J.N., Winder, L., Holland, J.M., Alston, R.D., 1999. Red–blue plots for clusters in count data. Ecol. Lett. 2, 106–113.

Platzer, E.G., 2007. Mermithid nematodes. Am. Mosq. Control Assoc. Bull.7, 58–64.

Pompanon, F., Deagle, B.E., Symondson, W.O.C., Brown, D.S., Jarman, S.N., Taberlet, P., 2012. Who is eating what: diet assessment using next generation sequencing. Mol. Ecol. 21, 1931–1950.

Renault, T., Le Deuff, R.M., Chollet, B., Cochennec, N., Gérard, A., 2000. Concomitant herpes-like virus infections in hatchery-reared larvae and nursery-cultured spat *Crassostrea giga*s, *Ostrea eduli*s. Dis. Aquat. Org. 42, 173–183.

Sanger, F., Nicklen, S., Coulson, A.R., 1977. DNA sequencing with chain terminating inhibitors. Proc. Natl. Acad. Sci. U.S.A., 74, 5463–5467.

Segers, R., Butt, T.M., Carder, J.H., Keen, J.N., Kerry, B.R., Peberdy, J.F., 1999. The subtilisins of fungal pathogens of insects, nematodes and plants: distribution and variation. Mycol. Res. 103, 395–402.

Shapiro, D.I., Fuxa, J.R., Braymer, H.D., Pashley, D.P., 1991. DNA restriction polymorphism in wild isolates of *Spodoptera frugiperda* nuclear polyhedrosis virus. J. Invertebr. Pathol. 56, 96–158.

Shapiro-Ilan, D.I., Gardner, W.A., Fuxa, J.R., Word, B.W., Nguyen, K.B., Adams, B.J., et al., 2003. Survey of entomopathogenic nematodes and fungi endemic to pecan orchards of the Southeastern United States and their virulence to the pecan weevil (Coleoptera: Curculionidae). Environ. Entomol. 32, 187–195.

Shapiro-Ilan, D.I., Bruck, D.J., Lacey, L.A., 2012a. Principles of epizootiology and microbial control, in: Vega, F.E., Kaya, H.K. (eds.), Insect Pathology, 2nd edn. Academic Press, San Diego, CA, pp. 29–72.

Shapiro-Ilan, D.I., Gardner, W.A., Wells, L., Wood, B.W., 2012b. Cumulative impact of a clover cover crop on the persistence and efficacy of *Beauveria bassiana* in suppressing the pecan weevil (Coleoptera: Curculionidae). Environ. Entomol. 41, 298–307.

Skovmand, O., Kerwin, J., Lacey, L. A., 2007. Microbial control of mosquitoes and black flies, in: Lacey, L.A., Kaya, H.K. (eds.), Field Manual of Techniques in Invertebrate Pathology: Application and Evaluation of Pathogens for Control of Insects and other Invertebrates, 2nd edn. Springer, Dordrecht, pp. 735–750.

Smithies, O., 1955a. Grouped variation in the occurrence of new protein components in normal human serum. Nature 175, 307–308.

Smithies, O., 1955b. Zone electrophoresis in starch gels: group variation in the serum proteins of normal human adults. Biochem. J. 61, 629–641.

Southwood, T.R.E., Henderson, P.A., 2000. Ecological Methods. Blackwell Scientific, Oxford.

St. Leger, R., May, B., Allee, L.L., Frank, D.C., Staples, R.C., Roberts, D.W., 1992. Genetic differences in allozymes and in formation of infection structures among isolates of the entomopathogenic fungus *Metarhizium anisopliae*. J. Invertebr. Pathol. 60, 89–101.

Steinkraus, D.C., 2007. Documentation of naturally occurring pathogens and their impact in agroecosystems, in: Lacey, L.A., Kaya, H.K. (eds.), Field Manual of Techniques in Invertebrate Pathology: Application and Evaluation of Pathogens for Control of Insects and Other Invertebrates, 2nd edn. Springer, Dordrecht, pp. 267–281.

Steinkraus, D.C., Hollingsworth, R.G., Boys, G.O., 1996. Aerial spores of *Neozygites fresenii* (Entomophthorales: Neozygitaceae): density, periodicity, and potential role in cotton aphid (Homoptera: Aphididae) epizootics. Environ. Entomol. 25, 48–57.

Stock, S.P., Goodrich-Blair, H., 2012. Nematode parasites, pathogens and associates of insects and invertebrates of economic importance, in: Lacey, L.A., Kaya, H.K. (eds.), Field Manual of Techniques in Invertebrate Pathology: Application and Evaluation of Pathogens for Control of Insects and Other Invertebrates, 2nd edn. Springer, Dordrecht, pp. 373–426.

Stock, S.P., Vandenburg, J., Glazer, I., Boemare, N.E., (eds.), 2009. Insect Pathogens Molecular Approaches and Techniques. CABI, Wallingford.

Stoltz, D.B., Hilsenhoff, W.L., Stitch, H.F., 1968. A virus disease in *Chironomus plumosus*. J. Invertebr. Pathol. 12, 118–128.

Sunagawa, S., DeSantis, T.Z., Piceno, Y.M., Brodie, E.L., DeSlavo, M.K., Voolstra, C.R., et al., 2009. Bacterial diversity and white plague disease-associated community changes in the Caribbean coral *Montastraea faveolata*. ISME J. 3, 512–521.

Tanada, Y., 1963. Epizootiology of infectious diseases, in: Steinhaus, E.A. (ed.), Insect Pathology and Advanced Treatise, Vol. 2. Academic Press, San Diego, CA, pp. 423–475.

Tautz, D., 1989. Hypervariability of simple sequences as a general source for polymorphic DNA markers. Nucleic Acid Res. 17, 6463–6471.

Torr, P., Spiridonov, S.E., Heritage, S., Wilson, M.J., 2007. Habitat associations of two entomopathogenic nematodes: a quantitative study using real-time quantitative polymerase chain reactions. J. Anim. Ecol. 76, 238–245.

Undeen, A.H., Avery, S.W., 1987. The isolation of microsporidia and other pathogens from concentrated ditch water. J. Am. Mosq. Control Assoc. 3, 54–58.

van der Heijden, M.G.A., Wagg, C., 2013. Soil microbial diversity and agro-ecosystem functioning. Plant Soil 363, 1–5.

Vittum, P.J., Villani, M.G., Tashiro, H., 1999. Turfgrass Insects of the United States and Canada. Cornell University Press, Ithaca, NY.

Vos, P., Hogers, R., Bleeker, M., Reijans, M., van de Leet, T., Hornes, M., et al., 1995. AFLPs: a new technique for DNA fingerprinting. Nucleic Acids Res. 23, 4407–4414.

Wang, Z., Gerstein, M., Snyder, M., 2009. RNA-Seq: a revolutionary tool for transcriptomics. Nat. Rev. Genet. 10, 57–63.

Wilson, K.H., Wilson, W.J., Radosevich, J.L., DeSantis, T.Z., Viswanathan, V.S., Kuczmarski, T.A., 2002. High-density microarray of small-subunit ribosomal DNA probes. Appl. Environ. Microbiol. 68, 2535–2541.

Wong, D., Bazopoulou, D., Pujol, N., Tavernarakis, N., Ewbank, J.J., 2007. Genome-wide investigation reveals pathogen-specific and shared signatures in the response of *Caenorhabditis elegans* to infection. Genome Biol. 8, R194.

Woods, S.A., Elkinton, J.S., Nurray, K.D., Liebhold, A.M., Gould, S.J.R., Podgwaite, J.D., 1991. Transmission dynamics of a nuclear polyhedrosis virus and predicting mortality in gypsy moth (Lepidoptera: Lymantriidae) populations. J. Econ. Entomol. 84, 423–430.

Xiao, G., Ying, S.H., Zheng, P., Wang, Z.L., Zhang, S., Xie, X.Q., et al., 2012. Genomic perspectives on the evolution of fungal entomopathogenicity in *Beauveria bassiana*. Sci. Rep. 2, 483.

Xue, Z., Li, H., Wang, X., Li, X., Liu, Y., Sun, J., Liu, C., 2015. A review of the immune molecules in the sea cucumber. Fish Shellfish Immunol. 44, 1–11.

Yousten, A., Fretzs, B., Jelleys, A., 1985. Selective medium for mosquito–pathogenic strains of *Bacillus sphaericus*. Appl. Environ. Microbiol.49, 1532–1533.

Zhu, Z., Jenkins, G., Zhang, W., Zhang, M., Guan, Z., James Yang, C., 2012. Single-molecule emulsion PCR in microfluidic droplets. Anal. Bioanal. Chem. 403, 2127–2143.

Zhang, S.-M., Zeng, Y., Loker, E.S., 2008. Expression profiling and binding properties of fibrinogen–related proteins (FREPs), plasma proteins from the schistosome snail host *Biomphalaria glabrata*. Innate Immun. 14, 175–189.

Zijlstra, C., van Hoof, R.A., 2006. A multiplex real-time polymerase chain reaction (TaqMan) assay for the simultaneous detection of *Meloidogyne chiwoodi* and *M. fallax*. Phytopathology 96, 1255–1262.

Zimmermann, G., 1986. The "Galleria bait method" for detection of entomopathogenic fungi in soil. J. Appl. Entomol. 102, 213–215.

Section II

The Basics of Invertebrate Pathogen Ecology

3

The Pathogen Population

Leellen F. Solter[1] and James J. Becnel[2]

[1] *Illinois Natural History Survey, Prairie Research Institute, University of Illinois, Champaign, IL, USA*
[2] *USDA ARS CMAVE, Gainesville, FL, USA*

3.1 Introduction

Pathogens are broadly defined as microbial agents that cause disease (departure from health) in animal and plant hosts and in other microbes. In the context of invertebrate animal hosts, the pathogens include viruses, entomopathogenic nematodes (EPNs), bacteria (including bacterial symbionts of EPNs), fungi, microsporidia and a large heterogenous group of protists. Prions have not been isolated from invertebrate animals. Species-specific characteristics of pathogens, their populations, and their genetic diversity determine their infectivity, pathogenicity, virulence and transmissibility. However, properties of the host are as important to the final outcome of infection as the pathogen itself (Pirofski and Casadevall, 2012), and the biotic and abiotic environments also strongly influence interactions. This section examines individually the properties of pathogens (this chapter), hosts (Chapter 4), the abiotic environment (Chapter 5), and the biotic environments (Chapter 6) in order to better understand the mechanisms of pathogen–host interactions.

Although pathogens are harmful to their hosts, they are evolutionarily integral components of ecosystems (Leighton, 2003), and, like all other reproducing organisms, they are species that undergo selection to optimally reproduce and maintain genetic stability (Timms et al., 2010; Brilli et al., 2013). They must compete for resources with their hosts (Budischak et al., 2015; Cressler et al., 2014), as well as with parasites and microbes, including other pathogens, facultative pathogens and commensals that utilize a host or host population. This chapter focuses on the characteristics that determine pathogen "success," as measured by reproduction in the host(s) and transmission to new host individuals, and that contribute to epizootics and impacts on host populations.

3.2 Characteristics of Pathogens

Pathogens have evolved myriad strategies to compete, survive and persist in populations of their hosts. Each species possesses the characteristics necessary to achieve exposure to hosts, colonization of a host, access to appropriate tissues for reproduction,

Ecology of Invertebrate Diseases, First Edition. Edited by Ann E. Hajek and David I. Shapiro-Ilan.

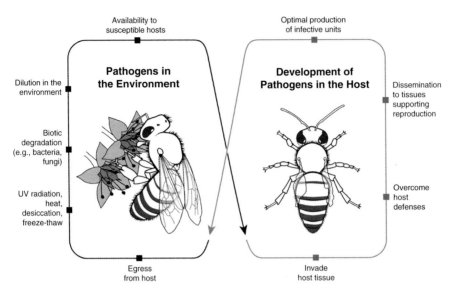

Fig. 3.1 Pathogen persistence in the host population depends on the survival of infective units in the environment, successful reproduction in individual hosts, and transmission to susceptible hosts. *Source*: Courtesy of D. Ruffatto and L. Solter.

avoidance of host immune responses, utilization of host resources for reproduction, egress from the host, and, in many cases, resistance to environmental insults before another host is encountered (Alberts et al., 2002, in part; Fig. 3.1). Some pathogen species obligately specialize on one host species, a few closely related host species, or on a primary and intermediate host species, while others are generalists with host ranges that vary widely in breadth.

3.2.1 Invasiveness and Infectivity

Pathogens are dispersed in the environment in the bodies of infected hosts or as infective stages that are released into the environment where they are available to susceptible hosts. Invasion of a potential host is necessary for infection but does not necessarily mean that an infection will occur. Before reproduction can begin, the pathogen must invade the host, overcome the host's immune defenses (Chapter 4), and access the appropriate tissues for reproduction, a species-specific characteristic known as tissue tropism (Katsuma et al., 2012). Tissue tropism can range in specificity from one type of tissue to multiple tissue types, while some pathogens utilize all host tissues to produce systemic infections. Some generalist pathogens may invade different tissues in different hosts, and some pathogens utilize different tissues at different times in their development within hosts.

The infectivity of a pathogen can be quantified in bioassays that measure the number of infective units needed to produce infection. The infective dosage (or pathogen concentration) producing infection in 50% of exposed hosts, the ID_{50} (IC_{50}), is calculated by regression analysis of a series of dosages. Infectivity bioassays can also provide a reasonable estimate of the minimum number of infective units needed to produce infection in an individual host, as well as the average number of units needed to produce infection in all exposed hosts (ID_{100}). The large differences in infectivity among pathogens have been predicted to be related to overcoming barriers to host entry and the biochemical mechanisms used to evade the host immune response; pathogens utilizing locally acting invasion "molecules" tend to be infective at low dosages, while those utilizing molecules that must diffuse into other tissues require higher dosages (Schmid-Hempel and Frank, 2007). Analysis of data for 43 human pathogens has validated this hypothesis (Legget et al., 2012).

3.2.1.1 Routes of Entry

Pathogen invasion involves entry into the potential host but requires that the pathogen becomes established in the host gut lumen or in tissues. The term "invasion" excludes passage through the gut in the food bolus or adherence to the exterior body structures with no subsequent infection occurring. Invasion mechanisms vary widely among pathogen groups. The following brief overview sets the stage for more detailed discussions in Section III.

Viruses are typically ingested by the host. Invasion of the host cells by the virions usually involves a specific gut environment (pH, enzymes) and recognition of receptor sites on the midgut epithelial cells. Breaching the peritrophic matrix of the insect midgut may be enhanced by viral proteins as well (Christian et al., 1993). Some viruses, including the nucleopolyhedroviruses (NPVs; Baculoviridae), cypoviruses (Reoviridae), and other virus families, have been shown to be transmitted from infected female to the offspring (Yue et al., 2007; Vilaplana et al., 2008; Antony, 2014; Section 3.2.5.2). Whether vertical transmission from infected females to offspring occurs via the embryo (transovarial transmission) or via egg surface contamination (transovum transmission) has been controversial, but molecular techniques are indicating that infection via embryos may be far more common for insect viruses than previously known (Chen et al., 2006; Virto et al., 2013).

Bacteria, many of which appear to be opportunistic pathogens in insects, are typically either ingested or invade the host body via wounds (Vallet-Gely et al., 2008). Bacterial epizootics do occur in invertebrates but appear to be less common than epizootics of viruses, fungi and microsporidia. Bacterial epizootics are frequently related to high host densities, such as European foulbrood and American foulbrood (caused by *Melissococcus plutonius* and *Paenibacillus larvae,* respectively) in honey bees (*Apis mellifera*) (Silvio et al., 2014; Müller et al., 2015). Although vertical transmission of opportunistic bacteria is unusual, *Spiroplasma* species and *Wolbachia* (rickettsial symbionts/pathogens that manipulate host fitness, reproduction and behavior) are transmitted from female to offspring (Vallet-Gely et al., 2008; Duplouy et al., 2015; Schuler et al., 2016).

Invasion of hosts by entomopathogenic fungi (EPF) is relatively unique because only a few taxa are known to be orally infective. The best known fungal clades that include invertebrate pathogens, the Hypocreales (Ascomycota) and Entomophthoromycotina, produce infective spores, usually called conidia, that attach to the cuticle of the host. A germ tube grows from the conidia and produces enzymes that allow the fungus to mechanically penetrate and grow through the cuticle to invade the insect body (Butt et al., 1995; Araújo and Hughes, 2016). One example of an orally infective fungus is *Ascosphaera apis* (Onygenales clade; Ascomycota), which produces chalkbrood in larval honey bees (*Apis* spp.) and bees in the family Megachilidae. The pathogen germinates in the closed gut of larvae (Vojvodic et al., 2011; Boomsma et al., 2014) and produces hyphae that penetrate the gut (Theantana and Chantawannakul, 2008), similar to penetration of the cuticle by other fungal species.

Unlike the fungi, to which they are related, most microsporidia infect their hosts by oral ingestion of spores. The infective spores germinate in the midgut lumen and a polar filament coiled within the spore everts and injects the contents of the spore into a host midgut cell. A significant number of microsporidian species, notably the type species *Nosema bombycis* infecting silk worms (*Bombyx mori*) and other closely related *Nosema* species, are also vertically transmitted (Solter, 2006).

Invasive mechanisms of protists vary, with some entering wounds or encysting on the host cuticle and creating openings, as reported for the parasitic ciliate *Lambornella clarki* (Washburn et al., 1988). Some taxa have infective stages that are ingested; for example, gregarine gametocysts (Detwiler and Janovy, 2008) and trypanosomatid promastigotes of *Crithidia mellificae* infecting honey bees (Schwartz et al., 2015).

Unlike the microbial pathogens, EPNs are metazoans, worms that actively forage for hosts using either ambush or cruiser behaviors (Chapter1). EPNs are included as pathogens of invertebrates because they carry species-specific symbiotic bacteria in the gut – *Photorhabdus* spp. in the nematode family Heterorhabditidae and *Xenorhabdus* spp. in the family Steinernematidae – that kill the host when released into the hemocoel. Different EPN species select specific body openings, including the mouth, anus, and spiracles, to invade the host, or they may bore through the cuticle or intersegmental membranes. Many EPN species invade the host through multiple routes (reviewed by Griffin et al., 2005). In addition to EPNs, facultatively parasitic nematodes and monoxenous nematodes without bacterial symbionts (e.g., mermithids) are sometimes considered with the entomopathogens (Chapter 11). These parasites also enter the host through the cuticle or via body openings.

3.2.1.2 Establishment of Infection and Tissue Tropism

A few pathogens, notably flagellate and eugregarine protists, become established in the gut lumen of the host when they are ingested and do not invade the host tissues. While they may anchor within the microvilli (Cook et al., 2001) or, in the case of eugregarines, briefly reside partially inside gut epithelial cells (Schrével and Desportes, 2013), these organisms usually do not cause severe pathologies unless populations build to extremely high densities (Bailey and Brooks, 1972; Lucarotti, 2000) or the host is stressed by other factors (Brown et al., 2000). Most pathogens, however, invade the hemocoel and/or cells of the host. Once entry to the host is gained, an infectious form must contact and invade the cells or tissues in which it can reproduce. For orally infective pathogens, the initial site of infection or access to target tissues is usually the midget epithelium. EPNs entering via the gut lumen bore through the peritrophic matrix and gut tissue directly to the hemocoel.

Systemic pathogens – those that invade and utilize most or all tissues in the body of the host – are found in all major taxa, including all species of EPF and bacteria, and the bacterial symbionts of EPNs. Many opportunistic bacteria produce toxins that destroy the midgut epithelium, opening the hemocoel to contamination (reviewed by Vallet-Gely et al., 2008). Bacteria and fungi initiate reproduction in the hemolymph and become systemic, often as host tissues are degraded by toxins, although toxins are not known to be involved in some systemic infections (e.g., the entomophthoralean fungi).

In contrast to systemic pathogens, many species of viruses, microsporidia, and some protists reproduce in specific host tissues, some only in the midgut cells. The mechanisms of dissemination to other target tissues are understood for some tissue-specific pathogen groups, such as alphabaculoviruses. These viruses "bud" through the basal membrane of the midgut cells, either as occlusion-derived virions (that have passed through midgut cells from the gut wall) or as budded virions produced in the midgut cell nuclei. Virions then move into the tracheal system, which is used as a conduit to infect other tissues (Engelhard et al., 1994; Passarelli, 2011; Harrison and Hoover, 2012). Many microsporidian species produce internally infective spores in the cytoplasm of the midgut epithelial cells that germinate through the gut basal membrane into the hemolymph (Johnson et al., 1997), and likely directly into tissues that are appressed to the gut (Solter and Maddox, 1998a). The mechanisms of tissue specificity or "tissue tropism" have been largely unknown, but recent molecular studies suggest that proteins produced by the host and/or the pathogen may determine which tissues support pathogen reproduction (Katsuma et al., 2012; Multeau et al. 2012). Tissues that are often specifically targeted include midgut epithelia (viruses, microsporidia), Malpighian tubules (microsporidia, neogregarines), fat body (viruses, neogregarines, microsporidia), nerve tissue (viruses, microsporidia), and epithelial tissue (viruses).

3.2.2 Pathogenicity, Virulence, and Pathogen Replication

The terms "virulence" and "pathogenicity" historically have been used interchangeably in the invertebrate pathology literature, but are not equivalent (also see Chapter 1). Shapiro-Ilan et al. (2005) reviewed these terms and gave precedence to definitions provided by Steinhaus and Martignoni (1970). Pathogenicity is a qualitative term representing an inherent quality; a microbe is either capable of producing disease in a host or it is not. However, a microbe may be pathogenic to its natural host(s) but not to a different, resistant species, or even to a different genotype of the same host species.

Virulence is the power to harm the host. Although it is generally understood that a "virulent pathogen" produces great harm and/or host death in a short time period, virulence is more appropriately used as a relative term to compare species or groups of pathogens, and is quantifiable. Virulence varies widely among species within all major insect pathogen groups, and a single pathogen species may vary in virulence when infecting different hosts or host populations. Some comparisons of virulence between pathogen species in the same major taxonomic groups are shown in Table 3.1.

For some pathogen species, very low initial dosages of infective units result in longer pathogen generation times and a generally slower progression of infection in the host, sometimes chronic, which may be the level of infection intensity that is most common for these species in the natural environment (e.g., Blaker et al., 2014). Adverse effects of chronic infections may be subtle, whereas higher dosages of the same pathogen (often used in laboratory studies to obtain infection in 100% of individuals) may result in rapid acute infection and high levels of mortality. However, some highly virulent pathogens for which lethality is characteristic can have an ID_{50} of only a few infective units and can cause nearly 100% host mortality within a short time period (Maddox, 1968; Opoku-Debrah et al., 2016).

In laboratory studies, virulence is usually evaluated by determining lethal time (LT) and/or mortality rates based on dosage (Onstad et al., 2006). The LT_{50} is the time

Table 3.1 Pathogens of low and high virulence, each compared to another species (or species complex) within each major taxon.

Pathogen species	Low virulence	High virulence
Bacteria[a]	*Rickettsiella melolonthae* (Legionellales) Chronic infection of beetle hosts, death in 4–6 mo. (Wille & Martignoni, 1952)	*Paenibacillus larvae* (Bacillaceae) Mortality in 3–6 days in honey bee workers (Behrens et al., 2010)
Fungi	*Hesperomyces* sp. (Laboulbeniales) Fungal thalli attach to beetle elytra, minor functional impairment, no mortality (Nalepa and Weir, 2007)	*Entomophaga maimaiga* (Entomophthorales) Epizootics in nature; IC_{100} as few as nine conidia; death in 4–6 days (Nielsen et al. 2005)
Microsporidia	*Nosema fumiferanae* (*Nosema* Clade) 30–50% prevalence has relatively little effect on host population; transovarially transmitted (Eveleigh et al. 2012)	*Vairimorpha disparis* (*Nosema* Clade) One spore fed to late-instar larvae = aborted pupation (Goertz and Hoch, 2008)
Nematodes	*Xenorhabdus bovienii* strain C S03 in *Steinernema weiseri* No virulence to host when injected; low virulence when in the nematode symbiont (Bisch et al. 2015)	*Xenorhabdus bovienii* strain SS 2004 in *Steinernema jollieti* Virulent alone when injected into host; virulent when in nematode symbiont (Bisch et al. 2015)
Viruses	Deformed wing virus (DWV) (Iflaviridae) Chronic and asymptomatic in honey bees in the absence of Varroa mites (de Miranda and Genersch, 2010)	Naturally occurring genotype DWV-B shortened adult bee lifespan by 53.5% compared to Genotype DWV-A at 38%; models show colony loss one annual cycle sooner (McMahon et al., 2016)

a) Most bacteria are facultative pathogens and strains of the same species may vary significantly in virulence. Most virulent bacteria produce highly insecticidal toxins.

post-inoculation at which 50% of the infected cohort exposed to a specified pathogen dosage have died. The LD_{50} (LD = lethal dosage) is the dosage that produces 50% mortality in a specified time period, and can vary depending on the age of the host at exposure. The LD_{50} of more virulent pathogens is expected to be lower than that of less virulent pathogens. However, the difficulty with using mortality to define virulence is that it does not take into account chronic pathogens that harm but do not necessarily kill the host (Casadevall and Pirofski, 2001) yet have significant impacts on host populations (e.g. Andreadis, 1984). In the case of chronic pathogens, determining differences in virulence becomes a much more subjective endeavor.

Reproduction of pathogens in the host is integral to pathogen-specific virulence and transmission to naïve hosts, as well as pathogen–host evolution. In general, the most virulent pathogens have the highest reproductive rates and are readily transmitted, while less virulent pathogens have lower reproductive rates and are less easily transmitted. Less virulent pathogens often require living hosts that spread infective units as they move in the environment and/or are vertically transmitted (Dimijian, 2000). However, the species-specific interactions that define the dynamics between pathogen and host vary widely for both virulent and chronic pathogens, as well as among and within the major pathogen groups. Details on the reproductive cycles for each pathogen group are covered in the relevant chapters.

3.2.2.1 Virulence Factors

Virulence is determined by factors that are contributed by the pathogen but which must be expressed in a susceptible host. Casadevall and Profski (1999, 2001) defined virulence factors as "microbial attributes that produce host damage." They included toxicity, aggressiveness, replication and transmission, adherence and attachment, antigenic variation, and immunologic reactions. Immunologic reactions are related to host immune response and factors possessed by the pathogen that suppress the response. Most of these attributes are interrelated; for example, aggressiveness, defined as the level of reproduction in the host, often depends on production of toxins and other metabolites to suppress the host immune system (Casadevall and Profski, 2001).

Toxins are the most obvious of the virulence factors. Proteins, peptides and secondary metabolites with different modes of action are produced by most entomopathogenic bacteria and many fungi. Crickmore et al. (2014) provide a current online list of ~900 crystal toxins (Cry or delta endotoxins), cytolytic proteins (Cyt), vegetative insecticidal proteins (Vip), and parasporins that are produced by the various strains of *Bacillus thuringiensis* (Bt) alone. Different bacterial toxins cause pathologies in invertebrates such as toxemia in the host gut lumen, paralyzed gut epithelia, invasion of the hemocoel, septicemia, and destruction of the internal organs (reviewed by Palma et al., 2014). Bt toxins kill the susceptible insect host, but Bt may or may not reproduce in the host, and it has been typed as an "environmental pathogen" because it can reproduce in the environment (Argôlo-Filho and Loguericio, 2014). Jurat-Fuentes and Jackson (2012) provide an excellent overview of Bt toxins and other soil-related bacteria that produce insecticidal toxins, including the honey bee pathogen *Paenibacillius larvae* (see also Fünfaus et al., 2013) and the grass grub pathogen *Serratia entomophila* (see also Hurst et al., 2000). *Vibrio tapetis*, one of a large group of aquatic environmental pathogens, has been shown to produce toxins in host clams (Borrego, 1996; reviewed by Paillard et al., 2004), as have *Vibrio* species that are toxic to or infect humans.

Toxins produced by the EPN-symbiotic bacteria *Photorhabdus* spp. and *Xenorhabdus* spp. include proteases, secondary metabolites and antimicrobials. Infective juvenile EPNs (IJs) release these bacteria into the hemolymph of the host, eliciting production by the host of antimicrobial proteins. The bacteria initially respond to the host by releasing proteases that inhibit the host's humoral and cellular immune responses, and then secrete toxins that destroy the midgut epithelium, causing cessation of host feeding, followed by host death. Additional toxins then liquefy the host tissues, which are used as a nutrient source for replication of the bacteria and the nematodes (Hinchliffe et al., 2010). The genome of *Photorhabdus luminescens* predicted more toxin-encoding genes than any other known bacterium (Duchaud et al., 2003).

Extracellular enzymes that degrade insect host cuticle are produced by some hypocrealean EPF. Proteases, chitinases and lipases solubilize the cuticle proteins and chitin, allowing the entry of the fungus through the layers of the cuticle and into the host hemolymph (Vega et al., 2012). Diverse secondary metabolites are produced by hypocrealean entomopathogens. The roles of these compounds in pathogenesis are unclear, although they have often been thought to potentially aid in disabling or harming hosts (Vega et al., 2012). These compounds may also suppress competing saprophytic micro-organisms after host death. Fewer secreted enzymes and toxins are known from species of entomophthoralean fungi, and the roles of the few compounds that are known remain unclear (De Fine Licht et al., 2016). Microsporidia are not known to produce toxins. The relatively high virulence of some species may be related to tissue tropism and disruption; species such as *Vairimorpha necatrix* and *Vairimorpha disparis*, pathogens of lepidopteran larvae, replicate in the fat body tissues where critical metabolic processes take place (Chapter 10).

3.2.2.2 Attenuation or Enhancement of Virulence

Virulence attenuation or enhancement is an outcome of interactions among pathogens, their hosts, and the environment (Casadeval and Pirofski, 2001). While the understanding of "optimal virulence" for pathogens is debated and difficult to evaluate, there is little doubt that various environmental and host factors serve to modulate the virulence of a population of pathogens. Most of what is known about virulence attenuation and enhancement is based on propagation in the laboratory of pathogens in natural hosts, laboratory hosts, cell culture, and axenic culture. Serial propagation using these culture methods frequently results in changes in virulence (Tanada and Fuxa, 1987). Laboratory experiments designed to compare virulence among isolates, evaluate attenuation, test conditions for pathogen survival, and increase pathogen virulence for use as biological control agents are elucidating the molecular and biochemical basis of virulence and the mechanisms used by pathogens to invade and reproduce in the host.

Attenuation in the laboratory or commercial culture commonly occurs when viruses, bacteria, EPNs and EPF are propagated on artificial media or in cell culture. Fungi were shown to lose enzymatic activity when grown on artificial media (Safavi, 2012), and fungal conidia had fewer pathogenicity-related genes when grown on media than in insect hosts (Shah et al., 2005). Baculoviruses in cell culture produced mutants with reduced production of polyhedra and increased production of the cell-to-cell infective budded virus (Slavicek et al., 1995). Conversely, serial passage of an opportunistic bacterium, *S. marcescens* strain db11, in its *Drosophila* host resulted in loss of virulence, while growth in the environment did not result in a reduction of virulence to the host (Mikonranta et al., 2015). Shapiro-Ilan and Raymond (2015) showed that virulence was reduced in heterorhabditid nematodes when IJs from large numbers of cadavers were combined to inoculate new hosts in serial passages, but was stable when low numbers of cadavers were combined. These results were predicted by kin-selection theory; more closely related nematodes continued to secrete virulence factors, while "cheaters" that did not secrete virulence factors were competitive when the population of nematodes was genetically more heterogenous. Attenuation in EPNs has been observed in both the nematodes and bacterial partners (Bai et al., 2005; Wang et al., 2007), and has also been attributed in part to inbreeding depression (Chaston et al., 2011).

The virulence of fungi, viruses, bacteria, and microsporidia can be attenuated when their host insects feed on specific host plants (reviewed by Cory and Hoover, 2006), and host plant effects on virulence have also been shown for EPNs (Hazir et al., 2016). Attenuation has been reported for microsporidia when they are propagated in hosts in which they do not naturally occur (Weiser, 1969), probably as a result of host immune responses or other physiological barriers to the development and maturation of infective spores (Solter and Maddox, 1998b).

Studies of enhanced virulence have also involved manipulation of the host–pathogen system. Chemicals and pathogen products (toxins, enzymes), as well as other pathogens, have been used to enhance infectivity, increase mortality rates and decrease LT. For example, Rangel et al. (2015) manipulated a variety of culture conditions and showed that fungal conidia germinated faster and were more virulent and stress-tolerant when conidia formed under conditions of environmental stress (minimal media, hypoxia, heat shock, osmotic stress, light), but under these conditions, fewer conidia were produced. Keyhani (2012) used transformation vectors to express mosquito-derived proteins, peptides, and a hormone in the entomopathogenic fungus *Beauveria bassiana* in order to disrupt the normal physiology of the host, rendering it more susceptible to the fungus. The molecules were selected with the criteria that they be critical to the host physiology, toxic when administered to the host, and have prior approval (previous studies) for use against insects. The LD_{50} was decreased by 5–10-fold for each of the four molecules used against *Manduca sexta*, *Galleria mellonella* and *Aedes aegypti*. Keyhani (2012) suggested that viruses, bacteria and EPNs could also be transformed for enhanced virulence.

3.2.3 Latency

Infection latency in invertebrate hosts has been defined in several ways, including: (i) survival in a host that does not cause recognizable symptoms but that is provokable to pathological activity (Bergold, 1958); (ii) a nonreproductive period of pathogens in host individuals or populations due to unknown factors (Tanada and Fuxa, 1987); and (iii) an initial period of infection in which the pathogen is developing but is not yet producing mature infective units (incubation time), which is also defined as the time from infection to infectiousness (Fine, 2003). In addition, latent infections have been included in the definition of covert infection, where the infection is asymptomatic and the pathogen is not horizontally transmitted (Sorrell et al., 2009). These definitions might be incorporated into a more general concept of latency as "inapparent infection" (Podgwaite and Mazzone, 1986). The pathogen may interact with the host during a latency period by actively suppressing the immune system or reproducing but not yet producing mature infectious stages; meanwhile, the host immune system may actively suppress the pathogen.

Tanada and Fuxa (1987) argued that incubation does not equal latency, and also that inapparent but inducible infections are not necessarily latent. However, although pathogens may well be inactive and noninfectious ("occult") in a host, it has been difficult to determine whether or not they are actively interacting with the host. Molecular techniques have provided new methodologies to detect latent infections in the strict sense (no known activity in the host). Murillo et al. (2011) used polymerase chain reaction (PCR) to detect infections by two latent baculoviruses, Spodoptera exigua multinucleopolyhedrovirus (SeMNPV) and Mamestra brassicae nucleopolyhedrovirus (MbNPV), in a laboratory colony of *Spodoptera exigua*. Likewise, latent infections of MbNPV were detected in all stages of colony-reared cabbage moth (*Mamestra brassicae*) (Hughes et al., 1993). Infections in both studies were inducible, becoming active and overt either spontaneously (Murillo et al, 2011) or upon exposure to other viruses (Hughes et al, 1993). In another study, latent white spot syndrome virus was detected in specific pathogen-free shrimp, and latency-related genes were predicted based on the presence of several virus proteins found to occur at higher levels in latent infections than in active ones (Khadijah et al., 2003).

Goertz et al. (2007), using the definition of latency that incorporates incubation time, calculated the average time from inoculation of the gypsy moth (*Lymantria dispar*) with the microsporidium *Nosema lymantriae* to shedding of mature infectious spores in the environment by living larvae. Although infectious spores were produced in the cells in as little as 4 days, infected larvae did not shed the spores until 7–8 days post-inoculation. This delay may have implications for the epizootiology of the pathogen; susceptible larvae in the host population with the potential for exposure to inoculum may have sufficient time to pupate, removing the possibility of infection.

Models predict that, in addition to impacts on host populations when diseases outbreak, sublethal, nondisseminating (latent) infections may be destabilizing and may have significant effects on host population dynamics, including reduction of the host population (Briggs and Godfray, 1996; Boots and Norman, 2000). Laboratory studies confirm these findings, particularly when stressors are added to the system (Boots and Begon, 1994; Sait et al., 1994).

3.2.4 Obligate, Opportunistic, and Facultative Pathogens

An obligate pathogen must reproduce in living host cells in order to complete its lifecycle. This definition is without controversy for viruses, microsporidia, and some protozoa. Although these pathogens often have infective stages that can survive in the environment for periods of time, they must infect a living host to reproduce. Many entomophthoralean fungi that are obligate pathogens in nature can be grown on rich media or in cell culture simulating insect hemolymph in the laboratory, although some species have never been grown *in vitro*.

Facultative pathogens are defined as pathogens that can reproduce in the environment (e.g., soil, phylloplane) but can also invade and reproduce in an animal or plant host, causing injury or disease. Jurat-Fuentes and Jackson (2012) differentiated facultative pathogens from opportunistic pathogens citing Bucher (1973), defining opportunistic pathogens as being capable of multiplying in an animal host but requiring a stress factor in the host to invade host tissues. Pathogens like the bacterium *S. marcescens* and the fungus *Aspergillus flavus* that are ubiquitous but rarely pathogenic in unstressed organisms are considered to be opportunistic.

The boundaries that separate facultative from obligate pathogens are not always clear. For example, *A. apis* is a specialist fungus found solely in association with honey bees and is often labeled as an obligate pathogen in the literature, but *A. apis* grows vigorously and can produce sexual ascospores in culture media (Takatori and Tanaka, 1982; Jenson et al., 2013). The bacterium *P. larvae* is the agent of American foulbrood and is also considered to be an obligate pathogen of honey bees. *Paenibacillus larvae* multiplies vegetatively in the gut lumen of larval bees and then invades the hemocoel, where it sporulates (Genersch, 2010). However, like *A. apis*, *P. larvae* can be cultured on specialized artificial media and will produce spores (Dingman and Stahly, 1983; Genersch et al., 2005). EPNs are interesting because they are obligate in nature (there is no evidence of their reproduction outside a host in nature) but they can be cultured *in vitro* in the laboratory or in industrial settings.

Bt is best known as an insect biopesticide and is nearly ubiquitous in the environment. Argôlo-Filho and Loguericio (2014) addressed the disparity between the viewpoints that Bt is an arthropod pathogen and that it is a saprophytic and opportunistic bacterium of the soil, water and phylloplane. While various isolates are highly infective to arthropods, Bt is easily cultured in artificial media, and Bt toxin genes are transformed into plants, which then produce pesticidal toxins. In addition, Bt is known to colonize, reproduce vegetatively and sporulate in the gut lumen of various vertebrate and invertebrate animals without causing disease. Argôlo-Filho and Loguericio (2014) put forward an argument that anthropogenic activities have resulted in restricted habitats for specific arthropod hosts, thereby creating selection for pathogenic Bt strains that are relatively host-specific and depend on the arthropod hosts for optimal sporulation. They suggest the term "environmental pathogen" to account for the enormous variety of niches and ecological interactions utilized by this highly variable and successful microbe.

3.2.5 Transmission

Once a pathogen reproduces in a host and mature infective stages are formed, barring the ability to indefinitely reproduce in the environment without a host, survival of the gene pool requires a mechanism for transmission to new hosts. Transmission of pathogens is "the driving force behind overall population dynamics" (Begon, 2009), and the mechanisms by which transmission occurs are broadly categorized as horizontal, vertical, or mechanical (see also Chapter 1). These are not mutually exclusive categories – many pathogens utilize a combination of transmission mechanisms.

3.2.5.1 Horizontal Transmission

Horizontal transmission occurs when a pathogen is passed from one individual host to another, excluding direct transmission from parent to offspring (Onstad et al, 2006). Horizontal transmission typically occurs within a host generation, but can also occur across generations. Exposure can be direct; for example, the microsporidium *Nosema apis* can be passed from an infected honey bee forager to a susceptible bee by trophallaxis (Naug and Gibbs, 2009). Pathogens can also be transmitted directly by cannibalism, as documented for *Paranosema locustae* in grasshoppers (Shi et al., 2009) and for a virus, Leptopilina boulardi filamentous virus, that is transmitted between larvae of the cannibalistic endoparasitoid *Leptopilina boulardi* within the bodies of superparasitized *Drosophila* hosts (Varaldi et al., 2012). Sadeh and Rosenheim (2016) and Sadeh et al. (2016) modeled cannibalism and disease, and concluded that cannibalism may also play an important role in the spread of otherwise vertically transmitted arthropod pathogens (see Section 3.2.5.2).

More commonly reported for terrestrial pathogens is indirect horizontal transmission of pathogens on contaminated surfaces, including food resources. Many, if not most, pathogen species have infectious stages that are environmentally resistant and can survive outside the host for periods of time that can, depending on the species, range from a few days to multiple years. Infectious stages may be disseminated into the environment in the air and water, in the feces of living infected hosts, in the decomposing carcasses of infected hosts, and potentially in the feces or regurgitated food of predators (e.g., Kring et al., 1988; Caceras et al., 2009; Goertz and Hoch, 2013a;

Gupta et al., 2013; Auld et al., 2014). In aquatic environments, invertebrates are exposed to pathogens in the water column (Hartikainen et al., 2014; Kough et al., 2015) and on underwater surfaces and sediments (e.g., Andreadis, 2002; Hewson et al., 2014).

Hosts are typically exposed to bacteria, viruses, microsporidia and some protists by ingesting pathogens as they feed, and numerous studies have elucidated means by which hosts come into contact with infectious pathogens. For example, Graystock et al. (2015) found that honey bees and bumble bees infected with several gut pathogens (*Nosema apis, Nosema ceranae, Nosema bombi, Crithidia bombi* and *Apicystis bombi*) contaminated floral resources in flight cages. The pathogens were then disseminated to other flowers during foraging by uninfected "vector" bees, which were also found to have ingested the pathogens.

Infective conidia produced by fungi in the Hypocreales and Entomophthorales emerge from the dead host and contaminate adjacent surfaces, but may also become airborne. *Lymantria dispar* larvae caged above the ground became infected with *Entomophaga maimaiga* (Weseloh and Andreadis, 1992; Hajek et al., 1999), and conidia were abundant in the air when conditions were sufficiently wet to induce conidiogenesis (Hajek et al., 1999). Conidia of EPF are found in soils and in water as well (reviewed by Vega et al., 2012), and fungi in endophytic phases within plant tissues, including *Metarhizium* and *Beauveria*, can be infective to invertebrates (reviewed by Barelli et al., 2016; Chapter 9).

Horizontal transmission can be transgenerational. The *L. dispar* baculovirus Lymantria dispar multinucleopolyhedrovirus (LdMNPV), like other lepidopteran baculoviruses, causes degradation of the host epithelium and subsequent spilling of virus polyhedra into the environment when the host dies. Murray and Elkinton (1989) showed that egg masses oviposited by female *L. dispar* moths collected from low-virus areas (thus, not likely vertically transmitted or surface-contaminated) and deployed in areas of known high virus prevalence became contaminated with virus. Virus prevalence in the progeny was significantly higher than in progeny from egg masses oviposited by females collected from high-prevalence sites (possibly infected or contaminated) and deployed in low-virus sites. Another example is the transmission of *Nosema pyrausta* from first-generation European corn borer (*Ostrinia nubilalis*) to second-generation larvae by fecal dissemination of infective spores in corn leaf whorls. Second-generation larvae also horizontally transmitted the pathogen to their generational cohort (Lewis et al., 2009).

Like pathogens of terrestrial animals, aquatic pathogens can be either horizontally or vertically transmitted. Coral reefs and other animal-rich aquatic environments are reported to be teaming with viruses, bacteria and other putative pathogens, and biofilms on underwater surfaces may draw pathogens into food webs (Shapiro et al., 2014). In water, pathogen lifecycles that include obligate intermediate hosts are commonly found. Examples are microsporidia in the genus *Amblyospora* that are maintained by horizontal transmission from mosquito larvae to copepod intermediate hosts. Spores produced by the copepods are infective to mosquito larvae (Andreadis, 2005).

3.2.5.2 Vertical Transmission

Pathogens are vertically transmitted when they are passed from an infected host directly to the offspring. Most research on vertical transmission has addressed transmission from infected female arthropods to offspring by "transovum" or "transovarial" means.

There is considerable confusion in the invertebrate pathology literature regarding these terms. Transovum transmission was defined by Onstad et al. (2006), after Steinhaus and Martignoni (1970), as transmission via the egg, and "transovarial" (or "transovarian") transmission was defined as a special case of transovum transmission whereby pathogens are transmitted within the tissues of the embryo. To simplify, and for consistency with Shapiro-Ilan et al. (2012), after Andreadis (1987), we retain this definition of transovarial transmission but define transovum transmission as transmission of pathogens by infected female hosts on the surface of the eggs.

In addition to female-to-offspring transmission, there are a few reports of venereal transmission by male arthropods to females and subsequently to offspring (de Miranda and Fries, 2008; Bolling et al., 2012; Virto et al., 2013), but direct male–to–offspring transmission via pathogens in the spermatophore is difficult to document. In laboratory studies, infected males may contaminate the environment where the eggs are oviposited, confounding a determination of transmission mechanisms (Solter et al., 1991).

Vertically transmitted pathogens usually express low virulence (Dunn and Smith, 2001) and cause chronic or covert infections that allow the host to survive through the post-infection developmental stages to the adult stage and reproduce (Solter, 2006; Myers and Cory, 2016). In some host–pathogen interactions, vertical transmission is part of a more complex lifecycle; for example, vertically transmitted microsporidia cause feminization in amphipods (Terry et al.- 2004) and male killing in *Aedes* mosquitoes (Andreadis, 2005). Female mosquitoes transovarially infected with *Amblyospora* spores survive to produce the next generation, in which infected males and a few females produce a different type of spore that is passed upon death to the copepod intermediate host (Andreadis, 2005).

Viruses, microsporidia and some bacteria are commonly vertically transmitted and, for a few species, no horizontal transmission mechanisms have been discovered. Two examples are *Nosema granulosis* in amphipods (Ironside et al., 2003) and *Nosema empoascae* in the potato leafhopper (*Empoasca fabae*) (Ni et al., 1997). With the exception of some mutualistic yeasts and *Coelomycidium simulii* in blackflies, vertical transmission has not been reported for fungi (Tanada and Kaya, 1993). However, in a twist on vertical transmission, *Coelomomyces* infects the ovaries of female mosquitoes, which results in castration of the mosquito and oviposition of fungal sporangia instead of eggs (Lucarotti and Andreadis, 1995).

Vertical transmission is common in aquatic invertebrates: amphipods infected with microsporidia transmit these pathogens to up to 100% of offspring (Terry et al., 2004), and many well-known and economically important viruses, bacteria and myxozoa are vertically transmitted in shrimp (viruses; Walker and Winton, 2010) and clams (*Vibrio* bacteria; Prado et al., 2014), as well as in bryozoans (myxozoa) that are intermediate hosts for salmon diseases (Okamura et al., 2011). Many of these pathogens are also horizontally transmitted.

3.2.5.3 Indirect Transmission

Indirect transmission (also "mechanical transmission" in invertebrate pathology literature), a form of horizontal transmission, occurs when pathogens are transferred from infected hosts to uninfected susceptible hosts by a vector that is not necessary to the lifecycle of the host and is not infected with the same pathogen (Foil and Gorham, 2000). For example, ovipositing hymenopteran parasitoids are capable of inoculating virus into susceptible larval hosts after ovipositing on an infected host (Young and Yearian, 1990). The efficiency of such transmission by parasitoids has not been shown to be high, but has been evaluated for only a few species (Hochberg, 1991). Genersch and Aubert (2010) reviewed the literature on the transmission of several virus diseases of honey bees by the parasitic Varroa mite. The mite apparently acts solely as a mechanical vector of acute bee paralysis virus, but as both a mechanical and a biological vector (the mite develops infection) of deformed wing virus (DWV). Mechanical transmission that does not involve inoculating the hemolymph of the host has also been demonstrated. Kistner et al. (2015) showed a 58% increase in mortality of grasshoppers in field experiments when ants vectored the conidia of the fungal pathogen *Entomophaga grylli*.

3.2.6 Genetic Variability and Potential for Coevolution with Hosts

Pathogens impose varying degrees of pressure on their hosts to evolve cellular and humoral immune responses, as well as evasive behaviors and morphological characteristics that resist pathogen invasion. In turn, host responses drive genetic changes in pathogens to overcome resistance. The ability to study coevolved traits at the genetic level has revolutionized studies of coevolution and is elucidating the importance of pathogen–host interactions in the web of life. The duration over which these evolutionary interactions occur can be as little as decades (Thompson, 2005). In addition to phylogenetic drivers of pathogen–host interactions, the biotic and abiotic environment and community ecology in which pathogens and their hosts interact are factors in coevolution and specificity (Birnbaum and Gerardo, 2016).

Pathogens and hosts both adapt to maximize reproduction and fitness (Roy et al., 2006), and pathogens must exploit their hosts without killing individuals before reproduction is optimized, even if pathogen progeny are produced after host death (Cory and Franklin, 2012). Pathogen-driven host extinction, although rare (McCallum, 2012), theoretically risks extinction of the pathogen population (Kapusinszky et al., 2015), but pathogen pressure also elicits genetic changes and variations in hosts that increase resistance and buffer host population declines (reviewed by Altizer and Pedersen, 2008).

3.2.6.1 Species and Strains

Differentiating a pathogen strain, or even a species, among populations of pathogens is often a difficult endeavor because of the inherent genetic variability within a species. Phylogenies based on specific genes may group together species and strains that have different capacities for infectivity, virulence, reproductive capacity and host specificity. Because pathogenicity and virulence depend on interactions with a specific host, pathogens that produce different responses may or may not be different strains (Ehrlich et al., 2008).

Identification of pathogen strains traditionally has relied on biological characteristics, including the host species from which the pathogen is isolated, phenotypic characteristics, pathogenicity, and relative virulence. Currently augmenting these still-important characteristics is an array of biochemical and genetic information that elucidates differences at the genome level. Techniques for obtaining these data include analysis of structural proteins, serological detection systems, nucleic acid hybridization, restriction endonuclease analysis, electrophoretic profiling of whole genomes, and sequencing of genome nucleic acids. These and other techniques are reviewed extensively for the different pathogen taxa in Stock et al. (2009).

Species and strains are defined somewhat differently for each pathogen group; for example, bacteria must have at least 70% DNA relatedness and possess at least one distinctive diagnostic phenotypic trait to be considered conspecific (Boemare and Tailleiz, 2009), while many microsporidian taxa appear to have few phenotypic apomorphies that stand up to molecular scrutiny, and isolates with up to 99% gene homology (ss-rDNA) can be designated as distinct species. Although there is high genetic diversity (polymorphisms) in the genomes of the few microsporidian species that have been evaluated, it is not yet known if these differences are represented in the same or different species and/or populations (Pombert et al., 2013).

Within genera of EPF, including *Beauveria*, *Metarhizium* and *Entomophaga*, there may be few phenotypic traits to distinguish strains or even species within a species complex. Therefore, host specificity and genetic characters have been critical to making taxonomic determinations at all levels. Genetic data are also critical for linking the anamorphic (asexual) and telomorpic (sexual) stages in the Hypocreales (Vega et al., 2012).

EPNs in the genera *Steinernema* and *Heterorhabditis* are paired with their bacterial symbionts *Xenorhabdus* spp. and *Photorhabdus* spp., respectively (Lewis and Clarke, 2012). These two nematode genera are not closely related, and the similarities in biology appear to be evolutionarily convergent (Adams et al., 1998). Both genera of symbiotic bacteria are in the family Enterobacteriaceae, but they have no common ancestor within the family (Boemare, 2002). The phylogenies of *Heterorhabditis* and its *Photorhabdus* symbiont are highly correlated (coevolved) (Maneesakorn et al., 2011), while those of *Steinernema* and *Xenorhabdus* are not (Lee and Stock, 2010). Tailliez et al. (2009) analyzed concatenated sequences from four genes to build a phylogeny for *Xenorhabdus* and *Photorhabdus*. *Xenorhabdus* isolates with less than 97% nucleotide identity were considered separate species, while *Photorhabdus* isolates with less than 97% identity were considered different subspecies.

Baculoviruses are divided into four clades, with relative relationships varying among clades. One NPV clade isolated from Lepidoptera and the clade containing granuloviruses (GVs) are highly divergent, while another lepidopteran NPV clade (with two subgroups) contains very closely related species with 97.0–99.9% identity (Jehle et al., 2006; Harrison and Hoover, 2012). Comparisons with less closely related NPVs show that genomic rearrangement is high (Harrison, 2009). A study of the 40 families of RNA viruses that infect insects found them to be taxonomically related but highly divergent, and a lack of understanding of their origin has resulted in difficulties in developing a phylogeny (Chen et al., 2012).

3.2.6.2 Host Specificity

Strict specificity of a pathogen to a single host is rare, but the availability of a limited number of potential host species in a given habitat may drive coevolution to higher levels of specificity, leading to new pathogen strains and host strains with changes in specific amino acids (Thompson, 2005). Pathogen specificity occurs as an adaptation to host resistance. Adaptations (e.g., changes in an enzyme that can overcome an immune response) can be highly specific (but not necessarily so) to the response of a particular host species to a pathogen (Obbard and Dudas, 2014).

The host ranges of even closely related pathogens can vary considerably and the gain or loss of genes and gene families appears to be involved in the specificity of pathogens. While loss of genes may reduce the number of host species a pathogen can utilize, acquisition of genes can increase the potential host range and allow pathogens to adapt to new environments (Wichadakul et al., 2015). An interesting comparison is between a microsporidian pathogen of vertebrates, *Encephalitozoon hellem*, and a congener, *Encephalitozoon romaleae* (a pathogen of the lubber grasshopper, *Romalea microptera*). These two species share a close evolutionary history: both genomes contain the same arthropod gene insert, which two other congeneric species – both vertebrate pathogens – do not (Pombert et al., 2012). *Encephalitozoon hellem* has been isolated from HIV-infected humans globally; from several bird taxa, including lovebirds, parrots, puffins, hummingbirds and an ostrich (Mathis et al., 2005); and from monkeys, chimpanzees and gorillas (Butel et al., 2015). *Encephalitozoon romaleae* has only been recovered from *R. microptera* (Lange et al., 2009). In addition to the arthropod gene acquired by a common ancestor of *E. hellem* and *E. romaleae*, a group of protein-coding genes for folate and purine metabolism from bacteria and animals was acquired that are not present in any other microsporidian genomes that have been sequenced. These genes are functional in *E. hellem*, which has a very broad vertebrate host range, but the folate-synthesizing genes have devolved to pseudogenes in the host specific *E. romaleae*. If *E. romaleae* moved to the grasshopper host from vertebrates after coevolving with *E. hellem* in insects, it may have lost folate synthesis in the context of a different host biochemistry (Pombert et al., 2012).

The role of host resistance in the evolution of a host-specific pathogen was demonstrated by Nagamine and Sako (2016) in studies of baculovirus infectivity in the silkworm (*Bombyx mori*). *Bombyx mori* is resistant to the lepidopteran virus Autographa californica multinucleopolyhedrovirus (AcMNPV), but not to the closely related Bombyx mori nucleopolyhedrovirus (BmNPV). BmNPV appears to take advantage of a mutation of the AcMNPV 143 helicase gene that is recognized as "foreign" by *B. mori* cells in its Ac-P143 form but, in its Bm-P143 form, leads to susceptibility. Thus, coevolution of AcMNPV with a single host, *B. mori*, that was apparently not highly susceptible to this virus led to speciation of the virus and specificity of BmNPV to the host.

3.3 Pathogen Effects on Host Development and Behavior

Both acutely virulent and chronic pathogens can impact their hosts by causing delayed or accelerated development, increased or decreased weight gain, decreased movement or increased irritability, wandering and climbing behaviors, reduced flight, mating and oviposition, and other effects. Hormonal and other physiological changes in hosts caused by pathogen manipulation or utilization of host tissues underpin many of these behaviors. For example, the microsporidium *N. ceranae* disrupts the balance of vitellogenin (Vg) and juvenile hormone (JH) that is necessary for the maturation of *A. mellifera* from the nurse/housekeeping stage to the foraging stage (temporal polyethism). Goblirsch et al. (2013) showed that Vg is suppressed and JH spikes in infected bees, resulting in early polyethism and a 9-day reduction in lifespan. In addition, seven different viruses were shown to accelerate polyethism in *A. mellifera* even more intensively than *N. ceranae* (Natsopoulou et al., 2016).

Summit disease (or treetop disease), the negative geotropism exhibited by hosts of some viral and fungal pathogens, is thought to enhance dissemination of a pathogen when the infected host dies in an elevated position, and has been documented to be genetically driven by some pathogen species. Hoover et al. (2011) determined that the *egt* gene of the *L. dispar* baculovirus, LdMNPV, encodes an enzyme that inactivates a molting hormone in *L. dispar*, resulting in climbing behavior. In tests using SeMNPV in *Spodoptera exigua*, Han et al. (2015) further suggested that the *egt* gene prolongs time to death, facilitating molting-related climbing behavior that would not occur earlier in the infection process. A particularly intriguing interaction producing behavior that favors the pathogen is that of the fungus *Ophiocordyceps unilateralis* infecting the arboreal ant, *Camponotus leonardi*. Infected ants develop tremors that cause them to fall to the forest floor, after which they move to vegetation on the north side of a tree and bite the rib on the underside of a leaf. The ants die attached to the leaf about 25 cm above the forest floor, where conditions are optimal for fungal growth and production of spores (Andersen et al., 2009). Host behaviors shown to be manipulated by pathogen genes are considered to be "extended phenotypes" of the pathogen that favor its own fitness (Dawkins, 1982).

While these behaviors are dramatic, myriad other, less obvious behaviors with unknown underlying mechanisms have been reported; for example, *Bombus* spp. infected with the trypanosomatid *Crithidia* sp. select floral resources that are high in iridoid glycosides, secondary metabolites that can reduce parasite loads (Richardson et al., 2015). This behavior also enhances pollination of the host plant, *Chelone glabra* (Richardson et al., 2016). Roy et al. (2006) provided an excellent review of the behaviors exhibited by insects infected with EPF, including changes in food consumption, changes in reproductive behaviors (e.g., increased attractiveness to mates and increased pheromone production and responses), and changes in social behaviors (e.g., grooming, defensive responses, and behavioral fever – a basking behavior that results in elevated body temperature and longer lifespan of infected hosts). Many of these behavioral changes are observed for other pathogen taxa, also.

3.4   Pathogen Populations

3.4.1   Density-Dependent Pathogens

General theory posits that pathogen density is positively correlated with host density, because increased numbers of hosts lead to increased interaction between infected and susceptible individuals and the production and dissemination of more infective units. Factors that determine the density of invertebrate pathogens in the environment and in the host population include infectivity, pathogen reproductive rate, pathogen survival capacity, and transmission of the pathogen, as well as the properties of the host population, with which the pathogen is inextricably tied (Tanada and Fuxa, 1987). In addition, spatial distribution of the pathogen in the environment, which includes distribution of infected hosts (Dwyer, 1991), can influence the local enzootic presence of disease and the potential for epizootics to occur (Shapiro-Ilan et al., 2012). Transmission is a key process in the relationship between host and pathogen, but it is the most difficult parameter to determine across host populations in different environments and of different densities (McCallum et al., 2001).

A high density of pathogens in the environment during and after an epizootic may increase the potential for host encounter and transmission, but contact by susceptible host individuals also depends on pathogen distribution, as well as host stage, behavior, and response to invasion. Although high host density theoretically results in host stress and increased susceptibility (May and Anderson, 1979), phenotypically plastic host resistance by Lepidoptera to viruses (Reeson et al., 1998) and by Coleoptera to fungal infections (Barnes and Siva-Jothy, 2000) has been shown to increase when host density and the risk of disease are high. However, Reilly and Hajek (2008) showed the opposite response with the virus LdMNPV in *L. dispar*. Classical examples of density-dependent pathogen–host interactions that have included virulence, transmission and host resistance are now understood to be complicated by the genetic diversity and plasticity of both the pathogen and the host. In addition, environmental factors, including the impacts of host self-regulation, such as starvation stress elicited by defoliation, may be involved (Bowers et al., 1993). In other words, the density-dependent host–pathogen cycle is "context-dependent" (Myers and Cory, 2016).

3.4.2 Density-Independent Pathogens

Pathogens that impact host populations regardless of host density may also be limiting (Myers and Cory, 2016). However, density-dependence and density-independence are relative: all obligate pathogens rely on sufficient numbers of available susceptible hosts for transmission and production of new infective units before they degrade in the environment (Weseloh, 2002). Like density-dependent pathogens, density-independent pathogens do not fall into a neat category of interactions with their hosts.

In order to be maintained in the host population, density-independent pathogens may survive by persisting in the environment. A well-studied example is the fungal pathogen *E. maimaiga*. Resting spores formed in late-stage host larvae can survive for years in leaf litter and soil and thus build up in the environment over many seasons (Liebhold et al., 2013). Epizootics in *L. dispar* can, therefore, occur even when host population levels are low – generally when the moisture level, an independent factor, is appropriate for spore germination and formation of infective conidia (Weseloh, 2002; Liebhold et al., 2013). However, under normal moisture levels, in very low host populations (as titers of resting spores in the soil decline; Hajek et al., 2004), larval survival is high, while at very high population levels, survival is low, indicating density dependence (Weseloh, 2002; Hajek et al. 2015; Hajek and van Nouhuys, 2016).

3.4.3 Pathogen Persistence in the Host Population

Pathogens are maintained in invertebrate populations within the bodies of their hosts and/or by producing infective stages that can tolerate conditions in the environment until encounter by a susceptible host. Strategies for persistence in the hosts include chronic infections that facilitate vertical transmission and long periods of pathogen release by living hosts (Duke et al., 2002; Lewis et al., 2009), persistence of low virulence stages in diapausing hosts (Siegel et al., 1988), alternative hosts (Jehle et al., 2006;

Roy et al., 2006), intermediate/alternate hosts (Andreadis, 2005), bodies of dead hosts (Brandenberg and Kennedy, 1981; Goertz and Hoch, 2008), mycorrhizal growth and reproduction in plants of the Hypocreales fungi (Bing and Lewis, 1991), and latency in host populations (see also Section 3.2.3).

3.4.3.1 Chronic Infections and Vertical Transmission
Viruses and microsporidia, as well as some protists, may produce low-intensity, chronic infections that allow not only survival of the host through subsequent life stages and potential vertical transmission, but also relatively long periods during which infective units are disseminated in the environment. Vertical transmission provides continuous host-to-host infections that may persist through multiple generations (Ni et al., 1997; Terry et al., 2004; Fries et al. 2006; Vilaplana et al., 2010), but high mortality rates in transovarially infected hosts are known for some pathogens, including microsporidia (van Frankenhuyzen et al., 2007; Lewis et al., 2009) and viruses (Barreau et al., 1997; Jiang et al., 2005; Gupta et al., 2010). Persistence in the host population of pathogens that have negative impacts on the host generally requires both vertical and horizontal transmission mechanisms (Agnew and Koella, 1999; Sadeh and Rosenheim, 2016).

3.4.3.2 Alternative and Alternate/Intermediate Hosts
All major pathogen groups include species that infect "alternative" hosts – those that are not required intermediate hosts in the life cycle. While these alternative hosts may simply represent a broad host range, any particular pathogen–host interaction is an evolutionary experiment, and a pathogen may or may not have a primary host (Casadevall, 2005). The ability to successfully infect multiple host species may guarantee that hosts are available even when populations of one or more hosts are low. Models predict that a modest amount of transmission among hosts of different species enhances pathogen persistence; however, low transmission among host species leads to epizootics within individual host species that resemble epizootics in the absence of multiple hosts, and high interspecific transmission is predicted to drive the most impacted species into (local) extinction (Dobson, 2004).

Alternate hosts are defined as two or more unrelated hosts required to complete the life cycle of a pathogen or parasite. The microsporidium *Amblyospora connecticus* is an example of a pathogen that uses alternate hosts to persist in its host populations (Andreadis, 1990). Two spore types are produced in the mosquito host *Ochlerotatus (Aedes) cantator*; one type is only infective to copepods and the other is vertically transmitted to the next *O. cantator* generation. Another spore type is produced in copepods and is only infective to the mosquitoes. In its salt marsh habitat, the pathogen overwinters in the copepod host, *Acanthocyclops vernalis*, and survives the summer vertically transmitted in *O. cantator*, when marsh pools alternate between dry and flooded conditions and the copepods are absent.

Several RNA viruses, including DWV, are infective to both *A. mellifera* and the Varroa mite parasite, but these viruses can also be transmitted among *A. mellifera* individuals and impact colony health in the absence of the mite (Highfield et al., 2009). The mite appears to increase the prevalence of DWV, but whether it increases persistence is unknown (Di Prisco et al., 2011). Many viruses and some bacteria, particularly wall-less bacteria, are vectored to plants or vertebrate animals. The ecology of these interactions is beyond the scope of this chapter, but it is interesting to note that vectors protect the pathogens while moving them between plant/vertebrate animal hosts, and the alternation between vectors and hosts appears to stabilize the genome of arboviruses (Moutailler et al., 2011).

3.4.3.3 Pathogen Survival in Cadavers and in Plant Tissues

Persistence of pathogens in the host population may be enhanced when mature infectious stages are protected from environmental harm within the bodies of dead hosts (Goertz and Hoch, 2008). Ecological theory suggests that cannibalism and intraspecific necrophagy (consuming dead conspecific individuals) does not promote pathogen persistence in host populations because only one susceptible individual typically consumes the infected host and, thus, the entire bolus of infective pathogen units (Rudolf and Antonovics, 2007). However, Sadeh and Rosenheim (2016) predicted that cannibalism in combination with vertical transmission can enhance pathogen spread. Necrophagy is well known in grasshoppers and appears to contribute to maintenance of the microsporidium *Paranosema locustae*, a pathogen that is also vertically and horizontally transmitted in the host populations (Shi et al., 2009, 2014). In addition, pathogens that are protected in the environment within the bodies of dead hosts for some period of time are not necessarily consumed as a bolus. Cadavers of gypsy moth larvae killed by *E. maimaiga* fall from trees and disintegrate, allowing release of the resting spores into the soil, where a reservoir can develop (Hajek et al., 1998).

Survival of EPNs can also be enhanced within host cadavers. Nematodes inside infected hosts are protected from desiccating (Koppenhöfer et al., 1997) and freezing (Lewis and Shapiro-Ilan, 2002) conditions. Furthermore, the symbiotic bacteria associated with EPNs produce compound(s) that deter various scavengers, including wasps, cockroaches, crickets, collembolans, and ants (Gulcu et al., 2012).

Plants may be a source of persistence for pathogens (Wang et al., 2005; Ortiz-Urquiza et al., 2015). *Beauveria bassiana* and *Metarhizium* spp. are both endophytic and entomopathogenic, infecting insect hosts that feed on their plants (Behie et al., 2015; Barelli et al., 2016; Chapter 9). *Metarhizium* appears to have evolved as a plant endophyte but uses different proteins by expressing different genes for invading plant tissues and insect cuticle (reviewed by Barelli et al., 2016). Interestingly, the death of insect hosts in proximity to the plants provides a source of nitrogen to the latter (Behie and Bidochka, 2014). Although there is no reproduction within plants, the undescribed bacterium "BEV" is deposited within plant tissues by the infected leafhopper (*Euscelidius variegatus*) host. Susceptible insect hosts feeding on the plant tissues become infected, demonstrating that plants may serve as reservoirs for insect pathogenic bacteria (Purcell et al., 1994). Entomopathogenic viruses may also be harbored, and even transported, in plants without invading plant tissues (Gildow and D'Arcy, 1988).

3.4.3.4 Latency in Host Populations

Pathogen latency may be an important component in the persistence of pathogen populations in host populations, allowing for vertical transmission of pathogens, particularly when environmental conditions and host densities are variable or hosts are migratory (Vilaplana et al., 2010). Using viral DNA and RNA detection, Vilaplana et al. (2010) found persistent NPV infections in field populations of symptomless African armyworm (*Spodoptera exempta*) larvae and adults, but very little overt disease in the host population. Laboratory colonies were symptomless, but adults were 100% infected (Vilaplana et al. 2008).

Burden et al. (2002) suggested that latent infections are most likely persistent suble-thal infections resulting from exposure to a low initial dosage of a viral pathogen. They demonstrated persistent infections of GV in the Indian meal moth (*Plodia interpunctella*), including in the offspring, after challenge with sublethal dosages of the pathogen. However, Vega Thurber et al. (2008) found herpes-like viral sequences in the genome of the coral *Porites compressa* and suggested that the virus was present and latent in the host population. The presence of the virus increased in field-collected, asymptomatic corals when they were stressed with pH changes, temperature, and increased nutrients. The role of environmental stress in eliciting overt disease and epizootics in inverte-brates with latent infections remains difficult to determine in field populations, due to environmental complexity and the interactions of multiple factors.

3.4.4 Persistence of Pathogen Stages in the Environment

Pathogens that are disseminated into the environment by infected hosts must survive until susceptible hosts are present in the appropriate stages; in temperate climates, this often means surviving the winter or summer host diapause period. Pathogens are vul-nerable to abiotic factors, including soil conditions and chemistry, sunlight, moisture, pH, and temperature extremes (see Chapter 5), as well as biotic factors, including pre-dation and degradation by bacteria and fungi (see Chapter 6; reviewed by Tanada and Fuxa, 1987; Shapiro-Ilan et al., 2012). Pathogen persistence in the environment has been difficult to measure because the introduction of freshly produced pathogens by infected hosts obfuscates the fate of pathogens released earlier, and infective forms of pathogens purified in the laboratory for evaluation may have significantly lower persis-tence than forms released by hosts into the natural habitat (Fuller et al., 2012).

Many invertebrate pathogens that are horizontally transmitted and exposed to the environment have persistent stages protected from environmental insults by morpho-logical structures. For example, entomophthorean resting spores (zygospores or azy-gospores) are enveloped in thick layers of chitin and protein–chitin complexes (Ohkawa and Aoki, 1980), allowing for survival over years in the soil and leaf litter (Weseloh and Andreadis, 2002). The spore walls of microsporidia consist of three distinct layers: an inner plasma membrane, a relatively thick endospore formed by a chitin–protein com-plex, and an outer exospore thought to be composed primarily of proteins (Cali and Takvorian, 2014). Terrestrial microsporidia generally survive in the environment over the winter in temperate climates, remaining infective for the next generation of hosts (Becnel and Andreadis, 1999). Third-stage infective EPN juveniles retain the second-instar sheath, which appears to aid in resistance to desiccation (Richert-Campbell and Gaugler, 1991). Virions of Baculoviridae, Reoviridae and Poxviridae (entomopox viruses) are embedded in occlusion bodies composed of protein matrices (Rohrmann, 2013) and may persist for years in protected environments such as soil (Grzywacz, 2016). Walled bacterial spores are well known for their ability to survive for many years in the environment, as well as for adapting to extreme and changeable conditions (Lambert and Kussell, 2014; Shen and Chou, 2016).

Ultraviolet (UV) radiation is the single most damaging abiotic factor – no pathogens are known to survive more than minimal exposure to direct sunlight (typically a few hours or less) – and survival in the environment generally requires protection from UV radiation under leaves and bark on plants, in leaf litter, in the soil, or in host feces and cadavers. Nevertheless, tolerances vary widely among pathogen groups, species, strains, and the conditions under which pathogens are developed. As an example of the tolerance range within a species, Rangel et al. (2015) grew the fungus *Metarhizium robertsi* on minimal media (nutrient stress), complex media + salts (osmotic stress), complex media and insect hosts. They recorded higher tolerance to UV-B radiation and heat for isolates that were nutrient-stressed and osmotically-stressed than for isolates grown on more complex media, including the host.

Temperature, humidity and moisture all affect pathogen survival and viability, with different combinations of environmental conditions that are either detrimental to or optimal for pathogen survival reported in an extensive literature, particularly for biological control agents (various pathogen groups reviewed in Vega and Kaya, 2012). Environmental conditions also serve as triggers for the infectivity of some pathogens, such as germination of fungi (e.g. Hajek et al., 2004; Filotas and Hajek, 2004) or the searching activity and establishment of EPNs (Yadev and Lalramliana, 2012) in the presence of moisture. Desiccation may be detrimental to pathogens, although many can tolerate dry conditions if protected from sunlight. While high humidity may be beneficial, it can also encourage growth of antagonists that degrade pathogens (Shapiro-Ilan et al., 2012).

Soil is an enormously complex environment, made up of biotic and abiotic factors with which invertebrate pathogens must contend, interacting primarily at the microenvironment scale (Klingen and Haukeland, 2006). While soil shields microbes from environmental extremes and UV radiation (Lacey et al., 2015) and serves as a reservoir for some pathogens, soil types, chemistry, moisture, and organic matter all figure into the persistence of pathogens (Koppenhöfer and Fuzy, 2006). Extensive studies on EPNs have shown that soil type, pH, and moisture affect the infectivity, virulence and persistence of EPNs, but effects vary by species. A soil that favors infectivity may not favor persistence, and vice versa (Shapiro et al., 2000; Koppenhöfer and Fuzy, 2006, 2007). Biotic factors in the soil may enhance persistence of pathogens; for example, plant roots and their associated rhizosphere may favor persistence of fungal pathogens, and pathogens may be dispersed or vectored by soil arthropods (reviewed by Shapiro-Ilan et al., 2012). Other factors, such as plant secondary compounds, may serve as pathogen inhibitors (Klingen et al., 2002), and antagonistic factors include viral phages and predators that feed on nematodes and microorganisms. In water, there is evidence that the impact of micropredators such as rotifers and ciliates can be significant in decreasing pathogen populations (Schmeller et al., 2014).

3.5 Dispersal and Spatial Distribution of Pathogens

Models developed by White et al. (2000) for the examination of insect–pathogen dynamics predicted four patterns of host–pathogen dispersal: (i) pathogen extinction followed by dispersal of uninfected hosts; (ii) host dispersal outstripping pathogen dispersal, with epizootics occurring well behind the wave of host population increase or spread; (iii) pathogen dispersal with the host, but slightly behind the wave of host population increase or spread, allowing a high-density host population only at a leading edge; and (iv) pathogen dispersal with the host, resulting in lower host populations across the habitat. Within the host population, the speed and efficacy of pathogen spread are strongly related to host movement and the rate of transmission to susceptible hosts (Dwyer and Elkinton, 1995), as well as to the spatial distribution of hosts in the environment (Dwyer, 1992) and the density and distribution of the pathogen itself (Carruthers et al., 1991). In addition to host dispersal and transmission, invertebrate pathogens spread within and among host populations by means of physical factors in the environment (wind, rain), parasites of infected hosts, and, occasionally, elimination of ingested infected hosts by predators (reviewed by Roy and Pell, 2000).

3.5.1 Physical Factors: Wind and Water Dispersal

In freshwater communities, pathogens released by living and dead infected hosts are moved in the water column, in currents, and on surfaces that are not stationary (plant, animal, and abiotic), where they are ingested by invertebrates. Some pathogens possess obvious adaptations: microsporidia infecting aquatic invertebrates may have mucoid coverings (Undeen and Becnel, 1992), keel-like structures (Hazard and Anthony, 1974), or tail-like structures (Adler et al., 2000) that allow them to remain suspended in the water column, where they are dispersed by water movement (Weiser, 1963). Iridescent virus prevalence in blackflies was correlated with increased covert infections occurring during high water flow in rivers (Williams, 1995), and "Marine snow" – aggregations in seawater of living and nonliving material in areas of high invertebrate densities – serves to move pathogens through the water column within large populations of benthic invertebrates (Lyons et al., 2005).

Rain splash may move fungal conidia and viruses to plant parts where insects feed (Bruck and Lewis, 2002). D'Amico and Elkinton (1995) placed dead, LdNPV-killed first-instar *L. dispar* larvae on branches and found higher NPV mortality in larvae on lower branches after artificial and natural rain events. Young and Yearian (1986) demonstrated that virus applied to the soil was splashed to lower soybean leaves by rainfall, and Fuxa and Richter (1996) showed that agricultural operations and rainfall appeared to have little or no effect on the presence of virus occlusion bodies on the soil surface, where they were available to contaminate soybean leaves.

Wind also plays an important role in the dispersal of some pathogens; notably, the conidia of EPF that are actively ejected from cadavers easily become airborne (Hajek et al., 1999; Steinkraus et al., 1999; Hemmati et al., 2001). Additionally, baculoviruses were reported to be dispersed in windblown dust (Olofsson, 1988), and microsporidia were recovered from windblown dust approximately 4300 km from the area of origin (Favet et al., 2013). While windborne microinvertebrate vectors of arboviruses (Jones et al., 1999; Bishop et al., 2000) and plant pathogens (Perotto et al., 2014) are better studied, long-distance transport of diseased invertebrates by wind is also likely. Ballooning infected invertebrates such as lepidopteran larvae (Dwyer and Elkinton, 1995) and aphids carried long distances on the wind (Feng et al., 2004) can disperse pathogens.

3.5.2 Biological Factors

In terms of targeted pathogen dispersal, the movement and distribution of infected hosts are the most important factors (Fuxa and Tanada, 1987). Infective stages of microsporidia and viruses, as well as some species of fungi and protists, are frequently released from living, mobile hosts. One of the most dramatic examples of a pathogen dispersed by a living host is *Massospora cicadina*, a fungal pathogen of periodical cicadas (*Magicicada* spp.). Resting spores germinate in the soil, infecting imagoes as they emerge. The infected cicadas produce conidia within their abdomens, which are dispersed during flight as segments of the abdomen fall off. Resting spores are produced within the abdomens of cicadas infected by conidia (the second cohort of infection during a season), and these are also dispersed from the posterior abdominal segments during flight (Roy et al., 2006). The newly produced resting spores are infective within 1 year, but can persist for the 13–17 years needed to infect the next generation of cicadas (Duke et al., 2002). More typically, pathogens that infect the midgut tissues and Malpighian tubules (e.g., viruses and microsporidia) are released in the feces, becoming dispersed as the host moves through the environment.

Most invertebrates in aquatic habitats, with the exception of flying adults of aquatic insects, appear to disperse passively. The presence of a marine oyster and its haplosporidian pathogen in both New Zealand and Chile suggests passive dispersal in ocean currents by rafting of adult oysters attached to floating objects (Foighil et al., 1999; Hill-Spanik et al., 2015). Additionally, models of the Panulirus argus lobster virus PaV1 suggested that spread of the pathogen, which is short-lived in open water, more likely occurred by passive dispersal of infected post-larvae in Caribbean water currents (Kough et al., 2015). Although dispersal in aquatic systems is difficult to study, more active dispersal in infected hosts to naïve populations may occur (Bilton et al., 2001).

In addition to dispersal by host(s), pathogens may be dispersed by movement of predators and parasites and by phoresy, such as dispersal of EPNs by earthworms (Campos-Herrera et al., 2006). However, although laboratory studies have shown the ability of some pathogens to survive the digestive tract of predators and retain infectivity (Section 3.2.5.1) or to be mechanically transmitted from infected host to susceptible host by parasites (Section 3.2.5.3), field data that demonstrate the importance of these mechanisms in dispersing pathogens within or among host populations are lacking.

Behaviors of infected invertebrates may also contribute to pathogen dispersal, including the negative geotropism exhibited by insects infected with baculoviruses and entomophthoralean fungi described in Section 3.3. Pathogen-induced changes in feeding location before death, changes in host movement, and host liquefaction may also benefit pathogen dispersal (Roy et al., 2006).

Invertebrate pathogens that have active dispersal mechanisms include motile bacteria (colony swarming), some protists (e.g., trypanosomes, gregarines), fungi (conidia expulsion), and EPNs. EPNs are unique among invertebrate pathogens in the extent to which active dispersal is part of their life history and impacts transmission to the host. EPN foraging is classified as a continuum from ambush to cruiser strategy (Chapter 11), and recently Fushing et al. (2008) and Shapiro-Ilan et al. (2014) reported EPN foraging success based on group movement and infection.

3.5.3 Spatial Distribution

While populations of density-dependent pathogens tend to be spatially associated with their hosts, pathogens that are not strongly host density dependent, and those that are spread in the environment by mobile infected hosts or, like some fungal conidia, are airborne (or are ubiquitous in the environment; Ormond et al., 2010), are likely to be somewhat evenly distributed, regardless of the patchiness of the host population. In certain situations, the pathogen load in the environment is sufficient to cause epizootics in low-density host populations. For example, persistent *E. maimaiga* resting spores may build up during widespread epizootics and continue to produce epizootics in pre-outbreak *L. dispar* populations for several years until low host densities and time reduce the pathogen below infection threshold (Solter and Hajek, 2009). However, dosages of more evenly distributed pathogens are generally less massive than those of pathogens that are clumped in a dead host (Dwyer, 1991). Dwyer (1991) demonstrated that LdMNPV, a density-dependent virus, is clumped in the environment in the bodies of dead hosts, and that late-instar *L. dispar* are more likely to become infected by LdMNPV than early instars, probably because of their higher mobility and feeding rates, which enhance the potential for pathogen encounter. Dwyer (1992) included spatial distributions of host and pathogen and host movement in a model of disease spread and predicted fairly accurately the rate of spread of a density-dependent NPV in the Douglas fir tussock moth (*Orygia pseudotsugata*).

Habitat heterogeneity may be directly involved in the patchiness of pathogen populations. EPNs tend to persist where soil moisture is adequate, which may result in patchy occurrence even in contiguous sites where other characteristics of the environment are similar (Lawrence et al., 2006). Raymond et al. (2010) reported that while the soil was a universal substrate for all Bt genotypes identified, only two of five clades, including dominant insect pathogens, were found on leaf material, and their presence was not dependent on the presence of insects.

The prevalence and patchiness of a pathogen population may change temporally. Grosholz (1993) found that the prevalence of an iridovirus varied inversely with the relative patchiness, inter-patch spacing, and within-patch density of the isopod host, all of which were seasonally dependent. Pathogen densities in the environment also vary with numbers of infected hosts in the environment, often related to host population density, season, and temperature. Temperature and humidity affect the production and release of infective entomophthoralean conidia with daily and seasonal patterns. Several studies have shown diel periodicity, with conidia produced in the early morning hours when humidity is high and sunlight is not present (e.g., Newman and Carner, 1974; Steinkraus et al., 1996), or during rainfall and other weather conditions (Reilly et al., 2014).

3.6 Pathogen Interactions

The survival of pathogens and pathogen populations requires that they interact, directly or indirectly, with other biotic factors (including those species with which they share hosts) and with abiotic factors in the environment. Biotic factors (Chapter 6) include other microbes utilizing the host and found in the environment, predators and parasites of the host, and sometimes other organisms in the host's microhabitat. Abiotic factors (Chapter 5) include temperature, moisture, solar radiation, and chemicals that are naturally occurring in the environment, as well as anthropogenic factors.

3.6.1 Interactions with other Biological Agents

Predators and parasites may directly compete with pathogens for hosts and, from the perspective of host population suppression, serve as synergists in mediating outbreaks

(Dwyer et al., 2004), but more subtle interactions also occur. The carabid beetle *Calosoma sycophanta* is a voracious predator of *L. dispar* larvae during outbreaks, but in laboratory tests it preferred larvae infected with the microsporidium *V. disparis* over healthy larvae and thus increased infection rates in test larvae when foraging (Goertz and Hoch, 2013a). However, predatory invertebrates may also avoid prey that is infected (Goertz and Hoch, 2013b), become infected by the same pathogen (Alma et al., 2010), or have higher predation success when hosts are infected (Johnson et al., 2006). Pathogen prevalence in prey may also be reduced by predation, most likely due to decreased prey density (Laws et al., 2009).

Similar to predators, parasitoids exhibit a range of behaviors, from preference to avoidance of infected hosts, but endoparasitoids generally die if they do not eclose before the host dies due to the infection. Hymenopteran parasitoids may become infected with pathogens (particularly microsporidia) of the host, resulting in reduced fecundity (Saito and Bjornson, 2013), increased larval development time, decreased adult longevity, and inability to distinguish infected from uninfected hosts (**Simões** et al., 2012; Oreste et al., 2016). Pathogens can also impair parasitoids without infecting them; emergence was lower, development slower, and mortality higher in parasitoids reared in GV-infected *Spodoptera litura* (Azam et al., 2016). Conversely, Hajek and van Nouhuys (2016) showed *L. dispar* larvae to be a shared resource for four parasitoid species and two lethal pathogens: a fungus and a virus. The parasitoids did not successfully co-infect with the fungus, but could survive co-infection with the virus.

Multiple pathogen species infect most arthropod populations, and co-infections of pathogens in a single host individual are not uncommon. In both individual hosts and host populations, pathogens may coexist with, compete with, synergize, or antagonize other pathogens and impact host mortality rates. Co-infections with two microsporidian species may result in competition between the pathogens, particularly if the same tissues are targeted, or two species may coexist if tissue tropism is different (Pilarska et al., 2006). The microsporidium *N. ceranae* appears to have outcompeted *N. apis* in most *A. mellifera* populations globally after host-switching from the Asian honey bee (*A. cerana*). Both pathogens infect the midgut tissues (Huang and Solter, 2013). Viral co-infections in honey bees may be exacerbated by the Varroa mite vector and appear to increase colony mortality (Pohorecka et al., 2014). Laboratory studies showed higher mortality in bees co-infected with virus and *N.* ceranae; *N. ceranae* infection increased the rate of virus production (Toplak et al., 2013). Co-infection with *E. maimaiga* and LdMNPV is common in *L. dispar* populations, with the expected co-infection rates due to chance at lower host densities but lower than expected rates at high host densities, suggesting that the fungus does suppress the virus at higher population levels (Malakar et al., 1999; Hajek and van Nouhuys, 2016).

Exposure to otherwise marginally infective pathogens may potentiate latent, naturally occurring pathogens, resulting in overt disease (Cory et al., 2000). In a study of co-occurring forest tent caterpillars (*Malacosoma disstria*) and western tent caterpillars (*Malacosoma californicum pluviale*) in British Columbia, Cooper et al. (2003) found that the virus Malacosoma californicum pluviale nucleopolyhedrovirus (McpINPV) was not highly infective to *M. disstria* but appeared to potentiate infection by a latent naturally occurring virus, Malacosoma disstria nucleopolyhedrovirus (MadiNPV). The authors suggested that the presence of the McpINPV virus in high-density *M. disstria* populations could initiate outbreaks of latent MadiNPV.

The role of microbial symbionts, particularly bacterial gut symbionts, has increasingly been shown to be important in antagonizing (Brownlie and Johnson, 2009) or synergizing (Chaston and Goodrich-Blair, 2010) invertebrate pathogens. Endosymbiotic bacteria, primarily *Wolbachia* or *Spiroplasma* in the hemolymph, may help regulate the host immune system to evade a response, but they have the effect of eliciting resistance to some pathogens and producing sensitivity to others (reviewed by Eleftherianos et al., 2013). Even ectosymbiotic fungi in the Laboulbeniales were shown to protect an ant host from infection by the entomopathogenic fungus *Metarhizium brunneum* (Konrad et al., 2015).

3.6.2 Interactions with Pesticides and Other Chemicals

The effects of chemicals on invertebrate pathogens vary widely depending on the species involved and the environment they inhabit. Laboratory experiments produce results that are often difficult to replicate under field conditions and may not accurately reflect the impacts (Lacey et al., 2015). For example, greenhouse studies of the effects of fungicides and herbicides showed little difference in the abundance of EPF, but organic fields supported a significantly higher abundance of the fungi than conventionally tilled ones (Clifton et al., 2015). Some studies suggest that use of chemical pesticides reduces epizootics of invertebrate pathogens applied as pesticides (Mascarin et al., 2016), while others have found synergistic pest control using combinations of fungal pathogens and synthetic chemicals (Oi et al., 2008; Zou et al., 2014), but there was no expectation of persistent microbial control in either case.

Of significant recent concern is the effect of chemicals on pathogens of *A. mellifera*. Collison (2015) reviewed studies showing a large number of different chemical residues, both agricultural chemicals and pesticides used to treat hives for parasites, in pollen, wax, and bee bread. Several studies associated pesticides, particularly neonicotinoids, with an increase in prevalence of *N. ceranae* (Pettis et al., 2012, 2013; Wu et al., 2012), and Huang et al. (2013) showed that diminishing levels of fumagillin, an antibiotic used to treat *A. mellifera* for nosema disease, can exacerbate *N. ceranae* infections. However, the hive environment is so complex that it is difficult to determine whether pesticides, parasites, or other pathogens are the primary factors causing colony failure, or if there are unidentified factors. A better understanding of the interactions of plants, animals, and microbial organisms under the impact of synthetic chemicals is needed in order to engage in knowledgeable decision-making regarding cropping systems and water supplies, as well as the protection of natural areas.

3.6.3 Enhancing Factors

Infections in invertebrate animals are sometimes enhanced in naturally occurring host populations by interacting natural enemies. Reports of epizootics and massive mortality resulting from two or more pathogens have been reported for lepidopteran insects (Zelinskaya, 1980; Hajek and van Nouhuys, 2016), and synergism has been demonstrated in numerous laboratory studies. As described in Section 3.6.1, the presence of a pathogen, even one that has marginal impacts on a host, may elicit acute infections of a latent pathogen species (Cooper et al., 2003), and some symbiotic microbes are also capable of synergizing specific pathogens.

Artificially produced enhancing factors for the manipulation of pathogens in biological control programs and remediation of disease in beneficial species have advanced considerably, from identification of chemicals such as the dyes Calcofluor White and Congo Red used to disrupt the insect peritrophic matrix (Sajjadian and Hosseininaveh, 2015), to studies of the molecular mechanisms of pathogenicity. Proteins have been identified, including granulin from GV, that can enhance NPV infections (Lepore et al., 1996). Molecular technologies are currently used to identify the genes and gene products in microbial organisms that, for example, serve as stressors to hosts, allowing changes in susceptibility to pathogens (Vega Thurber et al., 2009), and that are known enhancers but can potentially be modified to reduce viral infections in beneficial mass-reared insects (Li et al., 2015; Cao et al., 2016).

3.7 Conclusion

Research on invertebrate pathogens and their interactions with their hosts contributes to efforts to develop safe and effective biological control of food animals and crop pests, remediate diseases in edible and other mass-reared invertebrates, study and control zoonotic and arthropod-borne diseases, create models that inform the understanding of human, food and companion-animal diseases, and protect biodiversity. It also advances our understanding of the role of pathogens in plant and animal evolution and the ecological underpinnings of the web of life on earth. Invertebrate animals are incredibly numerous and diverse life forms, and many species can be inexpensively mass-reared to investigate infectious disease processes, both for pathogens that are phylogenetically related across vertebrate and invertebrate host taxa and for those that are unique to invertebrate hosts.

New research tools for discovery and analysis are providing exciting opportunities to elucidate the mechanisms by which pathogens exploit their hosts and interact with other microbes, as well as the roles they play in the ecology of specific habitats. Metagenomic data are now being analyzed more completely using high-throughput sequencing to produce entire genomes of interacting organisms. These studies have the potential to reveal how the ecosystem of pathogens within individual animals functions (Waldor et al., 2015). Identification of pathogen strains, the origins of invading pathogens, and transmission routes, as well as the evolution of populations, can also be studied with high-throughput sequencing (McAdam et al., 2014). Detection and abundance of pathogens in the environment or in host populations for ecological studies and biodiversity inventories is increasingly possible using eDNA (Huver et al., 2015) and quantitative polymerase chain reaction (qPCR) methods (Walker et al., 2015).

Mechanistic studies of pathogen metabolism, reproduction and interactions with hosts are burgeoning. Next-generation sequencing and transcriptome analysis allow researchers to evaluate pathogen gene expression in cells or tissues, the response of the host immune system, differences in gene expression between populations of pathogens and hosts, and responses to the environment (Wolf, 2013). Analytical tools, databasing, and data storage are proliferating and providing access to pathogen data (as well as the means to extract useful information from enormous data sets) at unprecedented rates.

These tools and methods, as well as sequencing methods and analyses, have become increasingly affordable and available to researchers in education, government, and industry laboratories for use in the study of the biology, ecology, evolution, diversity, and impacts of invertebrate pathogens. Understanding pathogen biology from the cellular level to the level of population dynamics provides the basis for understanding the ecology of disease in nature, as well as for manipulating or mitigating invertebrate diseases.

References

Adams, B.J., Burnell, A.M., Powers, T.O., 1998. A phylogenetic analysis of *Heterorhabditis* (Nemata: Rhabditidae) based on internal transcribed spacer 1 DNA sequence data. J. Nematol. 30, 22–39.

Adler, P.H., Becnel, J.J., Moser, B., 2000. Molecular characterization and taxonomy of a new species of Caudosporidae (Microsporidia) from black flies (Diptera: Simuliidae) with host-derived relationships of the North American caudosporids. J. Invertebr. Pathol. 75, 133–143. doi: 10.1006/jipa.1999.4902

Agnew, P., Koella, J.C., 1999. Constraints on the reproductive value of vertical transmission for a microsporidian parasite and its female-killing behavior. J. Animal Ecol. 68, 1010–1019. doi:10.1046/j.1365-2656.1999.00349.x

Alberts, B., Johnson, A., Lewis, J., Raff, M., Roberts, K., Walter, P., 2002. Molecular Biology of the Cell, 4th edn. New York: Garland Science.

Alma, C.R., Gillespie, D.R., Roitberg, B.D., Goettel, M.S., 2010. Threat of infection and threat-avoidance behavior in the predator *Dicyphus hesperus* feeding on whitefly nymphs infected with an entomopathogen. J. Insect Behav. 23, 90–99. doi:10.1007/s10905-009-9198-8

Altizer, S., Pedersen, A., 2008. Host-pathogen evolution, biodiversity and disease risk for natural populations, in: Carroll, S., Fox, C. (eds.), Conservation Biology: Evolution in Action. Oxford University Press, New York, New York, pp. 259–277.

Andersen, S.B., Gerritsma, S., Yusah, K.M., Mayntz, D., Hywel-Jones, N.L., Billen, J., et al., 2009. The life of a dead ant: the expression of an adaptive extended phenotype. Am. Nat., 174, 424–433. DOI:10.1086/603640

Andreadis, T.G., 1984. Epizootiology of *Nosema pyrausta* in field populations of the European corn borer (Lepidptera: Pyralidae). Environ. Entomol. 13, 882–887. doi: http://dx.doi.org/10.1093/ee/13.3.882

Andreadis, T.G., 1987. Transmission, in: Fuxa, J., Tanada, Y. (eds.), Epizootiology of Insect Diseases. Wiley-Interscience, New York, pp. 159–176.

Andreadis, T.G., 1990. Epizootiology of *Amblyospora connecticus* (Microsporida) in field populations of the saltmarsh mosquito, *Aedes Cantator*, and the cyclopoid copepod, *Acanthocyclops Vernalis*. J. Protozool. 37, 174–182. doi: 10.1111/j.1550-7408.1990.tb01123.x

Andreadis, T.G., 2002. Epizootiology of *Hyalinocysta chapmani* (Microsporidia: Thelohaniidae) infections in field populations of *Culiseta melanura* (Diptera: Culicidae) and *Orthocyclops modestus* (Copepoda: Cyclopidae): a three-year investigation. J. Invertebr. Pathol. 81, 114–121. doi:10.1016/S0022-2011(02)00154-4

Andreadis, T.G., 2005. Evolutionary strategies and adaptations for survival between mosquito–parasitic microsporidia and their intermediate copepod hosts: a comparative examination of *Amblyospora connecticus* and *Hylanocysta chapmani* (Microsporidia: Amblyosporidae). Folia Parasitol. 52, 23–35. doi: 10.14411/fp.2005.004

Antony, B., 2014. Detection of nucleopolyhedrosisviruses in the eggs and caterpillars of tea looper caterpillar *Hyposidra infexiaria* (Walk.) (Lepidoptera: Geometridae) as evidence of transovarial transmission. Arch. Phytopathol. Plant Protect. 47, 1426–1430. http://dx.doi.org/10.1080/03235408.2013.845469

Araújo, J.P.M., Hughes, D.P., 2016. Diversity of entomopathogenic fungi: which groups conquered the insect body? Adv. Gen. 94, 1–39. http://dx.doi.org/10.1016/bs.adgen.2016.01.001

Argôlo-Filho, R.C., Loguericio, L.L., 2014. *Bacillus thuringiensis* is an environmental pathogen and host-specificity has developed as an adaptation to human-generated ecological niches. Insects 5, 62–91. doi:10.3390/insects5010062

Auld, S.K.J.R., Hall, S.R., Ochs, J.H., Sebastian, M., Duffy, M.A., 2014. Predators and patterns of within-host growth can mediate both among-host competition and evolution of transmission potential of parasites. Am. Nat. 184(Suppl. 1), S77–S90. doi:10.1086/676927

Azam, A., Kunimi, Y., Inoue, M.N., Nakai, M., 2016. Effect of granulovirus infection of *Spodoptera litura* (Lepidoptera: Noctuidae) larvae on development of the endoparasitoid *Chelonus inanitus* (Hymenoptera: Braconidae). Appl. Entomol. Zool. 51, 479–488. doi:10.1007/s13355-016-0423-6Source

Bai, C., Shapiro-Ilan, D.I., Gaugler, R., Hopper, K.R., 2005. Stabilization of beneficial traits in *Heterorhabditis bacteriophora* through creation of inbred lines. Biol. Control. 32, 220–227. http://dx.doi.org/10.1016/j.biocontrol.2004.09.011

Bailey, C.H., Brooks, W.M., 1972. Effects of *Herpetomonas muscarum* on development and longevity of the eye gnat *Hippelates pusio* (Diptera: Chloropidae). J. Invertebr. Pathol. 20, 31–36. doi:10.1016/0022-2011(72)90077-8

Barelli, L., Moonjely, S., Behie, S.W., Bidochka, M.J., 2016. Fungi with multifunctional lifestyles: endophytic insect pathogenic fungi. Plant Mol. Biol. 90, 657–664. doi: 10.1007/s11103-015-0413-z

Barnes, A.I., Siva-Jothy, M.T., 2000. Density-dependent prophylaxis in the mealworm beetle *Tenebrio molitor* L. (Coleoptera: Tenebrionidae): cuticular melanization is an indicator of investment in immunity. P. Roy. Soc. B Biol. Sci. 267, 177–182. doi:10.1098/rspb.2000.0984

Barreau, C., Jousset, F-X., Bergoin, M., 1997. Venereal and vertical transmission of the *Aedes albopictus* parvovirus in *Aedes aegypti* mosquitoes. Am. J. Trop. Med. Hyg. 57, 126–131.

Becnel, J.J., Andreadis, T.G., 1999. Microsporidia in insects, in: Wittner, M., Weiss, L.M. (eds.), The Microsporidia and Microsporidiosis. American Society for Microbiology Press, Washington, DC, pp. 447–501.

Begon, M., 2009. Ecological epidemiology, in: Levin, S.A., Carpenter, S.R., Godfray, H.C.J., Kinzig, A.P., Loreau, M., Losos, J.B., et al. (eds.), The Princeton Guide to Ecology. Princeton University Press, Princeton, NJ, pp. 220–226.

Behie, S.W., Bidochka, M.J., 2014. Ubiquity of insect-derived nitrogen transfer to plants by endophytic insect pathogenic fungi: an additional branch of the soil nitrogen cycle. Appl. Environ. Microbiol. 80, 1553–1560. doi:10.1128/AEM.03338-13

Behie, S.W., Jones, S.J., Bidochka, M.J., 2015. Plant tissue localization of the endophytic insect pathogenic fungi *Metarhizium* and *Beauveria*. Fungal Ecol. 13, 112–119. DOI:10.1016/j.funeco.2014.08.001

Behrens, D., Forsgren, E., Fries, I., Moritz, R.F.A., 2010. Lethal infection thresholds of *Paenibacillus larvae* for honey bee drone and worker larvae (*Apis mellifera*). Environ. Microbiol. 12, 2838–2845. doi:10.1111/j.1462-2920.2010.02257.x

Bergold, G.H., 1958. Viruses of insects, in: Doerr, R., Hallauer, C. (eds.), Handbuch der Virusforschung. Springer-Verlag, Vienna, pp. 60–142.

Bilton, D.T., Freeland, J.R., Okamura, B., 2001. Dispersal in freshwater invertebrates. Annu. Rev. Ecol. Syst. 32, 159–181. doi: 10.1146/annurev.ecolsys.32.081501.114016

Bing, L.A., Lewis. L.C., 1991. Suppression of *Ostrinia nubilalis* (Hübner) (Lepidoptera: Pyralidae) by endophytic *Beauveria bassiana* (Balsamo) Vuillemin. Environ. Entomol. 20, 1207–1211.

Birnbaum, S.S.L., Gerardo, N.M., 2016. Patterns of specificity of the pathogen *Escovopsis* across the fungus-growing ant symbiosis. Am. Nat. 188, 52–65. doi:10.1086/686911

Bisch G., Pages, S., McMullen, J.G., Stock, S.P., Duvic, B., Givaudan, A., Gaudriault, S., 2015. *Xenorhabdus bovienii* CS03, the bacterial symbiont of the entomopathogenic nematode *Steinernema weiseri*, is a non-virulent strain against lepidopteran insects. J. Invertebr. Pathol. 124, 15–22. doi:10.1016/j.jip.2014.10.002

Bishop, A.L., Barchia, I.M., Spohr, L.J., 2000. Models for the dispersal in Australia of the arbovirus vector, *Culicoides brevitarsis* Kieffer (Diptera: Ceratopogonidae). Prev. Vet. Med. 47, 243–254. doi:10.1016/S0167-5877(00)00175-6

Blaker, E.A., Strange, J.P., James, R.R., Monroy, F.P., Cobb, N.S., 2014. PCR reveals high prevalence of non/low sporulating *Nosema bombi* (microsporidia) infections in bumble bees (*Bombus*) in Northern Arizona. J. Invertebr. Pathol. 123, 25–33. doi: 10.1016/j.jip.2014.09.001

Boemare, N., 2002. Biology, taxonomy and systematics of *Photorhabdus* and *Xenorhabdus*, in: Gaugler, R. (ed.), Entomopathogenic Nematology. CABI, Wallingford, pp. 35–56.

Boemare, N., Taillez, P., 2009. Molecular approaches and techniques for the study of entomopathogenic bacteria, in: Stock, S.P., Vandenberg, J., Glazer, Il, Boemare, N. (eds.), Insect Pathogens: Molecular Approaches and Techniques. CABI, Wallingford, pp. 32–49. doi: 10.1079/9781845934781.0000

Bolling, B.G., Olea-Popelka, F.J., Eisen, L., Morre, C.G., Blair, C.D., 2012. Transmission dynamics of an insect-specific flavivirus in a naturally infected *Culex pipiens* laboratory colony and effects of co-infection on vector competence for West Nile virus. Virology 427, 90–97. doi:10.1016/j.virol.2012.02.016

Boomsma, J.J., Jensen, A.B., Meyling, N.V., Eilenberg, J., 2014. Evolutionary interaction networks of insect pathogenic fungi. Ann. Rev. Entomol. 59, 467–485. doi: 10.1146/annurev-ento-011613-162054

Boots, M., Begon, M., 1994. Resource limitation and the lethal and sublethal effects of a viral pathogen in the Indian meal moth. Ecol. Entomol. 19, 319–326. doi: 10.1111/j.1365-2311.1994.tb00248.x

Boots, M., Norman, R., 2000. Sublethal infection and the population dynamics of host-microparasite interactions. J. Anim. Ecol. 69, 517–524. doi: 10.1046/j.1365-2656.2000.00417.x

Borrego, J.J., Luque, A., Castro, D., Santamaria, J.A., Martinez-Manzanares, E., 1996. Virulence factors of *Vibrio* P1, the causative agent of brown ring disease in the Manila clam, *Ruditapes philippinarum*. Aquat. Living Resour. 9, 125–136. doi: 10.1051/alr:1996016

Bowers, R.G., Begon, M., Hodgkinson, D. E., 1993. Host-pathogen population cycles in forest insects? Lessons from simple models reconsidered. Oikos 67, 529–538. doi: 10.2307/3545365

Brandenberg, R.L., Kennedy, G.G., 1981. Overwintering of the pathogen *Entomophthora floridana* and its host, the two-spotted spider mite. J. Econ. Entomol. 74, 428–431. doi: 10.1093/jee/74.4.428 428-431

Briggs, C.J., Godfray, H.C.J., 1996. The dynamics of insect-pathogen interactions in seasonal environments. Theor. Popul. Biol., 50, 149–177. doi: 10.1006/tpbi.1996.0027

Brilli, M., Lió, P., Lacroix, V., Sagot, M.-F., 2013. Short and long-term genome stability analysis of prokaryotic genomes. BMC Genomics 14, 309. doi: 10.1186/1471-2164-14-309

Brown, M.J.F., Loosli, R., Schmid-Hempel, P., 2000. Condition-dependent expression of virulence in a trypanosome infecting bumble bees. Oikos 91, 421–427. doi: 10.1034/j.1600-0706.2000.910302.x

Brownlie, J.C., Johnson, K.N., 2009. Symbiont-mediated protections in insect hosts. Trends Microbiol. 17, 348–354. http://dx.doi.org/10.1016/j.tim.2009.05.005

Bruck, D.J., Lewis, L.C., 2002. Rainfall and crop residue effects on soil dispersion and *Beauveria bassiana* spread to corn. Appl. Soil Ecol. 20, 183–190. http://dx.doi.org/10.1016/S0929-1393(02)00022-7

Bucher, G.E., 1973. Definition and identification of insect pathogens. Ann. N.Y. Acad. Sci. 217, 8–17.

Budischak, S.A., Sakamoto, K., Megow, L.C., Cummings, K.R., Urban, J.F., Ezenwa, V.O., 2015. Resource limitation alters the consequences of co-infection for both hosts and parasites. Int. J. Parasitol. 45, 455–463. doi: 10.1016/j.ijpara.2015.02.005

Burden, J.P., Griffiths, C.M., Cory, J.S., Smith, P., Sait, S.M., 2002. Vertical transmission of sublethal granulovirus infection in the Indian meal moth, *Plodia interpunctella*. Mol. Ecol. 11, 547–555. doi: 10.1046/j.1461-0248.2003.00459.x

Butel, C., Mundeke, S.A., Drakulovski, P., Krasteva, D., Ngole, E.M., Mallie, M., et al., 2015. Assessment of infections with microsporidia and *Cryptosporidium* spp. in fecal samples from wild primate populations from Cameroon and Democratic Republic of Congo. Int. J. Primatol. 36, 227–243. doi: 10.1007/s10764-015-9820-x

Butt, T.M., Ibrahim, L., Clark, S.J., Beckett, A., 1995. The germination behavior of *Metarhizium aniospliae* on the surface of aphid and flea beetle cuticles. Mycol. Res. 99, 945–950. doi: 10.1371/journal.pone.0081686

Caceras, C.E., Knight, C.J., Hall, S.R., 2009. Predator-spreaders: predation can enhance parasite success in a planktonic host-parasite system. Ecology 90, 2850–2858. doi: 10.1890/08-2154.1

Cali, A., Takvorian, P.M., 2014. Developmental morphology and life cycles of the microsporidia, in: Weiss, L.M., Becnel, J.J. (eds.), Microsporidia: Pathogens of Opportunity. John Wiley & Sons, Chichester, pp. 71–133.

Campos-Herrera, R., Trigo, D., Gutiérrez, C., 2006, Phoresy of the entomopathogenic nematode *Steinernema feltiae* by the earthworm *Eisenia fetida*. J. Invertebr. Pathol. 92, 50–54. doi: 10.1016/j.jip.2006.01.007

Cao, M.-Y., Kuang, X.-X., Li, H.-Q., Lei, X.-J., Xiao, W.-F., Dong, Z.-Q., et al., 2016. Screening and optimization of an efficient *Bombyx mori* nucleopolyhedrovirus inducible promoter. J. Biotechnol. 231, 72–80. doi:10.1016/j.jbiotec.2016.05.037Source

Carruthers, R.I., Sawyer, A.J., Hural, K., 1991. Use of fungal pathogens for biological control of insect pests, in: Rice, B.J. (ed.), Sustainable Agriculture Research and Education in the Field. National Academy Press, Washington, DC, pp. 336–372.

Casadevall, A., 2005. Host as the variable: model hosts approach the immunological asymptote. Infect. Immun. 73, 3829–3832. doi: 10.1128/IAI.73.7.3829-3832.2005

Casadevall, A., Pirofski, L., 1999. Host-pathogen interactions: redefining the basic concepts of virulence and pathogenicity. Infect. Immun. 67, 3703–3713.

Casadevall, A., Pirofski, L., 2001. Host-pathogen interactions: the attributes of virulence. J. Inf. Dis. 184, 337–344. doi:10.1086/322044

Chaston, J., Goodrich-Blair, H., 2010. Common trends in mutualism revealed by model associations between invertebrates and bacteria. FEMS Microbiol Rev. 34, 41–58. doi: 10.1111/j.1574-6976.2009.00193.x

Chaston, J.M., Dillman, A.R., Shapiro-Ilan, D.I., Bilgrami, A.L., Gaugler, R., Hopper, K.R., Adams, B.J., 2011. Outcrossing and crossbreeding recovers deteriorated traits in laboratory cultured *Steinernema carpocapsae* nematodes. Int. J. Parasitol. 41, 801–809.

Chen, Y.P., Pettis, J.S., Collins, A., Feldlaugher, M.F., 2006. Prevalence and transmission of honey bee viruses. Appl. Environ. Microbiol. 72, 606–611. doi: 10.1128/AEM.72.1.606-611.2006

Chen, Y.P., Becnel, J.J., Valles, S.M., 2012. RNA viruses infecting pest insects, in: Vega, F.E., Kaya, H.K. (eds.), Insect Pathology, 2nd edn. Academic Press, San Diego, CA, pp. 133–170. doi: 10.1016/B978-0-12-384984-7.00005-1

Christian, P.D., Hanzlik, T.N., Dall, D.J., Gordon, K.J., 1993. Insect viruses and pest control, in: Oakeshott, J., Whitten, M.J. (eds.), Molecular Approaches to Fundamental and Applied Entomology. Springer, New York, pp. 128–163.

Clifton, E.H., Jaronski, S.T., Hodgson, E.W., Gassmann, A.J., 2015. Abundance of soil-borne entomopathogenic fungi in organic and conventional fields in the midwestern USA with an emphasis on the effect of herbicides and fungicides on fungal persistence. PLoS ONE 10(7), e0133613. doi:10.1371/journal.pone.0133613

Collison, C., 2015. A closer look: bee health and pesticides. Available from: http://www.beeculture.com/a-closer-look-bee-health-and-pesticides/ (accessed May 8, 2017).

Cook, T.J.P., Janovy, J., Jr., Clopton, R.E., 2001. Epimerite-host epithelium relationships among eugregarines parasitizing the damselflies *Enallagma civile* and *Ischnura verticalis*. J. Parasitol. 87, 988–996. doi: 10.1645/0022-3395(2001)087[0988:EHERAE]2.0.CO;2

Cooper, D., Cory, J.S., Theilmann, D.A., Myers, J.H., 2003. Nucleopolyhedroviruses of forest and western tent caterpillars: cross-infectivity and evidence for activation of latent virus in high-density field populations. Ecol. Entomol. 28, 41–50. doi: 10.1046/j.1365-2311.2003.00474.x

Cory, J.S., Franklin, M.T., 2012. Evolution and the microbial control of insects. Evol. Appl. 5, 455–469. doi: 10.1111/j.1752-4571.2012.00269.x

Cory, J.S., Hoover, K., 2006. Plant mediated effects in insect-pathogen interactions. Trends Ecol. Evol. 21, 278–286. doi: 10.1016/j.tree.2006.02.005

Cory, J.S., Hirst, M.L., Sterling, P.H., Speight, M.R., 2000. Narrow host range nucleopolyhedrovirus for control of the browntail moth (Lepidoptera: Lymantriidae). Environ. Entomol. 29, 661–667. doi: 10.1603/0046-225X-29.3.661

Cressler, C.E., Nelson, W.A., Day, T., McCauley, E., 2014. Disentangling the interaction among host resources, the immune system and pathogens. Ecol. Lett. 17, 284–293. doi: 10.1111/ele.12229

Crickmore, N., Zeigler, D.R., Feitelson, J., Schnepf, E., Van Rie, J., Lereclus, D., et al., 2014. Bacillus thuringiensis toxin nomenclature. Available from: http://www.lifesci.sussex. ac.uk/Home/Neil_Crickmore/Bt/ (accessed May 8, 2017).

D'Amico, V., Elkinton, J.S., 1995. Rainfall effects on transmission of gypsy moth (Lepidoptera: Lymantriidae) nuclear polyhedrosis virus. Environ. Entomol. 24, 1144–1149. doi: http://dx.doi.org/10.1093/ee/24.5.1144

Dawkins, R. 1982. The Extended Phenotype. Oxford University Press, Oxford.

De Fine Licht, H., Hajek, A.E., Jensen, A.B., Eilenberg, J., 2016. Utilizing genomics to study entomopathogenicity in the fungal phylum Entomophthoromycota: a review of current genetic resources. Adv. Gen. 94, 41–65. http://dx.doi.org/10.1016/bs.adgen.2016.01.003

de Miranda, J.R., Fries, I., 2008. Venereal and vertical transmission of deformed wing virus in honey bees (*Apis mellifera* L.). J. Invertebr. Pathol. 98, 184–189. doi: 10.1016/j. jip.2008.02.004

de Miranda, J.R., Genersch, E., 2010. Deformed wing virus. J. Invertebr. Pathol. 103, S48–S61. doi: 10.1016/j.jip.2009.06.012

Detwiler, J., Janovy, J. Jr., 2008. The role of phylogeny and ecology in experimental host specificity: Insights from a eugregarine-host system. J. Parasitol. 94, 7–12. doi: 10.1645/ GE-1308.1

Dimijian, G.G. 2000. Pathogens and parasites: strategies and challenges. Proc. Baylor Univ. Med. Cent. 13, 19–29.

Dingman, D.W., Stahly, D.P., 1983. Medium promoting sporulation of *Bacillus larvae* and metabolism of medium components. Appl. Environ. Microbiol. 46, 860–869.

Di Prisco, G., Zhang, X., Pennacchio, F., Caprio, E., Li, J., Evans, J., et al., 2011. Dynamics of persistent and acute deformed wing virus infections in honey bees, *Apis mellifera*. Viruses 3, 2425–2441. doi: 10.3390/v3122425

Dobson, A., 2004. Population dynamics of pathogens with multiple hosts. Am. Nat. 164, S64–S78. doi: 10.1086/424681

Duchaud, E., Rusniok, C., Frangeul, L., Buchrieser, C., Givaudan, A., Taourit, S., Bocs, S., 2003. The genome sequence of the entomopathogenic bacterium *Photorhabdus luminescens*. Nat. Biotechnol. 21, 1307–1313. doi: 10.1038/nbt886

Duke, L., Steinkraus, D.C., English, J.E., Smith, K.G., 2002. Infectivity of resting spores of *Massospora cicadina* (Entomophthorales: Entomophthoraceae), an entomopathogenic fungus of periodical cicadas (*Magicicada* spp.) (Homoptera: Cicadidae). J. Invertebr. Pathol. 80, 1–6.

Dunn, A.M., Smith, J.E., 2001. Microsporidian life cycles and diversity: the relationship between virulence and transmission. Microbes Infect. 3, 381–388. doi: 10.1016/ S1286-4579(01)01394-6

Duplouy, A., Chouchoux, C., Hanski, I., van Nouhuys, S., 2015. *Wolbachia* infection in a natural parasitoid wasp population. PLoS ONE 10, 30134843. doi: 10.1371/journal.pone.0134843

Dwyer, G., 1991. The roles of density, stage and patchiness in the transmission of an insect virus. Ecology 72, 559–574. doi: 10.2307/2937196

Dwyer, G., 1992. On the spatial spread of insect pathogens: theory and experiment. Ecology 73, 479–494. doi: 10.2307/1940754

Dwyer, G., Elkinton, J.S., 1995. Host dispersal and the spatial spread of insect pathogens. Ecology 76, 1262–1275. doi: 10.2307/1940933

Dwyer, G., Dushoff, J., Yee, S.H., 2004. The combined effects of pathogens and predators on insect outbreaks. Nature 430, 341–345. doi:10.1038/nature02569

Ehrlich, G.D., Hiller, N.L., Hu, F.Z., 2008. What makes pathogens pathogenic? Genome Biol. 9, 225. doi: 10.1186/gb-2008-9-6-225

Eleftherianos, J., Atri, J., Acetta, J., Castillo, J.C., 2013. Endosymbiotic bacteria in insects: guardians of the immune system? Front Physiol. 4, 46. doi: 10.3389/fphys.2013.00046

Engelhard, E.K., Kam-Morgan, L.N.W., Washburn, J.O., Volkman, L.E., 1994. The insect tracheal system: a conduit for the systemic spread of *Autographa californica* M nuclear polyhedrosis virus. Proc. Natl. Acad. Sci. U.S.A. 91, 3224–3227. doi: 10.1073/pnas.91.8.3224

Eveleigh, E.S., Lucarotti, C.J., McCarthy, P.C., Morin, B., 2012. Prevalence, transmission and mortality associated with *Nosema fumiferanae* infections in field populations of spruce budworm *Choristoneura fumiferanae*. Agric. For. Entomol. 14, 389–398. doi: 10.1111/j.1461-9563.2012.00580.x

Favet, J., Lapanje, A., Giongo, A., Kennedy, S., Aung, Y.-Y., Cattaneo, A., et al., 2013. Microbial hitchhikers on intercontinental dust: catching a lift in Chad. ISME J. 7, 850–867.

Feng, M.-G., Chen, C., Chen, B., 2004. Wide dispersal of aphid-pathogenic Entomophthorales among aphids relies on migratory alates. Environ. Microbiol. 6, 510–516. doi: 10.1111/j.1462-2920.2004.00594.x

Filotas, M.J., Hajek, A.E., 2004. Influence of temperature and moisture on infection of forest tent caterpillars (Lepidoptera: Lasiocampidae) exposed to resting spores of the entomopathogenic fungus *Furia gastropachae* (Zygomycetes: Entomophthorales). Environ. Entomol. 33, 1127–1136.

Fine, P.E.M., 2003. The interval between successive cases of an infectious disease. Am. J. Epidemiol. 158, 1039–1047. doi: 10.1093/aje/kwg251

Foighil, D.O., Marshall, B.A., Hilbish, T.J., Pino, M.A., 1999. Trans-Pacific range extension by rafting is inferred for the flat oyster *Ostrea chilensis*. Biol. Bull. 196, 122–126.

Foil, L.D., Gorham, J.R., 2000. Mechanical transmission of disease agents by arthropods, in: Eldridge, B.F., Edman, J.D. (eds.), Medical Entomology. Kluwer Academic Publishers, The Netherlands, pp. 461–514. doi: 10.1007/978-94-011-6472-6_12

Fries, I., Lindström, A., Korpela, S., 2006. Vertical transmission of American foulbrood (*Paenibacillus larvae*) in honey bees (*Apis mellifera*). Vet. Microbiol. 114, 269–274. doi: 10.1016/j.vetmic.2005.11.068

Fuller, E., Elderd, B.D., Dwyer, G. 2012. Pathogen persistence in the environment and insect-baculovirus interactions: disease-density thresholds, epidemic burnout and insect outbreaks. Am. Nat. 179, E70–E96. doi: 10.1086/664488

Fünfaus, A., Poppingia, L., Genersch, E., 2013. Identification and characterization of two novel toxins expressed by the lethal honey bee pathogen *Paenibacillus larvae*, the causative agent of American foulbrood. Environ. Microbiol. 15, 2951–2965. doi: 10.1111/1462-2920.12229

Fushing, H., Zhu, L., Shapiro-Ilan, D.I., Campbell, J.F., Lewis, E.E., 2008. State-space based mass event-history model I: many decision-making agents with one target. Ann. Appl. Stat. 2, 1503–1522. doi: 10.1214/08-AOAS189

Fuxa, J.R., Richter, A.R., 1996. Effect of agricultural operations and precipitation on vertical distribution of a nuclear polyhedrosis virus in soil. Biol. Control 6, 324–329. doi: 10.1006/bcon.1996.0041

Fuxa, J.R., Tanada, Y., 1987. Epizootiology of Insect Diseases. Wiley-Interscience, New York.

Genersch, E., 2010. American foulbrood in honey bees and its causative agent, *Paenibacillus larvae*. J. Invertebr. Pathol. 103, S10–S19. doi: 10.1016/j.jip.2009.06.015

Genersch, E., Aubert, M., 2010. Emerging and re-emerging viruses of the honey bee (*Apis mellifera* L.). Vet. Res. 41, 54. doi: 10.1051/vetres/2010027

Genersch, E., Ashiralieva, A., Fries, I., 2005. Strain- and genotype-specific differences in virulence of *Paenibacillus larvae* subsp. larvae, a bacterial pathogen causing American foulbrood disease in honey bees. Appl. Environ. Microbiol. 71, 7551–7555. doi: 10.1128/AEM.71.11.7551-7555.2005

Gildow, F.E., D'Arcy, C.J., 1988. Barley and oats as reservoirs for an aphid virus and the influence on barley yellow dwarf virus transmission. Phytopathology 78, 811–816. doi: 10.1094/Phyto-78-811

Goblirsch, M., Huang, A.Y., Spivak, M., 2013. Physiological and behavioral changes in honey bees (*Apis mellifera*) induced by *Nosema ceranae* infection. PLoS ONE 8, e58165. doi: 10.1371/journal.pone.0058165

Goertz, D., Hoch, G., 2008. Horizontal transmission pathways of terrestrial microsporidia: a quantitative comparison of three pathogens infecting different organs in *Lymantria dispar* L. (Lep.: Lymantriidae) larvae. Biol. Control 44, 196–206. http://dx.doi.org/10.1016/j.biocontrol.2007.07.014

Goertz, D., Hoch, G., 2013a. Influence of the forest caterpillar hunter *Calosoma sycophanta* on the transmission of microsporidia in larvae of the gypsy moth *Lymantria dispar*. Agr. Forest Entomol. 15, 178–186. doi: 10.1111/afe.12000

Goertz, D., Hoch, G., 2013b. Effects of the ant *Formica fusca* on the transmission of microsporidia infecting gypsy moth larvae. Entomol. Exp. App. 147, 251–261. doi:10.1111/eea.12063

Goertz, D., Solter, L.F., and Linde, A., 2007. Horizontal and vertical transmission of a *Nosema* sp. (Microsporidia) from *Lymantria dispar* (L.) (Lepidoptera: Lymantriidae). J. Invertebr. Pathol. 95, 9–16. doi: 10.1016/j.jip.2006.11.003

Graystock, P., Goulson, D., Hughes, W.O.H., 2015. Parasites in bloom: flowers aid dispersal and transmission of pollinator parasites within and between bee species. Proc. Roy. Soc. Lond. B Biol. Sci. 282. doi: 10.1098/rspb.2015.1371

Griffin, C.T., Boemare, N.E., Lewis, E.E., 2005. Biology and behavior, in: Grewal, P.S., Ehlers, R.-U., Shapiro-Ilan, D.I. (eds.), Nematodes as Biocontrol Agents. CABI, Wallingford, pp. 47–64.

Grosholz, E.D., 1993. The influence of habitat heterogeneity on host-pathogen population dynamics. Oecologia 96, 347–353. doi:10.1007/BF00317504

Grzywacz, D., 2016. Basic and applied research: Baculovirus, in: Lacey, L.A. (ed.), Microbial Control of Insect and Mite Pests. Academic Press, San Diego, CA, pp. 27–46.

Gulcu, B., Hazir, S., Kaya, H.K., 2012. Scavenger deterrent factor (SDF) from symbiotic bacteria of entomopathogenic nematodes. J. Invertebr. Pathol. 110, 326–333.

Gupta, R.K., Amin, M., Bali, K., Monobrullah, M.D., Jasrotia, P., 2010. Vertical transmission of sublethal granulovirus infection in the tobacco caterpillar *Spodoptera litura*. Phytoparasitica 38, 209–216. doi: 10.1007/s12600-010-0090-z

Gupta, R.K., Gani, M., Jasrotia, P. Srivastava, K., 2013. Development of the predator *Eocanthecona furcellata* on different proportions of nucleopolyhedrovirus infected *Spodoptera litura* larvae and potential for predator dissemination of virus in the field. BioControl 58, 543–552. doi: 10.1007/s10526-013-9515-1

Hajek, A.E., van Nouhuys, S., 2016. Fatal diseases and parasitoids: from competition to facilitation in a shared host. Proc. Roy. Soc. Lond. B Biol. Sci. 283, 20160154. doi: 10.1098/rspb.2016.0154

Hajek, A.E., Tatman, K.M., Wheeler, M.M., 1998. Location and persistence of cadavers of gypsy moth, *Lymantria dispar*, containing *Entomophaga maimaiga* azygospores. Mycologia 90, 754–760. doi: 10.2307/3761315

Hajek, A.E., Olsen, C.H., Elkinton, J.S., 1999. Dynamics of airborne conidia of the gypsy moth (Lepidoptera: Lymantriidae) fungal pathogen *Entomophaga maimaiga* (Zygomycetes: Entomophthorales). Biol. Control 16, 111–117. doi: 10.1006/bcon.1999.0740

Hajek, A.E., Strazanac, J.S., Wheeler, M.M., Vermeylen, F.M., Butler, L., 2004. Persistence of the fungal pathogen *Entomophaga maimaiga* and its impact on native Lymantriidae. Biol. Control 30, 466–473. doi: 10.1016/j.biocontrol.2004.02.005

Hajek, A.E., Tobin, P.C., Haynes, K.J., 2015. Replacement of a dominant viral pathogen by a fungal pathogen does not alter the synchronous collapse of a forest insect outbreak. Oecologia 177, 785–797.

Han, Y., van Houte, S., Drees, G.F., van Oers, M.M., Ros, V.I.D., 2015. Parasitic manipulation of host behaviour: baculovirus SeMNPV EGT facilitates tree-top disease in *Spodoptera exigua* larvae by extending the time to death. Insects 6, 716–731. doi: 10.3390/insects6030716

Harrison, R.L., 2009. Structural divergence among genomes of closely related baculoviruses and its implications for baculovirus evolution. J. Invertebr. Pathol. 101, 181–186. doi: 10.1016/j.jip.2009.03.012

Harrison, R., Hoover, K., 2012. Baculoviruses and other occluded insect viruses, in; Vega, F.E., Kaya, H.K. (eds.), Insect Pathology, 2nd edn. Academic Press, San Diego, CA, pp. 73–132. doi: 10.1016/B978-0-12-384984-7.00004-X

Hartikainen, H., Ashford, O.S., Berney, C., Okamura, B., Feist, S.W., Baker-Austin, C., et al., 2014. Lineage-specific molecular probing reveals novel diversity and ecological partitioning of haplosporidians. ISME J. 8, 177–186. doi: 10.1038/ismej.2013.136

Hazard, E.I., Anthony, D.W., 1974. A redescription of the genus *Parathelohania* Codreanu 1966 (Microsporida: Protozoa) with a reexamination of previously described species of *Thelohania* Henneguy 1892 and descriptions of two new species of *Parathelohania* from Anopheline mosquitoes. USDA – ARS Tech. Bull. 1505.

Hazir, S., Shapiro-Ilan, D.I., Hazir, C., Leite, L.G., Cakmak, I., Olson, D., 2016. Multifaceted effects of host plants on entomopathogenic nematodes. J. Invertebr. Pathol. 135, 53–59. doi: 10.1016/j.jip.2016.02.004

Hemmati, F., Pell, J.K., McCartney, H.A., Deadman, M.L., 2001. Airborne concentrations of conidia of *Erynia neoaphidis* above cereal fields. Mycol. Res. 105, 485–489. doi: 10.1017/S0953756201003537

Hewson, I., Button, J.B., Gudenkauf, B.M., Miner, B., Newton, A.L., Gaydos, J.K., et al., 2014. Densovirus associated with sea-star wasting disease and mass mortality. P. Natl. Acad. Sci. U.S.A. 48, 17 278–17 283. doi: 10.1073/pnas.1416625111

Highfield, A.C., El Nagar, A., Mackinder, L.C.M., Nöel, L.M., Hall, M.J., Martin, S.J., Schroeder, D.C., 2009. Deformed wing virus implicated in overwintering honey bee colony losses. Appl. Environ. Microbiol. 75, 7212–7220. doi: 10.1128/AEM.02227-09

Hill-Spanik, K.M., McDowell, J.R., Stokes, N.A., Reece, K.S., Burreson, E.M., Carnegie, R.B., 2015. Phylogeographic perspective on the distribution and dispersal of a marine pathogen, the oyster parasite *Bonamia exitiosa*. Mar. Ecol. Prog. Ser. 536, 65–76. doi:10.3354/meps11425

Hinchliffe, S.J., Hares, M.C., Dowling, A.J., ffrench-Constant, R.H., 2010. Insecticidal toxins from the *Photorhabdus* and *Xenorhabdus* bacteria. Open Toxinol. J. 3, 83–100.

Hochberg, M.E., 1991. Extra-host interactions between a braconid endoparasitoid *Apanteles glomeratus,* and a baculovirus for larvae of *Pieris brassicae.* J. Anim. Ecol. 60, 65–77. doi: 10.2307/5445

Hoover, K., Grove, M., Gardner M., Hughes, D.P., McNeil, J., Slavicek, J., 2011. A gene for an extended phenotype. Science 333, 1401. doi: 10.1126/science.1209199

Huang, W.-F., Solter, L.F., 2013. Comparative development and tissue tropism in *Nosema apis* and *Nosema ceranae.* J. Invertebr. Pathol. 113, 35–41. doi:10.1371/journal.ppat.1003185

Huang, W.-F., Solter, L.F., Yau, P.M., Imai, B., 2013. *Nosema ceranae* escapes fumagillin control in honey bees. PLoS Pathog. 9(3), e1003185.

Hughes, D.S., Possee, R.D., King, L.A., 1993. Activation and detection of a latent baculovirus resembling *Mamestra brassicae* nuclear polyhedrosis virus in *M. brassicae* insects. Virology 194, 608–615. doi: 10.1006/viro.1993.1300

Hurst, M.R., Glare, T.R., Jackson, T.A., Ronson, C.W., 2000. Plasmid-located pathogenicity determinants of *Serratia entomophila,* the causal agent of amber disease of grass grub, show similarity to the insecticidal toxins of *Photorhabdus luminescens.* J. Bacteriol. 182, 5127–5138. doi: 10.1128/JB.182.18.5127-5138.2000

Huver, J.R., Koprivnikar, J., Johnson, P.T.J., Whyard, S., 2015. Development and application of an eDNA method to detect and quantify a pathogenic parasite in aquatic ecosystems. Ecol. Appl. 25, 991–1002. doi:10.1890/14-1530.1

Ironside, J.E., Dunn, A.M., Rollinson, D., Smith, J.E., 2003. Association with host mitochondrial haplotypes suggests that feminizing microsporidia lack horizontal transmission. J. Evol. Biol. 16, 1077–1083. DOI: 10.1046/j.1420-9101.2003.00625.x

Jehle, J.A., Lange, M., Wang, H., Hu, Z., Wang, Y., Hauschild, R., 2006. Molecular identification and phylogenetic analysis of baculoviruses from Lepidoptera. Virology 346, 180–193. doi: 10.1016/j.virol.2005.10.032

Jensen, A.B., Aronstein, K., Flores, J.M. Vojvodic, S., Palacio, M.A., Spivak, M., 2013. Standard methods for fungal brood disease research. J. Apic. Res. 52. doi: 10.3896/ IBRA.1.52.1.13

Jiang, J.-X., Zeng, A.-P., Ji, X.-Y., Jiang, Z.-R., 2005. Vertical transmission of *Spodoptera exigua* nuclear polyhedirosis virus in the beet armyworm, *Spodotera exigua.* Acta Entomol. Sin. 48, 922–927.

Jones, C.J., Isard, S.A., Cortinas, M.R., 1999. Dispersal of synanthropic Diptera: lessons from the past and technology for the future. Ann. Entomol. Soc. Am. 92, 829–839.

Johnson, M.A., Becnel, J.J., Undeen, A.H., 1997. A new sporulation sequence in *Edhazardia aedis* (Microsporidia: Culicosporidae), a parasite of the mosquito *Aedes aegypti* (Diptera: Culicidae). J. Invertebr. Pathol. 70, 69–75. doi: 10.1006/jipa.1997.4678

Johnson, P.T.J., Stanton, D.E., Preu, E.R., Forshay, K.J., Carpenter, S.R., 2006. Dining on disease: How interactions between infection and environment affect predation risk. Ecology 87, 1973–1980. doi: 10.1890/0012-9658(2006)87[1973:DODHIB]2.0.CO;2

Jurat-Fuentes, J.L., Jackson, T.A. 2012. Bacterial entomopathogens, in: Vega, F.E., Kaya, H.K. (eds.), Insect Pathology, 2nd edn. Academic Press, San Diego, CA, pp. 265–349.

Kapusinszky, B., Mulvaney, U., Jasinska, A.J., Deng, X., Freimer, N., Delwart, E., 2015. Local virus extinctions following a host population bottleneck. J. Virology 89, 8152–8161. doi: 10.1128/JVI.00671-15

Katsuma, S., Kobayashi, J., Koyano, Y., Matsuda-Imai, N., Kang, W., Shimada, T., 2012. Baculovirus-encoded protein BV/ODV-E26 determines tissue tropism and virulence in lepidopteran insects. J. Virol. 86, 2545–2555. doi: 10.1128/JVI.06308-11

Keyhani, N.O., 2012. Using host molecules to increase fungal virulence for biological control of insects. Virulence 3, 415–517. doi: 10.4161/viru.20956

Khadijah, S., Neo, S.Y., Hossain, M.S., Miller, L.D., Mathavan, S., Kwang, J., 2003. Identification of white spot syndrome virus latency-related genes in specific-pathogen-free shrimps by use of a microarray. J. Virol. 77, 10162–10167. doi: 10.1128/JVI.77.18.10162-10167.2003

Kistner, E.J., Saums, M., Belovsky, G.E., 2015. Mechanical vectors enhance fungal entomopathogen reduction of the grasshopper pest *Camnula pellucida* (Orthoptera: Acrididae). Environ. Entomol. 44, 144–152. doi:10.1093/ee/nvu004

Klingen, I., Eilenberg, J., Meadow, R., 2002. Effects of farming system, field margins and bait insect on the occurrence of insect pathogenic fungi in soils. Agric. Ecosyst. Environ. 91, 191–198. http://dx.doi.org/10.1016/S0167-8809(01)00227-4

Klingen, I., Haukeland, S., 2006. The soil as a reservoir for natural enemies of pest insects and mites with emphasis on fungi and nematodes, in: Eilenberg, J., Hokkanen, H. (eds.), An Ecological and Societal Approach to Biological Control. Springer, Dordrecht, pp. 145–211. doi: 10.1007/978-1-4020-4401-4_9

Konrad, M., Grasse, A.V., Tragust, S., Cremer, S., 2015. Anti-pathogen protection versus survival costs mediated by an ectosymbiont in an ant host. Proc. Roy. Soc. Lond. B Biol. Sci. 282, 20141976. http://dx.doi.org/10.1098/rspb.2014.1976

Koppenhöfer, A.M., Fuzy, E.M., 2006. Effect of soil type on infectivity and persistence of the entomopathogenic nematodes *Steinernema scarabaei, Steinernema glaseri, Heterorhabditis zealandica*, and *Heterorhabditis bacteriophora*. J. Invertebr. Pathol. 92, 11–22. doi: http://dx.doi.org/10.1016/j.jip.2006.02.003

Koppenhöfer, A.M., Fuzy, E.M., 2007. Soil moisture effects on infectivity and persistence of the entomopathogenic nematodes *Steinernema scarabaei, S. glaseri, Heterorhabditis zealandica*, and *H. bacteriophora*. Appl. Soil Ecol. 1, 128–139. doi: http://dx.doi.org/10.1016/j.apsoil.2006.05.007

Koppenhöfer, A.M., Baur, M.E., Stock, P.S., Choo, H.Y., Channasri, B., Kaya, H.K., 1997. Survival of entomopathogenic nematodes within host cadavers in dry soil. Applied Soil Ecol. 6, 231–240. doi: 10.1016/S0929-1393(97)00018-8

Kough, A.S., Paris, C.B., Behringer, D.C., Butler, M.J., 2015. Modelling the spread and connectivity of waterborne marine pathogens: the case of PaV1 in the Caribbean. ICES J. Mar. Sci. 72(Suppl. 1), 139–146. doi:10.1093/icesjms/fsu209

Kring, T.J., Young, S.Y., Yearian, W.C., 1988. The striped lynx spider *Oxyopes salticus* Hentz (Araneae: Oxyopidae) as a vector of a nuclear polyhedrosis virus in *Anticarsia gemmatalis* Huebner (Lepidoptera: Noctuidae). J. Entomol. Sci. 23, 394–398.

Lacey, L.A., Grzywacz, D., Shapiro-Ilan, D.I., Frutos, R., Brownbridge, M., Goettel, M.W., 2015. Insect pathogens as biological control agents: back to the future. J. Invertebr. Pathol. 132, 1–41. doi:10.1016/j.jip.2015.07.009

Lambert, G., Kussell, E., 2014. Memory and fitness optimization of bacteria under fluctuating environments. PLoS Genet.10, e1004556. doi: 10.1371/journal.pgen.1004556

Lange, C.E., Johny, S., Baker, M.D., Whitman, D.W., Solter, L.F., 2009. A new *Encephalitozoon* species (Microsporidia) isolated from the lubber grasshopper, *Romalea microptera* (Beauvois) (Orthoptera: Romaleidae). J. Parasitol. 95, 976–986. doi: 10.1645/GE-1923.1

Lawrence, J.L., Hoy, C.W., Grewal, P.S., 2006. Spatial and temporal distribution of endemic entomopathogenic nematodes in a heterogeneous vegetable production landscape. Biol. Control 37, 247–255. http://dx.doi.org/10.1016/j.biocontrol.2006.02.002

Laws, A.N., Frauendorf, T.C., Gómez, J.E., Algaze, I.M., 2009. Predators mediate the effects of a fungal pathogen on prey: an experiment with grasshoppers, wolf spiders, and fungal pathogens. Ecol. Entomol. 34, 702–708. doi: 10.1111/j.1365-2311.2009.01122.x

Lee, M.M., Stock, S.P., 2010. A multilocus approach to assessing co-evolutionary relationships between *Steinernema* spp. (Nematoda: Steinernematidae) and their bacterial symbionts *Xenorhabdus* spp. (Gamma-Proteobacteria: Entero- bacteriaceae). Syst. Parasitol. 77, 1–12. doi: 10.1007/s11230-010-9256-9

Leggett, H.C., Cornwallis, C.K., West, S.A., 2012. Mechanisms of pathogenesis, infective dose and virulence in human parasites. PLoS Pathog. 8, e1002512. doi: 10.1371/journal. ppat.1002512

Leighton, F.A., 2003. Pathogens and disease, in: Hoffman, D.J., Rattner, B.A., Burton, G.A., Cairns, J. (eds.), Handbook of Ecotoxicology. CRC Press, Boca Raton, FL, pp. 667–677.

Lepore, L.S., Roelvink, P.R., Granados, R.R., 1996. Enhancin, the granulosis virus protein that facilitates nucleopolyhedrovirus (NPV) infections, is a metalloprotease. J. Invertebr. Pathol. 68, 131–140. doi: 10.1006/jipa.1996.0070

Lewis, E.E., Clarke, D.J., 2012. Nematode parasites and entomopathogens, in: Vega, F.E., Kaya, H.K. (eds.), Insect Pathology, 2nd edn. Academic Press, San Diego, CA, pp. 395–424. doi: 10.1016/B978-0-12-384984-7.00011-7

Lewis, E.E., Shapiro-Ilan, D.I., 2002. Host cadavers protect entomopathogenic nematodes during freezing. J. Invertebr. Pathol. 81, 25–32.

Lewis, L.C., Bruck, D.J., Prasifka, J.R., Raun, E.S., 2009. *Nosema pyrausta*: its biology, history, and potential role in a landscape of transgenic insecticidal crops. Biol. Control 48, 223–231. http://dx.doi.org/10.1016/j.biocontrol.2008.10.009

Li, R., Zhang, R., Zhang, L., Zou, J., Xing, Q., Dou, H., et al., 2015. Characterizations and expression analyses of NF-kappa B and Rel genes in the Yesso scallop (*Patinopecten yessoensis*) suggest specific response patterns against Gram-negative infection in bivalves. Fish Shellfish Immunol. 44, 611–621. doi: 10.1016/j.fsi.2015.03.036

Liebhold, A.M., Plymale, R., Elkinton, J.S., Hajek, A.E., 2013. Emergent fungal entomopathogen does not alter density dependence in a viral competitor. Ecology 94, 1217–1222. doi: 10.1890/12-1329.1

Lucarotti, C.J., 2000. Cytology of *Leidyana canadensis* (Apicomplexa : Eugregarinida) in *Lambdina fiscellaria fiscellaria* larvae (Lepidoptera : Geometridae). J. Invertebr. Pathol. 75, 117–125.

Lucarotti, C.J., Andreadis, T.G., 1995. Reproduction strategies and adaptations for survival among obligatory microsporidian and fungal parasites of mosquitoes: a comparative analysis of *Amblyospora* and *Coelomomyces*. J. Am. Mosquito Contr. 11, 111–121. doi: 10.1006/jipa.1999.4911

Lyons, M.M., Ward, J.E., Smolowitx, R., Uhlinger, K.R., 2005. Lethal marine snow: pathogen of bivalve mollusc concealed in marine aggregates. Limnol. Oceanogr. 50, 1983–1988. doi: 10.4319/lo.2005.50.6.1983

Maddox, J.V., 1968. Generation time of the microsporidian *Nosema necatrix* in larvae of the armyworm, *Pseudaletia unipuncta*. J. Invertebr. Pathol. 11, 90–96. doi:10.1016/0022-2011(68)90057-8

Malakar, R., Elkinton, J.S., Carroll, S.D., D'Amico, V., 1999. Interactions between two gypsy moth (Lepidoptera: Lymantriidae) pathogens: Nucleopolyhedrovirus and *Entomophaga maimaiga* (Zygomycetes: Entomophthorales): Field studies and a simulation model. Biol. Control 16, 189–198. doi: 10.1006/bcon.1999.0751

Maneesakorn, P., An, R., Daneshvar, H., Taylor, K., Bai, X., Adams, B.J., et al., 2011. Phylogenetic and cophylogenetic relationships of entomopathogenic nematodes (*Heterorhabditis*: Rhabditida) and their symbiotic bacteria (*Photorhabdus*: Enterobacteriaceae). Mol. Phylogenet. Evol. 59, 271–280. doi: 10.1016/j.ympev.2011.02.012

Mascarin, G.M., Guarin-Molina, J.H., Arthurs, S.P., Humber, R.A., Moral, R. de A., Borges, C., et al., 2016. Seasonal prevalence of the insect pathogenic fungus *Colletotrichum nymphaeae* in Brazilian citrus groves under different chemical pesticide regimes. Fungal Ecol. 22, 43–51. doi: 10.1016/j.funeco.2016.04.005

Mathis, A., Weber, T., Deplazes, P., 2005. Zoonotic potential of the microsporidia. Clin. Microbiol. Rev. 18, 423–445. doi: 10.1128/CMR.18.3.423-445.2005

May, R.M., Anderson, R.M., 1979. Population biology of infectious diseases: part II. Nature 280, 455–461. doi: 10.1038/280455a0

McAdam, P.R., Richardson, E.J., Fitzgerald, J.R., 2014. High-throughput sequencing for the study of bacterial pathogen biology. Curr. Opin. Microbiol. 19, 106–113. doi: 10.1016/j.mib.2014.06.002

McCallum, H., 2012. Disease and the dynamics of extinction. Proc. Roy. Soc. Lond. B Biol. Sci. 367(1604), 2828–2839. doi: 10.1098/rstb.2012.0224

McCallum, H., Barlow, N., Hone, J., 2001. How should pathogen transmission be modelled? Trends Ecol. Evol. 16, 295–300. doi: 10.1016/S0169-5347(01)02144-9

McMahon, D.P., Natsopoulou, M.E., Doublet, V., Fuerst, M., Weging, S., Brown, M.J.F., et al., 2016. Elevated virulence of an emerging viral genotype as a driver of honey bee loss. Proc. Roy. Soc. Lond. B Biol. Sci. 283 (1833), 20160811. doi: 10.1098/rspb.2016.0811

Mikronranta, L., Mappes, J., Laakso, J., Ketola, R., 2015. Host evolution decreases virulence in an opportunistic bacterial pathogen. BMC Evol. Biol. 15, 165–173. doi: 10.1186/s12862-015-0447-5

Moutailler, S., Roche, G., Thiberge, J.-M., Caro, V., Rougeon, F., Failloux, A.-B., 2011. Host alternation is necessary to maintain the genome stability of Rift Valley Fever Virus. PloS Neglect. Trop. D. 5, e1156. doi: 10.1371/journal.pntd.0001156

Müller, S., Garcia-Gonzalez, E., Genersch, E., Sussmuth, R.D., 2015. Involvement of secondary metabolites in the pathogenesis of the American foulbrood of honey bees caused by *Paenibacillus larvae*. Nat. Prod. Rep. 32, 765–778.

Multeau, C., Froissart, R., Perrin, A., Castelli, I., Casartelli, M., Ogliastro, M., 2012. Four amino acids of an insect densovirus capsid that determine midgut tropism and virulence. J. Virol. 86, 5937–5941. doi: 10.1039/c4np00158c

Murillo, R., Hussey, M.S., Possee, R.D., 2011. Evidence for covert baculovirus infections in a *Spodoptera exigua* laboratory culture. J. Gen. Virol. 92, 1061–1070. doi: 10.1099/vir.0.028027-0

Murray, K.D., Elkinton, J.S., 1989. Environmental contamination of egg masses as a major component of transgenerational transmission of gypsy moth nuclear polyhedrosis virus (LdMNPV). J. Invertebr. Pathol. 53, 324–334. doi: 10.1016/0022-2011(89)90096-7

Myers, J., Cory, J., 2016. Ecology and evolution of pathogens in natural populations of Lepidoptera. Evol. Appl. 9, 231–247. doi: 10.1111/eva.12328

Nagamine, T., Sako, Y., 2016. A role for the anti-viral host defense mechanism in the phylogenetic divergence in baculovirus evolution. PLoS ONE 11(5), e0156394. doi: 10.1371/journal.pone.0156394

Nalepa, C.A., Weir, A., 2007. Infection of *Harmonia axyridis* (Coleoptera: Coccinellidae) by *Hesperomyces virescens* (Ascomycetes: Laboulbeniales): role of mating status and aggregation behavior. J. Invertebr. Pathol. 94, 196–203. http://dx.doi.org/10.1016/j.jip.2006.11.002

Natsopoulou, M.E., McMahon, D.P., Paxton, R.J., 2016. Parasites modulate within-colony activity and accelerate the temporal polyethism schedule of a social insect, the honey bee. Behav. Ecol. Sociobiol. 70, 1019–1031. doi: 10.1007/s00265-015-2019-5

Naug, D., Gibbs, A., 2009. Behavioral changes mediated by hunger in honey bees infected with *Nosema ceranae*. Apidologie 40, 595–599. doi: 10.1051/apido/2009039

Newman, G.G., Carner, G.R., 1974. Diel periodicity of *Entomophthora gammae* in the soybean looper. Environ. Entomol. 3, 888–890.

Ni, X., Backus, E., Maddox, J.V., 1997. Transmission mechanisms of *Nosema empoascae* (Microspora : Nosematidae) in *Empoasca fabae* (Homoptera: Cicadellidae). J. Invertebr. Pathol. 69, 269–275. doi: 10.1006/jipa.1996.4646

Nielsen, C., Keena, M., Hajek, A.E., 2005. Virulence and fitness of the fungal pathogen *Entomophaga maimaiga* in its host *Lymantria dispar*, for pathogens and host strains originating from Asia, Europe and North America. J. Invertebr. Pathol. 89, 232–242. doi: 10.1016/j.jip.2005.05.004

Obbard, D.J., Dudas, G., 2014. The genetics of host–virus coevolution in invertebrates. Curr. Opin. Virol. 8, 73–78. doi: 10.1016/j.coviro.2014.07.002

Ohkawa, A., Aoki, J., 1980. Fine structure of resting spore formation and germination in *Entomophthora virulenta*. J. Invertebr. Pathol. 35, 279–289. doi: 10.1016/0022-2011(80)90163-9

Okamura, B., O'Dea, A., Knowles, T., 2011. Bryozoan growth and environmental reconstruction by zooid size variation. Mar. Ecol. Prog. Ser. 430, 133–146. doi: 10.3354/meps08965

Oi, D.H., Williams, D.F., Pereira, R.M., Horton, P., Davis, T.S., Hyder, A.H., et al., 2008. Combining biological and chemical controls for the management of red imported fire ants (Hymenoptera: Formicidae). Am. Entomol. 54, 46–55. doi: http://dx.doi.org/10.1093/ae/54.1.46

Olofsson, E., 1988. Dispersal of the nuclear polyhedrosis virus of *Neodiprion sertifer* from soil to pine foliage with dust. Entomol. Exp. Appl. 46, 181–186. doi: 10.1111/j.1570-7458.1988.tb01109.x

Onstad, D.W., Fuxa, J.R., Humber, R.A., Oestergaard, J., Shapiro-Ilan, D.I., Gouli, V.V., et al., 2006. Abridged Glossary of Terms Used in Invertebrate Pathology, 3rd edn. Society for Invertebrate Pathology. Available from: http://www.sipweb.org/resources/glossary.html (accessed May 8, 2017).

Opoku-Debrah, J.K., Hill, M.P., Knox, C., Moore, S.D., 2016. Heterogeneity in virulence relationships between *Cryptophlebia leucotreta* granulovirus isolates and geographically distinct host populations: lessons from codling moth resistance to CpGV-M. BioControl 61, 449–459. doi: 10.1007/s10526-016-9728-1

Oreste, M., Bubici, G., Poliseno, M., Tarasco, E., 2016. Effects of entomopathogenic fungi on *Encarsia formosa* Gahan. (Hymenoptera: Aphelinidae) activity and behavior. Biol. Control 100, 46–53. doi: 10.1016/j.biocontrol.2016.05.011

Ormond, E.L., Thomas, A.P.M., Pugh, P.J.A., Pell, J.D., Roy, H.E., 2010. A fungal pathogen in time and space: the population dynamics of *Beauveria bassiana* in a conifer forest. FEMS Microbiol. Ecol. 74, 146–154. doi: 10.1111/j.1574-6941.2010.00939.x

Ortiz-Urquiza, A., Luo, Z., Keyhani, N.O., 2015. Improving mycoinsecticides for insect biological control. Appl. Microbiol. Biotechnol. 99, 1057. doi: 10.1007/s00253-014-6270-x

Paillard, C., Le Roux, F., Borrego, J.J., 2004. Bacterial disease in marine bivalves, a review of recent studies: trends and evolution. Aquat. Living Resour. 17, 477–498. doi: 10.1051/alr:2004054

Palma, L., Munoz, D., Berry, C., Murillo, J., Caballero, P., 2014. *Bacillus thuringiensis* toxins: an overview of their biocidal activity. Toxins 6, 3296–3325. doi: 10.3390/toxins6123296

Passarelli, A.L., 2011. Barriers to success: How baculoviruses establish efficient systemic infections. Virology 411, 383–392. doi: 10.1016/j.virol.2011.01.009

Perotto, M.C., Di Rienzo, J.A., Lanati, S., Panonto, S., Macchiavelli, R., Cafrune, E.E., Conci, V.C., 2014. Temporal and spatial spread of potyvirus infection and its relationship to aphid populations. Australas. Plant Path. 43, 623–630. doi: 10.1007/s13313-014-0312-9

Pettis, J.S., vanEngelsdorp, D., Johnson, J., Dively, G., 2012. Pesticide exposure in honey bees results in increased levels of the gut pathogen *Nosema*. Naturwissenschaften 99, 153–158. doi: 10.1007/s00114-011-0881-1

Pettis, J.S., Lichtenberg, E.M., Andree, M., Stitzinger, J., Rose, R., vanEngelsdorp, D., 2013. Crop pollination exposes honey bees to pesticides which alters their susceptibility to the gut pathogen *Nosema ceranae*. PLoS ONE 8(7), e70182. doi: 10.1371/journal.pone 0070182

Pilarska, D.K., Solter, L.F., Kereselidze, M., Linde, A., Hoch, G., 2006. Microsporidian infections in *Lymantria dispar* larvae: interactions and effects of multiple species infections on pathogen horizontal transmission. J. Invertebr. Pathol. 93, 105–113. doi: 10.1016/j.jip.2006.05.003

Pirofski, L., Casadevall, A., 2012. What is a pathogen? A question that begs the point. BMC Biol. 201210:6. doi: 10.1186/1741-7007-10-6

Podgwaite, J.D., Mazzone, H.M., 1986. Latency of insect viruses. Adv. Virus Res. 31, 293–320.

Pombert, J.-F., Selman, M., Burki, F., Floyd, B.P., Farinelli, K., Solter, L.F., et al., 2012. Gain and loss of multiple functionally-related horizontally transferred genes in the reduced genomes of two microsporidian parasites. Proc. Natl. Acad. Sci. U.S.A. 109, 12 638–12 643. doi: 10.1073/pnas.1205020109

Pombert, J.-F., Xu, J., Smith, D., Heiman, S.Y., Cuomo, C.A., Weiss, L.M., Keeling, P.J., 2013. Complete genome sequences from three genetically distinct strains reveal high intraspecies genetic diversity in the microsporidian *Encephalitozoon cuniculi*. Eukaryot. Cell. 12, 503–511. doi: 10.1128/EC.00312-12

Pohorecka, K., Bober, A., Skubida, M., Zdanska, D., Toroj, K., 2014. A comparative study of environmental conditions, bee management and the epidemiological situation in apiaries varying in the level of colony losses. J. Apic. Sci. 58, 107–132. doi: 10.2478/JAS-2014-0027

Prado, S., Dubert, J., da Costa, F., Martinez-Patino, D., 2014. Vibrios in hatchery cultures of the razor clam, *Solen marginatus* (Pulteney). J. Fish Dis. 37, 209–217. doi: 10.1111/jfd.12098

Purcell, A.H., Suslow, K.G., Klein, M., 1994. Transmission via plants of an insect pathogenic bacterium that does not multiply or move in plants. Microb. Ecol. 27, 19–26. doi: 10.1007/BF00170111

Raymond, B., Wyres, K.L., Sheppard, S.K., Ellis, R.J., Bonsall, M.B., 2010. Environmental factors determining the epidemiology and population genetic structure of the *Bacillus cereus* group in the field. PLoS Pathog. 6, e1000905. doi: 10.1371/journal.ppat.1000905

Rangel, D.E., Braga, G.U., Fernandes, E.K., Keyser, C.A., Hallsworth, J.E., Roberts, D.W., 2015. Stress tolerance and virulence of insect-pathogenic fungi are determined by environmental conditions during conidial formation. Curr. Genet. 61, 383–404. doi: 10.1007/s00294-015-0477-y

Reeson, A.F., Wilson, K., Gunn, A., Hails, R.S., Goulson, D., 1998. Baculovirus resistance in the noctuid *Spodoptera exempta* is phenotypically plastic and responds to population density. Proc. Roy. Soc. Lond. B Biol. Sci. 265, 1787–1791. doi: 10.1098/rspb.1998.0503

Reilly, J.R., Hajek, A.E., 2008. Density-dependent resistance of the gypsy moth *Lymantria dispar* to its nucleopolyhedrovirus, and the consequences for population dynamics. Oecologia 154, 691–701. doi: 10.1007/s00442-007-0871-3

Reilly, J.R., Hajek, A.E., Liebhold, A.M., Plymale, R.S., 2014. The impact of *Entomophaga maimaiga* on outbreak gypsy moth populations: the role of weather. Environ. Entomol. 43, 632–641. doi: http://dx.doi.org/10.1603/EN13194

Richardson, L.L., Adler, L.S., Leonard, A.S., Andicoechea, J., Regan, K.H., Anthony, W.E., et al., 2015. Secondary metabolites in floral nectar reduce parasite infections in bumble bees. Proc. Roy. Soc. Lond. B Biol. Sci. 282, 20142471. doi: 10.1098/rspb.2014.2471

Richardson, L.L., Bowers, M.D., Irwin, R.E., 2016. Nectar chemistry mediates the behavior of parasitized bees: consequences for plant fitness. Ecology 97, 325–337. doi: 10.1890/15-0263.1

Rickert-Campbell, L., Gaugler, R., 1991. Role of the sheath in desiccation tolerance of two entomopathogenic nematodes. Nematologica 37, 324–332. doi: 10.1163/187529291X00321

Rohrmann, G.F., 2013. Baculovirus Molecular Biology, 3rd edn. National Center for Biotechnology Information, Bethesda, MD.

Roy, H.E., Pell, J.K., 2000. Interactions between entomopathogenic fungi and other natural enemies: implications for biological control. Biocontrol Sci. Techn. 10, 737–752. doi: 10.1080/09583150020011708

Roy, H.E., Steinkraus, D.C., Eilenberg, J., Hajek, A.E., Pell, J.K., 2006. Bizzare interactions and endgames: entomopathogenic fungi and their arthropod hosts. Annu. Rev. Entomol. 51, 331–357. doi: 10.1146/annurev.ento.51.110104.150941

Rudolf, V.H.W., Antonovics, J., 2007. Disease transmission by cannibalism: rare event or common occurrence? Proc. Roy. Soc. Lond. B Biol. Sci. 274, 1205–1210. doi: 10.1098/rspb.2006.0449

Sadeh, A., Northfield, T.D., Rosenheim, J.A., 2016. The epidemiology and evolution of parasite transmission through cannibalism. Ecology 97, 2003–2011. doi: 10.1890/15-0884.1

Sadeh, A., Rosenheim, J.A., 2016. Cannibalism amplifies the spread of vertically transmitted pathogens. Ecology 97, 1994–2002. doi: 10.1890/15-0825.1

Safavi, S.A., 2012. Attenuation of the entomopathogenic fungus *Beauveria bassiana* following serial in vitro transfers. Biologia (Bratislava) 67, 1062–1068. doi: 10.2478/s11756-012-0120-z

Sait, S.M., Begon, M., Thompson, D.J., 1994. The effects of a sublethal baculovirus infection in the Indian meal moth, *Plodia interpunctella*. J. Anim. Ecol. 63, 541–550. doi: 10.2307/5220

Saito, T., Bjornson, S., 2013. The convergent lady beetle, *Hippodamia convergens* Guerin-Meneville and its endoparasitoid *Dinocampus coccinellae* (Schrank): the effect of a microsporidium on parasitoid development and host preference. J. Invertebr. Pathol. 113, 18–25. doi: 10.1016/j.jip.2013.01.003

Sajjadian, S., Hosseininaveh, V., 2015. Destruction of peritrophic membrane and its effect on biological characteristics and activity of digestive enzymes in larvae of the Indian meal moth, *Plodia interpunctella* (Lepidoptera: Pyralidae). Eur. J. Entomol. 112, 245–250. doi: 10.14411/eje.2015.046

Schmeller, D.S., Blooi, M., An, M., Trenton, W.J., Garner, M.C., Fisher, F.A., et al., 2014. Microscopic aquatic predators strongly affect infection dynamics of a globally emerged pathogen. Curr. Biol. 24, 176–180. doi: http://dx.doi.org/10.1016/j.cub.2013.11.032

Schmid-Hempel P., Frank S.A., 2007. Pathogenesis, virulence, and infective dose. PLoS Pathog. 83, 1372–1373. doi: 10.1371/journal.ppat.0030147

Schrével, J., Desporte, I., 2013. Introduction: gregarines among Apicomplexa, in: Desporte, I., Schrével, J. (eds.), Treatise on Zoology: Anatomy, Taxonomy, Biology, The Gregarines, Vol. 1. Brill, Leiden, pp. 7–24.

Schuler, H., Koeppler, K., Daxboeck-Horvath, S., Rasool, B., Krumboeck, S., Schwarz, D., et al., 2016. The hitchhiker's guide to Europe: the infection dynamics of an ongoing *Wolbachia* invasion and mitochondrial selective sweep in *Rhagoletis cerasi*. Mol. Ecol. 25, 1595–1609. doi: 10.1111/mec.13571

Schwartz, R.S., Bauchan, G.R., Murphy, C.A., Ravoet, J., de Graaf, D.C., Evans, J.D., 2015. Characterization of two species of Trypanosomatidae from the honey bee *Apis mellifera*: *Crithidia mellificae* Langridge and McGhee, and *Lotmaria passim* n. gen., n. sp. Eukaryot. Microbiol. 62, 567–583. doi: 10.1111/jeu.12209

Shah, F.A., Wang, C.S., Butt, T.M., 2005. Nutrition influences growth and virulence of the insect-pathogenic fungus *Metarhizium anisopliae*. FEMS Microbiol. Lett. 251, 259–266. http://dx.doi.org/10.1016/j.femsle.2005.08.010

Shapiro, D.I., McCoy, C.W., Fares, A., Obreza, T., Dou, H. 2000. Effects of soil type on virulence and persistence of entomopathogenic nematodes in relation to control of *Diaprepes abbreviatus* (Coleoptera: Curculionidae). Environ. Entomol. 29, 1083–1087. doi: http://dx.doi.org/10.1603/0046-225X-29.5.1083

Shapiro, K., Krusor, C., Mazzillo, F.F.M., Conrad, P.A., Largier, J.L., Mazet JAK, Silver, M.W., 2014. Aquatic polymers can drive pathogen transmission in coastal ecosystems. Proc. Roy. Soc. Lond. B Biol. Sci. 281, 20141287. doi: 10.1098/rspb.2014.1287

Shapiro-Ilan, D.I., Raymond, B., 2015. Limiting opportunities for cheating stabilizes virulence in insect parasitic nematodes. Evol. Applic. 9, 462–470. doi: 10.1111/eva.12348

Shapiro-Ilan, D.I., Fuxa, J.R., Lacey, L.A., Onstad, D.W., Kaya, H.K., 2005. Definitions of pathogenicity and virulence in invertebrate pathology. J. Invertebr. Pathol. 88, 1–7. doi: 10.1016/j.jip.2004.10.003

Shapiro-Ilan, D.I., Bruck, D.J., Lacey, L.A., 2012. Principles of epizootiology and microbial control, in: Vega, F.E., Kaya, H.K. (eds.), Insect Pathology, 2nd edn. Academic Press, San Diego, CA, pp. 29–72. doi: 10.1016/B978-0-12-384984-7.00003-8

Shapiro-Ilan, D.I., Lewis, E.E., Schliekelman, P., 2014. Aggregative group behavior in insect parasitic nematode dispersal. Int. J. Parasitol. 44, 49–54. doi: 10.1016/j.ijpara.2013.10.002

Shen, J.-P., Chou, C.-F., 2016. Morphological plasticity of bacteria – open questions. Biomicrofluidics 10, 031501. doi: 10.1063/1.4953660

Shi, W.-P., Wang, Y.-Y., Fen, L., Guo, C., Chent, X., 2009. Persistence of *Paranosema* (*Nosema*) *locustae* (Microsporidia: Nosematidae) among grasshopper (Orthoptera: Acrididae) populations in the Inner Mongolia Rangeland, China. BioControl 54, 77–84. doi: 10.1007/s10526-008-9153-1

Shi, W., Guo, Y., Xu, C., Tan, S., Miao, J., Feng, Y., et al., 2014. Unveiling the mechanism by which microsporidian parasites prevent locust swarm behavior. Proc. Natl. Acad. Sci. U.S.A. 111, 1343–1348. doi: 10.1073/pnas.1314009111

Siegel, J.P., Maddox, J.V., Ruesink, W.G., 1988. Seasonal progress of *Nosema pyrausta* in the European corn borer. J. Invertebr. Pathol. 52, 130–136. doi: 10.1016/0022-2011(88)90111-5

Silvio, E., Denner, A., Bobis, O., Forsgren, E., Robin, F.A., 2014. Diversity of honey stores and their impact on pathogenic bacteria of the honey bee, *Apis mellifera*. Ecol. Evol. 4, 3960–3967. doi: 10.1002/ece3.1252

Simões, R.A, Reis, L.G., Bento, J.M.S., Solter, L.F., Delalibera, I. Jr., 2012. Biological and behavioral parameters of the parasitoid *Cotesia flavipes* (Hymenoptera: Braconidae) are altered by the pathogen *Nosema* sp. (Microsporidia: Nosematidae). Biol. Control 63, 164–171. doi: 10.1016/j.biocontrol.2012.06.012

Slavicek, J.M, Hayesplazolles, N., Kelly, M.E., 1995. Rapid formation of few polyhedral mutants of *Lymantria dispar* multinucleocapsid nuclear polyhedrosis virus during serial passage in cell culture. Biol. Contr. 5, 251–261. doi: 10.1006/bcon.1995.1031

Solter, L.F., 2006. Transmission as a predictor of ecological host specificity with a focus on vertical transmission of microsporidia. J. Invertebr. Pathol. 92, 132–140. doi: 10.1016/j.jip.2006.03.008

Solter, L.F., Hajek, A.E., 2009. Control of gypsy moth, *Lymantria dispar*, in North America since 1878, in: Hajek, A.E., Glare, T.R., O'Callaghan, M. (eds.), Use of Microbes for Control and Eradication of Invasive Arthropods. Springer, New York, pp. 181–212. doi: 10.1007/978-1-4020-8560-4_11

Solter, L.F., Maddox, J.V., 1998a. Timing of an early sporulation sequence of microsporidia in the genus *Vairimorpha* (Microsporidia: Burenllidae). J. Invertebr. Pathol. 72, 323–329. doi: 10.1006/jipa.1998.4815

Solter, L.F., Maddox, J.V., 1998b. Physiological host specificity of microsporidia as an indicator of ecological host specificity. J. Invertebr. Pathol. 71, 207–216. doi: 10.1006/jipa.1997.4740

Solter, L.F., Maddox, J.V., Onstad, D.W., 1991. Transmission of *Nosema pyrausta* in adult European corn borers. J. Invertebr. Pathol. 57, 220–226. doi: 10.1016/0022-2011(91)90120-F

Sorrell, I., White, A., Pedersen, A.B., Hails, R.S., Boots, M., 2009. The evolution of covert, silent infection as a parasite strategy. Proc. Roy. Soc. Lond. B Biol. Sci. 276, 2217–2226. doi: 10.1098/rspb.2008.1915

Steinhaus, E.A., Martignoni, M.E., 1970. An Abridged Glossary of Terms Used in Invertebrate Pathology, 2nd edn. USDA Forest Service, PNW Forest and Range Experiment Station. Available from: http://www.sipweb.org/resources/glossary.html (accessed May 8, 2017).

Steinkraus, D.C., Hollingsworth, R.G., Boys, G.O., 1996. Aerial spores of Neozygites fresenii (Entomophthorales: Neozygitaceae): density, periodicity, and potential role in cotton aphid (Homoptera: Aphididae) epizootics. Environ. Entomol. 25, 48–57. doi: http://dx.doi.org/10.1093/ee/25.1.48

Steinkraus, D.C., Howard, M.N., Hollingsworth, R.G., Boys, G.O., 1999. Infection of sentinel cotton aphids (Homoptera: Aphididae) by aerial conidia of *Neozygites fresenii* (Entomophthorales: Neozygitaceae). Biol. Control 14, 181–185. doi: 10.1006/bcon.1998.0690

Stock, P., Vandenberg, J., Boemare, N., Glazer, I. (eds.), 2009. Insect Pathogens: Molecular Approaches and Techniques. CABI, Wallingford.

Tailliez, P., Laroui, C., Ginibre, N. Boemare, N., 2009. Phylogeny of *Photorhabdus* and *Xenorhabdus* based on universally conserved protein-coding sequences and implications for the taxonomy of these two genera. Proposal of new taxa: *X. vietnamensis* sp. nov., *P. luminescens* subsp. *caribbeanensis* subsp. nov., *P. luminescens* subsp. *hainanensis* subsp. nov., *P. temperata* subsp. *khanii* subsp. nov., *P. temperata* subsp. *tasmaniensis* subsp. nov. Int. J. Syst. Evol. Micr. 60, 1921–1937. doi: 10.1099/ijs.0.014308-0

Takatori, K., Tanaka, I., 1982. *Ascosphaera apis* isolated from chalk brood in honey bees. Nihon Chikusan Gakk. 53, 89–92. doi: 10.1016/j.jip.2009.06.018

Tanada, Y., Fuxa, J.R. 1987. The pathogen population, in: Fuxa, J.R., Tanada, Y. (eds.), Epizootiology of Insect Diseases. John Wiley & Sons, New York.

Tanada, Y., Kaya, H.K., 1993. Insect Pathology. Academic Press, San Diego, CA.

Terry, R.S., Smith, J.E., Sharpe, R.G., Rigaud, R., Littlewood, D.T.J., Ironside, J.E., et al., 2004. Widespread vertical transmission and associated host sex-ratio distortion within the eukaryotic phylum Microspora. Proc. Roy. Soc. Lond. B Biol. Sci. 271, 1783–1789. doi: 10.1098/rspb.2004.2793

Theantana, T., Chantawannakul, P., 2008. Protease and β-N-acetylglucosaminidase of honey bee chalkbrood pathogen *Ascosphaera apis*. J. Apicult. Res. 47, 68–76. doi: 10.1080/00218839.2008.11101426

Thompson, J.N., 2005. The Geographic Mosaic of Coevolution. University of Chicago Press, Chicago, IL.

Timms, A.R., Cambray-Young, J., Scott, A.E., Petty, N.K., Connerton, P.L., Clarke, L., et al., 2010. Evidence for a lineage of virulent bacteriophages that target *Campylobacter*. BMC Genomics 11, 214. doi: 10.1186/1471-2164-11-214

Toplak, I., Ciglenecki, U.J., Aronstein, K., Gregorc, A., 2013. Chronic bee paralysis virus and *Nosema ceranae* experimental co-infection of winter honey bee workers (*Apis mellifera* L.). Viruses 5, 2282–2297. doi: 10.3390/v5092282

Undeen, A.H., Becnel, J.J., 1992. Longevity and germination of *Edhazardia aedis* Microspora: Amblyosporidae) spores. Biocontrol Sci. Techn. 2, 247–256. doi: 10.1080/09583159209355238

Vallet-Gely, I., Lemaitre, B., Boccard, F., 2008. Bacterial strategies to overcome insect defences. Nature Rev. Microbiol. 6, 302–313. doi: 10.1038/nrmicro1870

van Frankenhuyzen, K., Nystrom, C., Liu, Y., 2007. Vertical transmission of *Nosema fumiferanae* (Microsporidia: Nosematidae) and consequences for distribution, post-diapause emergence and dispersal of second-instar larvae of the spruce budworm, *Choristoneura fumiferana* (Clem.) (Lepidoptera: Tortricidae). J. Invertebr. Pathol. 96, 173–182. http://dx.doi.org/10.1016/j.jip.2007.03.017

Varaldi, J., Martinez, J., Patot, S., Lepetit, D., Fleury, F., Gandon, S., 2012. An inherited virus manipulating the behavior of its parasitoid host: epidemiology and evolutionary consequences, in: Beckage, N.E., Drezen, J.M. (eds.), Parasitoid Viruses: Symbionts and Pathogens. Elsevier, London, pp. 203–214. doi: 10.1016/B978-0-12-384858-1.00017-5

Vega, F.E., Kaya, H.K., 2012. Insect Pathology, 2nd edn. Academic Press, San Diego, CA.

Vega, F.E., Meyling, N.V., Luangsa-ard, J.J., Blackwell, M., 2012. Fungal entomopathogens, in: Vega, F.E., Kaya, H.K. (eds.), Insect Pathology, 2nd edn. Academic Press, San Diego, CA, pp. 171–220.

Vega Thurber, R.L., Barott, K.L., Hall, D., Liu, H., Rodrigues-Mueller, B., Desnues, C., et al., 2008. Metagenomic analysis indicates that stressors induce production of herpes-like viruses in the coral *Porites compressa*. Proc. Nat. Acad. Sci. U.S.A. 105, 18413–18418. doi: 10.1073/pnas.0808985105

Vega Thurber, R., Willner, H.D., Rodriguez-Mueller, B., Desnues, C., Edwards, R.A., Angly, F., et al., 2009. Metagenomic analysis of stressed coral holobionts. Environ Microbiol. 11, 2148–2163. doi: 10.1111/j.1462-2920.2009.01935.x

Vilaplana, L., Redman, E., Wilson, K., Cory, J.S., 2008. Density-related variation in vertical transmission of a virus in the African armyworm. Oecologia 155, 237–246. doi: 10.1007/s00442-007-0914-9

Vilaplana, L., Wilson, K., Redman, E.M., Cory, J.S., 2010. Pathogen persistence in migratory insects: high levels of vertically-transmitted virus infection in field populations of the African armyworm. Evol. Ecol. 24, 147–160. doi: 10.1007/s10682-009-9296-2

Virto, C., Zarate, C.A., Lopez-Ferber, M., Murillo, R., Caballero, P., Williams, T., 2013. Gender-mediated differences in vertical transmission of a nucleopolyhedrovirus. PLOS ONE 8, e70932. doi: 10.1371/journal.pone.0070932

Vojvodic, S., Jensen, A.B., Markussen, B., Eilenberg, J., Boomsma, J.J., 2011. Genetic variation in virulence among chalkbrood strains infecting honey bees. PLoS ONE 6, e25035. doi: 10.1371/journal.pone.0025035

Waldor, M.K., Tyson, G., Borenstein, E., Ochman, H., Moeller. A., Finlay, B.B., et al., 2015. Where next for microbiome research? PLOS Biol. 13(1), e1002050. doi: 10.1371/journal. pbio.1002050

Walker, M., Basáñez, M.G., Ouédraogo, A.L., Hermsen, C., Bousema, T., Churcher, T.S., 2015. Improving statistical inference on pathogen densities estimated by quantitative molecular methods: malaria *gametocytaemia* as a case study. BMC Bioinformatics 16, 5. doi: 10.1186/s12859-014-0402-2

Walker, P.J., Winton, J.R., 2010. Emerging viral diseases of fish and shrimp. Vet Res. 41, 51. doi: 10.1051/vetres/2010022

Wang, C.S., Hu, G., St. Leger, R.J., 2005. Differential gene expression by *Metarhizium anisopliae* growing in root exudate and *Metarhizium anisopliae* in the rhizosphere host (*Manduca sexta*) cuticle or hemolymph reveals mechanisms of physiological adaptation. Fungal Genet. Biol. 42, 704–718. http://dx.doi.org/10.1016/j.fgb.2005.04.006

Wang, Y., Bilgrami, A.L, Shapiro-Ilan, D., Gaugler, R., 2007. Stability of entomopathogenic bacteria, *Xenorhabdus nematophila* and *Photorhabdus luminescens*, during in vitro culture. J. Industr. Microbiol. Biotechnol. 34, 73–81.

Washburn, J.O., Egerter, D.E., Anderson, J.R., Saunders, J.A., 1988. Density reduction in larval mosquito (Diptera: Culicidae) populations by interactions between a parasitic cilate (Ciliophora: Tetrahymenidae) and an opportunistic fungal (Oomycetes: Pythiaceae) parasite. J. Med. Entomol. 25, 307–314. doi: http://dx.doi.org/10.1093/jmedent/25.5.307

Weiser, J., 1963. Sporozoan infections, in: Steinhaus, E.A. (ed.), Insect Pathology: An Advanced Treatise. Academic Press, San Diego, CA, pp. 291–334.

Weiser, J., 1969. Immunity of insects to protozoa, in: Jackson, G.J., Herman, R., Singer, I. (eds.), Immunity to Parasitic Animals, Vol I. Appleton-Century-Crofts, New York, pp. 129–147.

Weseloh, R.M., 2002. Modeling the impact of the fungus *Entomophaga maimaiga* (Zygomycetes: Entomophthorales) on gypsy moth (Lepidoptera: Lymantriidae): incorporating infection by conidia. Environ. Entomol. 3, 1071–1084. doi: http://dx.doi. org/10.1603/0046-225X-31.6.1071

Weseloh, R.M., Andreadis, T.G., 1992. Epizootiology of the fungus *Entomophaga maimaiga*, and its impact on gypsy moth populations. J. Invertebr. Pathol. 59, 133–141. doi: 10.1016/0022-2011(92)90023-W

Weseloh, R.M., Andreadis, T.G., 2002. Detecting the titer in forest soils of spores of the gypsy moth (Lepidoptera: Lymantriidae) fungal pathogen, *Entomophaga maimaiga* (Zygomycetes: Entomophthorales). Can. Entomol. 134, 269–279. doi: http://dx.doi. org/10.4039/Ent134269-2

White, A., Watt, A.D., Hails, R.S. and Hartley, S.E., 2000. Patterns of spread in insect-pathogen systems: the importance of pathogen dispersal. Oikos, 89: 137–145. doi: 10.1034/j.1600-0706.2000.890115.x

Wichadakul, D., Kobmoo, M., Ingsriswang, S., Tangphatsornruang, S., Chantasingh, D., Luangsa-ard, J.J., Eurwilaichitr, L., 2015. Insights from the genome of *Ophiocordyceps polyrhachis-furcata* to pathogenicity and host specificity in insect fungi. BMC Genomics 16, 881. doi: 10.1186/s12864-015-2101-4

Wille, H., Martignoni, M.E., 1952. Vorläufige Mitteilung über einen neuen Krankheitstypus beim Engerling von *Melolontha vulgaris* F. Schwiez. A. All. Pathol. Bakteriol. 15, 470–473.

Williams, T., 1995. Patterns of covert infection by invertebrate pathogens: iridescent viruses of blackflies. Molec. Ecol. 4, 447–457. doi: 10.1111/j.1365-294X.1995.tb00238.x

Wolf, J.B.W., 2013. Principles of transcriptome analysis and gene expression quantification: an RNA-seq tutorial. Mol. Ecol. Resour. 13, 559–572. doi: 10.1111/1755-0998.12109

Wu, J.Y., Smart, M.D., Anelli, C.M., Sheppard, W.S., 2012. Honey bees (*Apis mellifera*) reared in brood combs containing high levels of pesticide residues exhibit increased susceptibility to *Nosema* (Microsporidia) infection. J. Invertebr. Pathol. 109, 326–329. http://dx.doi.org/10.1016/j.jip.2012.01.005

Yadav, A.K., Lalramliana, 2012. Soil moisture effects on the activity of three entomopathogenic nematodes (Steinernematidae and Heterorhabditidae) isolated from Meghalaya, India. J. Parasit. Dis. 36, 94–98. doi: 10.1007/s12639-011-0076-x

Young, S.Y., Yearian, W.C., 1986. Movement of a nuclear polyhedrosis virus from soil to soybean and transmission in *Anticarsia gemmatalis* (Hübner) (Lepidoptera: Noctuidae) populations on soybean. Environ. Entomol. 15, 573–580. doi: http://dx.doi.org/10.1093/ee/15.3.573

Young, S.Y., Yearian, W.C., 1990. Transmission of nuclear polyhedrosis virus by the parasitoid *Microplitis croceipes* (Hymenoptera: Braconidae) to *Heliothis virescens* (Lepidoptera: Noctuidae) on soybean. Environ. Entomol. 19, 251–256. doi: http://dx.doi.org/10.1093/ee/19.2.251

Yue, C., Schroeder, M., Gisder, S., Genersch, E., 2007. Vertical-transmission routes for deformed wing virus in honey bees (*Apis mellifera).* J. Gen. Virol. 88, 2329–2336. doi: 10.1099/vir.0.83101-0

Zelinskaya, L.M., 1980. Role of microsporidia in the abundance dynamics of the gypsy moth (*Porthetria dispar*) in forest plantings along the lower Dnepr River (Ukrainian Republic, USSR). Vestnik Zoology (Zool. Bull.) 1, 57–62.

Zou, C., Li, L., Dong, T., Zhang, B., Hu, Q., 2014. Joint action of the entomopathogenic fungus *Isaria fumosorosea* and four chemical insecticides against the whitefly *Bemisia tabaci*. Biocontrol Sci. Techn. 24, 315–324. http://dx.doi.org/10.1080/09583157.2013.860427

4

The Host Population

Louela A. Castrillo

Department of Entomology, Cornell University, Ithaca, NY, USA

4.1 Introduction

Insects have been around since the Early Ordovician Period (Misof et al., 2014), and a current estimate suggests that today there are up to 8.7 million species (Mora et al., 2011). Insects inhabit diverse terrestrial habitats, with some using aquatic habitats as part of their life cycle. Some of the terrestrial habitats represent the more extreme habitats on the planet (e.g., polar regions or arid environments), where insects have evolved morphological, physiological, and behavioral adaptations to optimize their survival. With each distinct habitat come not only abiotic environmental threats, but also a range of potential biotic ones, including natural enemies such as predators, parasites, and microbial pathogens.

The microorganisms insects encounter are often rich in diversity and high in number, and those found associated with insects can be transient or symbiotic (i.e., in close association). The transient microorganisms, mostly bacteria and fungi, can be found on the integument or in the gut, reflecting the microbial flora present in the environment or in the food they ingest (e.g., Staudacher et al., 2016). Symbiotic microorganisms may be commensals or mutualists, and insects have developed ways to acquire and maintain some of the latter (e.g., termites and wood-feeding cockroaches with protistans in their gut, ambrosia beetles and *Sirex* woodwasps with symbiotic fungi in their mycangia). Entomopathogenic microorganisms (in this review, these include multicellular nematodes) are also in close association with insects, but with negative consequences to their hosts. Because of the adverse effects of these pathogens on host populations, these microorganisms have been commercialized or are being developed as control agents against insect pests of agricultural, structural, and medical importance. Most studies to elucidate insect (host)–microorganism (pathogen) interactions have focused on improving the efficacy of these pathogens for pest control (Chapter 13) and on controlling microorganisms attacking beneficial insects (e.g., pollinators, biological control agents, and insects reared as food and as feed) (Chapter 14). But the development of insect resistance to some pathogens, the increasing awareness of the diversity and importance of microorganisms other than pathogens to host biology, the need to conduct studies in the context of the natural ecology of insects, and the use of insects as

Ecology of Invertebrate Diseases, First Edition. Edited by Ann E. Hajek and David I. Shapiro-Ilan.
© 2018 John Wiley & Sons Ltd. Published 2018 by John Wiley & Sons Ltd.

models of immune responses (e.g., *Drosophila melanogaster, Apis mellifera*, and *Tenebrio molitor*) have resulted in very active research on insect defenses against microbial pathogens in the last few decades.

Insects are not passive players in the infection process. Microbial pathogens encountering potential insect hosts are confronted with barriers and defenses that insects possess or deploy to prevent or counter microbial infection. Insects have a multilayered and multifaceted defense system that includes structural barriers, an innate immune system, and behavioral responses that determine whether an invading microorganism becomes established to grow and multiply, invading one or more organs or overcoming the host in general, and potentially reproducing to infect other insects. Whether insect–microorganism interactions initially lead to infections depends also on a variety of intrinsic host factors and pathogen factors (Chapter 3), which are compounded by external ones, such as suboptimal environmental conditions and food sources (e.g., plants species for phytophagous insects) (Fig. 4.1; Chapter 5). These same interacting factors also determine the severity of impact of a pathogen on a given insect host: whether it dies from or survives the infection.

This chapter presents an overview of our current knowledge of insect defenses against microbial infection and offers readers an insight into the multiple and complex ways by which insects defend themselves, allowing them success in exploiting diverse habitats where they come in contact with various microorganisms, some of which are potentially pathogenic. The interplay between insect hosts and microbial pathogens is complex, and this chapter is by no means a complete nor a comprehensive review of all relevant topics related to the host population, but it summarizes our knowledge about how hosts influence their interactions with pathogens.

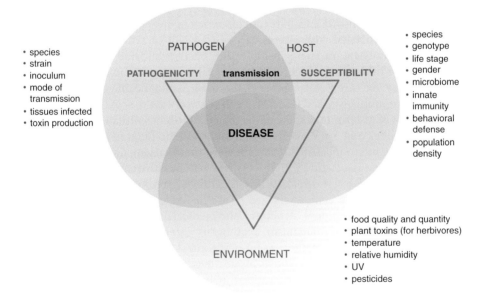

Fig. 4.1 Some of the interacting host, pathogen, and environmental factors determining the development of insect disease. (*See color plate section for the color representation of this figure.*)

4.2   General Host Factors

4.2.1   Routes of Pathogen Acquisition

Insects acquire pathogens via multiple ways. The most common route is via ingestion, when insects consume food contaminated with bacterial spores, viral occlusion bodies, microsporidian spores, or protozoan cysts. This route of acquisition also includes cannibalism, in insects with cannibalistic tendencies, when conspecifics weakened by infection are preyed upon, or when insects under stressful conditions, such as nutritional stress or overcrowding, become cannibalistic (e.g., Chapman et al., 1999; Williams and Hernandez, 2006). In contrast to other pathogens, fungi initiate infection by penetrating through the insect cuticle by use of enzymes (i.e., proteases, chitinases, lipases) and physical pressure (St. Leger, 1994). Some nematodes (e.g., Heterorhabditidae, Steinernematideae) are also able to breach the cuticle (Chapter 13), in addition to penetrating through natural body openings such as spiracles, the mouth, and the anus (e.g., Lewis et al, 2006; Koppenhöfer et al., 2007). Some pathogens are carried or injected directly into the hemocoel. These include bacterial symbionts of entomopathogenic nematodes (e.g., Burnell and Stock, 2000) and viruses or microsporidia injected by parasitoid ovipositors (e.g., Hamm et al., 1983; Own and Brooks, 1986; Tillman et al., 2004). The primary mode of pathogen acquisition generally sets up the first stage of direct insect–pathogen interaction, and also determines the initial insect defenses employed. After the pathogen gains entry into an insect host, the host employs additional defenses to thwart the development of infection.

4.2.2   Insect Species, Life Stage, Age, and Gender

Variations in insect biology, as defined by a combination of factors such as species, life stage, age, and sex, determine insect development type, morphology, physiology, and behavior. These are intrinsic host factors that, along with habitat type and environmental conditions, determine the type and diversity of microorganisms that insects are exposed to and their likely susceptibility to those that are potentially pathogenic. For example, foliage-feeding lepidopteran larvae with chewing mouthparts are susceptible to pathogens that are transmitted orally, while insects with sucking mouthparts (e.g., homopterans) are generally resistant because their feeding strategy precludes oral acquisition of most infective propagules. In lepidopterans or other holometabolous insects, the actively feeding larval stage is more likely to be infected than the adult stage with siphoning mouthparts, unless the infected larva survives and infection is carried into the adult stage. Differences in morphology, behavior, and reproductive activities between sexes, specifically in the adult stage of many insects, also result in variations in exposure to pathogens and, consequently, in susceptibility to different types of pathogens.

  Variations in insect susceptibility to different pathogens, taken in context of their life histories, could also result from differences in immune responses. Studies on immune function, based on measures of cellular and humoral immune factors, have revealed variations within and between genders, life stages, and ages (i.e., newly emerged versus sexually mature adults). These variations have been explained as being due to genetic and physiological differences and, from an evolutionary point of view, to variable risks from changing environments and costs associated with maintaining or mounting immune defenses.

Early studies on insect defense have focused on variations in host–pathogen interactions from the conceptual approach of specific interactions. But insect defense against infections comes at a cost and results in trade-offs with other components of fitness and survival (e.g., Cotter et al., 2010; Ardia et al., 2012). More recent studies have taken an evolutionary approach, focusing on the fitness costs of insect defense and the resulting variation in responses under different environmental conditions (i.e., ecological immunology) (Sadd and Schmid-Hempel, 2009). Ecological immunology posits that given an insect's limited resources and the costs of maintaining and mounting immune defenses, trade-offs are made that affect other life histories, in particular mating and reproduction (e.g., Adamo et al., 2001; Gwynn et al, 2005; Leman et al., 2009; Gershman et al., 2010; McNamara et al., 2014a; Schwenke et al., 2016). But studies have now shown that the immune system can be plastic, with trade-offs mediated by environmental factors, in particular diet quality and quantity (e.g., Lee et al., 2006; Povey et al., 2009; Cotter et al., 2011) (Section 4.4.3.3.).

Changes in immune responses during various stages of an insect's life cycle have been observed in the larval, pupal, and adult stages of the burying beetle *Nicrophorus vespilloides* (Urbanski et al., 2014). Measures of cellular and humoral defense effectors showed increasing activities during larval development, followed by a drop in phagocytic hemocytes (Section 4.4.1.1) and an increase in phenoloxidase during the pupal stage, and a low level of both cellular and humoral effectors in the adult stage. Similarly, studies on *Manduca sexta* and *Carabus lefebvrei* also showed different patterns in cellular and humoral activity at different life stages (Eleftherianos et al., 2008; Giglio and Giulianini, 2013). In these studies, the observed correlation between changes in immune activity for developmental stages that exhibited different morphologies, physiologies, and behaviors suggests immune adaptations to different risk levels from exposure to different microorganisms in different environments during an insect's life history (Schmid-Hempel and Ebert, 2003).

As insects age, the efficiency of their immune defenses can also decline (immunosenescence), as has been shown in some species (e.g., Adamo et al., 2001; Doums et al., 2002; Armitage and Boomsma, 2010; Mackenzie et al., 2011; Roberts and Hughes, 2014). Older honeybees (*Apis mellifera*) exposed to *Nosema ceranae* were found to have more spores of the microsporidian in their gut compared to younger bees (Roberts and Hughes, 2014). Intensity of infection was negatively correlated with levels of prophenoloxidase in the hemolymph, suggesting that older bees were less able to replace activated prophenoloxidase (Roberts and Hughes, 2014). A study of *Bombus* spp. workers also showed that cellular immune response declined with age (Doums et al., 2002). The degree of encapsulation/melanization response following immune challenge (i.e., implantation of a nylon filament) was lower in older workers compared to younger ones (Doums et al., 2002). In *D. melanogaster*, older flies were found to be less efficient in clearing bacterial infections, due to decreased phagocytic activity (Mackenzie et al., 2011).

4.2.3 Population Density

Insect population density can be a critical factor in pathogen transmission and disease resistance. The likelihood of pathogen transmission (i.e., horizontal transmission) increases with higher host densities, as the rate of encounter between infected and uninfected conspecifics rises, and as the number of infected insects increases, so does the titer of pathogen inocula in the environment. For example, epizootics caused by baculoviruses have been observed following and during population explosions of forest lepidopterans, including gypsy moth (*Lymantria dispar*) and the western tent caterpillar (*Malacosoma californicum pluviale*) (e.g., Elkinton, 1990; Cory and Myers, 2009). Also, higher insect density or crowding can lead to intraspecific competition and induce physiological stress, resulting in increased susceptibility to infection (Lindsey et al., 2009). Overcrowding has been shown to result in the conversion of resident covert baculoviruses into an overt and lethal state (e.g., Cooper et al., 2003; Opoku-Debrah et al., 2013). In confined spaces (i.e., mass rearing containers), stress caused by crowding results in an increased likelihood of infection from opportunistic pathogens such as *Serratia marcescens* (e.g., King et al., 1975; Lighthart et al., 1988).

With the increased risk of diseases in high-density insect populations, it has been hypothesized that those insects living in crowded conditions invest more in their immune defense than those in low-density populations or those with solitary habits (i.e., density-dependent prophylaxis) (Wilson and Reeson, 1998). The selective advantage to individuals with increased resistance to pathogens in response to higher conspecific density would also result in build-up of more resistant populations (Wilson and Reeson, 1998). The oriental armyworm (*Mythimna* (formerly *Pseudaletia*) *separata*), exposed to diet contaminated with a granulovirus, had higher survival rate when reared at higher densities than when reared singly, indicating that virus resistance in this species is density-dependent (Kunimi and Yamada, 1990). The mechanism for density-dependent resistance could be a combination of physiological defenses (i.e., elevated antimicrobial activity, cuticular melanism) and behavioral defenses (e.g., Reeson et al., 1998; Wilson et al., 2002; Parker et al., 2010). In the phase polyphenic desert locust (*Schistocerca gregaria*), recently molted adults reared under crowded conditions (*gregaria* phase) were more resistant than the *solitaria* phase to the fungal pathogen *Metarhizium acridum* (formerly *M. anisopliae* var. *acridum*) (Wilson et al., 2002). The enhanced resistance was found to be associated with increased antimicrobial activity (Wilson et al., 2002). Other studies reporting density-dependent prophylaxis showed larval crowding associated with cuticular melanism (e.g., Goulson and Cory, 1995; Barnes and Siva-Jothy, 2000).

4.3 Barriers to Microbial Infection

4.3.1 Insect Integument

The outermost layer of an insect is the integument, which determines the insect shape and serves structural (exoskeleton), sensory, and protective functions. It is composed of a single layer of epidermal cells, a basement membrane, and a cuticle covering the epidermis (Merzendorfer, 2013). The cuticle covers the entire insect body and also lines the fore- and hindgut and trachea. It is composed of three layers: a thin epicuticle (with an inner layer consisting mostly of tanned lipoproteins and an outer waxy layer), an exocuticle, and an endocuticle. The latter two are collectively called the procuticle and are made of protein and chitin together with lipids and quinones. Variations in protein, chitin, and water content determine, in part, the mechanical properties of insect cuticles, which can be hard and rigid, membranous, or elastic (Merzendorfer, 2013). These variations allow for various functions, from muscle attachment to elastic intersegmental membranes between sclerites to allow for movement. Sclerotization of the integument stabilizes and hardens it, giving it high mechanical strength and resistance to chemical and enzymatic digestion (Hopkins and Kramer, 1992).

The integument provides an effective barrier against entrance by microbial pathogens, except by entomopathogenic fungi and some nematodes. As mentioned earlier, fungal pathogens employ a combination of enzymatic and physical means to breach potential host cuticles. But even for these fungi, attachment to the integument, a prerequisite to infection, can be affected by surface structures and the topography of the exoskeleton (e.g., St. Leger et al., 1991; Sosa-Gomez et al., 1997). Further, adhesion does not always lead to formation of germ tubes that initiate entry into the hemocoel, since conditions on the integument, such as available nutrients or the presence of antagonistic microorganisms, toxic lipids, and phenolic compounds, can inhibit spore or conidial germination (St. Leger, 1991). Penetration of the cuticle by fungi requires passage of the germ tube through the epicuticle and the procuticle, and this often occurs through the membranous intersegmental regions, rather than the hard and rigid sclerites. Similarly, entomopathogenic fungi such as *Beauveria bassiana* proliferating inside infected adult cadavers with hardened sclerites, like the Colorado potato beetle (*Leptinotarsa decemlineata*), have been observed emerging through the softer intersegmental membranes while the larval stage with mostly soft cuticle becomes completely covered with fungal mycelia (personal observation).

The cuticle can also be impregnated with antimicrobial lipids and peptides. Cuticular compounds, including aldehydes, have been shown to have antimycotic effects against *B. bassiana* and *Metarhizium anisopliae* (e.g., Sosa-Gomez et al., 1997). Breaks in the cuticle also result in clotting and melanization (Section 4.4.1.1.), which not only close the wound but also trap and inactivate invading microorganisms (Merzendorfer, 2013). Some insects, such as ants and termites, also produce exocrine gland secretions with antimicrobial properties during grooming, which also prevent infection (e.g., Schlüns and Crozier, 2009; Graystock and Huges, 2011; Hamilton et al., 2011). Exocrine glands (i.e., glands releasing their secretions to the exterior) are epidermal cells with specialized glandular functions. There are two major types of exocrine gland cells, corresponding to types I and III of the three general classes of insect gland cells, with type III involved in defense (Billen and Morgan, 1998; Merzendorfer, 2013). Exocrine glands and their secretions are relatively well studied in social insects for their role in group living (i.e., communication) and, more importantly, in defense against potentially pathogenic microorganisms. For example, some ants possess a pair of metapleural glands, located at the posterolateral margin of the thorax, that produce secretions containing a diversity of carboxylic acids and proteinaceous compounds with antimicrobial properties (Ortius-Lechner et al., 2000; Fernandez-Marin et al., 2006). These secretions spread passively over the cuticle, but can also be regulated by ant activity (i.e., acquisition of small amounts by use of front legs during grooming), especially after exposure to fungal conidia (Fernandez-Marin et al., 2006). Further, ants adjust the amount of metapleural gland secretions to the virulence of the fungi present, with more secretions produced after exposure to the pathogens *B. bassiana* and *M. brunneum* compared to nonpathogenic *Aspergillus niger* or *Escovopsis weberi* (Yek et al., 2012).

Solitary insects also use exocrine gland secretions for defense. Willow-feeding chryso-melid larvae utilize phenolic compounds from their host plants as precursors in the production of salicylaldehyde, which is the main component of their exocrine gland secretions (Gross and Hilker, 1995). The larvae emit these secretions when disturbed, resulting in a cloud of volatiles surrounding them. These secretions have been shown to be effective against predators and exhibit antibacterial and antifungal activities against a wide range of microorganisms (Gross et al., 2002). Because of the dorsal location of these glands and their distribution along the thorax and the abdomen, the cloud of volatiles can provide protection over a large part of the insect integument against fungal pathogens such as *M. anisopliae* (Gross et al., 2002).

4.3.2   Tracheae

The tracheae are invaginations of the insect exoskeleton, from large primary tubes into fine distal branches (tracheoles), that allow gaseous exchange to all parts of the body. Tracheal tubes are lined with an outer cuticle, with a protein or chitin layer beneath it (Harrison and Waserthal, 2013). The external openings of the tracheal system are the spiracles, located in the insect thorax and abdomen, usually with one pair per segment. Although spiracles have a closing mechanism in most insects, regulated in response to oxygen and carbon levels, these openings provide entry points for some pathogens (i.e., fungi and nematodes). The fungal pathogens *B. bassiana* and *M. anisopliae* have been observed initiating infections through the spiracles, but with variable results (e.g., Lacey et al, 1988; Fuguet and Vey, 2004). This is likely due to the induction of a fast and intense melanization response against invading fungal structures, as has been observed in response to fungal penetration of the insect integument (e.g., Fuguet and Vey, 2004). In contrast to entomopathogenic fungi, dauer larvae of *Steinernema* nematodes can reach the hemocoel by actively seeking entry through the spiracles and other natural body openings (e.g., Burnell and Stock, 2000; Lewis et al., 2006). Some species of scarab larvae, however, have spiracles that are cribiform, with plates punctuated with aeropyles (small holes) that serve as the only openings to the tracheae (e.g., Forschler and Gardner, 1991).

In addition to its cuticular lining serving as a barrier to infection, the tracheal epidermal cells beneath the cuticular intima are able to launch immune responses to bacteria, as has been shown in *D. melanogaster*. Wagner et al. (2008) showed that the fly's tracheal system has a peculiar organization that allows for combating airborne bacterial pathogens. The epithelial cells possess a variety of pathogen-recognition receptors and effector molecules, such as antimicrobial peptides (AMPs), but signaling occurs through only a single transduction pathway (Wagner et al., 2008). The insect tracheal cells, however, can be susceptible to viruses. In *Autographa californica*, these cells are secondary sites of infection by the Autographa californica M nucleopolyhedrosis virus (AcMNPV). Infection starts in the midgut cells, with the release of virions following the dissociation of viral polyhedra by alkaline digestive enzymes. Virus multiplication in the midgut nuclei produces budded viruses that target primarily the tracheolar cells, which then facilitate a systemic infection, resulting in inevitable host death (Englehard et al., 1994). The direct infection of the tracheal cells via the spiracles is also possible and may be a source for horizontal transmission of the virus (Kirkpatrick et al., 1994).

4.3.3 Insect Gut

4.3.3.1 Peritrophic Membrane and Basal Lamina

Following ingestion of pathogen propagules, the insect midgut serves as the primary site of entry for bacteria, viruses, microsporidia, and protists, since both the foregut and hindgut are of ectodermal origin and are lined with cuticle. In contrast, the midgut is of endodermal origin. It has a layer of columnar epithelial cells responsible for secretion of digestive enzymes and absorption of nutrients. Between the gut lumen and the epithelial cells is the peritrophic membrane, which is present in all insects except for those that feed on plant sap and blood. The peritrophic membrane, also known as the peritrophic matrix, is noncellular in structure and consists of an organized lattice of chitin fibrils held together by chitin-binding proteins (Hegedus et al., 2009). The membrane serves as a sieve, permeable only to small molecules, and a physical barrier between the midgut epithelial cells and digestive enzymes, ingested food materials, and some pathogens. Variations in the thickness of the peritrophic membrane could account, in part, for variations in insect susceptibility to some pathogens. Dauer larvae of nematodes *Steinernema carpocapsae* and *Heterorhabditis bacteriophora* entering through the mouth penetrated the midgut of *Galleria mellonella* faster than the midgut of the scarab *Phyllophaga hirticula*, which has a thicker peritrophic membrane (Forschler and Gardner, 1991). The peritrophic membrane also serves as a biochemical barrier, able to sequester and inactivate some ingested toxins (Hegedus et al., 2009).

On the basal side of the midgut epithelium is a basal lamina with multiple structural and filtration functions. It serves as a scaffold for regeneration of epithelial cells after sloughing off and separates the midgut from the hemocoel (Passarelli, 2011). Thus, the midgut basal lamina could also serve as a barrier, preventing ingested pathogen propagules from reaching and infecting other tissues and organs in the hemocoel. The basal laminae are thin, flexible, noncellular sheets composed of proteins with openings of only 15 nm between their protein layers (Reddy and Locke, 1990), precluding virus passage (Passarelli, 2011). Despite its small pore size, however, the basal lamina could still allow viruses passage when gaps are present due to its dynamic nature (i.e., when it is formed, degraded, and reformed) (Passarelli, 2011).

4.3.3.2 Conditions in the Gut Lumen

Pathogens in the gut lumen are subject to conditions that could be detrimental to their growth and survival. These conditions include the presence of digestive enzymes, gut redox potential, ionic strength and pH, and the presence of resident gut microbiota (e.g., Dillon and Dillon, 2004). Insect gut pH varies along its length, and in the midgut this depends on the optimum range for digestive enzyme activity and nutrient absorption that still limits the harmful effects of some ingested compounds. The optimal range of gut pH for digestive enzymes in most insects is pH 6–7, but it can range from highly acidic (pH 3–4) in some dipterans to alkaline (pH 10–11) in some lepidopterans (Douglas, 2013a). In those with low pH, the acidic condition is associated with bacteriostatic or bacteriocidal activity against microorganisms present in decomposing food sources. Overend et al. (2016) showed that the gut region of *D. melanogaster* with low pH protects against colonization by pathogenic bacteria

(*Pseudomonas* spp.) and regulates the number of nonpathogenic bacteria present (*Acetobacter* spp. and *Lactobacillus* spp.). In herbivorous lepidopteran larvae, the alkaline condition in their midgut allows uncoupling of tannins present in foliage, making plant nutrients available for absorption (Berenbaum, 1980). The high pH, however, also allows dissolution of ingested *Bacillus thuringiensis* protein crystals, which results in the release of insecticidal toxins (de Maagd et al., 2001) or dissolution of viral occlusion bodies, releasing virions (e.g. Washburn et al., 1995). The enzymes secreted by epithelial cells, like proteases and lipases, for breakdown of ingested food may also have antimicrobial activity, as has been shown in larval digestive juices from *Bombyx mori* against Bombyx mori nucleopolyhedrovirus (Ponnuvel et al. 2003; Nakasawa et al., 2004).

The time of passage of ingested pathogen propagules through the gut could vary depending on the amount and type of food ingested and the host physiology. A longer time in the gut could allow fungal conidia to attach to the cuticle-lined foregut and germinate, initiating infection, although studies on several insect species show no fungal germination or infection initiating from ingestion (e.g., Dillon and Charnley, 1986, 1988; Chouvenc, et al., 2009). Studies by Dillon and Charnley (1986, 1988) showed that the passage of *M. anisopliae* conidia through the gut of the desert locust (*S. gregaria*) resulted in a 50% reduction in viability, primarily from the fungistatic effects of resident gut bacteria. These bacteria produced antifungal phenolic compounds that were active against multiple strains of *M. anisopliae*, as well as other insect- and plant-pathogenic fungi. Further testing with other insects also revealed antifungal activity in their guts (Dillon and Charnley, 1988), suggesting that these phenols may be a common defense against entomopathogenic fungi.

4.3.3.3 Sloughing of Infected Epidermal Cells

Variations in the susceptibility of some insects to various orally transmitted pathogens have been linked, in part, to the sloughing of infected midgut cells, which are then replaced by new, uninfected cells that prolong host survival or allow the host to survive infection. This appears to be a general mechanism of resistance and has been associated with insect defense against bacteria (e.g., *B. thuringiensis*, *Paenibacillus popilliae*), viruses (e.g., AcMNPV, silkworm viruses), and microsporidia (e.g., *Nosema acridophagus*) (Henry, 1967; Inoue and Miyagawa, 1978; Splittstoesser et al., 1978; Chiang et al., 1986; Englehard and Volkman 1995; Hoover et al., 2000). In rice moth (*Corcyra cephalonica*) larvae infected with *B. thuringiensis*, Chiang et al. (1986) observed the following sequence of cellular responses in the midgut region: swelling and sloughing of affected columnar cells accompanied by growth of new cells from the basal area of the epithelium, followed by merocrine secretions from goblet cells producing a mucus layer protecting the newly developed cells. These cellular activities limited bacterial damage to the gut and prolonged the survival of infected insects. In *Drosophila*, the renewal of gut epithelial cells in combination with innate immune defenses is required for fighting bacterial infection (Buchon et al., 2009). In *Trichoplusia ni* infected with AcMNPV, the timing (host larval age) and speed of sloughing of infected midgut cells are also critical in determining resistance to infection, since infected cells can produce budded virus that later infects tracheal cells, resulting in the development of a systemic infection (Englehard and Volkman, 1995; Washburn et al., 1998).

4.4   Defenses against Microbial Infection

4.4.1   Innate Immune System

Pathogens capable of breaching the cuticle or the insect midgut can gain access to other tissues and organs, as well as to the hemolymph in the hemocoel. Here, the host's innate immune system comes into action, providing defense against microbial infection. Although insects do not possess an adaptive immune system (i.e., acquired through exposure to specific pathogens during the organism's lifetime), they have an innate immune system consisting of constitutive and induced defenses. Constitutive immunity is always present and is capable of defense without previous contact or exposure, while induced immunity is activated after the pathogen has been detected and the effectors are synthesized *de novo* (e.g., Schmid-Hempel and Ebert, 2003). The former offers the advantage of a rapid or "front-line" response to microorganisms that multiply rapidly within the host and inflict damage (Shudo and Ywasa, 2001; Matova and Anderson, 2006), especially in environments or conditions with high a likelihood of pathogen exposure (Graham et al., 2015). Cellular response (i.e., mediated directly by hemocytes) and rapidly activated enzyme cascades (i.e., phenoloxidase and lysozymes) are constitutive defenses, while AMPs and reactive oxygen species (ROS) are induced defenses. The AMPs, ROS, and activated-enzyme cascades (i.e., the phenoloxidase cascade) are often referred to as humoral defenses. The insect innate immune system is sometimes separated into cellular and humoral defenses, but this is mostly for convenience, since hemocytes are involved in both (i.e., hemocyte-mediated defense responses and hemocyte-mediated effector responses, respectively) (Strand, 2008). In combination, the constitutive and induced defenses provide an effective way of combating invading pathogens. A study by Haine et al. (2008) showed a fast and effective microbial clearance (>99.5%) of *Staphylococcus aureus* in the hemolymph of female *T. molitor* adults within less than an hour after injection. This was followed by induced antimicrobial activity that started to increase only after most of the bacteria were cleared, peaking at 24 hours post-injection (Haine et al., 2008).

As mentioned earlier, maintenance and induction of these defenses can be costly and may require trade-offs with other life-history traits. Additional costs include metabolic costs and possible self-harm (e.g., Poulsen et al., 2002; Sadd and Siva-Jothy, 2006). In *T. molitor* adults with implanted nylon tubing, the resulting phenoloxidase enzyme cascade, which catalyzed production of melanin that was deposited on the implants, resulted in the melanization of the adjacent malphigian tubules (Sadd and Siva-Jothy, 2006). Because these tubules extend throughout the body cavity, are intimately associated with the hemolymph, and are devoid of an impermeable protective membrane, they can be subject to the cytotoxic effects of upregulated immune effectors (Sadd and Siva-Jothy, 2006). In this case, the phenotype and function of malphigian tubules adjacent to sites of insertion were negatively impacted by the resulting immune response.

4.4.1.1   Constitutive Innate Immunity: Cellular Immunity

Insects have an open circulatory system, with the hemolymph consisting of liquid plasma and cellular hemocytes circulating freely throughout the body cavity. The hemolymph has multiple functions, including serving as a medium that bathes cells, transfers substances to and from these cells, and provides cellular and humoral defenses (Douglas and Siva-Jothy, 2013). Hemocytes circulating in the hemolymph are produced during embryogenesis from procephalic mesodermal cells and during the larval or nymphal stages from hematopoietic organs (Strand, 2008). In lepidopterans, these organs are located in the meso- and metathorax near the imaginal wing disks (Lavine and Strand, 2002), while in *Drosophila* they are formed along the anterior part of the dorsal vessel (where they are called lymph glands) (Jung et al., 2005). The prohemocytes are proposed to be the stem cells that develop into one of the differentiated hemocytes, but the lineage relationship varies among lepidopterans, *Drosophila*, and mosquitoes (Strand, 2008). Differentiated hemocytes are identified based on a combination of different characters (i.e., morphological, histochemical, and functional), and current nomenclature differs among hemocytes in lepidopterans, *Drosophila*, and mosquitoes (Lavine and Strand, 2002; Ribeiro and Brehelin, 2006; Strand, 2008). For this review, the nomenclature for hemocytes in lepidopterans will be used, but see Table 4.1 for a comparison with the nomenclature for *Drosophila*. These differentiated hemocytes play different roles in insect immune responses (Table 4.1).

Table 4.1 Classification of differentiated insect hemocytes from lepidopteran larvae and *Drosophila* and their defense-related function(s) (Lavine and Strand, 2002; Ribeiro and Brehelin, 2006; Strand, 2008; Douglas and Siva-Jothy, 2013).

Insect	Hemocyte type	Properties	Immune function
Lepidopteran larvae	plasmatocytes	adherent cell type; spherical or oval cells, with no granular inclusions	capsule formation around foreign bodies too large to phagocytose; nodule formation
	granulocytes (or granular hemocytes)	adherent cell type; spherical cells with large number of granular inclusions	phagocytosis, first cells in contact with foreign bodies at the beginning of capsule or nodule formation
	oenocytes (or oenocytoids)	nonadherent cell type	contain prophenoloxidase
	spherulocytes (or spherule cells)	nonadherent cell type; rounded cells with large inclusions (spherules)	expression of lysozymes
Drosophila	plasmatocytes	adherent cell type; spherical cells, may contain granules	phagocystosis, first cells in contact with foreign bodies at the beginning of capsule formation
	crystal cells	nonadherent cell type; similar in structure and function to lepidopteran oenocytes	contain prophenoloxidase
	lamellocytes	adherent cell type; similar to lepidopteran plasmatocytes, but flattened, with only a few in circulating hemolymph of healthy larvae	capsule or nodule formation

Hemocyte-mediated defenses include phagocytosis, nodule formation, encapsulation, and clotting. In phagocytosis, individual cells engulf and destroy small targets such as bacteria, fungi, and protozoa (Strand, 2008). The process starts with receptor-mediated recognition and binding of the small target to a hemocyte, followed by phagosome formation and engulfment of the target (Strand, 2008). Phagocytosis, however, is not limited to biotic targets but also includes synthetic beads, as have been used in studies on variability in immune responses. In lepidopterans, phagocytosis is carried out by granular cells and plasmatocytes. These two types of hemocytes together make up more than 50% of circulating hemocytes in lepidopteran larvae (Strand and Pech, 1995).

For targets too numerous or too big be engulfed, insects deploy hemocytes that surround these targets in multiple layers. For bacterial aggregates, the response is called nodulation (or nodule formation), while for larger targets like parasitoids eggs, fungal germ tubes, and nematodes, it is called encapsulation (or capsule formation) (Lavine and Strand, 2002). In either case, an overlapping sheath of hemocytes is formed around the target and, as in phagocytosis, granular cells and plasmatocytes are involved. Both processes start with granular cells binding to the target, followed by layers of plasmatocytes forming a multicellular sheath, and finally a single layer of granular cells (Pech and Strand, 1996). Nodules may or may not be melanized, while capsules are often ultimately melanized (Marmaras and Lampropoulou, 2009).

In insect larvae at an early stage of fungal infection, the penetration sites are often easily detected by the presence of melanotic spots formed in response to the invading fungal germ tubes. The intensity of melanization, however, can vary with insect host and fungal pathogen and can affect host survival. A study by Golkar et al. (1993) showed that intense and diffuse melanization in *Anopheles gambiae* larvae, following encystment of *Lagenidium giganteum* zoospores, resulted in reduced mortality among infected larvae, while a weak and localized melanization in *Aedes aegypti* larvae resulted in near-100% mortality even though a lower number of encysted zoospores were observed compared to *A. gambiae.*

Hemocytes also coagulate to form clots at sites of external wounds. These clots, formed primarily by granulocytes in lepidopterans and by plasmatocytes in *Drosophila*, not only contribute to wound healing but also serve as scaffolds for repair and the prevention of microbial infection (Theopold et al., 2004; Dushay, 2009). Clots first form as soft clots, which become hardened following crosslinking of proteins and deposition of melanin (Theopold et al., 2004).

Activation of insect cellular defense depends on the recognition of self versus nonself. Conserved pathogen-associated molecular patterns (PAMPs), including peptidoglycans and lipopolysaccharide in bacterial cell walls and β-1,3-glucans in fungal cell walls, are recognized by insect humoral pathogen recognition receptors (PRRs) that function as opsonins or by PRRs on the surfaces of hemocytes (Strand, 2008). In the absence of these distinct PAMPs, some pathogens are able to avoid detection by the host. For example, *B. bassiana* hyphal bodies, without a well-defined cell wall, can elude the defensive response in *Spodoptera exigua* larvae (Hung and Boucias, 1992). The absence of a well-defined cell wall, with its elicitors of detection, allows hyphal bodies to evade detection and encapsulation even in the presence of immunocompetent hemocytes (Pendland and Boucias, 1993).

4.4.1.2 Constitutive Innate Immunity: Phenoloxidase

Deposition of insoluble melanin, or melanization, is a defensive response triggered by cuticular damage or wounding caused by abiotic factors or biotic agents such as parasitoids and pathogens, in particular fungi. It can block entry of fungal germ tubes by walling off breaks in the cuticle, and, as mentioned earlier, is a component of cellular immunity against other pathogens, sequestering them inside melanotic nodules or capsules. Melanization is the result of phenoloxidase activity, initiated by the cleavage of prophenoloxidase (proPO) to its active form, phenoloxidase (PO), via limited proteolysis by serine proteinases (Cerenius and Söderhall, 2004; Pham and Schneider, 2008). PO then oxidizes tyrosine derivatives into quinones that polymerize to form melanin (Pham and Schneider, 2008). ProPO is synthesized in oenocytes in lepidopterans or crystal cells in *Drosophila* (Iwama and Ashida, 1996; Waltzer et al., 2002) and is stored in hemocytes. It is also present in the plasma of some insects (Cerenius and Söderhall, 2004). Hemolymph proPO is actively transported transepithelially to the cuticle, as shown in a study of *B. mori* by Asano and Ashida (2001).

The process of melanization includes four steps: (i) recognition of the pathogen via specific pathogen-associated patterns; (ii) recruitment of adherent plasmatocytes to surround the pathogen; (iii) production of melanin by internal layers of plasmatocyte walls; and (iv) release of melanin by cellular lysis to encapsulate the pathogen (González-Santoyo and Cordóba-Aguilar, 2011). Like cellular defense, activation of the proPO system is initiated by pathogen detection (Cerenius and Söderhall, 2004), but it can also be triggered in the absence of PAMPs (Cerenius et al., 2008). For example, melanotic capsules are formed in *Drosophila* upon detection of endogenous factors from disrupted basement membrane (Brennan and Anderson, 2004). Once encapsulated, growth and development of the pathogen are prevented, and blocking of its absorption of nutrients can eventually lead to its death (e.g., Chen and Chen, 1995). The efficacy of melanization against fungi and other pathogens, however, varies depending on host conditions (i.e., intrinsic factors affecting PO levels), pathogen virulence, and the mechanisms the pathogen has for evading host defenses.

4.4.1.3 Induced Innate Immunity: Reactive Oxygen Species

In *Drosophila*, ingestion of bacterial pathogens results in the generation of ROS (e.g., Ha et al., 2005a), followed by induction of AMPs (Vallet-Gely et al., 2008). Production of ROS constitutes the first line of induced defenses in *D. melanogaster*, and the amounts generated can be adequate to kill ingested pathogens (Vallet-Gely et al., 2008). Ingestion of the microsporidian *N. ceranae* by honeybees is also followed by immediate production of ROS, indicating that this is a general defense mechanism among insects (Dussaubat et al., 2012). ROS production, however, did not prevent honeybee mortality from *N. ceranae* infection due to gut tissue degeneration and impaired gut epithelial renewal (Dussaubat et al., 2012). This observation concurs with the concept that in some insects, a combination of innate immune defenses and gut epithelial renewal to maintain gut homeostasis is required for effective host recovery from some infections (Buchon et al., 2009, 2013). ROS have also been found concentrated at sites of melanin production during cellular defense and can facilitate the killing of the pathogen. Thus, these molecules participate in insect immunity in multiple ways (Nappi et al., 1995; Nappi and Vass, 1998; Kumar et al., 2003).

ROS are oxygen-derived molecules that readily oxidize other molecules, and an excess of ROS results in oxidative stress (i.e., damages to diverse cellular macromolecules such as DNA, lipids, and proteins) and cell injuries (Zug and Hammerstein, 2015b). During an infection, residual ROS is eliminated concurrent with its generation to protect the host and to maintain reduction–oxidation balance (Ha et al., 2005b). This balance is mediated by antioxidant enzymes such as an extracellular immune-regulated catalase found in *Drosophila* (Ha et al., 2005b). Given the contrasting roles of oxidative stress and innate immune response, it has been suggested that ROS could be key factors underlying physiological mechanisms in trade-offs between insect immunity and life-history traits (Zug and Hammerstein, 2015b). ROS could also be key players in the interplay between the symbiont *Wolbachia* and its insect hosts, determining host immune responses leading to resistance to or tolerance of the bacterium (Zug and Hammerstein, 2015b; Monnin et al., 2016).

4.4.1.4 Induced Innate Immunity: Antimicrobial Peptides

AMPs are considered the second line of inducible defense against ingested bacterial pathogens in *D. melanogaster*, after generation of ROS (Vallet-Gely, et al., 2008). AMPs are a diverse class of evolutionary conserved molecules that are typically cationic and with fewer than 100 amino acids. They are produced by prokaryotes and eukaryotes for defense (Zhang and Gallo, 2016). Insect AMPs, of which over 150 have been described, are active mostly against Gram-negative and Gram-positive bacteria and fungi. Insect AMPs have been classified into four groups or families based on their structural motif and their high content of specific residues: (i) linear and amphipathic α-helical peptides without cysteine residues (e.g., cecropins and cecropin-like molecules); (ii) cysteine-rich peptides (e.g., defensins); (iii) proline-rich peptides (e.g., apidaecin, abaecin, formaecins and metchnikowins); and (iv) glycine-rich peptides (e.g., attacins) (Bulet and Stöcklin, 2005) (Table 4.2).

Most of the insect AMPs discovered to date come from members of the orders Coleoptera, Diptera, Hemiptera, Hymenoptera, and Lepidoptera, but distribution of the different AMPs varies considerably among these groups (Mylonakis et al., 2016). Defensins, cecropins, and attacins are widely distributed, but some AMPs have been identified from only one insect order (e.g., gallerimycin and gloverin in lepidopterans, coleoptericins in coleopterans, apidaecin and hymenoptericins in hymenopterans) or genus (e.g., metchnikowin in *Drosophila*) (Mylonakis et al., 2016) (Table 4.2). In the majority of these insects, the AMPs are produced mostly by fat bodies and secreted into the hemolymph (so these are systemic), and locally by epithelial cells (e.g., Tzou et al., 2000). In the termite *Pseudacanthotermes spiniger*, however, two AMPs have been found constitutively expressed in the salivary glands and granulocytes of uninfected insects, from where they can be released into the hemolymph following an immune challenge (Lamberty et al., 2001). This finding is in marked contrast to other described AMPs and shows an alternative means of AMP production by insects.

Most AMPs have antibacterial activities and their mode of action has been fairly well studied. Because of their net positive charge and high ratio of hydrophobic amino acids, AMPs can bind selectively to negatively charged bacterial membranes, which leads to membrane disruption and cell lysis, or enter into bacterial cells, which inhibits cellular functions (i.e., inhibition of nucleic acid and protein synthesis and blocking enzyme activities) (e.g., Zhang and Gallo, 2016). Some AMPs target both bacteria and fungi,

Table 4.2 Examples of insect antimicrobial peptides (AMPs).

Classification	AMP	Insect source(s): Species (order)[a]	Target(s)[b]	Reference
α-helical peptides without cysteine residues	cecropins	*Aedes aegypti* (D)	G+, G−, F	Lowenberger et al. (1999)
		Bombyx mori (L)	G+, G−	Taniai et al. (1995)
		Drosophila melanogaster (D)	G−, F[c]	Lemaitre et al. (1997)
		Hyalophora cecropia (L)	G+, G−	Hultmark et al. (1982)
cysteine-rich peptides	defensins	*Aedes aegypti* (D)	G+, G−	Lowenberger et al. (1995)
		Bombus pascuorum (H)	G+, G−, F	Rees et al. (1997)
		Drosophila melanogaster (D)	G+	Imler and Bulet (2005)
		Galleria mellonella (L)	F	Lee et al. (2004)
		Oryctes rhinoceros (C)	G+	Ishibashi et al. (1999)
	drosomycin	*Drosophila melanogaster* (D)	F	Lemaitre et al. (1997)
	gallerimycin	*Galleria mellonella* (L)	F	Schuhmann et al. (2003)
	hymenoptaecin	*Apis mellifera* (H)	G+, G−	Casteels et al. (1993)
	termicin	*Pseudacanthotermes spiniger* (I)	G+, F	Lamberty et al. (2001)
proline-rich peptides	abaecin	*Apis mellifera* (H)	G+, G−	Casteels et al. (1990)
	apidaecin	*Apis mellifera* (H)	G−	Casteels et al. (1989)
	drosocin	*Drosophila melanogaster* (D)	G−, F[c]	Lemaitre et al. (1997)
	formaecin	*Myrmecia gulosa* (H)	G−	Mackintosh et al. (1998)
	metchnikowin	*Drosophila melanogaster* (D)	G+, G−, F	Lemaitre et al. (1997)
glycine-rich peptides	attacin	*Hyalophora cecropia* (L)	G−	Carlsson et al. (1991)
		Drosophila melanogaster (D)	G−, F[c]	Lemaitre et al. (1997)
	diptericin	*Drosophila melanogaster* (D)	G−, F[c]	Lemaitre et al. (1997)
	gloverin	*Manduca sexta* (L)	G+, F	Xu et al. (2012)

a) C (Coleoptera); D (Diptera); H (Hymenoptera); I (Isoptera); L (Lepidoptera).
b) G+ (Gram-positive bacteria); G− (Gram negative bacteria); F (fungi).
c) A relatively lower inducibility of the AMP genes by fungus *B. bassiana* was detected after pricking the fly integument, in contrast to a strong response to the Gram-negative bacterium *Escherichia coli* (Lemaitre et al., 1997).

and a few are limited to fungi, but information on antifungal mechanisms is relatively limited (Hegedus and Marx, 2013). Among cysteine-rich peptides and defensins, only four examples with antifungal properties are known: drosomycin, gallerimycin, heliomycin, and termicin (Fehlbaum et al., 1994; Lamberty et al., 1999, 2001; Schuhmann et al., 2003). Information on the mode of action is available only from drosomycin and termicin. Drosomycin lyses fungal hyphae, resulting in membrane permeabilization and pore formation (Fehlbaum et al., 1994). Termicin, in combination with antifungal glucanases (these break down β-1,3-glucans of fungal cell walls), also produced by termites is spread on the cuticle during self- and allogrooming (Section 4.4.3.2.). Termicin inhibits growth of *M. anisopliae*, and, in combination with glucanases, also prevents attachment of the fungus to the insect cuticle (Hamilton et al., 2011; Hamilton and Bulmer, 2012). The mode of action of metchnikowin, a proline-rich AMP, is unclear in insects, but expression of the metchnikowin gene from *Drosophila* in barley resulted in resistance to a fungal pathogen via modulation of different cascades of the plant immune system (Rahnamaeian and Vilcinskas, 2012).

In *Drosophila*, induction of AMP production ensues from the activation of the signaling pathways Toll and immune deficiency (Imd) following fly exposure to Gram-positive bacteria or fungi and Gram-negative bacteria, respectively (de Gregorio et al., 2002; Bulet and Stöcklin, 2005). Detection of these microorganisms results from direct contact between receptor proteins on host cells (such as peptidoglycan recognition proteins and Gram-negative bacteria-binding proteins) and microbial PAMPs (e.g., Gottar et al., 2002). *Drosophila melanogaster* produces 20 AMPs, grouped into seven classes (Lemaitre and Hoffmann, 2007) (Table 4.2). Flies can discriminate among different pathogen types, resulting in selective activation of the appropriate signaling pathway to trigger production of the applicable AMPs (Lemaitre et al., 1997). The Toll pathway is activated upon detection of either Gram-positive bacteria or fungal pathogens, initiating a cascade of reactions that results in the translocation of specific proteins to the nucleus of fat body cells where the target AMP genes are activated, such as *Drosomycin* for fungi (Bulet and Stöcklin, 2005). In response to Gram-negative bacteria, the Imd pathway is activated, targeting AMP production of genes such as *Diptericin* (Lemaitre and Hoffman, 2007). The activation of some inducible immune genes may be dependent on one pathway only, but some can be induced by both cascades. For example, systemic drosomycin production in *Drosophila* results from both pathways, but local production (gut epithelia) of diptericin results from Imd activation only (Tzou et al., 2000; Ferrandon et al., 2007). Induction of the Imd pathway can be rapid, occurring within minutes, in order to combat fast-replicating bacteria (e.g., Paquette et al., 2010). This is in contrast to the Toll pathway, which is activated within hours, with gene transcription potentially lasting for days (Lemaitre et al., 1997).

Another pathway, the Janus kinase signal transducer and activator of transcription (JAK-STAT) signaling pathway, is also involved in insect immune defenses, and is activated in response to septic injury and bacterial and viral pathogens (Morin-Poulard et al., 2013; Myllymäki and Rämet, 2014). Studies on *Drosophila* infected with the Drosophila C virus showed that this pathway was involved in controlling virus titers and was required but not sufficient for the activation of antiviral genes (Dostert et al., 2005). The JAK-STAT pathway is also involved in the control of fly hematopoiesis and gut homeostasis, and is highly conserved in eukaryotes (Jiang et al., 2009; Morin-Poulard et al., 2013). The numerous studies on *Drosophila*-induced innate defenses have provided a wealth of information critical to our current understanding of the mechanisms underlying insect immunity. For more information on the activation and regulation of the mentioned signaling pathways and host pathogen recognition receptor proteins or the overall *Drosophila* innate immune defenses, the following references are suggested: Ferrandon et al. (2007), Lemaitre and Hoffmann (2007), and Buchon et al. (2014).

4.4.2 Microbiome-Based Defenses

4.4.2.1 Gut Microbiota

The insect gut is home to various microorganisms, mostly bacteria but also fungi, protists, and archaea (e.g., Engel and Moran, 2013). A survey of gut bacteria from 218 insect species, representing 21 orders, found 18 bacterial phyla, composed primarily of members of Proteobacteria and Firmicutes, followed by Bacteroidetes, Actinobacteria, and Tenericutes (Yun et al., 2014). The number and diversity of microorganisms in the insect gut could vary significantly between insect species, and also among members of the same species and during an insect's life cycle, as has also been shown in other studies (e.g., Chu et al., 2013; Hroncova et al., 2015; Chen et al., 2016). Between species, variations arise from differences in habitat, food, and morphology (i.e., mouthparts and gut structure), among others things (e.g., Engel and Moran, 2013; Yun et al., 2014; Douglas, 2015). Among members of the same species, variations can arise directly from differences in habitat or food consumption, or in response to available food sources (e.g., Cox and Gilmore, 2007; Robinson et al., 2010; Chandler et al., 2011; Chu et al., 2013). A survey of the gut microbiota of crop rotation-resistant populations of the western corn rootworm (*Diabrotica virgifera virgifera*) showed shifts in bacterial composition when only the nonpreferred soybean was available (Chu et al., 2013). The change in prevalence of resident bacteria resulted in a physiological adaptation that provided the insect tolerance to antiherbivory defenses in soybeans (Chu et al., 2013). A few other studies also point to the importance of gut microbiota in facilitating insect consumption of plants with toxic compounds, suggesting that variation in herbivory of chemically defended plants may be caused, in part, by the resident microorganisms (i.e., gut microbial facilitation hypothesis) (Hammer and Bowers, 2015; but see Douglas, 2013b).

Other benefits insects derive from their associations with gut microorganisms include defensive mechanisms that preclude the establishment of ingested pathogens or prevent the development of infection. These defensive mechanisms include competition for nutrients and space, secretion of compounds antagonistic to pathogens, and enhancement of the host's innate immune system (Douglas, 2015) (Section 4.4.1.2.). Early studies on honeybee gut microbiota identified a mixture of fungi (e.g., *Mucor* sp., *Rhizopus* sp.) and bacteria (*Bacillus* spp.), some of which inhibit the fungus *Ascosphaera apis*, the causal organism of chalkbrood in honeybees (Gilliam, 1978; Gilliam et al., 1988). Bee colonies resistant to chalkbrood were found to harbor higher numbers of these microorganisms in their guts, and to add these microorganisms to the pollen they collected and stored, reducing disease transmission (Gilliam et al., 1988). This negative effect of resident gut microorganisms (mediated by production of antimycotic compounds) against fungal pathogens is similar to the gut microbe–pathogen interactions observed by Dillon and Charnley (1986, 1988) in desert locusts. Later studies on honeybee gut microbiota also identified lactic acid bacteria *Lactobacillus* and *Bifidobacterium* of multiple phylotypes with antibiotic properties (Olofson and Vasquez, 2008; Vasquez et al., 2012). Inhibition assays with *Paenibacillus larvae* (the causal agents of American foulbrood) against individual lactic acid bacteria phylotypes showed variable results, whereas the combination of all phylotypes led to complete *P. larvae* inhibition (Forsgren et al., 2010). In bumblebees, a high percentage of gut bacteria, specifically the Betaproteobacteria, was found to be inversely correlated with the likelihood of infection by the trypanosomatid gut parasite *Crithidia bombi* (Koch and Schmid-Hempel, 2011). The mechanism by which this resistance is produced is unclear, however.

Colonization resistance could also be based on nutrient competition or niche occupation and has been shown or inferred in a number of studies (e.g., Jarosz, 1979; Veivers et al., 1982; Dillon et al., 2005). In desert locusts, the diversity of symbiotic gut flora was negatively correlated with the density of the pathogenic bacterium *S. marcescens* (Dillon et al., 2005). The authors also found an inverse relationship between diversity of the bacterial flora and the proportion of locusts harboring *Serratia*, indicating that species-rich gut communities are more resistant to pathogen invasion than those that are species-poor.

4.4.2.2 Intracellular Symbionts

Insects also harbor intracellular symbiotic bacteria and fungi, characterized as either bacteriocyte symbionts (if bacteria: they are characterized as mycetocytes if yeasts) or "guest microbes." Both are maternally transmitted via transovarial transmission (Ishikawa, 2003). The first group is housed in specialized cells called bacteriocytes (or mycetocytes) that may be assembled as a specialized organ called a bacteriosome (or mycetome) (Buchner, 1965). Most of these symbionts are bacteria and have been found in the insect orders Blattodea, Coleoptera, Hemiptera, and Phthiraptera, (Douglas, 2015). Fungal symbionts, identified as being yeastlike, have been found in homopterans (planthoppers), hymenopterans (carpenter ants), and coleopterans (anobiids) (Noda, 1974, 1977; Noda and Kodama, 1995, 1996). The main function of these intracellular symbionts is nutritional provisioning or augmentation of host viability and fecundity, and their relationship with their insect host is specific (Engel and Moran, 2013; Douglas, 2015). Recent full genome sequencing data for several insect hosts and their symbionts showed coevolution, with gene loss-acquisition and integration of host–symbiont metabolic functions (e.g., Wernegreen, 2002; Wilson and Duncan, 2015). Within the bacteriocyte (or mycetocyte) symbiont group are two subgroups: the primary symbionts and the secondary symbionts (Douglas, 2015). Primary symbionts are restricted to bacteriocytes (or mycetocytes) and are required by their hosts for survival; secondary symbionts may be housed in bacteriocytes or sheath cells bounding bacteriocytes or found in the hemolymph. They are not always present in all individuals of a host species and may be transmitted both horizontally and vertically (Douglas, 2015). Among the benefits these secondary symbionts provide is host resistance to parasitoids, fungal pathogens, and other bacterial symbionts. Studies on the aphid secondary symbiont *Regiella insecticola* have shown that it confers its host resistance to the fungal pathogen *Pandora neoaphidis* (Ferrari et al., 2004; Moran et al., 2005; Scarborough et al., 2005). *Regiella* not only improved the ability of its host to survive fungal infection but also reduced spore production in infected cadavers (Scarborough et al., 2005). In *Anopheles* spp., the naturally occurring intracellular bacteria *Asaia* inhibited transmission of artificially introduced *Wolbachia* (see later), likely from competition, and this suggests one means by which interspecies horizontal transfer of intracellular symbionts could be affected (Hughes et al., 2014).

The second group of intracellular symbionts, referred to as "guest microbes," includes members of the genus *Wolbachia* (phylum Proteobacteria, order Rickettsiales). Only one species, *W. pipientis*, is recognized, but numerous strains have been characterized based on genotype, phenotype, and host range (Stouthamer et al., 1999; Werren et al., 2008). *Wolbachia pipientis* is an obligate symbiont identified in arthropods (insects and spiders) and nematodes. Among the bacteria identified by Yun et al. (2014) from a survey of 21 insect orders, *Wolbachia* formed the most represented genus, which is not surprising given that it is estimated to infect >20% (up to 76%) of all insect species (e.g., Jeyaprakash and Roy, 2000; Werren and Windsor, 2000; Zug and Hammerstein, 2012). *Wolbachia* always infects the female germline (i.e., transovarial transmission) (Werren et al., 2008), but somatic tissues may also be infected (Saridaki and Bourtzis, 2010). Further, *Wolbachia* can be transmitted horizontally between species, as evidenced by incongruence between *Wolbachia* and host phylogenies (Werren et al., 1995). Strains associated with insects are reproductive parasites, manipulating their host biology to facilitate their transmission (i.e., via cytoplasmic incompatibility and female-biased sex-ratio distortion) (Stouthamer et al., 1999; Werren et al., 2008). This parasitism, however, may sometimes be accompanied by benefits such as increased host fecundity, nutritional provisioning, and host protection against pathogens.

In insects naturally infected with *Wolbachia*, such as *Drosophila* spp., the symbiont has been shown to protect its hosts from pathogenic bacteria and viruses (e.g., Hedges et al., 2008; Teixera et al., 2008; Osborne et al., 2009; Wong et al., 2011). *Wolbachia*-infected *D. melanogaster* adults were more resistant to two naturally occurring viral pathogens, Drosophila C virus (DCV) and the Nora virus, and one introduced virus, the flock house virus; they showed lower titers of the first two (Teixera et al., 2008). The reduction in DCV titers was 10 000 less than in *Wolbachia*-free flies and resulted in longer host lifespans (Teixera et al., 2008). Protection against viruses, however, appears to be limited to RNA viruses and could vary with *Drosophila* species and *Wolbachia* strain (Osborne et al., 2009; Graham et al, 2012). The mechanisms by which antiviral protection is conferred in naturally infected *Drosophila* spp. are unclear, but these are likely to be different from those underlying antipathogen protection (viruses, bacteria, *Plasmodium*, and nematodes) conferred to insects artificially infected with *Wolbachia*, such as *Aedes* spp. and *Anopheles* spp. (Wong et al., 2011; Zug and Hammerstein, 2015a, 2015b). In the latter group, protection is likely due to immune enhancement via recent artificial introduction of the symbiont, with the upregulation of host immune responses – in particular the Toll and IMD pathways (e.g., Kambris et al., 2009; Wong et al., 2011). Further, in *Drosophila* naturally infected with *Wolbachia*, the mechanisms for antibacterial and antiviral protection appear to be independent, since the former is not as common as the latter (Rottschaefer and Lazzaro, 2012; Wong et al., 2011). Also worth noting is that naturally occurring *Wolbachia* does not always produce beneficial effects and could increase host susceptibility to pathogens (e.g., Wong et al., 2011; Graham et al., 2012; Rottschaefer and Lazzaro, 2012).

4.4.3 Behavioral Defenses

Some insects perform evasive and defensive behaviors against pathogens as part of their repertoire of defense mechanisms. These behaviors are sometimes considered first-line defense against infection. Unlike mechanical barriers (i.e., cuticle, peritrophic membrane, and basal lamina) and the innate immune system in insects, however, behavioral defenses are not ubiquitous. Further, the behavioral defenses insects employ include a range of activities that not only prevent contact with pathogens but also increase tolerance to infection (de Roode and Lefèvre, 2012). Pathogen resistance results in reduced pathogen growth or reduced development of infection, while tolerance solely alleviates the detrimental effects of the pathogen or the costs of infection (Schneider and Ayres, 2008; Kutzer and Armitage, 2016). Behavioral defenses, also called behavioral immunity, can be grouped into three major defensive mechanisms that offer stepwise protection: (i) pathogen avoidance or anti-infection resistance; (ii) pathogen clearance or anti-growth resistance; and (iii) tolerance (de Roode and Lefèvre, 2012).

Behavioral immunity has been described in both solitary and social insects. In the latter group, the likelihood of pathogen transmission is relatively higher because of their frequent social contacts within and between generations and due to group living with close relatives (Schmid-Hempel, 1998; Fefferman et al., 2007). As such, these insects have developed a variety of social and behavioral defenses, termed social immunity, to avoid contact with or transmission of pathogens and to limit or suppress infection (Cremer et al., 2007; Wilson-Rich et al., 2009). A study on the trade-offs between increased susceptibility to the fungal pathogen *M. anisopliae* from group living and the benefits of social immunity in leaf-cutting *Acromyrmex* ants showed that group living conferred better resistance than would be predicted in individuals at high population densities (Hughes et al., 2002). The division of labor in social insects also allows for employment of multiple behavioral defenses to collectively protect the colony (Chapuisat, 2010). Social immunity is achieved through the collective actions of individuals, involving not only the risk of being exposed to pathogens (e.g., workers grooming or removing infected adults or brood; Hart and Ratnieks, 2002; Wilson-Rich et al., 2009) but can also require self-exclusion. For example, *Temnothorax unifasciatus* ants infected with *M. anisopliae* left their nest, which further reduced their chances of survival (Heinze and Walter, 2010). The evolutionary success of social insects indicates that the benefits derived from social living outweigh the risks of disease susceptibility shared among nestmates and the transmission of these pathogens in the nest (Wilson-Rich et al., 2009). The various behavioral defenses observed among social insects can be unique to group living (e.g., hygienic behavior such as removal of diseased brood from nests) or can be similar to those of solitary insects (e.g. protection by use of antimicrobial compounds). For this review, behavioral responses by social and solitary insects will be grouped together by mechanism of defense.

4.4.3.1 Avoidance/Evasion

Avoidance of pathogens can be accomplished in multiple ways (e.g., spatial, temporal, or trophic), which may be used singly or in combination (de Roode and Lefèvre, 2012). Behaviors aimed at avoiding contact with pathogens occur following detection of the pathogen (i.e., via volatiles from pathogens) or of cadavers of conspecifics (Parker et al., 2010). Gypsy moth larvae avoid consuming leaf disks with virus-infected cadavers and even uninfected cadavers, indicating that avoidance may be triggered not by the pathogen but by components of the cadaver (Capinera et al., 1976; Parker et al., 2010). Laboratory studies (Petri dish assays) of unrestrained larvae of the Japanese beetle (*Popillia japonica*) exhibited a 50% increase in crawling in the presence of *H. bacteriophora* dauer larvae, although this did not result in a significant difference in mortality compared to unrestrained beetle larvae (Gaugler et al., 1994). Crawling, however, may be more effective in the field in moving the potential host away from pathogenic nematodes (Schroeder et al., 1993; Gaugler et al., 1994).

Another way to avoid pathogens is by sexual selection, where female mate choice is influenced by male health (e.g., Knell and Webberley, 2004), since infected males can transfer pathogens to females during copulation (e.g., Yue et al., 2007; Amiri et al., 2016). This is especially critical in social insects, since the queen's decision on who to mate with affects the success of later generations (Baer, 2015). Optimal mating strategies, however, are also determined by factors other than pathogen avoidance, and thus females (or queens) do not always show preference for healthy or uninfected males (e.g., Webberley et al., 2002; Luong and Kaya, 2005; Rosengaus et al., 2011). Further, some pathogens have the ability to manipulate host behavior to facilitate their transmission, which could result in attraction to infected sexual partners (e.g., Møller, 1993) (see Chapter 3).

4.4.3.2 Grooming and Hygienic Behaviors

Grooming and hygienic behaviors also fall into the pathogen-avoidance behaviors but involve removal of the pathogen or infected kin, respectively, in the nest. Grooming can be accomplished in different ways and generally varies with insect mouthparts (Zhukovskaya et al., 2013): (i) dipterans and lepidopterans with piercing-sucking or siphoning mouthparts use their legs to clean their head, thorax, and abdomen and clean the legs either by rubbing them against each other (e.g., Walker and Archer, 1988; Wuellner et al., 2002) or by coiling and uncoiling the proboscis or moving it sideways (e.g., Hikl and Krenn, 2011); (ii) orthopterans and coleopterans with chewing mouthparts use their mouthparts to clean, sometimes ingesting pathogens (e.g., Valentine, 2007); (iii) many hymenopterans use a combination of both strategies (e.g., Basibuyuk and Quicke, 1999), for example cleaning the forelegs with mouthparts and rubbing other body parts to remove surface debris.

Ants (e.g., *Formica selysi*) returning to their nests after foraging remove debris and contaminants (including pathogens and parasites) from their own body surface (personal or self-grooming), and nestmates also groom returning individuals (e.g., Reber et al, 2011). Grooming one another (allogrooming), another benefit of social living in insects, permits more efficient mechanical removal of pathogens and parasites not easily accessed during personal grooming. In *F. selysi*, allogrooming for 10 minutes resulted in a reduction by half the number of fungal conidia on contaminated ants versus those kept in isolation (Reber et al., 2011). Grooming frequency can vary depending on the presence (and detection) of pathogens and previous colony experience. *F. selysi* show more frequent self-grooming in the presence of *M. anisopliae*, indicating that these ants detected this fungal pathogen. Further, allogrooming of reintroduced nestmates, whether contaminated or not, was more frequent if the colony had previously been exposed to the fungus (Reber et al., 2011). In the dampwood termite (*Zootermopsis angusticollis*), allogrooming also increased in frequency during and after exposure to *M. anisopliae* conidia (Rosengaus et al., 1998). A separate study of termites by Yanagawa et al. (2011) showed that the musty odor of the fungus *M. anisopliae* and other fungi (*Isaria fumosorosea* and *Beauveria* spp.) triggered allogrooming of treated termites by their nestmates. Exposure of these termites to dauer larvae of the nematode *S. carpocapsae* also resulted in increased allogrooming (Wilson-Rich et al., 2007).

Contact with infected nestmates poses the risk of disease transmission to groomers unless the pathogen propagules are inactivated, as in the case of ants that apply formic acid during allogrooming. This equates to surface disinfection of individuals being groomed and inhibits fungal conidia collected in the mouths of groomers (Tragust et al., 2013). In addition, other social insects inactivate the pathogen spores removed during allogrooming by destroying them mechanically via compression in the infrabuccal pocket or by separating them under nest materials, resulting in low rates of pathogen transmission (e.g., Febvay et al., 1984; Rosengaus et al., 1998; Hughes et al., 2002; Cremer et al., 2007).

In honeybees, the hygienic behavior of nurse workers (i.e., uncapping of cells and removal of diseased brood) is their primary resistance mechanism against chalkbrood (Gilliam et al., 1988, 1983). Indeed, bees can be bred or selected for resistance on the basis of elevated hygienic behavior (Gilliam, 1997). Compounds released by the fungal pathogen *A. apis* are detected by honeybees, triggering induction of hygienic behavior (Swanson et al., 2009). Bees also remove diseased or dead adults from the nest, a defense behavior termed "undertaking" (Wilson-Rich et al., 2009). Together with grooming, these behaviors decrease disease transmission in the colony

4.4.3.3 Diet-Based Prophylactic and Therapeutic Defenses

Some insect herbivores consume plants containing toxins: secondary metabolites that may deter feeding but can also act as attractants or feeding stimulants (Feeny, 1992). These toxins may have antimicrobial properties, which reduce insect susceptibility to infection (i.e., nutritional immunology). Host plants provide not only the nutrients required for insect growth and development but also an array of toxic compounds such as alkaloids, cardenolides, oxalic acids, terpenoids, and tannins that can increase insect survival against parasites and pathogens (e.g. Nishida, 2002; de Roode et al., 2008). This defensive feeding behavior is prophylactic and is exhibited by both uninfected and infected individuals.

In contrast to prophylactic feeding, therapeutic feeding behavior is exhibited by infected insects only, which are induced to consume plants containing compounds that reduce or clear infections. One variation of prophylactic defense was shown by monarch butterflies (*Danaus plexippus*) infected with the neogregarine *Ophryocystis elektroscirrha*. Although infected insects could not cure themselves of infection by consuming milkweeds with antiparasitic properties (i.e., with higher levels of cardenolides), infected adults preferentially laid their eggs on such milkweeds, resulting in reduced infection risk and lower pathogen load in their offspring (Lefrevre et al., 2010, 2012).

Also considered therapeutic feeding is increased food consumption to offset the cost of the immune response. Host nutrition has been shown to modulate the insect innate defenses, with dietary nutrients (in particular, proteins) affecting components of the innate immune system (Lee et al., 2006, 2008; Strand, 2008; Povey et al., 2009). A study on the immune system of the Australian plague locust (*Chortoices terminifera*) in the field showed that levels of constitutive defenses (i.e., hemocyte density and lysozyme-like antimicrobial activity) were positively correlated with protein levels in the hemolymph, suggesting that protein levels and dietary protein intake can affect defensive responses (Graham et al., 2015, 2014). To offset the protein cost of resisting infection, some insects have been shown to alter their feeding behavior by increasing consumption of macronutrients (e.g., Lee et al., 2006; Povey et. al., 2009; Shikano et al., 2016). *Spodoptera littoralis* larvae surviving nucleopolyhedrovirus infection were observed to have consumed a diet with a higher protein to carbohydrate ratio compared to infected larvae dying of infection and uninfected larvae (Lee et al., 2006). Similarly, survival of *S. exempta* larvae infected with the bacterium *B. subtilis* increased with increasing protein-to-carbohydrate ratio in the diet, and infected larvae increased their protein intake compared to controls (Povey et al., 2009).

4.4.3.4 Thermoregulation

The insect body temperature is generally dependent on ambient temperature, but with the adaptation of behavioral changes and physiological adjustments, insects can regulate their body temperature to near optimal for enzymatic and tissue functions. In some insects, behavioral thermoregulation is also employed to suppress development of pathogens, resulting in increased host survival (or, at least, longevity) due to prolonged disease development (Carruthers et al., 1992; Blanford and Thomas, 1999). Insects that seek warm locations or bask, aligning their body's long axis to the sun, can raise their body temperature above their normal set point (i.e., behavioral fever) (Blanford and Thomas, 1999). Although costly to uninfected individuals, this behavior can suppress fungal (including microsporidia) and bacterial pathogens. Behavioral fever has been observed in houseflies infected with fungus *Entomophthora muscae* (Watson et al., 1993), grasshoppers infected with fungus *B. bassiana* (Inglis et al., 1996) or the microsporidium *N. acridophagus* (Boorstein and Ewald, 1987), and locusts infected with the fungus *B. bassiana* or *M. acridum* (Blanford and Thomas, 1999). Behavioral fever can also be achieved by social behaviors. Honeybee colonies generate broodcomb fever following infection with the pathogen *A. apis* and before larvae are killed, suggesting a preventative defense behavior against chalkbrood (Starks et al., 2000). The preponderance of examples against fungal pathogens suggests that this defensive behavior may be more effective against a pathogen group with a slower developmental time for infection and host kill. Defensive thermoregulation could also be effected by chilling, or by seeking environments with cooler temperatures (which would impact the development of some parasites, such as the acantocephalan *Moniliformis moniliformis* in cockroaches *Supella longipalpa* and *Blatta orientalis*) (Moore and Freehling, 2002).

4.4.3.5 Deposition of Antimicrobial Compounds

Some insects deposit antimicrobial compounds on their food, incorporate antimicrobial compounds into their nests, or deposit microorganisms secreting antimicrobial compounds to limit potential pathogen inocula or their susceptibility to infection. The burying beetle (*Nicrophorus vespilloides*) lays its eggs on small animal carcasses and deposits oral and anal secretions with antimicrobial activity on the carcass (Arce et al., 2012). *Nicrophorus* spp. secretions help preserve the carcass from bacterial and fungal decomposers, prolonging the availability of the food resource for the brood (Suzuki, 2001; Rozen et al., 2008). The secretions also protect the brood from infection by potential pathogens that might be present (Jacobs et al., 2014). In the absence of these compounds, experimentally isolated eggs exposed to soilborne bacteria had reduced hatching (Rozen et al., 2008).

The wood ant (*Formica paralugubris*) deliberately collects pieces of solidified conifer resin and incorporates these into its nests (Castella et al., 2008). Resin contains antimicrobial compounds, and the presence of resin in ant nests inhibits the growth of bacteria (e.g., *Pseudomonas* spp.) and fungi, especially xerophilic fungi (Christie et al., 2003). Moreover, the presence of resin significantly increased survival of adult and larval *P. paralugubris* exposed to the bacterium *Pseudomonas fluorescens*, as well as survival of larvae but not adults exposed to the fungus *M. anisopliae* (Chapuisat et al., 2007). *In vitro* tests also showed inhibitory effects of resin on both microorganisms (Chapuisat et al., 2007). The resin deposits are likely to confer protection against other microorganisms, given that resin is rich in terpenes and oleic compounds that have a broad spectrum of antibacterial and antifungal activities (Phillips and Croteau, 1999).

The European beewolf (*Philanthus triangulum*), a solitary digger wasp, cultivates *Streptomyces* bacteria in specialized antennal glands and applies them to the brood cell prior to oviposition (Kaltenpoth et al., 2005). The bacterium is ingested by larvae and is present on the walls of the cocoon, providing protection from fungi that develop on the remains of honeybee prey, likely by production of antibacterial compounds (Kaltenpoth et al., 2005).

4.5   Resistance via Priming

The mechanisms by which insects develop resistance to pathogens can be as varied and as numerous as the interacting factors that determine the development of infection and the steps leading to it. Although insects have different layers and facets of defense against pathogens, these microorganisms are continuously developing ways to overcome these defenses, resulting in an unceasingly evolving combat between these two groups. It is beyond the scope of this chapter to cover the various ways resistance is achieved against each pathogen group, species, or strain. Instead, a brief discussion will be given on the subject of disease resistance conferred by immune priming, a relatively recent area of inquiry.

Immune priming is defined as enhanced protection against a pathogen following exposure to a sublethal dose of the same pathogen (homologous challenge) (Little and Kraaijeveld, 2004; Kurtz, 2005; Contreras-Garduño et al., 2016). This improved protection can occur within generations (within-generational priming) or across generations (transgenerational priming) (e.g., Sadd et al., 2005; Moret, 2006; Sadd and Schmid-Hempel, 2006; Lopez et al., 2014). Transgenerational priming can be derived from either parent individually or from both parents (Little et al., 2003; Roth et al., 2010; Freitak et al., 2014), but it can be limited to the egg stage of offspring in some insects (e.g., Dubuffet et al., 2015). It can also incur costs to the offspring, such as reduced sperm viability in sons as a result of trade-off by the parent(s) (McNamara et al., 2014b). Transgenerational priming may also be constrained by the presence of another pathogen (nonhomologous) during the initial challenge (e.g., presence of gregarines during infection with the bacterium *B. thuringiensis* in *Tribolium*) (Tate and Graham, 2015).

Contreras-Garduño et al. (2016) listed three criteria for immune priming: (i) the immune resistance is specific; (ii) the protection is long-lasting; and (iii) the immune response is biphasic. In a biphasic response, the immune reaction increases after the first challenge before returning to a basal level, then increases again to a greater degree after the second challenge (e.g., Brehelin and Roch, 2008; Roth et al., 2009; Contreras-Garduño et al., 2016). Immune priming differs from immune enhancement in that the latter is activated by consumption of probiotics, exposure to or presence of non-harming immune stimulants such as gut microbiota, or a previous exposure to a different pathogenic microorganism (heterologous challenge) (Kurtz, 2005; Contreras-Garduño et al., 2016). Thus, immune enhancement is nonspecific, and the response is not as strong or as fast as in priming. Enhancement also has no memory, although a sustained immune response can result in enhancement across generations (Contreras-Garduño et al., 2016).

Mechanisms of within- and transgenerational immune priming can include upregulated constitutive immunity (i.e., phenoloxidase activity, hemocyte differentiation) and AMPs (e.g., Moret, 2006; Sadd and Schmid-Heimpel 2007; Rodrigues et al., 2010). Most studies on priming have focused on bacterial pathogens, however, with limited data on other pathogens (e.g., Rodrigues et al., 2010; Tidbury et al., 2011). A few studies have shown evidence of priming against fungal pathogens (e.g., Dubovskiy et al., 2013; Fisher and Hajek, 2015). Dubovskiy et al. (2013) showed transgenerational priming to the fungus *B. bassiana* in *G. mellonella* larvae after multiple generations were subjected to constant selective pressure. *B. bassiana*-resistant larvae showed higher cuticular phenoloxidase activity, but not hemolymph phenoloxidase, and higher expression of a fungal enzyme inhibitor gene; the defense mechanism thus focused on protecting the integument (Dubovskiy et al., 2013). There is also evidence to indicate that immune priming is not universal in insects, however, and that it could vary by insect host and pathogen (e.g., Reber and Chapuisat, 2012; Dubuffet et al., 2015; Shikano et al., 2016).

4.6   Conclusion

The diversity and number of insects inhabiting variable habitats that are also rich with microorganisms, some of which are pathogenic, provide proof of the active and effective immune defenses insects employ. Studies on how insects defend themselves against microbial pathogens reveal multiple layers and facets of complementary mechanisms, ranging from structural barriers preventing pathogen entry to induction of AMPs against pathogens in the gut and hemolymph to behavioral defenses that prevent exposure to pathogens or reduce pathogen inocula in the environment. The maintenance and induction of these defenses come at a cost, requiring trade-offs with insect life histories. But these defenses can be plastic, and insects adjust their behavior and physiological responses to offset the accompanying costs of immunity.

Considerable progress has been made in the last few decades in our understanding of insect immunity, with full genome sequence data for some insects providing the basis for examining mechanisms underlying physiological and behavioral components of immune responses to pathogens. The comprehension of the role of insect microbiome in insect defense also contributes to a holistic approach in these studies. While much still needs to be learned, the combination of two complementary approaches, studying host–pathogen interactions from specific and evolutionary points of view, will further facilitate studies of insect defenses, whether to enhance our basic understanding or for applied purposes.

Acknowledgments

This chapter is dedicated to my former graduate adviser, Wayne M. Brooks (retired, Department of Entomology, North Carolina State University), for the classical training and instruction in insect pathology he provided.

References

Adamo, S.A., Jensen, M., Younger, M., 2001. Changes in lifetime immunocompetence in male and female *Gryllus texensis* (formerly *G. integer*): trade-offs between immunity and reproduction. Anim. Behav. 62, 417–425.

Amiri, E., Meixner, M.D., Kryger, P. 2016. Deformed wing virus can be transmitted during natural mating in honey bees and infect the queens. Sci. Rep. 6, 33065.

Arce, A.N., Johnston, P.R., Smiseth, P.T., Rozen, D.E., 2012. Mechanisms and fitness effects of antibacterial defense in a carrion beetle. J. Evol. Biol. 25, 930–937.

Ardia, D.R., Gantz, J.E., Schneider, B.C., Strebel, S., 2012. Costs of immunity: an induced immune response increases metabolic rate and decreases antimicrobial activity. Funct. Ecol. 26, 732–739.

Armitage, S.A.O., Boomsma, J.K., 2010. The effects of age and social interactions on innate immunity in a leaf-cutting ant. J. Insect Physiol. 56, 78–787.

Asano, T., Ashida, M., 2001. Cuticular prophenoloxidase of the silk worm *Bombyx mori*: Purification and demonstration of its transport from hemolymph. J. Biol. Chem. 276, 11 100–11 112.

Baer, B., 2015. Female choice in social insects, in: Peretti, A.V., Aisenberg, A. (eds.), Cryptic Female Choice in Arthropods. Springer, Cham, pp. 461–477.

Barnes, A.I., Siva-Jothy, M.T., 2000. Density-dependent prophylaxis in the mealworm beetle *Tenebrio molitor* L. (Coleoptera: Tenebrionidae): cuticular melanization is an indicator of investment in immunity. Proc. R. Soc. B 267, 177–182.

Basibuyuk, H.H., Quicke, D.J.L., 1999. Grooming behaviors in Hymenoptera (Insecta): potential phylogenetic significance. Zool. J. Linn. Soc. 125, 3490382.

Berenbaum, M., 1980. Adaptive significance of midgut pH in larval Lepidoptera. Am. Nat. 115, 138–146.

Billen, J., Morgan, E.D., 1998. Pheromone communication in social insects: sources and secretions, in: Vander Meer, R.K., Breed, M.D., Espelie, K.E., Winston, M.L. (eds.), Pheromone Communication in Social Insects: Ants, Wasps, Bees, and Termites. Westview Press, Boulder, CO, pp. 3–33.

Blanford, S., Thomas, M.B., 1999. Host thermal biology: the key to understanding host-pathogen interactions and microbial pest control? Agric. For. Entomol. 1, 195–202.

Boorstein, S.M., Ewald, P.W., 1987. Costs and benefits of behavioural fever in *Melanoplus sanguinipes* infected by *Nosema acridophagus*. Physiol. Zool. 60, 586–595.

Brehelin, M., Roch, P., 2008. Specificity, learning and memory in the innate immune response. Invertebr. Survival J. 5, 103–109.

Brennan, C.A., Anderson, K.V., 2004. *Drosophila*: the genetics of innate immune recognition and response. Annu. Rev. Immunol. 22, 457–483.

Buchner, P., 1965. Endosymbiosis of Animals with Plant Microorganisms. Interscience, New York.

Buchon, N., Broderick, N.A., Poidevin, M., Pradervand, S., Lemaitre, B., 2009. *Drosophila* intestinal response to bacterial infection: activation of host defense and stem cell proliferation. Cell Host Microbe 5, 200–211.

Buchon, N., Broderick, N.A., Lemaitre, B., 2013. Gut homeostasis in a microbial world: insights from *Drosophila melanogaster*. Nat. Rev. Microbiol. 11, 615–626.

Buchon, N., Silverman, N., Cherry, S., 2014. Immunity in *Drosophila melanogaster* – from microbial recognition to whole-organism physiology. Nat. Rev. Immunol. 14, 796–810.

Bulet, P., Stöcklin, R., 2005. Insect antimicrobial peptides: structures, properties and gene regulation. Protein Pept. Lett. 12, 3–11.

Burnell, A.M., Stock, S.P., 2000. *Heterorhabditis*, *Steinernema* and their bacterial symbionts- lethal pathogens of insects. Nematology 2, 31–42.

Capinera, J.L., Kirouac, S.P., Barbosa, P., 1976. Phagodeterency of cadaver components to gypsy moth larvae, *Lymantria dispar*. J. Invertebr. Pathol. 28, 277–279.

Carlsson, A., Engstrom, P., Palva, E.T., Bennich, H., 1991. Attacin, an antibacterial protein from *Hyalophora cecropia*, inhibits synthesis of outer membrane proteins in *Escherichia coli* by interfering with *omp* gene transcription. Infect. Immun. 59, 3040–4045.

Carruthers, R.I., Larkin, T.S., Firstencel, H., Feng, Z., 1992. Influence of thermal ecology on the mycosis of rangeland grasshoppers. Ecol. 73, 193–204.

Casteels, P., Ampe, C., Jacobs, F., Vaeck, M., Tempst, P., 1989. Apidaecins: antibacterial peptides from honeybees. EMBO J. 8, 2387–2391.

Casteels, P., Ampe, C., Riviere, L., Damme, J.V., Elicone, C., Fleming, M., et al., 1990. Isolation and characterization of abaecin, a major antibacterial response peptide in the honeybee (*Apis mellifera*). Eur. J. Biochem. 187, 381–386.

Casteels, P., Ampe, C., Jacobs, F., Tempst, P., 1993. Functional and chemical characterization of hymenoptaecin, an antibacterial polypeptide that is infection-inducible in the honeybee (*Apis mellifera*). J. Biol. Chem. 268, 7044–7054.

Castella, G., Chapuisat, M., Christie, P., 2008. Prophylaxis with resin in wood ants. Anim. Behav. 75, 1591–1596.

Cerenius, L., Söderhäll, K., 2004. The prophenoloxidase-activating system in invertebrates. Immunol. Rev. 198, 116–128.

Cerenius, L., Lee, B.L.,Söderhäll, K., 2008. The proPO-system: pros and cons for its role in invertebrate immunity. Trends Immunol. 29, 263–271.

Chandler, J.A., Lang, J.M., Bhatnagar, S., Eisen, J.A., Kopp, A., 2011. Bacterial communities of diverse *Drosophila* species: ecological context of a host-microbe model system. PLoS Genet. 7, e1002272.

Chapman, J.W., Williams, T., Escribano, A., Caballero, P., Cave, R.D., Goulson, D., 1999. Age-related cannibalism and horizontal transmission of a nuclear polyhedrosis virus in larval *Spodoptera frugiperda*. Ecol. Entomol. 24, 268–275.

Chapuisat, M., 2010. Social evolution: sick ants face death alone. Curr. Biol. 20, R104–R105.

Chapuisat, M., Oppliger, A., Magliano, P., Christie, P., 2007. Wood ants use resin to protect themselves against pathogens. Proc. R. Soc. B 274, 2013–2017.

Chen, B., Teh, B.S., Sun, C., Hu, S., Lu, X., Boland, W., Shao, Y., 2016. Biodiversity and activity of the gut microbiota across the life history of the insect herbivore *Spodoptera littoralis*. Nat. Sci. Rep. 6, 29505.

Chen, C.C., Chen, C.S., 1995. *Brugia pahangi*: effects of melanization on the uptake of nutrients by microfilariae in vitro. Exp. Parasitol. 81, 72–78.

Chiang, A.S., Yen, D.F., Peng, W.K., 1986. Defense reaction of midgut epithelial cells in the rice moth larva (*Corcyra cephalonica*) infected with *Bacillus thuringiensis*. J. Invertebr. Pathol. 47, 333–339.

Chouvenc, T., Su, N.Y., Robert, A., 2009. Inhibition of *Metarhizium anisopliae* in the alimentary tract of the eastern subterranean termite *Reticulitermes flavipes*. J. Invertebr. Pathol. 101, 130–136.

Christie, P., Oppliger, A., Bancala, F., Castella, G., Chapuisat, M., 2003. Evidence for collective medication in ants. Ecol. Lett. 6, 19–22.

Chu, C.C., Spencer, J.L., Curzi, M.J., Zavala, J.A., Seufferheld, M.J., 2013. Gut bacteria facilitate adaptation to crop rotation in the western corn rootworm. Proc. Natl. Acad. Sci. U.S.A. 110, 11 917–11 922.

Contreras-Garduño, J., Lanz-Mendoza, H., Franco, B., Nava, A., Pedraza-Reyes, M., Canales-Lazcano, J., 2016. Insect immune priming: ecology and experimental evidences. Ecol. Entomol. doi:10.1111/eee.12300.

Cooper, D., Cory, G.S., Theilman, D.A., Myers, J.H., 2003. Nucleopolyhedroviruses of forest and western tent caterpillars: cross-infectivity and evidence for activation of latent virus in high density field populations. Ecol. Entomol. 28, 41–50.

Cory, J.S., Myers, J.H., 2009. Within and between population variation in disease resistance in cyclic populations of the western tent caterpillar: a test of the disease defence hypothesis. J. Anim. Ecol. 78, 646–655

Cotter, S.C., Topham, E., Price, A.J.P., Kilner, R.M., 2010. Fitness costs associated with mounting a social immune response. Ecol. Lett. 13, 1114–1123.

Cotter, S.C., Simpson, S.J., Raubenheimer, D., Wilson, K., 2011. Macronutrient balance mediates trade-offs between immune function and life history traits. Funct. Ecol. 25, 186–198.

Cox, C.R., Gilmore, M.S., 2007. Native microbial colonization of *Drosophila melanogaster* and its use as a model of *Enterococcus faecalis* pathogenesis. Infect. Immun. 75, 1565–1567.

Cremer, S., Armitage, S.A.O., Schmid-Hempel, P., 2007. Social immunity. Curr. Biol. 17, R693–R702.

de Gregorio, E., Spellman, P.T., Tzou, P., Rubin, G.M., Lemaitre, B., 2002. The Toll and Imd pathways are the major regulators of the immune response in *Drosophila*. EMBO J. 21, 2568–2579.

de Maagd, R.A., Bravo, A., Crickmore, N., 2001. How *Bacillus thuringiensis* has evolved specific toxins to colonize the insect world. Trends Genet. 17, 193–199.

de Roode, J.C., Lefèvre, T., 2012. Behavioral immunity in insects. Insects 3, 789–820.

de Roode, J.C., Pedersen, A.B., Hunter, M.D., Altizer, S., 2008. Host plant species affects virulence in monarch butterfly parasites. J. Anim. Ecol. 77, 120–126.

Dillon, R.J., Charnley, A.K., 1986. Inhibition of *Metarhizium anisopliae* by the gut bacterial flora of the desert locust, *Schistocerca gregaria*: evidence for an antifungal toxin. J. Invertebr. Pathol. 47, 350–360.

Dillon, R.J., Charnley, A.K., 1988. Inhibition of *Metarhizium anisopliae* by the gut bacterial flora of the desert locust: characterization of antifungal toxins. Can. J. Microbiol. 34, 1075–1082.

Dillon, R.J., Dillon, V.M., 2004. The gut bacteria of insects: Nonpathogenic interactions. Annu. Rev. Entomol. 49, 71–92.

Dillon, R.J., Vennard, C.T., Buckling, A., Charnley, A.K., 2005. Diversity of locust gut
  bacteria protects against pathogen invasion. Ecol. Lett. 8, 1291–1298.
Dostert, C., Jouanguy, E., Irving, P., Troxler, L., Galiana-Arnoux, D., Hetru, C., et al., 2005.
  The Jak-STAT signaling pathway is required but not sufficient for the antiviral response
  of drosophila. Nature Immunol. 6, 946–953.
Douglas, A.E., 2013a. Alimentary canal, digestion and absorption, in: Simpson, S.J.,
  Douglas, A.E. (eds.), The Insects: Structure and Function, 5th edn. Cambridge
  University Press, New York, pp. 46–80.
Douglas, A.E., 2013b. Microbial brokers of insect-plant interactions revisited. J. Chem.
  Ecol. 39, 952–961.
Douglas, A.E., 2015. Multiorganismal insects: Diversity and function of resident
  microorganisms. Annu. Rev. Entomol. 60, 17–34.
Douglas, A.E., Siva-Jothy, M.T., 2013. Circulatory system, blood and the immune system,
  in: Simpson, S.J., Douglas, A.E. (eds.), The Insects: Structure and Function, 5th edn.
  Cambridge University Press, New York, pp. 107–131.
Doums, C., Moret, Y., Benelli, E, Schmid-Hempel, P., 2002. Senescence of immune defense
  in *Bombus* workers. Ecol. Entomol. 27. 138–144.
Dubovskiy, I.M., Whitten, M.M.A., Yaroslavtseva, O.N., Greig, C., Kryukov, V.Y.,
  Grizanova, E.V., et al., 2013. Can insects develop resistance to insect pathogenic fungi?
  PLoS ONE 8, e60248.
Dubuffet, A., Zanchi, C., Boutet, G., Moreau, J., Teixera, M., Moret, Y., 2015. Trans-
  generational immune priming protects the eggs only against gram-positive bacteria.
  PLoS Pathog. 11, e1005178.
Dushay, M.S., 2009. Insect hemolymph clotting. Cell. Mol. Life Sci. 66, 2643–2650.
Dussaubat, C., Brunet, J.C., Higes, M., Colbourne, J.K., Lopez, J., Choi, J.H., et al., 2012.
  Gut pathology and responses to the microsporidium *Nosema ceranae* in the honey bee
  Apis mellifera. PLoS ONE 7, e3707.
Eleftherianos, I., Baldwin, H., Ffrench-Constant, R.H., Reynolds, S.E., 2008. Developmental
  modulation of immunity: changes within the feeding period of the fifth larval stage in
  the defence reactions of *Manduca sexta* to infection by *Photorhabdus*. J. Insect Physiol.
  54, 309–318.
Elkinton, J.S., 1990. Population dynamics of gypsy moth in North America. Annu. Rev.
  Entomol. 35, 571–596.
Engel, P., Moran, N.A., 2013. The gut microbiota of insects- diversity in structure and
  function. FEMS Microbiol. Rev. 37, 699–735.
Engelhard, E.K., Volkman, L.E., 1995. Developmental resistance within fourth instar of
  Trichoplusia ni orally inoculated with *Autographa californica* M nuclear polyhedrosis
  virus. Virology 209, 384–389.
Engelhard, E.K., Kam-Morgan, L.N.W., Washburn, O.J., Volkman, L.E., 1994. The insect
  tracheal system: a conduit for the systemic spread of *Autographa californica* M nuclear
  polyhedrosis virus. Proc. Natl. Acad. Sci. U.S.A. 91, 3224–3227.
Febvay, G., Decharme, M., Kermarrec, A., 1984. Digestion of chitin by the labial glands
  of *Acromyrmex octospinosus* Reich (Hymenoptera, Formicidae). Can. J. Zool. 62,
  229–234.
Feeny, P., 1992. The evolution of chemical ecology: contributions from the study of
  herbivorous insects, in: Rosenthal, G., Berenbaum, M., (eds.), Herbivores: Their

Interactions with Secondary Plant Metabolites. Academic Press, San Diego, CA, pp. 1–44.

Fefferman, N.H., Traniello, J.F.A., Rosengaus, R.B., Calleri, D.V., 2007. Disease prevention and resistance in social insects: modelling the survival consequences of immunity, hygienic behavior, and colony organization. Behav. Ecol. Sociobiol. 61, 565–577.

Fehlbaum, P., Bulet, P., Michaut, L., Lagueux, M., Broekaert, W.F., Hetru, C., Hoffmann, J.A., 1994. Insect immunity. Septic injury of *Drosophila* induces the synthesis of a potent antifungal peptide with sequence homology to plant antifungal peptides. J. Biol. Chem. 269, 33 159–33 163.

Fernadez-Marin, H., Zimmerman, J.K., Rehner, S.A., Wcislo, W.T., 2006. Active use of the metapleural glands by ants in controlling fungal infection. Proc. Roy. Soc. Lond. B Biol. Sci. 273, 1689–1695.

Ferrandon, D., Imler, J.L., Hetru, C., Hoffmann, J.A., 2007. The *Drosophila* systemic immune response: sensing and signalling during bacterial and fungal infections. Nature Rev. Immunol. 7, 862–874.

Ferrari, J., Darby, A.C., Daniell, T.J., Godfray, H.C. Jr., Douglas, A.E., 2004. Linking the bacterial community in pea aphids with host-plant use and natural enemy resistance. Ecol. Entomol. 29, 60–65.

Fisher, J.J., Hajek, A.E., 2015. Maternal exposure of a beetle to pathogens protects offspring against fungal disease. PLoS ONE 10, e0125197.

Forschler, B., Gardner, W., 1991. Parasitism of *Phyllophaga hirticula* (Coleoptera: Scarabaeidae) by *Heterorhabditis heliothidis* and *Steinernema carpocapsae*. J. Invertebr. Pathol. 58, 396–407.

Forsgren, E., Olofsson, T.C., Vasquez, A., Fries, I., 2010. Novel lactic acid bacteria inhibiting *Paenibacillus* larvae in honey bee larvae. Aphidologie 41, 99–108.

Freitak, D., Schmidtberg, H., Dickel, F., Lochnit, G., Vogel, H., Vilcinskas, A. 2014. The maternal transfer of bacteria can mediate trans-generational immune priming in insects. Virulence 5, 547–554.

Fuguet, R., Vey, A., 2004. Comparative analysis of the production of insecticidal and melanizing macromolecules by strains of *Beauveria* spp. in vivo studies. J. Invertebr. Pathol. 85, 152–167.

Gaugler, R., Wang, Y., Campbell, J.F., 1994. Aggressive and evasive behaviors in *Popillia japonica* (Coleoptera: Scarabaeidae) larvae: defenses against entomopathogenic nematode attack. J. Invertebr. Pathol. 64, 193–199.

Gershman, S.N., Barnett, C.A., Pettinger, A.M., Weddle, C.B., Hunt, J., Sakaluk, S.K., 2010. Give 'til it hurts: trade-offs between immunity and male reproductive effort in the decorated cricket, *Gryllodes sigillatus*. J. Evol. Biol. 23, 829–839.

Giglio, A., Giulianini, P.G., 2013. Phenoloxidase activity among developmental stages and pupal cell types of the ground beetle *Carabus (Chaetocarabus) lefebvrei* (Coleoptera: Carabidae). J. Insect Physiol. 54, 309–318.

Gilliam, M., 1978. Bacteria belonging to the genus *Bacillus* isolated from selected organs of queen honey bees, *Apis mellifera*. J. Invertebr. Pathol. 31, 389–391.

Gilliam, M., 1997. Identification and roles of non-pathogenic microflora associated with honey bees. FEMS Microbiol. Lett. 155, 1–10.

Gilliam, M., Taber III, S., Richardson, G.V., 1983. Hygienic behavior of honey bees in relation to chalkbrood disease. Apidologie 14, 29–39.

Gilliam, M., Taber III, S., Lorenz, B.J., Prest, D.B., 1988. Factors affecting development of chalkbrood disease in colonies of honey bees, *Apis mellifera*, fed pollen contaminated with *Ascosphaera apis*. J. Invertebr. Pathol. 52, 314–325.

Golkar, L., LeBrun, R.A., Ohayon, H., Gounon, P., Papierok, B., Brey, P.T., 1993. Variation of larval susceptibility to *Lagenidium giganteum* in three mosquito species. J. Invertebr. Pathol. 62, 1–8.

González-Santoyo, I., Córdoba-Aguilar, A., 2012. Phenoloxidase: a key component of the insect immune system. Ent. Expt. Appl. 142, 1–16.

Gottar, M., Gobert, V., Michel, T., Belvin, M., Duyk, G., Hoffmabb, J.A., et al., 2002. The *Drosophila* immune response against gram-negative bacteria is mediated by a peptidoglycan recognition protein. Nature, 416, 640–644.

Goulson, D., Cory, J.S., 1995. Responses of *Mamestra brassicae* (Lepidoptera, Noctuidae) to crowding – interactions with disease resistance, color phase and growth. Oecologia 104, 416–423.

Graham, R.I., Grzywacz, D., Mushoboli, W.L., Wilson, K., 2012. *Wolbachia* in a major African crop pest increases susceptibility to viral disease rather than protects. Ecol. Lett. 15, 993–1000.

Graham, R.I., Deacutis, J.M., Pulpitel, T., Ponton, F., Simpson, S.J., Wilson, K., 2014. Locusts increase carbohydrate consumption to protect against a fungal biopesticide. J. Insect Physiol. 69, 27–34.

Graham, R.I., Deacutis, J.M., Simpson, S.J., Wilson, K., 2015. Body condition constrains immune function in field populations of female Australian plague locust *Chortoicetes terminifera*. Parasitol. Immunol. 37, 233–241.

Graystock, P., Huges, W.O.G., 2011. Disease resistance in a weaver ant, *Polyrhachis dives*, and the role of antibiotic-producing glands. Behav. Ecol. Sociobiol. 65, 2319–2327.

Gross, J., Hilker, M., 1995. Chemoecological studies of the exocrine grandular larval secretions of two chrysomelid species (Coleoptera): *Phaedon cochleariae* and *Chrysomela lapponica*. Chemoecol. 5/6, 185–189.

Gross, J., Podsiadlowski, L., Hilker, M., 2002. Antimicrobial activity of exocrine glandular secretion of *Chrysomela* larvae. J. Chem. Ecol. 28, 317–331.

Gwynn, D.M., Callaghan, A., Gorham, J., Walters, K.F.A., Fellowes, M.D.E., 2005. Resistance is costly: trade-offs between immunity, fecundity and survival in the pea aphid. Proc. Roy. Soc. Lond. B Biol. Sci. 272, 1803–1808.

Ha, E.M., Oh, C.T., Bae, Y.S., Lee, W.J., 2005a. A direct role for dual oxidase in *Drosophila* gut immunity. Science 310, 847–850.

Ha, E.M., Oh, C.T., Ryu, J.H., Bae, Y.S., Kang, S.W., Jang, I., et al., 2005b. An antioxidant system required for host protection against gut infection in *Drosophila*. Dev. Cell 8, 125–132.

Haine, E.R., Moret, Y., Siva-Jothy, M.T., Rolff, J., 2008. Antimicrobial defense and persistent infection in insects. Science 322, 1257–1259.

Hamilton, C., Bulmer, M.S., 2012. Molecular antifungal defenses in subterranean termites: RNA interference reveals in vivo roles of termicins and GNBPs against a naturally encountered pathogen. Dev. Comp. Immunol. 36, e372–e377.

Hamilton, C., Lay, F., Bulmer, M.S., 2011. Subterranean termite prophylactic secretions and external antifungal defenses. J. Insect Physiol. 57, 1259–1266.

Hamm, J.J., Nordlund, D.A., Mullinix, B.G. Jr., 1983. Interaction of the microsporidium *Vairimorpha* sp. with *Microplitis croceipes* (Cresson) and *Cotesia marginiventris*

(Cresson) (Hymenoptera: Braconidae), two parasitoids of *Heliothis zea* (Boddie) (Lepidoptera: Noctuidae). Environ. Entomol. 12, 1547–1550.

Hammer, T.J., Bowers, M.D., 2015. Gut microbes may facilitate insect herbivory of chemically defended plants. Oecologia 179, 1–14.

Harrison, J.F., Wasserthal, L.T., 2013. Gaseous exchange, in: Simpson, S.J., Douglas, A.E. (eds.), The Insects: Structure and Function, 5th edn. Cambridge University Press, New York, pp. 501–545.

Hart, A.G., Ratnieks, F.L.W., 2002. Waste management in the leaf-cutting ant *Atta colombica*. Behav. Ecol. 13, 224–231.

Hedges, L.M., Brownlie, J.C., O'Neill, S.L., Johnson, K.N., 2008. *Wolbachia* and virus protection in insects. Science 322, 702.

Hegedus, D., Erlandson, M., Gillot, C., Toprak, U., 2009. New insights into peritrophic matrix synthesis, architecture, and function. Annu. Rev. Entomol. 54, 285–302.

Hegedus, N., Marx, F., 2013. Antifungal proteins: more than microbials. Fungal Biol, Rev. 26, 132–145.

Heinze, J., Walter, B., 2010. Moribund ants leave their nests to die in social isolation. Curr. Biol. 20, 249–252.

Henry, J.E., 1967. *Nosema acridophagus* sp. n., a microsporidian isolated from grasshoppers. J. Inverterbr. Pathol. 9, 331–334.

Hikl, A.L., Krenn, H.W., 2011. Pollen processing behavior of *Heliconius* butterflies: a derived grooming behavior. J. Insect Sci. 11, 99.

Hoover, K., Washburn, J.O., Volkman, L.E., 2000. Midgut-based resistance of *Heliothis virescens* to baculovirus infection mediated by phytochemicals in cotton. J. Insect Physiol. 46, 999–1007.

Hopkins, T.L., Kramer, K.J., 1992. Insect cuticle sclerotization. Annu. Rev. Entomol. 37, 273–302.

Hroncova, Z., Havlik, J., Killer, J., Doskocil, I., Tyl, J., Kamler, M., et al., 2015. Variation in honey bee gut microbial diversity affected by ontogenetic stage, age and geographic location. PLoS ONE 10, e0118707.

Hughes, G.L., Dodson, B.L., Johnson, R.M., Murdock, C.C., Tsujimoto, H., Suzuki, Y., et al., 2014. Native microbiome impedes vertical transmission of *Wolbachia* in *Anopheles* mosquitoes. Proc. Natl. Acad. Sci. U.S.A. 111, 12498–12503.

Hughes, W.O.H., Eilenberg, J., Boomsma, J.J., 2002. Trade-offs in group living: transmission and disease resistance in leaf cutting ants. Proc. Biol. Sci. 269, 1811–1819.

Hultmark, D., Engstrom, A., Bennich, H., Kapur, R., Boman, H.G., 1982. Insect immunity: isolation and structure of cecropin D and four minor antibacterial components from *Cecropia* pupae. Eur. J. Biochem. 127, 207–217.

Hung, S.Y., Boucias, D.G., 1992. Influence of *Beauveria bassiana* on the cellular defense response of the beet armyworm, *Spodoptera exigua*. J. Invertebr. Pathol. 60, 152–158.

Imler, J.L., Bulet, P. 2005. Antimicrobial peptides in *Drosophila*: structures, activities and gene regulation. Chem. Immunol. Allergy 86, 1–21.

Inglis, G.D., Johnson, D.L., Goettel, M.S., 1996. Effects of temperature and thermoregulation on mycosis by *Beauveria bassiana* in grasshoppers. Biol. Control 7, 131–139.

Inoue, H., Miyagawa, M., 1978. Regeneration of midgut epithelial cell in the silkworm, *Bombyx mori*, infected with viruses. J. Invertebr. Pathol. 32, 373–380.

Ishibashi, J., Saido-Sakanaka, H., Yang, J., Sagisaka, A., Yamakawa, M., 1999. Purification, cDNA cloning and modification of a defensin from the coconut rhinoceros beetle, *Oryctes rhinoceros*. Eur. J. Biochem. 266, 616–623.

Ishikawa, H., 2003. Insect symbiosis, an introduction, in: Bourtzis, K., Miller, T.A. (eds.), Insect Symbiosis. CRC Press, Boca Raton, FL, pp. 1–22.

Iwama, R., Ashida, M., 1996. Biosynthesis of prophenoloxidase in hemocytes of larval hemolymph of the silk worm, Bombyx mori. Insect Biochem. 16, 547–555.

Jacobs, C.G.C., Wang, Y., Vogel, H., Vilcinskas, A., van der Zee, M., Rozen, D.E., 2014. Egg survival is reduced by grave-soil microbes in the carrion beetle, *Nicrophorus vespillloides*. BMC Evol. Biol. 14, 208.

Jarosz, J., 1979. Gut flora of *Galleria mellonella* suppressing ingested bacteria. J. Invertebr. Pathol. 34, 192–198.

Jeyaprakash, A., Hoy, M.A., 2000. Long PCR improves *Wolbachia* DNA amplification: *wsp* sequences found in 76% of sixty-three arthropod species. Insect Mol. Biol. 9, 393–405.

Jiang, H., Patel, P.H., Kohlmaier, A., Grenley, M.O., McEwen, D.G., Edgar, B.A., 2009. Cytokine/Jak/Stat signaling mediates regeneration and homeostasis in the *Drosophila* gut. Cell 137, 1343–1355.

Jung, S.H., Evans, C.J., Uemura, C., Banerjee, U., 2005. The *Drosophila* lymph gland as a developmental model of hematopoiesis. Development 132, 2521–2533.

Kaltenpoth, M., Göttler, W., Herzner, G., Strohm, E., 2005. Symbiotic bacteria protect wasp larvae from fungal infestation. Curr. Biol. 15, 475–479.

Kambris, Z., Cook, P.E., Phuc, H.K., Sinkins, S.P., 2009. Immune activation by life-shortening *Wolbachia* and reduced filarial competence in mosquitoes. Science 326, 134–136.

King, E.G., Bell, J.V., Martin, D.F., 1975. Control of bacterium *Serratia marcescens* in an insect host-parasite rearing program. J. Invertebr. Pathol. 26, 35–40.

Kirkpatrick, B.A., Washburn, J.O., Engelhard, E.K., Volkman, L.E., 1994. Primary infection of insect trachea by *Autographa californica* M nuclear polyhedrosis virus. Virology 203, 184–186.

Knell, R.J., Webberley, K.M., 2004. Sexually transmitted diseases of insects: distribution, evolution, ecology and host behavior. Biol. Rev. 79, 557–581.

Koch, H., Schmid-Hempel, P., 2011. Socially transmitted gut microbiota protect bumble bees against an intestinal parasite. Proc. Natl. Acad. Sci. U.S.A. 108, 19 288–19 292.

Koppenhöfer, A.M., Grewal, P.S., Fuzy, E.M., 2007. Differences in penetration routes and establishment rates of four entomopathogenic nematode species into four white grub species. J. Invertebr. Pathol. 94, 184–195.

Kumar, S., Christophides, G.K., Cantera, R., Charles, B., Han, Y.S., Meister S., et al. 2003. The role of reactive oxygen species on *Plasmodium* melanotic encapsulation in *Anopheles gambiae*. Proc. Natl. Acad. Sci. U.S.A. 100, 14 139–14 144.

Kunimi, Y., Yamada, E., 1990. Relationship of larval phase and susceptibility of the armyworm *Pseudaletia separata* Walker (Lepidoptera: Noctuidae) to a nuclear polyhedrosis virus and a granulosis virus. Appl. Entomol. Zool. 25, 289–297.

Kurtz, J. 2005. Specific memory within innate immune systems. Trends Immunol. 26, 186–192.

Kutzer, M.A.M., Armitage, S.A.O., 2016. Maximizing fitness in the face of parasites: a review of host tolerance. Zoology 119, 281–289.

Lacey, C.M., Lacey, L.A., Roberts, D.R., 1988. Route of invasion and histopathology of *Metarhizium anisopliae* in *Culex quinquefasciatus*. J. Invertebr. Pathol. 52, 108–118.

Lamberty, M., Ades, S., Uttenweiler-Joseph, S., Brookhart, G., Bushey, D., Hoffmann, J.A., Bulet, P., 1999. Insect immunity. Isolation from the lepidopteran *Heliothis virescens* of a novel insect defensin with potent antifungal activity. J. Biol. Chem. 274, e9320–e9326.

Lamberty, M., Zachary, D., Lanot, R., Bordereau, C., Robert, A., Hoffmann, J.A., Bulet, P., 2001. Insect immunity. Constitutive expression of a cysteine-rich antifungal and a linear antibacterial peptide in a termite insect. J. Biol. Chem. 276, e4085–e4092.

Lavine, M.D., Strand, M.R., 2002. Insect hemocytes and their role in immunity. Insect Biochem. Mol. Biol. 32, 1295–1309.

Lee, K.P., Cory, S.J., Wilson, K., Raubenheimer, D., Simpson, S. J., 2006. Flexible diet choice offsets protein costs of pathogen resistance in a caterpillar. Proc. R Soc. B 273, 823–829.

Lee, K.P., Simpson, S.J., Wilson, K., 2008. Dietary protein-quality influences melanization and immune function in insects. Funct. Ecol. 22, 1052–1061.

Lee, Y.S., Yun, E.K., Jang, W.S., Kim, I., Lee, J.H., Park, S.Y., et al., 2004. Purification, cDNA cloning and expression of an insect defensin from the great wax moth, *Galleria mellonella*. Insect Mol. Biol. 13, e65–e72.

Lefèvre, T., Oliver, L., Hunter, M.D., de Roode, J.C., 2010. Evidence for trans-generational medication in nature. Ecol. Lett. 13, 1485–1493.

Lefèvre, T., Chiang, A., Kelavkar M., Li, H., de Castillejo, C.L.F., Oliver, L., et al., 2012. Behavioral resistance against protozoan parasite in the monarch butterfly. J. Anim. Ecol. 81, 70–79.

Lemaitre, B., Hoffman, J.A., 2007. The host defense of *Drosophila melanogaster*. Annu. Rev. Immunol. 25, 697–743.

Lemaitre, B., Reichhart, J.M., Hoffman, J.A., 1997. *Drosophila* host defense: differential induction of antimicrobial peptide genes after induction by various classes of organisms. Proc. Natl. Acad. Sci. U.S.A. 94, 14614–14619.

Leman, J.C., Weddle, C.B., Gershman, S.N., Kerr, A.M., Ower, G.D., St. John, J.M., et al., 2009. Lovesick: immunological costs of mating to male sagebrush crickets. J. Evol. Biol. 22, 163–171.

Lewis, E.E., Campbell, J., Griffin, C., Kaya, H., Peter, A., 2006. Behavioral ecology of entomopathogenic nematodes. Biol. Control 36, 66–79.

Lighthart, B., Sewall, D., Thomas, D.R., 1988. Effect of several stress factors on the susceptibility of the predatory mite, *Metaseiulus occidentalis* (Acari: Phytoseiidae), to the weak bacterial pathogen *Serratia marcescens*. J. Invertebr. Pathol. 54, 33–42.

Lindsey, E., Altizer, S., 2009. Sex differences in immune defenses and response to parasitism in monarch butterflies. Evol. Ecol. 23, 607–620.

Lindsey, E., Mehta, M., Dhulipala, V., Oberhauser, K., Altizer, S., 2009. Crowding and disease: effects of host density in response to infection in a butterfly-parasite interaction. Ecol. Entomol. 34, 551–561.

Little, T.J., Kraaijeveld, A.R., 2004. Ecological and evolutionary implications of immunological priming in invertebrates. Trends Ecol. Evol. 19, 58–60.

Little, T.J., O'Connor, B., Colegrave, N., Watt, K., Read, A.F., 2003. Maternal transfer of strain-specific immunity in an invertebrate. Curr. Biol. 13, 489–492.

Lopez, J.H., Schuehly, W., Crailsheim, K., Riessberger-Gallé, U., 2014. Trans-generational immune priming in honeybees. Proc. Roy. Soc. Lond. B Biol. Sci. 281, 20140454.

Lowenberger, C., Bulet, P., Charlet, M., Hetru, C., Hodgeman, B., Christensen, B.M., Hoffmann, J.A., 1995. Insect immunity: isolation of three novel inducible defensins from the vector mosquito, *Aedes aegypti*. Insect Biochem. Mol. Biol. 25, 867–873.

Lowenberger, C., Charlet, M., Vizioli, J., Kamal, S., Richman, A., Christensen, B.M., Bulet, P., 1999. Antimicrobial activity spectrum, cDNA cloning, and mRNA expression of a newly isolated member of the cecropin family from the mosquito vector *Aedes aegypti*. J. Biol. Chem. 274, 20092–20097.

Luong, L.T., Kaya, H.K., 2005. Sexually transmitted parasites and host mating behavior in the decorated cricket. Behav. Ecol. 16, 794–799.

Mackenzie, D.K., Bussiere, L.F., Tinsley, M.C., 2011. Senescence of the cellular immune response in *Drosophila melanogaster*. Exp. Gerontol. 46, 853–859.

Mackintosh, J.A., Veal, D.A., Beattie, A.J., Gooley, A.A., 1998. Isolation from an ant *Myrmecia gulosa* of two inducible O-glycosylated proline-rich antibacterial peptides. J. Biol. Chem. 273, 6139–6143.

Marmaras, V.J., Lampropoulou, M., 2009. Regulators and signalling in insect hemocyte immunity. Cell. Signal. 21, 186–195.

Matova, N., Anderson, K.V., 2006. Rel/NF-κB double mutants reveal that cellular immunity is central to *Drosophila* host defense. Proc. Natl. Acad. Sci. U.S.A. 103, 16424–16429.

McNamara, K.B., van Lieshout, E., Simmins, L.W., 2014a. Females suffer a reduction in the viability of stored sperm following an immune challenge. J. Evol. Biol. 27, 133–140.

McNamara, K.B., van Lieshout, E., Simmons, L.W., 2014b. The effect of maternal and paternal immune challenge on offspring immunity and reproduction in a cricket. J. Evol. Biol. 27, 1020–1028.

Merzendorfer, H., 2013. Integument, in: Simpson, S.J., Douglas, A. E. (eds.), The Insects: Structure and Function, 5th edn. Cambridge University Press, New York, pp. 463–498.

Misof, B., Liu, S., Meusemann, K., Peters, R.S., Donath, A., Mayer, C., et al., 2014. Phylogenomics resolves the timing and pattern of insect evolution. Science 346, 763–767.

Møller, A.P., 1993. A fungus infecting domestic flies manipulates sexual behavior of its host. Behav. Ecol. Sociobiol. 33, 403–407.

Monnin, D., Kremer, N., Berny, C., Henri, H., Dumet, A., Voituron, Y., et al., 2016. Influence of oxidative stress on bacterial density and cost of infection in *Drosophila-Wolbachia* symbioses. J. Evol. Biol. 29, 1211–1222.

Moore, J., Freehling, M., 2002. Cockroach hosts in thermal gradients suppress parasite development. Oecologia 133, 261–266.

Mora, C., Tittensor, D.P., Adl, S., Simpson, A.G.B., Worm, B., 2011. How many species are there on earth and in the ocean. PLoS Biol. 9, e1001127.

Moran, N.A., Russell, J.A., Koga, R., Fukatsu, T., 2005. Evolutionary relationships of three new species of *Enterobacteriaceae* living as symbionts of aphids and other insects. Appl. Environ. Entomol. 71, 3302–3310.

Moret, Y., 2006. "Trans-generational immune priming": specific enhancement of the antimicrobial immune response in the mealworm beetle, *Tenebrio molitor*. Proc. Roy. Soc. Lond. B Biol. Sci. 273, 1399–1405.

Morin-Poulard, I., Vincent, A., Crozatier, M., 2013. The *Drosophila* JAK-STAT pathway in blood cell formation and immunity. JAK-STAT 2(3), e25700.

Myllymäki, H., Rämet, M., 2014. JAK/STAT pathway in *Drosophila* immunity. Scand. J. Immunol. 79, 377–385.

Mylonakis, E., Podsiadlowski, L., Muhammed, M., Vilcinskas, A., 2016. Diversity, evolution and medical applications of insect antimicrobial peptides. Proc. Roy. Soc. Lond. B Biol. Sci. 371, 20150290.

Nakazawa, H., Tsuneishi, E., Ponnuvel, K.M., Furukawa, S., Asaoka, A., Tanaka, H., et al., 2004. Anti-viral activity of a serine protease from the digestive juice of *Bombyx mori* larvae against nucleopolyhedrovirus. Virology 321, 154–162.

Nappi, A.J., Vass, E., 1998. Hydrogen peroxide production in immune-reactive *Drosophila melanogaster*. J. Parasitol. 84, 1150–1157.

Nappi, A.J., Vass, E., Frey, F., Carton, Y., 1995. Superoxide anion generation in *Drosophila* during melanotic encapsulation of parasites. Eur. J. Cell Biol. 68, 450–456.

Nishida, R., 2002. Sequestration of defensive substances from plants by Lepidoptera. Annu, Rev. Entomol. 47, 57–92.

Noda, H., 1974. Preliminary histological observation and population dynamics of intracellular yeast-like symbiotes in the smaller brown planthopper, *Laodelphax striatellus* (Homoptera: Delphacidae). Appl. Entomol. Zool. 9, 275–277.

Noda, H., 1977. Histological and histochemical observation of intracellular yeastlike symbiotes in the fat body of the smaller brown planthopper, *Laodelphax striatellus* (Homoptera: Delphacidae). Appl. Entomol. Zool 12, 134–141.

Noda, H., Kodama, K., 1995. Phylogenetic position of yeast-like symbiotes of rice planthoppers based on partial 18S rDNA sequences. Insect Biochem. Mol. Bio. 25, 639–646.

Noda, H., Kodama, K., 1996. Phylogenetic position of yeast-like endosymbionts of anobiid beetles. Appl. Environ. Microbiol. 62, 162–167.

Olofsson, T.C., Vásquez A., 2008. Detection and identification of a novel lactic acid bacterial flora within the honey stomach of the honeybee *Apis mellifera*. Curr. Microbiol. 57, 356–363.

Opoku-Debrah, J.K., Hill, M.P., Knox, C., Moore, S.D., 2013. Overcrowding of false codling moth, *Thaumatotibia leucotreta* (Meyrick) leads to the isolation of five new *Cryptophlebia leucotreta* granulovirus (CrleGV-SA) isolates. J. Invertebr. Pathol. 112, 219–228.

Orthius-Lechner, D., Maile, R., Morgan, E.D., Boomsma, J.J., 2000. Metapleural gland secretion of the leaf-cutter ant *Acromyrmex octospinosus*: new compounds and their functional significance. J. Chem. Ecol. 26, 1667–1683.

Osborne, S.E., Leong, Y.S., O'Neill, S.L., Johnson, K.N., 2009. Variation in antiviral protection mediated by different *Wolbachia* strains in *Drosophila simulans*. PLoS Pathogens 5, e10006556.

Overend, G., Luo, Y., Henderson, L., Douglas, A.E., Davies, S.A., Dow, J.A.T., 2016. Molecular mechanism and functional significance of acid generation in the *Drosophila* midgut. Sci. Rep. 6, 27242.

Own, O.S., Brooks, W.M., 1986. Interactions of the parasite *Pediobius foveolatus* (Hymenoptera: Eulophidae) with two *Nosema* spp. (Microsporida: Nosematidae) of the Mexican bean beetle (Coleoptera: Coccinellidae). Environ. Entomol. 15, 32–39.

Paquette, N., Broemer, M., Aggarwal, K., Chen, L., Husson, M., Ertürk-Hasdemir, D., et al., 2010. Caspase-mediated cleavage, IAP binding, and ubiquitination: linking three mechanisms crucial for *Drosophila* NF-κB Signaling. Mol. Cell 37, 172–182.

Parker, B.J., Elderd, B.D., Dwyer, G., 2010. Host behavior and exposure risk in an insect-pathogen interaction. J. Anim. Ecol. 79, 863–870.

Passarelli, A.L., 2011. Barriers to success: How baculoviruses establish efficient systemic infections. Virology 411, 383–392.

Pech, L.L., Strand, M.R., 1996. Granular cells are required for encapsulation of foreign targets by insect haemocytes. J. Cell Sci. 109, 2053–2060.

Pendland, J.C., Boucias, D.G., 1993. Evasion of host defense by in vivo produced protoplast-like cells of the insect mycopathogen *Beauveria bassiana*. J. Bacteriol. 175, 5962–5969.

Pham, L.N., Schneider, D.S., 2008. Evidence for specific memory in the insect innate immune response, in: Beckage, N. (ed.), Insect Immunology. Academic Press, San Diego, CA, pp. 97–126.

Phillips, M.A., Croteau, R.B., 1999. Resin-based defenses in conifers. Trends Plant Sci. 4, 184–190.

Ponnuvel, K.M., Nakazawa, H., Furukawa, S., Asaoka, A., Ishibashi, J., Tanaka, H., Yamakawa, M., 2003. A lipase isolated from the silkworm *Bombyx mori* shows anti-viral activity against nucleopolyhedrovirus. J. Virol. 77, 10725–10729.

Poulsen, M., Bot, A.N.M., Nielsen, M.G., Boomsma, J.J., 2002. Experimental evidence for the costs and hygienic significance of the antibiotic metapleural gland secretion in leaf-cutting ants. Behav. Ecol. Sociobiol. 52, 151–157.

Povey, S., Cotter, S.C., Simpson, S.J., Lee, K.P., Wilson, K., 2009. Can the protein costs of bacterial resistance be offset by altered feeding behavior? J. Anim. Ecol. 78, 437–446.

Rahnamaeian, M., Vilcinskas, A., 2012. Defense gene expression is potentiated in transgenic barley expressing antifungal peptide metchnikowin throughout powdery mildew challenge. J. Plant Res. 125, 115–124.

Reber, A., Chapuisat, M., 2012. No evidence for immune priming in ants exposed to a fungal pathogen. PLoS ONE 7, e35372.

Reber, A., Purcell, J., Buechel, S.D., Buri, P., Chapuisat, M., 2011. The expression and impact of antifungal grooming in ants. J. Evol. Biol. 24, 954–964.

Reddy, J.T., Locke, M., 1990. The size limited penetration of gold particles through insect basal laminae. J. Insect Physiol. 36, 397–408.

Rees, J.A., Moniatte, M., Bulet, P., 1997. Novel antibacterial peptides isolated from a European bumblebee, *Bombus pascuorom* (Hymenoptera, Apoidea). Insect Biochem. Mol. Biol. 27, 413–422.

Reeson, A.F., Wilson, K., Gunn, A., Hails, R.S., Goulson, D., 1998. Baculovirus resistance in the noctuid *Spodoptera exempta* is phenotypically plastic and responds to population density. Proc. Roy. Soc. Lond. B Biol. Sci. 265, 1787–1791.

Ribeiro, C., Brehelin, M., 2006. Insect hemocytes: what type of cell is that? J. Insect Physiol. 52, 417–429.

Roberts, K.E., Hughes, W.O.H., 2014. Immunosenescence and resistance to parasite infection in the honeybee, *Apis mellifera*. J. Invertebr. Pathol. 121, 1–6.

Robinson, C.J., Schloss, P., Ramos, Y., Raffa, K., Handelsman, J., 2010. Robustness of the bacterial community in the cabbage whitefly larval midgut. Microb. Ecol. 59, 199–211.

Rodrigues, J., Brayner, F.A., Alves, L.C., Dixit, R., Barillas-Mury, C., 2010. Hemocyte differentiation mediates innate immune memory in *Anopheles gambiae* mosquitoes. Science 329, 1353–1355.

Rosengaus, R.B., Maxmen, A.B., Coates, L.E., Traniello, J.F.A., 1998. Disease resistance: a benefit of sociality in the dampwood termite *Zootermopsis angusticollis* (Isoptera: Termopsidae). Behav. Ecol. Sociobiol. 44, 125–134.

Rosengaus, R.B., James, L.T., Hartke, T.R., Brent, C.S., 2011. Male preference and disease risk in *Zootermopsis angusticollis* (Isoptera: Termopsidae). Environ. Entomol. 40, 1554–1565.

Roth, O., Sadd, B.M., Schmid-Hempel, P., Kurtz, J., 2009. Strain-specific priming of resistance in the red flour beetle, *Tribolium castaneum*. Proc. R. Soc. B 276, 145–151.

Roth, O., Joop, G., Eggert, H., Hilbert, J., Daniel, J., Schmid-Hempel, P., Kurtz, J., 2010. Paternally derived immune priming for offspring in the red flour beetle, *Tribolium castaneum*. J. Anim. Ecol. 79, 403–413.

Rottshaeffer, S.M., Lazzaro, B.P., 2012. No effect of *Wolbachia* on resistance to intracellular infection by pathogenic bacteria in *Drosophila melanogaster*. PLoS ONE 7, e40500.

Rozen, D.E., Engelmoer, D.J.P., Smiseth, P.T., 2008. Antimicrobial strategies in burying beetles breeding on carrion. Proc. Natl. Acad. Sci. U.S.A. 105, 17890–17895.

Sadd, B.M., Schmid-Hempel, P., 2006. Insect immunity shows specificity in protection upon secondary pathogen exposure. Curr. Biol. 16, 1206–1210.

Sadd, B.M., Schmid-Hempel, P., 2007. Facultative but persistent trans-generational immunity via the mother's eggs in bumblebees. Curr. Biol. 17, R1046.

Sadd, B.M., Schmid-Hempel, P., 2009. Perspective: principles of ecological immunology. Evol. Appl. 2, 113–121.

Sadd, B.M., Siva-Jothy, M.T., 2006. Self-harm caused by an insect's innate immunity. Proc. Roy. Soc. Lond. B Biol. Sci. 273, 2571–2574.

Sadd, B.M., Kleinlogel, Y., Schmid-Hempel, R., Schmid-Hempel, P., 2005. Trans-generational priming in a social insect. Biol. Lett. 1, 386–388.

Saridaki, A., Bourtzis, K., 2010. *Wolbachia*: more than just a bug in insect genitals. Curr. Opin. Microbiol. 13, 67–72.

Scarborough, C.L., Ferrari, J., Godfray, H.C., 2005. Aphid protected from pathogen by endosymbiont. Science 310, 1781.

Schmid-Hempel, P., 1998. Parasites in Social Insects. Princeton University Press, Princeton, NJ.

Schmid-Hempel, P., Ebert, D., 2003. On the evolutionary ecology of specific immune defence. Trends Ecol. Evol. 18, 27–32.

Schlüns, H., Crozier, R.H., 2009. Molecular and chemical immune defenses in ants (Hymenoptera: Formicidae). Myrmecol. News 12, 237–249.

Schneider, D.S., Ayres, J.S., 2008. Two ways to survive infection: what resistance and tolerance can teach us about treating infectious diseases. Nat. Rev. Immunol. 8, 889–895.

Schroeder, P.C., Villani, M.G., Ferguson, C.S., Nyrop, J.P., Shields, E.J., 1993. Behavioral interactions between Japanese beetle (Coleoptera: Scarabaeidae) grubs and an entomopathogenic nematode (Nematoda: Heterorhabditidae) within turf microcosms. Environ. Entomol. 22, 595–600.

Schuhmann, B., Seitz, V., Vilcinskas, A., Podsiadlowski, L., 2003. Cloning and expression of gallerimycin, an antifungal peptide expressed in immune response of greater wax moth larvae, *Galleria mellonella*. Arch. Insect Biochem. Physiol. 53, e125–e133.

Schwenke, R.A., Lazzaro, B.P., Wolfner, M. F., 2016. Reproduction-immunity trade-offs in insects. Annu. Rev. Entomol. 61, 239–256.

Shikano, I., Hua, K.N., Cory, J.S., 2016. Baculovirus-challenge and poor nutrition inflict within-generation fitness costs without triggering transgenerational priming. J. Invertebr. Pathol. 136, 35–42.

Shudo, E., Iwasa, Y., 2001 Inducible defense against pathogens and parasites: optimal choice among multiple options. J. Theor. Biol. 209, 233–247.

Splittstoesser, C.M., Kawanishi, C.Y., Tashiro, H., 1978. Infection of the European chafer, *Amphimallon majalis* by *Bacillus popilliae*: light and electron microscope observations. J. Invertebr. Pathol. 31, 84–89.

Sosa-Gomez, D.R., Boucias, D.G., Nation, J.L., 1997. Attachment of *Metarhizium anisopliae* to the southern green stink bug *Nezara viridula* cuticle and fungistatic effect of cuticular lipids and aldehydes. J. Invertebr. Pathol. 69, 31–39.

St. Leger, R.J., 1991. Integument as a barrier to microbial infections, in: Binnington, K., Retnakaran, A. (eds.), Physiology of the Insect Epidermis. CSIRO Publishing, Clayton, pp. 284–306.

St. Leger, R.J., 1994. The role of cuticle degrading proteases in fungal pathogenesis of insects. Can. J. Bot. 73, S1119–S1125.

St. Leger, R.J., Goettel, M., Roberts, D.W., Staples, R.C., 1991. Prepenetration events during infection of host cuticle by *Metarhizium anisopliae*. J. Invertebr. Pathol. 58, 168–179.

Starks, P.T., Blackie, C.A., Seeley, T.D., 2000. Fever in honeybee colonies. Naturwissenschaften 87, 229–231.

Staudacher, H., Kaltenpoth, M., Breeuwer, J.A.J., Menken, S.B.J., Heckel, D.G., Groot, A.T., 2016. Variability of bacterial communities in the moth *Heliothis virescens* indicates transient associations with the host. PLoS ONE 11, e0154514.

Stouthamer, R., Breeuwer, J.A.J., Hurst, G.D.D., 1999. *Wolbachia pipientis*: microbial manipulator of arthropod reproduction. Annu. Rev. Microbiol. 53, 71–102.

Strand, M.R., 2008. The insect cellular immune response. Insect Sci. 15, 1–14.

Strand, M.R., Pech, L.L, 1995. Immunological basis for compatibility in parasitoid-host relationships. Annu. Rev. Entomol. 40, 31–56.

Suzuki, S., 2001. Suppression of fungal development on carcasses by the burying beetle *Nicrophorus quadripunctatus* (Coleoptera: Silphidae) Ent. Sci. 4, 403–405.

Swanson, J.A.I., Torto, B., Kells, S.A., Mesce, K.A., Tumlinson, J.H., Spivak, M., 2009. Odorants that induce hygienic behavior in honeybees: identification of volatile compounds in chalkbrood-infected honeybee larva. J. Chem. Ecol. 35, 1108–1116.

Taniai, K., Kadono-Okuda, K., Kato, Y., Yamamoto, M., Shimabukuro, M., Chowdhury, S., et al., 1995. Structure of two cecropin B-encoding genes and bacteria-inducible binding proteins which bind to the 5′-upstream regulatory region in the silkworm, *Bombyx mori*. Gene 163, 215–219.

Tate, A.T., Graham, A.L., 2015. Trans-generational priming of resistance in wild flour beetles reflects the primed phenotypes of laboratory populations and is inhibited by co-infection by a common parasite. Funct. Ecol. 29, 1059–1069.

Teixera, L., Ferreira, A., Ashburner, M., 2008. The bacterial symbiont *Wolbachia* induces resistance to RNA viral infections in *Drosophila melanogaster*. PLoS Biol. 6, e1000002.

Theopold, U., Schmidt, O., Söderhäll, K., Dushay, M.S., 2004. Coagulation in arthropods: defense, wound closure and healing. Trends Immunol. 25, 289–295.

Tidbury, H.J., Pedersen, A.B., Boots, M., 2011. Within and transgenerational immune priming in an insect to a DNA virus. Proc. Roy. Soc. Lond. B Biol. Sci. 278, 871–876.

Tillman, P.G., Styler, E.L., Hamm, J.J., 2004. Transmission of ascovirus from *Heliothis virescens* (Lepidoptera: Noctuidae) by three parasitoids and effects of virus on the survival of parasitoid *Cardiochiles nigriceps* (Hymenoptera: Braconidae). Environ. Entomol. 33, 633–643.

Tragust, S., Mitteregger, B., Barone, V., Konrad, M., Ugelvig, L.V., Cremer, S., 2013. Ants disinfect fungus exposed brood by oral uptake and spread of their poison. Curr. Biol. 23, 76–82.

Tzou, P., Ohresser, S., Ferrandon, D., Capovilla, M., Reichhart, J.M., Lemaitre, B., et al. 2000. Tissue-specific inducible expression of antimicrobial peptide genes in *Drosophila* surface epithelia. Immun. 13, 737–748.

Urbanski, A., Czarniewska, E., Baraniak, E., Rosinski, G., 2014. Developmental changes in cellular and humoral responses of the burying beetle *Nicrophorus vespilloides* (Coleoptera, Silphidae). J. Insect. Physiol. 60, 98–103.

Valentine, B., 2007. Mutual grooming in cucujoid beetles (Coleoptera: Silvanidae). Insecta Mundi 2, 1–3.

Vallet-Gely, I., Lemaitre, B., Boccard, F., 2008. Bacterial strategies to overcome insect defences. Nat. Rev. Microbiol. 6, 302–313.

Vasquez, A., Forsgren, E., Fries, I., Paxton, R.J., Flaberg, E., Szekely, L., Olofsson, T.C., 2012. Symbionts as major modulators in insect health: Lactic acid bacteria and honeybees. PLos ONE 7, e33188.

Veivers, P.C., O'Brien, R.W., Slaytor, M., 1982. Role of bacteria in maintaining the redox potential in the hindgut of termites and preventing entry of foreign bacteria. J. Insect Physiol. 28, 947–951.

Wagner, C., Isermann, K., Fehrenbach, H., Roeder, T., 2008. Molecular architecture of the fruit fly's airway epithelial immune system. BMC Genomics 9, 446–457.

Walker, E.D., Archer, W.E., 1988. Sequential organization of grooming behaviors of the mosquito *Aedes triseriatus*. J. Insect Behav. 1, 97–109.

Waltzer, L., Bataillé, L., Peyrefitte, S., Haenlin, M., 2002. Two isoforms of serpent containing either one or two GATA zinc fingers have different roles in *Drosophila* haematopoiesis. EMBO J. 21, 5477–5486.

Washburn, J.O., Kirkpatrick, B.A., Volkman, L.E., 1995. Comparative pathogenesis of *Autographa californica* M nuclear polyhedrosis virus in larvae of *Trichoplusia ni* and *Heliothis virescens*. Virology 209, 561–568.

Washburn, J.O., Kirkpatrick, B.A., Hass-Stapleton, E. Volkman, L.E., 1998. Evidence that the stilbene-derived optical brightener M2R enhances *Autographa californica* M nucleopolyhedrosis virus infection of *Trichoplusia ni* and *Heliothis virescens* by preventing sloughing of infected midgut epithelial cells. Biol. Control 11, 58–69.

Watson, D.W., Mullens, B.A., Petersen, J.J., 1993. Behavioral fever response of *Musca domestica* (Diptera: Muscidae) to infection by *Entomophthora muscae* (Zygomycetes: Entomophthorales). J. Invertebr. Pathol. 61, 10–16.

Webberley, K.M., Hurst, G.D.D., Buszko, J., Majerus, M.E.N., 2002. Lack of parasite-mediated sexual selection in a ladybird/sexually transmitted disease system. Anim. Behav. 63, 131–141.

Werren, J.H., Windsor, D.M., 2000. *Wolbachia* infection frequencies in insects: evidence of a global equilibrium? Proc. Roy. Soc. Lond. B Biol. Sci. 267, 1277–1285.

Werren J.H., Zhang, W., Guo, L.R., 1995. Evolution and phylogeny of *Wolbachia*: reproductive parasites of arthropods. Proc. Biol. Sci. 261, 55–63.

Werren, J.H., Baldo, L., Clark, M.E., 2008. *Wolbachia*: master manipulators of invertebrate biology. Nat. Rev. Microbiol. 6, 741–751.

Wernergreen, J.J., 2002. Genome evolution in bacterial endosymbionts of insects. Nat. Rev. Genet. 3, 850–861.

Williams, T., Hernandez, O., 2006. Costs of cannibalism in the presence of an iridovirus pathogen of *Spodoptera frugiperda*. Ecol. Entomol. 31, 106–113.

Wilson, A., Duncan, R.P., 2015. Signatures of host/symbiont genome coevolution in insect nutritional symbioses. Proc. Natl. Acad. Sci. U.S.A. 112, 10255–10261.

Wilson, K., Reeson, A.F., 1998. Density-dependent prophylaxis: evidence from Lepidoptera-baculovirus interactions? Ecol. Entomol. 23, 100–101.

Wilson, K., Thomas, M.B., Blanford, S., Doggett, M., Simpson, S.J., Moore, S.L., 2002. Coping with crowds: density-dependent disease resistance in desert locusts. Proc. Nat. Acad. Sci. U.S.A. 99, 5471–5475.

Wilson-Rich, N., Stuart, R.J., Rosengaus, R.B., 2007. Susceptibility and behavioral responses of the dampwood termite *Zootermopsis angusticollis* to the entomopathogenic nematode *Steinernema carpocapsae*. J. Invertebr. Pathol. 95, 17–25.

Wilson-Rich, N., Spivak, M., Fefferman, N.H., Starks, P.T., 2009. Genetic, individual, and group facilitation of disease resistance in insect societies. Annu, Rev. Entomol. 54, 405–423.

Wong, Z.S., Hedges, L.M., Brownlie, J.C., Johnson, K.N., 2011. *Wolbachia*-mediated antibacterial protection and immune gene regulation in *Drosophila*. PLoS ONE 6, e25430.

Wuellner, C.T., Porter, S.D., Gilbert, L.E., 2002. Eclosion, mating and grooming behavior of the parasitoid fly *Pseudacteon curvatus* (Diptera: Phoridae). Florida Entomol. 84, 563–566.

Xu, X.X., Zhong, X., Yi, H.Y., Yu, X.Q., 2012. *Manduca sexta* gloverin binds microbial components and is active against bacteria and fungi. Dev. Comp. Immunol. 38, 255–284.

Yanagawa, A., Fujiwara-Tsuji, N., Akino, T., Yoshimura, T., Yanagawa, T., Shimizu, S., 2011. Musty odor of entomopathogens enhances disease-prevention behaviors in the termite *Coptotermes formosanus*. J. Invertebr. Pathol. 108, 1–6.

Yek, S.H., Nash, D.R., Jensen, A.B., Boomsma, J.J., 2012. Regulation and specificity of antifungal metapleural gland secretion in leaf cutting ants. Proc. Roy. Soc. Lond. B Biol. Sci. 279, 4215–4222.

Yue, C., Schroder, M., Gisder, S., Genersch, E., 2007. Vertical-transmission routes for deformed wing virus of honeybees (*Apis mellifera*). J. Gen. Virol. 88, 2329–2336

Yun, J.H., Roh, S.W., Whon, T.W, Jung, M.J., Kim, M.S., Park, D.S., et al., 2014. Insect gut bacterial diversity determined by environmental habitat, diet, developmental stage, and phylogeny of host. Appl. Environ. Microbiol. 80, 5254–5264.

Zhang, L., Gallo, R.L., 2016. Antimicrobial peptides. Curr. Biol. 26, R1–R21.

Zhukovskaya, M., Yanagawa, A., Forschler, B.T., 2013. Grooming behavior as a mechanism of insect disease defense. Insects 4, 609–630.

Zug, R., Hammerstein, P., 2012. Still a host of hosts for *Wolbachia*: analysis of recent data suggests that 40% of terrestrial arthropods species are infected. PLoS ONE 6, e38544.

Zug, R., Hammerstein, P., 2015a. Bad guys turned nice: a critical assessment of *Wolbachia* mutualisms in arthropod hosts. Biol. Rev. 90, 89–111.

Zug, R., Hammerstein, P., 2015b. *Wolbachia* and the insect immune system: what reactive oxygen species can tell us about the mechanisms of *Wolbachia*-host interactions. Front. Microbiol. 6, 1201.

5

Abiotic Factors

Dana Ment[1], Ikkei Shikano[2] and Itamar Glazer[1]

[1] Department of Entomology, ARO, Volcani Centre, Rishon LeZion, Israel
[2] Department of Entomology and Center for Chemical Ecology, Pennsylvania State University, University Park, PA, USA

5.1 Introduction

Pathogens of invertebrates, along with their hosts, inhabit all niches throughout the world, in ecosystems ranging from sub-arctic to arid, temperate, and tropical. The abiotic factors that affect pathogen survival, reproduction, distribution and pathogenic effects on hosts consist of environmental elements such as temperature, moisture, and ultraviolet (UV) radiation, habitat characteristics including soil texture, soil type, and pH, as well as chemical inputs such as fertilizers, pesticides, and pollutants. In the present chapter, we will provide an updated overview of what has been learned regarding abiotic effects on different entomopathogen groups during the past 30 years. We focus our discussion on entomopathogens that have demonstrated potential roles in microbial control. The abiotic factors discussed complement other influences (biological and genetic, which are described in other chapters) that dictate the prevalence and activity of pathogenic organisms.

5.2 The Surviving Unit

We define the surviving units as the life stages or propagules of the organism that persist in the environment and are transmitted to the next host. While another way to think of these stages would be as stages for persistence and transmission, we are emphasizing the fact that they are the stages exposed to abiotic conditions and that they must survive in order to encounter a new host.

5.2.1 Nematodes

Among several families within the order Rhabditida, environmental stress (lack of food, high population density, heat, etc.) induces the development of a unique juvenile stage, the "dauer juvenile" (Riddle, 1988). This stage is adapted morphologically and physiologically to remain in the environment without feeding while it searches

Ecology of Invertebrate Diseases, First Edition. Edited by Ann E. Hajek and David I. Shapiro-Ilan.
© 2018 John Wiley & Sons Ltd. Published 2018 by John Wiley & Sons Ltd.

for a new food source. Among entomopathogenic nematodes (EPNs) (Steinernematidae and Heterorhabditidae), the dauer stage is also the infective stage that locates and invades insect hosts (Forst and Clarke, 2002). This infective stage also provides protection to symbiotic bacteria carried in its intestine (Forst and Clarke, 2002; Stock, 2015). It is equipped with two layers of external membrane: the cuticles of the third stage and of the second molt (which is retained to provide additional protection) (Timper and Kaya, 1989; Campbell and Gaugler, 1991; Rickert et al., 1991). The cuticle is suggested as playing an important role in EPN desiccation survival (Patel and Wright, 1998), as survival and the rate of water loss during desiccation at 80% relative humidity (RH) between *Steinernema carpocapsae* and three other species were associated with the cuticle ultrastructure.

Infective juveniles (IJs) are dependent on internal sources for energy until a host is located. These energy reserves are critical for juvenile survival, as several studies have demonstrated that nematode infectivity declines as energy reserves are depleted (Lewis et al., 1995; Patel and Wright, 1997a,b; Patel et al., 1997b). Quantitative and qualitative analysis of biochemical reserves of EPNs have shown that lipid content and fatty acid composition are highly dependent on media components (Selvan et al., 1993; Abu Hatab et al., 1998), and these lipid reserves are essential to fitness and persistence in nature.

The other significant energy reserve found in nematodes is glycogen (Barrett and Wright, 1998; Wright, 1998). This storage compound occurs in appreciable amounts in some EPNs (Selvan et al., 1993), but its functional significance has been less clear. Utilization of glycogen by IJs of *S. carpocapsae, Steinernema glaseri,* and *Heterorhabditis bacteriophora* can be inferred from the data of Lewis et al. (1995). Subsequent work confirmed that the glycogen content of *S. carpocapsae, S. glaseri, Steinernema feltiae,* and *Steinernema riobrave* declined during storage, suggesting that glycogen may play a significant if secondary (to neutral lipids) role in the maintenance of infectivity in these species (Patel and Wright, 1997b). A physiological basis for the differences in longevity observed in this study is not clear. Lipid content and energy reserves have been related to EPN longevity (Selvan et al., 1993; Abu Hatab and Gaugler, 2001), but a correlation between EPN persistence and other factors, such as the activity of antagonists in nonsterile soil, has also been observed (Hass et al. 2002). Shapiro-Ilan et al. (2006) compared the longevity of 29 EPN strains from 11 species in soil. The variation in longevity was also attributed to some physiological differences between the strains, as well as environmental factors.

5.2.2 Fungi

Lifecycles of entomopathogenic fungi involve an infective stage, usually conidia in Ascomycetes and Entomophthorales, which germinate on the host cuticle and penetrate via appressoria. Once in the host, the fungus multiplies and fungal hyphae inhabit the cadaver. Under humid conditions, the hyphae penetrate outward through the host cuticle, conidiophores are formed, and sporulation occurs. Entomophthorales also form resting structures, depending on environmental conditions, which possess a thick cell wall and oil storage droplets and are dormant after being formed. Conditions breaking dormancy are species-dependent and occur under specific environmental conditions, which also involve climatic conditions and host cues. For both Entomophthorales and

Ascomycetes, the surviving units of the fungus depend on physical factors for dispersal, such as rainfall, wind (for terrestrial pathogens), and water currents (for pathogens of aquatic hosts).

In contrast, Microsporidia, which were classified until recently as protozoa and now are classified as fungi, can reproduce only inside living cells. The infective forms are environmental spores. The spore wall of most species is an endospore layer comprising a chitin–protein matrix. A unique feature is the polar filament, which is attached to the spore and is everted during germination. Spore germination occurs solely within the host hemoceol, and for most species in the gut after ingestion. Once in the host's cells, the sporoplasm from the spore reproduces. The lifecycle is complex and depends on the microsporidian species (Solter and Becnel, 2007).

Longevity of Ascomycete conidia is best under conditions of low temperatures (Hong et al., 1997), low moisture content (specifically, conidial moisture content) (Hedgecock et al., 1995), and low humidity (Clerk and Madelin, 1965; Daoust and Roberts, 1983). Hong et al. (1997) reported variation in the sensitivity of spore longevity to temperature and equilibrium RH, which was strain- and species-dependent for both entomopathogenic and phytopathogenic fungi. Whether in aqueous suspension or a dry state, temperature has different effects on the surviving unit. Post-application temperature can be more detrimental to the conidia than the arthropod host, as it affects germination, mycelial growth, spore production, and pathogenicity.

Once germinated, Entomophthorales discharge short-lived, generally sticky primary conidia, relying on forcible and physical factors for dispersal (Gilbert and Gill, 2010). Primary conidia are sensitive to unfavorable abiotic conditions, especially low humidity, and conidial dispersal is modulated by temperature, humidity, light, and host factor (Boucias and Pendland, 2012).

Persistence of conidia in the environment depends on habitat architecture, host behavior, and environmental factors. In soil, *Metarhizium anisopliae* conidia persisted for several years at the same levels at which the fungus was originally applied (Inglis et al., 2001; Milner et al., 2003). However, in the crop canopy, persistence was only 16–28 days for *Beauveria bassiana* conidia (Inglis et al., 1993). In recent years, *Metarhizium* spp. were found to be rhizosphere associates, and root colonization is a prerequisite for beneficial effects on plants (St. Leger, 2008; Liao et al., 2014). However, the survival units that promote these rhizosphere interactions are not known. One theory is that microsclerotia (compact hyphal aggregates functioning as overwintering structures) are produced, as these are also produced by plant pathogenic fungi. *Metarhizium* microsclerotia were obtained in artificial liquid culture as a novel approach to the production of these fungi at higher concentrations, where they possess desiccation tolerance. Under favorable conditions, microsclerotia germinate and sporulation occurs (Jackson and Jaronski, 2009).

5.2.3 Viruses

Baculoviruses are the most well studied terrestrial invertebrate viruses. Their virions have two structurally and biochemically distinct phenotypes: the occlusion-derived viruses (ODVs), which transmit the virus to new hosts, and the budded viruses (BVs), which spread the virus infection within a host through cell-to-cell transmission (Summers and Volkman, 1976). The ODVs are protected from the environment within a paracrystalline matrix formed by viral occlusion proteins, which occur in two distinct types: occlusion bodies (OBs), produced by lepidopteran, dipteran, and hymenopteran nucleopolyhedroviruses (NPVs), and granules, produced by the granuloviruses (GVs). OBs and granules consist of a single viral protein, called polyhedrin and granulin, respectively (Harrison and Hoover, 2012). The larval body tissue of infected lepidopteran hosts is converted into millions of OBs or granules, which are released into the environment upon host death and cadaver liquefaction. In other hosts, such as sawflies and mosquitoes, baculovirus infections are restricted to the midgut, and OBs are shed continually with the feces. These viral occlusions can persist in the environment for considerable periods of time (Cory and Myers, 2003).

Far less is known about how aquatic invertebrate viruses survive in the environment. The structure of the virus particle, primarily the internal lipid membrane, is thought to provide stability to iridescent viruses in water (Kelly, 1985). Similarly, the virions of another aquatic virus, white spot syndrome virus (WSSV), has a multilayered structure comprising a nucleocapsid surrounded by a thick, lipid-containing envelope that is thought to provide environmental protection (Zhou et al., 2008).

5.2.4 Bacteria

The dormant stage of *Bacillus* spp. bacteria is the endospore. It contains the proteinacious crystalline inclusion bodies. This stage is highly resistant to environmental factors, including heat, UV, desiccation, and oxidizing agents. The endospore is sensitive to changes in environmental cues and responds with germination followed by vegetative growth when conditions are amenable (Gilbert and Gill, 2010). Endospores were reported to be 5–50 times more resistant to UV radiation than vegetative cells (Setlow, 2001). UV resistance is attributed to several spore-specific attributes, including a thick spore coat layer, a DNA damage-protective mechanism, a DNA repair pathway, accumulation of dipicolinic acid (DPA) as the Ca^{2+} chelate, and pigmentation (Nicholson et al., 2000; Nicholson, 2002; Moeller et al., 2005). Prolonged activity and persistence in the host environment is seldom observed and thus continued presence of the bacteria is often a result of recycling. Natural recycling can prolong the persistence of microbial control applications through periods as long as 6 months (de Melo-Santos et al., 2009).

5.3 Abiotic Factors Affecting Invertebrate Pathogens

5.3.1 Temperature

Extreme temperature conditions (freezing or extreme heat, such as >38 $^{\circ}$C) can adversely influence the survival of entomopathogens. Additionally, temperature has a tremendous effect on host–pathogen interactions. Pathogenicity and virulence depend on temperature, which has differential effects for each pathogen group and for each species within each group.

For cold tolerance, there are two main strategies used by cold-tolerant organisms to survive freezing temperatures: freeze tolerance – surviving ice formation in their tissues; and freeze susceptibility – avoiding ice nucleation and maintaining their body fluids in liquid phase at temperatures well below their freezing point, by supercooling. Supercooling is defined as the ability of an organism to maintain its body fluids in liquid phase at temperatures below the freezing point (Wharton and Block, 1993). Sugars and polyols, such as trehalose, maltose, glucose and dextrose, and polyethylene glycol have roles as cryoprotectants or antifreezes. The production of natural cryoprotectants, such as trehalose in nematodes (Wharton et al., 1984) and trehalose, mannitol, and arabitol in fungi (Weinstein et al., 2000; Tibbett et al., 2002; Tibbett and Cairney, 2007), is correlated with enhanced supercooling abilities.

5.3.1.1 Nematodes

Nematodes inhabiting the soil are exposed to subzero temperatures in temperate, arctic, and sub-arctic regions, as well as at high altitudes (Wharton, 1986). Isolation of EPNs from such cold regions indicates that they are capable of withstanding subzero conditions. However, little is known about their cold-tolerance mechanisms.

Brown and Gaugler (1990) demonstrated that *S. feltiae*, *Steinernema anomali*, and *H. bacteriophora* were all freeze-tolerant, with lower lethal temperatures of –22, –14, and –19 °C, respectively. Wharton and Surry (1994) showed that *H. zealandica* is freeze-avoiding: its sheath prevented inoculative freezing (formation of internal ice), allowing extensive supercooling to –32 °C, whereas exsheathed IJs froze above –6 °C did not survive.

Shapiro-Ilan et al. (2014) showed that cold tolerance differs among different strains of steinernematids and heterorhabditids. Intraspecies variation in freeze tolerance was observed among *H. bacteriophora* and *S. riobrave* strains, yet within-species variation was not detected among *S. carpocapsae* strains. In interspecies comparisons, poor freeze tolerance was observed in *H. indica*, *S. glaseri*, *S. rarum*, and *S. riobrave*, whereas *Heterorhabditis georgiana* and *S. feltiae* exhibited the highest freeze tolerance.

Heat imposes severe stress on living systems, but evolutionary processes have equipped many organisms with unique mechanisms to withstand high temperatures and to repair resultant damage (Morimoto, 2008). High temperatures (>32 °C) have an adverse effect on the reproduction, growth, and survival of nematodes (Zervos et al., 1991; Grewal et al., 1994). Survival of EPNs at high temperatures (37 °C) was substantially increased when they were pre-exposed to mild heat (30 °C), rather than directly exposed to heat stress (Selvan et al., 1993; Grewal et al., 1994). These results indicate the need for an adaptation period, perhaps to induce gene regulation and a shift in metabolism for the stress response. Heat shock proteins (HSPs) are known to be involved in organismal survival at elevated temperatures (Schlesinger, 1990), and production of *hsp*70 has been detected in EPNs (Selvan et al., 1993). Production of *hsp*70 was also detected in *H. bacteriophora* HP88 (Selvan et al., 1993), even though temperatures above 32 °C hamper the reproduction, activity, and viability of this strain (Grewal et al., 1994).

Species and strains of EPNs differ in their ability to withstand heat stress. Mukuka et al. (2010a,b,c) characterized the diversity of this trait among 36 populations of *H. bacteriophora* isolated from diverse environments across the globe, 18 hybrid or inbred strains of this species, 5 strains of *H. indica*, and 1 of *H. megidis*. Nematodes were tested with or without prior adaptation to heat at 35 °C for 3 hours. The mean tolerated temperature ranged from 33.3 to 40.1 °C for nonadapted populations and from 34.8 to 39.2 °C for adapted ones. *H. indica* was the most tolerant species, followed by *H. bacteriophora* and *H. megidis*. Pre-exposure to high temperatures (adaptation) caused an increase in the tolerance for all species tested. Correlation between a strain's heat tolerance and the mean annual temperature in its place of origin was weak.

It is noteworthy that Lee et al. (2016) found that EPN IJs exhibited extreme plasticity of olfactory behavior as a function of cultivation temperature. Many odorants that were attractive for IJs grown at lower temperatures were repulsive for IJs grown at higher temperatures, and vice versa. In this study, EPNs also showed temperature-dependent changes in their host-seeking strategies: IJs cultured at lower temperatures appeared to more actively cruise for hosts than IJs cultured at higher temperatures.

5.3.1.2 Fungi

The activity of fungi in an insect population is greatly influenced by both temperature and humidity. While high humidity is crucial for spore germination and disease progression in insects in aerial habitats, low humidity and temperature are prerequisite for conidial longevity in the environment (Daoust and Roberts, 1983; Benz, 1987; Milner et al., 1997; Arthurs and Thomas, 2001; Dimbi et al., 2004). In some cases, environmental temperature is favorable for disease progression while host body temperature is not, as with behavioral fever in grasshoppers. Grasshoppers infected with either *B. bassiana* or *Metarhizium flavoviride* that were allowed to bask for different periods of time were less susceptible to infection as basking periods increased (Inglis et al., 1997b). When allowed, the grasshoppers moved to heat sources, using thermoregulation as a behavioral regulation to inhibit fungal disease progression. *Locusta migratoria* inoculated with *Metarhizium acridum* that did not thermoregulate during blastospore development in the hemolymph had reduced hemocyte and increased blastospore concentrations, while thermoregulating insects had similar hemocyte concentration as uninfected insects (Ouedraogo et al., 2003).

Environmental temperature is known to influence the virulence of many entomopathogenic fungi and can determine their effectiveness as biological control agents under natural conditions (Feng et al., 1999; Dimbi et al., 2004; Ment et al., 2011). Numerous studies have demonstrated temperature effects on fungal efficacy under simulated field regimes (Arthurs and Thomas, 2001; Chandler et al., 2005; Polar et al., 2005; Lysyk, 2008; Ment et al., 2011). The study of thermal characteristics is used as a tool to select suitable fungal pathogens for control given the host habitat. In most studies, the upper limit for germination was 35 °C for *Metarhizium* spp., while the lowest was 5 °C for *Lecanicillium* spp. (Kope et al., 2008). Nonfavorable temperatures hinder other characteristics of *M. anisopliae* strains, including the ability to survive, grow, and infect arthropods at temperatures above (McCammon and Rath, 1994; Brooks and Wall, 2005) and below (Fernandes et al., 2008) the optimum temperature range. For example, in Polar et al. (2005), temperature tolerance of *M. anisopliae* strains was a critical factor when spores were applied to the surface of mammals to control ticks. In Ment et al. (2011), two *Metarhizium* species were characterized: *Metarhizium brunneum* tolerated thermal shock as would occur under field application conditions and *Metarhizium pingshaense* tolerated continuous exposure up to 35 °C as would occur on the mammalian body surface. These results suggest a combination of strains in a single commercial product may be beneficial, but the odds of registering such a product are low.

For *Metarhizium* and *Beauveria* spp., the latitude of origin correlates with thermo-tolerance or cold activity. *B. bassiana* isolates from higher latitudes were more cold-active, but there was not a similar correlation for heat (Fernandes et al., 2008). For *Metarhizium* spp., the correlation with latitude was only demonstrated for thermotol-erance (Fernandes et al., 2008; Rangel et al., 2005). In addition to selection for thermo-tolerant isolates by specific screening assays, thermotolerance of *M. anisopliae* was increased by applying directed evolution approaches using continuous culture (De Crecy et al., 2009). When applying this methodology, the selected variants displayed robust growth at 36.5 °C and maintained virulence parameters.

Favorable environmental conditions are crucial during pre-penetration events, germination, and appressorium differentiation, when the fungus is exposed to the envi-ronment (Dimbi et al., 2004; Kope et al., 2008; Ment et al., 2011). However, during post-penetration, growth in the infected host under adverse environmental conditions is not necessarily inhibitory. For example, exposing tick eggs infected with *M. brun-neum* to unfavorable environmental conditions resulted in the development of unique resting structures – chlamydospores – originating from the fungal hyphae (Ment et al., 2010). In this pathosystem, the formation of chlamydospores under unfavorable envi-ronmental conditions during the process of host colonization allowed the fungus to survive and recover when favorable conditions returned. Therefore, once the penetra-tion process occurs, environmental conditions are no longer a factor in the success of fungal activity: after growth within the host, the fungus can shift into another develop-mental form – chlamydospores – and consequently survive under unfavorable condi-tions. In the case of Entomophthorales, while the fungus is more protected once it is growing within a host, high temperatures can cure infections at this time too (e.g., when gypsy moth (*Lymatnria dispar*) larvae infected by *Entomophaga maimaiga* are held constantly at 30 °C, the infection is cured; Hajek et al., 1990). Once hosts die of entomophthoralean infections, the resting spores that are produced are the important stage for persistence of the pathogen in the habitat, but resting spore production varies by species and conditions (Boucias and Pendland, 2012). For example, *Zoophthora radicans* produces more resting spores when hosts are held at low temperatures, while *E. maimaiga* produces more resting spores when hosts are held at high temperatures (Hajek, 1997; Milner and Lutton, 1983).

Similarly, for most microsporidia, it is only during the infection process of the host that temperatures should lie between 20 and 30 °C (Benz, 1987). In laboratory assay on *Gammarus pulex* naturally infected with *Microsporidium* spp., temperature had a sig-nificant effect on mortality rates. At 15 °C no mortality occurred, while at 25 °C 42.9% of gammarids died from infections (Grabner et al., 2014). Air-dried spores of *Nosema apis*, a pathogen of bees, lost viability after 3 or 5 days at 40, 45, or 49 °C (Malone et al., 2001). While *N. apis* was as virulent as *N. ceranae* at 33 °C, it was less infective than *N. ceranae* at lower and higher temperatures of 25 and 37 °C. The lowest infectivity for honeybees was recorded for *N. ceranae* kept at 37 °C (Martín-Hernández et al., 2009). The temperature tolerance of aquatic species of microsporidia varies between species, but usually the pathogen will be adapted to the temperature range of its host's habitat (Lacey and Kaya, 2013).

5.3.1.3 Viruses

The vast majority of studies on the direct impact of ambient temperature on virus stability outside of their hosts have been conducted on insect baculoviruses. Basic temperature stability studies on baculoviruses, briefly described in this paragraph, were conducted in the 1960s–80s and have been summarized in detail in Benz (1987). Virions in their OBs retain activity through freezing and thawing and can remain active at 4 °C for decades if kept in a sealed container or for years at room temperature. At higher temperatures (>40 °C), such as those reached in compost piles and some agroecosystems, baculovirus activity is gradually lost. At 40 °C, some active baculovirus particles are still present after 5 days, or in some cases at up to 20 days. Some baculoviruses can retain activity at 60 °C after 24 hours of exposure, although the activity of most baculoviruses is greatly reduced after 40 minutes. Besides baculoviruses, the temperature tolerances of some other invertebrate viruses have also been characterized. Invertebrate iridescent virus 6 (IIV-6) occurs primarily in damp and aquatic habitats and has a potential host range (based on field collections and laboratory inoculations) that includes at least 100 species of insects, 2 terrestrial species of Crustacea, and 1 species of Chilopoda (Devauchelle et al., 1985). Its infectivity outside of the host steadily decreased over time at 4 and 25 °C and at 27 ± 0.2 °C, 50 cm below the surface of an artificial pond (Marina et al., 2000). The loss of activity tended to be greater at higher temperatures, but this effect was masked by high variability in virus infectivity among replicates after storage at the higher temperatures (Marina et al., 2000). Under laboratory conditions, IIV-6 lost little activity after 60 minutes at 50 °C but was almost completely inactivated by heat treatment at 70 °C for 60 minutes or at 80 °C for 30 minutes (Martínez et al., 2003). The capsids of the nimavirus WSSV retained morphological integrity at temperatures <45 °C but became denatured at >60 °C (Chen et al., 2012). Whether WSSV retains infectivity after exposure to temperatures as high as 45 °C was not tested.

As viruses are obligate intracellular parasites, their replication is generally dependent on the development of their host and the impact of environmental effects on host physiology. Thus, the measured impact of temperature on virus lethality and replication can depend on the range of temperatures tested. For example, the production of progeny virus of *Autographa californica* multinucleopolyhedrovirus (AcMNPV) produced in cabbage looper (*Trichoplusia ni*) larvae increased linearly with temperature if the maximum temperature tested was optimal for *T. ni* development (29 °C) (van Beek et al., 2000). However, increasing the temperature beyond optimal to one that is stressful for the host can result in substantial decreases in the amount of virus progeny produced (Day and Dudzinski, 1966; Johnson et al., 1982; Ribeiro and Pavan, 1994). Temperature effects on the mortality of infected hosts are mixed. At temperature ranges that support host survival and development, temperature usually has little or no impact on baculovirus-induced mortality rate (Ignoffo, 1966; Frid and Myers, 2002; Sporleder et al., 2008; Shikano and Cory, 2015). In contrast, increasing mortality was observed in sugarcane borer (*Diatraea saccharalis*) larvae with increasing temperatures (up to 37 °C) after infection by four different baculovirus strains (Ribeiro and Pavan, 1994). Pupae of *Bombyx mori* injected with an unidentified NPV had higher survival at 35 °C than at 25 °C (Kobayashi et al., 1981). Infections by aphid lethal paralysis virus or *Rhopalosiphum padi* virus, which are transovarially transmitted in the aphids

Rhopalosiphum padi and *Diuraphis noxia*, were absent at low temperatures and increased with rearing temperature (Laubscher and von Wechmar, 1992).

The impact of temperature on the susceptibility of some hosts can vary depending on the identity of the virus. When infected by Galleria mellonella nucleopolyhedrovirus (GmNPV), the greater wax moth (*G. mellonella*) died faster as temperatures increased from 18 to 40 °C (Stairs, 1978). In contrast, when infected by *Tipula* iridescent virus, which is a virus collected from the marsh crane fly (*Tipula oleracea*), its mortality decreased dramatically as temperature was increased from 24 to 30 °C (Witt and Stairs, 1976). Since the optimal temperature for survival and development of *G. mellonella* is about 32 °C, the differential impact of temperature on the two viruses likely reflects the evolutionary relationship between the host and the viruses.

Few field studies have investigated the influence of temperature on the severity and rate of transmission of viral disease. Since low temperatures do not affect the lethality of most baculoviruses, these viruses could be applied for area-wide management to reduce pest populations early or before the start of the growing season, as demonstrated against *Helicoverpa zea* and *Heliothis virescens* (Bell and Hayes, 1994). Increasing temperatures in the field might also enhance pest control, as was demonstrated for NPV in *Spodoptera frugiperda* (Elderd and Reilly, 2014). Faster *S. frugiperda* larval development was associated with a higher fraction of infected larvae, suggesting that the low-risk individuals in a population are those that eat less leaf tissue under cooler temperatures; as temperature increases, so does the likelihood of encountering and ingesting a lethal viral dose in this group (Elderd and Reilly, 2014).

Temperature also has a profound effect on viral disease outbreaks in aquatic environments. Many countries producing Pacific oysters (*Crassostrea gigas*) have reported periodic mass mortalities during the summer months (Glude, 1974; Koganezawa, 1974; Renault et al., 1994; Cheney et al., 2000). Infection by ostreid herpes virus 1 has been linked with summer mortality (Sauvage et al., 2009). Oyster broodstocks from some geographical origins exhibited higher and more rapid mortality at 25–26 °C than at 22–23 °C, and herpes virus particles were observed in oysters reared at high but not low temperatures (Le Deuff et al., 1996). WSSV, which infects a wide range of crustacean hosts, is one of the most devastating and virulent viral agents threatening the penaeid shrimp culture industry. The prevalence of WSSV in grow-out ponds and hatcheries was reported to increase when water temperatures exceeded 25 °C in China (Zhan et al., 1998) but to decrease in the warm season in tropical countries such as Ecuador and Thailand (Rodriguez et al., 2003; Withyachumnarnkul et al., 2003). Temperature changes appear to have an overwhelmingly strong effect on the proliferation of WSSV in Chinese shrimp compared to changes in salinity (15–35%) and pH (6.5–9.0) (Gao et al., 2011). WSSV-infected juvenile Pacific white shrimp (*Litopenaeus vannamei*) and Kuruma shrimp (*Marsupenaeus japonicas*) died rapidly at moderate temperatures (27–28 °C), but mortality was reduced at lower temperatures (15–18 °C) (Guan et al., 2003; Moser et al., 2012). WSSV infection in crayfish (*Pacifastacus leniusculus* and *Procambarus clarkii*) caused 100% mortality at temperatures between 18 and 24 °C but no mortality at 4–12 °C (Jiravanichpaisal et al., 2004; Du et al., 2008). Mortality of infected shrimp was also substantially reduced or eliminated when the temperature was elevated from moderate (27–28 °C) to high (31–33 °C) (Vidal et al., 2001; Guan et al., 2003; Rahman et al., 2006; You et al., 2010), even though infected shrimp maintained at hyperthermic (i.e., having a body temperature greatly above normal) temperatures had detectable lymphoid organ spheroids suggestive of a chronic viral infection (Vidal et al., 2001). When the WSSV-infected shrimp were transferred from 32.3 ± 0.8 to $25.8 \pm 0.7\,°C$ (ambient temperature), ensuing mortality from WSSV was 100% (Vidal et al., 2001). Thus, daily fluctuations in temperature can increase shrimp survival and delay mortality, especially if the temperature is held longer at an extreme (Rahman et al., 2007a). Raising the water temperature to $33\,°C$ during the early stages of infection reduced mortality, but this effect was not seen later in the infection (Rahman et al., 2007b). Improved survival of WSSV-infected crustaceans at high and low temperature extremes has been attributed to low viral replication (Vidal et al., 2001; Granja et al., 2006; Jiravanichpaisal et al., 2006; You et al., 2010; Gao et al., 2011).

Some fungi-infected insects, such as locusts, grasshoppers, and houseflies, survive longer after infection at warmer temperatures. In the only test of therapeutic behavioral fever in an invertebrate–virus system, Western tent caterpillars infected by Malacosoma californicum pluviale nucleopolyhedrovirus (McpINPV) showed no change in temperature preference compared to uninfected caterpillars (Frid and Myers, 2002). This is not surprising, given that temperature did not influence the survival of infected individuals. Whether virus-infected hosts engage in behavioral fever is likely to vary with the host–virus system. Behavioral fever could be particularly important in aquatic systems, where temperature has a strong influence on virus lethality. The vertical distribution of the shrimp *Pandalus borealis* changes with developmental stage in thermally stratified water columns (Ouellet and Allard, 2006), suggesting that this species may select developmentally optimal temperatures, although it could also be due to the distribution of developmentally appropriate prey within the water column. Demonstration of behavioral fever could lead to changes in rearing practices, such as increasing the depth of the water column, to provide WSSV-infected crustaceans with a wider range of temperatures.

5.3.1.4 Bacteria

Studies of temperature effects on bacterial pathogens of invertebrates, including species of Rickettsiales, are scarce. Few studies have examined temperature effects on the activity of *Bacillus* spp. However, for *Bacillus thuringiensis israelensis* (Bti) R-153-78 infecting second-, third-, and fourth-instar *Culex quinquefasciatus* larvae, temperatures ranging from 15 to $30\,°C$ did not affect its activity. For *B. subtilis*, sporulation at an elevated temperature invariably resulted in spores with increased heat resistance. Inactivation of *Bacillus* spores occurs under wet heat conditions at temperatures higher than observed for vegetative cells ($40\,°C$) (Nicholson et al., 2000). In a bacterial epizootic that occurred in a Mediterranean commercial sponge population (*Hippospongia communis*), increases in water temperature were probably the primary factor underlying the outbreak (Peters, 1993), but the exact factors responsible for the sponge mortalities remain unidentified.

5.3.2 Moisture and Humidity

Natural environments can vary significantly in water availability during different seasons and at different locations. Continuous changes from dry to wet states can impose considerable physiological constraints on pathogens, but many have evolved effective mechanisms to tolerate environmental fluctuations in water availability. As water is essential to the survival and development of the hosts of invertebrate pathogens, pathogen infectivity and replication within the host can also be significantly affected by water availability.

5.3.2.1 Nematodes

Several nematode species are capable of anhydrobiosis and can survive in a desiccated state (Glazer, 2002; Gal et al., 2004). Anhydrobiosis is usually reached following a slow rate of water loss (Cooper and van Gundy, 1971; Crowe and Madin, 1975; Wharton, 1986; Crowe, 2014). Some nematodes form tight coils when exposed to desiccation, which slows the rate of water loss by reducing the area of cuticle exposed to air (Glazer, 2002).

Large variation in the capability to withstand dry conditions was recorded among different EPNs. Early studies showed that steinernematids are more desiccation-tolerant than heterorhabditids (Liu and Glazer, 2000; Glazer, 2002). Mukuka et al. (2010a,b,c) screened the desiccation tolerance of 43 strains of *Heterorhabditis* spp. and 18 hybrid/inbred strains of *H. bacteriophora*, showing significant variation between nematode strains and species. Shapiro-Ilan et al. (2014) showed that *S. carpocapsae* exhibited the highest level of desiccation tolerance among species, followed by *S. feltiae* and *S. rarum*; the heterorhabditid species exhibited the least desiccation tolerance, and *S. riobrave* and *S. glaseri* were intermediate.

Solomon et al. (1999, 2000) studied physiological changes occurring in the EPN *S. feltiae* during desiccation. Changes included a decrease in glycogen levels (Gal et al., 2001) and increased levels of trehalose in the adaptation period during gradual dehydration (24 hours at 97% RH) (Solomon et al., 1999, 2000; Glazer, 2002). Trehalose is an osmoprotectant believed to play a major role in protecting cell membrane structure and integrity during dehydration (Crowe and Crowe, 1992; Rolim et al., 2003; Crowe, 2014). Gal et al. (2001) identified elevated production of a known desiccation-related protein from the late embryonic abundant (LEA) group LEA3 in *S. feltiae* IS-6 IJs during desiccation stress. Proteomic analysis of *S. feltiae* following desiccation stress showed that out of more than 400 protein spots identified by 2D-gel electrophoresis, 10 showed changes in abundance upon desiccation, one being the chaperonin HSP60 (Chen and Glazer, 2004; Chen et al. 2005).

Yaari et al. (2016) analyzed transcriptome expression in *Steinernema* species that differed in their tolerance to stress. Analyzing gene expression patterns for the stress response showed a large fraction of downregulated genes in the desiccation-tolerant nematode *S. riobrave* and a larger fraction of upregulated genes in the desiccation-susceptible *S. feltiae* strains Carmiel and Gvulot. Metabolic pathways and the expression of specific stress-related genes were compared. This study revealed many genes and metabolic cycles that are differentially expressed in the water-stressed nematodes.

Removal of water from the cell body can also occur as a result of osmotic pressure. Being soil inhabitants, nematodes may have to tolerate considerable variation in the concentrations of salts in the environment and thus be subject to osmotic stress (Evans and Perry, 1976; Wharton, 1986). The greatest osmotic problem is the influx of water under hypoosmotic conditions. Some nematode species possess specific mechanisms for the removal of excess water, such as having restricted cuticular permeability. Some actively remove excess water under hypotonic conditions through the intestine. Influx of water under hypotonic conditions may be reduced by excreting salts and thus decreasing the osmotic gradient (Wharton, 1986). The mechanisms of osmoregulation among steinernematids and heterorhabditis are not known, but studies have demonstrated their capability to withstand osmotic stress (Thurston et al., 1994; Finnegan et al., 1999). Glazer and Salame (2000) evaluated the effect of different osmolytes on the viability of *S. carpocapsae*. Exposing fresh IJs to a high-temperature assay (45 °C) resulted in a rapid reduction in viability, yet under the same assay the mortality of evaporatively and osmotically desiccated nematodes showed enhanced ability to withstand high-temperature stress.

Numerous laboratory studies have examined the effect of soil moisture on the efficacy and survival of EPNs (Gaugler and Kaya, 1990; Kaya and Gaugler, 1993; Glazer, 2002). In the laboratory, the virulence of *H. bacteriophora*, *S. glaseri*, *S. feltiae*, and *S. carpocapsae* increased with increasing soil moisture content in sandy loam soils, ranging from below the permanent wilting point to near saturation (Grant and Villani, 2003). Hudson and Nguyen (1989) tested the infectivity of *S. scapterisci* in mole crickets, and found that soil moisture that varied from 5 to 15% had no effect on infection. Salame and Glazer (2015) compared the effect of soil moisture gradient on the downward movement of *S. carpocapsae* and *H. bacteriophora*. *H. bacteriophora* IJs abandoned the upper soil layers as dryness intensified, with >80% found in the bottom (20–25 cm) layer of soil. In contrast, >70% of *S. carpocapsae* IJs remained in the upper layer. The results provide initial evidence of a possible stress-avoidance strategy in *H. bacteriophora* under natural dry conditions.

In a survey of Spanish soils for EPNs, out of 150 different sites, EPNs were recovered from 35; significantly higher occurrence was reported for steinernematids than for heterorhabditis (Garcia del Pino and Palomo, 1996). The study concluded that soil moisture and temperature regimes are more important than other factors in determining the distribution and prevalence of EPNs in cold, moist soils. In conventional-till and no-till maize in North Carolina, there was a quadratic relationship between soil moisture content and numbers of sentinel *G. mellonella* infected by *S. carpocapsae* but not by *S. riobrave* or *H. bacteriophora* (Millar and Barbercheck, 2002). Reduced virulence of EPNs in low-moisture conditions can be increased by rehydrating the soil to simulate rainfall or irrigation (Grant and Villani, 2003). Moisture also played an important role in pest suppression by EPNs in a coastal prairie in California (Preisser and Strong, 2004), and it might be a major factor impacting EPN efficacy after applications in peaty versus mineral soils in Ireland (Williams et al., 2013).

5.3.2.2 Fungi

Moisture is probably the most widely recognized environmental factor that determines transmission of fungal entomopathogens (Hall and Papierok, 1982; Carruthers and Soper, 1987; Hajek, 1997); therefore, it regulates critical steps in the dynamics of insect epizootics caused by fungal pathogens. Moisture and humidity both impact spore production, survival, and infection. Hong et al. (1997) reported variation in the sensitivity of spore longevity to temperature, moisture content, and equilibrium RH, which is strain- and species-dependent for both entomopathogenic and phytopathogenic fungi. Specifically, for *M. acridum*, the combination of low moisture content (5.0–9.4%) and temperature at the range of 25–38 °C resulted in greater longevity than storage at 55 °C

(Hedgecock et al., 1995). For *M. anisopliae*, any further decrease of RH below 12% did not provide further increase in longevity (Hong et al., 1997). Similarly, *Nomuraea rileyi* survived long storage periods under dry conditions and low temperatures, but quickly (<7 days) lost its viability when wet (Ignoffo, 1992). Contradictory to this, most microsporidia require wet conditions for survival. Exposure to dry conditions at 30 °C for 30 days reduced insecticidal activity of microsporidia spores of *Nosema plodiae* by 90% (Ignoffo, 1992).

Spore germination *in vitro* and on hosts requires high RH or free water. The humidity threshold depends on the fungus, and it may vary depending on the host, its ecology, and the microclimate (Ibrahim et al., 1999; Jaronski, 2010). In a study on termites, *M. anisopliae* germination was increasingly delayed at water activities equivalent to 99, 98, and 96% RH, and completely inhibited at 94, 92, and 90% RH (Milner et al., 1997). Successful sporulation is also dependent on RH. In a fungus–tick pathosystem, the natural environment in which ticks oviposit is relatively favorable for fungal infection. Most female ixodid ticks lay eggs in large cohorts in protected, sheltered areas (e.g., under stones, under leaves, deep in sand, in cracks in soil, etc.). This behavior provides an environment of relatively high humidity and "buffered" temperature around the eggs. Eggs of many tick species require RH of over 85% for their normal development (Dusbabek, 1994), which is ideal for mycosis development. Modification of RH after egg infection by *M. brunneum* from 100 to 55–75% resulted in inhibition of conidial production and, instead, in chlamydospore development inside the eggs (Ment et al., 2010). Similarly, in *Aedes aegypti* eggs, infection by *M. anisopliae* was successful only after a minimal period of incubation at near-saturated conditions (Santos et al., 2009). Humidity did not affect total mortality or mycosis in *M. acridum*-infected adults of *Schistocerca gregaria* (Fargues et al., 1997), but high environmental humidity associated with rainfall was required for its sporulation on *S. gregaria* cadavers, and the likelihood of sporulation differed between microsites (Arthurs et al., 2001). Vast fluctuations in minimum and maximum RH favor the development of *N. bombycis* on silkworm larvae, probably due to an increased multiplication rate of the pathogen (Chakrabarti and Manna, 2008). Favorable conditions for *Nosema* sp. I sporulation while infecting the coleopteran *Lophocateres pusillus* were 31.7–33.8 °C and 97–100% RH (Ghosh and Saha, 1995).

5.3.2.3 Viruses

High humidity and moisture have been reported to accompany viral epizootics in insects (Benz, 1987). While increased host stress under these conditions is typically assumed to increase virus-induced mortality (Benz, 1987), another key factor may be rainfall. The physical action of water droplets from simulated rainfall hitting baculovirus-filled cadavers was demonstrated to spread *Lymantria dispar* nucleopolyhedrovirus (LdNPV) from the upper branches to the lower branches of red oak trees (D'Amico and Elkinton, 1995). Since the NPV is washed off the upper branches, virus-induced host mortality on the upper branches was reduced, but mortality on the lower branches increased (D'Amico and Elkinton, 1995). Similarly, mortality induced by *Orgyia psuedotsugata* nucleopolyhedrovirus (OpNPV) on Douglas fir (*Psuedotsuga menziesii*) branches increased from 12% one day before a light rain to 100% two days after (Thompson, 1978). This downward movement of NPVs within the canopy by rainfall has also been suggested to increase the protection of NPVs from sunlight by moving them to more shaded areas (Young and Yearian, 1989). The movement of NPVs by rainfall in trees may be mostly limited to branches and foliage, since rainfall on trees with high prevalence of virus-killed cadavers on their stems and branches did not contaminate egg masses on the trunks (Murray and Elkinton, 1989). Higher prevalence of virus particles on the lower canopy of low-lying plants after rainfall can also occur via splash contamination of viruses from the soil reservoir, as application of NPV of the velvet bean caterpillar (*Anticarsia gemmatalis*) to soil caused larval mortality on soybean plants only after rainfall (Young and Yearian, 1989, 1986).

Baculovirus OBs can persist for long periods in water when protected from sunlight and extreme high temperatures. NPVs of *S. frugiperda* and *H. zea* applied to maize crops through irrigation water demonstrated successful protection of the crops (Hamm and Hare, 1982; Valicente and Costa, 1995). Lepidopteran NPVs can even be detected in freshwater and marine aquatic environments (Holmes et al., 2008; Hewson et al., 2011). However, when water is combined with soil, the soil moisture can inactivate NPVs, although the effect may vary with soil type (Peng et al., 1999). Persistence of *Anticarsia gemmatalis* nucleopolyhedrovirus (AgNPV) was significantly lower in marsh soil than in agricultural soil, and inactivation was faster in soil at −0.3 bar water potential than in soil at water potentials of 0 or −5 bar (Peng et al., 1999). Additionally, AgNPV lost 99% of its activity within 64 days when stored in unautoclaved marsh water, but remained 98–99% active in distilled water or autoclaved marsh water. This suggests that heat-sensitive agent(s), perhaps microorganisms, can inactivate NPVs in some moist soil types (Peng et al., 1999). Similarly, dew that forms on the plant surface can inactivate OBs through exposure to phytochemicals, such as isoflavonoids, on the surface of chickpea leaves (Stevenson et al., 2010).

5.3.2.4 Bacteria

Long-term exposure to free water and humidity causes losses of propagules of bacteria (Benz, 1987; Ignoffo, 1992). Short-term exposure to water for 7–30 days will have no effect on bacterial spores or crystals. Both *B. sphaericus* and *B. thuringiensis* subsp. *israelensis* have good persistence in water habitats (Mulla et al., 1990; Regis et al., 2001; de Melo-Santos et al., 2009). Under desiccated conditions obtained artificially using vacuum systems, *Bacillus* spores did not exhibit any detectable mortality. Even prolonged desiccation of $<10^{-4}$ Pa for 80 hours was not lethal, but it did cause significant mutagenesis (Nicholson et al., 2000).

Mediterranean flour moth (*Ephestia kuehniella*) larvae fed with spores and crystals of *B. thuringiensis kurstaki* (Btk) had higher nodule concentrations (i.e., an immune response) at 32 °C than at 15 and 23 °C, and higher nodule concentrations at 85% RH than at 43% RH, indicating a better effect on pathogenicity in warmer, more humid environments (Mostafa et al., 2005).

5.3.3 Ultraviolet Radiation

UV radiation inactivates all groups of pathogens. Sunlight, specifically UV radiation between 290 and 400 nm, is the most destructive environmental factor affecting the persistence of all entomopathogens (Shapiro-Ilan et al., 2012). "The half-life of different types of inoculum (conidia, spores, occluded virions) exposed to natural sunlight is estimated at about one hour for the most sensitive entomopathogen to about 96 h for the most resistant entomopathogen" (Ignoffo, 1992).

5.3.3.1 Nematodes

Shapiro-Ilan et al. (2015) indicated significant variation in UV tolerance among EPN strains and species. Overall, several steinernematids exhibited higher levels of UV tolerance than other strains or species, with *S. carpocapsae* strains generally exhibiting the highest level. Previous comparisons among EPNs for UV tolerance only included one to four species or strains (Gaugler and Boush, 1978; Gaugler et al., 1992; Mason and Wright, 1997; Fujiie and Yokoyama, 1998; Grewal et al., 2002; Wang and Grewal, 2002; Jagdale and Grewal, 2007). Gaugler et al. (1992) compared the two nematodes *H. bacteriophora* (NC1) and *S. carpocapsae* (All) and concluded that the lower UV tolerance observed in the *H. bacteriophora* strain was indicative of a broader lack of environmental tolerance inherent in heterorhabditids compared with steinernematids. The impact of UV radiation on EPNs depends on wavelength. Gaugler and Boush (1978) reported that longer UV wavelengths, such as 366 nm, did not impact the virulence of *S. carpocapsae*, whereas shorter wavelengths, such as 254 nm, negatively affected the nematode. Confirming this finding, Fujiie and Yokoyama (1998) reported sunlight and UV wavelengths of 310 and 254 nm affected *Steinernema kushidai* virulence whereas 350 nm had no effect. Jagdale and Grewal (2007) applied a wavelength of 340 nm for 3 hours and observed that warm- and cold-stored *S. carpocapsae* and *S. riobrave* maintained significantly higher original virulence than when stored at culture temperature. Shapiro-Ilan et al. (2015) applied 254 nm and suggested the reduced impact and differential effects observed in the previous study were likely due to the longer wavelength used.

5.3.3.2 Fungi

The half-life of conidia of most species of entomopathogenic fungi directly exposed to simulated sunlight ranged from about 1 to 4 hours (Zimmermann, 1982; Ignoffo, 1992). Several studies have investigated the effect of UV-A and UV-B on the survival of *Metarhizium* and *Beauveria* (Alves et al., 1998; Braga et al. , 2001, 2002; Rangel et al., 2004; Fernandes et al., 2007). All studies indicated that a few hours of direct exposure to sunlight, specifically UV-B, fully inactivated conidia of Hypocrealen fungi. In addition, UV radiation has been demonstrated to delay the germination of surviving conidia (Zimmermann, 1982; Moore et al., 1993; Braga et al., 2001; Filotas et al., 2006). For both *Metarhizium* and *Beauveria*, a significant inverse correlation between UV-B tolerances and the latitudes of origin of the isolates was observed (Braga et al., 2001; Fernandes et al., 2007). To further show the effect of UV, formulations incorporating UV protectants such as vegetable oils, adjuvant oils, oil-soluble sunscreens, optical brighteners of the stilbene class, and clay increased survival of conidia under field conditions (Alves et al., 1998; Moore et al., 1993; Inglis et al., 1995; Braga et al., 2015).

The effect of UV on microsporidia was extensively studied in several species and systems. Since these microorganisms inhabit variable habitats, including terrestrial and aquatic habitats possessing variable concentrations of salts, these parameters have an effect on the actual UV radiation to which the spores are exposed. Microsporidian spores germinate after digestion, and a previous exposure to UV reduces infectivity. While direct exposure of spores of the microsporidia *Nosema* spp. to 254 nm for 20 minutes completely inhibited their germination (Undeen and Meer, 1990), mixing with various substrate increased their survival up to 28 hours (Mandava and Morgan, 1985).

5.3.3.3 Viruses

Sunlight is one of the most effective agents in the inactivation of many viruses, and is the most important factor influencing the field persistence of baculoviruses. NPVs and GVs can be completely inactivated by sunlight in less than 24 hours, with a mean half-life varying from 2 to 5 days (Bullock, 1967; David et al., 1968; Jaques, 1972, 1985; Broome et al., 1974; Ignoffo et al., 1977; McLeod et al., 1977). However, field applications of baculoviruses are not completely inactivated at a constant rate. While *Cydia pomonella* granulovirus (CpGV) applied to orchard trees lost 78% of its activity after 7 days, a small proportion (4%) persisted for 3 weeks (Glen and Payne, 1984). The longer persistence of some OBs could be due to inherently greater stability in the environment or to protection from sunlight offered by shade on the undersides of leaves (Brassel and Benz, 1979; Glen and Payne, 1984; Huber and Lüdcke, 1996). AgNPV sprayed on upper leaf surfaces lost ≥60% of its activity within 2 days, compared to ≤13% when sprayed on the undersides of leaves. Also, cytoplasmic polyhedrovirus (CPV)-induced mortality of *H. virescens* rapidly decreased from 98 to just 7% if the CPV was exposed to direct sunlight for 8 hours prior to inoculation (Ali and Sikorowski, 1986). Aquatic viruses are also vulnerable to sunlight, as IIV-6 in a tray with water was almost completely inactivated after 36 hours of direct sun exposure (Hernández et al., 2005).

UV radiation in regions B (UV-B; 280–310 nm) and A (UV-A; 320–400 nm) are the most critical in deactivating viruses (Ignoffo, 1992; Moscardi, 1999). Baculovirus, CPV, and entomopox virus were all highly sensitive to a UV source combining two ranges of spectra (160–215 and 290–400 nm); half-lives were approximately 2 hours for all viruses (Ignoffo et al., 1977). The complete inactivation of virus takes longer under direct sunlight than under artificial UV lights or simulated sunlight (Shapiro et al., 2002), likely because the optimum exposure to the degrading effects of UV radiation in sunlight is only about 4 hours/day (Ignoffo, 1992). Due to the extreme sensitivity to UV, efforts are under way to protect NPV-based biopesticides from UV radiation through microencapsulation with different polymers (Villamizar et al., 2010; Gómez et al., 2013; Gifani et al., 2015).

5.3.3.4 Bacteria

As reviewed by Ignoffo (1992), the half-life of bacteria spores exposed to sunlight ranged from 0.5 hours for bacteria to about 4 hours for the endotoxin. Btk remained active after 3 hours of exposure to UV, but over time significant loss of virulence toward *Helicoverpa armigera* larvae was detected (Cohen et al., 1991). In general, bacterial spores are 10–50 times more resistant to UV than vegetative cells, depending on the species examined and whether spores were examined in dry state or in aqueous suspension (Nicholson et al., 2000). In a study on *B. sphaericus*, viability was lost and toxicity was impaired after 6 hours of exposure to UV-B (Hadapad et al., 2009). UV tolerance is attributed to pigments existing in the endospores in different types and concentrations (de Melo-Santos et al., 2009). Mutants of *Bacillus thuringiensis* (Bt), producing melanin, were significantly more resistant to UV radiation at 254 and 366 nm (Saxena et al., 2002). In addition, increased UV resistance of the endospore is caused by differences in the UV photochemistry of DNA in spores and the efficient and relatively error-free repair of the novel photoproduct formed by UV light in spore DNA (Nicholson et al., 2000).

5.3.4   Chemical Inputs

The effects of combining microbial control agents with chemical pesticides on pest mortality depend on the mode of action of the chemical. Most chemical insecticides mixed with microbial control agents have an additive effect, but some may inhibit important characteristics of invertebrate pathogens, such as movement, replication, and survival. The compatibility of pathogens with chemical pesticides can be achieved through genetic selection. Other chemicals to consider are fertilizers, metals, and chemical additives in the formulation of the microbial control agents.

5.3.4.1   Nematodes

Steinernematid and heterorhabditid nematodes can survive exposure to many chemical pesticides (Hara and Kaya, 1982; Rovesti et al., 1988; Rovesti and Deseö, 1990). However, IJs are highly susceptible to several nematicides likely to be found in the agroecosystem (Rovesti and Deseö, 1990, 1991). Glazer et al. (1997) genetically improved the resistance of *H. bacteriophora* strain HP88 to several nematicides: nematodes were 8–9-fold more resistant to fenamiphos (an organophosphate) and avermectin (a biologically derived product), and 70-fold more resistant to oxamyl (a carbamate).

It has been shown that some insecticides have additive or synergistic effects on insect mortality when applied together with EPNs. Combinations of *H. bacteriophora* and the anthranilic diamide chloranlraniliprole resulted in mostly synergistic but also additive mortality of third-instar *Anomala orientalis*, *Popillia japonica*, and *Cyclocephala borealis* in greenhouse and field experiments (Koppenhöfer and Fuzy, 2008). Imidacloprid mostly interacted synergistically in combinations with several EPN species *(S. glaseri, H. bacteriophora, H. marelata, H. megidis)* in third instars of *Cyclocephala borealis, C. hirta, Cyclocephala pasadenae, P. japonica*, and *A. orientalis* (Koppenhöfer et al., 2000, 2002; Koppenhöfer and Fuzy, 2008). The EPN–imidacloprid interaction is primarily based on reduced defensive and evasive larval behaviors resulting in increased host attachment and penetration (Koppenhöfer et al., 2000). Neonicotinoid–EPN combination generally had no negative effect on EPN reproduction in hosts. It resulted in higher IJ densities in the soil following application, due to the greater number of infected white grubs (Koppenhöfer et al., 2003).

Soil chemistry has an important effect on EPN activity. At high concentrations, NaCl, KCl, and $CaCl_2$ inhibited the ability of *S. glaseri* to move through a soil column and to locate and infect a susceptible host (Thurston et al., 1994). $CaCl_2$ and KCl had no effect on *H. bacteriophora* survival, infection efficiency, or movement through a soil column, but moderate concentrations of these salts enhanced *H. bacteriophora* virulence. NaCl at high salinities (>16 dS/m) adversely affected all of these parameters (Thurston et al., 1994). A field survey in an Ohio vegetable production area indicated that both biotic and abiotic factors were associated with EPN abundance; these factors included increased enrichment and food web structure, as well as lower P, higher K, and a lower C : N ratio (Hoy et al., 2008).

Shapiro et al. (1996) evaluated the effects of three fertilizers (fresh cow manure, composted manure, and urea) on the virulence of *S. carpocapsae* against *G. mellonella*. Urea and fresh manure decreased nematode virulence in laboratory experiments. In field experiments, however, only the fresh manure treatment reduced nematode virulence. Composted manure did not affect nematode virulence.

Other chemicals in agricultural soil have been shown to affect nematode persistence and efficacy. Sun et al. (2016) assessed the impact of heavy metal accumulation on EPNs (*S. carpocapsae* and *H. bacteriophora*) and plant parasitic nematodes (*Meloidogyne incognita*). Generally, pretreatment with Cu, Zn, or Cr solutions for 24 hours caused direct mortality of the IJs and lowered the virulence of *H. bacteriophora* against *G. mellonella*. In addition, pretreatment of *H. bacteriophora* with Cu, Zn, or Cr salts resulted in increased penetration of *M. incognita* into cucumber roots compared with controls. The results indicate that heavy metals can have a direct negative effect on EPNs (via direct mortality), as well as an indirect effect through disrupting biocontrol efficacy against plant parasitic nematodes.

5.3.4.2 Fungi

The effect of chemical pesticides (particularly fungicides) on entomopathogenic Ascomycetes fungi was studied mainly in *in vitro* assays (Alizadeh et al., 2007; Cuthbertson et al., 2008; Shah et al., 2009; Asi et al., 2010). For *B. bassiana*, noncompatible insecticides included the insect growth regulators (IGRs) Flufenoxuron and Teflubenzuron, the formamidin Amitraz, and the cyclodiene Endosulfanate (Alizadeh et al., 2007). Lorsban (chlorpyriphos) was most toxic to both *M. anisopliae* and *Isaria fumosorosea*, while IGRs were less toxic and Tracer (spinosad) was nontoxic (Asi et al., 2010). *In vitro* effects of fungicides at recommended doses and at doses 10 time higher than recommended inhibited germination of *M. anisopliae*, *B. bassiana*, *I. fumosorosea*, and *Lecanillium longisporum* conidia. *Beauveria bassiana* mycelial growth was inhibited, but the virulence of *B. bassiana* and *L. longisporum* was not affected. However, the fungicides tolylfluanid and azoxystrobin reduced the virulence of *M. anisopliae* and *I. fumosorosea*, respectively (Shah et al., 2009). In a joint pesticide test program of all 10 insecticides, Hostaquick (heptenophos, organophosphate) was the only one toxic for the entomopathogenic fungi *Verticllium* and *Beauveria* spp. Most fungicides were toxic to the entomopathogenic fungi (Sterk et al., 1999).

Fungicide resistance can be enhanced through artificial selection, as was achieved for the commercial strain *B. bassiana*-GHA. Removal of selection pressure for three passages and subculturing with exposure to fungicides did not reduce the new trait of enhanced fungicide resistance (Shapiro-Ilan et al., 2002). However, endemic strains isolated from pecan weevil (*Curculio caryae*) populations were more resistant to the fungicides, although they could not be selected for enhanced resistance. Fungicide exposure increased their virulence. Selection for fungicide resistance was achieved for *M. brunneum* as well. However, for both fungi, removing the selection pressure had negative effects on other beneficial traits, such as reducing germination or growth and reducing virulence to *G. mellonella* (Shapiro-Ilan et al., 2011).

Similarly, pesticide effects on microsporidia (e.g., *N. pyrausta*) in combination with microbial (Bt) and conventional insecticides have been examined. An additive effect in decreasing numbers of surviving European corn borer (*Ostrinia nubilalis*) larvae was obtained when aqueous *N. pyrausta* applications were combined with granular applications of *B. thuringiensis*, carbaryl, or carbofuran (Lewis et al., 2009).

5.3.4.3 Viruses

The effect of combining viruses with chemical insecticides on mortality and virus replication depends on the mode of action of the chemical. Most chemical insecticides mixed with NPVs have an additive effect (Mohamed et al., 1983a,b; Harper, 1986). The chitin synthesis inhibitor chlorfluazuron acted synergistically with AcMNPV against third-instar *S. exigua* by disrupting the structure of the peritrophic membrane (Guo et al., 2007). Chlordimeform and benomyl were demonstrated to act synergistically with *Heliothis* NPV against *H. virescens* larvae, while methoprene acted antagonistically (Mohamed et al., 1983a,b). The effects varied with insect age at the time of infection, and the mechanisms behind the synergistic and antagonistic interactions were not investigated. Methoprene is a juvenile hormone analog (JHA) and may interfere with virus-induced mortality by inducing larval molt. Insects that were treated with both JHA (methoprene or fenoxycarb) and NPV gained more weight before death compared to those treated with NPV alone, and consequently yielded more virus progeny (Longworth and Singh, 1980; Glen and Payne, 1984; Kolodny-Hirsch et al., 1995; Lasa et al., 2007a). In contrast, mixtures of *Lymantria dispar* multinucleopolyhedrovirus (LdMNPV) or *Spodoptera litura* nucleopolyhedrovirus (SpltNPV) with extracts of neem seeds or its insecticidal component, azadirachtin (which have strong antifeedant, insect growth regulatory, and reproductive effects), induced faster mortality of gypsy moth (*L. dispar*) and oriental leafworm (*S. litura*) larvae, although killed hosts were smaller and were likely to yield less progeny virus (Shapiro et al., 1994; Cook et al., 1996; Nathan and Kalaivani, 2005).

The addition of stilbene optical brighteners to baculovirus formulations has been extensively studied and has been demonstrated to synergistically reduce the virus doses needed to kill insects, accelerate speed of kill, and even expand virus host range to marginally permissive hosts (Hamm and Shapiro, 1992; Shapiro, 1992, 2000; Shapiro and Robertson, 1992; Shapiro and Dougherty, 1994; Argauer and Shapiro, 1997; Boughton et al., 2001; Okuno et al., 2003; Lasa et al., 2007b). Studies using the optical brightener Calcofluor White M2R have shown that it does not increase the number of infected primary target cells in the midgut, but instead inhibits sloughing of infected primary target cells, thereby permitting the establishment and spread of the virus (Washburn et al., 1998).

Studies examining the effects of chemical inputs on virus efficacy in other invertebrate systems are rare. Shrimp that were exposed to PCBs in their aquatic environment were significantly more susceptible to baculovirus infection (Couch and Courtney, 1977).

5.3.4.4 Bacteria

In general, due to their anatomical characteristics, *Bacillus* spores are highly resistant to various chemical compounds (reviewed in Nicholson et al., 2000). According to Ravensberg (2011), no reports exist concerning the integration of bacteria with chemical pesticides, even when tank-mixed. However, in a recent study, Spiromesifen reduced the development of Bt by up to 65% and Bifenthrin with and without carbosulfan was toxic to Bt and reduced vegetative growth (Agostini et al., 2014). In a study on Btk compatibility with 27 insecticides, including organophosphates, carbamates, urea derivatives, and antifeedants, carbamates were generally more compatible than were the other groups tested, and technical formulations were less harmful to the bacteria than wettable powders, which were less harmful than emulsifiable concentrates (Morris, 1977). Another Bt-compatible pesticide is the neonicotinoid Thiamethoxam (Batista Filho et al., 2001).

Synergistic interactions were observed for Bt applied against the *S. littoralis* when combined with pyrethroids and organophosphates. The carbamates, diflubenzuron, and a combination of methomyl and diflubenzuron all showed an additive effect when applied jointly with Bt varieties. Chemical pesticides that cause antifeedant effects are not compatible with Bt, since the chemical pesticide reduces Bt spore ingestion, indicating that spatial separation for these applications should be considered (Navon, 2000).

5.3.5 Other Habitat Characteristics

Survival and persistence of entomopathogenss in various habitats are affected by numerous abiotic factors in addition to the factors discussed so far. This chapter has focused on the principal and most studied abiotic factors (temperature, moisture, UV, chemical inputs). In this section, we expand upon other habitat characteristics, including oxygen level, soil type, soil texture, soil depth, pH, cation exchange capacity, and organic matter content in the soil habitat. Other factors apart from soil, such as wind, plant height, and erosion, can also affect the efficacy and persistence of entomopathogens.

5.3.5.1 Nematodes

Since nematodes are aerobic organisms, low oxygen availability can reduce their survival (Evans and Perry, 1976; Wharton, 1986). Oxygen becomes a limiting factor in clay soils, water-saturated soil, or soils with high levels of organic matter. In laboratory studies, it has been shown that *S. carpocapsae* survives oxygen tensions of as low as 0.5% saturation at 20 °C (Burman and Pye, 1980). In sandy soil, survival of *S. carpocapsae* and *S. glaseri* decreased as oxygen concentration decreased from 20 to 1% (Kung et al., 1990). Reduction of oxygen concentrations induces a dormancy state (anabiosis) in several free-living nematode species (Wharton, 1986), but this phenomenon has not been reported with steinernematids or heterorhabditids.

The chemistry and pH of the soil solutions affect EPNs, but nematodes tolerate a wide range of soil pH. Kung et al. (1990) found reduced survival of steinernematid nematodes at pH 10 but no differences from pH 4 to 8. Mortality of the cotton leafworm (*S. littoralis*) from *H. bacteriophora* and *S. carpocapsae* was higher and more rapid at pH 6.9 and 8.0 than at pH 5.6 (Ghally, 1995). Acid deposition may be a limiting factor for nematodes in some areas (Sharpe and Drohan, 1999), but no studies appear to have documented such effects on EPNs.

Qiu and Bedding (2000) investigated the effect of anaerobic conditions on survival, infectivity, and physiological changes in IJs of *S. carpocapsae*. Under aerobic conditions, the survival rate of IJs decreased slightly to 91% in the first 6 weeks and then dropped sharply to about 78% at week 7 and 55% at week 8 of storage. Analysis of changes in key energy reserve materials under anaerobic conditions indicates that levels of glycogen and trehalose declined rapidly over time from initial levels of about 5.0 and 1.5 ng per nematode to 1.33 and 0.3 ng per nematode at day 6, respectively. When anaerobically incubated nematodes were returned to an aerobic environment, both glycogen and trehalose levels increased sharply, while the lipid and lactate levels decreased correspondingly. This indicates that, like most other animals, IJs of *S. carpocapsae* depend on carbohydrate reserves to provide energy under anaerobic conditions.

EPNs are differentially affected by soil texture and structure (Kung et al., 1990a; Barbercheck and Kaya, 199la; Barbercheck, 1993). Movement is more restricted in soils with restrictive pore spaces (heavy or poorly structured soils) than in those with a more porous structure. In the laboratory, survival and movement of *H. bacteriophora*, *S. carpocapsae*, and *S. glaseri* varied with soil texture and bulk density (Portillo-Aguilar et al., 1999). All three species moved more in sandy loam than in loam or silty clay loam. Movement generally decreased as soil bulk density increased. However, the degree to which soils of high bulk density reduced movement differed among species and soil textures: *H. bacteriophora* was the least restricted, whereas *S. carpocapsae* was the most. In general, rates of movement and infection were strongly correlated with the amount of soil pore space having dimensions similar to or greater than the diameter of the EPNs.

Under natural conditions in the field, soil type might have more influence on heterorhabditids than on steinernematids, since heterorhabditids tend to dwell deeper in the soil (Horninick, 2002). However, varied results are observed between the different studies. In no-till and conventional-till maize fields in North Carolina, no significant relationships were detected between the occurrence of endemic *S. carpocapsae* or *H. bacteriophora* and soil organic matter, pH, or soil texture (Millar and Barbercheck, 2002). In Florida citrus groves, soil type was not correlated with infection of citrus root weevils (*Diaprepes abbreviates*) by *S. carpocapsae* (Beavers et al., 1983), but suppression of root weevils by *S. riobrave* was affected by it (Shapiro et al., 2000), being greater in coarse, sandy soils than in fine-textured soils (Duncan et al., 2001).

5.3.5.2 Fungi

Abiotic factors affecting the efficacy of fungal entomopathogens in the soil are predominantly texture, soil type, temperature, and moisture, as well as pH, cation exchange capacity, and inorganic salts (Jaronski, 2010). Jaronski (2007) gives a thorough overview of the soil ecology of entomopathogenic Ascomycota, including habitat and soil characteristics, the effects of different soil types on fungistasis, soil pH, texture, agricultural inputs, and microbes as effectors of the persistence and infectivity of entomopathogenic fungi.

Soil texture and type are important factors in the inundative application of spores as they affects the spores' distribution and contact with the target host. Moisture content determines fungus infectivity and persistence (see earlier). However, both soil texture and moisture interact in a complex relationship, in which the host, its size, and its movement behavior all affect efficacy. In general, independent of soil type, increased moisture resulted in increased efficacy of *M. anisopliae* in killing third-instar sugarbeet root maggot (*Tetanops myopaeformis*). However, introducing the fungus to sandy loam, in which clay content was the lowest, resulted in high efficacy independent of moisture content (Jaronski, 2010). *Beauveria bassiana* predominated in soils with high clay content, higher pH, and low organic matter content, whereas *M. anisopliae* predominated in soils with high organic matter content and in sand (Bidochka et al., 1998; Quesada-Moraga et al., 2007; Garrido-Jurado et al., 2011). Latitude was associated with *B. bassiana* occurrence, whereas longitude was associated with *M. anisopliae* occurrence. These data indicate the complex interaction of the mycopathogens with the abiotic factors of the soil, without encompassing complex interactions with soil-inhabiting microorganisms.

In cultivated soils, as reviewed by Barbercheck (1992), tillage treatments did not disturb conidia of *B. bassiana* in the upper 5 cm. In more recent studies conducted in Canada and Spain, *M. anisopliae* was more closely associated with cultivated soil than *B. bassiana*, especially in annual field crops (Bidochka et al., 1998; Quesada-Moraga et al., 2007).

The effect of oxygen levels has been understudied as an effector of the soil habitat, but *in vitro* studies indicate that conidial yield of *B. bassiana* is maximized under hypoxia (16% O_2), although germination was impaired due to oxidative stress (Garza-López et al., 2011). On the other hand, oxygen-enriched atmosphere increased the conidia yield of *M. anisopliae* (Tlecuitl-Beristain et al., 2009).

Entomophthorales fungi have several means of persistence in the soil: usually as resting spores, and for a few species as hyphal bodies or conidia. For each form, survival depends on both the temperature and the moisture of the soil and the environment. Active inoculum of *Pandora neoaphidis* and *Conidiobolus obscurus* was found in field-collected soil in early spring in both natural soil habitats and cultivated soil, but most frequently in the latter. Quiescent inoculum of both of these fungal species from soil samples became activated only in the presence of its aphid host (Nielsen et al., 2003).

5.3.5.3 Viruses

While NPVs can degrade rapidly in sunlight, they can remain stable for long periods in soil, albeit at lower concentrations. In agroecosystems, OBs of AgNPV from contaminated plants leached to depths of 25.0–37.5 cm and persisted for more than 1 year, even though the soil was mechanically disturbed (disked) during that time (Fuxa and Richter, 1996). In a forest environment, where there is less soil disturbance, 25–50% of active OBs of OpNPV are estimated to persist in the soil 10 years after an epizootic; enough OBs to infect Douglas fir tussock moth larvae persisted for 41 years in some sheltered locations (Thompson and Scott, 1981). Wind and rainfall are key abiotic factors responsible for transporting OBs from the soil to the aboveground surfaces of plants, where they can be ingested by host insects. Their effectiveness depends not only on wind velocity and rainfall intensity, but also on plant height, soil types, soil depth, soil moisture, soil NPV concentrations, and distances of OB reservoirs to the plants (Young and Yearian, 1986, 1989; Olofsson, 1988; Fuxa and Richter, 1996, 2001, 2007; Fuxa et al., 2007; Fuxa, 2008). For instance, transport of OBs by simulated rain was significantly greater from sandy soil than from clay, but transport by wind was greater from clay than from sandy soil (Fuxa et al., 2007).

Additionally, certain soil-forming minerals have been shown to alter the infectivity of HaSNPV and IIV-6, but not the activity of cricket paralysis virus (CrPV), although the mechanism is unclear (Christian et al., 2006). For example, kaolinite and talc reduced HaSNPV activity but increased IIV-6 activity. Patterns of adsorption to the minerals also differed for each type of virus. HaNPV and IIV-6 showed moderate to high affinity for all the minerals tested, except of IIV-6 to bentonites, while CrPV showed generally low affinity for most of the minerals but very high affinity for ferric oxide and for one type of kaolinite (Christian et al., 2006). Thus, it has been suggested that reservoirs of insect viruses may be greater in soils with a high content of iron oxides or kaolinite (Christian et al., 2006).

Baculovirus and CPV OBs can also be susceptible to degradation from high pH in the dew (i.e., water on the leaf surface) of some plants, such as cotton (Gudauskas and Canerday, 1968; Andrews and Sikorowski, 1973; McLeod et al., 1977; Young et al., 1977; Ali and Sikorowski, 1986). Among dew samples collected from cotton, only dew that contained high pH (9.3) inactivated baculovirus OBs (McLeod et al., 1977), while OBs were not inactivated in dew on soybean leaves, which has a more neutral pH (7.7) (Young et al., 1977). Dew on cotton plants that were higher in pH (8.9) inactivated OBs if it was air-dried and resuspended in deionized water daily (mimicking the natural formation and evaporation of dew), because the drying of dew increased the pH (to 9.6) and the ionic strength in the dew (Young et al., 1977). High pH in dew was associated with the dissolution of viral OBs and premature liberation of the infective virions (Young et al., 1977).

Low pH in plant tissue or from acid rain on the leaf surface can also reduce NPV infectivity (Keating et al., 1988; Neuvonen et al., 1990; Saikkonen and Neuvonen, 1993). Since NPV OBs require high pH in the insect midgut in order to release infective virions, it has been hypothesized that insect feeding on acidic foliage decreases midgut pH and inhibits the release of virions (Keating et al., 1988). Data on pH effects on other invertebrate viruses are scarce. Capsids of WSSV exhibit marked stability at a broad pH range (1–10), but degradation occurs at pH 10.5 (Chen et al., 2012). High salinity can cause dissociation of nucleocapsid proteins (VP15 and VP95), but does not cause noticeable alterations in the morphology and ultrastructure of the nucleocapsid and capsid (Chen et al., 2012). Thus, changes in salinity (15–35%) and pH (6.5–9.0) had little effect on the proliferation of WSSV in Chinese shrimp, especially compared to the strong effect of temperature (Gao et al., 2011).

5.3.5.4 Bacteria

Spore-forming bacteria such as Bt inhabit the soil. The effect of soil characteristics such as texture, type, and moisture content on bacterium virulence and persistence is not well known. Organic matter, water availability, and aeration affect bacterial germination, efficacy, and persistence, depending on the actual bacterial strain (Vilas-Bôas et al., 2000; Tetreau et al., 2012), but data on this subject are limited. Studies on Bt Cry toxins indicate that Cry1Ac toxicity was maintained in soil but that it was lost after a 2-week incubation at 25 °C and that time to host death was slower at 4 °C, with no effect of soil sterilization (Hung et al., 2016). The observed decline in detectable protein may have been due to the germination of Bt spores, which was favored by higher temperatures (Hung et al., 2016).

In long-term persistence studies, soil treated for 11 years with Bti showed no differential abundance between the various sampling sites, which differed in soil characteristics. Also, Bti abundance was significantly higher in treated sites than in untreated sites and the number of applications did not influence abundance (Hendriksen, 2015). Similarly, in studies on Btk, long-term persistence was demonstrated in the topsoil (0–2 cm), with a half-life exceeding 100 days, and after 1 year, analysis of the top 15 cm of soil showed that 77% of Btk remained in the topsoil layer. The bacterium survived at relatively low densities 13 years after application, possibly due to growth in hosts (Hendriksen and Hansen, 2002; Hendriksen and Carstensen, 2013; Hendriksen, 2015).

5.4 Mechanisms of Survival

5.4.1 Nematodes

When the environment is unfavorable, some nematode populations may be able to avoid stress by moving into protective niches. This behavioral pattern might explain the findings of Glazer et al. (1995), which indicated a higher number of heterorhabditid nematodes in deeper soil layers (35–40 cm) during summer seasons as compared with the upper soil layer (5–10 cm depth).

Salame and Glazer (2015) compared the effect of a soil moisture gradient on downward movement of a highly desiccation-tolerant EPN (*S. carpocapsae*) and a poorly desiccation-tolerant one (*H. bacteriophora*). *Heterorhabditis bacteriophora* IJs abandoned the upper soil layers as dryness intensified, with >80% of IJs found in the bottom (20–25 cm) layer. In contrast, >70% *S. carpocapsae* IJs remained in the upper layer. In covered buckets, with 10% moisture throughout the experiment, heterorhabditid IJs were equally distributed between the 10–15 and 20–25 cm layers; only 7% remained in the upper layer. Again, >70% *S. carpocapsae* IJs remained in the upper layer throughout. Soil type influenced *H. bacteriophora* IJs' downward migration.

The insect host cadaver can also serve as a protective niche by providing protection from adverse conditions such as extreme temperatures (Lewis and Shapiro-Ilan, 2002) and low humidity (Brown and Gaugler, 1997; Koppenhöfer et al., 1997).

5.4.2 Fungi

In recent years, the life histories of hypocrealean and basidiomycetes EPF were found to include plant endophytes and rhizosphere colonizers, and as such they are involved in promoting plant growth and antagonists of plant diseases and arthropod pests (Roy et al., 2010). For example, as was discovered in natural populations of the citrus rust mite (*Phyllocoptruta oleivora*), the occurrence of the fungi *Acaromyces ingoldii* and *Meira* spp. in grapefruit flowers and in the flavedo of the fruits' peel caused mortality of the mites. However, these fungi were not parasitizing the mites, but rather secreted toxic metabolites (Gerson et al., 2008). As endophytes, abiotic factors may display less challenge to EPF persistence, but knowledge on this is lacking.

Even though metagenomic approaches are commonly used as tools to investigate processes in microorganisms and a vast knowledge exists concerning the effects of abiotic factors on entomopathogenic fungi, there is a lack of knowledge about specific processes involved in stress response and tolerance. Delayed germination after thermal shock and exposure to UV are probably associated with the need to repair damages before germination occurs (Fernandes et al., 2008). Conidia with higher concentrations of melanin were more tolerant to UV radiation (Ignoffo et al., 1977; Singaravelan et al., 2008). In *M. anisopliae*, exposure to UV resulted in upregulation of enzymes involved in oxidative stress, such as catalase, superoxide dismutase (SOD), and glutathione reductase, but their expression could not be correlated with greater tolerance to UV (Miller et al., 2004). In Entomophthorales, the cadaver and the conidia halo (high-density deposition) provide some protection for conidia from UV radiation, but when not protected, the primary conidia are inactivated 3 minutes after exposure to UV (Pell et al., 2001). The main means for longer-term survival by Entomophthorales is the dormant thick-walled resting spores, with only specific environmental cues for germination. In *E. maimaiga*, the resting spore reservoir accumulates and remains in the top soil surfaces at the bases of forest trees, so that it is present in a dispersal path for larval hosts (*L. dispar*) in spring (Hajek et al., 1998).

5.4.3 Viruses

Infective virions of NPVs and GVs are protected from the environment within viral OBs and granules, respectively. These baculoviruses can persist for long periods in soil and water and on plant surfaces, but only if protected from UV radiation (see earlier). One of the fascinating features of baculoviruses is that they can manipulate host behavior following infection to increase the likelihood of their transmission. The ecdysteroid UDP-glucosyl transferase (*egt*) gene from LdMNPV was demonstrated to induce climbing behavior in infected *L. dispar* larvae; larvae subsequently died in the upper canopy (termed "Wipfelkrankheit" or "treetop disease") (Hoover et al., 2011). AcMNPV induced treetop disease in two different hosts, *Trichoplusia ni* and *Spodoptera exigua* larvae, although the effect of *egt* on climbing behavior was only observed in larvae that molted after infection (Ros et al., 2015). Additionally, *Mamestra brassicae* larvae infected by MbNPV tended to die on the apex of cabbage leaves (Vasconcelos et al., 1996). While it may seem counterintuitive for NPV-infected larvae to die and release OBs at higher elevations or on the apex of leaves, where UV intensity would be greater, liquefaction of host tissues induced by virus-encoded chitinase and cathepsin genes (Hawtin et al., 1997) causes OBs to be dispersed to lower positions in the canopy, where uninfected hosts feed. Dissemination of OBs through liquefaction to the lower canopy can also increase protection from UV by shade from the upper canopy.

In abandoned shrimp ponds, where shrimp populations were completely decimated by a WSSV outbreak, the WSSV genome was detected by polymerase chain reaction (PCR) in at least 90% of the water samples collected in the ponds 3 months after the outbreak, and in more than 10% of samples after more than 1.5 years (Quang et al., 2009). However, detection of WSSV genomes may not necessarily represent infective virions, as not enough viable WSSV particles remain to infect shrimp after 40 days in seawater or after 21 days of sun-drying of virus-contaminated sediments (Satheesh Kumar et al., 2013). WSSV persist in decomposing dead shrimp, and can become suspended in the water column if disturbed. Cannibalism of dead shrimp is also a key route of transmission. Other important characteristics of WSSV transmission include its ability to vertically transmit to the offspring of infected individuals and its broad host range of 98 known species, including crabs, copepods, polychaete worms, and zooplankton (Escobedo-Bonilla et al., 2008). The wide range of hosts suggests that WSSV may not need to survive in the environment for long periods of time, as new hosts are encountered quickly.

5.4.4 Bacteria

Endospores of bacteria are highly resistant to abiotic stress. The spore coat is involved in resistance to chemical compounds, oxidative stress, heat, and UV radiation. It also restricts entry of hydrophilic compounds (Nicholson et al., 2000). In addition, the low water content of the core is related to wet heat resistance. Spores are far more resistant to various chemical compounds than vegetative cells for a number of reasons, including their impermeability, low water content, and DNA protection (Nicholson et al., 2000).

5.5 Conclusion

Abiotic environmental factors have diverse effects on entomopathogen survival, efficacy, and dispersal in the ecosystem. Extreme temperatures, desiccation, and UV radiation have the most detrimental effects on pathogen survival despite the existence of durable resting stages for most groups. In contrast, air and water can facilitate pathogen dispersal and transmission. With the exception of the powerful effect of sunlight, individual abiotic factors at levels found in normal field conditions have little influence on longer-term pathogen stability and persistence in the environment. However, fluctuations in a single abiotic factor rarely occur independently of other abiotic factors. For example, entomopathogens in habitats that are exposed to direct sunlight experience higher UV radiation and have a greater likelihood of experiencing higher temperature and lower RH, whereas those in shade may experience lower UV radiation, lower temperature, and higher RH. Thus, it is increasingly clear that the influence of abiotic factors on the stability of pathogens must be examined in light of other ecologically relevant factors. For instance, while NPV OBs can persist in soil for long periods of time (Fuxa and Richter, 1996; Thompson and Scott, 1981), soil type, soil moisture and water potential, and soil microorganisms can alter their stability (Peng et al., 1999). Similarly, NPV OBs are highly stable in water, but water on the surface of plants can mix with plant exudates that have high pH, resulting in the premature dissolution of the OBs and release of unprotected virions (McLeod et al., 1977; Young et al., 1977). Interactions between the moisture and pH of soil and leaf surfaces can also interfere with virulence factors in fungi and degrade bacterial endotoxins (West et al., 1985; Inyang et al., 1999; Garrido-Jurado et al., 2011; Sanahuja et al., 2011). The abundance of studies examining single abiotic factors oversimplifies the complexity of pathogen persistence in the field. The lack of attention to interactions among abiotic factors is a critical inadequacy in invertebrate host–pathogen studies.

As viruses are obligate intracellular pathogens, abiotic (and biotic) factors experienced by their hosts can have profound effects on their ability to establish infections and replicate. For instance, complex interactions among temperature, food quality, and larval density significantly influenced the immunocompetence of the Indian meal moth (*Plodia interpunctella*) (Triggs and Knell 2012). The outcome of host–pathogen interactions within these complex yet realistic interactions among environmental factors can be difficult to predict, and examples are rare. In the mosquito *Anopheles stephensi*, dietary supplementation, which boosted midgut immunity, reduced infection by the rodent malarial parasite *Plasmodium yoelii* at intermediate temperatures but not at low or high temperature extremes (Murdock et al., 2014). Although this is an example using a vector–parasite interaction, where the parasite may not necessarily reduce mosquito survival, the complexity of the outcome inflicted by two interacting environmental factors should still apply to pathogens and their invertebrate hosts. Interactive effects among environmental factors on host–pathogen interactions can also differ depending on the identity of the infecting pathogen. The interactive effects of temperature and dietary nutrient ratios (a proxy for host plant quality) on the survival, growth, and performance of cabbage looper (*Trichoplusia ni*) larvae differed depending on the identity of the infecting virus (*Trichoplusia ni* single nucleopolyhedrovirus (TnSNPV) or AcMNPV) (Shikano and Cory 2015).

Biotic interactions may even enhance protection of entomopathogens against abiotic factors. For example, iflavirus was found co-occluded in baculovirus OBs, which enhanced the protection of iflavirus against UV light (Jakubowska et al., 2016). Understanding how complex interactions among environmental factors (abiotic by abiotic and/or abiotic by biotic) influence the efficacy of entomopathogens is critical for improving their use as biological control agents in agriculture, for protecting beneficial invertebrate species against pathogens, and for predicting epizootics in natural systems. Assessing the impact of a multitude of environmental factors as a whole, rather than as single factors, is a key challenge in better understanding the ecology of invertebrate pathogens.

References

Abu Hatab, M., Gaugler, R., 2001. Diet composition and lipids of in vitro-produced *Heterorhabditis bacteriophora*. Biol. Control 20, 1–7.

Agostini, L.T., Duarte, R.T., Volpe, X.L., Agostini, T.T., de Carvalho, G.A., Abrahão, Y.P., Polanczyk, R.A., 2014. Compatibility among insecticides, acaricides, and *Bacillus thuringiensis* used to control *Tetranychus urticae* (Acari: Tetranychidae) and *Heliothis virescens* (Lepidoptera: Noctuidae) in cotton fields. Afr. J. Agric. Res. 9, 941–949.

Ali, S., Sikorowski, P.P., 1986. Effects of sunlight, cotton foliage surface, and temperature on the infectivity of cytoplasmic polyhedrosis virus to *Heliothis virescens* larvae (Lepidoptera: Noctuidae). J. Econ. Entomol. 79, 364–367.

Alizadeh, A., Samih, M.A., Khezri, M., Riseh, R.S., 2007. Compatibility of *Beauveria bassiana* (Bals.) Vuill. with several pesticides. Int. J. Agric. Biol. 9, 31–34.

Alves, R.T., Bateman, R.P., Prior, C., Leather, S.R., 1998. Effects of simulated solar radiation on conidial germination of *Metarhizium anisopliae* in different formulations. Crop Prot. 17, 675–679.

Andrews, G.L., Sikorowski, P.P., 1973. Effects of cotton leaf surfaces on the nuclear polyhedrosis virus of *Heliothis zea* and *Heliothis virescens* (Lepidoptera: Noctuidae). J. Invertebr. Pathol. 22, 290–291.

Argauer, R., Shapiro, M., 1997. Fluorescence and relative activities of stilbene optical brighteners as enhancers for the gypsy moth (Lepidoptera: Lymantriidae) baculovirus. Biol. Microb. Control 90, 416–420.

Arthurs, S., Thomas, M.B., 2001. Effect of dose, pre-mortem host incubation temperature and thermal behaviour on host mortality, mycosis and sporulation of *Metarhizium anisopliae* var. acridum in *Schistocerca gregaria*. Biocontrol Sci. Technol. 11, 411–420.

Arthurs, S.P., Thomas, M.B., Lawton, J.L., 2001. Seasonal patterns of persistence and infectivity of *Metarhizium anisopliae* var. acridum in grasshopper cadavers in the Sahel. Entomol. Exp. Appl. 100, 69–76.

Asi, M.R., Bashir, M.H., Afzal, M., Ashfaq, M., Sahi, S.T., 2010. Compatibility of entomopathogenic fungi, *Metarhizium anisopliae* and *Paecilomyces fumosoroseus* with selective insecticides. Pak. J. Bot. 42, 4207–4214.

Barbercheck, M.E., 1992. Effect of soil physical factors on biological control agents of soil insect pests. Fla. Entomol. 75, 539–548.

Batista Filho, A., Almeida, J.E.M., Lamas, C., 2001. Effect of thiamethoxam on entomopathogenic microorganisms. Neotrop. Entomol. 30, 437–447.

Bell, M.R., Hayes, J.L., 1994. Areawide management of cotton bollworm and tobacco budworm (Lepidoptera: Noctuidae) through application of a nuclear polyhedrosis virus on early-season alternate hosts. J. Econ. Entomol. 87, 53–57.

Benz, G., 1987. Environment, in: Fuxa, J.R., Tanada, Y. (eds.), Epizootiology of Insect Diseases. John Wiley & Sons, New York, pp. 177–214.

Bidochka, M.J., Kasperski, J.E., Wild, G.A., 1998. Occurrence of the entomopathogenic fungi *Metarhizium anisopliae* and *Beauveria bassiana* in soils from temperate and near-northern habitats. Can. J. Bot. 76, 1198–1204.

Boucias, D., Pendland, J.C., 2012. Principles of Insect Pathology. Springer Science & Business Media, Berlin.

Boughton, A.J., Lewis, L.C., Bonning, B.C., 2001. Potential of *Agrotis ipsilon* nucleopolyhedrovirus for suppression of the black cutworm (Lepidoptera: Noctuidae) and effect of an optical brightener on virus efficacy. J. Econ. Entomol. 94, 1045–1052.

Braga, G.U., Flint, S.D., Messias, C.L., Anderson, A.J., Roberts, D.W., 2001. Effect of UV-B on conidia and germlings of the entomopathogenic hyphomycete *Metarhizium anisopliae*. Mycol. Res. 105, 874–882.

Braga, G.U., Rangel, D.E., Fernandes, É.K., Flint, S.D., Roberts, D.W., 2015. Molecular and physiological effects of environmental UV radiation on fungal conidia. Curr. Genet. 61, 405–425.

Braga, G.U., Rangel, D.E., Flint, S.D., Miller, C.D., Anderson, A.J., Roberts, D.W., 2002. Damage and recovery from UV-B exposure in conidia of the entomopathogens *Verticillium lecanii* and *Aphanocladium album*. Mycologia 94, 912–920.

Brassel, J., Benz, G., 1979. Selection of a strain of the granulosis virus of the codling moth with improved resistance against artificial ultraviolet radiation and sunlight. J. Invertebr. Pathol. 33, 358–363.

Broome, J.R., Sikorowski, P.P., Neel, W.W., 1974. Effect of sunlight on the activity of nuclear polyhedrosis virus from *Malacosoma disstria*. J. Econ. Entomol. 67, 135–136.

Brown, I.M., Gaugler, R., 1997. Temperature and humidity influence emergence and survival of entomopathogenic nematodes. Nematologica 43, 363–375.

Brown, I.M., Gaugler, R., 1996. Cold tolerance of steinernematid and heterorhabditid nematodes. J. Therm. Biol. 21, 115–121.

Bullock, H.R., 1967. Persistence of Heliothis nuclear-polyhedrosis virus on cotton foliage. J. Invertebr. Pathol. 9, 434–436.

Burman, M., Pye, A.E., 1980. *Neoaplectana carpocapsae*: respiration of infective juveniles. Nematologica 26, 214–219.

Campbell, L.R., Gaugler, R., 1991. Mechanisms for exsheathment of entomopathogenic nematodes. Int. J. Parasitol. 21, 219–224.

Carruthers, R. I., and Soper, R. S., 1987. Fungal diseases, in: Fuxa, J.R., Tanada, Y. (eds.), Epizootiology of Insect Diseases. John Wiley & Sons, New York, pp. 357–416.

Chakrabarti, S., Manna, B., 2008. Influence of temperature and relative humidity in infection of *Nosema bombycis* (Microsporidia: Nosematidae) and cross-infection of *N. mylitta* on growth and development of mulberry silkworm, *Bombyx mori*. Int. J. Ind. Entomol. 17, 173–180.

Chandler, D., Davidson, G., Jacobson, R.J., 2005. Laboratory and glasshouse evaluation of entomopathogenic fungi against the two-spotted spider mite, *Tetranychus urticae* (Acari: Tetranychidae), on tomato, *Lycopersicon esculentum*. Biocontrol Sci. Technol. 15, 37–54.

Chen, S. and Glazer, I., 2004. The effect of direct and gradual increase osmotic pressure to survival of entomopathogenic nematodes. Phytoparasitica 32, 486–497.

Chen S., Glazer, I., Gollop, N., Cash, P., Argo, E., Innes, A., et al., 2005. Proteomic analysis of the entomopathogenic nematode *Steinernema feltiae* IS-6 IJs under evaporative and osmotic Stresses. Molec. Biochem. Parasitol. 145, 195–204.

Chen, W., Zhang, H., Gu, L., Li, F., Yang, F., 2012. Effects of high salinity, high temperature and pH on capsid structure of white spot syndrome virus. Dis. Aquat. Organ. 101, 167–171.

Cheney, D.P., Macdonald, B.F., Elston, R.A., 2000. Summer mortality of Pacific oysters, *Crassostrea gigas* (Thunberg): initial findings on multiple environmental stressors in Puget Sound, Washington, 1998. J. Shellfish Res. 19, 353–359.

Christian, P.D., Richards, A.R., Williams, T., 2006. Differential adsorption of occluded and nonoccluded insect-pathogenic viruses to soil-forming minerals. Appl. Environ. Microbiol. 72, 4648–4652.

Clerk, G.C., Madelin, M.F., 1965. The longevity of conidia of three insect-parasitizing hyphomycetes. Trans. Br. Mycol. Soc. 48, 193–209.

Cohen, E., Rozen, H., Joseph, T., Braun, S., Margulies, L., 1991. Photoprotection of *Bacillus thuringiensis kurstaki* from ultraviolet irradiation. J. Invertebr. Pathol. 57, 343–351.

Cook, S.P., Webb, R.E., Thorpe, K.W., 1996. Potential enhancement of the gypsy moth (Lepidoptera: Lymantriidae) nuclear polyhedrosis virus with the triterpene azadirachtin. Environ. Entomol. 25, 1209–1214.

Cooper, A.F., Van Gundy, S.D., 1971. Senescence, quiescence, and cryptobiosis, in: Zuckerman, B.M., Mai, W.F., Rhode, R.A. (eds.), Plant Parasitic Nematodes, Vol. II. Academic Press, New York, pp. 297–318.

Cory, J.S., Myers, J.H., 2003. The ecology and evolution of insect baculoviruses. Annu. Rev. Ecol. Evol. Syst. 34, 239–272.

Couch, J.A., Courtney, L., 1977. Interaction of chemical pollutants and virus in a crustacean: a novel bioassay system. Ann. N. Y. Acad. Sci. 79, 497–504.

Crowe, J.H., 2014. Anhydrobiosis: an unsolved problem. Plant Cell Environ. 37, 1491–1493.

Crowe, J.H., Crowe, L.M., 1992. Membrane integrity in anhydrobiotic organisms: toward a mechanism for stabilizing dry cells. Water Life, 87–103.

Crowe, J.H., Madin, K.A.C., 1975. Anhydrobiosis in nematodes: evaporative water loss and survival. J. Exp. Zool. 193, 323–333.

Cuthbertson, A.S., Blackburn, L., Northing, P., Luo, W., Cannon, R.C., Walters, K.A., 2008. Further compatibility tests of the entomopathogenic fungus *Lecanicillium muscarium* with conventional insecticide products for control of sweetpotato whitefly, *Bemisia tabaci* on poinsettia plants. Insect Sci. 15, 355.

Wharton, D. A., 1986. A Functional Biology of Nematodes. Croom Helm, London.

D'Amico, V., Elkinton, J.S., 1995. Rainfall effects on transmission of gypsy moth (Lepidoptera: Lymantriidae) nuclear polyhedrosis virus. Environ. Entomol. 24, 1144–1149.

Daniel, M., Dusbabek, F., 1994. Micrometeorological and microhabitats factors affecting maintenance and dissemination of tick-borne diseases in the environment. In: Sonenshine, D.E. (ed.), Ecological Dynamics of Tick-borne Zoonoses. Oxford University Press, New York, pp. 91–138.

Daoust, R.A., Roberts, D.W., 1983. Studies on the prolonged storage of *Metarhizium anisopliae* conidia: effect of temperature and relative humidity on conidial viability and virulence against mosquitoes. J. Invertebr. Pathol. 41, 143–150.

David, W.A.L., Gardiner, B.O.C., Woolner, M., 1968. The effects of sunlight on a purified granulosis virus of *Pieris brassicae* applied to cabbage leaves. J. Invertebr. Pathol. 11, 496–501.

Day, M.F., Dudzinski, M.L., 1966. The effect of temperature on the development of *Sericesthis iridescent* virus. Aust. J. Biol. Sci. 19, 481–493.

De Crecy, E., Jaronski, S., Lyons, B., Lyons, T.J., Keyhani, N.O., 2009. Directed evolution of a filamentous fungus for thermotolerance. BMC Biotechnol. 9, 74.

Devauchelle, G., Attias, J., Monnier, C., Barray, S., Cerutti, M., Guerillon, J., Orange-Balange, N., 1985. Chilo iridescent virus, in: Willis, D.B. (ed.), Iridoviridae. Springer, Philadelphia, PA, pp. 37–48.

Dimbi, S., Maniania, N.K., Lux, S.A., Mueke, J.M., 2004. Effect of constant temperatures on germination, radial growth and virulence of *Metarhizium anisopliae* to three species of African tephritid fruit flies. BioControl 49, 83–94.

Du, H., Dai, W., Han, X., Li, W., Xu, Y., Xu, Z., 2008. Effect of low water temperature on viral replication of white spot syndrome virus in *Procambarus clarkii*. Aquaculture 277, 149–151.

Elderd, B.D., Reilly, J.R., 2014. Warmer temperatures increase disease transmission and outbreak intensity in a host-pathogen system. J. Anim. Ecol. 83, 838–849.

Escobedo-Bonilla, C.M., Alday-Sanz, V., Wille, M., Sorgeloos, P., Pensaert, M.B., 2008. A review on the morphology, molecular characterization, morphogenesis and pathogenesis of white spot syndrome virus. J. Fish Dis. 31, 1–18.

Evans, A.A.F., Perry, R.N., 2009. Survival mechanisms, in: Perry, R.N., Moens, M., Starr, J.L. (eds.), Root-Knot Nematodes. CABI, Wallingford, pp. 201–222.

Fargues, J., Ouedraogo, A., Goettel, M.S., Lomer, C.J., 1997. Effects of temperature, humidity and inoculation method on susceptibility of *Schistocerca gregaria* to *Metarhizium flavoviride*. Biocontrol Sci. Technol. 7, 345–356.

Feng, M.G., Poprawski, T.J., Nowierski, R.M., Zeng, Z., 1999. Infectivity of *Pandora neoaphidis* (Zygomycetes: Entomophthorales) to *Acyrthosiphon pisum* (Hom., Aphididae) in response to varying temperature and photoperiod regimes. J. Appl. Entomol. 123, 29–35.

Fernandes, E.K., Rangel, D.E., Moraes, A.M., Bittencourt, V.R., Roberts, D.W., 2007. Variability in tolerance to UV-B radiation among *Beauveria* spp. isolates. J. Invertebr. Pathol. 96, 237–243.

Fernandes, É.K.K., Rangel, D.E.N., Moraes, Á.M.L., Bittencourt, V.R.E.P., Roberts, D.W., 2008. Cold activity of *Beauveria* and *Metarhizium*, and thermotolerance of *Beauveria*. J. Invertebr. Pathol. 98, 69–78.

Filotas, M.J., Vandenberg, J.D., Hajek, A.E., 2006. Concentration-response and temperature-related susceptibility of the forest tent caterpillar (Lepidoptera: Lasiocampidae) to the entomopathogenic fungus *Furia gastropachae* (Zygomycetes: Entomophthorales). Biol. Control 39, 218–224.

Finnegan, M., Griffin, C., O'Regan, M., Downes, M., 1999. Effect of salt and temperature stresses on survival and infectivity of *Heterorhabditis* spp. IJs. Nematology 1, 69–78.

Frid, L., Myers, J.H., 2002. Thermal ecology of western tent caterpillars *Malacosoma californicum* pluviale and infection by nucleopolyhedrovirus. Ecol. Entomol. 27, 665–673.

Fujiie, A., Yokoyama, T., 1998. Effects of ultraviolet light on the entomopathogenic nematode, *Steinernema kushidai* and its symbiotic bacterium, *Xenorhabdus japonicus*. Appl. Entomol. Zool. 33, 263–269.

Fuxa, J.R., 2008. Threshold concentrations of nucleopolyhedrovirus in soil to initiate infections in *Heliothis virescens* on cotton plants. Microb. Ecol. 55, 530–539.

Fuxa, J.R., Richter, A.R., 1996. Effect of agricultural operations and precipitation on vertical distribution of a nuclear polyhedrosis virus in soil. Biol. Control 6, 324–329.

Fuxa, J.R., Richter, A.R., 2001. Quantification of soil-to-plant transport of recombinant nucleopolyhedrovirus: effects of soil type and moisture, air currents, and precipitation. Appl. Environ. Microbiol. 67, 5166–5170.

Fuxa, J.R., Richter, A.R., 2007. Effect of nucleopolyhedrovirus concentration in soil on viral transport to cotton (*Gossypium hirsutum* L.) plants. BioControl 52, 821–843.

Fuxa, J.R., Richter, A.R., Milks, M.L., 2007. Threshold distances and depths of nucleopolyhedrovirus in soil for transport to cotton plants by wind and rain. J. Invertebr. Pathol. 95, 60–70.

Gal, T.Z., Solomon, A., Glazer, I., Koltai, H., 2001. Alterations in the levels of glycogen and glycogen synthase transcripts during desiccation in the insect-killing nematode *Steinernema feltiae* IS-6. J. Parasitol. 87, 725–732.

Gal, T.Z., Glazer, I., Koltai, H., 2004. An LEA group 3 family member is involved in survival of *C. elegans* during exposure to stress. FEBS Lett. 577, 21–26.

Gao, H., Kong, J., Li, Z., Xiao, G., Meng, X., 2011. Quantitative analysis of temperature, salinity and pH on WSSV proliferation in Chinese shrimp *Fenneropenaeus chinensis* by real-time PCR. Aquaculture 312, 26–31.

Garcia del Pino, F., Palomo, A., 1996. Natural occurrence of entomopathogenic nematodes (Rhabditida: Steinernematidae and Heterorhabditidae) in Spanish soils. J. Invertebr. Pathol. 68, 84–90.

Garrido-Jurado, I., Torrent, J., Barrón, V., Corpas, A., Quesada-Moraga, E., 2011. Soil properties affect the availability, movement, and virulence of entomopathogenic fungi conidia against puparia of *Ceratitis capitata* (Diptera: Tephritidae). Biol. Control 58, 277–285.

Garza-López, P.M., Konigsberg, M., Gómez-Quiroz, L.E., Loera, O., Garza-López, P.M., Gómez-Quiroz, L.E., Konigsberg, M., 2011. Physiological and antioxidant response by *Beauveria bassiana*. World J. Microbiol. Biotechnol. 28, 353–359.

Gaugler, R., Boush, G.M., 1978. Effects of ultraviolet radiation and sunlight on the entomogenous nematode, *Neoaplectana carpocapsae*. J. Invertebr. Pathol. 32, 291–296.

Gaugler, R., Bednarek, A., Campbell, J.F., 1992. Ultraviolet inactivation of heterorhabditid and steinernematid nematodes. J. Invertebr. Pathol. 59, 155–160.

Gerson, U., Gafni, A., Paz, Z., Sztejnberg, A., 2008. A tale of three acaropathogenic fungi in Israel: *Hirsutella*, *Meira* and *Acaromyces*. Exp. Appl. Acarol. 46, 183–194.

Ghosh, S., Saha, K., 1995. Impact of temperature and relative humidity on the incidence of a microsporan parasite, *Nosema* sp. I, infecting coleopteran stored grain pest, *Lophocateres pusillus*. J. Entomol. Res. 19, 207–214.

Gifani, A., Marzban, R., Safekordi, A., Ardjmand, M., Dezianian, A., 2015. Ultraviolet protection of nucleopolyhedrovirus through microencapsulation with different polymers. Biocontrol Sci. Technol. 25, 814–827.

Gilbert, L.I., Gill, S.S., 2010. Insect Control: Biological and Synthetic Agents. Academic Press, New York.

Glazer, I., 2002. Survival biology, in: Gaugler, R. (ed.), Entomopathogenic Nematology. CABI, Wallingford, pp. 169–187.

Glazer, I., Salame, L., 2000. Osmotic survival of the entomopathogenic nematode *Steinernema carpocapsae*. Biol. Control 18, 251–257.

Glazer, I., Salame, L., Segal, D., 1997. Genetic enhancement of nematicide resistance in entomopathogenic nematodes. Biocontrol Sci. Technol. 7, 499–512.

Glen, D., Payne, C., 1984. Production and field evaluation of codling moth granulosis virus for control of *Cydia pomonella* in the United Kingdom. Ann. Appl. Biol. 104, 87–98.

Glude, A., 1974. Summary report of the Pacific coast oyster mortality investigations 1965–1972. Proceedings of the Third US–Japan Meeting on Aquaculture.

Gómez, J., Guevara, J., Cuartas, P., Espinel, C., Villamizar, L., 2013. Microencapsulated *Spodoptera frugiperda* nucleopolyhedrovirus: insecticidal activity and effect on arthropod populations in maize. Biocontrol Sci. Technol. 23, 829–846.

Grabner, D.S., Schertzinger, G., Sures, B., 2014. Effect of multiple microsporidian infections and temperature stress on the heat shock protein 70 (hsp70) response of the amphipod *Gammarus pulex*. Parasit. Vectors 7, 170.

Granja, C.B., Vidal, O.M., Parra, G., Salazar, M., 2006. Hyperthermia reduces viral load of white spot syndrome virus in *Penaeus vannamei*. Dis. Aquat. Organ. 68, 175–180.

Grewal, P.S., Selvan, S., Gaugler, R., 1994. Thermal adaptation of entomopathogenic nematodes: niche breadth for infection, establishment, and reproduction. J. Therm. Biol. 19, 245–253.

Grewal, P.S., Wang, X., Taylor, R.A.J., 2002. Dauer juvenile longevity and stress tolerance in natural populations of entomopathogenic nematodes: is there a relationship? Int. J. Parasitol. 32, 717–725.

Guan, Y., Yu, Z., Li, C., 2003. The effects of temperature on white spot syndrome infections in *Marsupenaeus japonicus*. J. Invertebr. Pathol. 83, 257–260.

Gudauskas, R.T., Canerday, D., 1968. The effect of heat, buffer salt and H-ion concentration, and ultraviolet light on the infectivity of *Heliothis* and *Trichoplusia* nuclear-polyhedrosis viruses. J. Invertebr. Pathol. 12, 405–411.

Guo, H.-F., Fang, J.-C., Liu, B.-S., Wang, J.-P., Zhong, W.-F., Wan, F.-H., 2007. Enhancement of the biological activity of nucleopolyhedrovirus through disruption of the peritrophic matrix of insect larvae by chlorfluazuron. Pest Manag. Sci. 63, 68–74.

Hadapad, A.B., Hire, R.S., Vijayalakshmi, N., Dongre, T.K., 2009. UV protectants for the biopesticide based on *Bacillus sphaericus* Neide and their role in protecting the binary toxins from UV radiation. J. Invertebr. Pathol. 100, 147–152.

Hajek, A.E., 1997. Ecology of terrestrial tungal entomopathogens, in: Jones, J.G. (ed.), Advances in Microbial Ecology. Springer, Philadelphia, PA, pp. 193–249.

Hajek, A.E., Carruthers, R.I., Soper, R.S., 1990. Temperature and moisture relations of sporulation and germination by *Entomophaga maimaiga* (Zygomycetes: Entomophthoraceae), a fungal pathogen of *Lymantria dispar* (Lepidoptera: Lymantriidae). Environ. Entomol. 19, 85–90.

Hall, R.A., Papierok, B., 1982. Fungi as biological control agents of arthropods of agricultural and medical importance. Parasitology 84, 205–240.

Hamm, J.J., Hare, W.W., 1982. Application of entomopathogens in irrigation water for control of fall armyworms and corn earworms (Lepidoptera: Noctuidae) on corn. J. Econ. Entomol. 75, 1074–1079.

Hamm, J.J., Shapiro, M., 1992. Infectivity of fall armyworm (Lepidoptera: Noctuidae) nuclear polyhedrosis virus enhanced by a fluorescent brightener. J. Econ. Entomol. 85, 2149–2152.

Hara, A.H., Kaya, H.K., 1982. Effects of selected insecticides and nematicides on the in vitro development of the entomogenous nematode *Neoaplectana carpocapsae*. J. Nematol. 14, 486–491.

Harper, J.D., 1986. Interactions between baculoviruses and other entomopathogens, chemical pesticides, and parasitoids, in: Granados, R.R., Federici, B.A. (eds.), The Biology of Baculoviruses, Vol. 2, Practical Applications for Insect Control. CRC Press, Boca Raton, FL, pp. 133–156.

Harrison, R., Hoover, K., 2012. Baculoviruses and other occluded insect viruses, in: Vega, F., Kaya, H. (eds.), Insect Pathology, 2nd edn. Elsevier, Amsterdam, pp.73–131.

Hawtin, R.E., Zarkowska, T., Arnold, K., Thomas, C.J., Gooday, G.W., King, L.A., et al., 1997. Liquefaction of *Autographa californica* nucleopolyhedrovirus-infected insects is dependent on the integrity of virus-encoded chitinase and cathepsin genes. Virology 238, 243–253.

Hedgecock, S., Moore, D., Higgins, P.M., Prior, C., 1995. Influence of moisture content on temperature tolerance and storage of *Metarhizium flavoviride* conidia in an oil formulation. Biocontrol Sci. Technol. 5, 371–378.

Hendriksen, N.B., 2016. Influence of multi-year *Bacillus thuringiensis* subsp. *israelensis* on the abundance of *B. cereus* group populations in Swedish riparian wetland soils. New Challenges in Biological Control, 15.

Hendriksen, N.B., Carstensen, J., 2013. Long-term survival of *Bacillus thuringiensis* subsp. *kurstaki* in a field trial. Can. J. Microbiol. 59, 34–38.

Hendriksen, N.B., Hansen, B.M., 2002. Long-term survival and germination of *Bacillus thuringiensis* var. *kurstaki* in a field trial. Can. J. Microbiol. 48, 256–261.

Hernández, A., Marina, C.F., Valle, J., Williams, T., 2005. Persistence of invertebrate iridescent virus 6 in tropical artificial aquatic environments. Arch. Virol. 150, 2357–2363.

Hewson, I., Brown, J.M., Gitlin, S.A., Doud, D.F., 2011. Nucleopolyhedrovirus detection and distribution in terrestrial, freshwater, and marine habitats of Appledore Island, Gulf of Maine. Microb. Ecol. 62, 48–57.

Holmes, S.B., Fick, W.E., Kreutzweiser, D.P., Ebling, P.M., England, L.S., Trevors, J.T., 2008. Persistence of naturally occurring and genetically modified *Choristoneura fumiferana* nucleopolyhedroviruses in outdoor aquatic microcosms. Pest Manag. Sci. 64, 1015–1023.

Hong, T.D., Ellis, R.H., Moore, D., 1997. Development of a model to predict the effect of temperature and moisture on fungal spore longevity. Ann. Bot. 79, 121–128.

Hoover, K., Grove, M., Gardner, M., Hughes, D.P., McNeil, J., Slavicek, J., 2011. A gene for an extended phenotype. Science 333, 1401.

Huber, J., Lüdcke, C., 1996. UV-inactivation of baculoviruses: the bisegmented survival curve. IOBC WPRS Bull. 19, 253–256.

Hudson, W.G., Nguyen, K.B., 1989. Effects of soil moisture, exposure time, nematode age, and nematode density on infection of *Scapteriscus vicinus* and *S. acletus* (Orthoptera: Gryllotalpidae) by a Uruguayan *Neoaplectana* sp. (Rhabditida: Steinernematidae). Environ. Entomol. 18, 719–722.

Hung, T.P., Truong, L.V., Binh, N.D., Frutos, R., Quiquampoix, H., Staunton, S., 2016. Persistence of detectable insecticidal proteins from *Bacillus thuringiensis* (Cry) and toxicity after adsorption on contrasting soils. Environ. Pollut. 208(Part B), 318–325.

Hung, T.P., Truong, L.V., Binh, N.D., Frutos, R., Quiquampoix, H., Staunton, S., 2016. Fate of insecticidal *Bacillus thuringiensis* Cry protein in soil: differences between purified toxin and biopesticide formulation. Pest Manag. Sci. 72, 2247–2253.

Ibrahim, L., Butt, T.M., Beckett, A., Clark, S.J., 1999. The germination of oil-formulated conidia of the insect pathogen, *Metarhizium anisopliae*. Mycol. Res. 103, 901–907.

Ignoffo, C.M., 1966. Effects of temperature on mortality of *Heliothis zea* larvae exposed to sublethal doses of a nuclear-polyhedrosis virus. J. Invertebr. Pathol. 8, 290–292.

Ignoffo, C.M., 1992. Environmental factors affecting persistence of entomopathogens. Fla. Entomol. 75, 516–525.

Ignoffo, C., Hostetter, D., Sikorowski, P., Sutter, G., Brooks, W., 1977. Inactivation of representative species of entomopathogenic viruses, a bacterium, fungus and protozoan by an ultraviolet light source. Environ. Entomol. 6, 411–441.

Inglis, G.D., Goettel, M.S., Johnson, D.L., 1993. Persistence of the entomopathogenic fungus, *Beauveria bassiana*, on phylloplanes of crested wheatgrass and alfalfa. Biol. Control 3, 258–270.

Inglis, G.D., Goettel, M.S., Johnson, D.L., 1995. Influence of utraviolet Light protectants on persistence of the entomopathogenic fungus, *Beauveria bassiana*. Biol. Control 5, 581–590.

Inglis, G.D., Johnson, D.L., Goettel, M.S., 1996. Effects of temperature and thermoregulation on mycosis by *Beauveria bassiana* in grasshoppers. Biol. Control 7, 131–139.

Inglis, G.D., Johnson, D.L., Cheng, K.J., Goettel, M.S., 1997a. Use of pathogen combinations to overcome the constraints of temperature on entomopathogenic hyphomycetes against grasshoppers. Biol. Control 8, 143–152.

Inglis, G.D., Johnson, D.L., Cheng, K.-J., Goettel, M.S., 1997b. Use of pathogen combinations to overcome the constraints of temperature on entomopathogenic Hyphomycetes against grasshoppers. Biol. Control 8, 143–152.

Inglis, G.D., Goettel, M.S., Butt, T.M., Strasser, H., 2001. Use of hyphomycetous fungi for managing insect pests., in: Butt, T.M., Jackson, C., Magan, N. (eds.), Fungi as Biocontrol Agents: Progress, Problems and Potential. CABI, Wallingford, pp. 23–69.

Inyang, E.N., Butt, T.M., Beckett, A., Archer, S., 1999. The effect of crucifer epicuticular waxes and leaf extracts on the germination and virulence of *Metarhizium anisopliae* conidia. Mycol. Res. 103, 419–426.

Jackson, M.A., Jaronski, S.T., 2009. Production of microsclerotia of the fungal entomopathogen *Metarhizium anisopliae* and their potential for use as a biocontrol agent for soil-inhabiting insects. Mycol. Res. 113, 842–850.

Jagdale, G.B., Grewal, P.S., 2007. Storage temperature influences desiccation and ultra violet radiation tolerance of entomopathogenic nematodes. J. Therm. Biol. 32, 20–27.

Jakubowska, A.K., Murillo, R., Carballo, A., Williams, T., Lent, J.W.M. Van, Caballero, P., Herrero, S., 2016. Iflavirus increases its infectivity and physical stability in association with baculovirus. Peer J. 4, e1687.

Jaques, R.P., 1985. Stability of insect viruses in the environment, in: Moramorasch, K., Sherman, K.E. (eds.), Viral Insecticides for Biological Control. Academic Press, New York, pp. 285–369.

Jaques, R.P., 1972. The inactivation of foliar deposits of viruses of *Trichoplusla ni* (Lepidoptera: Noctuidae) and *Pieris rapae* (Lepidoptera: Pieridae) and tests on protectant additives. Can. Entomol. 104, 1985–1994.

Jaronski, S.T., 2007. Soil ecology of the entomopathogenic Ascomycetes: a critical examination of what we (think) we know, in: Ekesi, S. Maniania, N.K. (eds.), Use of Entomopathogenic Fungi in Biological Pest management. Research Signposts, Trivandrum, pp. 91–144.

Jaronski, S.T., 2010. Ecological factors in the inundative use of fungal entomopathogens. BioControl 55, 159–185.

Jiravanichpaisal, P., Söderhäll, K., Söderhäll, I., 2006. Characterization of white spot syndrome virus replication in in vitro-cultured haematopoietic stem cells of freshwater crayfish, *Pacifastacus leniusculus*. J. Gen. Virol. 87, 847–854.

Jiravanichpaisal, P., Söderhäll, K., Söderhäll, I., 2004. Effect of water temperature on the immune response and infectivity pattern of white spot syndrome virus (WSSV) in freshwater crayfish. Fish Shellfish Immunol. 17, 265–275.

Johnson, D.W., Boucias, D.B., Barfield, C.S., Allen, G.E., 1982. A temperature-dependent developmental model for a nucleopolyhedrosis virus of the velvetbean caterpillar, *Anticarsia gemmatalis* (Lepidoptera: Noctuidae). J. Invertebr. Pathol. 40, 292–298.

Keating, S.T., Yendol, W.G., Schultz, J.C., 1988. Relationship between susceptibility of gypsy moth larvae (Lepidoptera: Lymantriidae) to a baculovirus and host plant foliage constituents. Environ. Entomol. 17, 952–958.

Kelly, D.C., 1985. Insect iridescent viruses. Curr. Top. Microbiol. Immunol. 116, 23.

Kobayashi, M., Inagaki, S., Kawase, S., 1981. Effect of high temperature on the development of nuclear polyhedrosis virus in the silkworm, *Bombyx mori*. J. Invertebr. Pathol. 38, 386–394.

Koganezawa, A., 1974. Present status of studies on the mass mortality of cultured oysters in Japan and its prevention. Proceedings of the Third US–Japan Meeting on Aquaculture.

Kolodny-Hirsch, D.M., Curtis, W., Nelson, J., 1995. Dietary effects of methoprene on *Lymantria dispar* (Lepidoptera: Lymantriidae) growth and nuclear polyhedrosis virus yield. Biol. Microb. Control 88, 825–829.

Kope, H.H., Alfaro, R.I., Lavallée, R., 2008. Effects of temperature and water activity on *Lecanicillium* spp. conidia germination and growth, and mycosis of *Pissodes strobi*. BioControl 53, 489–500.

Koppenhöfer, A.M., Fuzy, E.M., 2008. Early timing and new combinations to increase the efficacy of neonicotinoid-entomopathogenic nematode (Rhabditida: Heterorhabditidae) combinations against white grubs (Coleoptera: Scarabaeidae). Pest Manag. Sci. 64, 725–735.

Koppenhöfer, A.M., Baur, M.E., Stock, S.P., Choo, H.Y., Chinnasri, B., Kaya, H.K., 1997. Survival of entomopathogenic nematodes within host cadavers in dry soil. Appl. Soil Ecol. 6, 231–240.

Koppenhöfer, A.M., Brown, I.M., Gaugler, R., Grewal, P.S., Kaya, H.K., Klein, M.G., 2000. Synergism of entomopathogenic nematodes and imidacloprid against white grubs: greenhouse and field evaluation. Biol. Control 19, 245–251.

Koppenhöfer, A.M., Cowles, R.S., Cowles, E.A., Fuzy, E.M., Baumgartner, L., 2002. Comparison of neonicotinoid insecticides as synergists for entomopathogenic nematodes. Biol. Control 24, 90–97.

Koppenhöfer, A.M., Cowles, R.S., Cowles, E.A., Fuzy, E.M., Kaya, H.K., 2003. Effect of neonicotinoid synergists on entomopathogenic nematode fitness. Entomol. Exp. Appl. 106, 7–18.

Kung, S.-P., Gaugler, R., Kaya, H.K., 1990. Influence of soil pH and oxygen on persistence of Steinernema spp. J. Nematol. 22, 440–445.

Lacey, L.A., Kaya, H.K., 2013. Field Manual of Techniques in Invertebrate Pathology: Application and Evaluation of Pathogens for Control of Insects and Other Invertebrate Pests. Springer Science & Business Media, Philadelphia, PA.

Lasa, R., Caballero, P., Williams, T., 2007a. Juvenile hormone analogs greatly increase the production of a nucleopolyhedrovirus. Biol. Control 41, 389–396.

Lasa, R., Ruiz-Portero, C., Alcázar, M.D., Belda, J.E., Caballero, P., Williams, T., 2007b. Efficacy of optical brightener formulations of *Spodoptera exigua* multiple nucleopolyhedrovirus (SeMNPV) as a biological insecticide in greenhouses in southern Spain. Biol. Control 40, 89–96.

Laubscher, J.M., von Wechmar, M.B., 1992. Influence of aphid lethal paralysis virus and *Rhopalosiphum padi* virus on aphid biology at different temperatures. J. Invertebr. Pathol. 60, 134–140.

Le Deuff, R.., Renault, T., Gerard, A., 1996. Effects of temperature on herpes-like virus detection among hatchery-reared larval Pacific oyster *Crassostrea gigas*. Dis. Aquat. Organ. 24, 149–157.

Lee, J.H., Dillman, A.R., Hallem, E.A., 2016. Temperature-dependent changes in the host-seeking behaviors of parasitic nematodes. BMC Biol. 14, 1–17.

Leger, R.J.S., 2008. Studies on adaptations of *Metarhizium anisopliae* to life in the soil. J. Invertebr. Pathol. 98, 271–276.

Lewis, E.E., Shapiro-Ilan, D.I., 2002. Host cadavers protect entomopathogenic nematodes during freezing. J. Invertebr. Pathol. 81, 25-32.

Lewis, E.E., Gaugler, R., Harrison, R., 1992. Entomopathogenic nematode host finding: response to host contact cues by cruise and ambush foragers. Parasitology 105, 309.

Lewis, L.C., Bruck, D.J., Prasifka, J.R., Raun, E.S., 2009. *Nosema pyrausta*: its biology, history, and potential role in a landscape of transgenic insecticidal crops. Biol. Control 48, 223–231.

Liao, X., O'Brien, T.R., Fang, W., Leger, R.J.S., 2014. The plant beneficial effects of *Metarhizium* species correlate with their association with roots. Appl. Microbiol. Biotechnol. 98, 7089–7096.

Liu, Q.Z., Glazer, I., 2000. Desiccation survival of entomopathogenic nematodes of the genus *Heterorhabditis*. Phytoparasitica 28, 331–340.

Longworth, J.F., Singh, P., 1980. A nuclear polyhedrosis virus of the light brown apple moth, *Epiphyas postvittana* (Lepidoptera: Tortricidae). J. Invertebr. Pathol. 35, 84–87.

Lysyk, T.J., 2008. Effects of ambient temperature and cattle skin temperature on engorgement of *Dermacentor andersoni*. J. Med. Entomol. 45, 1000–1006.

Malone, L.A., Gatehouse, H.S., Tregidga, E.L., 2001. Effects of time, temperature, and honey on *Nosema apis* (Microsporidia: Nosematidae), a parasite of the honeybee, *Apis mellifera* (Hymenoptera: Apidae). J. Invertebr. Pathol. 77, 258–268.

Mandava, N.B., Morgan, E.D., 1985. Handbook of Natural Pesticides. CRC Press, Boca Raton, FL.

Marina, C.F., Feliciano, J.M., Valle, J., Williams, T., 2000. Effect of temperature, pH, ion concentration, and chloroform treatment on the stability of invertebrate iridescent virus 6. J. Invertebr. Pathol. 75, 91–94.

Martín-Hernández, R., Meana, A., García-Palencia, P., Marín, P., Botías, C., Garrido-Bailón, E., et al., 2009. Effect of temperature on the biotic potential of honeybee Microsporidia. Appl. Environ. Microbiol. 75, 2554–2557.

Martínez, G., Christian, P., Marina, C., Williams, T., 2003. Sensitivity of invertebrate iridescent virus 6 to organic solvents, detergents, enzymes and temperature treatment. Virus Res. 91, 249–254.

Mason, J.M., Wright, D.J., 1997. Potential for the control of *Plutella xylostella* larvae with entomopathogenic nematodes. J. Invertebr. Pathol. 70, 234–242.

McLeod, P.J., Yearian, W.C., Young, S.Y., 1977. Inactivation of Baculovirus heliothis by ultraviolet irradiation, dew, and temperature. J. Invertebr. Pathol. 30, 237–241.

de Melo-Santos, M.A.V., de Araújo, A.P., Rios, E.M.M., Regis, L., 2009. Long lasting persistence of *Bacillus thuringiensis* serovar. israelensis larvicidal activity in *Aedes aegypti* (Diptera: Culicidae) breeding places is associated to bacteria recycling. Biol. Control 49, 186–191.

Ment, D., Gindin, G., Glazer, I., Perl, S., Elad, D., Samish, M., 2010. The effect of temperature and relative humidity on the formation of *Metarhizium anisopliae* chlamydospores in tick eggs. Fungal Biol. 114, 49–56.

Ment, D., Iraki, N., Gindin, G., Rot, A., Glazer, I., Abu-Jreis, R., Samish, M., 2011. Thermal limitations of *Metarhizium anisopliae* efficacy: selection for application on warm-blooded vertebrates. BioControl 56, 81–89.

Millar, L.C., Barbercheck, M.E., 2002. Effects of tillage practices on entomopathogenic nematodes in a corn agroecosystem. Biol. Control 25, 1–11.

Miller, C.D., Rangel, D., Braga, G.U.L., Flint, S., Kwon, S.-I., Messias, C.L., et al., 2004. Enzyme activities associated with oxidative stress in Metarhizium anisopliae during germination, mycelial growth, and conidiation and in response to near-UV irradiation. Can. J. Microbiol. 50, 41–49.

Milner, R.J., Lutton, G.G., 1983. Effect of temperature on *Zoophthora randican*s (Brefeld) Batko: an introduced microbial control agent of the spotted alfafa aphid, *Therioaphis trifolii* (Monell) F. Maculata. Aust. J. Entomol. 22, 167–173.

Milner, R.J., Staples, J.A., Lutton, G.G., 1997. The effect of humidity on germination and infection of termites by the hyphomycete, *Metarhizium anisopliae*. J. Invertebr. Pathol. 69, 64–69.

Milner, R.J., Samson, P., Morton, R., 2003. Persistence of conidia of *Metarhizium anisopliae* in sugarcane fields: effect of isolate and formulation on persistence over 3.5 years. Biocontrol Sci. Technol. 13, 507–516.

Moeller, R., Horneck, G., Facius, R., Stackebrandt, E., 2005. Role of pigmentation in protecting *Bacillus* sp. endospores against environmental UV radiation. FEMS Microbiol. Ecol. 51, 231–236.

Mohamed, A.I., Young, S.Y., Yearian, W.C., 1983a. Susceptibility of *Heliothis virescens* (F.) (Lepidoptera: Noctuidae) larvae to microbial agent-chemical pesticide mixtures on cotton foliage. Environ. Entomol. 12, 1403–1405.

Mohamed, A.I., Young, S.Y., Yearian, W.C., 1983b. Effects of microbial agent-chemical pesticide mixtures on *Heliothis virescens* (F.) (Lepidoptera: Noctuidae). Environ. Entomol. 12, 478–481.

Moore, D., Bridge, P.D., Higgins, P.M., Bateman, R.P., Prior, C., 1993. Ultra-violet radiation damage to *Metarhizium flavoviride* conidia and the protection given by vegetable and mineral oils and chemical sunscreens. Ann. Appl. Biol. 122, 605–616.

Morimoto, R.I., 2008. Proteotoxic stress and inducible chaperone networks in neurodegenerative disease and aging. Genes Dev. 22, 1427–1438.

Morris, O.N., 1977. Compatibility of 27 chemical insecticides with *Bacillus thuringiensis* var *kurstaki*. Can. Entomol. 109, 855–864.

Moscardi, F., 1999. Assessment of the application of baculoviruses for control of Lepidoptera. Annu. Rev. Entomol. 44, 257–289.

Moser, J.R., Álvarez, D.A.G., Cano, F.M., Garcia, T.E., Molina, D.E.C., Clark, G.P., et al., 2012. Water temperature influences viral load and detection of White Spot Syndrome Virus (WSSV) in *Litopenaeus vannamei* and wild crustaceans. Aquaculture 326–329, 9–14.

Mostafa, A.M., Fields, P.G., Holliday, N.J., 2005. Effect of temperature and relative humidity on the cellular defense response of *Ephestia kuehniella* larvae fed *Bacillus thuringiensis*. J. Invertebr. Pathol. 90, 79–84.

Mukuka, J., Strauch, O., Ehlers, R.-U., 2010a. Variability in desiccation tolerance among different strains of the entomopathogenic nematode *Heterorhabditis bacteriophora*. Nematology 12, 711–720.

Mukuka, J., Strauch, O., Hoppe, C., Ehlers, R.-U., 2010b. Improvement of heat and desiccation tolerance in *Heterorhabditis bacteriophora* through cross-breeding of tolerant strains and successive genetic selection. BioControl 55, 511–521.

Mukuka, J., Strauch, O., Hoppe, C., Ehlers, R.-U., 2010c. Fitness of heat and desiccation tolerant hybrid strains of *Heterorhabditis bacteriophora* (Rhabditidomorpha: Heterorhabditidae). J. Pest Sci. 83, 281–287.

Mulla, M.S., Darwazeh, H.A., Zgomba, M., 1990. Effect of some environmental factors on the efficacy of *Bacillus sphaericus* 2362 and *Bacillus thuringiensis* (H-14) against mosquitoes. Bull. Soc. Vector Ecol. 15, 166–175.

Murray, K.D., Elkinton, J.S., 1989. Environmental contamination of egg masses as a major component of transgenerational transmission of gypsy-moth nuclear polyhedrosis-virus (LdMNPV). J. Invertebr. Pathol. 53, 324–334.

Nathan, S.S., Kalaivani, K., 2005. Efficacy of nucleopolyhedrovirus and azadirachtin on *Spodoptera litura* Fabricius (Lepidoptera: Noctuidae). Biol. Control 34, 93–98.

Navon, A., 2000. *Bacillus thuringiensis* insecticides in crop protection – reality and prospects. Crop Prot., XIVth International Plant Protection Congress 19, 669–676.

Neuvonen, S., Saikkonen, K., Haukioja, E., 1990. Simulated acid rain reduces the susceptibility of the European pine aawfly (*Neodiprion sertifer*) to its nuclear polyhedrosis virus. Oecologia 83, 209–212.

Nicholson, W.L., 2002. Roles of Bacillus endospores in the environment. Cell. Mol. Life Sci. 59, 410–416.

Nicholson, W.L., Munakata, N., Horneck, G., Melosh, H.J., Setlow, P., 2000. Resistance of *Bacillus* endospores to extreme terrestrial and extraterrestrial environments. Microbiol. Mol. Biol. Rev. 64, 548–572.

Nielsen, C., Hajek, A.E., Humber, R.A., Bresciani, J., Eilenberg, J., 2003. Soil as an environment for winter survival of aphid-pathogenic Entomophthorales. Biol. Control 28, 92–100.

Okuno, S., Takatsuka, J., Nakai, M., Ototake, S., Masui, A., Kunimi, Y., 2003. Viral-enhancing activity of various stilbene-derived brighteners for a *Spodoptera litura* (Lepidoptera: Noctuidae) nucleopolyhedrovirus. Biol. Control 26, 146–152.

Olofsson, E., 1988. Dispersal of the nuclear polyhedrosis virus of *Neodiprion sertifer* from soil to pine foliage with dust. Entomol. Exp. Appl. 46, 181–186.

Ouedraogo, R.M., Cusson, M., Goettel, M.S., Brodeur, J., 2003. Inhibition of fungal growth in thermoregulating locusts, *Locusta migratoria*, infected by the fungus *Metarhizium anisopliae* var acridum. J. Invertebr. Pathol. 82, 103–109.

Ouellet, P., Allard, J.-P., 2006. Vertical distribution and behaviour of shrimp *Pandalus borealis* larval stages in thermally stratified water columns: laboratory experiment and field observations. Fish. Oceanogr. 15, 373–389.

Patel, M.N., Wright, D.J., 1998. The ultrastructure of the cuticle and sheath of infective juveniles of entomopathogenic steinernematid nematodes. J. Helminthol. 72, 257.

Pell, J.K., Eilenberg, J., Hajek, A.E., Steinkraus, D.C., 2001. Biology, ecology and pest management potential of Entomophthorales, in: Butt, T.M., Jackson, C., Magan, N. (eds.), Fungi as Biocontrol Agents: Progress, Problems and Potential. CABI, Wallingford, pp. 71–153.

Peng, F., Fuxa, J.R., Richter, A.R., Johnson, S.J., 1999. Effects of heat-sensitive agents, soil type, moisture, and leaf surface on persistence of *Anticarsia gemmatalis* (Lepidoptera: Noctuidae) nucleopolyhedrovirus. Environ. Entomol. 28, 330–338.

Peters, E.C., 1993. Diseases of other invertebrate phyla: porifera, cnidaria, ctenophora, annelida, echinodermata. Pathobiol. Mar. Estuar. Org. 393–449.

Polar, P., de Muro, M.A., Kairo, M.T., Moore, D., Pegram, R., John, S.A., Roach-Benn, C., 2005. Thermal characteristics of *Metarhizium anisopliae* isolates important for the development of biological pesticides for the control of cattle ticks. Vet. Parasitol. 134, 159–167.

Preisser, E.L., Strong, D.R., 2004. Climate affects predator control of an herbivore outbreak. Am. Nat. 163, 754–762.

Qiu, L., Bedding, R., 2000. Energy metabolism and its relation to survival and infectivity of infective juveniles of *Steinernema carpocapsae* under aerobic conditions. Nematology 2, 551–559.

Quang, N.D., Hoa, P.T.P., Da, T.T., Anh, P.H., 2009. Persistence of white spot syndrome virus in shrimp ponds and surrounding areas after an outbreak. Environ. Monit. Assess. 156, 69–72.

Quesada-Moraga, E., Navas-Cortés, J.A., Maranhao, E.A.A., Ortiz-Urquiza, A., Santiago-Álvarez, C., 2007. Factors affecting the occurrence and distribution of entomopathogenic fungi in natural and cultivated soils. Mycol. Res. 111, 947–966.

Rahman, M.M., Escobedo-Bonilla, C.M., Corteel, M., Dantas-Lima, J.J., Wille, M., Sanz, V.A., et al., 2006. Effect of high water temperature (33 °C) on the clinical and virological outcome of experimental infections with white spot syndrome virus (WSSV) in specific pathogen-free (SPF) *Litopenaeus vannamei*. Aquaculture 261, 842–849.

Rahman, M.M., Corteel, M., Dantas-Lima, J.J., Wille, M., Alday-Sanz, V., Pensaert, M.B., et al., 2007a. Impact of daily fluctuations of optimum (27 °C) and high water temperature (33 °C) on *Penaeus vannamei* juveniles infected with white spot syndrome virus (WSSV). Aquaculture 269, 107–113.

Rahman, M.M., Corteel, M., Wille, M., Alday-Sanz, V., Pensaert, M.B., Sorgeloos, P., Nauwynck, H.J., 2007b. The effect of raising water temperature to 33 °C in *Penaeus vannamei* juveniles at different stages of infection with white spot syndrome virus (WSSV). Aquaculture 272, 240–245.

Rangel, D.E., Braga, G.U., Flint, S.D., Anderson, A.J., Roberts, D.W., 2004. Variations in UV-B tolerance and germination speed of *Metarhizium anisopliae* conidia produced on insects and artificial substrates. J. Invertebr. Pathol. 87, 77–83.

Rangel, D.E., Braga, G.U., Anderson, A.J., Roberts, D.W., 2005. Variability in conidial thermotolerance of *Metarhizium anisopliae* isolates from different geographic origins. J. Invertebr. Pathol. 88, 116–125.

Ravensberg, W.J., 2011. A Roadmap to the Successful Development and Commercialization of Microbial Pest Control Products for Control of Arthropods. Springer, Dordrecht.

Regis, L., Silva-Filha, M.H., Nielsen-LeRoux, C., Charles, J.-F., 2001. Bacteriological larvicides of dipteran disease vectors. Trends Parasitol. 17, 377–380.

Renault, T., Cochennec, N., Le Deuff, R.M., Chollet, B., 1994. Herpes-like virus infecting Japanese oyster (*Crassostrea gigas*) spat. Bull. Eur. Assoc. Fish Pathol. 14, 64–66.

Ribeiro, H., Pavan, O., 1994. Effect of temperature on the development of baculoviruses. J. Appl. Entomol. 118, 316–320.

Rickert Campbell, L., Gaugler, R., 1991. Role of the sheath in desiccation tolerance of two entomopathogenic nematodes. Nematologica 37, 324–332.

Riddle, D.L., 1988. The dauer larva, in: Wood, W.B. (ed.), The Nematode *Caenorhabditis elegans*. Cold Spring Harbor Laboratory, Cold Spring Harbor, NY, pp. 393–412.

Rolim, M.F., de Araujo, P.S., Panek, A.D., Paschoalin, V.M.F., Silva, J.T., 2003. Shared control of maltose and trehalose utilization in *Candida utilis*. Braz. J. Med. Biol. Res. 36, 829–837.

Ros, V.I.D., van Houte, S., Hemerik, L., van Oers, M.M., 2015. Baculovirus-induced tree-top disease: how extended is the role of egt as a gene for the extended phenotype? Mol. Ecol. 24, 249–258.

Rovesti, L., Deseö, K.V., Heinzpeter, E.W., Tagliente, F., 1988. Compatibility of pesticides with the entomopathogenic nematode *Heterorhabditis bacteriophora* Poinar (Nematoda: Heterorhabditidae). Nematologica 34, 462–476.

Rovesti, L., Deseö, K.V., 1990. Compatibility of chemical pesticides with the entomopathogenic nematodes, *Steinernema carpocapsae* Weiser and *S. feltiae* Filipjev (Nematoda: Steinernematidae). Nematologica 36, 237–245.

Rovesti, L., Deseö, K.V., 1991. Compatibility of pesticides with the entomopathogenic nematode, *Heterorhabditis heliothidis*. Nematologica 37, 113–116.

Roy, H.E., Vega, F.E., Chandler, D., Goettel, M.S., Pell, J., Wajnberg, E., 2010. The Ecology of Fungal Entomopathogens. Springer, Philadelphia, PA.

Saikkonen, K.T., Neuvonen, S., 1993. Effects of larval age and prolonged simulated acid rain on the susceptibility of European pine sawfly to virus infection. Oecologia 95, 134–139.

Salame, L., Glazer, I., 2015. Stress avoidance: vertical movement of entomopathogenic nematodes in response to soil moisture gradient. Phytoparasitica 43, 647–655.

Sanahuja, G., Banakar, R., Twyman, R.M., Capell, T., Christou, P., 2011. *Bacillus thuringiensis*: a century of research, development and commercial applications. Plant Biotechnol. J. 9, 283–300.

Santos, A.H., Tai, M.H.H., Rocha, L.F.N., Silva, H.H.G., Luz, C., 2009. Dependence of *Metarhizium anisopliae* on high humidity for ovicidal activity on *Aedes aegypti*. Biol. Control 50, 37–42.

Satheesh Kumar, S., Ananda Bharathi, R., Rajan, J.J.S., Alavandi, S.V., Poornima, M., Balasubramanian, C.P., Ponniah, A.G., 2013. Viability of white spot syndrome virus (WSSV) in sediment during sun-drying (drainable pond) and under non-drainable pond conditions indicated by infectivity to shrimp. Aquaculture 402–403, 119–126.

Sauvage, C., Pepin, J.F., Lapegue, S., Boudry, P., Renault, T., 2009. Ostreid herpes virus 1 infection in families of the Pacific oyster, *Crassostrea gigas*, during a summer mortality outbreak: differences in viral DNA detection and quantification using real-time PCR. Virus Res. 142, 181–187.

Saxena, D., Ben-Dov, E., Manasherob, R., Barak, Z., Boussiba, S., Zaritsky, A., 2002. A UV tolerant mutant of B*acillus thuringiensis* subsp. *kurstaki* producing melanin. Curr. Microbiol. 44, 25–30.

Schlesinger, M.J., 1990. Heat shock proteins. J. Biol. Chem. 265, 12111–12114.

Selvan, S., Gaugler, R., Lewis, E.E., 1993. Biochemical energy reserves of entomopathogenic nematodes. J. Parasitol. 79, 167.

Setlow, P., 2001. Resistance of spores of Bacillus species to ultraviolet light. Environ. Mol. Mutagen. 38, 97–104.

Shah, F.A., Ansari, M.A., Watkins, J., Phelps, Z., Cross, J., Butt, T.M., 2009. Influence of commercial fungicides on the germination, growth and virulence of four species of entomopathogenic fungi. Biocontrol Sci. Technol. 19, 743–753.

Shapiro, D.I., Tylka, G.L., Lewis, L.C., 1996. Effects of fertilizers on virulence of *Steinernema carpocapsae*. Applied Soil Ecol. 3, 27–34.

Shapiro, D.I., McCoy, C.W., Fares, A., Obreza, T., Dou, H., 2000. Effects of soil type on virulence and persistence of entomopathogenic nematodes in relation to control of *Diaprepes abbreviatus*. Environ. Entomol. 29, 1083–1087.

Shapiro, M., 1992. Use of optical brighteners as radiation protectants for gypsy moth (Lepidoptera: Lymantriidae) nuclear polyhedrosis virus. J. Econ. Entomol. 85, 1682–1686.

Shapiro, M., Dougherty, E.M., 1994. Enhancement in activity of homologous and heterologous viruses against the gypsy moth (Lepidoptera: Lymantriidae) by an optical brightener. J. Econ. Entomol. 87, 361–365.

Shapiro, M., Robertson, J.L., 1992. Enhancement of gypsy moth (Lepidoptera: Lymantriidae) baculovirus activity by optical brighteners. J. Econ. Entomol. 85, 1120–1124.

Shapiro, M., Farrar, R.R. Jr., Domek, J., Javaid, I., 2002. Effects of virus concentration and ultraviolet irradiation on the activity of corn earworm and beet armyworm (Lepidoptera: Noctuidae) nucleopolyhedroviruses. J. Econ. Entomol. 95, 243–249.

Shapiro, M., Robertson, J.L., Webb, R.E., 1994. Effect of neem seed extract upon the gypsy moth (Lepidoptera: Lymantriidae) and its nuclear polyhedrosis virus. J. Econ. Entomol. 87, 356–360.

Shapiro-Ilan, D.I., Reilly, C.C., Hotchkiss, M.W., Wood, B.W., 2002. The potential for enhanced fungicide resistance in *Beauveria bassiana* through strain discovery and artificial selection. J. Invertebr. Pathol. 81, 86–93.

Shapiro-Ilan, D.I., Stuart, R.J., McCoy, C.W., 2006. A comparison of entomopathogenic nematode longevity in soil under laboratory conditions. J. Nematol. 38, 119–129.

Shapiro-Ilan, D.I., Reilly, C.C., Hotchkiss, M.W., 2011. Comparative impact of artificial selection for fungicide resistance on *Beauveria bassiana* and *Metarhizium brunneum*. Environ. Entomol. 40, 59–65.

Shapiro-Ilan, D.I., Bruck, D.J., Lacey, LA., 2012. Principles of epizootiology and microbial control, in: Vega, F.E., Kaya, H.K. (eds.), Insect Pathology, 2nd edn. Elsevier, Amsterdam, pp. 29–72.

Shapiro-Ilan, D.I., Brown, I., Lewis, E.E., 2014. Freezing and desiccation tolerance in entomopathogenic nematodes: diversity and correlation of traits. J. Nematol. 46, 27–34.

Shapiro-Ilan, D.I., Hazir, S., Lete, L., 2015. Viability and virulence of entomopathogenic nematodes exposed to ultraviolet radiation. J. Nematol. 47, 184–189.

Shikano, I., Cory, J.S., 2015. Impact of environmental variation on host performance differs with pathogen identity: implications for host-pathogen interactions in a changing climate. Sci. Rep. 5, 15351.

Singaravelan, N., Grishkan, I., Beharav, A., Wakamatsu, K., Ito, S., Nevo, E., 2008. Adaptive melanin response of the soil fungus *Aspergillus niger* to UV radiation stress at "Evolution Canyon," Mount Carmel, Israel. PLoS ONE 3, e2993.

Solomon, A., Paperna, I., Glazer, I., 1999. Desiccation survival of the entomopathogenic nematode *Steinernema feltiae*: induction of anhydrobiosis. Nematology 1, 61–68.

Solomon, A., Salomon, R., Paperna, I., Glazer, I., 2000. Desiccation stress of entomopathogenic nematodes induces the accumulation of a novel heat-stable protein. Parasitology 121, 409–416.

Solter, L.F., Becnel, J.J., 2007. Entomopathogenic microsporidia, in: Lacey, L.A., Kaya, H.K. (eds.), Field Manual of Techniques in Invertebrate Pathology. Springer, Amsterdam, pp. 199–221.

Sporleder, M., Zegarra, O., Cauti, E.M.R., Kroschel, J., 2008. Effects of temperature on the activity and kinetics of the granulovirus infecting the potato tuber moth *Phthorimaea operculella* Zeller (Lepidoptera: Gelechiidae). Biol. Control 44, 286–295.

Stairs, G.R., 1978. Effects of a wide range of temperatures on the development of *Galleria mellonella* and its specific baculovirus. Environ. Entomol. 7, 297–299.

Sterk, G., Hassan, S.A., Baillod, M., Bakker, F., Bigler, F., Blümel, S., et al., 1999. Results of the seventh joint pesticide testing programme carried out by the IOBC/WPRS-Working Group "Pesticides and Beneficial Organisms." BioControl 44, 99–117.

Stevenson, P.C., D'Cunha, R.F., Grzywacz, D., 2010. Inactivation of baculovirus by isoflavonoids on chickpea (*Cicer arietinum*) leaf surfaces reduces the efficacy of nucleopolyhedrovirus against *Helicoverpa armigera*. J. Chem. Ecol. 36, 227–235.

Summers, M.D., Volkman, L.E., 1976. Comparison of biophysical and morphological properties of occluded and extracellular non-occluded baculovirus from in vivo and in vitro host systems. J. Virol. 17, 962–972.

Sun, Y., Bai, G., Wang, Y.-X., Zhang, Y., Pan, J., Cheng, W., et al., 2016. The impact of Cu, Zn and Cr salts on the relationship between insect and plant parasitic nematodes: a reduction in biocontrol efficacy. Appl. Soil Ecol. 107, 108–115.

Tetreau, G., Alessi, M., Veyrenc, S., Périgon, S., David, J.-P., Reynaud, S., Després, L., 2012. Fate of *Bacillus thuringiensis* subsp. israelensis in the field: evidence for spore recycling and differential persistence of toxins in leaf litter. Appl. Environ. Microbiol. 78, 8362–8367.

Thompson, C.G., 1978. Nuclear polyhedrosis epizootiology, in: Brookes, M.H., Stark, R.W., Campbell, R.W. (eds.), The Douglas-Fir Tussock Moth: A Synthesis. USDA Technical Bulletin 1585, Washington, DC, pp. 124–138.

Thompson, C.G., Scott, D.W., 1981. Long-term persistence of the nuclear polyhedrosis virus of the Douglas-Fir tussock moth, *Orgyia pseudotsugata* (Lepidoptera: Lymantriidae), in forest soil. J. Invertebr. Pathol. 10, 254–255.

Thurston, G.S., Ni, Y., Kaya, H.K., 1994. Influence of salinity on survival and infectivity of entomopathogenic: nematodes. J. Nematol. 26, 345–351.

Tibbett, M., Cairney, J.W., 2007. The cooler side of mycorrhizas: their occurrence and functioning at low temperatures. Botany 85, 51–62.

Tibbett, M., Sanders, F., Cairney, J., 2002. Low-temperature-induced changes in trehalose, mannitol and arabitol associated with enhanced tolerance to freezing in ectomycorrhizal basidiomycetes (Hebeloma spp.). Mycorrhiza 12, 249–255.

Timper, P., Kaya, H.K., 1989. Role of the second-stage cuticle of entomogenous nematodes in preventing infection by nematophagous fungi. J. Invertebr. Pathol. 54, 314–321.

Tlecuitl-Beristain, S., Viniegra-González, G., Díaz-Godínez, G., Loera, O., 2009. Medium selection and effect of higher oxygen concentration pulses on *Metarhizium anisopliae* var. lepidiotum conidial production and quality. Mycopathologia 169, 387–394.

Triggs, A., Knell, R.J., 2012. Interactions between environmental variables determine immunity in the Indian meal moth *Plodia interpunctella*. J. Anim. Ecol. 81, 386–394.

Undeen, A.H., Meer, R.K.V., 1990. The effect of ultraviolet radiation on the germination of *Nosema algerae* Vávra and Undeen (Microsporida: Nosematidae) spores. J. Protozool. 37, 194–199.

Valicente, F.H., Costa, E.F., 1995. Controle da lagarta do cartucho, *Spodoptera frugiperd*a (JE Smith), com o *Baculovirus spodoptera*, aplicado via água de irrigação. An. Soc. Entomológica Bras. 24, 61–67.

van Beek, N., Hughes, P.R., Wood, H.A., 2000. Effects of incubation temperature on the dose-survival time relationship of *Trichoplusia ni* larvae infected with *Autographa californica* nucleopolyhedrovirus. J. Invertebr. Pathol. 76, 185–190.

Vasconcelos, S.D., Cory, J.S., Wilson, K.R., Sait, S.M., Hails, R.S., 1996. Modified behavior in baculovirus-infected lepidopteran larvae and its impact on the spatial distribution of inoculum. Biol. Control 7, 299–306.

Vidal, O.M., Granja, C.B., Aranguren, F., Brock, J.A., Salazar, M., 2001. A profound effect of hyperthermia on survival of *Litopenaeus vannamei* juveniles infected with white spot syndrome virus. J. World Aquac. Soc. 32, 364–372.

Vilas-Bôas, L.A., Vilas-Bôas, G.F., Saridakis, H.O., Lemos, M.V. Lereclus, D., Arantes, O.M., 2000. Survival and conjugation of *Bacillus thuringiensis* in a soil microcosm. FEMS Microbiol. Ecol. 31, 255–259.

Villamizar, L., Barrera, G., Cotes, A.M., Martínez, F., 2010. Eudragit S100 microparticles containing *Spodoptera frugiperda* nucleopolyehedrovirus: physicochemical characterization, photostability and in vitro virus release. J. Microencapsul. 27, 314–324.

Wang, X., Grewal, P.S., 2002. Rapid genetic deterioration of environmental tolerance and reproductive potential of an entomopathogenic nematode during laboratory maintenance. Biol. Control 23, 71–78.

Washburn, J.O., Kirkpatrick, B.A., Haas-Stapleton, E., Volkman, L.E., 1998. Evidence that the stilbene-derived optical brightener M2R enhances *Autographa californica* M nucleopolyhedrovirus infection of *Trichoplusia ni* and *Heliothis virescens* by preventing sloughing of infected midgut epithelial cells. Biol. Control 11, 58–69.

Weinstein, R.N., Montiel, P.O., Johnstone, K., 2000. Influence of growth temperature on lipid and soluble carbohydrate synthesis by fungi isolated from fellfield soil in the maritime Antarctic. Mycologia 92, 222–229.

West, A.W., Burges, H.D., Dixon, T.J., Wyborn, C.H., 1985. Survival of *Bacillus thuringiensis* and *Bacillus cereus* spore inocula in soil: effects of pH, moisture, nutrient availability and indigenous microorganisms. Soil Biol. Biochem. 17, 657–665.

Wharton, D.A., Block, W., 1993. Freezing tolerance in some Antarctic nematodes. Funct. Ecol. 7, 578.

Wharton, D.A., Young, S.R., Barrett, J., 1984. Cold tolerance in nematodes. J. Comp. Physiol. B 154, 73–77.

Witt, D.J., Stairs, G.R., 1976. Effects of different temperatures on *Tipula iridescent* virus infection in *Galleria mellonella* larvae. J. Invertebr. Pathol. 28, 151–152.

Yaari, M., Doron-Faigenboim, A., Koltai, H., Salame, L., Glazer, I., 2016. Transcriptome analysis of stress tolerance in entomopathogenic nematodes of the genus *Steinernema*. Int. J. Parasitol. 46, 83–95.

You, X.-X., Su, Y.-Q., Mao, Y., Liu, M., Wang, J., Zhang, M., Wu, C., 2010. Effect of high water temperature on mortality, immune response and viral replication of WSSV-infected *Marsupenaeus japonicus* juveniles and adults. Aquaculture 305, 133–137.

Young, S.Y., Yearian, W.C., 1986. Movement of a nuclear polyhedrosis virus from soil to soybean and transmission in *Anticarsia gemmatalis* (Hübner)(Lepidoptera: Noctuidae) populations on soybean. Environ. Entomol. 15, 573–580.

Young, S.Y., Yearian, W.C., 1989. Persistence and movement of nuclear polyhedrosis virus on soybean plants after death of infected *Anticarsia gemmatalis* (Lepidoptera: Noctuidae). Environ. Entomol. 18, 811–815.

Young, S.Y., Yearian, W.C., Kim, K.S., 1977. Effect of dew from cotton and soybean foliage on activity of *Heliothis* nuclear polyhedrosis virus. J. Invertebr. Pathol. 29, 105–111.

Zervos, S., Johnson, S.C., Webster, J.M., 1991. Effect of temperature and inoculum size on reproduction and development of *Heterorhabditis heliothidis* and *Steinernema glaseri* (Nematoda: Rhabditoidea) in *Galleria mellonella*. Can J Zool 69, 1261–1264.

Zhan, W.-B., Wang, Y.-H., Fryer, J.L., Yu, K.-K., Fukuda, H., Meng, Q.-X., 1998. White spot syndrome virus infection of cultured shrimp in China. J. Aquat. Anim. Health 10, 405–410.

Zhou, Q., Li, H., Qi, Y.P., Yang, F., 2008. Lipid of white-spot syndrome virus originating from host-cell nuclei. J. Gen. Virol. 89, 2909–2914.

Zimmermann, G., 1982. Effect of high temperatures and artificial sunlight on the viability of conidia of *Metarhizium anisopliae*. J. Invertebr. Pathol. 40, 36–40.

6

The Biotic Environment

Jenny S. Cory and Pauline S. Deschodt

Department of Biological Sciences, Simon Fraser University, Burnaby, BC, Canada

6.1   Introduction

Invertebrate pathology has traditionally focused on single host–single pathogen inter-actions on a limited number of pathogen species, with a limited number of groups with high virulence and insect pest control potential. However, from an ecological perspective, it is clear that both insects and their pathogens are rarely found in isolation in natural situations. Multiple pathogens are often isolated from individual species, and the use of sensitive molecular tools and next-generation sequencing has made it clear that communities of microorganisms with varying impacts on their hosts can be detected in single organisms. Recently, interest in microorganisms has extended to the host microbiome and how it interacts with host processes, including defense against parasites and pathogens. In addition, host–pathogen interactions can also be modulated by trophic interactions, and in particular, the host plant that the insect feeds on can alter the outcome. Thus, biotic interactions are likely to play a major role in how invertebrate pathogens impact their host populations. However, there is still a lack of studies on more complex interactions within pathogen – or, more broadly, natural enemy – communities at all spatial scales. This information is important to the development and use of microbial insecticides and to the prevention of disease among managed invertebrates, in addition to our knowledge about the roles played by pathogens in natural communities. Entomopathogens are increasingly being used together in pest control programs, and yet there are no clear guidelines as to whether this is likely to enhance or impede biological control and successful pest management. Understanding the mechanisms underlying pathogen–natural enemy interactions would allow us to develop a framework for mixed microbial applications. In this chapter, we discuss three areas of biotic interactions involving invertebrate pathogens: (i) tritrophic interactions among invertebrates, pathogens, and plants; (ii) competition within the natural enemy complex; and (iii) microbe-mediated defense by the host microbiome.

Ecology of Invertebrate Diseases, First Edition. Edited by Ann E. Hajek and David I. Shapiro-Ilan.
© 2018 John Wiley & Sons Ltd. Published 2018 by John Wiley & Sons Ltd.

6.2 Tritrophic Interactions

Many insect pathogens need to be ingested to initiate infection and, for herbivorous hosts, both host plant species and food quality can alter the likelihood of infection and many other interaction traits. Most research on tritrophic interactions involving entomopathogens has focused on obligate pathogens, such as baculoviruses, where the link to host plant identity is likely to be strong. Many studies have shown that host plant secondary chemicals can interact with baculoviruses in the midgut, reducing their infectivity and altering traits such as speed of kill and pathogen productivity (reviews in Cory and Hoover, 2006; Cory, 2010).

There is also good evidence that other entomopathogens, such as the bacterium *Bacillus thuringiensis* (Bt), can be affected by host plants (Cory and Hoover, 2006; Jafary et al., 2016), although most research on Bt and host plant effects tends to be in relation to the cost and development of resistance.

Entomopathogenic fungi primarily infect through the invertebrate cuticle, and interaction with secondary plant chemicals in the gut is less likely; however, there are other routes by which plants can have a direct impact on fungi, such as via plant volatiles or plant surface chemistry (review in Cory and Ericsson, 2010). The possible impact of plant volatiles is one of the most intriguing areas under investigation, particularly given their role in the attraction of insect parasitoids; however, their impact on entomopathogenic fungi has been equivocal, with both positive and negative effects on conidiation and germination being recorded and no effects being found on infection rate (Cory and Ericsson, 2010). One issue might be the difference between the impact of green leaf volatiles and herbivore-induced volatiles. However, a recent study by Lin et al. (2016) suggests that the pathogenicity of the fungus *Lecanicillium lecanii* for the aphid *Lipaphis erysimi* was enhanced after exposure to herbivore-induced volatiles from *Arabidopsis thaliana*, and that this response was related to insect density.

Entomopathogenic nematodes have received less attention in terms of tritrophic effects. Nematodes are less intimately associated with the host plants of their target insects, as they are mobile. They also spend more time in the soil, where they attack insects that feed on the roots (which have fewer phytochemicals). They enter their insect hosts through various orifices and are not reliant on being ingested. Thus, direct effects of plant chemicals are less likely, although the nematodes (or their symbiotic bacteria; see Chapter 11) can still be affected by insect host quality, host plant quality, and immunocompetence (Barbercheck et al., 1995; Shapiro-Ilan et al., 2008; Gassmann et al., 2010; Hazir et al., 2016), and plant chemistry also influences whether nematodes are attracted to plants (e.g., in response to "call-for-help" signals: Rasmann et al., 2005; Hiltpold et al., 2010).

Host plants can also impact the insect–pathogen interaction in other indirect ways, including via morphological and architectural differences, which can affect the persistence of the pathogen by shading it from ultraviolet (UV) degradation, changing the microclimate, or modifying the behavior of the insect, thereby altering the likelihood of pathogen encounter and thus transmission rate (Cory and Hoover, 2006).

Most research on tritrophic interactions has focused on holometabolous insects and pathogens that need to be ingested and only infect the nonreproducing larval stages. This is the stage where most, if not all, feeding takes place for many herbivorous insects; however, there is less information on hemimetabolous insects and the impact of changes in plant quality and infection on feeding adult stages. In experimental studies, the timing of tritrophic – or, more broadly, nutritional – studies can make significant differences to the outcome (Cory and Hoover, 2006). For example, the direct (likely negative) impact of particular phytochemicals, such as phenolics, on a pathogen in the larval midgut is likely to be different if an insect has been continuously fed on a plant containing high levels of phenolics prior to challenging them with a pathogen. Feeding on foliage with high levels of secondary chemicals will reduce the insect growth rate for many species, which can change insect conditions, result in smaller insects, or decrease host health – conditions that often increase susceptibility to pathogens. The likelihood of infection is condition-dependent, and teasing apart the impact of nutrition at different stages in the infection process will aid our understanding of the factors that modulate the virulence of pathogens. When cabbage looper (*Trichoplusia ni*) larvae were reared on two host plants, broccoli and cucumber, broccoli was clearly superior in terms of growth rate and survival. Larvae reared on cucumber were also more susceptible to Trichoplusia ni single nucleopolyhedrovirus (TnSNPV) than those reared on broccoli by almost fourfold, and larval weight had no impact on the results (Shikano et al., 2010). In this case, the insects were only reared on the host plants prior to virus challenge; they were then infected on artificial diet and reared through artificial diet post-infection, to remove the direct effect of phytochemicals. More recent work has started to investigate the more general impact of nutrition on invertebrate susceptibility to disease in a range of study systems. Several studies have shown that parents who have experienced nutritional stress earlier in life produce offspring with increased disease resistance (e.g., Mitchell and Read, 2005; Boots and Roberts, 2012), although this is not always the case (Valtonen et al., 2012; Shikano et al., 2016). For example, *T. ni* larvae fed a dilute (less nutritious) diet (with increased cellulose) produced offspring that were more resistant to both *B. thuringiensis* and TnSNPV (Shikano et al., 2015), whereas the same insect culture fed a low protein diet showed no change in resistance to Autographa californica multinucleopolyhedrovirus (AcMNPV) in the offspring (Shikano et al., 2016). This suggests that the type of nutritional manipulation is important.

One interesting question that has been posed is whether the diet that an insect feeds on after pathogen challenge could alter the outcome of infection. A novel study on the lepidopteran *Spodoptera littoralis* demonstrated that insects that were fed a diet that was higher in protein after challenge with a baculovirus (Spodoptera littoralis multinucleopolyhedrovirus, SlMNPV) had higher survival (ranging from 20 to 75% survival), suggesting a protein cost to resisting baculovirus challenge. More intriguingly, when the larvae were left to select their own diet after virus challenge, surviving virus-challenged insects – but not those that eventually died – selected a higher ratio of protein to carbohydrate (Lee et al., 2006), a surprising result for such virulent pathogens as baculoviruses. There appeared to be a small cost to feeding on a high-protein diet for uninfected insects, so these data support the hypothesis that insects can self-medicate in response to pathogen challenge (Abbott, 2014). A study on *Spodoptera exempta* larvae, which were injected with the opportunistic bacterial pathogen *Bacillus subtilis*, showed a similar pattern (Povey et al., 2008). Alternative explanations to self-medication are that the insect is showing compensatory feeding to make up the nutrients used in fighting off infection or that the pathogen is manipulating feeding to enhance its own fitness. A recent study with *T. ni* and AcMNPV and TnSNPV compared these hypotheses and found that increased protein consumption only occurred in response to challenge by one of the viruses (AcMNPV) but did not result in increased survival or increased pathogen fitness. The authors concluded that in the AcMNPV–*T. ni* system at least, changes in feeding behavior are more likely to be a result of compensatory feeding, and therefore responses to changing ratios of macronutrients are likely to vary with species and conditions (Shikano and Cory, 2016).

More traditionally, self-medication involves the use of plant secondary chemicals to reduce infection. For example, the coniferous tree resin often incorporated into the nests of wood ants (*Formica paralugubris*) can reduce infection by the bacterium *Pseudomonas fluorescens* and the fungus *Metarhizium anisopliae*, creating a form of social immunity (Chapuisat et al., 2007). This behavior is often prophylactic rather than in response to infection (therapeutic). However, honeybees (*Apis mellifera*) have been shown to increase their rate of resin collecting in response to challenge by the fungal agent that causes chalkbrood (*Ascophaera apis*), a result more in agreement with a self-medication response, although the costs of this behavior to unchallenged individuals have not been demonstrated (Simone-Finstrom and Spivak, 2012). Ants (*Formica fusca*) have also been shown to enhance their survival by selecting diets with enhanced reactive oxygen species (ROS; in this case, added hydrogen peroxide), which aids against challenge by the fungus *Beauveria bassiana* (Bos et al., 2015). Extending the concept of self-medication further, studies on the monarch butterfly (*Danaus plexippus*) suggest that it can protect its kin from the protozoan parasite *Ophryocystis elektroscirrha*. Infected caterpillars showed no evidence of self-medication, but infected females preferentially oviposited on species of milkweed (*Asclepias* spp.) that contained higher levels of toxic cardenolides (Lefèvre et al., 2010). Cardenolides reduce the fitness costs of *O. elektroscirrha* infection, so this behavior has been termed transgenerational self-medication.

6.2.1 Further Complexity

Plants, through either their secondary chemicals or macronutrients, can clearly alter numerous aspects of the insect–pathogen relationship both within and between generations. Whether the observed effects scale up to the population level or modulate the relationship over time is less clear. Constitutive phytochemicals, and particularly herbivore-induced chemicals, can reduce the impacts of pathogens such as baculoviruses, but can they alter population dynamics? It is likely that at higher pathogen concentrations, plant-based effects will be swamped, but at lower concentrations plant variation and quality could have an impact on disease prevalence. However, recent mathematical modeling studies on gypsy moth (*Lymantria dispar*) have suggested that induced host plant effects (due to host feeding) and forest composition could have an impact on insect outbreaks (Elderd et al., 2013). A further level of complexity with the facultative hypocrealean fungi, such as *B. bassiana* and *Metarhizium* spp., is that they can have multiple roles, acting as endophytic symbionts in plants, where they can both suppress plant diseases and improve plant growth (Ownley et al., 2010; Moonjely et al., 2016). How do these different functions interact in different nutritional contexts? From an evolutionary viewpoint, plants could modulate the insect–pathogen interaction over time (Biere and Tack, 2013). The toxicity of Bt varies with host plant, and this affects the evolution of resistance, the costs of resistance, as well as its dominance (Janmaat and Myers, 2007; Raymond et al., 2007). With obligate pathogens that need to be ingested, host plants could potentially exert a strong selective force; in the pine beauty moth (*Panolis flammea*), different nucleopolyhedrovirus (NPV) genotypes behaved differently on different host plant species (Hodgson et al., 2004). Thus, there is the potential that different genotypes could be selected when the insects feed on host plants with different chemistries. Evidence to support this hypothesis has been found in the western tent caterpillar (*Malacosoma californicum pluviale*), where viral isolates collected from larvae feeding on different host plant species had higher virulence (speed of kill) on the host plant from which they were isolated (Cory and Myers, 2004). There are still many questions remaining about the roles of plants and nutrition in insect–pathogen interactions. It is hoped that future studies will provide an answer as to whether host plant identity can influence the frequency or intensity of epizootics, and how nutrition might influence the outcomes when multiple natural enemies interact at an individual, population, and community level.

6.3 Pathogen–Natural Enemy Interactions

Invertebrates are attacked by a wide range of natural enemies, including numerous species of predators, in addition to many groups of pathogens and parasitoids. These can interact with a host at the individual level (within-host dynamics) and at the population level through intraguild interactions (Rosenheim et al., 1995) and potentially via other members of the community in which these species are embedded. Different natural enemies can compete with one another either directly (interference) or indirectly via competition for resources or by coopting the immune system (apparent competition) (Mideo, 2009) (Fig. 6.1.). Alternatively, they can act independently or even enhance the susceptibility of one natural enemy to another via trait-mediated effects, whereby a host's behavior, physiology, or morphology is changed, making it more vulnerable to other natural enemies (Sih et al., 1998). In addition to invertebrate pathogens interacting with other entomopathogens, parasitoids, and invertebrate predators, there is also an array of biotic antagonists, primarily pathogens, that attack invertebrate pathogens directly (particularly fungi and nematodes); these relationships will not be explored within this chapter, but reviews can be found elsewhere (e.g., Kaya, 2002).

6.3.1 Entomopathogen–Entomopathogen Interactions

The study of mixed entomopathogen infections extends back many decades, particularly for baculoviruses (Harper et al., 1986). The main focus of the earlier mixed-infection studies was on the overall impact on host mortality and mortality rate of challenging an insect with two pathogens, in order to determine whether synergism, interference, or additive effects occurred. All of these outcomes have been demonstrated (Cory et al., 1997; Mantzoukas et al., 2013). However, co-exposing a host to two pathogens does not necessarily mean that they will co-infect (both infect and potentially develop and reproduce within the host, at least to some measurable extent), although the second pathogen could still have an impact on the first, even if one dominates at the end of the infection process. In terms of pathogen fitness, it is the production of transmission stages which

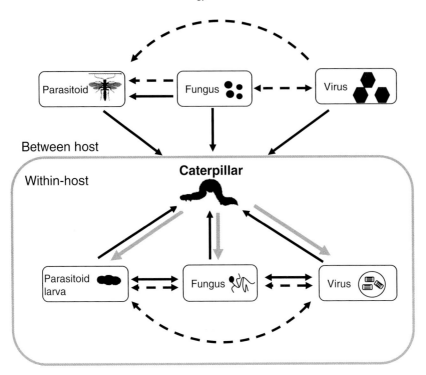

Fig. 6.1 Natural enemies have a negative interaction with their host as they parasitize or infect individuals and consume host resources (solid black arrow). Interactions can take place at two levels: within the host and at the population level. Within the host, pathogens and parasitoids will have an indirect impact on each other by competing for host resources (exploitation competition). Alternatively, they could interact directly, such as by the production of toxins (interference competition) (dashed line). In addition, the host immune system will have a direct negative effect on a pathogen or parasitoid (gray arrow), and this response could also indirectly influence co-infecting organisms, either positively (by making the host more vulnerable) or negatively (by fighting off the second infection with an enhanced immune response; immune-mediated competition). At the host population level, a generalist pathogen, such as some fungi, can directly infect parasitoids or predators (intraguild predation; solid black arrow) or remove potential susceptible hosts from the population. All natural enemies compete for uninfected hosts in the population (dashed line), and population persistence may depend on the nature of pathogen transmission (whether or not it is density-dependent) and the critical population threshold.

is important, but this has rarely been measured. Laboratory studies have shown that the timing and order of infection are important for the outcome of mixed infections, as are the relative concentrations of different pathogens. For example, co-infection of a "fast" Helicoverpa armigera single nucleopolyhedrovirus (HearSNPV) (average speed of kill 5–6 days) with a slow strain of Helicoverpa armigera granulovirus (HearGV) (average speed of kill 16 days) showed that the two viruses competed and that HearGV appeared to inhibit HearSNPV replication, even when the challenge with HearGV came 36 hours after HearSNPV infection (Hackett et al., 2000). Migratory locusts (*Locusta migratoria*) infected with a microsporidian (*Paranosema locustae*) followed by *Metarhizium acridum* died more rapidly than expected, but only when the second pathogen was added 9 days after the initial challenge, which corresponded with microsporidian spore maturation (Tokarev et al., 2011). Similarly, interactions among entomopathogenic nematodes and other entomopathogens may be antagonistic, additive, or synergistic, depending on the pathogen and host species involved, as well as the relative timing of infection (Koppenhöfer and Grewal, 2005).

Although the majority of mixed-infection studies have focused on host mortality or speed of kill, some (but not many) have measured impacts on pathogen fitness, in particular the production of transmission stages. Laboratory investigations of co-infection of the lepidopteran *Galleria mellonella* with two nematode species, *Steinernema carpocapsae* and *S. glaseri*, demonstrated that co-infections were common and that *S. glaseri* was the superior competitor (in terms of production of progeny) due to its faster development and more flexible association with its bacterial symbionts (Koppenhöfer et al., 1995). Another study with the same nematode species and host also showed that *S. carpocapsae* infection was increased by the presence of *S. glaseri* (Wang and Ishibashi, 1999). Co-infection of two entomophthoralean fungi, *Pandora blunckii* and *Zoophthora radicans*, in the diamondback moth (*Plutella xylostella*) showed that co-infection (resulting in reproduction by both fungi) was relatively uncommon (around 20%) and tended to occur only when the species were introduced simultaneously. The last species to be introduced was most likely to benefit, and in some scenarios overall mortality was lower in mixed compared to single infections (Sandoval-Aguilar et al., 2015). Thus, co-infecting species can clearly affect one another's fitness. However, questions such as how frequently co-challenge results in a mixed (co-)infection (at the individual level, i.e., with reproduction by both pathogens) and whether the first pathogen to establish always wins are unlikely to have simple answers. The outcome of experiments with multiple pathogen species appears to be context- and species-dependent, and a more systematic approach, which addresses underlying mechanisms, is needed in order to begin to predict the outcome of mixed infections.

When a single host is challenged with two pathogens, the pathogens are clearly competing for the same resource, so it is likely that one or both species will have reduced fitness compared to when they are infecting singly, unless they produce synergistic effects on host mortality. In trying to decide whether it will be beneficial to combine two or more entomopathogens for pest control, it will be useful to have a framework to indicate the direction that the outcome of the mixed infection is likely to take. One approach is to try and understand the competitive mechanisms behind the interactions. For example, Staves and Knell (2010) studied the competitive interactions among different strains of a fungal pathogen, *M. anisopliae* (intraspecific competition), and between *M. anisopliae* and the entomopathogenic nematode *Steinernema feltiae* (interspecific competition), using the wax moth (*G. mellonella*) as their target host. They found that competition among the fungal strains depended on strain virulence (speed of kill in this case), with the more virulent strains being better competitors, although this also depended on dose. However, in competition with the nematode, less virulent strains of the fungus were more successful. They suggested that this difference is due to the nature of the competition, with the fungi engaging in indirect competition via host exploitation and the fungus and nematode engaging in direct competition involving fungal toxin production (interference competition). In another study using two previously undescribed species of *Steinernema* in *G. mellonella*, Bashey et al. (2013) characterized the different strains of *Xenorhabdus* spp. bacteria carried by the nematodes and examined which produced bacteriocins that were capable of inhibiting the growth of another strain of the same bacterium. They found that where there was no interference between *Xenorhabdus* strains, the speed with which the host was killed was a predictor of competitive success. However, where there was interference via bacteriocins, this overcame any superiority in terms of killing speed. Thus, in some host–pathogen systems at least, it may be possible to predict the outcome of mixed infections, given sufficient information on the nature of the competition taking place.

Studies on mixed pathogen infections are moving away from a focus on host (pest) mortality and toward a more fundamental understanding of the processes and consequences of mixed infections, addressing these issues in a broad range of invertebrate hosts. For example, it has long been known that honeybees are infected by a wide range of pathogens, and studies are now investigating how they interact and what the consequences are for bee biology (e.g., Doublet et al., 2015; Klinger et al., 2015). A key focus is the evolution of virulence and why pathogens evolve to have a particular level of virulence (here defined as harm to the host). A full discussion of the evolution of virulence is beyond the scope of this review, but one of the dominant models is the trade-off model (a trade-off between virulence and transmission) (Alizon et al., 2008). Mixed infections are predicted to increase virulence, depending on the nature of the competitive interaction (Alizon et al., 2013). For example, an experimental evolution study in which two strains of Bt – one pathogenic to the host (the diamondback moth, *P. xylostella*) and the other not – were passaged for several generations in hosts showed that virulence in the mixed infection (compared to single passage of the pathogenic strain) decreased over time. The reduction in virulence was accompanied by an increased capacity to suppress the growth of competitors (antagonism) (Garbutt et al., 2011). This outcome impacts transmission to new hosts, and potentially host–pathogen dynamics. These studies have relevance for the use of microbial control agents in situations where secondary cycling of the pathogen occurs (Chapter 1) and are important for ongoing pest suppression within and between generations. Both the yield (production of transmission stages) and the virulence of the pathogen that results from a mixed infection affect the outcome. However, empirical studies examining whether or how mixed infections alter virulence and transmission are rare.

Generally, biodiversity is expected to improve natural pest control (Cardinale et al., 2006). There is support for this in studies with predators (e.g., Snyder et al., 2006). A lab experiment examining host mortality from an entomopathogen diversity perspective found that increasing pathogen species richness (combinations of *B. bassiana* and the nematodes *Heterorhabditis megidis*, *Steinernema carpocapsae*, and *S. feltiae*) increased mortality in the Colorado potato beetle (*Leptinotarsa decemlineata*) and wax moth (*G. mellonella*), primarily due to a fungus–nematode synergism (Jabbour et al., 2011). At a slightly larger scale, a greenhouse experiment demonstrated that a combination of the nematode *Steinernema kraussei* and the fungus *M. anisopliae* resulted in a synergistic effect in controlling overwintering larvae of the black vine weevil (*Otiorhynchus sulcatus*) (Ansari et al., 2010).

Larger-scale population-level experiments on the impact of mixed pathogen applications on pest suppression or insect population dynamics are very rare. An exception is a recent study looking at the impact of combining *Bacillus thuringiensis kurstaki* (Btk) with an NPV application on cabbage moth (*Mamestra brassicae*) mortality, using small-scale ($1\,m^2$) field plots. Focusing on the consequences for the NPV, the study showed that co-inoculation resulted in proportionally more cadavers with observable NPV rep-lication (compared to the virus alone), even when the virus was applied 4 days after the Btk. However, secondary transmission of virus to a second generation of *M. brassicae* was lower, as co-exposure with Btk reduced the size of the cadavers and thus the amount of NPV inoculum produced (Hesketh and Hails, 2015).

These few studies illustrate that there might be promise in the use of mixed pathogen applications for pest management.

A long-term field study of gypsy moth (*L. dispar*) populations in the United States, where it is invasive, showed that a newly emerging specific fungal pathogen, *Entomo-phaga maimaiga*, was now a dominant pathogen, taking over from a species-specific baculovirus (Lymantria dispar multinucleopolyhedrovirus, LdMNPV) that had previ-ously caused epizootics in outbreak populations. It also found that the fungus seems to outcompete parasitoids; the authors hypothesized that this was because the fungus developed more quickly and both the fungus and the parasitoids needed to kill their host to complete their development (parasitoid) or produce transmission stages (fungus). Virus and fungus were found to co-infect in the field, which supports the idea that they are only competing for resources (Hajek and van Nouhuys, 2016). However, there is a need for more and larger-scale experiments on mixed entomopathogen appli-cations to examine whether laboratory findings scale up to the field level. At a more fundamental level, we need to understand how pathogen interactions impact transmis-sion to new hosts and examine the dynamics of pathogen communities over time and how these interact with host resistance.

6.3.2 Entomopathogen–Parasitoid Interactions

The majority of studies of multispecies interactions with entomopathogens involve parasitoids, and the outcome of the interaction is usually viewed from the perspective of its impact on the parasitoid, with little consideration for pathogen fitness. Both the pathogen and the parasitoid are competing for host resources, so the outcome is pri-marily an issue of timing. Like pathogen–pathogen interactions, population-level inter-actions involving parasitoids and pathogens have received less attention.

6.3.2.1 Effects of Pathogens on Parasitoids

From the parasitoid's perspective, the main challenge in sharing a host with another parasite is whether its larvae can complete their development and emerge successfully from the host. In other words, the costs of the pathogen–parasitoid interaction are likely to be higher for the parasitoid, because it will be a matter of life or death for its offspring (Hochberg, 1991; Cossentine, 2009). Regardless of whether the infection is due to bacteria, fungi, microsporidia, viruses, or nematodes, parasitoids will be vulner-able to the early death of the infected host (Brooks, 1993; Escribano et al., 2000a; Furlong, 2004; Harvey, 2005; Cossentine, 2009; Mbata and Shapiro-Ilan, 2010; Martins et al., 2014). Overall, entomopathogen infections are largely detrimental to adult para-sitoid emergence (Atwood et al., 1997; Nakai and Kumini, 1998; Escribano et al., 2002; Furlong, 2004; Martins et al., 2014). Two main factors will alter the outcome of parasi-toid–pathogen interactions: the initial pathogen density and the time lag between infec-tion and oviposition in the shared host. The survival rate and emergence of the adult parasitoids generally decrease when they are in competition with an entomopathogen, especially at high pathogen population density, regardless of whether the parasitoids are interacting with viruses (Cai et al., 2012), fungi (Aqueel and Leather, 2011), or bacteria (Atwood et al., 1997). For example, Guo et al. (2013) showed that the emergence of the solitary braconid endoparasitoid *Meteorus pulchricornis* was dependent on the concentration of Spodoptera exigua multinucleopolyhedrovirus (SeMNPV) in *Spodoptera exigua* larvae when challenged simultaneously with the virus; there was no effect up to a concentration of 10^5 OBs/ml (OBs = occlusion bodies that contain virions; see Chapter 7), but emergence declined markedly as concentrations of inoculum were increased. The interval between oviposition and infection is probably the most crucial factor in determining the outcome (Cory et al., 1997). If oviposition precedes infection, the parasitoids have more time to develop and can emerge before the pathogen uses up the available resources and kills the host (e.g. Escribano et al., 2000a; Furlong, 2004; Guo et al., 2013; Martins et al., 2014). However, timing appears to be more crucial when oviposition occurs before pathogen infection. Rashki et al. (2009) found that application of *Beauveria bassiana* 24 hours after parasitization of the green peach aphid (*Myzus persicae*) by the aphid parasitoid *Aphidius matricariae* resulted in a reduction in emergence of almost two-thirds, indicating that young larvae might be particularly sensitive to competition. Results are variable when oviposition takes place at the same time as or after pathogen challenge (Rashki et al., 2009; Aqueel and Leather, 2011; Potrich et al., 2014). Nevertheless, the longer the pathogen has to develop and replicate, the more host quality will decline for ovipositing female parasitoids. For example, the production of mummies in the Russian wheat aphid (*Diuraphis noxia*) by the parasitoid *Aphelinus asychis* did not differ when aphids were treated simultaneously with a concentration of 5.2×10^4 conidia/cm^2 of the entomopathogenic fungus *Isaria fumosoroseea* (= *Paecilomyces fumosoroseus*), but steadily decreased as the time between infection and oviposition increased up to 3 days (Mesquita and Lacey, 2001).

While it is assumed that the competition between a parasitoid and a pathogen is indirect and based on accessing the resources needed to grow or replicate, earlier studies suggested that some strains of an NPV isolated from the armyworm (*Pseudaletia unipuncta*) produced a toxin that could kill the braconid parasitoid *Apanteles militaris* (Kaya and Tanada, 1973), indicating a more direct form of interference competition; however, this appears not to have been followed up. A more clear example of direct effects was found in a study assessing interactions between the nematode *Heterorhabditis indica* and the parasitoid *Habrobracon hebetor* (the host insect was Indian meal moth, *Plodia interpunctella*); the nematodes infected and killed the larvae of the parasitoid (Mbata and Shapiro-Ilan, 2010).

While a large number of studies have examined the emergence success of parasitoid offspring and their survival time in a host that is shared with an entomopathogen, little is known about the fitness of the newly emerged generation, particularly in terms of fecundity or mating behavior. Both the size and longevity of female parasitoids are strongly linked to their fitness (fecundity) (West et al., 1996; Legner et al., 1997), and any competition taking place within the host could affect both factors. Survival of the emerging parasitoid first depends on whether it is susceptible to the co-infecting pathogen and how this is influenced by life stage. Whereas parasitoids are not susceptible to baculoviral infection (Cossentine, 2009; Guo et al., 2013), there are reports of some of the broader host-range fungi, such as *B. bassiana*, infecting non-target beneficial insects, including parasitoids (e.g., Shipp et al., 2003). However, these experiments are often carried out with high concentrations of spores under very artificial conditions, and sometimes with unlikely routes of entry, making it unclear whether microbial control agents actually impact beneficial species under field conditions. Adult parasitoid longevity after emergence is also influenced by pathogen load during infection (Aiuchi et al., 2012; Potrich et al., 2014). Generally, parasitoid longevity decreased when parasitoid and pathogen were sharing the same host (Martins et al., 2014; Potrich et al., 2014). Nevertheless, in terms of biological control and integrated pest-management programs, parasitoids showed a lower susceptibility to pathogens at concentrations that were used in the field (Ruiu et al., 2007; Aiuchi et al., 2012; Potrich et al., 2014). Aiuchi et al. (2012) showed that the longevity of F1-generation adult *Aphidius colemani* parasitoids was not negatively affected by a concentration of 10^6 conidia/ml of the entomopathogenic fungi *Lecanicillium* used to control the cotton aphid (*Aphis gossypii*), whereas a higher concentration of 10^7 conidia/ml did reduce adult longevity. Similarly, Dean et al. (2012) found that the susceptibility of two hymenopteran parasitoids of the emerald ash borer (*Agrilus planipennis*) to the entomopathogenic fungus *B. bassiana* was minimal when they were exposed to the fungus in a bioassay that mimicked field conditions (twigs dipped in a fungal suspension of 2×10^7 spores/cm^2).

6.3.2.2 Effects of Parasitoids on Pathogens

Studies that have considered the impact of sharing a host with a parasitoid on the fitness of an entomopathogen are rare. Unlike the impact of an entomopathogen on a parasitoid's survival, which can be easily measured by the emergence or death of the parasitoid larva, the effect of a mixed infection on a pathogen's fitness is less all-or-nothing. A pathogen could still replicate within a host that contains a parasitoid larva and produce transmission stages, although likely with a decreased yield compared to a single infection. Generally, research exclusively focuses on the pathogen density used to infect the host, the speed of kill, and the comparison between host susceptibility to pathogens in parasitized and nonparasitized hosts (Furlong, 2004; Cossentine, 2009; Guo et al., 2013).

One prediction is that a parasitized host might be more susceptible to pathogen infection. Numerous species of braconid and ichneumonid parasitoids contain polydnaviruses, which are intimately involved with the wasps and persist as proviruses in both the somatic cells and the germline. At oviposition, the virus is injected into the host and expresses genes that suppress the host immune system and alter host growth, enabling the wasp larva to develop (Strand and Burke, 2013). Thus, it is suggested that these immunocompromised hosts might be more susceptible to subsequent pathogen challenge. More broadly, an insect host fighting any parasitoid can either be immunosuppressed (and thus not able to fight natural enemies on multiple fronts) or have an upregulated immune system (which might be more reactive to a second invader). The data are equivocal: in some studies, parasitism has been found to improve host susceptibility to pathogens (Hochberg, 1991; Guo et al., 2013), but Escribano et al. (2000a) found that prior parasitization did not affect host susceptibility to a pathogen (NPV) in any larval instar of *Spodoptera frugiperda* when insects were challenged with the virus at least 12 hours after parasitization by the braconid *Chelonus insularis*. In addition, parasitization can also induce other physiological changes that are likely to affect host susceptibility to pathogen infection. For instance, in the interaction between the aphid parasitoid *Aphidius colemani* and the entomopathogenic fungus *Lecanicillium* spp., the mummification process that follows oviposition induced physical and chemical changes within the host that inhibited pathogen penetration 8 days after parasitization (Aiuchi et al., 2012). Parasitism may also impact pathogen behavior in entomopathogenic nematodes. Mbata and Shapiro-Ilan (2010) observed increased host preference of *H. indica* for infecting *P. interpunctella* that were parasitized by *H. hebetor* relative to nonparasitized hosts; the authors hypothesized that the observation supports a risk-sensitive foraging model (the path of least resistance), as described by Fushing et al. (2008). A more systematic approach to analyzing these interactions is needed.

The impact on pathogen infection of the parasitoid larvae within a host is also dependent on timing and pathogen concentrations, as well as host age (Hoverman et al., 2013, Escribano et al., 2000b). However, only a few studies take into account important pathogen fitness criteria such as the production and viability of transmission stages in parasitized and nonparasitized hosts. Furlong (2004) showed that fungal reproduction within the diamondback moth (*P. xylostella*) was significantly reduced when parasitoid development was well advanced at the time of infection (at least 3 days after parasitism). These results are supported by other studies on baculoviruses, which tend to show a decrease in virus occlusion body production or yield per unit wet weight in parasitized later-instar larvae compared to earlier instars with the same time lag between oviposition and pathogen challenge (Escribano et al., 2000b; Cai et al., 2012). This could relate to the greater resistance of later-instar host larvae; for example, host susceptibility to baculovirus infection decreases with increasing larval stages (Cory and Myers, 2003). Increasing time between parasite oviposition and pathogen inoculation will also result in there being fewer resources for pathogen replication (Escribano et al., 2000b; Cossentine, 2009). However, pathogens rarely lose out completely in mixed infections; for example, when *Pieris brassicae* larvae were parasitized by *Apanteles glomeratus* and infected with Pieris brassicae granulovirus (PbGV), OBs were produced even when hosts died after the parasitoid emerged successfully (Hochberg, 1991).

As with the pathogen–pathogen mixed infections discussed earlier, co-infection could potentially select for increase pathogen virulence over time (Cossentine, 2009). A study on the interaction between an NPV in *S. frugiperda* larvae and the ichneumonid parasitoid *Campoletis sonorensis* showed no change in biological activity after the virus had been amplified in parasitized – as compared to unparasitized – larvae for 10 generations. However, examination of DNA restriction profiles for different generations showed changes in virus population structure in both parasitized and nonparasitized hosts (Escribano et al. 2000b). Co-infection of an NPV with the parasitoid *Chelonus insularis* in the same host again showed changes in virus population structure, as well as increases in speed of kill and infectivity; however, these appeared in both parasitized and unparasitized treatments (Escribano et al., 2001). Evolution of pathogen virulence in parasitoid–entomopathogen interactions still needs to be explored. A better understanding of the effect of mixed natural-enemy interactions within the host on pathogens in terms of virulence would help us determine how the within-host relationship between pathogen and parasitoid could evolve in the longer term, and how we could manage this interaction in order to maximize pest suppression.

6.3.2.3 Population Level Effects

Only a few studies have scaled up from small-scale laboratory studies to population-level or field experiments. Microcosms have been used to investigate host–parasitoid–pathogen dynamics in the stored product insect *P. interpunctella*. This host species system exhibits generational cycles, and authors have investigated whether and how the addition of the ichneumonid parasitoid *Venturia canescens* combined with Plodia interpunctella granulovirus (PiGV) altered these cycles. Begon et al. (1996) found that the shift from a two- to a three-species system (i.e., adding the virus to the host and parasitoid) moved the population cycles from a single generation to multigenerational, indicating the potential impact that multispecies interactions involving pathogens could have in field populations. At a larger scale, mixed microbial–parasitoid applications in which parameters related to the success of either natural enemy are assessed are rare. Labbé et al. (2009) examined whether *B. bassiana* would disrupt control of the whitefly *Trialeurodes vaporariorum* by the parasitoid *Encarsia formosa* and the predator *Dicyphus hesperus* in the greenhouse. They did not find a decrease in either parasitism or predation, suggesting that there were no significant effects of adding the fungus to this system.

6.3.3 Pathogen–Predator Interactions

Interactions between entomopathogens and predators tend to be focused on two areas: ingestion of an infected host by a predator (a form of intraguild predation) and predator-induced changes in behavior that alter stress levels and the likelihood of predation. The exceptions is when a pathogen infects an invertebrate predator – a situation mainly limited to fungi (differential intraguild predation: Cottrell and Shapiro-Ilan, 2003; Shipp et al., 2003; Roy et al., 2008b). Predators in general are not affected by the consumption of insects infected by most entomopathogens; however, the quality of diseased prey could impact predator fitness if it was already suboptimal. A recent meta-analysis indicates that consumption of infected organisms can reduce invertebrate predator longevity, fecundity, and survival by up to 30% (Flick et al., 2016). Eating infected insects could reduce the inoculum available if it was broken down in the predator gut (Roy et al., 2008a); alternatively, predators feeding on infected insects (living or dead) could be seen as a means of enhancing spread in the field. Pathogens such as baculoviruses require an alkaline gut to break them down, whereas many predators have more acidic guts, so that those feeding on cadavers of pathogen-killed insects disperse infective pathogens. A range of invertebrate and vertebrate predators, including birds and earthworms, have been recorded as potential dispersal agents of baculoviruses and entomopathogenic nematodes in both the lab and the field (Epsky et al., 1988; Entwistle et al., 1993; Vasconcelos et al. 1996; Reilly and Hajek, 2012; Shapiro-Ilan and Brown, 2013; Infante-Rodriguez et al., 2016), which could be important for pathogen spread in natural situations, although this is difficult to estimate empirically.

Predators can also change the behavior of their prey, and in this way increase the likelihood of infection. Most research on this topic is on aphids and their fungus, *Pandora neoaphidis*. Transmission of *P. neoaphidus* to the pea aphid (*Acyrthosiphon pisum*) was increased in the presence of both the lady beetle (*Coccinella septempunctata*) and the parasitoid *Aphidius ervi* in microcosms, although this did not increase aphid suppression (Baverstock et al., 2009, 2010). In an interesting study that extends the impact of predators on entomopathogen infection, it was found that exposing the larval stage of the Colorado potato beetle (*Leptinotarsa decemlineata*) to predators (predatory beetles and bugs) resulted in increased susceptibility to pathogens (*B. bassiana* and nematodes) when they pupated in the soil, due to a weakened immune response (Ramirez and Snyder, 2009). In another interesting study, it was suggested that entomopathogenic nematodes within an infected host avoid consumption by predators or scavengers by producing compounds (called scavenger deterrent factor) in their symbiotic bacteria (Gulcu et al., 2012).

6.3.4 Conclusion

The use of natural enemies to suppress pest populations in agriculture, horticulture, and forestry means that it is important to understand how interactions between different control agents affect their ability to attack, kill the target host, and spread within the pest population. Laboratory studies involving mixed infections with entomopathogens are common at the within-host level, but tend to focus solely on host mortality or parasitoid emergence. The cost of the interaction for parasitoids is mainly directed by the ability of the parasitoid larvae to complete their development and emerge from the host in the presence of a competitor, whereas the cost of the interaction for pathogens is more a matter of the degree of replication achieved. Nevertheless, in both situations, the final status of each parasite is dependent on pathogen density at infection and the gap in time between infection and infection (pathogen–pathogen) or infection and oviposition (pathogen–parasitoid), although the replication/reproduction times of both players will modulate the interaction. However, there are clearly differences between species combinations; some of these might relate to methodological differences, but a more systematic approach is needed to elucidate whether the outcomes can be predicted by the organisms or species involved.

Pathogen–pathogen and pathogen–parasitoid interactions need to be further investigated, especially in terms of virulence evolution and the nature of within-host competition. Few studies have assessed whether there are consequences for the parasitoid in terms of offspring quality or reproductive success. Laboratory studies are not representative of the field and do not take into account behavioral changes, pathogen persistence and transmission, host plant effects, or spatial structure. They also do not factor in evolution. Parasitoid females could develop abilities to detect infected hosts and choose healthier host patches to improve the likelihood of offspring emergence and reduce competition (Sait et al., 1996; Stark et al., 1999). Co-infection might select for more rapidly developing pathogens and parasitoids, as suggested by the evolution of virulence literature, but this would have to be traded off against any negative effects in transmission, such as reduced yield or smaller females (West et al., 1996; Harvey, 2005). Alternatively, selection by the entomopathogen for larger parasitoids might result in a synergism between the two players, as parasitoid fecundity would increase (Y. Norouzi, D.R. Gillespie, and J.S. Cory, unpublished data).

6.4 Microbe-Mediated Defense

One of the newest and potentially most important areas to emerge in disease ecology is the role of nonpathogenic microorganisms, the microbiome, in the modulation of disease resistance. Broadly defined as a collection of genomes of microorganisms in a system (the system being an organism, the host), the microbiome has been shown to influence a variety of functions in invertebrates, including host plant use and defense against parasitoids, in addition to altering interactions with pathogens (Oliver and Martinez, 2014). Many of these interactions are described as symbiotic, usually facultative, and they cover a spectrum of relationships ranging from true mutualism to commensalism to parasitism, all with varying degrees of transience (Shapira, 2016). Many insects have been shown to host secondary (or facultative) bacterial symbionts. The best studied are the maternally transmitted symbionts, which ensure their persistence by manipulating host reproduction or increasing host fitness in other ways, such as by defending against natural enemies, including pathogens.

The impact of symbionts in defense against insect pathogens can potentially take two forms: where host defense is modified by the symbiont and where the symbiont alters other demographic factors (e.g., host sex ratio, age structure, or population density), which in turn alter the risk of infection and disease prevalence (e.g., Ryder et al., 2014). Only the former will be covered here.

6.4.1 Heritable Symbionts

Defense mediated by microbes has been best studied in aphids (Zytynska and Weisser, 2016) and *Drosophila* spp. (Hamilton and Perlman, 2013), with much of the research concentrating on defense against parasitoids by maternally transmitted secondary endosymbionts. One of the earliest studies on defense against entomopathogens was the demonstration of the protective role of the bacterial symbiont *Regiella insecticola* in the pea aphid (*Acyrthosiphon pisum*) against the aphid-specialist entomophthoralean fungus *Pandora neoaphidis*, where both mortality and the likelihood of sporulation were reduced (Scarborough et al., 2005). Protection against *P. neoaphidis* has since been demonstrated by bacterial endosymbionts in three additional genera: *Rickettsia*, *Rickettsiella*, and *Spiroplasma* (Łukasik et al., 2013). *Regiella insecticola* has also been shown to protect *A. pisum* against another specialist fungus, *Zoophthora occidentalis*, but not the broader host-range fungus, *B. bassiana*, suggesting that symbiont-mediated protection has evolved in response to particular fungal species, or else is related to aphid-specialization and host-range breadth (Parker et al., 2013). It has been shown that a species of *Drosophila*, *D. neotestacea*, is protected against a sterilizing nematode, *Howardula aoronymphium*, by a maternally transmitted *Spiroplasma* (Jaenike et al., 2010). Field studies on symbionts are rare; however, this *Spiroplasma* has spread rapidly across North America in the last few decades (Cockburn et al., 2013), demonstrating the powerful impact that defensive symbionts can have on host populations, and potentially also on communities.

There is also some evidence for symbiont interactions with insect viruses. *Wolbachia* are vertically transmitted intracellular bacteria that have been isolated from a wide range of invertebrates. They are best known as reproductive parasites that frequently distort the sex ratio in favor of females by a range of mechanisms that increase their transmission and persistence. *Wolbachia* can delay mortality in adult *Drosophila melanogaster* caused by a range of RNA viruses (both a naturally occurring *Drosophila* virus and other non-*Drosophila* viruses) introduced by injection, but not by a heterologous DNA virus (Hedges et al., 2008, Teixeira et al., 2008). However, a male-killing strain of *Wolbachia* did not provide any protection against two of the same RNA viruses in

D. bifasciata (Longdon et al., 2012), and anti-RNA virus protection varied with different *Wolbachia* strain–*D. simulans* fly line combinations (Osborne et al., 2009), indicating that the antiviral traits are not universal. Although *Wolbachia* is found in a diverse array of invertebrates, its defensive role against pathogens has been almost exclusively studied in Diptera. An exception is a study on the African armyworm (*Spodoptera exempta*), where *Wolbachia* was implicated in the enhancement of baculovirus infection, rather than in protection (Graham et al., 2012). The role of *Wolbachia* as a defensive symbiont, while fascinating, is clearly equivocal, and requires more detailed study in a wider range of non-model taxa and under a variety of conditions.

The mechanisms underlying the defensive behavior of these maternally inherited symbionts are not clear. The interactions between pathogen and symbiont could be direct, such as through antagonistic interactions via toxins or antimicrobial substances, or could be the result of competition for limiting resources. Alternatively, the entomopathogen could be affected indirectly via symbiont impacts on the host immune system (Haines, 2008; Mideo, 2009). These different mechanisms are likely to have different implications for the evolutionary relationship between pathogen, symbiont, and host (Ford and King, 2016). Protection against parasitoids in the pea aphid by the bacterium *Hamiltonella defensa* is linked to the lysogenic, toxin-producing bacteriophage that it carries (Oliver et al., 2009). As yet, data on the defensive mechanisms of secondary symbionts against other natural enemies are limited. However, transcriptomic studies on the *Spiroplasma*–nematode system found no support for resource competition between symbiont and parasite, and no strong support for immune priming; but there was evidence for the upregulation of a novel toxin in response to nematode challenge (Hamilton et al., 2014).

6.4.2 Do Gut Microflora Influence Pathogen Susceptibility?

In addition to the heritable symbionts, insect guts host a wide range of other microbes. The gut microbiota has been shown to influence a range of activities, including food digestion, herbivore host range, and the production of pheromones (Oliver and Martinez, 2014). Recent data from the gypsy moth (*Lymantria dispar*) indicate that although egg masses host a diverse bacterial community in the wild, this is strongly filtered in the larval stage, initially being heavily influenced by the host plant, but eventually merging into very similar gut communities in different gypsy moth populations through larval development (Mason and Raffa, 2014). However, this is likely to be related to the breadth of host plant chemistries used (Broderick et al., 2004), and it would suggest that the bacteria on the phyllosphere are more important than any transmitted vertically from the parent, although the dynamics of the gut microbial community and the impact of any transient effects from ingested bacteria have received little attention.

Information on the gut microbiomes of a wide range of invertebrates is increasing at a rapid rate, and recently this has expanded to the potential role of the gut microbiome in the modulation of defense against insect pathogens and parasites. This is not a new area for insect pathology (Jarosz, 1979; Dillon and Charnley, 1986). Earlier studies on the desert locust (*Schistocerca gregaria*) suggested that an antifungal toxin produced by the gut microbiota inhibited infection by the fungus *M. anisopliae* (Dillon and Charnley,

1988). Interestingly, the diversity of this gut community made a difference, with more diverse communities being more resilient (Dillon et al., 2005). However, it is not clear whether the host microbiota plays any role in fungal inhibition when hosts are infected via the cuticle, which is considered the more common route of infection.

Recent information on the bee microbiome indicates that both bumblebees and honeybees support a gut microbiota with relatively low diversity (Martinson et al., 2011). However, data from the bumblebee (*Bombus terrestris*) show that the gut microbiota is acquired by exposure to the feces of nestmates, and more importantly that this reduces the prevalence of a trypansosomatid gut parasite, *Crithidia bombi*, indicating a benefit of sociality. Correlations between *C. bombi* presence and the proportion of dominant Gammaproteobacteria and Betaproteobacteria in the gut community in field-collected individuals support this conjecture, as well as indicating that the presence of another common parasite, the intracellular microsporidian *Nosema bombi*, is not affected by the gut bacterial community (Koch and Schmid-Hempel, 2011). Moreover, specificity within the *C. bombi–B. terrestris* system (i.e., genotype × genotype interactions between host and parasite) also appears to be primarily driven by the gut microbiota, rather than the host genotype (Koch and Schmid-Hempel, 2012).

One of the more controversial issues has been the role of gut bacteria in the pathogenicity of Bt. Several studies have suggested that Bt requires an indigenous gut microbiota to cause mortality via septicemia; when the gut microbiota was removed with antibiotic treatment, the ability of Bt to kill lepidopteran larvae was lost (Broderick et al., 2006, 2009). However, follow-up studies suggested that this was not the case, as the effect of that specific antibiotic cocktail on Bt had not been taken into consideration, and prior exposure to antibiotics – and not the absence of gut bacteria – resulted in reduced mortality when larvae were exposed to Bt (Johnson and Crickmore, 2009; Raymond et al., 2009, 2010). Recently, a different approach was taken to addressing this question: RNAi-mediated silencing of immune genes, rather than antibiotic treatment to remove the gut microbiota. When an immune-related gene was suppressed in the lepidopteran *Spodoptera littoralis*, immunosuppression was documented, together with increased larval mortality, resulting from Bt and a Cry toxin alone (Caccia et al., 2016). This was accompanied by a proliferation of midgut bacteria and a change in the bacterial species found in the hemolymph, which became dominated by bacteria usually found in the gut. These data strongly implicate the role of gut bacteria in Bt-mediated mortality, but also highlight the difficulties of studying the role of the gut microbiota through broad-scale antibiotic treatment.

Studies on the role of the host microbiota in defense against insect pathogens have lagged behind in other systems. However, there are some intriguing data on baculoviruses that indicate a possible role for the gut microbiota. In the lepidopteran *Spodoptera exigua*, infection with its baculovirus, SeMNPV, resulted in the downregulation of immune-related genes and an increase in the density of gut microbiota (but not in the hemolymph). When larvae without a gut microbiota (fed on a diet with streptomycin) were challenged with NPV, they died more slowly and produced fewer transmission stages (occlusion bodies) (Jakubowska et al., 2013). This result is intriguing, as there is often a trade-off between speed of kill and yield of NPV occlusion bodies, such that a slower death results in more transmission stages; however, this result suggests that this relationship might be altered by the presence of the gut microbiota.

6.4.3 Future Directions

Clearly, there are benefits to carrying symbionts; however, the trade-offs between the costs and benefits of maintaining populations of microorganisms and how these are modulated by the conditions that an insect encounters in nature still need to be examined in more detail and in more systems. Bacterial endosymbionts have been shown to have negative effects on various aphid fitness metrics, but these tend to be context-dependent (Zytynska and Weisser, 2016). More broadly, we still have little understanding of the role of symbionts in natural populations and how they influence dynamics and communities. Are symbionts and other microbiota likely to be important for microbial pest control? As the broader host range entomopathogens tend to be selected for insect pest control, the limited current information suggests that these are less likely to be affected by the heritable facultative symbionts. The gut microbiota, in contrast, appears to have more general effects in terms of its interactions with entomopathogens, although there is the suggestion that reducing the immunocompetence of insects could increase their vulnerability to infection via opportunistic bacteria. However, information on symbionts, and the host microbiota more broadly, is restricted to a very small number of insect groups and a limited number of pathogen genera. Examination of a broader range of taxa is crucial if we are to understand the role of the host microbiome in insect–pathogen interactions. This should include information on the dynamics of microbiome, in addition to its role in natural populations and communities.

6.5 Conclusion

Multispecies interactions involving entomopathogens are in their infancy. Although there is a legacy of studies investigating mixed infections in laboratory experiments, their scope has been limited and has focused primarily on host mortality. This is not surprising, given that research on insect pathogens has until recently been motivated by their development as microbial insecticides and disease prevention in managed insects. However, there is a need to scale up studies to a more realistic population level and to broaden our outlook and address questions relating to pathogen, as well as host, fitness. Disease ecology and evolution are expanding research areas. Many fundamental questions relating to issues such as the role of pathogens in population and community dynamics, the impact of biodiversity on disease, and the factors that modulate virulence are being investigated using invertebrate models. Understanding these issues should feed back into the use of insect pathogens as microbial insecticides, but should almost promote entomopathogens and their hosts as systems for addressing important issues relating to our understanding of disease and its management.

Acknowledgments

JSC would like acknowledge the support of an NSERC discovery grant and PSD the support of the Simons Foundation Doctoral Entrance fellowship.

References

Abbott, J., 2014. Self-medication in insects: current evidence and future perspectives. Ecol. Entomol. 39, 273–280.

Aiuchi D., Saito Y., Tone J., Kanazawa M., Tani M., Koike, M., 2012. The effect of entomopathogenic *Lecanicillium spp.* (Hypocreales: Cordycipitaceae) on the aphid parasitoid *Aphidius colemani* (Hymenoptera: Aphidiinae). Appl. Entomol. Zool. 47, 351–357.

Alizon, S., de Roode, J., Michalakis, Y., 2013. Multiple infections and the evolution of virulence. Ecol. Lett. 16, 556–567.

Alizon S., Hurford, A., Mideo, N., Van Baalen N., 2008. Virulence evolution and the trade-off hypothesis: history, current state of affairs and the future. J. Evol. Biol. 22, 245–259.

Ansari, M.A., Shah, F.A., Butt, T.M., 2010. The entomopathogenic nematode *Steinernema kraussei* and *Metarhizium anisopliae* work synergistically in controlling overwintering larvae of the black vine weevil, *Otiorhynchus sulcatus*, in strawberry growbags. Biocontr. Sci. Technol. 20, 99–105.

Aqueel, M.A., Leather, S.R., 2011. Virulence of *Verticillium lecanii* (Z.) against cereal aphids; does timing of infection affect the performance of parasitoids and predators? Pest. Manag. Sci. 69, 493–498.

Atwood, D.W., Young III, S.Y., Kring, T.J., 1997. Development of *Cotesia marginiventris* (Hymenoptera: Braconidae) in tobacco budworm (Lepidoptera: Noctuidae) larvae with *Bacillus thuringiensis* and thiodicarb. J. Econ. Entomol. 90, 751–756.

Barbercheck, M.E., Wang, J., Hirsh, I.S., 1995. Host plant effects on entomopathogenic nematodes. J. Invertebr. Pathol. 66, 169–177.

Bashey, F., Hawlena, H., Lively, C.M., 2013. Alternative paths to success in a parasite community: within-host competition can favor higher virulence or direct interference. Evolution 67, 900–907.

Baverstock, J., Clark, S.J., Alderson, P.G., Pell, J.K., 2009. Intraguild interactions between the entomopathogenic fungus *Pandora neoaphidus* and an aphid predator and parasitoid at the population scale. J. Invertebr. Pathol. 102, 167–172.

Baverstock, J., Roy, H.E., Pell, J.K., 2010. Entomopathogenic fungi and insect behaviour: from unsuspecting hosts to targeted vectors. BioControl 55, 89–102.

Begon, M., Sait, S.M., Thompson, D.J., 1996. Predator-prey cycles with period shifts between two- and three-species systems. Nature 381, 311–315.

Biere, A., Tack, A.J.M., 2013. Evolutionary adaptation in three-way interactions between plants, microbes and arthropods. Funct. Ecol. 27, 646–660.

Boots, M., Roberts, K.E., 2012. Maternal effects in disease effects: poor maternal environment increases offspring resistance to an insect virus. Proc. R. Soc. B 279, 4009–4014.

Bos, N., Sundström, L., Fuchs, S., Freitak, D., 2015. Ants medicate to fight disease. Evolution 69, 2979–2984.

Broderick, N.A., Raffa, K.F., Handelsman, J., 2006. Midgut bacteria required for *Bacillus thuringiensis* insecticidal activity. Proc. Natl. Acad. Sci. U.S.A. 103, 15196–15199.

Broderick, N.A., Robinson, C.J., McMahon, M.D., Holt, J., Handelsman, J., Raffa, K.F., 2009. Contributions of gut bacteria to *Bacillus-thuringiensis* induced mortality vary across a range of Lepidoptera. BMC Biology 7, 11.

Brooks, W.M., 1993. Host-parasitoid-pathogen interactions, in: Beckage, N.E., Thompson, S.N., Federici, B.A. (eds.), Parasites and pathogens of Insects, Vol. 2, Academic Press, San Diego, CA, pp. 231–272.

Caccia, S., Lelio, I.D., La Storia, A., Marinelli, A., Varricchio, P., Franzetti, E., et al., 2016. Midgut microbiota and host immunocompetence underlie *Bacillus thuringiensis* killing mechanism. Proc. Natl. Acad. Sci. U.S.A. 113, 9486–9491.

Cai, Y., Fan, J., Sun, S., Wang, F., Yang, K., Li, G., Pang, Y., 2012. Interspecific interaction between *Spodoptera exigua* multiple nucleopolyhedrovirus and *Microplitis bicoloratus* (Hymenoptera: Braconidae: Microgastrinae) in *Spodoptera exigua* (Lepidoptera: Noctuidae) larvae. J. Econ. Entomol. 105, 1503–1508.

Cardinale, B.J., Srivastava, D.S., Duffy, J.E., Wright, J.P., Downing, A.L., Sankaran, M., Jouseau, C., 2006. Effects of biodiversity on the functioning of trophic groups and ecosystems. Nature 443, 989–992.

Chapuisat, M., Oppliger, A., Magliano, P., Christe, P., 2007. Wood ants use resin to protect themselves against pathogens. Proc. R. Soc. B 274, 2013–2017.

Cockburn, S.N., Haselkorn, T.S., Hamilton, P.T., Landzberg, E., Jaenike, J., Perlman, S.J., 2013. Dynamics of the continent-wide spread of a *Drosophila* defensive symbiont. Ecol. Lett. 16, 609–616.

Cory, J.S., 2010. The biology of baculoviruses, in: Asgari, S., Johnson, K.N. (eds.), Insect Virology. Caister Academic Press, Poole, pp. 405–421.

Cory, J.S., Ericsson, J.D., 2010. Fungal entomopathogens in a tritrophic context. BioControl 55, 75–88.

Cory, J.S., Hoover, K., 2006. Plant-mediated effects in insect–pathogen interactions. Trends Ecol. Evol. 21, 278–286.

Cory, J.S., Hails, R.S., Sait, S.M., 1997. Baculovirus ecology, in: Miller, L.K. (ed.), The Baculoviruses. Plenum Press, New York, pp. 301–339.

Cory, J.S., Myers, J.H., 2003. The ecology and evolution of insect baculoviruses. Annu. Rev. Ecol. Evol. Syst. 34, 239–272.

Cory, J.S., Myers, J.H., 2004. Adaptation in an insect host-plant pathogen interaction. Ecol. Lett. 7, 632–639.

Cossentine, J.E., 2009. The parasitoid factor in the virulence and spread of lepidopteran baculoviruses. Virol. Sin. 24, 305–314.

Cottrell, T.E., Shapiro-Ilan, D.I., 2003. Susceptibility of a native and an exotic lady beetle (Coleoptera: Coccinellidae) to *Beauveria bassiana*. J. Invertebr. Pathol. 84, 137–144.

Dean, K.M., Vandenberg, J.D., Griggs, M.H., Bauer, L.S., Fierke, M.K., 2012. Susceptibility of two hymenopteran parasitoids of *Agrilus planipennis* (Coleoptera: Buprestidae) to the entomopathogenic fungus *Beauveria bassiana* (Ascomycota: Hypocreales). J. Invertebr. Pathol. 109, 303–306.

Dillon, R.J., Charnley, A.K., 1986. Inhibition of *Metarhizium anisopliae* by the gut bacterial flora of the desert locust *Schistocerca gregaria*: evidence for an antifungal toxin. J. Invertebr. Pathol. 47, 350–360.

Dillon, R.J., Charnley, A.K., 1988. Inhibition of *Metarhizium anisopliae* by the gut bacterial flora of the desert locust: characterization of antifungal toxins. Can. J. Microbiol. 34, 1075–1082.

Dillon, R.J., Vennard, C.T., Buckling, A., Charnley, A.K., 2005. Diversity of locust gut bacteria protects against pathogen invasion. Ecol. Lett. 8, 1291–1298.

Doublet, V., Natsopoulou, M.E., Zschiesche, L., Paxton, R.J., 2015. Within-host competition among honey bee pathogens *Nosema ceranae* and Deformed wing virus is asymmetric and to the disadvantage of the virus. J. Invertebr. Pathol. 124, 31–34.

Elderd, B.D., Rehill, B.J., Haynes, K.J., Dwyer, G., 2013. Induced plant defences, host-pathogen interactions and forest insect outbreaks. Proc. Natl. Acad. Sci. U.S.A. 110, 14978–14983.

Entwistle, P.F., Forkner, A.C., Green, B.M., Cory, J.S., 1993. Avian dispersal of nuclear polyhedrosis viruses after induced epizootics in the pine beauty moth, *Panolis flammea* (Lepidoptera: Noctuidae). Biol. Contr. 3, 61–69.

Epsky, N.D., Walter, D.E., Capinera, J.L., 1988. Potential role of nematophagous microarthropods as biotic mortality factors of entomogenous nematodes (Rhabditida: Steinernematidae, Heterorhabditidae). J. Econ. Entomol. 81, 821–825.

Escribano, A., Williams, T., Goulson, D., Cave, R.D., Chapman, J.W., Caballero, P., 2000a. Parasitoid-pathogen-pest interactions of *Chelonus insularis*, *Campoletis sonorensis*, and a nucleopolyhedrovirus in *Spodoptera frugiperda* larvae. Biol. Contr. 19, 265–273.

Escribano, A., Williams, T., Goulson, D., Cave, R.D., Chapman, J.W., Caballero, P., 2000b. Effect of parasitism on a nucleopolyhedrovirus amplified in *Spodoptera frugiperda* larvae parasitized by *Campoletis sonorensis*. Entomol. Exp. Appl. 97, 257–264.

Escribano, A., Williams, T., Goulson, D., Cave, R.D., Chapman, J.W., Caballero, P., 2001. Consequences of interspecific competition on the virulence and genetic composition of a nucleopolyhedrovirus in *Spodoptera frugiperda* larvae parasitized by *Chelonus insularis*. Biocontr. Sci. Technol. 11, 649–662.

Flick, A.J., Acevedo, M.A., Elderd, B.D., 2016. The negative effects of pathogen-infected prey on predators: a meta-analysis. Oikos 125, 1554–1560.

Ford, S.A., King, K.C., 2016. Harnessing the power of defensive microbes: evolutionary implications in nature and disease control. PLoS Pathog. 12(4), e1005465.

Furlong, M.J., 2004. Infection of the immature stages of *Diadegma semiclausum*, an endolarval parasitoid of the diamondback moth, by *Beauveria bassiana*. J. Invertebr. Pathol. 86, 52–55.

Fushing, H., Zhu, L., Shapiro-Ilan, D.I., Campbell, J.F., Lewis, E.E., 2008. State-space based mass event-history model I: many decision making agents with one target. Ann. Appl. Stat. 2, 1503–1522.

Garbutt, J., Bonsall, M.B., Wright, D.J., Raymond, B., 2011. Antagonistic competition modulates virulence in *Bacillus thuringiensis*. Ecol. Lett. 14, 765–772.

Gassmann, A.J., Stock, S.P., Tabashnik, B.E., Singer, M.S., 2010. Tritrophic effects of host plants on an herbivore-pathogen interaction. Ann. Entomol. Soc. Am. 103, 371–378.

Graham, R.I., Grzywacz, D., Mushobozi, W.L., Wilson, K., 2012. *Wolbachia* in a major African crop pest increases susceptibility to a viral disease rather than protects. Ecol. Lett. 15, 993–1000.

Gulcu, B., Hazir, S., Kaya, H.K., 2012. Scavenger deterrent factor (SDF) from symbiotic bacteria of entomopathogenic nematodes. J. Invertebr. Pathol. 110, 326–333.

Guo, H.-F., Fang, J.-C., Zhong, W.-F., Liu, B.-S., 2013. Interactions between *Meteorus pulchricornis* and *Spodoptera exigua* multiple nucleopolyhedrovirus. J. Insect Sci. 13, 12.

Hackett, K.J., Boore, A., Deming, C., Buckley, E., Camp, M., Shapiro, M., 2000. *Helicoverpa armigera* granulovirus interference with progression of *H. zea* nucleopolyhedrovirus disease in *H. zea* larvae. J. Invertebr. Pathol. 75, 99–106.

Haines, E.R., 2008. Symbiont-mediated protection. Proc. R. Soc. B 275, 353–361.

Hajek, A.E., van Nouhuys, S., 2016. Fatal diseases and parasitoids: from competition to facilitation in a shared host. Proc. R. Soc. B 283, 20160154.

Hamilton, P.T., Peng, F., Boulanger, M.J., Perlman, S.J., 2016. A ribosome-inactivating protein in a *Drosophila* defensive symbiont. Proc. Natl. Acad. Sci. U.S.A. 113, 350–355.

Hamilton, P.T., Perlman, S.J., 2013. Host defense via symbiosis in *Drosophila*. PLoS Pathog. 9(12), e1003808.

Harper, J.D., 1986. Interactions between baculoviruses and other entomopathogens, chemical pesticides and parasitoids, in: Granados, R.R., Federici, B.A. (eds.), The Biology of Baculoviruses, Vol. 1. CRC Press, Boca Raton, FL, pp. 133–135.

Hazir, S., Shapiro-Ilan, D.I., Hazir, C., Leite, L.G., Cakmak, I., Olson, D., 2016. Multifaceted effects of host plants on entomopathogenic nematodes. J. Invertebr. Pathol. 135, 53–59.

Hedges, L.M., Brownlie, J.C., O'Neill, S.L., Johnson, K.N., 2008. *Wolbachia* and virus protection in insects. Science 322, 702.

Hesketh, H., Hails, R.S., 2015. *Bacillus thuringiensis* impacts on primary and secondary baculovirus transmission dynamics in Lepidoptera. J. Invertebr. Pathol. 132, 171–181.

Hiltpold, I., Toepfer, S., Kuhlmann, U., Turlings, T.C.J., 2010. How maize root volatiles influence the efficacy of entomopathogenic nematodes against the western corn rootworm? Chemoecology. 20, 155–162.

Hochberg, M.E., 1991. Intra-host interactions between a braconid endoparasitoid, *Apanteles glomeratus*, and a baculovirus for larvae of *Pieris brassicae*. J. Anim. Ecol. 60, 51–63.

Hodgson, D.J., Hitchman, R.B., Vanbergen, A.J., Hails, R.S., Possee, R.D., Cory, J.S., 2004. Host ecology determines the relative fitness of virus genotypes in mixed-genotype nucleopolyhedrovirus infections. J. Evol. Biol. 17, 1018–1025.

Hoverman, J.T., Hoye, B.J., Johnson, P.T.J., 2013. Does timing matter? How priority effects influence the outcome of parasite interactions within hosts. Oecologia. 173, 1471–1480.

Infante-Rodriguez, D.A., Berber, J.J., Mercado, G., Valenzeula-Gonzalez, J., Munoz, D., Williams, T., 2016. Earthworm mediated dispersal of baculovirus occlusion bodies: experimental evidence from a model system. Biol. Contr. 100, 18–24.

Jabbour, R., Crowder, D.W., Aultman, E.A., Snyder, W.E., 2011. Entomopathogen biodiversity increases host mortality. Biol. Contr. 59, 277–283.

Jaenike, J., Unckless, R., Cockburn, S.N., Boelio, L.M., Perlman, S.J., 2010. Adaptation via symbiosis: recent spread of a *Drosophila* defensive symbiont. Science 329, 212–215.

Jafary, M., Karimzadeh, J., Farazmand, H., Rezapanah, M., 2016. Plant-mediated vulnerability of an insect herbivore to *Bacillus thuringiensis* in a plant-herbivore-pathogen system. Biocontr. Sci. Technol. 26, 104–115.

Jakubowska, A.K., Vogel, H., Herrero, S., 2013. Increase in gut microbiota after immune suppression in baculovirus-infected larvae. PLoS Pathog. 9(5), e1003379.

Janmaat, A.F., Myers, J.H., 2007. Host-plant effects the expression of resistance to *Bacillus thuringiensis kurstaki* in *Trichoplusia ni* (Hubner): an important factor in resistance evolution. J. Evol. Biol. 20, 62–69.

Jarosz, J., 1979. Gut flora of *Galleria mellonella* suppressing ingested bacteria. J. Invertebr. Pathol. 34, 192–198.

Johnstone, P.R., Crickmore, N., 2009. Gut bacteria are not required for the insecticidal activity of *Bacillus thuringiensis* toward the tobacco hornworm, *Manduca sexta*. Appl. Environ. Microb. 75, 5094–5099.

Kaya, H.K., 2002. Natural enemies and other antagonists, in: Gaugler, R. (ed.), Entomopathogenic Nematology, CABI, Wallingford, pp. 189–204.

Kaya, H.K., Tanada, Y., 1973. Hemolymph factor in armyworm larvae infected with a nuclear polyhedrosis virus toxic to *Apanteles miltaris*. J. Invertebr. Pathol. 21, 211–14.

Klinger, E.G., Vojvodic, S., DeGrandi-Hoffman, G., Welker, D.L., James, R.J., 2015. Mixed infections reveal virulence differences between host-specific bee pathogens. J. Invertebr. Pathol. 129, 28–35.

Koch, H., Schmid-Hempel, P., 2011. Socially transmitted gut microbiota protect bumble bees against an intestinal parasite. Proc. Natl. Acad. Sci. U.S.A. 108, 19 288–19 292.

Koch, H., Schmid-Hempel, P., 2012. Gut microbiota instead of host genotype drive the specificity in the interaction of a natural host-parasite system. Ecol. Lett. 15, 1095–1103.

Koppenhöfer, A.M., Grewal, P.S., 2005. Compatibility and interactions with agrochemicals and other biocontrol agents, in: Grewal, P.S., Ehlers, R.-U., Shapiro-Ilan, D.I. (eds.), Nematodes as Biological Control Agents. CABI, Wallingford, pp. 363–381.

Koppenhöfer, A., Kaya, H.K., Shanmugam, S., Wood, G.L., 1995. Interspecific competition between steinernematid nematodes within an insect host. J. Invertebr. Pathol. 66, 99–103.

Lee, K.P., Cory, J.S., Wilson, K., Raubenheimer, D., Simpson, S.J., 2006. Flexible diet choice offsets protein costs of pathogen resistance in a caterpillar. Proc. R. Soc. B 273, 823–829.

Lefèvre, T., Oliver, M., Hunter, M.D., de Roode, J.C., 2010. Evidence for trans-generational medication in nature. Ecol. Lett. 13, 1485–1493.

Legner, E.F., Gerling, D., 1967. Host-feeding and oviposition on *Musca domestica* by *Spalangia cameroni*, *Nasonia vitripennis*, and *Muscidifurax raptor* (Hymenoptera: Pteromalidae) influences their longevity and fecundity. Ann. Entomol. Soc. Am. 60, 678–691.

Lin, Y., Hussain, M., Avery, P.B., Qasim, M., Fang, D., Wang, L., 2016. Volatiles from plants induced by multiple aphid attacks promote conidial performance of *Lecanicillium lecanii*. PLoS ONE 11(3), e0151844.

Longdon, B., Fabian, D.K., Hurst, G.D.D., Jiggins, F.M., 2012. Male-killing *Wolbachia* do not protect *Drosophila bifasciata* against viral infection. BMC Microbiol. 12, 58.

Łukasik, P., van Asch, M., Guo, H., Ferrari, J., Godfray, H.C.J., 2013. Unrelated facultative endosymbionts protect aphids against a fungal pathogen. Ecol. Lett. 16, 214–218.

Mantzoukas, S., Milonas, P., Kontodimas, D., Angelopoulos, K., 2013. Interaction between the entomopathogenic bacterium *Bacillus thuringiensis* subsp. *kurstaki* and two entomopathogenic fungi in bio-control of *Sesamia nonagrioides* (Lefevbre) (Lepidoptera: Noctuidae). Ann. Microbiol. 63, 1083–1091.

Martins, I.C.F., Silva, R.J., Alencar, J.R.D.C.C., Silva, K.P., Cividanes, F.J., Duarte, R.T., et al., 2014. Interactions between the entomopathogenic fungi *Beauveria bassiana* (Ascomycota: Hypocreales) and the aphid parasitoid *Diaeretiella rapae* (Hymenoptera: Braconidae) on *Myzus persicae* (Hemiptera: Aphididae). J. Econ. Entomol. 107, 933–938.

Martinson, V.G., Danforth, B.N., Minckley, R.L., Rueppell, O., Tingek, S., Moran, N.A., 2011. A simple and distinctive microbiota associated with honey bees and bumble bees. Mol. Ecol. 20, 619–628.

Mbata, G.N., Shapiro-Ilan, D.I., 2010. Compatibility of *Heterorhabditis indica* (Rhabditida: Heterorhabditidae) and *Habrobracon hebetor* (Hymenoptera: Braconidae) for biological control of *Plodia interpunctella* (Lepidoptera: Pyralidae). Biol. Contr. 54, 75–82.

Mesquita, A.L.M., Lacey L.A., 2001. Interactions among the entomopathogenic fungus, *Paecilomyces fumosoroseus* (Deuteromycotina: Hyphomycetes), the parasitoid, *Aphelinus asychis* (Hymenoptera: Aphelinidae), and their aphid host. Biol. Contr. 22, 51–59.

Mideo, N., 2009. Parasite adaptations to within host competition. Trends Parasitol. 25, 261–268.

Mitchell, S.E., Read, A.R., 2005. Poor maternal environment enhances offspring disease resistance in an invertebrate. Proc. R. Soc. B 272, 2601–2607.

Moonjely, S., Barelli, L., Bidochka, M.D., 2016. Insect pathogenic fungi as endophytes. Adv. Genet. 94, 107–135.

Nakai, M., Kumini, Y., 1998. Effects of the timing of entomopoxvirus administration to the smaller tea tortrix, *Adoxophyes sp.* (Lepidoptera: Tortricidae) on the survival of the endoparasitoid, *Ascogaster reticulatus* (Hymenoptera: Braconidae). Biol. Contr. 13, 63–69.

Oliver, K.M., Degnan, P.H., Hunter, M.S., Moran, N., 2009. Bacteriophages encode factors required for protection in a symbiotic mutualism. Science 325, 992–994.

Oliver, K.M., Martinez, A.J., 2014. How resident microbes modulate ecologically-important traits of insects. Curr. Opin. Ins. Sci. 4, 1–7.

Osborne, S.E., Leong, Y.S., O'Neill, S.L., Johnson, K.N., 2009. Variation in anti-viral protection mediated by different *Wolbachia* strains in *Drosophila simulans*. PLoS Pathog. 5(11), e1000656.

Ownley, B.H., Gwinn, K.D., Vega, F.E., 2010. Endophytic fungal pathogens with activity against plant pathogens: ecology and evolution. Biocontrol 55, 113–128.

Parker, B.J., Spragg, C.J., Altincicek, B., Gerardo, N.M., 2013. Symbiont-mediated protection against fungal pathogens in the pea aphid: a role for pathogen specificity? Appl Environ. Microb. 79, 2455–2458.

Potrich, M., Alves, L.F.A., Lozano, E., Roman, J.C., Pietrowski, V., Neves, P.M.O., 2015. Interactions between *Beauveria bassiana* and *Trichogramma pretiosum* under laboratory conditions. Entomol. Exp. Appl. 154, 213–221.

Povey, S., Cotter, S.C., Simpson, S.J., Lee, K.P, Wilson, K., 2008. Can the protein costs of bacterial resistance be offset by altered feeding behaviour. J. Anim. Ecol. 78, 437–446.

Ramirez, R.A., Snyder, W.E., 2009. Scared sick? Predator-pathogen facilitation enhances exploitation of a shared resource. Ecology 90, 2832–2839.

Rashki, M., Kharazi-pakdel, A., Allahyari, H., van Alphen, J.J.M., 2009. Interactions among the entomopathogenic fungus, *Beauveria bassiana* (Ascomycota: Hypocreales), the parasitoid, *Aphidius matricariae* (Hymenoptera: Braconidae), and its host, *Myzus persicae* (Homoptera: Aphididae). Biol. Contr. 50, 324–328.

Rasmann, S., Köllner, T.G., Degenhardt, J., Hiltpold, I., Toepfer, S., Kuhlmann, U., et al., 2005. Recruitment of entomopathogenic nematodes by insect-damaged maize roots. Nature 434, 732–737.

Raymond, B., Sayyed, A.H., Wright, D.J., 2007. Host plant and population determine the fitness costs of resistance to *Bacillus thuringiensis*. Biol. Lett. 3, 82–85.

Raymond, B., Johnston, P.R., Wright, D.J., Ellis, R.J., Crickmore, N., Bonsall, M.B., 2009. A mid-gut microbiota is not required for the pathogenicity of *Bacillus thuringiensis* to diamondback moth larvae. Environ. Microbiol. 11, 2556–2563.

Raymond, B., Johnston, P.R., Nielsen-LeRoux, C., Lereclus, D., Crickmore, N., 2010. *Bacillus thuringiensis*: an impotent pathogen? Trends Microbiol. 18, 189–194.

Reilly, J.R., Hajek, A.E., 2012. Prey-processing by avian predators enhances virus transmission in the gypsy moth. Oikos 121, 1311–1316.

Rosenheim, J.A., Kaya, H.K., Ehler, L.E., Marois, J.J., Jaffee, B.A., 1995. Intraguild predation among biological-control agents: theory and evidence. Biol. Contr. 5, 303–335.

Roy, H.E., Baverstock, J., Ware, R.L., Clark, S.J., Majerus, M.E.N., Baverstock, K.E., Pell, J.K., 2008a. Intraguild predation of the aphid pathogenic fungus *Pandora neoaphidus* by the invasive coccinellid *Harmonia axyridis*. Ecol. Entomol. 33, 175–182.

Roy, H.E., Brown, P.M.J., Rothery, P., Ware, R.L., Majerus, M.E.N., 2008b. Interactions between the fungal pathogen *Beauveria bassiana* and three species of coccinellid: *Harmonia axyridis*, *Coccinella septempunctata* and *Adalia bipunctata*. BioControl 53, 265–276.

Ruiu, L., Satta, A., Floris, I., 2007. Susceptibility of the house fly pupal parasitoid *Muscidifurax raptor* (Hymenoptera: Pteromalidae) to the entomopathogenic bacteria *Bacillus thuringiensis* and *Brevibacillus laterosporus*. Biol. Contr. 43, 188–194.

Ryder, J.J., Hoare, M.-J., Pastok, D., Bottery, M., Boots, M., Fenton, A., et al., 2014. Disease epidemiology in arthropods is altered by the presence of nonprotective symbionts. Am. Nat. 183, 89–104.

Sait, S.M., Begon, M., Thompson, D.J., Harvey, J.A., 1996. Parasitism of baculovirus-infected *Plodia interpunctella* by *Venturia canescens* and subsequent virus transmission. Funct. Ecol. 10, 586–591.

Sandoval-Aguilar, J.A., Guzmán-Franco, A.W., Pell, J.K., Clark, S.J., Alatorre-Rosas, R., Santillán-Galicia, M.A., Valdovinos-Ponce, G., 2015. Dynamics of competition and co-infection between *Zoophthera radicans* and *Pandora blunckii* in *Plutella xylostella* larvae. Fung. Ecol. 17, 1–9.

Scarborough, C.L., Ferrari, J., Godfray, H.C.J., 2005. Aphid protected from pathogen by endosymbiont. Science 310, 1781.

Shapira, M., 2016. Gut microbiotas and host evolution: scaling up symbiosis. Trends Ecol. Evol. 7, 539–549.

Shapiro-Ilan, D.I., Brown, I., 2013. Earthworms as phoretic hosts for *Steinernema carpocapsae* and *Beauveria bassiana*: Implications for enhanced biological control. Biol. Control. 66, 41–48.

Shapiro-Ilan, D.I., Rojas, M.G., Morales-Ramos, J.A., Lewis, E.E., Tedders, W.L., 2008. Effects of host nutrition on virulence and fitness of entomopathogenic nematodes: lipid and protein based supplements in *Tenebrio molitor* diets. J. Nematol. 40, 13–19.

Shikano, I., Cory, J.S., 2016. Altered nutrient intake by baculovirus-challenged insects: self-medication or compensatory feeding? J. Invertebr. Pathol. 139, 25–33.

Shikano, I., Ericsson, J.D., Cory, J.S., Myers, J.H., 2010. Indirect plant-mediated effects on insect immunity and disease resistance in a tritrophic system. Basic Appl. Ecol. 11, 15–22.

Shikano, I., Oak, M.C., Halpert-Scanderbeg, O., Cory, J.S., 2015. Trade-offs between trans-generational transfer of nutritional stress tolerance and immune priming. Funct. Ecol. 29, 1156–1164.

Shikano, I., Hua, K., Cory, J.S., 2016. Baculovirus-challenge and poor nutrition inflict within-generation fitness costs without triggering transgenerational immune priming. J. Invertebr. Pathol. 136, 35–42.

Shipp, J.L., Zhang, Y., Hunt, D.W.A., Ferguson, G., 2003. Influence of humidity and greenhouse microclimate on the efficacy of *Beauveria bassiana* (Balsamo) for control of greenhouse arthropod pests. Environ. Entomol. 32, 1151–1163.

Sih, A., Englund, G., Wooster, D., 1998. Emergent impacts of multiple predators on prey. Trends Ecol. Evol. 13, 350–355.

Simone-Finstrom, M.D., Spivak, M., 2012. Increased resin collection after parasite challenge: a case of self-medication in honey bees? PLoS ONE 7(3), e34601.

Snyder, W.E., Snyder, G.B., Finke, D.L., Straub, D.S., 2006. Predator diversity strengthens herbivore suppression. Ecol. Lett. 9, 789–796.

Stark, D.M., Mills, N.J., Purcell, A.H., 1999. Interactions between the parasitoid *Ametadoria misella* (Diptera: Tachinidae) and the granulovirus of *Harrisina brillians* (Lepidoptera: Zygaenidae). Biol. Contr. 14, 146–151.

Staves, P.A., Knell, R.J., 2010. Virulence and competitiveness: testing the relationship during inter- and intraspecific mixed infections. Evolution 64, 2643–2652.

Strand, M.R., Burke, G.R., 2013. Polydnavirus-wasp associations: evolution, genome association and function. Curr. Opin. Virol. 3, 587–594.

Teixeira, L., Ferreira, A., Ashburner, M., 2008. The bacterial symbiont *Wolbachia* induces resistance to RNA viral infections in *Drosophila melanogaster*. PLoS Biol. 6(12), e1000002.

Tokarev, Y.S., Levchenko, M.V., Naumov, M.V., Senderskiy, I.V., Lednev, G.R., 2011. Interactions of two insect pathogens, *Paranosema locustae* (Protista: Microsporidia) and *Metarhizium acridum* (Fungi: Hypocreales), during a mixed infection of *Locusta migratoria* (Insects: Orthoptera) nymphs. J. Invertebr. Pathol. 106, 336–338.

Valtonen, T.M., Kangassalo, K., Pölkki, M., Rantala, M.J., 2012. Transgenerational effects of parental larval diet on offspring development time, adult body size and pathogen resistance in *Drosophila melanogaster*. PLoS ONE 7(2), e31611.

Vasconcelos, S.D., Williams, T., Hails, R.S., Cory, J.S., 1996. Prey selection and baculovirus dissemination by carabid predators of Lepidoptera. Ecol. Entomol. 21, 98–104.

Wang, X.D., Ishibashi, N., 1999. Infection of the entomopathogenic nematode *Steinernema carpocapsae*, as affected by the presence of *Steinernema glaseri*. J. Nematol. 31, 207–211.

West, S.A., Flanagan, K.E., Godfray, H.C.J., 1996. The relationship between parasitoid size and fitness in the field, a study of *Achrysocharoides zwoelferi* (Hymenoptera: Eulophidae). J. Anim. Ecol. 65, 631–639

Zytynska, S.E., Weisser, W., 2016. The natural occurrence of secondary bacterial symbionts in aphids. Ecol. Entomol. 41, 13–26.

Section III

Ecology of Pathogen Groups

7

Viruses

Trevor Williams

Instituto de Ecologia AC (INECOL), Xalapa, Veracruz, Mexico

7.1   Introduction

The most commonly studied invertebrate viruses are those that frequently cause overt disease in their hosts. As such, many of the examples presented in this chapter involve insect pests of crops or forests that have been studied in the search for effective biological control agents. The majority of these approaches have focused on the use of viruses, mostly baculoviruses (*Baculoviridae*), as the active ingredient in biological insecticides. These types of products are usually applied in an inundative strategy of biological control in order to infect and kill a high proportion of pest insects in a short period of time. An alternative approach involves an inoculative strategy, in which small amounts of pathogen are released into the pest population. The pathogen multiplies over several transmission cycles until the pathogen population is sufficiently large to effectively control the pest population through the development of epizootics of disease.

The interest generated in invertebrate viruses largely depends on whether the host is considered to be of benefit, or not, to humans. Viruses that kill pests and vectors are generally viewed favorably, and considerable information has been obtained on the ecology of these diseases. In contrast, viruses of beneficial or commercially valuable invertebrates such as insect pollinators or shellfish are studied primarily when disease has a tangible economic impact on their populations. The same applies to insect mass-rearing facilities that produce massive numbers of insects for use in pest or vector-control programs involving the sterile insect technique (Kariithi et al., 2013). The foremost example of viruses infecting beneficials, however, is that of pathogenic viruses of honeybees, which have attracted a great deal of attention over the past decade in the search for the causative agent(s) of colony collapse disorder, which has been decimating bee populations in many parts of North America and Europe (Cox-Foster et al., 2007; Martin et al., 2012). Pathogens of beneficial invertebrates in terrestrial and aquatic ecosystems are considered elsewhere in this book (see Chapters 14 and 15), and are only mentioned briefly here.

The use of the terms pathogenicity and virulence often varies across the literature on invertebrate viruses. This is because ecologists, evolutionary biologists, and invertebrate pathologists have applied different definitions depending on the focus of their

Ecology of Invertebrate Diseases, First Edition. Edited by Ann E. Hajek and David I. Shapiro-Ilan.
© 2018 John Wiley & Sons Ltd. Published 2018 by John Wiley & Sons Ltd.

studies, or have used different combinations of metrics to define each concept (see a discussion of these issues by Thomas and Elkinton, 2004; Shapiro-Ilan et al., 2005). To avoid confusion, and because I have drawn examples from across all of the disciplines involving host–virus interactions, I have opted mostly to avoid these terms in favor of the metrics that were employed, such as infectivity (the capacity to infect), dose or concentration–mortality relationships, and speed of kill.

In addition to baculoviruses, many other invertebrate viruses are known to infect invertebrates in terrestrial or aquatic habitats (Rybov 2016; Williams et al., 2016). However, this chapter has restricted its focus to the better-known virus families for which most information is available. That said, a world of opportunities remains available for any researcher wishing to study the ecology of the better- and lesser-known viruses (tetraviruses, nodaviruses, birnaviruses, idnoreoviruses, herpesviruses, nidoviruses, etc.) that infect insects and other invertebrates.

7.2 Diversity of Invertebrate Pathogenic Viruses

The virus pathogens of invertebrates are classified in orders, families, genera, and species based on multiple criteria related to the physical characteristics of the virus particle, the genome properties (e.g., type of viral nucleic acid, genome organization and gene content, deduced phylogenetic relationships, the replication cycle within the host cell), and ecological characteristics (e.g., types of host infected and nature of virus disease (pathology)). Although virus species is a recognized concept and has an established definition, isolates of viruses are given names that are not italicized, even if they include the name of the host species (Kuhn and Jahrling, 2010). I have adopted this practice here. Virus family names, in contrast, are italicized.

One key characteristic that determines the ecology of these pathogens is the presence or absence of an occlusion body (OB) (Table 7.1.). This is a crystalline matrix of protein that surrounds the virus particle (virion) and protects it during periods outside the host. This structure is particularly important in the transmission of viruses that infect invertebrates in terrestrial habitats, as it allows the virus to persist on plant surfaces, where plant secondary chemicals and solar ultraviolet (UV) radiation can inactivate it, or in the soil, where enzymes released by microorganisms may otherwise degrade viral proteins and nucleic acids.

The baculoviruses, entomopoxviruses, and cypoviruses are all characterized by forming large OBs, typically 0.5–4.0 μm in diameter, that can be visualized using a phase contrast microscope. The nonoccluded viruses, such as the densoviruses, nudiviruses, iflaviruses, hytrosaviruses, and iridescent viruses, tend to exploit routes of transmission that do not involve extended periods in the environment. The ascoviruses and polydnaviruses have intimate relationships with parasitoid wasps. These viruses are carried between hosts by the parasitoids and are never exposed, or are exposed only very briefly, to environmental conditions outside the host insect.

As most of our understanding of invertebrate virus ecology comes from baculoviruses, it is worth briefly mentioning here the baculovirus transmission cycle. When a susceptible lepidopteran larva consumes OBs, these break down in the alkaline insect midgut and release occlusion-derived virions (ODVs). These virions have to cross the peritrophic membrane, which is a tube of chitin and glycoproteins that lines the midgut

Table 7.1 Main virus pathogens of invertebrates mentioned in this chapter.

Virus (family)	Genome[a]	Occlusion body	Hosts[b]	Main tissues infected	Route of horizontal infection[c]	References
Ascoviruses (*Ascoviridae*)	dsDNA	No	Lep., Hym.	Fat body	Vectored by parasitoid wasps	Bideshi et al. (2010)
Baculoviruses[d] (*Baculoviridae*)	dsDNA	Yes	Lep., Hym., Dip.	All tissues	Ingestion of occlusion bodies	Rohrmann (2013)
Cypoviruses (*Reoviridae*)	dsRNA	Yes	Lep., Dip.	Midgut	Ingestion of occlusion bodies	Mori and Metcalf (2010)
Densoviruses (*Parvoviridae*)	ssDNA	No	Lep., Dip., Hem., Ort., Bla., crustaceans	All tissues	Ingestion of virions	Bergoin and Tijssen (2010)
Dicistroviruses (*Dicistroviridae*)	ssRNA	No	Hem., Hym., Ort., Dip., Acari, crustaceans	Variable	Ingestion or injection of virions, vertical transmission	Bonning and Miller (2010)
Entomopoxviruses (*Poxviridae*)	dsDNA	Yes	Col., Lep., Ort., Dip.	Fat body, hemocytes	Ingestion of spheroid occlusion bodies	Thézé et al. (2013)
Hytrosaviruses (*Hytrosaviridae*)	dsDNA	No	Dip.	Salivary gland	Ingestion of virus-contaminated saliva	Abd-Alla et al. (2010)
Iflaviruses (*Iflaviridae*)	ssRNA	No	Lep., Hym., Hem., Acari	Gut and other tissues	Ingestion of virus particles	van Oers (2010)
Invertebrate iridescent viruses (*Iridoviridae*)	dsDNA	No	Many insect orders, crustaceans	All tissues	Cannibalism, wounding, parasitism	Williams and Ward (2010)
Nudiviruses (*Nudiviridae*)	dsDNA	No	Col., Lep., Ort., Dip., crustaceans	Variable (gut, fat body, reproductive tissues)	Ingestion or sexual reproduction	Jehle (2010)

(Continued)

Table 7.1 (Continued)

Virus (family)	Genome[a]	Occlusion body	Hosts[b]	Main tissues infected	Route of horizontal infection[c]	References
Polydnaviruses (*Polydnaviridae*)	dsDNA	No	Hym. (Lep. but without replication)	Female wasp reproductive tissues (but causes immune suppresion in parasitized lepidopteran hosts)	Only vertical transmission during oviposition by parasitoid wasps (lepidopteran hosts are infected, but without virus replication)	Strand and Burke (2015)

a) ds, double stranded; ss, single stranded.
b) Bla., Blattodea; Col., Coleoptera; Dip., Diptera; Hem., Hemiptera; Hym., Hymenoptera; Lep., Lepidoptera; and Ort., Orthoptera.
c) The majority of the invertebrate viruses can also be transmitted vertically from parents to offspring (see Section 7.5.2).
d) Baculoviruses can be divided into NPVs (genera: *Alphabaculovirus* in Lepidoptera, *Deltabaculovirus* in Diptera, *Gammabaculovirus* in phytophagous Hymenoptera) or GVs (genus *Betabaculovirus*) based on morphology and genetic factors.

and protects it from abrasion and pathogens. The ODVs then infect midgut epithelial cells, where they undergo replication to produce virions that bud through the basal membrane of the cell into the hemolymph. The budded virions disperse in the hemolymph to infect other cells during the systemic phase of infection (Rohrmann, 2013). Following multiple rounds of systemic infection, the infected cells accumulate large numbers of OBs, which are then released into the environment, often following the death of the insect, for transmission to other susceptible larvae (see Chapter 3).

Of the virus families listed in Table 7.1., only the cypoviruses, dicistroviruses, and iflaviruses have an RNA genome; all the others are DNA viruses. The type, organization, and quantity of nucleic acid in the viral genome all have implications for the infection strategy, replication scheme, and ability to carry supplementary, nonessential genes that improve aspects of virus fitness. Thus, with the exception of the densoviruses, the genomes of DNA viruses tend to be far larger than the genomes of RNA viruses. This reflects the diversity of "survival strategies" that viruses can adopt, ranging from structurally simple particles with a small, compact genome to large complex particles with an extensive array of genes that have structural, replication, and auxiliary functions. The ecological consequences of this diversity will be explored in the course of this chapter.

7.3   Distribution of Invertebrate Pathogenic Viruses

Invertebrate pathogenic viruses are present on all continents of the world, in terrestrial, freshwater, and marine habitats. A recent metagenomic study even reported the presence of dicistroviruses, iflaviruses, and iridoviruses in a remote Antarctic lake that was frozen for most of the year, although the host species were not identified (López-Bueno et al., 2015). As obligate intracellular microparasites, the primary factor that determines the presence of pathogenic viruses in a particular locality is, of course, the presence of the invertebrate host. The principal factors that determine the presence of the host are suitable climatic conditions and the availability of a suitable food supply, be it a plant in the case of phytophagous insects, a vertebrate host in the case of hematophagous arthropods, or plankton, algae, or organic particles in the case of marine crustaceans or mollusks (see Chapters 4 and 6). Even for occluded viruses than can persist in the environment for extended periods, the periodic presence of the host population is required to maintain a viable pathogen population.

Our current understanding of the diversity and distribution of invertebrate pathogens has less to do with the geographical distribution of pathogens and much more to do with the geographical distribution of invertebrate pathologists and the availability of scientific infrastructure for the study of diseased insects and other invertebrates. This was clearly reflected in a qualitative analysis of the development of virus-based biological insecticides in different geographical regions, in which North America and Europe were developing more viruses for pest control than the countries of Africa, Central and South America, Oceania, and the Indian subcontinent (Entwistle, 1998). That said, the rapid growth in the study of insect pathogenic viruses in China over the past 2 decades has resulted in significant advances in the use of these pathogens in pest control (Sun, 2015). As a result of the use of viruses in biological control, particularly baculoviruses, many of the following examples are from viruses of insect pests in forest and agricultural ecosystems.

7.4 Key Aspects of Pathogen Ecology

The survival of a pathogen in a particular host population depends on a complex set of interactions that modulate transmission. Transmission itself is the process by which a pathogen or parasite is passed from an infected host to a susceptible host of the same or subsequent generations (see Chapter 1). Because of this, the mechanistic aspects of virus transmission (i.e., the route by which the virus leaves one host and gains entry to a new host to achieve infection) are highly influential in the ecology of virus diseases.

The majority of invertebrate viruses employ direct transmission (Tanada and Kaya, 1993). This means that the virus passes directly from one host to another through reproduction or sexual contact. Alternatively, some common viruses have an intermediate step in which they leave the infected host and wait in the environment until encountered by a new susceptible host. In contrast, indirectly transmitted pathogens require an intermediate (secondary) host or a vector organism in order to pass from one primary host to another.

In the following sections, it will become clear that pathogen survival involves the interplay of transmission with persistence in the environment (in the case of occluded viruses) and dispersal across a range of spatial scales, from local movement between plants and soil to regional dispersal, usually via host-mediated migration. Virus transmission, persistence, and dispersal are modulated by host-related factors such as foraging behavior, and by biotic factors, often involving the host plant in the case of phytophagous insects, and abiotic factors that reflect specific characteristics of the environment. It is therefore important to bear in mind in the following sections that transmission, persistence, and dispersal should not be viewed in isolation but rather as a set of interacting and interdependent processes.

With the development of molecular tools over the past 2 decades, we are slowly becoming aware that many invertebrate species harbor covert (inapparent) infections by viral pathogens that can affect different aspects of their development or reproductive capacity in the absence of clear signs of disease. However, our understanding of the ecology of nonlethal viruses lags many years behind that of lethal virus pathogens. There are several reasons for this: obvious diseases tend to attract the attention of researchers that study these organisms, studies are more easily targeted at individuals showing specific signs of morbidity or mortality in a population, the massive proliferation of the virus in lethally infected individuals and the associated pathological changes in tissues and organs simplify the correct identification of the causative agent, and standard laboratory techniques have usually been developed and verified for the detection and identification of the most serious invertebrate diseases. That said, advances in molecular detection techniques now allow the screening of large numbers of organisms in the search for particular pathogens, including viruses from insects (Zwart et al., 2008; Virto et al., 2014; Zhou et al., 2015) or other invertebrates (Ren et al., 2010; Panichareon et al., 2011). Alternatively, transcriptome studies and metagenomics approaches are proving highly informative in the discovery of nonlethal viruses in ants (Valles et al., 2012), bees (Cox-Foster et al., 2007), mosquitoes (Cook et al., 2013), dragonflies (Rosario et al., 2011), and moths (Pascual et al., 2012; Jakubowska et al., 2014, 2015), among others.

Serendipity has also played an important role in the discovery of nonlethal viruses. The detection of nonlethal viruses has frequently been accidental during the study of apparently healthy individuals (Lacey and Brooks, 1997), when working with apparently healthy cell lines (Carrillo-Tripp et al., 2014), or during the study of lethal viruses in which nonlethal viruses can appear as contaminants (Wagner et al., 1974; Jakubowska et al., 2016). As a result, most of the examples in the following sections focus on lethal viruses of insects, particularly baculoviruses, which are by far the best understood insect–virus pathosystems (Cory, 2010). However, when working on virus ecology, it is important to bear in mind that just because an experimental individual or population appears to be healthy, this is not evidence that it is not infected by one or more pathogenic viruses.

7.5   Transmission

Transmission is described as **horizontal** when the pathogen leaves an infected host and passes to a susceptible host (other than the host's offspring). This involves a spatial component in transmission, even for viruses that adopt a sit-and-wait strategy during the environmental phase of transmission. Virus particles that remain infectious outside the host can infect individuals from the same generation or subsequent generations. Alternatively, **vertical** transmission occurs when infected parents reproduce and pass the pathogen to their offspring. As such, vertical transmission is a mechanism for transgenerational transmission in pathogens that do not kill their host prior to reproduction. In fact, many pathogenic invertebrate viruses adopt a mixed strategy involving both horizontal and vertical transmission, depending on the conditions within the infected host and the relative probability of successful transmission by either route.

7.5.1   Horizontal Transmission

For lethal viruses, the death of the host is usually followed by the release of massive numbers of virus particles. This is characteristic of baculoviruses, such as nucleopolyhedroviruses (NPVs) and granuloviruses (GVs), that infect lepidopteran larvae. Many baculoviruses have genes for cathepsin and chitinase enzymes that rapidly break down the host tissues and liquefy the virus-killed insect (Ishimwe et al., 2015). This facilitates the release of virus OBs, which are then spread over the surfaces of the leaves and stems of the host plant by gravity or through the action of wind and rain. A single infected late-instar larva can release enormous numbers ($\sim 10^6$–10^9) of OBs (Shapiro, 1986). As susceptible larvae may become infected following the consumption of a single or a few OBs, depending on species and the growth stage of the larva, the death of a single infected insect can have the potential to transmit the infection to many other larvae that consume OB-contaminated foliage.

In agricultural settings, in which baculoviruses are used as insecticides, OBs are usually applied to the whole crop, resulting in a near-uniform distribution of OBs. Similarly, viruses applied as insecticides are present at high densities on the crop so that pest insects rapidly acquire a lethal infection during periods of feeding in the hours following the application of OBs (Lasa et al., 2007). In natural settings, insects acquire infections from OBs in the environment that likely have a random or clumped distribution (Dwyer, 1991). A clumped distribution of OBs reflects the local distribution of recent deaths of infected insects from which viral OBs have been released (Vasconcelos et al.,

1996b; D'Amico et al., 2005). As such, we would expect the patterns of transmission in areas where natural populations of viruses exist to be quite different to those in crops treated with virus-based insecticides, although the basic principles related to transmission remain the same.

An insect's susceptibility to virus infection usually decreases markedly as it grows. For example, in baculoviruses, the 50% lethal dose of OBs increases by 10^3–10^5-fold between the first and final instars in many species of Lepidoptera (Briese, 1986). The mechanism of this developmental resistance likely involves three main factors: (i) a decreasing surface area–volume ratio in the gut of growing larvae, which means that in order to reach midgut cells, virus particles must pass through an increasing volume of food bolus as the larvae age (Hochberg, 1991a); (ii) a stage-related increase in the thickness and reduction in the porosity of the peritrophic membrane through which virions must pass to reach and infect midgut epithelial cells (Wang and Granados, 2000; Levy et al., 2012); and (iii) an increase in the rate of sloughing of infected midgut cells in later, compared to earlier, instars (Kirkpatrick et al., 1998). Cell sloughing reduces the period available for the virus to replicate in infected midgut cells and produce the budded virions that establish a systemic infection (McNeil et al., 2010). As the peritrophic membrane represents a major barrier to pathogens infecting through the gut, several GVs and NPVs produce mucin glycoprotein-degrading enzymes, which they carry in the OBs or the virions, to degrade the membrane and facilitate access to midgut cells (Peng et al., 1999b; Slavicek and Popham, 2005; Hoover et al., 2010). Similarly, a chitinase domain in the fusolin protein of entomopoxviruses is activated in the insect midgut to degrade the chitin component of the peritrophic membrane and facilitate access of the large entomopoxvirus virions to midgut cells (Mitsuhashi and Miyamoto, 2003; Chiu et al., 2015).

As a result, the probability of horizontal transmission depends on complex interactions among the density of inoculum OBs in the environment, the spatial distribution of the inoculum (uniform, random, or clumped), the host density, the feeding behavior, and the susceptibility of insects to infection (Dwyer, 1991; Goulson et al., 1995; D'Amico et al., 1996; Reeson et al., 2000; Parker et al., 2010). As many of these variables differ markedly between distinct species of insects and their viruses, it is clear that quantitative estimates of transmission require an understanding of the behavior of healthy and infected insects, the rate of decay of OBs in the environment, and variation generated through heterogeneity in host susceptibility and host plant effects (for herbivorous hosts), in addition to the usual estimates of host and pathogen densities in each pathosystem.

A clear example of how density affects transmission comes from comparative studies on lepidopteran larvae that live alone or in groups. Solitary species usually exist at relatively low densities, as dispersal behavior or cannibalism reduces the numbers of individuals in a particular locality. In contrast, gregarious species experience high local densities of conspecifics within each group of individuals. In the case of solitary species, resistance to infection depends mainly on larval weight, whereas in gregarious larvae, resistance to infection increases faster than body weight gain, as the risks of virus transmission for each individual in a group-living species increase with age (Hochberg, 1991a).

Two additional routes of horizontal transmission are frequently found in invertebrate viruses: cannibalism of infected individuals and transmission during sexual contact. As both of these routes involve specific behaviors, they are considered in Sections 7.9.2 and 7.9.3.

7.5.1.1 Estimating Horizontal Transmission

Initial attempts to quantify the transmission process in baculoviruses adopted the principle of mass action, in which transmission is directly proportional to the density of susceptible and infected individuals (or infectious OBs in the case of baculoviruses) in the local population (Anderson and May, 1981). The mass action principle assumes that the efficiency of transmission is a constant (the transmission coefficient), reflecting a fixed probability of infection following contact between a susceptible individual and a pathogen particle in the environment (McCullum et al., 2001). For the host population, this can be written as $dS/dt = -\nu SP$, where S is the density of susceptible hosts, P is the density of virus particles in the environment, and ν is a constant describing the probability of transmission.

However, a series of field studies with NPVs has demonstrated that the efficiency of transmission is not constant but varies with the density of susceptible insects and the pathogen (D'Amico et al., 1996), insect growth stage, area of foliage consumed (Goulson et al., 1995), pathogen clumping (Dwyer 1991; D'Amico et al., 2005), heterogeneity in host susceptibility (Dwyer et al., 1997; Reeson et al., 2000; Hudson et al., 2016), duration of exposure to the pathogen, and density-dependent variation in insect behavior (Reeson et al., 2000). Similar findings were reported in laboratory studies on a GV of the meal moth, *Plodia interpunctella*, that was transmitted via cannibalism of infected individuals. In this case, the transmission coefficient increased with the density of susceptible hosts and decreased with the density of infected cadavers (Knell et al., 1998). The transmission efficiency also declined over time as infected cadavers were rapidly consumed by cannibalistic larvae in the first few hours of the experiments, resulting in a reduction in the overall pathogen density.

Behavioral, physiological, and environmental factors may affect both the probability of contact between pathogen particles and susceptible insects and the probability of successful infection once contact has occurred. Consequently, an alternative approach has been developed in which a proportion of the host population is considered to occupy a pathogen-free refuge, the size of which can vary according to the size of the pathogen or of susceptible insect populations (Hails et al., 2002). This approach proved useful for comparison of transmission risks in lepidopteran populations exposed to wild-type and recombinant baculoviruses (Hails et al., 2002) and in mosquito larvae exposed to an iridescent virus (Marina et al., 2005), highlighting the versatility of the procedure. The value of other formal approaches to the study of virus transmission in insect populations is discussed in detail in Chapter 12.

7.5.2 Vertical Transmission

Vertical transmission of invertebrate viruses, from parents to offspring, is a route that is only available to pathogens that do not invariably kill their hosts prior to reproduction. The presence of vertically transmitted infections in insects has been suspected since early observations that the offspring of seemingly healthy insects could spontaneously succumb to virus diseases even under clean laboratory conditions (Kukan, 1999). Vertical transmission is also an issue of concern in laboratory colonies of insects that are required to be pathogen-free for use in a diversity of experimental settings (Helms and Raun 1971; Fuxa et al., 1999) or in insect mass-production facilities, where the impact of vertically-transmitted pathogens can be devastating (Greenberg, 1970; Boucias et al., 2013; Morales-Ramos et al., 2014).

Molecular studies have demonstrated that vertical transmission is a common feature of viruses in natural populations of insects (Carpenter et al., 2007; Virto et al., 2014; Cory, 2015), including beneficial insects such as honeybees (de Miranda and Fries, 2008) and other invertebrates (Cowley et al., 2002; Barbosa-Solomieu et al., 2005). Similarly, electron microscopy studies have provided evidence for the presence of virus particles in the ovarian tissues or developing eggs of adult female Lepidoptera infected with NPV (Smith-Johannsen et al., 1986), densovirus (Garzon and Kurstak, 1968), or nudiviruses (Raina et al., 2000; Rallis and Burand, 2002), as well as for a hytrosavirus in the tsetse fly (*Glossina pallidipes*) (Jura et al., 1989) and an ascovirus and iflavirus in parasitoid wasps (Bigot et al., 1997; Reineke and Asgari, 2005), among many other examples. In the case of rhabdoviruses or entomopoxviruses of parasitoid wasps, the viruses may replicate in the poison or accessory glands of the infected female wasp before being injected into host insects together with the parasitoid egg(s), which are subsequently infected by the virus (Lawrence and Akin, 1990; Lawrence and Matos, 2005).

Studies focusing on male involvement in vertical transmission are less common than studies on females. Nevertheless, evidence in favor of male involvement includes observations on the presence of virus in the testes for NPVs, GVs, and nudiviruses in Lepidoptera (Lewis et al., 1977; Burden et al., 2002; Pereira et al., 2008), reovirus in Coleoptera (Kitajima et al., 1985), and rhabdoviruses in Diptera (Longdon et al., 2011), among others. Indeed, male involvement in transmission during mating has been reported in densoviruses (Barik et al., 2016), sigmaviruses (Longdon and Jiggins, 2012), NPVs (Knell and Webberley, 2004), iflaviruses (Yue et al., 2007), nudiviruses (Zelazny, 1976; Burand, 2009), and iridescent viruses (Marina et al., 1999; Adamo et al., 2014), among others.

To determine whether the virus particles responsible for vertical transmission are present inside or on the surfaces of eggs, experimental egg masses are often subjected to surface decontamination using formalin or hypochlorite solutions. If no difference is observed in the incidence of virus infection in the progeny from surface-decontaminated versus untreated eggs, it is usually concluded that the virus is likely to have been transmitted in the developing embryos within eggs, which is known as transovarial transmission (see Chapters 1 and 3). In contrast, if surface decontamination of eggs markedly reduces the incidence of infection in the offspring, it is likely that most infections are acquired by ingestion of virus particles on the exterior egg surface, which the hatching larvae consume as they chew their way out. If the contaminating virus particles were deposited by the female during oviposition, this is known as transovum transmission. The parental origin of the contaminating virus (versus environmental sources of inoculums) defines vertical transmission (Murray and Elkinton, 1989). In the case of baculoviruses, scanning electron microscopy (SEM) has been used to confirm the presence of viral OBs on the exterior egg chorion (Hamm and Young, 1974; Nordin et al., 1990).

In a few cases, viruses have developed a symbiotic, mutualistic relationship with their hosts that depends on vertical transmission. One example is seen in the ascoviruses that infect lepidopteran larvae (Bideshi et al., 2010). Horizontal transmission in ascoviruses is normally achieved when a female parasitoid wasp carries the virus on her ovipositor from an infected to a susceptible caterpillar. The Diadromus pulchellus ascovirus 4a (DpAV-4a) differs from other ascoviruses in that it replicates in the parasitoid ovary (Bigot et al., 1997). During oviposition, the wasp injects the virus into pupae of the leek moth (*Acrolepiopsis assectella*). The virus suppresses the leek moth immune response, allowing development of the parasitoid progeny, which themselves acquire the virus for future cycles of vertical transmission (Renault et al., 2002).

The symbiotic relationship with ichneumonid or braconid parasitoid wasps has evolved further in another family of viruses, the polydnaviruses (Strand, 2010). These are transmitted to offspring as so-called proviruses, which replicate in a wasp's reproductive tract, producing encapsidated particles that are injected into the wasp's host (usually a caterpillar) during oviposition. Once injected, the polydnavirus particles enter the cells of different caterpillar tissues but do not replicate. Instead, they carry and express a series of genes from the wasp's genome that favor the survival of the wasp's offspring. The most important effect of polydnaviruses is to protect the developing wasp egg or larva by suppressing the caterpillar immune response, which would otherwise encapsulate the developing parasitoid within hemocytes, leading to melanization and parasitoid death (Gundersen-Rindal et al., 2013). Therefore, the virus is essential for the successful development of the parasitoid, and the virus – which is in reality an extension of the genome of the wasp – achieves continuous cycles of vertical transmission (Strand, 2012).

7.6 Persistence

Viruses can persist in two quite different ways: (i) as a covert infection within the host, aimed at achieving vertical transmission through host reproduction; or (ii) in the environment as long-lived infective stages. Whether a virus adopts a lethal or a nonlethal strategy of host exploitation depends largely on the relative opportunity for horizontal versus vertical transmission, which is the clearest indicator of virus fitness (Cory and Franklin, 2012). So-called mixed-mode transmission strategies are observed in many of the invertebrate viruses.

7.6.1 Persistence within the Host

Inapparent infections are common in a wide range of invertebrates, but they have not been quantified because of a lack of interest in infections that do not cause immediate patent disease, because their detection usually requires considerable knowledge of invertebrate pathology and molecular techniques and an appreciation of the complexity of host–virus relationships, and because of difficulties in identifying novel viruses below the level of virus family (Okamura, 2016).

Publications on this topic variously refer to covert, inapparent, silent, occult, persistent, or sublethal infection, in which the virus replicates at a low level without killing the host. In this sense, covert infection differs from latent infection, in which the virus genome either is integrated into the host genome or persists in an inactive state in host cells with minimal replication (Lin et al., 1999; Fang et al., 2016). The latent state is poorly understood in invertebrate viruses, although it may involve the suppression of cell epigenetic silencing and the production of viral miRNAs that inhibit the expression of lytic viral genes (Wu et al., 2011; Hussain and Asgari, 2014).

For studies in which the presence of the pathogen has been confirmed, sublethal disease is characterized by decreased body weight and adult eclosion, decreased reproduction and longevity (Marina et al., 2003; Sood et al., 2010; Cabodevilla et al., 2011b), and adverse effects on sperm production (Sait et al., 1998). Due to the association between covert infections and reduced reproduction, sublethal disease has been implicated as a potentially important factor modulating the population dynamics of insect populations (Boots et al., 2003; Bonsall et al., 2005; Myers and Cory, 2016), although empirical evidence for this is sparse.

Sublethal effects have three possible origins: (i) as a direct result of the pathological effects of the virus within the host; (ii) due to the metabolic costs incurred from mounting an immune response to suppress the pathogen; or (iii) as a result of host traits that are corrected with disease resistant phenotypes (Myers and Kuken, 1995; Rothman and Myers, 1996; Bouwer et al., 2009). Fortunately, molecular techniques now allow covertly infected individuals to be identified with a high degree of confidence. These individuals can be differentiated from those who were exposed to viral inoculum but did not become infected and from those who became infected but managed to rid themselves of the infection.

Covert infections were initially detected using DNA hybridization techniques (Christian, 1992; Kukan and Myers, 1995), but these were superseded by polymerase chain reaction (PCR) and multiplex PCR, which detect viral genomic DNA in host tissues (Williams, 1993; Hughes et al., 1997; Lupiani et al., 1999; Arzul et al., 2002; Abd-Alla et al., 2007; Kemp et al., 2011), or reverse transcriptase PCR, which detects gene transcription as an indicator of virus replication (Burden et al., 2002; Martínez et al., 2005; Vilaplana et al., 2010). The use of expressed sequence tag (EST) libraries based on mRNA sequences purified from the host has also proved useful, although only viruses with high titers are likely to be detected using this method (Liu et al., 2011). Quantitative PCR (qPCR) now allows researchers to detect very low numbers of gene copies in experimental samples (Yue et al., 2007; Murillo et al., 2011; Blanchard et al., 2014), as do recently developed amplification techniques (Xia et al., 2014, 2015), transcriptomics, and next generation sequencing (Liu et al., 2011; Ma et al., 2011; Kolliopoulou et al., 2015; Webster et al., 2015).

Interestingly, the amounts of NPV present in covertly infected adult Lepidoptera differed markedly in different parts of the adult body, with particularly high virus loads in the head, legs, and wings; previously researchers have tended to focus on tissues and organs within the insect abdomen, such as the fat body. The whole-body virus load also differed with life stage, being highest in eggs and neonate larvae and lowest in final-instar and adult insects (Graham et al., 2015).

7.6.2 Persistence Outside of the Host

Viruses vary markedly in their ability to persist outside the host (Ignoffo, 1992). These differences reflect the importance of environmental persistence in their transmission cycle. As mentioned in Section 7.2, the virions of occluded viruses, namely the baculoviruses, entomopoxviruses, and cypoviruses, are protected by the protein matrix that forms the OB. This structure allows virions to persist for months or years in protected environments (Jaques, 1985). Some nonoccluded viruses, such as densoviruses (*Parvoviridae*), are also capable of extended periods of survival outside the host (Kawase and Kurstak, 1991), whereas the Oryctes nudivirus is inactivated within a few days in the environment (Zelazny, 1972). That said, with the exception of a number of baculoviruses, the persistence of invertebrate pathogenic viruses has not been the subject of systematic study or quantification, so information on environmental persistence in many virus families is limited.

One important factor to take into account in studies on virus persistence is that research performed prior to the development of PCR-based detection almost invariably employed bioassay techniques or serological reactions to estimate the quantities of infectious virus present in a particular sample. In contrast, molecular techniques are used to detect or quantify viral DNA or RNA in environmental samples (Hewson et al., 2011; Krokene et al., 2012), which is not necessarily equivalent to a measure of the quantity of virus that retains infectivity for the target host. Where possible, the results of molecular analyses should, therefore, be verified using biological assays.

7.6.2.1 Persistence on Plants

The sit-and-wait strategy of transmission of baculovirus, cypovirus, and entomopoxvirus pathogens of phytophagous insects necessitates that these viruses persist in an infective state on the food plant until consumed by a suitable host. However, virus persistence on plants involves a series of complex interactions of virus particles with plant architecture, leaf epidermal structure, leaf surface chemistry, and plant phenology that usually has to be studied as a set of variables rather than as individual factors. Environmental factors also interact with plant-related variables to influence virus persistence. For example, following the release of baculovirus OBs from an infected insect, OBs may be washed by rainfall on to the upperside or underside of leaves or plant stems. Each of these locations will differ in the presence, density, and physical characteristics of surface hairs (trichomes), surface wrinkles, and pits, stomata, glandular structures, and epicuticular waxes that are likely to affect OB adhesion and retention. Laboratory studies on the forces involved in OB attachment to hydrocarbons, such as those present in leaf waxes, have indicated that strong hydrophobic interactions are probably important in maintaining OB adhesion (Small et al., 1986). The upper and lower leaf surfaces and stems will also differ in their exposure to solar UV radiation; a major factor in the inactivation of virus pathogens in the environment (see Section 7.11.1). The presence of plant exudates can also result in different chemical environments being present at these sites. For example, in the case of plants of the family Malvaceae, leaf-surface pH values are high (pH 8–11) and can differ markedly between the upper and lower phylloplanes, depending on plant species. Phylloplane pH also tends to increase as the leaf ages (Harr et al., 1984). The alkalinity of the leaf surfaces is due to the presence of glandular trichomes that secrete carbonates and bicarbonates of magnesium, calcium, and potassium (Elleman and Entwistle, 1982). This contrasts with the phylloplane of many other plants, which tends to be slightly acidic (e.g., pH 5–6 in the case of maize) (Derridj, 1996).

When applied to cotton leaves, NPV OBs were inactivated within 24 hours, even when plants were not exposed to solar radiation (Young and Yearian, 1974; Elleman and Entwistle, 1985a). The presence of dew droplets on the leaves appears to solubilize the exudates and likely speeds the inactivation of OBs (Young et al., 1977). Although OBs exposed to cotton exudates retain their polyhedral structure, exposure to metal cations in leaf exudate may have reduced the solubility of these OBs in the insect midgut (Elleman and Entwistle, 1985a,b).

OBs on bark or plant stems represent an important pathogen reservoir in populations of the gypsy moth (*Lymantria dispar*). OBs on bark can infect neonate larvae as they search for suitable foliage (Woods et al., 1989) or can be washed by rainfall and contaminate egg masses prior to hatching (Murray and Elkinton, 1989).

Plant phenology will often influence OB persistence, both in terms of the types of plant structures available (leaves, flowers, fruits, etc.) and in terms of changes in plant architecture during growth and development. For example, the leaf whorl of maize plants is the preferred feeding site of larvae of the fall armyworm (*Spodoptera frugiperda*). The leaf whorl also provides a natural cuplike structure that protects OBs from solar radiation as developing leaves expand and grow out of the whorl (Castillejos et al., 2002). Similarly, the ability of OBs to persist on the surfaces of fruit can be quite different from that on leaves, which can have important implications for the effectiveness of OBs applied as biological insecticides, such as the NPV used to control *Heliothis virescens* on cotton (Fuxa, 2008) and the GVs used to control larvae of the codling moth (*Cydia pomonella*) and lepidopteran pests of citrus (Ballard et al., 2000; Moore et al., 2015). In contrast, although root-feeding invertebrates are infected by a number of occluded and nonoccluded viruses, the effect of root structure, root exudates, and root microbiota on virus persistence on and around root systems remains unknown.

Adopting a formal approach, Fuller et al. (2012) have argued that OB persistence on foliage cannot be estimated accurately unless virus decay is measured independently of infectiousness. To do so, they varied the density of infected *L. dispar* cadavers and the exposure time. Following different periods of decay, cadavers on oak branches were enclosed in gauze bags with healthy larvae to estimate transmission. Importantly, OB-contaminated foliage contained within gauze bags was protected from additional decay during the period in which transmission (infectiousness) was determined. The best estimates of average virus persistence in this system varied from 2.5 days in 2008 to 14.3 days in 2007, although the 2007 estimates were judged unreliable due to predation of experimental larvae. In contrast, data taken from studies by other authors using purified OB suspensions indicated average persistence times of 0.9–1.5 days for Lymantria dispar nucleopolyhedrovirus (LdNPV) on oak (Webb et al., 1999, 2001), 0.5–1.3 days for Trichoplusia ni nucleopolyhedrovirus (TnNPV) on cabbage (e.g., Jaques, 1972), and 0.4–0.9 days (half-life) for Helicoverpa armigera nucleopolyhedrovirus (HearNPV) on cotton (Sun et al., 2004). These findings provide support for previous observations that OBs released from infected cadavers persist on trees and field crops for significantly longer than purified OBs applied as biological insecticides (Magnoler, 1968; Evans and Entwistle, 1982; Young and Yearian, 1989; Pessoa et al., 2014). This is probably because the debris and substances from the insect cadaver provide improved adhesion to plant surfaces and/or protection from UV radiation (see Section 7.11.1).

A special case of virus persistence on plants is that of viruses of insect pollinators on flowers. Several recent studies have implicated flowers as sites that can become contaminated with parasites or viruses (e.g., deformed wing virus, *Iflaviridae*) when infected insects visit flowers. These parasites and pathogens can infect other susceptible pollinators that subsequently visit contaminated flowers, such as honeybees, bumblebees, and wasps (Singh et al., 2010; Evison et al., 2012; Fürst et al., 2014; McMahon et al., 2015). Consequently, the role of pollen, nectar, or other flower traits in the persistence and transmission of pollinator viruses has begun to generate interest among researchers concerned about recent global pollinator declines (McArt et al., 2014). The high visitation rates to flowers by pollinators and the ability of non-host pollinators to disperse pathogens and parasites from contaminated to noncontaminated flowers (Graystock et al., 2015) suggest that the persistence of pollinator viruses on flowers may play a significant, but poorly understood, role in the ecology of these pathogens.

7.6.2.2 Persistence in Soil

The soil is the most important environmental reservoir for occluded viruses. Rain splash, surface water, and windblown dust can move OB-contaminated soil particles from the soil on to plants, where they can be consumed by susceptible insects (Hochberg, 1989; Fuxa et al., 2007). When infected plant-feeding insects die, they fall on to the soil surface or remain on the plant and release large numbers of OBs, which are subsequently washed from leaf surfaces on to the soil. Alternatively, OB-contaminated leaves and stems senesce, fall to the soil, and are subsequently incorporated into the soil by agricultural practices such as tillage (Fuxa and Richter, 1996) or by the soil fauna. In forests, the leaf litter is also an abundant virus reservoir, in which OBs can persist for extended periods with little loss of infectivity (Podgwaite et al., 1979; Thompson and Scott, 1979).

In a systematic study on tillage and precipitation, viable NPV OBs were detected in the soil of soybean fields at depths of 0–25 cm, but were far less abundant at depths of 25–50 cm (Fuxa and Richter, 1996). Soil microcosm experiments indicated that NPV OBs released from virus-killed larvae or applied to the soil in water underwent a 3-logarithm reduction in viable OBs over a 17-month period (Fuxa et al., 2001). This may seem like a large reduction, but the enormous quantity of OBs produced in each infected insect means that even after extended periods in the soil, many OBs remain viable and have the potential to infect and replicate if consumed by susceptible insects. Indeed, the abundance of OBs in soil closely reflects the prevalence of infection in the host insect population in forests (Thompson et al., 1981), field crops (Fuxa and Richter, 2001), greenhouse crops (Murillo et al., 2007), and pastures attacked by soil-dwelling pests (Kalmakoff and Crawford, 1982). Such is the stability of the OB structure that viable OBs of a forest pest, *Orgyia pseudotsugata*, have been detected in soils several decades after the forest was cleared (Thompson et al., 1981).

Recognizing that soil represents a major environmental reservoir of OBs also means that it represents a unique resource for the discovery of novel virus isolates. Studies on open-field agricultural soils and greenhouse substrates have proved that soils contain a high diversity of NPVs and GVs (Murillo et al., 2007; Rios-Velasco et al., 2011; Gómez-Bonilla et al., 2012). These can be isolated using a simple bioassay technique in which soil samples are mixed with artificial diet and fed to early-instar larvae that succumb to virus disease if sufficient OBs are present (Richards and Christian, 1999). Viable isolates of Spodoptera exigua multiple nucleopolyhedrovirus (SeMNPV) were obtained from 29–38% of the greenhouse soil substrate samples tested using this technique (Murillo et al., 2007).

There is an intimate association between OBs and soil particles. However, as soil is one of the most heterogeneous habitats on earth, the findings on OB populations in one type of soil may not be readily extrapolated to other types. The clay component of soil is particularly important, and OB retention in soil depends on the relative abundance and type of clay component. Indeed, once bound to clay, baculovirus OBs can be very difficult to recover. In a study on seven different clays, HearNPV OBs bound strongly to all the clays tested, whereas two nonoccluded viruses – cricket paralysis virus (*Dicistroviridae*) and an iridescent virus (*Iridoviridae*) –preferentially bound to certain clay types but not to others (Christian et al., 2006). OB binding to soil components is likely to be affected by the cation-exchange capacity of the soil, which is determined largely by soil pH, the presence of clay minerals, and organic matter (Hunter-Fujita et al., 1998). The presence of iron-based minerals is likely to be a good indicator of soils that are suitable for baculovirus OB populations (Christian et al., 2006).

Baculoviruses are not the only invertebrate viruses that persist in soils. The occluded cypoviruses (Tanada et al., 1974) and entomopoxviruses (Hurpin and Robert, 1976) and the nonoccluded densoviruses (Watanabe and Shimizu, 1980), iridescent viruses (Reyes et al., 2004), and nodaviruses (Felix et al., 2011) have all been found to persist in soil. That said, the relationship between the soil virus populations and the prevalence of infection in host invertebrates, such as that observed in soil-dwelling lepidopteran pests, remains poorly understood in general (Kalmakoff and Crawford, 1982; Bourner et al., 1992; Prater et al., 2006).

7.6.2.3 Persistence in Water

Viruses are often stored in water in laboratory refrigerators for periods of months or years. However, there are no systematic studies of the persistence of occluded viruses in natural water bodies, probably because of a lack of interest in the use of these pathogens for the control of aquatic insects. The viruses that naturally infect hosts in aquatic habitats might be expected to be stable in water, but this is not always the case. The infectious titer of abalone herpesvirus fell markedly following 1–5 days of incubation in seawater at 15 °C (Corbeil et al., 2012). Similarly, qPCR-based studies indicated a >99.9% reduction in the number of genomes of the oyster herpesvirus (OsHV-1) in seawater over a 24-hour period, but the virus appeared to persist at high titers in the tissues of dead, infected oysters over a 7-day period (Hick et al., 2016).

Mosquito larvae are susceptible to NPVs and cypoviruses, the infectivity of which is modulated by calcium and magnesium ions present in solution (Becnel, 2006). The ability of these viruses to persist in the aquatic environment has not been studied in detail, although it is likely that they have retained the OB structure in order to persist in the soil habitat of pools that undergo periods of drying when rainfall is scarce. Storage of invertebrate iridescent virus 3 in water at 27 °C resulted in a near-exponential reduction in the infectious titer as determined by bioassay in larvae of *Aedes taeniorhynchus*. The virus persisted approximately twice as long in brackish water as in freshwater, possibly reflecting an adaptation to the brackish water habitat of the mosquito host (Linley and Nielsen, 1968).

The persistence of NPV OBs of the spruce budworm (*Choristoneura fumiferana*), a terrestrial lepidopteran, was monitored over 3 years in aquatic microcosms that had been inoculated with a large quantity of OBs ($>10^{10}$) in a forested area of Ontario, Canada. Viral DNA was detected in 8–9 out of 12 microcosms after 1 year, but only samples taken close to the bottom sediment proved positive by PCR after 3 years (Holmes et al., 2008). qPCR analysis of environmental samples from an island off the coast of Maine, USA, indicated the presence of NPV in soil under chokecherry trees (*Prunus virginiana*) infested by webworms (*Hyphantria* spp.) and in the sediment of freshwater pools, sea foam, and marine plankton samples. The widespread presence of OBs was attributed to the runoff from infected webworms and webworm feces during a period of frequent rainfall on the island (Hewson et al., 2011).

7.7 Dispersal

7.7.1 Host-Mediated Dispersal

Probably one of the most important yet least understood mechanisms of virus dispersal involves the movement of infected hosts. For insects, this usually occurs on two broad scales: (i) local movement on or among food plants by infected larvae that die and release OBs at a site different from the site where they acquired the infection; and (ii) flight of adult insects carrying covert infections that are transmitted vertically to their offspring at an oviposition site distant from the original site of infection of the parent.

The first issue of local movement by infected larvae has mainly been examined in relation to behavioral manipulation of the host insect by the virus (Section 7.9.4). This results in increased vertical and horizontal dispersal by infected insects, which improves the dispersal and transmission of the pathogen. Infected *Mamestra brassicae* larvae moved twice as far as healthy larvae in cabbage plots; an effect that was particularly evident during the 2–3-day period prior to death. The virus was also effectively dispersed by infected insects, which crawled over a distance up to 45 cm from the initial point of release (Vasconcelos et al., 1996b).

An excellent example of host-mediated dispersal comes from the Oryctes nudivirus, which infects the gut of both larval and adult rhinoceros beetles and has been successfully used for biological control of this pest (Hochberg and Waage, 1991). Adult beetles are good fliers and spend alternating periods feeding in the apices of coconut palms and reproducing beneath decomposing palm trunks. Infected adults live for ~4 weeks, and during this period they excrete large quantities of virus as they move between feeding and breeding sites, thereby contaminating both types of habitat and transmitting the virus to developing larvae or other adult conspecifics (Jackson, 2009). The rate of spread of this virus through the dispersal of infected adult beetles was estimated at between ~1 and 3 km/month on different islands in the Pacific and at 4 km/month in the Seychelles (Bedford, 1980; Lomer, 1986).

The dispersal of covertly infected adults of the African armyworm (*Spodoptera exempta*) is likely to represent the principal means by which its NPV (SpexNPV) travels along migration routes over distances of hundreds of kilometers during periodic outbreaks of this pest (Vilaplana et al., 2010). The prevalence of covert infection in adult moths collected during an outbreak in Tanzania ranged between 60 and 97%, depending on the PCR detection technique used. Infections were naturally efficiently transmitted to the progeny of infected parents. Outbreaks of this pest in Kenya, Tanzania, and other parts of East Africa tend to terminate in epizootics of virus disease. That said, the influence of covert infection on the dispersal of infected *S. exempta* adults and the factors that trigger the activation of lethal disease in their offspring have yet to be determined.

In a study on the invasion of forests in Wisconsin by *L. dispar*, the rate of dispersal of the pest was estimated at ~12 km/year, but, having arrived in a new section of the forest, the insect population required several generations to reach densities at which transmission of its NPV (LdMNPV) was likely to occur (Hajek and Tobin, 2011). This meant that the virus began to regulate the pest population some 4 years after the pest had established in a particular location, compared to a 3-year delay in the case of a fungal pathogen.

The dispersal of nonoccluded viruses that infect highly mobile insects, such as the viruses of crickets, drosophilids, and other dipterans, remains largely unstudied from an ecological perspective. An example of host-mediated dispersal of a nonoccluded virus comes from terrestrial isopods (woodlice, pillbugs) infected by an iridescent virus, which was influenced by the distance between suitable patches of habitat (Grosholz, 1993). The probability of dispersal decreased as interpatch distances increased. Habitat patchiness was also influential in the prevalence of virus disease: low levels of patchiness during the wet spring months were associated with a high prevalence of infection, which decreased as the dispersal of infected isopods became more restricted during the dry summer and fall months.

7.7.2 Environmental Factors Involved in Dispersal

There are many anecdotal accounts of virus dispersal through the action of rainfall and windblown dust. For example, the contamination of egg masses on foliage by OBs of the Douglas fir tussock moth (*O. pseudotsugata*) increased from 12 to 100% as the remains of infected cadavers were washed over foliage by a day of light rain (Brookes et al., 1978). Virus-decontaminated branches became contaminated by a sawfly NPV washed down by rain from infected cadavers on the upper branches of spruce trees (Evans and Entwistle, 1982). Indeed, the presence of OBs in raindrops hanging from pine needles beneath diseased sawfly (*Neodiprion sertifer*) colonies was quantified at 10^8 OBs/ml by direct counting under a microscope (Olofsson, 1989). This effect was confirmed in experiments using simulated and natural rainfall applied to infected cadavers of *L. dispar* on oak trees, in which branches below cadavers became contaminated by OBs washed down from higher branches (D'Amico and Elkinton, 1995). Simulated rainfall also strongly influenced the vertical distribution of NPV OBs on cabbage plants and in the soil of field plots (Goulson, 1997). Irrigation water may also be an effective means of virus dispersal in crops that are routinely irrigated (Young, 1990). Virus-contaminated dust was implicated in the dispersal of OBs from the soil to colonies of sawfly larvae feeding on pine at varying distances from a forest dirt track (Olofsson, 1988a).

Quantitative studies of local virus dispersal are rare. Simulated rainfall transported between 56 and 226 OBs of Helicoverpa zea single nucleopolyhedrovirus (HzSNPV) from different types of soil on to cotton plants in a greenhouse experiment. OB transport increased with increasing speed of air currents, and more OBs were transported from dry compared to wet soils. Of the three soils tested, OB retention was lowest in sandy soil and highest in clay soil. No OB transport was detected in the absence of simulated rainfall (Fuxa and Richter, 2001). In subsequent experiments, simulated rainfall was capable of transporting soil OBs distances of 30–75 cm to cotton plants, whereas air currents transported OBs 60–80 cm, irrespective of soil type. Transport from soil was detected for OBs at depths of up to 2 cm. In all cases, the lower portions of cotton plants were more heavily contaminated than the upper portions by wind- and rain-transported OBs (Fuxa et al., 2007).

7.7.3 Biotic Factors that Assist the Dispersal of Viruses

7.7.3.1 Predators

Numerous species of insect predators have been demonstrated to act as potential agents for the dispersal of baculovirus OBs. This is because most predators have an acidic gut, and the OBs in infected larvae pass through the gut without dissolution to be excreted in the predator's feces, sometimes for several days following the consumption of an infected prey item.

Birds appear to be particularly effective agents of dispersal of baculoviruses, not only because they are important predators of insect larvae, but also because of the large distances they can fly and disperse OBs between feeding sites. In a study on bird species trapped in and around pine forests treated with NPVs to control larvae of the Pine beauty moth (*Panolis flammea*), a total of nine bird species, representing 11–77% of birds captured, were found to produce viable OBs in their feces. Each bird dropping contained between 5×10^4 and 5×10^7 OBs, which represented a great many lethal doses of the viruses for early-instar larvae of *P. flammea* (Entwistle et al., 1993). Studies in field-crop and pasture systems have reported similar quantities of OBs in bird drop-pings (Crawford and Kalmakoff, 1978; Hostetter and Bell, 1985). In a separate study on sawfly control using NPVs applied to spruce trees, 90% of bird droppings collected from the trees contained viable OBs. Virus-contaminated bird droppings were collected at distances up to 6 km from sawfly infestations (Entwistle et al., 1977). Similarly, the feces of birds foraging for earthworms in OB-contaminated soil also tested positive for sawfly NPV OBs (Olofsson, 1989). Birds can additionally spread virus by processing infected larvae prior to consumption (Reilly and Hajek, 2012). In an aviary study, chickadees (*Poecile atricapilla*) consumed most infected larvae and excreted most OBs, but larvae were usually swallowed whole. In contrast, vireos (*Vireo olivaceus*) beat the urticating hairs off *L. dispar* larvae before eating them – an act that sprayed droplets of liquefied larval tissues on to nearby foliage. As a result, virus transmission due to rigorous prey processing by vireos exceeded transmission through the passage of OBs in feces (Reilly and Hajek, 2012).

Small mammals have been reported to be common dispersal agents for baculoviruses in forest ecosystems, and up to 75% of fecal samples may contain important quantities of OBs (Hostetter and Bell, 1985). However, most studies on agricultural pests have focused on predatory arthropods that consume moribund and virus-killed lepidopteran larvae. These studies have implicated carabids (Vasconcelos et al., 1996a), predatory hemipterans (Young and Yearian, 1987), earwigs (Dermaptera) (Castillejos et al., 2001), neuropterans (Boughton et al., 2003), spiders (Fuxa and Richter, 1994), crickets, and scavenging flies (Lee and Fuxa, 2000a) in the dissemination of OBs in their feces over periods of several days. The nests of paper wasps of the genus *Polistes* were found to contain OBs of several different NPVs, cypoviruses, and entomopoxviruses, reflecting the diseases of their lepidopteran prey (Morel and Fouillaud, 1994).

In soybean plots, virus dispersal was estimated at 80–120 cm/day and occurred in all directions from plots treated with AgMNPV. Virus dispersal was significantly corre-lated with the presence of predatory arthropods that tested positive for OBs in their feces. In greenhouse microcosms of collard plants, dispersal by larvae of the cabbage looper (*Trichoplusia ni*) infected with a wild-type NPV (AcMNPV) averaged 22–45 cm/day, but the virus dispersal rate increased to 38–71 cm/day in the presence of predators and a scavenging fly (Lee and Fuxa, 2000b). The susceptibility of diseased larvae to predation may be greater (Young and Kring, 1991), similar (Vasconcelos et al., 1996a), or less (Castillejos et al., 2001) than that of healthy conspecifics, depending on the predator–prey system and the severity of the disease.

7.7.3.2 Parasitoids

The abundance and high mobility of insect parasitoids means that they can also be highly effective agents for the dispersal of invertebrate viruses. Indeed, one family of viruses, the ascoviruses, depends almost entirely on endoparasitoid wasps as vectors for transmission of infections to healthy noctuid hosts (Stasiak et al., 2005). Virus dispersal and transmission via endoparasitoid wasps is often highly efficient because the ovipositor becomes contaminated with virus during oviposition into an infected insect and virions are injected directly into the host hemolymph during subsequent acts of oviposition in healthy insects (Brooks, 1993). There are numerous examples of studies on endoparasitoid wasps that demonstrate that wasps can vector viruses between hosts infected by baculoviruses (Cossentine, 2009), an entomopoxvirus (Lawrence, 2002), iridescent virus (López et al., 2002), and densovirus (Kurstak and Vago, 1967). Limited evidence from field experiments supports the idea that female endoparasitoids effectively disperse viruses under natural conditions (Hochberg, 1991b; Fuxa and Richter, 1994; López et al., 2002). However, due to the difficulties in tracking individual wasps, the rate of parasitoid-mediated dispersal of invertebrate viruses in the field has not been quantified. In some cases, ectoparasitoids may also be efficient vectors of viruses (Stoianova et al., 2012).

7.7.3.3 Other Organisms

Earthworms are normally abundant in agricultural and forest soils, and laboratory studies indicate that they are capable of moving NPV OBs from the soil surface to lower depths, where they are protected from exposure to UV radiation and high temperatures. Earthworms are capable of transporting OBs because their guts are slightly acidic and OBs can pass through them without loss of infectivity (Infante-Rodríguez et al., 2016).

Livestock were implicated in the dispersal of NPVs, GVs, and entomopoxviruses of the soil-dwelling pest complex *Wiseana* spp. in New Zealand. The transport of these viruses on the hooves of the animals resulted in the spread and increased prevalence of virus diseases in pastures (Kalmakoff and Crawford, 1982). Similarly, by moving virus from the soil on to pasture grasses, the presence of cattle increased the prevalence of NPV disease in *S. frugiperda* in the United States (Fuxa, 1991).

7.7.4 Agricultural Practices that Affect Dispersal

Agronomic practices are likely to have a major influence on virus populations in the environment. The avoidance of tillage in the production of soybean over a 2-year period was shown to increase soil populations of NPV OBs to the point where natural epizootics were initiated in velvetbean caterpillar (*Anticarsia gemmatalis*) populations in Brazil (Moscardi, 1989). In contrast, following the application of the same virus in the United States, tillage moved virus from the soil on to plants and resulted in an elevated prevalence of disease in the pest population (Young and Yearian, 1986). In another study in soybean, most agricultural operations did not influence the vertical distribution of established soil OB populations, although a decline in soil OBs was observed following the removal of crop refuse on the soil surface by disking (Fuxa and Richter, 1996). Other types of practices, such as the use of herbicidal or hormonal defoliants that cause near-total loss of foliage immediately prior to the harvest of cotton crops, are also likely to contribute a large influx of occluded viruses of cotton pests into the soil reservoir, although studies are lacking.

The movement of honeybee hives by commercial apiaries across large areas of the United States in response to demands for pollination services, in combination with an increase in the average size of apiaries, intensification of honey production, and the international trade in queens and bee semen, provides opportunities for the transmission and dispersal of honeybee viruses and parasitic vectors of bee diseases (Mutinelli, 2011; Smith et al., 2013). The intensive movement of hives also affords opportunities for the exchange of viruses and other pathogens between wild and commercial bees (Fürst et al., 2014). These are issues of major concern given the recent declines in natural and managed bee populations in the United States and Europe.

7.7.5 Spatial Patterns of Dispersal

Patterns of disease dispersal from the initial epicenter of an epizootic are often modeled as a reaction–diffusion model borrowed from chemical reaction kinetics with a diffusion component to describe spatial movement (White et al., 2000). This appears as a moving wavefront of infection, which is effectively a spatial transition zone, on one side of which the prevalence of disease is high (closer to the epicenter) and on the other side of which it is low. The speed of the wave is determined by a combination of individual-level processes involving transmission, production of progeny virus particles, virus decay in the environment, and movement of host insects. Evidence from studies on a spruce-feeding sawfly (*Gilpinia hercyniae*) and its NPV in Wales revealed that after traveling approximately 1000 m through the forest, the wave of infection began to break down as other minor waves of disease traveled outwards from the periphery of the major wave. These secondary epicenters were likely initiated by biotic vectors of the disease, such as birds (Entwistle et al., 1983), although this interpretation has been challenged in favor of seasonal effects on wave behavior (White et al., 1999).

The traveling wave model performed well in describing small-scale dispersal of Orgyia pseudotsugata nucleopolyhedrovirus (OpNPV) infection in an experimental system of fir seedlings (Dwyer, 1992). On a larger scale, ballooning of *L. dispar* larvae on silk threads was found to contribute to the initial wavefront of disease during a period of several weeks over a distance of ~100 m, but subsequently the model showed poor match to the observed spatial distribution of infections, possibly due to parasitoid vectoring of the virus (Dwyer and Elkinton, 1995). A detailed description of the use of models to understand the spatial spread of these pathogens is given in Chapter 12.

7.8   Genetic Diversity in Viruses

7.8.1   Genetic Diversity is Pervasive in Virus Populations

Genetic diversity is present in all invertebrate virus populations but has been particularly studied in baculoviruses. The fact that genetic diversity is maintained and transmitted between host generations indicates that this variation is selectively advantageous to each virus. Estimates of genetic diversity in baculoviruses depend largely on the techniques employed. Studies using restriction endonuclease enzymes, beginning in the 1980s, started to characterize diversity within and between baculovirus isolates from the same and different host species. It also became apparent that many natural isolates comprised mixtures of genotypes that could be separated by cloning using *in vitro* (cell culture) or *in vivo* techniques. *In vivo* cloning involves serial inoculation of larvae with very low doses of OBs or injection of larvae with low concentrations of budded virions from the hemolymph of an infected insect. These techniques can give quite different results in terms of the diversity and characteristics of the genotypes isolated, due to the divergent conditions required for replication and transmission in insects compared to *in vitro* systems (Erlandson, 2009). An alternative approach now involves deep sequencing and metagenomic analyses to determine the diversity of genotypic variants present in natural virus populations that have not been subjected to prior cloning steps (Baillie and Bouwer, 2012b; Chateigner et al., 2015).

NPVs and GVs tend to differ in their genetic diversity characteristics. NPVs tend to be genetically heterogeneous, with many variants occluded within a single OB. Therefore, most NPV infections involve mixtures of genotypes. In contrast, each GV OB contains a single virion with a single genome, and infections tend to involve very little diversity (Eberle et al., 2009; Erlandson, 2009), although mixed infections in GVs may be adaptive under certain circumstances, such as when the host population is resistant to one genotype (Graillot et al., 2016).

The diversity present in baculovirus genomes consists of single nucleotide polymorphisms (SNPs) and indels (insertions and deletions), which are often located around putative origins of replication (*hrs*) and baculovirus repeated open reading frames (*bro*), which have multiple functions in baculoviruses (Erlandson, 2009). Large deletions, sometimes representing over 10% of the genome, are present in a quarter or a third of all genotypes in some NPV populations (Simón et al., 2004a; Redman et al., 2010). These deletion variants can only persist in the presence of complete genotypes, which provide the missing gene products in cells that are simultaneously infected by complete and deletion genotypes, a process known as complementation. This is possible because, during the final phase of systemic NPV infection, each cell is infected by approximately four budded virions, each of which contains a single genome, which may be a deletion variant or a complete genotype (Bull et al., 2001). Co-infection of cells by multiple genotypes also provides numerous opportunities to generate diversity through recombination among genotypic variants, a mechanism known to be important for generating novel genotypes in baculoviruses (Kondo and Maeda, 1991; Kamita et al., 2003).

Ultradeep sequencing has revealed that the diversity in an isolate of Autographa californica multiple nucleopolyhedrovirus (AcMNPV) is astonishing, with every possible combination of variants present, albeit at different frequencies. Each genotypic variant was found to comprise an average of 94 SNPs scattered across the genome, and 25% of variants had large deletions (Chateigner et al., 2015). Other studies have identified important variation in genes encoding proteins that are located in the envelope of ODVs. These proteins include *per os* infection factors (PIFs) and ODV-E66, which are critical for primary infection of midgut cells (Simón et al., 2011; Craveiro et al., 2013; Thézé et al., 2014). Variation has also been identified in core genes involved in replication (Baillie and Bouwer 2012a; Chateigner et al., 2015) and in auxiliary genes such as *chitinase*, *egt*, and *enhancin* that improve transmission (D'Amico et al., 2013b; Harrison 2013; Martemyanov et al., 2015a). Additional diversity may also arise from the presence of mobile genetic elements such as transposons, indicating that invertebrate viruses can act as vectors for these elements (Gilbert et al., 2014). As might be expected, this notable diversity at the nucleic acid, gene, and genomic levels is reflected in numerous phenotypic traits that modulate virus fitness within and between host insects and their populations.

7.8.2   Genetic Diversity Favors Virus Survival

The genetic diversity in NPV populations is selectively advantageous and has clear ecological and evolutionary benefits to these viruses. When individual genotypic variants are examined, each variant usually exhibits a particular combination of phenotypic characteristics, which are often presented in terms of OB dose–mortality metrics, speed of kill, OB production in each insect, and OB production per milligram of insect tissue. These traits are clearly important for virus transmission, because they determine the likelihood of acquiring an infection, the time taken between initial infection and the release of progeny OBs that can infect other susceptible larvae, the total number of OBs released from each insect, and the efficiency with which the virus converts host resources into virus progeny. However, it is not possible to maximize all these traits simultaneously, as many involve correlations and tradeoffs imposed by biological constraints.

One of the best-characterized tradeoffs is that of speed of kill and total OB production. Slow-killing variants allow infected insects to continue feeding and growing during virus replication, thereby providing additional resources for the production of virus progeny. Fast-killing variants kill the host shortly following infection, so that each host represents a near-fixed resource to be exploited for progeny production. Additional evidence for the ecological role of the tradeoff between speed of kill and OB production in baculoviruses comes from a study of the *L. dispar*–LdMNPV system. Field isolates of LdMNPV varied in their tendency to kill larvae rapidly without producing progeny OBs and in the period during which infected larvae could grow prior to death, as indicated by post mortem body size (Fleming-Davies and Dwyer, 2015). Cadaver size was positively correlated with the prevalence of infection in neonate larvae exposed to bark pieces that had overwintered under natural conditions, indicating that rapid speed of kill was costly to virus environmental persistence and transmission to the following generation in this pathosystem.

As most NPV infections involve mixtures of genotypes, the analysis of individual clonal genotypes is unlikely to provide ecologically useful information. Instead, the role of mixed-genotype infections and the interactions among genotypes are likely to approximate natural virus populations to a far greater degree. For example, mixtures of genotypes present in natural (wild-type) isolates can increase the virus' ability to establish lethal infections (López-Ferber et al., 2003; Bernal et al., 2013; Redman et al., 2016) and increase the total production of OBs in infected insects (Barrera et al., 2013; Bernal et al., 2013). In these cases, the infectivity and OB production values of the wild-type population exceed those of the component variants, indicating a degree of cooperation among genotypes that, in the case of one NPV, is known to be mediated through the expression of a *pif* gene (Simón et al., 2013).

Genotypic heterogeneity in virus populations may also provide preadaptation, which allows the pathogen to exploit novel hosts, new host genotypes, or food plants with novel chemical defenses. In such cases, rare genotypes are likely to be favored over common ones in a given population, through a process of negative frequency-dependent selection.

Finally, diversity provides opportunities for risk-spreading by the virus in response to environmental stochasticity. Examples include the presence of genotypes with divergent tendencies for vertical or horizontal transmission, which may be differentially favored as host densities fluctuate (Cabodevilla et al., 2011a). Similarly, viral *chitinase* and *cathepsin* genes are responsible for post mortem melting of infected cadavers, which increases the rate of transmission (Goulson et al., 1995) but also exposes OBs to UV inactivation. Therefore, for a given speed-of-kill phenotype, variation in the frequency of genotypes lacking the viral *chitinase* gene (Vieira et al., 2012; D'Amico et al., 2013b) could determine the probabilities of transmission within and between host generations as a bet-hedging strategy in response to variation in opportunities for horizontal transmission over time.

7.8.3 What Generates So Much Genetic Diversity?

Genetic diversity is present within individual hosts, between different host insects, and between populations that are segregated by geographical, behavioral, or ecological factors. Genetic variation within and between virus populations will arise from host-related and other ecological processes, of which I consider three here: heterogeneity in host susceptibility to infection, the roles of food plants, and host species-mediated selection.

Heterogeneity in susceptibility to infection in *L. dispar* promotes polymorphism in the infectivity phenotype of LdMNPV (Fleming-Davies et al., 2015). This arises from a tradeoff between transmission rate and variation in susceptibility in this insect. High variation in susceptibility results in a fraction of the host population that acquires an infection at very low pathogen densities and a fraction that is resistant to infection even at high pathogen densities. In contrast, when variation in susceptibility is low, the probably of infection gradually increases with increasing pathogen density in the environment (Fleming-Davies et al., 2015). Similarly, heterogeneity in host susceptibility promotes mixed-genotype infections when the inoculum comprises mixtures of genotypes, as the probability of each viral variant establishing infection varies with host genotype (van der Werf et al., 2011). Additional support for this concept comes from a study demonstrating differential susceptibility of sibling groups of *L. dispar* larvae to infection by different strains of LdMNPV. From this, it is apparent that virus variants differ in their capacity to produce lethal infection of family groups and that susceptibility to infection also varies across families, indicating that the outcome of exposure to LdMNPV inoculum depends on viral genotype × host genotype interactions (Hudson et al., 2016).

Food plants can have major effects on the transmission of baculoviruses (Section 7.12.1). Individual genotypic variants and mixed-genotype populations can differ markedly in dose–mortality responses, speed of kill, and total OB yields when host larvae feed on different species of food plant, so that the transmission of one variant is favored over that of other variants on a given plant (Hodgson et al., 2002; Raymond et al., 2002). In such situations, the genotypic composition of the virus population is expected to vary according to the local species composition of available food plants. In another plant–insect–virus system, different species of crop and pasture grasses had no significant role in modulating the genetic composition of the virus population (Shapiro et al., 1991).

In the case of NPVs with an extended host range, laboratory studies indicate that infection of other host species with differing susceptibility can result in selection for certain genotypes or mixtures of genotypes within particular hosts, which represents an additional mechanism for maintaining genetic diversity in sympatric species that share a common virus pathogen (Hitchman et al., 2007; Zwart et al., 2009).

7.8.4   How Is Genetic Diversity Transmitted?

More than 20 genotypic variants can be present within individual NPV-infected insects (Cory et al., 2005; Baillie and Bouwer, 2012b). So how is this diversity transmitted? For vertically transmitted infections, virus variants have to be transmitted from the parental reproductive organs to the egg surface, or within the developing embryo of the offspring. This is likely to represent a bottleneck in the transmission of many variants, although studies have yet to address this issue.

For horizontally transmitted infections, the key to variant transmission is the OB. NPV OBs usually occlude groups of 30–80 ODVs, and each ODV typically contains between 1 and 10 nucleocapsids enveloped within the virion. Moreover, various nucleocapsids each containing a different genotype can be enveloped within an ODV (Clavijo et al., 2010). Therefore, when a larva consumes a single OB, numerous ODVs are released into the midgut lumen, and each ODV that infects a midgut cell can transmit between 1 and 10 genomes of potentially different genotypes. ODVs act independently in highly susceptible hosts, each with a certain (albeit low) probability of establishing a productive infection. This situation changes in less susceptible hosts, in which high doses of OBs are required to establish infection. In this case, ODVs appear to require a critical threshold before a productive infection can be established, possibly due to intrinsic host defense mechanisms (Zwart et al., 2009).

If deletion genotypes and complete genotypes are present in the same virion (i.e., if both genotypes replicated together in a co-infected cell prior to being occluded in an OB), then these genotypes will have shared the common pool of proteins necessary for peroral transmission and both variants will be transmitted together (Clavijo et al., 2009, 2010). When the dose of OBs consumed is very low, the opportunities for co-infection and complementation with complete genotypes are reduced. In such cases, primary infection in the insect midgut represents an important bottleneck to diversity (Zwart and Elena, 2015).

In contrast to the situation where larvae consume low doses of OBs, the density of OBs in the environment increases greatly during the development of an epizootic. In consequence, the average number of inoculum OBs consumed by the insect and the probability of the transmission of genotypic diversity both increase accordingly, so that *a priori* we may expect diversity in virus infections to increase during epizootics (Hodgson et al., 2003). However, this idea has been contradicted by observations on a nudivirus that infects shrimp, in which mixed-genotype infections were less prevalent during disease outbreaks (Hoa et al., 2011).

Finally, the possibility that NPVs are capable of generating genetic diversity *de novo* has come from a study on HearNPV, in which larvae inoculated with a high dose of OBs (95% mortality) were found to produce progeny OBs with a similar diversity of genetic variants as that present in the inoculum. In contrast, larvae inoculated with a single OB (5% mortality) produced OBs with a significantly higher number of variants: variants that were less similar compared to those present in the virus sample from which the inoculum OB was obtained (Baillie and Bouwer, 2013). The processes behind these intriguing observations have yet to be elucidated.

7.9 Role of Host Behavior in Virus Ecology

Despite the critical importance of host behavior in the transmission of most invertebrate virus pathogens, this aspect is often neglected in studies of pathogen ecology or during field testing of virus-based insecticides. For invertebrates that acquire infections by ingestion of contaminated food, the choice of what and where to eat will clearly influence the survival and reproduction of the host, but it will also determine the probability of infection and the host's ability to resist or overcome the pathogen.

7.9.1 Foraging Decisions: What and Where to Eat

Decisions made by ovipositing females of polyphagous species regarding their food plant species or the position of eggs laid on the plant can influence the prevalence of disease in their offspring. For example, in *S. frugiperda*, the prevalence of NPV disease at field sites was positively correlated with the presence of signalgrass (*Brachiaria platyphylla*) and negatively correlated with two other grasses (Fuxa and Geaghan, 1983), although this did not reflect the quantities of virus OBs required for acquisition of lethal disease on each type of plant (Richter et al., 1987). Similarly, the food plant species was clearly demonstrated to affect virus fitness in the winter moth (*Operophtera brumata*) (Raymond et al., 2002) and immune system function and disease resistance in *T. ni* larvae (Shikano et al., 2010).

The choice of host plant can also determine the virulence of the virus strain that an insect is at risk of acquiring. In the case of the Western tent caterpillar (*Malacosoma californicum pluviale*), virus isolates present on particular plant species killed their hosts faster when larvae consumed inoculum OBs on foliage of the same plant (Cory and Myers, 2004). However, whereas phytopathogenic viruses can increase oviposition by insect vectors on infected plants (Chen et al., 2013), it is not clear whether any species of adult female invertebrates is capable of detecting the presence of invertebrate pathogenic viruses or the remains of virus-killed individuals on plants and modifying their ovipositional decisions accordingly. The only exception to this comes from observations that females of *C. pomonella* reduced oviposition on apple cultivars that had been treated with a GV-based insecticide, probably due to components in the product formulation that altered leaf-surface metabolites, rather than a response to the presence of GV OBs (Lombarkia et al., 2013).

For phytophagous insects, decisions on where to feed on the plant can also have implications for the transmission of their pathogens. For example, in *L. dispar*, late instars avoid feeding on leaves contaminated by the remains of diseased cadavers but do not avoid foliage contaminated by purified NPV OBs (Capinera et al., 1976). In a later study, *L. dispar* larvae appeared able to detect and avoid virus-infected cadavers from a distance of at least 5 mm; this behavior varied between family groups, indicating a significant degree of heritability in this trait (Parker et al., 2010). A model developed from these observations suggested that the ability to detect infected cadavers from a close distance (<1 mm) would result in a decrease in the prevalence of infection of 4–7%, which appears modest in a single round of transmission but may have a significant impact on the risk of infection during the multiple cycles of transmission that occur during the development of an epizootic (Eakin et al., 2015). In contrast, *Spodoptera exigua* larvae showed no preference to avoid contact or consumption of leaf disks contaminated by infected cadavers (Rebolledo et al., 2015).

7.9.2   The Risks of Cannibalism

Insects from several orders show cannibalistic behavior, particularly during the final larval stages, or in situations of low food availability or high population density. The ecological and evolutionary consequences of this behavior have been reviewed elsewhere (Richardson et al., 2010). Cannibalism is also an efficient route for the transmission for certain pathogens, including baculoviruses, when healthy larvae consume moribund infected conspecifics prior to death (Dhandapani et al., 1993; Boots, 1998; Chapman et al., 1999). Cannibalism has been shown to result in transmission of densovirus and entomopoxviruses in Orthoptera (Streett and McGuire, 1990; Weissman et al., 2012) and iridescent viruses in Lepidoptera, Orthoptera, Diptera, and terrestrial isopods (Crustacea) (Williams, 2008). This is because diseased individuals often become lethargic in the final stages of infection and are unable to defend themselves from aggressive conspecifics. When this behavior occurs – infected cadavers are eaten – it should more correctly be described as conspecific necrophagy.

7.9.3   Sexually Transmitted Viral Diseases

For sexually transmitted viruses, the choice of sexual partner will often affect reproductive success and may directly affect survival of the individual or their offspring. Indeed, in a review, Knell and Webberley (2004) identified 17 pathosystems in which sexual transmission of a virus resulted in reduced insect reproduction and/or reduced offspring survival. Recent examples include healthy *S. exigua* females that mated with NPV-infected male moths. A quarter of the offspring from these matings were covertly infected by an NPV (Virto et al., 2013). Similarly, healthy females that mated with iflavirus-infected males efficiently transmitted the infection to their offspring (Virto et al., 2014). The semen of infected male honeybees was found to contain deformed wing virus (*Iflaviridae*), and transmission to the offspring of healthy queens was 100% efficient (Yue et al., 2007), while densovirus-infected male *Anopheles* mosquitoes inseminated healthy females with semen containing over 10^6 genomes of the virus (Barik et al., 2016). Sexual transmission of nudiviruses is a particularly well characterized example of the efficiency of transmission during mating (see Section 7.5.1). The application of molecular tools to the study of these pathosystems is greatly improving our understanding of the role of mating systems in the transmission of invertebrate pathogens in general.

7.9.4 Ecological Consequences of Host Manipulation by Viruses

There are many reports of changes in the behavior of virus-infected invertebrates. These can be related to pathological effects or they can be adaptive. In the latter case, behavioral changes can be classified into three broad groups:

1) The extended phenotype. In this group of examples, the virus manipulates the host to improve its own fitness through the expression of viral genes: the so-called "extended phenotype" of the virus. Indeed, several viruses have the capacity to manipulate invertebrate hosts in order to improve their transmission (Han et al., 2015b). Baculovirus-infected insects often show enhanced locomotory activity, or hyperactivity, midway through the course of the infection, which increases their rate of dispersal (Kamita et al., 2005). A closely related, but apparently distinct behavior is baculovirus-induced climbing behavior, which occurs shortly before death in many species of Lepidoptera. This behavior was first described in NPV-infected larvae of the nun moth (*Lymantria monacha*) over a century ago, and was named treetop disease due to the tendency of larvae to die and hang suspended from branches and foliage in the highest parts of coniferous trees. Since then, the behavior has been reported in *Mythimna* (*Pseudaletia*) *separata* (Ohbayashi and Iwabuchi, 1991), *L. dispar* (Murray and Elkinton, 1992), *M. brassicae* (Goulson, 1997), *Orgyia antigua* (Richards et al., 1999), *S. exigua* (van Houte et al., 2014b), and *T. ni* (Ros et al., 2015), among other lepidopteran species. Downward movement of infected larvae has been reported in *O. brumata* (Raymond et al., 2005). Although common among NPV-infected insects, climbing behavior has occasionally been reported in GV-infected individuals (Moore et al., 2011), but apparently not in other occluded insect viruses.

This behavior appears to be adaptive for the virus, as infected larvae that die on the uppermost parts of plants are likely to improve the transmission and dispersal of the pathogen by: (i) releasing large quantities of OBs, which subsequently contaminate foliage lower in the plant canopy by liquefaction or the action of rainfall; and (ii) being more susceptible to predation by birds, which are efficient agents for virus dispersal. The climbing behavior clearly depends on a positive phototactic response, although the molecular basis for this remains unknown (van Houte et al., 2014b). In the case of *S. exigua* larvae, virus-induced climbing behavior increased the probability of encounters with healthy conspecific larvae that became infected following necrophagy of virus-killed cadavers (Rebolledo et al., 2015).

Invertebrate viruses are also capable of manipulating the sexual activity of their hosts. Infection of cotton bollworm (*Helicoverpa zea*) adults by the nudivirus Hz-2v often resulted in malformation of the reproductive system and sterility, although sexual activity was enhanced. Infected females called for mates more frequently, produced more mating pheromone, and attracted more mates than healthy females. During copulation, the genitals of males became contaminated by virions in a waxy vaginal plug. Contaminated males then transmitted the virus to other females during subsequent matings. Moreover, a portion of the female population was fertile and asymptomatic and produced infected sterile offspring, through which the virus was transmitted. The molecular basis for these complex behavioral changes has yet to be elucidated (Burand, 2009). Similarly, iridescent virus infection of the cricket *Gryllus texensis* causes sterility in both sexes. However, infected individuals continue to engage in mating behavior, and infected males are quicker than uninfected males to court females. The virus is transmitted to healthy crickets through sexual activity, leading to its description as a "viral aphrodisiac" (Adamo et al., 2014).

2) Adoption of behaviors that reduce the costs of infection. Examples include self-medication involving a specific therapeutic and adaptive change in behavior in response to disease. For example, larvae of *S. exempta* infected with a NPV immediately adjusted their diet to reduce carbohydrate intake, while protein consumption gradually increased over time, resulting in markedly improved survival compared to insects that could not adjust their diet. The high protein/low carbohydrate diet was associated with higher levels of antimicrobial activity in the hemolymph and improved immune system function. This provides clear evidence of an adaptive self-medication response in this species (Lee et al., 2006; Povey et al., 2014).

3) A shared phenotype arising from the expression of both virus and host genes. In this case, a virus may elicit a particular behavior in one host species but not in another. The outcome and magnitude of shared-phenotype behavioral changes are therefore likely to depend on combinations of host, virus, and environment interactions (van Houte et al., 2013). For example, in contrast to the self-medication response of *S. exempta* mentioned earlier, modified feeding responses in infected *T. ni* larvae were dependent on both temperature and the identity of the virus (Shikano and Cory, 2015, 2016).

7.9.4.1 Molecular Basis for Host Manipulation

Behavioral changes observed in infected individuals are likely to be caused by pathological effects on the nervous system (Wang et al., 2015) or changes in metabolism or physiology (Thompson and Sikorowski, 1980; Chen et al., 2014). In a small number of cases, the molecular basis for viral manipulation of host behavior has been elucidated. The first to be identified was that of the ecdysteroid UDP-glucosyltransferase (*egt*) gene in baculoviruses (O'Reilly and Miller, 1991). The EGT enzyme inactivates ecdysteroid hormones by conjugation with sugars, which delays molting to the following instar. This results in continued feeding and growth, a higher production of OBs, and improved probability of transmission compared to gene-deletion viruses (Cory et al., 2004). The fact that *egt*-deletion variants are present in natural populations of NPVs suggests that variation in speed of kill and OB production may be selectively advantageous under certain circumstances (Harrison et al., 2008; Simón et al., 2012). The *egt* gene has also attracted attention in the development of baculovirus insecticides, as its deletion results in an improved speed of kill compared to wild-type virus (Popham et al., 2016).

The *egt* gene has also been implicated in the climbing behavior of infected *L. dispar* larvae prior to death (Hoover et al., 2011). However, *egt* is not responsible for this behavior in all insect–NPV pathosystems (Ros et al., 2015). In *S. exigua* larvae infected by AcMNPV, climbing behavior was dependent on molting during the infection period, and larvae that died without molting tended to move downwards rather than upwards. In contrast, in *S. exigua* larvae infected by the homologous virus (SeMNPV), the *egt* gene was shown to extend the lifespan of the infected host and facilitate climbing behavior (Han et al., 2015a).

Enhanced locomotory activity was linked to the expression of a baculoviral protein tyrosine phosphatase (*ptp*) gene in the silkworm (*Bombyx mori*) in a process that was light-activated (Kamita et al., 2005). In BmNPV-infected silkworms, the PTP protein does not require enzymatic activity to elicit enhanced locomotory activity but appears to be a viral structural protein present in the envelope of budded virions that interacts with a protein that modulates the actin cytoskeleton of the infected cell and manipulates host behavior via infection of larval brain tissues (Katsuma, 2015a). In the case of AcMNPV, phosphatase activity is required for enhanced locomotory activity, although the target substrate is presently unknown and appears to exert its effect independently of the processes that govern climbing behavior (van Houte et al., 2014a; Katsuma, 2015b). Understanding the molecular basis for host manipulation by pathogens is an issue that is currently generating excitement among invertebrate pathologists and evolutionary biologists.

7.10 Dynamics of Viruses in Host Populations

7.10.1 Pathogenic Viruses Can Regulate Populations

Viruses can be major mortality factors in populations of some invertebrates, particularly forest-feeding Lepidoptera (Erebiidae and Lasiocampidae) and sawflies (Hymenoptera, Diprionidae), as well as in beneficial insects such as silkworms and honeybees. Outbreaks of baculovirus diseases have been associated with the cyclic dynamics of some forest pests, with a periodicity of several generations (typically 5–15 years), particularly in temperate regions. In these systems, the density of the host population increases until it exceeds a threshold value, at which point the pathogen can spread rapidly through the population. This leads to dramatic declines in the host population, which falls back below the threshold density. The virus persists as a vertically transmitted covert infection or in environment reservoirs until the threshold density is exceeded once more and sustained transmission is possible again (Briggs et al., 1995). Following the seminal population model (model G) of Anderson and May (1981), significant advances have been made in developing models that accurately describe population dynamics of insect–virus pathosystems (Elderd, 2013). The *L. dispar*–LdMNPV system has been particularly well characterized in this respect (see Chapter 12 for a detailed review).

Another excellent example of insect–virus dynamics is that of the Western tent caterpillar (*M. c. pluviale*) and its NPV in British Columbia, Canada (Myers and Cory, 2016). This univoltine gregarious insect feeds on a variety of host trees and has cyclic populations with a periodicity of 8–11 years. The prevalence of virus infection over a 24-year period closely tracked host population density, as indicated by the density of the silk tents inhabited by families of larvae (Fig. 7.1a). During epizootics, 80–100% of larval families within tents were diseased and host population densities fell rapidly. A significant negative correlation between the rate of population change between one year and the next and the percentage of virus-diseased families was detected (Fig. 7.1b), which is indicative of population regulation by the pathogen. A lag in the recovery of the host population is required for cyclic dynamics, and reduced fecundity following an epizootic of disease was identified as the most probable cause for delayed recovery in the host population (Cory and Myers, 2009). Potential causes of the reduced fecundity were considered, including reduced food quality or availability following defoliation of trees, costs incurred from resistance to infection or from maintaining an immune response in

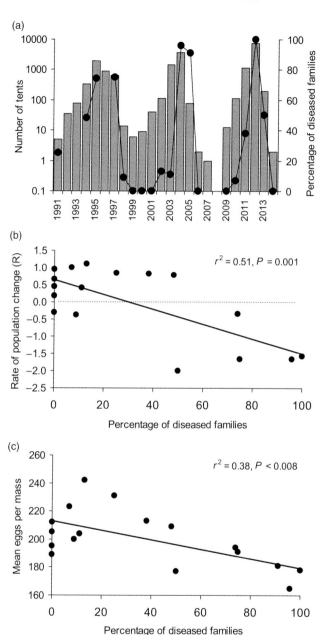

Fig. 7.1 Cyclic population dynamics of the Western tent caterpillar (*Malacosoma californicum pluviale*) and its NPV in British Columbia, Canada over a 24-year period. (a) Fluctuations in numbers of tents (families) on Galiano Island (columns) and percentages of families that contain diseased insects (black dots). (b) Negative correlation between rate of population change (R) and percentage of families containing diseased individuals, where R = log(n + 1/n), indicating the change in population size on Galiano Island from one year (n) to the next (n + 1). (c) Negative correlation between female fecundity (expressed as mean number of eggs in each egg mass) and the percentage of families containing infected insects. *Source*: Myers and Cory (2016), http://onlinelibrary.wiley.com/doi/10.1111/eva.12328/full. Used under CC BY 4.0 https://creativecommons.org/licenses/by/4.0/.

order to control a covert infection, and the pathological effects of sublethal disease in covertly infected insects. Sublethal infection is often associated with reduced fecundity or fertility in other species of Lepidoptera (Rothman and Myers, 1996). In the case of *M. c. pluviale* a significant negative correlation was detected between the prevalence of infection and the numbers of eggs laid in egg masses (Fig. 7.1c). The low fecundity observed during years with disease epizootics was followed by reduced population density in the following years, a prerequisite for cyclic dynamics. Myers and Cory (2016) therefore concluded that this delayed density-dependent effect on fecundity was most likely due to sublethal disease in *M. c. pluviale* populations, although testing for this using molecular methods was problematic during periods between outbreaks, due to the paucity of insects present in low-density populations.

7.10.2 Ecosystem Characteristics that Favor Virus Transmission

Epizootics of infection in high-density pest populations have been reported in a number of agricultural systems and have led researchers to consider developing baculoviruses as biological insecticides, sometimes with great success (Moscardi, 1999). Fuxa (2004) proposed that agroecosystems could be classified as permissive or nonpermissive based on their propensity to sustain a high prevalence of baculovirus infection in agricultural pests. Examples of such systems include the Spodoptera frugiperda multiple nucleopolyhedrovirus (SfMNPV), which is capable of sustained annual epizootics in a permissive system involving pasture grasses, but not in crops such as maize or sorghum (non-permissive systems). This was attributed to the behavior of larvae that are solitary and do not disperse between maize and sorghum plants but move readily between low-growing grasses. The pasture grasses are also very likely to become contaminated from the soil OB reservoir (Fuxa, 1982).

Plant–pest systems involving *A. gemmatalis* larvae on soybean and *T. ni* larvae on collards were also considered permissive to epizootics due to plant contamination by soil OB populations and a high prevalence of dispersal of their respective viruses by predatory arthropods. In contrast, the cotton–*Helicoverpa* spp. system was considered nonpermissive, as introductions of the virus in one season were never reflected in lethal disease the following season, possibly due to the poor persistence of OBs on cotton foliage (Fuxa, 2004). Forest ecosystems, in contrast, have a combination of factors that favor virus persistence, including an undisturbed soil reservoir, physical protection for OBs in tree bark crevices on twigs and branches, the dispersal of OBs by biotic and abiotic factors, and opportunities for horizontal and vertical transmission in resident insect populations, which do not exist (or exist only to a limited degree) in ephemeral crop habitats.

Of the host-related factors that favor epizootics, in addition to heterogeneity in resistance to infection (see Section 7.8.3), it has become increasingly recognized that the risk of disease is likely to increase with population density, as transmission is density-dependent (Section 7.5). As such, it would be advantageous for species that experience large fluctuations in density to be able to increase their investment in immune defenses to match the threat posed by pathogens in high-density populations – a response known as density-dependent prophylaxis (Cotter et al., 2004). This risk is particularly relevant to gregarious feeding species, in which the death from virosis of a single individual can rapidly spread to other group members (Hochberg, 1991a).

An additional physiological response to the presence of high densities of conspecifics is phase polyphenism. Numerous species of Lepidoptera and some other taxa exhibit a switch to a dark larval phase when reared at high densities, due to the melanization of the cuticle. Dark-phase individuals often differ from their pale conspecifics in metabolic rate, growth rate, and other variables that affect fitness, including their susceptibility to NPV disease (Goulson and Cory, 1995).

Interestingly, species that show phase polyphenism also have the capacity for density-dependent prophylaxis. In *S. exempta*, both rearing density and phase polyphenism were positively associated with increased phenoloxidase levels in the hemolymph and with reduced susceptibility to NPV disease. Under field conditions, transmission and mortality were also lower in larvae that had been reared in crowded environments (reviewed in Wilson and Cotter, 2009). Similarly, the armyworm (*M. separata*) became highly resistant to NPV and GV infection when reared at high densities. When virions were injected directly into the hemolymph, however, no difference was observed in the susceptibility of larvae that had been reared singly or in groups, suggesting that resistance to infection may be related to primary infection in the insect midgut (Kunimi and Yamada, 1990).

In *L. dispar*, which is not a phase-polyphenic species, rearing at high densities did not result in density-dependent prophylaxis. In fact, the ability to resist lethal infection and the survival times of infected insects were reduced at high rearing densities (Reilly and Hajek, 2008). In contrast, *A. gemmatalis* appears to represent an intermediate species, in which larvae are phase-polyphenic and can experience high-density populations but are not gregarious. In *A. gemmatalis*, the switch to the melanic form is an all-or-nothing response to the presence of one or more conspecifics. Larvae reared at high densities had a stronger encapsulation response, possessed higher numbers of hemocytes, and survived longer when infected, but no significant differences were detected among phase phenotypes for disease resistance (Silva et al., 2013). Nonetheless, the dark phenotype was observed to have a thicker and more robust peritrophic membrane than that of pale conspecifics (Silva et al., 2016), providing support for the idea that the midgut barrier may provide an important contribution to disease resistance in melanic phase insects. Rearing temperature was highly influential in determining the prevalence of each phase phenotype, immune response, and the survival of infected insects (Silva and Elliot, 2016). In light of climate change, these findings have potential implications for insect–virus population dynamics in phase-polyphenic species, as insects may become less susceptible to pathogens in a warmer climate.

7.10.3 Climate Change and Insect–Virus Population Dynamics

Our understanding of the likely impact of global warming on invertebrate–virus population dynamics is extremely limited, mainly due to a paucity of empirical testing of theoretical models. It is clear, however, that rising temperatures are expected to influence disease dynamics in agricultural, forest, and aquatic ecosystems. Two studies have indicated that warmer temperatures may increase the impact of virus pathogens on insect populations. The prevalence of lethal infection by SfMNPV in *S. frugiperda* larvae increased at higher temperatures, possibly due to increased feeding rates. The transmission rate itself was not temperature-sensitive, but heterogeneity in the risk of disease decreased with increasing temperature, resulting in higher mortality than observed at lower temperatures (Elderd and Reilly, 2014). As climate warming and increased CO_2 are associated with a decreased protein–carbohydate ratio in plants, Shikano and Cory (2015) compared the effects of temperature and nutrition on the growth and survival of *T. ni* larvae infected by NPVs. The survival of virus-challenged insects was positively correlated with dietary protein–carbohydrate ratio, suggesting that these pathogens could exacerbate the negative effects of reduced protein (nitrogen) availability in plants in a global warming scenario. Conversely, increasing rearing temperatures did not influence the susceptibility of *M. c. pluviale* larvae to NPV infection or the estimated yield of progeny OBs, although survival times of infected larvae declined rapidly with increasing temperature (Frid and Myers, 2002). One particular feature of this species is that it appears capable of temperature regulation through sun-basking behavior, possibly rendering it less susceptible to climatic variation than nonregulating insects.

7.11   Influence of Abiotic Factors on Viruses

Viruses in the environment are subjected to a series of abiotic challenges that they do not face within the host. Their ability to retain infectivity during periods in the environment varies widely depending on the ecosystem and the presence of the OB structure. In this section, I consider the effects of the most influential abiotic factors: UV radiation, seasonality, temperature, precipitation, and pH.

7.11.1   Effect of Ultraviolet Light on Viruses

The environmental factor that has attracted most attention from researchers is solar UV radiation. This is because UV rapidly inactivates viruses applied to plants as biological insecticides, thereby limiting their effectiveness for pest control. As such, the half-life of occluded viruses exposed to direct sunlight can often be measured in terms of hours (Ignoffo et al., 1997; Sajap et al., 2007), such that only a small fraction of the original inoculum remains viable a few days after applying OBs as a biological insecticide (Sun et al., 2004). As mentioned previously (Section 7.6.2.1), the remains of infected cadavers appear to provide protection against UV degradation so that OBs released from cadavers persist approximately twice as long as purified OBs on foliage (Fuller et al., 2012). In contrast, a large fraction (~50%) of the OBs applied to plants grown under the UV-protective structure of a plastic greenhouse can still be viable a week after the initial application (Bianchi et al., 1999; Lasa et al., 2007).

Exposure to solar UV is highest in tropical regions and at high altitudes and decreases with increasing latitude. Temperate regions also experience marked seasonal changes in UV irradiation and climatic conditions, particularly due to cloud cover and precipitation. UV-B (280–315 nm wavelength) is the most biologically harmful part of the solar radiation spectrum that arrives at the earth's surface. It has the ability to cause breaks in strands of nucleic acids or, more frequently, to fuse adjacent thymine bases in DNA strands, forming a cyclobutane thymine dimer that blocks normal DNA synthesis and often results in mutations. In viruses with a dsRNA genome, uracil dimers may also accumulate as a result of UV radiation.

Of the occluded viruses, entomopoxviruses were classified as the most resistant to UV-B in laboratory conditions and GVs were the most sensitive to UV-B inactivation, whereas NPVs and cypoviruses were intermediate in their susceptibility (Ignoffo et al., 1977). Light with a longer wavelength and a lower energy, such as UV-A (315–400 nm) and visible light, can also inactivate viruses given extended periods of exposure (Shapiro and Domek, 2002).

To counter UV-induced damage, several insect pathogenic viruses encode class II photolyase enzymes that use the energy of visible light to repair pyrimidine dimers and return them to their original state. To date, photolyase genes have been identified in several entomopoxviruses, a GV, and a growing number of NPVs (van Oers et al., 2008; Rabalski et al., 2016). Phylogenetic evidence indicates that baculoviruses probably obtained their photolyase genes by horizontal gene transfer from an ancestral lepidopteran host (Biernat et al., 2011). Other families of invertebrate DNA viruses have a selection of genes that allow them to repair different types of damage to their genomes, including strand breaks and base or nucleotide excision (Blanc-Mathieu and Otata, 2016).

For OBs on plant foliage, stems, or bark, the degree of shading provided by the upper layers of foliage or adjacent plants will reduce exposure to solar UV (Jaques, 1985). Consequently, OBs on foliage at the middle or lower parts of the plant canopy will receive a lower dose of UV and tend to persist longer than those in the top part. For example, compared to the upper and middle sections of plants, the density of OBs was markedly higher on the lower parts of soybean plants and pine trees (*Pinus cortata*) that were shaded from direct sunlight by upper-canopy foliage (Young and Yearian, 1989; Richards et al., 1999). Viral OBs were most abundant on heather (*Calluna vulgaris*) growing under pine trees, which represented the most shaded habitat in a pine plantation and formed a major environmental reservoir of a lymantriine NPV (Richards et al., 1999).

OBs persisted longer on the undersides of cotton, cabbage, and soybean leaves compared to the upper surfaces (Young and Yearian, 1974; Biever and Hostetter, 1985; Peng et al., 1999a). Similarly, because of the angle of incidence of solar radiation, NPV OBs on south-facing foliage of pine trees received an approximately fivefold higher dose of UV than those on north-facing foliage, for trees growing in the northern hemisphere (Killick and Warden, 1991); the opposite effect was observed for GV OBs applied to citrus trees growing in the southern hemisphere (Moore et al., 2015). OBs in the crevices at the bases of pine needles or on the bark of twigs, branches, and trunks may also be protected from UV radiation, so that they can remain viable during the winter period, when host larvae are absent (Kaupp, 1983; Olofsson, 1988b). As such, tree surfaces can constitute a reservoir of OBs that retain their infectivity, unlike those on plants in unshaded locations. For example, OBs of the winter moth (*O. brumata*) NPV retained infectivity on oak (*Quercus robur*) and sitka spruce (*Picea sitchensis*) in forested areas, whereas OBs on heather in unshaded habitats were rapidly inactivated (Raymond et al., 2005).

The use of UV lamps has been evaluated for the inactivation of pathogens such as white spot syndrome virus (WSSV), a whispovirus that can persist in water used for shrimp farming (Chang et al., 1998). However, the influence of natural sunlight on virus persistence in water has been little studied. One exception is the study on particles of an iridescent virus in trays of fresh water, which lost 97% of their infectious titer over a 60-hour period under shaded tropical conditions but 99.99% over a 36-hour period when exposed to direct sunlight. The persistence of the virus was negatively correlated with the accumulated dose of solar UV radiation (Hernández et al., 2005). Of course, viruses in tropical habitats are likely to experience a more severe combination of insolation and elevated temperatures compared to those in temperate zones.

7.11.2 Seasonal Effects on Viruses

Seasonality affects virus persistence in tropical and temperate regions by way of seasonal fluctuations in biotic and abiotic factors such as UV radiation, precipitation, and temperature, as well as the presence of the host plant foliage and plant phenology. That said, early studies on baculovirus ecology recognized that the seasonality of virus dynamics in soil and on the foliage of plants was driven by the seasonal characteristics of the host lifecycle. Following an initial application of NPV OBs to the soil of cabbage plots, the density of soil OBs reached a peak during the fall (500–1000 OBs/mg soil), remained high or fell very slightly during the winter, then fell due to tillage of plots in the spring (5–30 OBs/mg soil), before returning to peak levels during the summer and fall. This cyclic pattern was repeated during the 5 years of the study (Jaques, 1974). In each of these years, the density of OBs on cabbage leaves increased rapidly, from 1–10 OBs/cm^2 following planting, as plants became contaminated with OBs from the soil reservoir, to 100–1000 OBs/cm^2 in the summer and fall, when *Trichoplusia ni* larvae had become infected and OBs were washed by rainfall over cabbage leaves and on to the soil (Jaques, 1985). Similar patterns were seen in studies of soybean fields following the introduction of an NPV to control *A. gemmatalis*, where OB densities in soil and on plants showed synchronous seasonal fluctuations over a 3-year period, with the main loss of OBs in the soil occurring over the winter months (Fuxa and Richter, 1994, 1996).

Plants die or lose leaves during the fall (temperate regions) or dry season (tropics), and this is a mechanism by which OB-contaminated foliage can become incorporated into the soil reservoir. For example, the horizontal distribution of the virus population in the soil reflected the spatial pattern of fallen leaves at distances of up to 15 m from large poplar and plane trees contaminated with an NPV of the fall webworm, *Hyphantria cunea* (Hukuhara, 1973).

7.11.3 Effect of Temperature on Viruses

The influence of temperature on the stability of occluded viruses has been summarized by others (Jaques, 1985; Benz, 1987; Ouellette et al., 2010). At normal environmental temperatures (10–40 °C), these occluded viruses can usually retain infectivity for weeks, months, or even years (David and Gardiner, 1967). However, studies have often been complicated by environmental factors, including different levels of moisture and the presence of contaminants such as enzymes and other microorganisms. Exposure to elevated temperatures (≥60 °C) results in loss of infectivity within a few minutes (Ribeiro and Pavan, 1994).

Nonoccluded viruses differ in their sensitivity to heat. The infectivity of NPV budded virions was significantly reduced following exposure to temperatures exceeding 45 °C (Michealsky et al., 2008). Purified particles of iridescent virus were rapidly inactivated above 50 °C but gradually lost between 0.5 and 1.0 logarithm of infectivity over a 50-day period at temperatures of 4 or 25 °C, or the ambient temperature of a tropical pond (Marina et al., 2000; Martínez et al., 2003). The nodavirus Flock house virus lost 1–2 logarithms of infectivity following 10 minutes of exposure to 53–58 °C (Scotti et al., 1983). In contrast, densoviruses and iflaviruses are stable at high temperatures (50–60 °C), although they are inactivated above 70 °C (Seki et al., 1986; Jakubowska et al., 2016). In general, all the invertebrate viruses are stable for periods of years when frozen.

7.11.4 Humidity, Moisture and Precipitation

Several viruses that infect aquatic or soil-dwelling invertebrates are sensitive to desiccation. An iridescent virus lost 2 logarithms of infectivity in 24 hours in dry soil (6.4% moisture, −1000 kPa matric potential), whereas its half-life was 4.9 days in a natural soil and 6.3 days in sterilized soil. The moisture content of the natural and sterilized soils was 17–37% (−114 to −9.0 kPa), but this moisture range did not significantly influence virus stability (Reyes et al., 2004). Similarly, the shrimp whispovirus (WSSV) was totally inactivated following 48 hours of dry conditions (LeBlanc and Overstreet, 1991).

Occluded viruses are not generally affected by humidity (David et al., 1971; Ignoffo, 1992). However, wetted deposits of NPV or GV OBs, such as occur following rainfall or early-morning dew, were more rapidly inactivated by UV radiation than dry deposits (Ignoffo and Garcia, 1992). Periods of damp weather have been reported to increase the prevalence of NPV infection in *L. monacha* and the armyworm (*S. exempta*) (Persson, 1981; Benz, 1987) but not in the gypsy moth (*L. dispar*) (Hajek and Tobin, 2011). However, as high humidity is a stressor, it is possible that the reports reflect the activation of covert infections, triggering lethal disease in natural populations of these insects (Fuxa et al., 1999).

The persistence of OBs of Anticarsia gemmatalis nucleopolyhedrovirus (AgMNPV) over 28 months in an agricultural soil was affected by moisture, with the highest persistence in saturated soil (45% moisture, 0 kPa matric potential), the lowest in damp soil (30% moisture, −30 kPa), and intermediate persistence in the driest soil treatment (15% moisture, −500 kPa). A very different pattern was observed in soil taken from a marsh, possibly due to differences in physicochemical properties and the presence of microorganisms (Peng et al., 1999a).

The role of precipitation in the persistence of viruses has little to do with inactivation and more to do with transport and dispersal in the environment. Early studies recognized that it was difficult to wash baculovirus or cypovirus OBs from plant foliage exposed to natural or simulated rainfall (Burgerjon and Grison, 1965; David and Gardiner, 1966; Bullock, 1967). In a controlled field study, natural or simulated rainfall applied to LdMNPV-infected *L. dispar* cadavers on oak trees resulted in a reduction in the prevalence of infection of conspecific larvae from approximately 42–60% prior to rainfall to 20–35% following rainfall, indicating a significant loss of inoculum from oak foliage through the action of rain (values estimated from figures in D'Amico and Elkinton, 1995). Clearly, humidity and precipitation are factors that tend to interact with other environmental variables in each particular habitat and across different types of ecosystem.

7.11.5 Effect of pH on Viruses

The pH of the environment can be highly influential in the stability and persistence of invertebrate viruses. As mentioned in Section 7.6.2.1, the alkaline pH of cotton leaf surfaces can rapidly inactivate NPV OBs (Young and Yearian, 1974; Elleman and Entwistle, 1985a). In contrast, simulated acid rain (pH 3) reduced the mortality of infected sawfly larvae feeding on pine foliage, not due to a direct effect on the virus, but likely due to a pH-mediated change in the quality of plant foliage that improved larval survival (Neuvonen et al., 1990).

Most studies on the effect of pH in the environment have focused on the pH of the soil. As OBs break down in the presence of alkaline pH and release virions, the persistence of baculoviruses in high-pH soils is poor compared to that in soils of neutral or slightly acid pH (Jaques, 1985). In a study on SeMNPV isolated from soil substrates across four zones of horticultural greenhouses in southern Spain (Murillo et al., 2007), the pH of the soil substrate was found to vary seasonally (Fig. 7.2a), with pH values in the fall (mean pH 8.6) significantly higher than those during other periods of the year (pH 7.8–8.1). This was probably due to the application of alkaline substrate disinfection treatments at the end of the summer. Soil substrate pH differed between greenhouse zones in this area, but, more significantly, the prevalence of mortality in bioassays (an indicator of the abundance of SeMNPV OBs in substrate samples) was negatively correlated with substrate pH (Fig. 7.2b). Moreover, some genotypes were associated with soil substrates with higher pH and others were associated with lower pH, suggesting that certain genotypes may be better able to withstand high-pH conditions (Fig. 7.2c). Finally, soil substrate pH affected the probability of isolating single- or mixed-genotype isolates (Fig. 7.2d), suggesting that soil substrates with lower pH harbored larger and more diverse OB populations (Murillo et al., 2007). Although thought-provoking, the possibility that genotypic variants differ in their ability to persist in the environment, perhaps due to the size or robustness of their OBs, has not been the subject of systematic study.

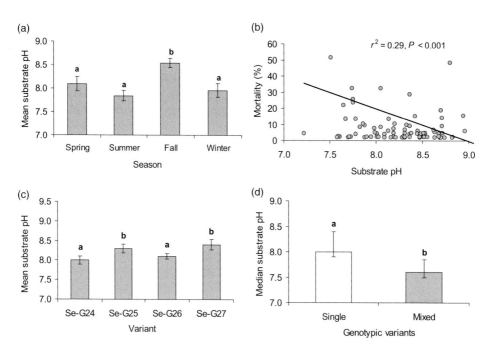

Fig. 7.2 Persistence of NPV (SeMNPV) occlusion bodies in greenhouse soil substrate in southern Spain. (a) Seasonal fluctuations in substrate pH. (b) Influence of substrate pH on the mortality of *Spodoptera exigua* larvae in bioassays – an indicator of the abundance of OBs in substrate samples. (c) Mean substrate pH from which four genotypic variants of SeMNPV (Se-G24 to Se-G27) were isolated in bioassays. (d) Median substrate pH from which single or mixed genotype infections were isolated in bioassays. Vertical bars indicate 95% confidence interval of means or interquartile range about the median. Columns headed by identical letters do not differ significantly (P > 0.05). *Source*: Murillo et al. (2007). Reproduced with permission of Elsevier.

7.12   Biotic Factors that Interact with Virus Populations

7.12.1   Plant Phenology, Structure, and Nutritional Value

Plant species and phenology have marked effects on the persistence of baculovirus OBs on plant surfaces (Section 7.6.2.1), but they also influence the nutritional quality of foliage and the physical toughness and other physical defenses against herbivory. As a result, the lethality of baculoviruses is often observed to vary markedly depending on the food plant species (Farrar and Ridgway, 2000; Wan et al., 2016). In some cases, this may be due to compensatory feeding, through which the insect consumes greater quantities of foliage on poor-quality food plants in an attempt to acquire sufficient nutrients and energy to achieve growth and development. Studies on virus transmission in different species or cultivars of plants should take differential feeding into account in order to accurately assess the role of the food plant on the relationship between OB density and prevalence of infection.

Studies on plant nutritional quality usually focus on key indicators of compounds required for insect growth and development, such as proteins (nitrogen content), carbohydrates, fatty acids, and defensive compounds. Food plant quality can affect different indicators of immune function, including hemocyte numbers, phenoloxidase levels, and encapsulation responses (Klemola et al., 2007; Shikano et al., 2010). Insects that feed on poor-quality plants appear to be at increased risk of infection by baculoviruses compared to those that feed on high-quality plants (Raymond and Hails, 2007; Shikano et al., 2010). The plant can also have a marked effect on the fitness of the pathogen. Winter moth (*O. brumata*) larvae feeding on oak, a good-quality food plant, were at lower risk of infection than conspecific larvae that fed on sitka spruce. However, each oak-fed larva produced larger numbers of OBs, and the overall production of OBs in each cohort of insects was also significantly greater on oak- compared to spruce-fed insects (Raymond and Hails, 2007). Furthermore, phenology-related reductions in the leaf nitrogen and phenolic content of silver birch (*Betula pendula*) affected hemocyte numbers in *L. dispar* larvae and the prevalence of lethal NPV disease, although not in a systematic manner. There was some evidence that larvae feeding on older nitrogen-depleted foliage were more likely to succumb to spontaneous virus disease that was triggered in covertly infected individuals (Martemyanov et al., 2015b).

In the case of physical defensive structures, very little attention has been paid to their role in modulating the primary infection process in baculoviruses. In a comparative study on armyworm (*Mythimna unipuncta*), the prevalence of NPV infection and the speed of kill of the virus were similar on spiny and smooth fescue grasses, and the peritrophic membrane remained undamaged despite the presence of grass fragments with spines in the food bolus (Keathley et al., 2012).

7.12.2   Phytochemical–Virus Interactions

Plants produce an enormous diversity of defensive compounds, many of which are designed to reduce feeding by insects or other herbivores. Studies on plant-mediated effects on insect–virus interactions have focused on baculoviruses (reviewed by Duffey et al., 1995). Plant chemistry effects on insect viruses are usually examined following observations that a baculovirus-based insecticide has poor efficacy on a particular type of crop (Stevenson et al., 2010), or else in order to examine the ecological effects of defoliation on disease dynamics in forest-feeding pests (Elderd et al., 2013).

Compounds on plant surfaces will interact with OBs from insect cadavers or other environmental reservoirs, whereas compounds within plant tissues only interact with OBs in the insect gut following consumption of virus-contaminated foliage. An interesting example of leaf-surface chemical effects is that of chickpea (*Cicer arietinum*), which has leaf trichomes that produce abundant organic acids, leading to a leaf-surface pH <3. However, following the application of NPV OBs, two phenolic compounds (isoflavonoids) are released on to the leaf surface and rapidly inactivate OBs, although details regarding the chemical mechanism for inactivation are uncertain (Stevenson et al., 2010).

As foliage is consumed and moves through the insect gut, a series of interactions occur between phytochemicals, digestive tract enzymes, and virus particles (reviewed by Cory and Hoover, 2006). Immediately following ingestion of plant material, salivary gland secretions containing enzymes such as glucose oxidase can produce reactive oxygen species (ROS), such as hydrogen peroxide, that can inactivate the bacterial pathogen *Bacillus thuringiensis* (Musser et al., 2005) and, possibly, viruses. Following this, exposure to plant phenolic compounds in the midgut can result in aggregation of baculovirus OBs, which then fail to release the ODVs that infect midgut cells (Duffey et al., 1995). ODVs are also likely to be damaged or inactivated by exposure to phenolic compounds or free radicals. For example, the lethality of NPV OBs decreased in the presence of two phenolic compounds, rutin and chlorogenic acid, in *H. zea* larvae (Felton et al., 1987). In a detailed study on the role of induced hydrolyzable tannins in oak, the presence of dietary tannins reduced the susceptibility of *L. dispar* larvae to their NPVs, as well as the risk of transmission (Elderd et al., 2013). Tannins are hydrolyzed in the insect gut to release phenolic compounds. The results of this study were used to develop a model to demonstrate that the prevalence of oak trees in mixed forests can effectively predict the severity and periodicity of *L. dispar* outbreaks (see Chapter 12).

The integrity of the peritrophic membrane lining the midgut may be compromised by interactions with phytochemicals and plant enzymes (Pechan et al., 2002) or physically altered in response to the consumption of different types of foliage, resulting in a negative correlation between peritrophic membrane thickness and the probability of lethal baculovirus infection (Plymale et al., 2008). Midgut cells can also suffer oxidative stress from the presence of reactive species, and they respond by sloughing off from the midgut wall before the virus has established a systemic infection in the insect (Hoover et al., 2000).

Phytochemicals, or phytochemical-induced signals, may cross the gut and modify immune function or host physiology, resulting in increased or decreased susceptibility to baculovirus disease in relation to food plant quality, as described in Section 7.12.1 (Klemola et al., 2007; Shikano et al., 2010). However, several studies on induced plant defenses have failed to detect altered susceptibility to NPVs (Plymale et al., 2007; Martemyanov et al., 2012; Sarfraz et al., 2013).

7.12.3 Virus Interactions with Alternative Hosts

The host range of invertebrate viruses varies widely from highly host-specific viruses, such as SeMNPV (Simón et al., 2004b), to viruses such as invertebrate iridescent virus 6 that can replicate in a wide range of arthropods and even ectothermic vertebrate hosts

(Ohba and Aizawa, 1979; Stöhr et al., 2016). This means that viruses with an extended host range can exploit alternative hosts and are less dependent on the presence and density of particular host species than are highly host-specific viruses. Nonetheless, the ability to exploit a range of host species is likely to come at the cost of reduced fitness for the pathogen in terms of the capacity to infect, replicate, and produce progeny particles in less than optimal host species. That said, very little attention has been paid to determining the tradeoff between the use of alternative host species and the fitness of invertebrate viruses in nature. There are several virus–invertebrate pathosystems that could be suitable for examining the ecological roles of alternative hosts in agricultural or natural ecosystems. In the case of baculoviruses, the extended host range of viruses such as AcMNPV and Anagrapha falcifera multiple nucleopolyhedrovirus (AnfaMNPV) means they can be used as biological insecticides to control multispecies complexes of noctuid looper pests (Vail et al., 1999). This system offers the intriguing possibility of determining the role of each of the looper species in the persistence, transmission, and genetic diversity of a multi-host baculovirus on *Brassica* spp. or other shared food plants.

Alternative host species also facilitate the survival of viruses in other systems. An iridescent virus that was transmitted through acts of cannibalism and interspecific predation persisted in two species of terrestrial isopods (woodlice) in a grassland ecosystem (Grosholz, 1992). Survival decreased and the prevalence of disease increased in mixed populations compared to the single-species population, apparently due to an increased frequency of interspecific aggression in mixed populations. Additionally, ascoviruses that are transmitted between hosts by endoparasitoid wasps appear capable of infecting numerous species of noctuid larvae, reflecting the oviposition preferences of their parasitoid vectors (Hamm et al., 1998).

Many viruses may be presumed to be host-specific because they have not been looked for in species other than the host from which they were initially isolated (Roy et al., 2009). Given that there is growing evidence that insects frequently harbor covert infections by viruses (Kemp et al., 2011; Virto et al., 2013), it is clear that disease-based estimates of the prevalence of virus pathogens in invertebrate populations represent a major underestimate of their true prevalence. As such, even when opportunities for horizontal transmission are scarce, sublethal disease and the activation of covert infections into lethal viroses (Cooper et al., 2003; Burden et al., 2006) are likely to have an impact on the populations of rare host species and non-pest species that, by definition, exist at low densities. Moreover, covert infections in populations of "unexpected" hosts are likely to be overlooked unless systematic surveys are performed (Roy et al., 2009). Some support for the unexpected-host hypothesis comes from a molecular study on the presence of an NPV in populations of the winter moth (*O. brumata*), which fortuitously identified the virus in two sympatric heather-feeding geometrids: the July highflyer (*Hydriomena furcata*) and the grey mountain carpet (*Entephria caesiata*) (Graham, 2005). However, the precise role of these species in the ecology of the winter moth–virus pathosystem remains unknown.

7.12.4 Competition and Facilitation in Virus Interactions with Other Organisms

7.12.4.1 Virus Interactions with Parasitoids

Of the interactions of invertebrate viruses with other parasites and pathogens, those related to parasitoid wasps have attracted by far the most attention, probably due to the abundance and conspicuousness of parasitoids in natural and agricultural ecosystems. However, despite the number and diversity of studies on virus–parasitoid interactions in the laboratory, few consensus principles have emerged on the competitive interactions among these natural enemies. This is probably due to three main factors: (i) a frequent focus on examining the "compatibility" of parasitoids and viruses used in the biological control of pests; (ii) complexity in the outcomes of virus–parasitoid co-infection studies arising from marked interspecific diversity in parasitoid biology; and (iii) complicating issues arising from the presence of different polydnaviruses in many of the most common braconid and ichneumonid parasitoids, which suppress host immune function.

Parasitoids, particularly endoparasitoids, which oviposit within the body of lepidopteran hosts, can detect physiological changes related to infection status and discriminate against infected hosts in favor of healthy individuals (Kyei-Poku and Kunimi, 1997; Matthews et al., 2004; Jiang et al., 2014). In many cases, parasitism must occur hours or days prior to virus infection for the parasitoid progeny to have any chance of developing in baculovirus-infected insects, otherwise the host is usually killed by the pathogen before wasp larvae can complete their development (Cossentine, 2009). In some baculoviruses and entomopoxviruses, developing wasp larvae are killed by virus-encoded toxic factors that eliminate the competitor (Okuno et al., 2002; Cossentine, 2009), whereas parasitoid larvae in iridescent virus-infected hosts themselves become infected by the virus and die (López et al., 2002).

The outcomes of virus–parasitoid interactions, from the perspective of the pathogen, range from facilitation to competition. The virus gains opportunities for transmission because parasitized hosts are often more susceptible to infection than are nonparasitized ones (Santiago-Alvarez and Caballero, 1990; Gou et al., 2013), likely due to immune suppression by parasitoid polydnaviruses (Washburn et al., 2000; Rivkin et al., 2006; although see D'Amico et al., 2013a). The presence of the virus is also frequently associated with reduced growth of developing wasp larvae, arising from the stunted growth of the host insect and the rapid sequestering of host resources for virus replication (Nakai et al., 1997; Azam et al., 2016).

Given that viruses replicating in parasitized insects face direct competition for host resources, it is not surprising that such conditions are appropriate for the triggering of patent disease in covertly infected insects (Stoltz and Makkay, 2003), or for selection for highly virulent strains of virus with rapid speed of kill (Escribano et al., 2001). Moreover, reduced host growth and the segregation of host resources by the developing parasitoids can result in a reduction in the number of progeny virus OBs produced in each infected and parasitized host (Escribano et al., 2001; Cai et al., 2012). This can impact directly on virus transmission, although a fraction of the parasitoids that emerge from virus-infected hosts may be contaminated and capable of transmitting the virus to other hosts during acts of oviposition (Brooks, 1993).

In a long-term study on *L. dispar* populations across different US states, four species of parasitoids, an NPV (LdMNPV), and a fungal pathogen were quantified in samples taken over a 17-year period (Hajek and van Nouhuys, 2016). At sites with outbreak populations of *L. dispar*, one braconid parasitoid was found in association with LdMNPV infection far more frequently than expected by chance, possibly due to a polydnavirus-mediated reduction in host resistance to virus infection, while other parasitoids appeared to avoid virus-infected gypsy moth larvae or were killed by the fungus before they could complete their development.

7.12.4.2 Virus Interactions with Other Pathogens

Interactions among different types of invertebrate viruses, or between viruses and other pathogens, depend on the route or mechanism of infection and the temporal sequence of infection by each entity. Studies focused on biological control tend to consider mixtures of pathogens administered simultaneously to pest insects, with the aim of identifying potentiation or antagonism in the insecticidal characteristics of the pathogens (Harper, 1986). Such studies have identified virus factors, such as enhancin in baculoviruses and spheroidin in entomopoxviruses, that degrade the larval peritrophic membrane and facilitate the primary infection of midgut cells by these viruses (Wang and Granados, 1997; Mitsuhashi et al., 1998).

In co-infected hosts, studies have focused on the outcome of within-host competition, which largely depends on the replication rate and speed of kill of each pathogen, or on the ability of one virus to disrupt the replication of another. Examples come from studies on co-infecting baculoviruses (Hacket et al., 2000; Wennmann et al., 2015) and on the interactions between baculoviruses and other viruses (Ishii et al., 2002), fungi (Malakar et al., 1999b), and entomopathogenic nematodes (Agra-Gothama et al., 1995).

An alternative approach, and one that is generally more applicable to an ecological context, is an examination of how virus-infected insects respond to superinfection by another pathogen. Superinfection occurs when an individual that is already infected by one pathogen is then infected by a second pathogen. For example, when NPV-infected larvae of the diamondback moth (*Plutella xylostella*) were challenged with different concentrations of *B. thuringiensis*, lower than expected mortality occurred at low doses of *B. thuringiensis*, suggesting a protective effect of virus infection (Raymond et al., 2006). Within-host interactions include the ability of a virus to induce cells to become refractive to superinfection some hours after the initial infection, thereby blocking the systemic infection process for the second virus (Beperet et al., 2014).

At the population level, the gypsy moth NPV (LdMNPV) and a fungal pathogen did not interact to influence the mortality of this insect (Malakar et al., 1999a). The virus continued to act in a density-dependent manner independent of the presence of the fungus, the prevalence of which was not affected by host density (Liebhold et al., 2013). In a long-term study, co-infection by LdMNPV and the fungal pathogen decreased with increasing host density. The reasons for this remain unclear but may be related to the speed of kill and propagule production by each type of pathogen, or to host responses to population density (Hajek and van Nouhuys, 2016).

The presence of a densovirus in natural populations of *Helicoverpa armigera* reduced the susceptibility of larvae to infection by NPVs (HearNPV) and to low doses of Bt toxin. HearNPV replication was also reduced in co-infected hosts, indicating an important protective effect by the densovirus (Xu et al., 2014). Similarly, the symbiont *Wolbachia* protected *Drosophila melanogaster* from several RNA viruses but not from a DNA virus (i.e., an iridescent virus) (Hedges et al., 2008; Teixeira et al., 2008). Subsequently, increased susceptibility to another DNA virus, an NPV, was observed in *Wolbachia*-infected *S. exempta* (Graham et al., 2012). Although the responses of *Wolbachia*-infected insects differ for RNA and DNA viruses, given the high incidence of *Wolbachia* in many insect populations, the ecological consequences of symbiont-mediated shifts in susceptibility to infection by viruses are likely to provide novel insights into individual and population-level processes in *Wolbachia*-infected insect virus pathosystems. This issue has particular implications for the use of *Wolbachia*-infected mosquitoes for the suppression of arbovirus transmission in tropical regions (Lambrechts et al., 2015).

In a more extreme case of facilitation between viruses, the nonoccluded Spodoptera exigua iflavirus 1 became intimately associated with the OBs of an NPV (SeMNPV) when both viruses replicated in co-infected larvae of *S. exigua* (Jakubowska et al., 2016). Iflavirus particles appeared to be incorporated into the OB protein matrix and were transmitted efficiently in OBs, which also protected the iflavirus from high temperatures and UV radiation during periods in the environment. In essence, the iflavirus became a hitchhiker, using the OBs as a vehicle to improve its survival and transmission opportunities.

Within-host interactions of viruses and other pathogens can be common and are highly influential to the evolution of virulence (Alizon et al., 2013). However, most studies are limited by their focus on the outcome of mixed infections in individuals, rather than population-level effects. Clearly, it is necessary to coordinate within-host and between-host studies over multiple cycles of transmission in order to obtain a useful perspective on the role of pathogen interactions and multiple infections in the evolution of virulence.

7.12.4.3 Virus Interactions with Microbiota

Finally, the role of microbiota in disease susceptibility is an issue that has begun to attract a great deal of attention following the advent of metagenomics techniques. Nonpathogenic bacteria on the phylloplane of different host plants did not elicit a host immune response and did not affect the susceptibility of *T. ni* larvae to NPV infection (Shikano et al., 2015). In contrast, when the gut microbiota of *S. exigua* larvae was controlled or eliminated using antibiotics, survival of SeMNPV-infected larvae increased and OB production decreased almost threefold in insects lacking gut microbiota (Jakubowska et al., 2013). Phylloplane organisms and the host plant can markedly influence the gut microbiota, but the implications of these findings for virus transmission in nature have yet to be determined.

7.13 Conclusion

Despite being simple replicating entities, devoid of life *per se*, viruses interacting with their host provide a wealth of challenging ecological and evolutionary questions. The diversity of the ecological processes described in this chapter bears testament to the complexity of invertebrate–virus relationships, which range from gene- and genome- to population- and species-level. With the recent growth in virus detection, sequence determination, and particle visualization techniques, our ability to examine invertebrate virus processes at the cell and organism levels is set to provide extraordinary opportunities to increase our understanding of host–virus relationships. At the other extreme, the development of population models supported by empirical observations continues to advance and offer opportunities to understand the fundamental ecological processes that regulate insect populations and drive particular patterns of population dynamics, best exemplified in phytophagous insects in forest ecosystems.

Of the key issues that need to be addressed over the coming decade, from a personal perspective the following stand out. (i) As we can screen large numbers of insects for covert virus infections, quantifying the impact of sublethal disease on invertebrate–virus population dynamics should become increasingly achievable. (ii) As we recognize that many virus populations are highly diverse, determining the selective nature of this diversity in transmission and persistence in natural ecosystems will doubtless provide a profitable line of research. (iii) Given the uncertainty that nations will be able to check global climate change, studies on the effects of predicted rises in temperature on fundamental ecological processes, including invertebrate disease dynamics, will become increasingly relevant.

Additional intriguing issues relate to the role of epigenetic mechanisms in modulating infectious disease processes in invertebrates and unraveling the relationship between virus genotype and phenotype, such as that observed in the case of *egt* and *ptp* manipulation of host behavior. Such studies are likely to provide original and exciting insights at the individual and population levels.

Acknowledgments

I thank Jenny Cory, Judy Myers, and Rosa Murillo for kind permission to reproduce figures, Miguel López-Ferber, Primitivo Caballero, James Becnel, Sean Moore, and Rodrigo Lasa for helpful discussions, and Gabriel Mercado for logistical support.

References

Abd-Alla, A., Bossin, H., Cousserans, F., Parker, A., Bergoin, M., Robinson, A., 2007. Development of a non-destructive PCR method for detection of the salivary gland hypertrophy virus (SGHV) in tsetse flies. J. Virol. Meth. 139, 143–149.

Abd-Alla, A.M., Boucias, D.G., Bergoin, M., 2010. Hytrosaviruses: structure and genomic properties, in: Assgari, S., Johnson, K. (eds.), Insect Virology. Caister Academic Press, Norfolk, pp. 103–121.

Adamo, S.A., Kovalko, I., Easy, R.H., Stoltz, D., 2014. A viral aphrodisiac in the cricket *Gryllus texensis*. J. Exp. Biol. 217, 1970–1976.

Agra-Gothama, A.A., Sikorowski, P.P., Lawrence, G.W., 1995. Interactive effects of *Steinernema carpocapsae* and *Spodoptera exigua* nuclear polyhedrosis virus on *Spodoptera exigua* larvae. J. Invertebr. Pathol. 66, 270–276.

Alizon, S., de Roode, J.C., Michalakis, Y., 2013. Multiple infections and the evolution of virulence. Ecol. Lett. 16, 556–567.

Anderson, R.M., May, R.M., 1981. The population dynamics of microparasites and their invertebrate hosts. Philos. Trans. R. Soc. Lond. B Biol. Sci. 291, 451–524.

Arzul, I., Renault, T., Thébault, A., Gérard, A., 2002. Detection of oyster herpesvirus DNA and proteins in asymptomatic *Crassostrea gigas* adults. Virus Res. 84, 151–160.

Azam, A., Kunimi, Y., Inoue, M.N., Nakai, M., 2016. Effect of granulovirus infection of *Spodoptera litura* (Lepidoptera: Noctuidae) larvae on development of the endoparasitoid *Chelonus inanitus* (Hymenoptera: Braconidae). Appl. Entomol. Zool. 51, 479–489.

Baillie, V.L., Bouwer, G., 2012a. High levels of genetic variation within core Helicoverpa armigera nucleopolyhedrovirus genes. Virus Genes 44, 149–162.

Baillie, V.L., Bouwer, G., 2012b. High levels of genetic variation within Helicoverpa armigera nucleopolyhedrovirus populations in individual host insects. Arch. Virol. 157, 2281–2289.

Baillie, V.L., Bouwer, G., 2013. The effect of inoculum dose on the genetic diversity detected within Helicoverpa armigera nucleopolyhedrovirus populations. J. Gen. Virol. 94, 2524–2529.

Ballard, J., Ellis, D.J., Payne, C.C., 2000. Uptake of granulovirus from the surface of apples and leaves by first instar larvae of the codling moth *Cydia pomonella* L. (Lepidoptera: Olethreutidae). Biocontr. Sci. Technol. 10, 617–625.

Barbosa-Solomieu, V., Dégremont, L., Vazquez-Juarez, R., Ascencio-Valle, F., Boudry, P., Renault, T., 2005. Ostreid herpesvirus 1 (OsHV-1) detection among three successive generations of Pacific oysters (*Crassostrea gigas*). Virus Res. 107, 47–56.

Barik, T.K., Suzuki, Y., Rasgon, J.L., 2016. Factors influencing infection and transmission of Anopheles gambiae densovirus (AgDNV) in mosquitoes. Peer J. 4, e2691.

Barrera, G., Williams, T., Villamizar, L., Caballero, P., Simón, O., 2013. Deletion genotypes reduce occlusion body potency but increase occlusion body production in a Colombian *Spodoptera frugiperda* nucleopolyhedrovirus population. PloS ONE 8, e77271.

Becnel, J.J., 2006. Transmission of viruses to mosquito larvae mediated by divalent cations. J. Invertebr. Pathol. 92, 141–145.

Bedford, G.O., 1980. Control of the rhinoceros beetle by baculovirus, in: Burges, H.D. (ed.), Microbial Control of Insects, Mites and Plant Diseases, Vol. 2. Academic Press, London, pp. 409–426.

Benz, G., 1987. Environment, in: Fuxa, J.R., Tanada, Y. (eds.), Epizootiology of Insect Diseases. John Wiley & Sons, New York, pp. 177–214.

Beperet, I., Irons, S., Simón, O., King, L.A., Williams, T., Possee, R.D., et al., 2014. Superinfection exclusion in alphabaculovirus infections is concomitant with actin reorganization. J. Virol. 88, 3548–3556.

Bergoin, M., Tijssen, P., 2010. Densoviruses: a highly diverse group, in: Assgari, S., Johnson, K. (eds.), Insect Virology. Caister Academic Press, Norfolk, pp. 59–82.

Bernal, A., Simón, O., Williams, T., Muñoz, D., Caballero, P., 2013. A *Chrysodeixis chalcites* single-nucleocapsid nucleopolyhedrovirus population from the Canary Islands is genotypically structured to maximize survival. Appl. Env. Microbiol. 79, 7709–7718.

Bianchi, F.J.J.A., Joosten, N.N., Gutierrez, S., Reijnen, T.M., Werf, W.V.D., Vlak, J.M., 1999. The polyhedral membrane does not protect polyhedra of AcMNPV against inactivation on greenhouse chrysanthemum. Biocontr. Sci. Technol. 9, 523–527.

Bideshi, D.K., Bigot, Y., Federici, B.A., Spears, T., 2010. Ascoviruses, in: Assgari, S., Johnson, K. (eds.), Insect Virology. Caister Academic Press, Norfolk, pp. 3–34.

Biernat, M.A., Ros, V.I.D., Vlak, J.M., van Oers, M.M., 2011. Baculovirus cyclobutane pyrimidine dimer photolyases show a close relationship with lepidopteran host homologues. Ins. Mol. Biol. 20, 457–464.

Biever, K.D., Hostetter, D.L., 1985. Field persistence of *Trichoplusia ni* (Lepidoptera: Noctuidae) single-embedded nuclear polyhedrosis virus on cabbage foliage. Env. Entomol. 14, 579–581.

Bigot, Y., Rabouille, A., Doury, G., Sizaret, P.Y., Delbost, F., Hamelin, M.H., Periquet, G., 1997. Biological and molecular features of the relationships between *Diadromus pulchellus* ascovirus, a parasitoid hymenopteran wasp (*Diadromus pulchellus*) and its lepidopteran host, *Acrolepiopsis assectella*. J. Gen. Virol. 78, 1149–1163.

Blanchard, P., Guillot, S., Antùnez, K., Köglberger, H., Kryger, P., de Miranda, J.R., et al., 2014. Development and validation of a real-time two-step RT-qPCR TaqMan assay for quantitation of Sacbrood virus (SBV) and its application to a field survey of symptomatic honey bee colonies. J. Virol. Meth. 197, 7–13.

Blanc-Mathieu, R., Ogata, H., 2016. DNA repair genes in the Megavirales pangenome. Curr. Opin. Microbiol. 31, 94–100.

Bonning, B.C., Miller, W.A., 2010. Dicistroviruses. Annu. Rev. Entomol. 55, 129–150.

Bonsall, M.B., Sait, S.M., Hails, R.S., 2005. Invasion and dynamics of covert infection strategies in structured insect–pathogen populations. J. Anim. Ecol. 74, 464–474.

Boots, M., 1998. Cannibalism and the stage-dependent transmission of a viral pathogen of the Indian meal moth, *Plodia interpunctella*. Ecol. Entomol. 23, 118–122.

Boots, M., Greenman, J., Ross, D., Norman, R., Hails, R., Sait, S., 2003. The population dynamical implications of covert infections in host–microparasite interactions. J. Anim. Ecol. 72, 1064–1072.

Boucias, D.G., Kariithi, H.M., Bourtzis, K., Schneider, D.I., Kelley, K., Miller, W.J., et al., 2013. Transgenerational transmission of the *Glossina pallidipes* hytrosavirus depends on the presence of a functional symbiome. PLoS ONE 8, e61150.

Boughton, A.J., Obrycki, J.J., Bonning, B.C., 2003. Effects of a protease-expressing recombinant baculovirus on nontarget insect predators of *Heliothis virescens*. Biol. Control 28, 101–110.

Bourner, T.C., Vargas-Osuna, E., Williams, T., Santiago-Alvarez, C., Cory, J.S., 1992. A comparison of the efficacy of nuclear polyhedrosis and granulosis viruses in spray and bait formulations for the control of *Agrotis segetum* (Lepidoptera: Noctuidae) in maize. Biocontr. Sci. Technol. 2, 315–326.

Bouwer, G., Nardini, L., Duncan, F.D., 2009. *Helicoverpa armigera* (Lepidoptera: Noctuidae) larvae that survive sublethal doses of nucleopolyhedrovirus exhibit high metabolic rates. J. Ins. Physiol. 55, 369–374.

Briese, D.T., 1986. Insect resistance to baculovirus, in: Granados R.R., Federici B.A. (eds.), The Biology of Baculoviruses, Vol. 2. CRC Press, Boca Raton, FL, pp. 237–263.

Briggs, C.J., Hails, R.S., Barlow, N.D., Godfray, H.C.J., 1995. The dynamics of insect–pathogen interactions, in: Grenfell, B.T., Dobson, A.P. (eds.), Ecology of Infectious Diseases in Natural Populations. Cambridge University Press, Cambridge, pp. 295–326.

Brookes, M.H., Stark, R.W., Campbell, R.W. (eds.), 1978. The Douglas-fir tussock moth: a synthesis. Forest Service, US Department of Agriculture. Technical Bulletin No. 1585, Washington, DC.

Brooks, W.M., 1993. Host-parasitoid-pathogen interactions, in: Beckage, N.E., Thompson, S.N., Federici, B.A. (eds.), Parasites and Pathogens of Insects, Vol. 2: Pathogens. Academic Press, New York, pp. 231–272.

Bull, J.C., Godfray, H.C.J., O'Reilly, D.R., 2001. Persistence of an occlusion-negative recombinant nucleopolyhedrovirus in *Trichoplusia ni* indicates high multiplicity of cellular infection. Appl. Env. Microbiol. 67, 5204–5209.

Bullock, H.R., 1967. Persistence of *Heliothis* nuclear polyhedrosis virus on cotton foliage. J. Invertebr. Pathol. 9, 434–436.

Burand, J.P., 2009. The sexually transmitted insect virus, Hz-2V. Virol. Sinica 24, 428–435.

Burden, J.P., Griffiths, C.M., Cory, J.S., Smith, P., Sait, S.M., 2002. Vertical transmission of sublethal granulovirus infection in the Indian meal moth, *Plodia interpunctella*. Mol. Ecol. 11, 547–555.

Burden, J.P., Possee, R.D., Sait, S.M., King, L.A., Hails, R.S., 2006. Phenotypic and genotypic characterisation of persistent baculovirus infections in populations of the cabbage moth (*Mamestra brassicae*) within the British Isles. Arch. Virol. 151, 635–649.

Burgerjon, A., Grison, P., 1965. Adhesiveness of preparations of *Smithiavirus pityocampae* Vago on pine foliage. J. Invertebr. Pathol. 7, 281–284.

Cabodevilla, O., Ibañez, I., Simón, O., Murillo, R., Caballero, P., Williams, T., 2011a. Occlusion body pathogenicity, virulence and productivity traits vary with transmission strategy in a nucleopolyhedrovirus. Biol. Control 56, 184–192.

Cabodevilla, O., Villar, E., Virto, C., Murillo, R., Williams, T., Caballero, P., 2011b. Intra-and intergenerational persistence of an insect nucleopolyhedrovirus: adverse effects of sublethal disease on host development, reproduction, and susceptibility to superinfection. Appl. Env. Microbiol. 77, 2954–2960.

Cai, Y., Fan, J., Sun, S., Wang, F., Yang, K., Li, G., Pang, Y., 2012. Interspecific interaction between *Spodoptera exigua* multiple nucleopolyhedrovirus and *Microplitis bicoloratus* (Hymenoptera: Braconidae: Microgastrina) in *Spodoptera exigua* (Lepidoptera: Noctuidae) larvae. J. Econ. Entomol. 105, 1503–1508.

Capinera, J.L., Kirouac, S.P., Barbosa, P., 1976. Phagodeterrency of cadaver components to gypsy moth larvae, *Lymantria dispar*. J. Invertebr. Pathol. 28, 277–279.

Carpenter, J.A., Obbard, D.J., Maside, X., Jiggins, F.M., 2007. The recent spread of a vertically transmitted virus through populations of *Drosophila melanogaster*. Mol. Ecol. 16, 3947–3954.

Carrillo-Tripp, J., Krueger, E.N., Harrison, R.L., Toth, A.L., Miller, W.A., Bonning, B.C., 2014. Lymantria dispar iflavirus 1 (LdIV1), a new model to study iflaviral persistence in lepidopterans. J. Gen. Virol. 95, 2285–2296.

Castillejos, V., Garcia, L., Cisneros, J., Goulson, D., Cave, R.D., Caballero, P., Williams, T., 2001. The potential of *Chrysoperla rufilabris* and *Doru taeniatum* as agents for dispersal of *Spodoptera frugiperda* nucleopolyhedrovirus in maize. Entomol. Exp. Appl. 98, 353–359.

Castillejos, V., Trujillo, J., Ortega, L.D., Santizo, J.A., Cisneros, J., Penagos, D.I., et al., 2002. Granular phagostimulant nucleopolyhedrovirus formulations for control of *Spodoptera frugiperda* in maize. Biol. Control 24, 300–310.

Chang, P.S., Chen, L.J., Wang, Y.C., 1998. The effect of ultraviolet irradiation, heat, pH, ozone, salinity and chemical disinfectants on the infectivity of white spot syndrome baculovirus. Aquaculture 166, 1–17.

Chapman, J.W., Williams, T., Escribano, A., Caballero, P., Cave, R.D., Goulson, D., 1999. Age-related cannibalism and horizontal transmission of a nuclear polyhedrosis virus in larval *Spodoptera frugiperda*. Ecol. Entomol. 24, 268–275.

Chateigner, A., Bézier, A., Labrousse, C., Jiolle, D., Barbe, V., Herniou, E.A., 2015. Ultra deep sequencing of a baculovirus population reveals widespread genomic variations. Viruses, 7, 3625–3646.

Chen, G., Pan, H., Xie, W., Wang, S., Wu, Q., Fang, Y., et al., 2013. Virus infection of a weed increases vector attraction to and vector fitness on the weed. Sci. Rep. 3, 2253.

Chen, Y.R., Zhong, S., Fei, Z., Gao, S., Zhang, S., Li, Z., et al., 2014. Transcriptome responses of the host *Trichoplusia ni* to infection by the baculovirus *Autographa californica* multiple nucleopolyhedrovirus. J. Virol. 88, 13781–13797.

Chiu, E., Hijnen, M., Bunker, R.D., Boudes, M., Rajendran, C., Aizel, K., et al., 2015. Structural basis for the enhancement of virulence by viral spindles and their in vivo crystallization. Proc. Nat. Acad. Sci. U.S.A. 112, 3973–3978.

Christian, P.D., 1992. A simple vacuum dot-blot hybridisation assay for the detection of Drosophila A and C viruses in single *Drosophila*. J. Virol. Meth. 38, 153–165.

Christian, P.D., Richards, A.R., Williams, T., 2006. Differential adsorption of occluded and non-occluded insect pathogenic viruses to soil forming minerals. Appl. Env. Microbiol. 72, 4648–4652.

Clavijo, G., Williams, T., Muñoz, D., López-Ferber, M., Caballero, P., 2009. Entry into midgut epithelial cells is a key step in the selection of genotypes in a nucleopolyhedrovirus. Virol. Sinica 24, 350–358.

Clavijo, G., Williams, T., Muñoz, D., Caballero, P., López-Ferber, M., 2010. Mixed genotype transmission bodies and virions contribute to the maintenance of diversity in an insect virus. Proc. R. Soc. Lond. B Biol. Sci. 277, 943–951.

Cook, S., Chung, B.Y.W., Bass, D., Moureau, G., Tang, S., McAlister, E., et al., 2013. Novel virus discovery and genome reconstruction from field RNA samples reveals highly divergent viruses in dipteran hosts. PLoS ONE 8, e80720.

Cooper, D., Cory, J.S., Theilmann, D.A., Myers, J.H., 2003. Nucleopolyhedroviruses of forest and western tent caterpillars: cross-infectivity and evidence for activation of latent virus in high-density field populations. Ecol. Entomol. 28, 41–50.

Corbeil, S., Williams, L.M., Bergfeld, J., Crane, M.S.J., 2012. Abalone herpes virus stability in sea water and susceptibility to chemical disinfectants. Aquaculture 326, 20–26.

Cory, J.S., 2010. The ecology of baculoviruses, in: Asgari, S., Johnson, K. (eds.), Insect Virology. Caister Academic Press, Norfolk, pp. 411–427.

Cory, J.S., 2015. Insect virus transmission: different routes to persistence. Curr. Opin. Ins. Sci. 8, 130–135.

Cory, J.S., Franklin, M.T., 2012. Evolution and the microbial control of insects. Evol. Appl. 5, 455–469.

Cory, J.S., Hoover, K., 2006. Plant-mediated effects in insect–pathogen interactions. Trends Ecol. Evol. 21, 278–286.

Cory, J.S., Myers, J.H., 2004. Adaptation in an insect host–plant pathogen interaction. Ecol. Lett. 7, 632–639.

Cory, J.S., Myers, J.H., 2009. Within and between population variation in disease resistance in cyclic populations of western tent caterpillars: a test of the disease defence hypothesis. J. Anim. Ecol. 78, 646–655.

Cory, J.S., Clarke, E.E., Brown, M.L., Hails, R.S., O'Reilly, D.R., 2004. Microparasite manipulation of an insect: the influence of the egt gene on the interaction between a baculovirus and its lepidopteran host. Func. Ecol. 18, 443–450.

Cory, J.S., Green, B.M., Paul, R.K., Hunter-Fujita, F., 2005. Genotypic and phenotypic diversity of a baculovirus population within an individual insect host. J. Invertebr. Pathol. 89, 101–111.

Cossentine, J.E., 2009. The parasitoid factor in the virulence and spread of lepidopteran baculoviruses. Virol. Sinica 24, 305–314.

Cotter, S.C., Hails, R.S., Cory, J.S., Wilson, K., 2004. Density-dependent prophylaxis and condition-dependent immune function in lepidopteran larvae: a multivariate approach. J. Anim. Ecol. 73, 283–293.

Cowley, J.A., Hall, M.R., Cadogan, L.C., Spann, K.M., Walker, P.J., 2002. Vertical transmission of gill-associated virus (GAV) in the black tiger prawn *Penaeus monodon*. Dis. Aquat. Org. 50, 95–104.

Cox-Foster, D.L., Conlan, S., Holmes, E.C., Palacios, G., Evans, J.D., Moran, N.A., et al., 2007. A metagenomic survey of microbes in honey bee colony collapse disorder. Science 318, 283–287.

Craveiro, S.R., Melo, F.L., Ribeiro, Z.M.A., Ribeiro, B.M., Báo, S.N., Inglis, P.W., Castro, M.E.B., 2013. *Pseudoplusia includens* single nucleopolyhedrovirus: genetic diversity, phylogeny and hypervariability of the *pif-2* gene. J. Invertebr. Pathol. 114, 258–267.

Crawford, A.M., Kalmakoff, J., 1978. Transmission of *Wiseana* spp. nuclear polyhedrosis virus in the pasture habitat. N.Z. J. Agric. Res. 21, 521–526.

D'Amico, V., Elkinton, J.S., 1995. Rainfall effects on transmission of gypsy moth (Lepidoptera: Lymantriidae) nuclear polyhedrosis virus. Env. Entomol. 24, 1144–1149.

D'Amico, V., Elkinton, J.S., Dwyer, G., Burand, J.P., Buonaccorsi, J.P., 1996. Virus transmission in gypsy moths is not a simple mass action process. Ecology 77, 201–206.

D'Amico, V., Elkinton, J.S., Podgwaite, J.D., Buonaccorsi, J.P., Dwyer, G., 2005. Pathogen clumping: an explanation for non-linear transmission of an insect virus. Ecol. Entomol. 30, 383–390.

D'Amico, V., Podgwaite, J.D., Zerillo, R., Taylor, P., Fuester, R., 2013a. Interactions between an injected polydnavirus and *per os* baculovirus in gypsy moth larvae. J. Invertebr. Pathol. 114, 158–160.

D'Amico, V., Slavicek, J., Podgwaite, J.D., Webb, R., Fuester, R., Peiffer, R.A., 2013b. Deletion of *v-chiA* from a baculovirus reduces horizontal transmission in the field. Appl. Env. Microbiol. 79, 4056–4064.

David, W.A.L., Gardiner, B.O.C., 1966. Persistence of a granulosis virus of *Pieris brassicae* on cabbage leaves. J. Invertebr. Pathol. 8, 180–183.

David, W.A.L., Gardiner, B.O.C., 1967. The effect of heat, cold, and prolonged storage on a granulosis virus of *Pieris brassicae*. J. Invertebr. Pathol. 9, 555–562.

David, W.A.L., Ellaby, S.J., Taylor, G., 1971. The stability of a purified granulosis virus of the European cabbageworm, *Pieris brassicae*, in dry deposits of intact capsules. J. Invertebr. Pathol. 17, 228–233.

de Miranda, J.R., Fries, I., 2008. Venereal and vertical transmission of deformed wing virus in honeybees (*Apis mellifera* L.). J. Invertebr. Pathol. 98, 184–189.

Derridj, S., 1996. Nutrients on the leaf surface, in: Morris, C.E., Nicot, P.C., Nguyen-The, C. (eds.), Aerial Plant Surface Microbiology. Plenum, New York, pp. 25–42.

Dhandapani, N., Jayaraj, S., Rabindra, R.J., 1993. Cannibalism on nuclear polyhedrosis-virus infected larvae by *Heliothis armigera* (Hubn.) and its effect on viral-infection. Ins. Sci. Appl. 14, 427–430.

Duffey, S.S., Hoover, K., Bonning, B., Hammock, B.D., 1995. The impact of host plant on the efficacy of baculoviruses. Rev. Pestic. Toxicol. 3, 137–275.

Dwyer, G., 1991. The roles of density, stage, and patchiness in the transmission of an insect virus. Ecology 72, 559–574.

Dwyer, G., 1992. On the spatial spread of insect pathogens: theory and experiment. Ecology 73, 479–494.

Dwyer, G., Elkinton, J.S., 1995. Host dispersal and the spatial spread of insect pathogens. Ecology 76, 1262–1275.

Dwyer, G., Elkinton, J.S., Buonaccorsi, J.P., 1997. Host heterogeneity in susceptibility and disease dynamics: tests of a mathematical model. Am. Nat. 150, 685–707.

Eakin, L., Wang, M., Dwyer, G., 2015. The effects of the avoidance of infectious hosts on infection risk in an insect-pathogen interaction. Am. Nat. 185, 100–112.

Eberle, K.E., Sayed, S., Rezapanah, M., Shojai-Estabragh, S., Jehle, J.A., 2009. Diversity and evolution of the *Cydia pomonella* granulovirus. J. Gen. Virol. 90, 662–671.

Elderd, B.D., 2013. Developing models of disease transmission: insights from ecological studies of insects and their baculoviruses. PLoS Pathog. 9, e1003372.

Elderd, B.D., Reilly, J.R., 2014. Warmer temperatures increase disease transmission and outbreak intensity in a host–pathogen system. J. Anim. Ecol. 83, 838–849.

Elderd, B.D., Rehill, B.J., Haynes, K.J., Dwyer, G., 2013. Induced plant defenses, host–pathogen interactions, and forest insect outbreaks. Proc. Nat. Acad. Sci. U.S.A. 110, 14 978–14 983.

Elleman, C.J., Entwistle, P.F., 1982. A study of glands on cotton responsible for the high pH and cation concentration of the leaf surface. Ann. Appl. Biol. 100, 553–558.

Elleman, C.J., Entwistle, P.F., 1985a. Inactivation of a nuclear polyhedrosis virus on cotton by the substances produced by the cotton leaf surface glands. Ann. Appl. Biol. 106, 83–92.

Elleman, C.J., Entwistle, P.F., 1985b. The effect of magnesium ions on the solubility of polyhedral inclusion bodies and its possible role in the inactivation of the nuclear polyhedrosis virus of *Spodoptera littoralis* by the cotton leaf gland exudate. Ann. Appl. Biol. 106, 93–100.

Entwistle, P.F., 1998. A world survey of virus control of insect pests, in: Hunter-Fujita, F.R., Entwistle, P.F., Evans, H.F., Crook, N.E. (eds.), Insect Viruses and Pest Management. John Wiley & Sons, Chichester, pp. 189–200.

Entwistle, P.F., Adams, P.H.W., Evans, H.F., 1977. Epizootiology of a nuclear-polyhedrosis virus in European spruce sawfly (*Gilpinia hercyniae*): the status of birds as dispersal agents of the virus during the larval season. J. Invertebr. Pathol. 29, 354–360.

Entwistle, P.F., Adams, P.H.W., Evans, H.F., Rivers, C.F., 1983. Epizootiology of a nuclear polyhedrosis virus (Baculoviridae) in European spruce sawfly (*Gilpinia hercyniae*): spread of disease from small epicentres in comparison with spread of baculovirus diseases in other hosts. J. Appl. Ecol. 20, 473–487.

Entwistle, P.F., Forkner, A.C., Green, B.M., Cory, J.S., 1993. Avian dispersal of nuclear polyhedrosis viruses after induced epizootics in the pine beauty moth, *Panolis flammea* (Lepidoptera: Noctuidae). Biol. Control 3, 61–69.

Erlandson, M.A., 2009. Genetic variation in field populations of baculoviruses: mechanisms for generating variation and its potential role in baculovirus epizootiology. Virol. Sinica 24, 458–469.

Escribano, A., Williams, T., Goulson, D., Cave, R.D., Chapman, J.W., Caballero, P., 2001. Consequences of interspecific competition on the virulence and genetic composition of a nucleopolyhedrovirus in *Spodoptera frugiperda* larvae parasitized by *Chelonus insularis*. Biocontr. Sci. Technol. 11, 649–662.

Evans, H.F., Entwistle, P.F., 1982. Epizootiology of the nuclear polyhedrosis virus of European spruce sawfly with emphasis on persistence of virus outside the host, in: Kurstak, E. (ed.), Microbial and Viral Pesticides. Marcel Dekker, New York, pp. 449–461.

Evison, S.E., Roberts, K.E., Laurenson, L., Pietravalle, S., Hui, J., Biesmeijer, J.C., et al., 2012. Pervasiveness of parasites in pollinators. PLoS ONE 7, e30641.

Fang, Z., Shao, J., Weng, Q., 2016. *De novo* transcriptome analysis of *Spodoptera exigua* multiple nucleopolyhedrovirus (SeMNPV) genes in latently infected Se301 cells. Virol. Sinica 31, 425–436.

Farrar, R.R., Ridgway, R.L., 2000. Host plant effects on the activity of selected nuclear polyhedrosis viruses against the corn earworm and beet armyworm (Lepidoptera: Noctuidae). Env. Entomol. 29, 108–115.

Felix, M.A., Ashe, A., Piffaretti, J., Wu, G., Nuez, I., Belicard, T., et al., 2011. Natural and experimental infection of *Caenorhabditis* nematodes by novel viruses related to nodaviruses. PLoS Biol. 9, e1000586.

Felton, G.W., Duffey, S.S., Vail, P.V., Kaya, H.K., Manning, J., 1987. Interaction of nuclear polyhedrosis virus with catechols: potential incompatibility for host-plant resistance against noctuid larvae. J. Chem. Ecol. 13, 947–957.

Fleming-Davies, A.E., Dwyer, G., 2015. Phenotypic variation in overwinter environmental transmission of a baculovirus and the cost of virulence. Am. Nat. 186, 797–806.

Fleming-Davies, A.E., Dukic, V., Andreasen, V., Dwyer, G., 2015. Effects of host heterogeneity on pathogen diversity and evolution. Ecol. Lett. 18, 1252–1261.

Frid, L., Myers, J.H., 2002. Thermal ecology of western tent caterpillars *Malacosoma californicum pluviale* and infection by nucleopolyhedrovirus. Ecol. Entomol. 27, 665–673.

Fuller, E., Elderd, B.D., Dwyer, G., 2012. Pathogen persistence in the environment and insect-baculovirus interactions: disease-density thresholds, epidemic burnout and insect outbreaks. Am. Nat. 179, 70–96.

Fürst, M.A., McMahon, D.P., Osborne, J.L., Paxton, R.J., Brown, M.J.F., 2014. Disease associations between honeybees and bumblebees as a threat to wild pollinators. Nature 506, 364–366.

Fuxa, J.R., 1991. Release and transport of entomopathogenic microorganisms, in: Levin, M., Strauss, H. (eds.), Risk Assessment in Genetic Engineering. McGraw-Hill, New York, pp. 83–113.

Fuxa, J.R., 1982. Prevalence of viral infections in populations of fall armyworm, *Spodoptera frugiperda*, in southeastern Louisiana. Env. Entomol. 11, 239–242.

Fuxa, J.R., 2004. Ecology of insect nucleopolyhedroviruses. Agric. Ecosyst. Env. 103, 27–43.

Fuxa, J.R., 2008. Threshold concentrations of nucleopolyhedrovirus in soil to initiate infections in *Heliothis virescens* on cotton plants. Microb. Ecol. 55, 530–539.

Fuxa, J.R., Geaghan, J.P., 1983. Multiple-regression analysis of factors affecting prevalence of nuclear polyhedrosis virus in *Spodoptera frugiperda* (Lepidoptera: Noctuidae) populations. Env. Entomol. 12, 311–316.

Fuxa, J.R., Richter, A.R., 1994. Virus released for long-term suppression of velvetbean caterpillar in soybeans. Louisiana Agric. 37, 8–11.

Fuxa, J.R., Richter, A.R., 1996. Effect of agricultural operations and precipitation on vertical distribution of a nuclear polyhedrosis virus in soil. Biol. Control 6, 324–329.

Fuxa, J.R., Richter, A.R., 2001. Quantification of soil-to-plant transport of recombinant nucleopolyhedrovirus: effects of soil type and moisture, air currents, and precipitation. Appl. Env. Microbiol. 67, 5166–5170.

Fuxa, J.R., Sun, J.Z., Weidner, E.H., LaMotte, L.R., 1999. Stressors and rearing diseases of *Trichoplusia ni*: evidence of vertical transmission of NPV and CPV. J. Invertebr. Pathol. 74, 149–155.

Fuxa, J.R., Matter, M.M., Abdel-Rahman, A., Micinski, S., Richter, A.R., Flexner, J.L., 2001. Persistence and distribution of wild-type and recombinant nucleopolyhedroviruses in soil. Microb. Ecol. 41, 222–231.

Fuxa, J.R., Richter, A.R., Milks, M.L., 2007. Threshold distances and depths of nucleopolyhedrovirus in soil for transport to cotton plants by wind and rain. J. Invertebr. Pathol. 95, 60–70.

Garzon, S., Kurstak, E., 1968. Infection des cellules des gonades et du système nerveux de *Galleria mellonella* par le virus de la densonucléose. Natur. Can. 95, 1125–1129.

Gilbert, C., Chateigner, A., Ernenwein, L., Barbe, V., Bézier, A., Herniou, E.A., Cordaux, R., 2014. Population genomics supports baculoviruses as vectors of horizontal transfer of insect transposons. Nat. Comm. 5, 3348.

Gómez-Bonilla, Y., López-Ferber, M., Caballero, P., Léry, X., Muñoz, D., 2012. Costa Rican soils contain highly insecticidal granulovirus strains against *Phthorimaea operculella* and *Tecia solanivora*. J. Appl. Entomol. 136, 530–538.

Goulson, D., 1997. Wipfelkrankheit: modification of host behaviour during baculoviral infection. Oecologia 109, 219–228.

Goulson, D., Cory, J.S., 1995. Responses of *Mamestra brassicae* (Lepidoptera: Noctuidae) to crowding: interactions with disease resistance, colour phase and growth. Oecologia 104, 416–423.

Goulson, D., Hails, R.S., Williams, T., Hirst, M.L., Vasconcelos, S.D., Green, B.M., et al., 1995. Transmission dynamics of a virus in a stage-structured insect population. Ecology 76, 392–401.

Graham, R.I., 2005. The impact of viral pathogens on host Lepidoptera population: the winter moth and its natural enemies. Unpublished PhD thesis, Oxford Brookes University, Oxford.

Graham, R.I., Grzywacz, D., Mushobozi, W.L., Wilson, K., 2012. *Wolbachia* in a major African crop pest increases susceptibility to viral disease rather than protects. Ecol. Lett. 15, 993–1000.

Graham, R.I., Tummala, Y., Rhodes, G., Cory, J.S., Shirras, A., Grzywacz, D., Wilson, K., 2015. Development of a real-time qPCR assay for quantification of covert baculovirus infections in a major African crop pest. Insects 6, 746–759.

Graillot, B., Bayle, S., Blachere-Lopez, C., Besse, S., Siegwart, M., Lopez-Ferber, M., 2016. Biological characteristics of experimental genotype mixtures of *Cydia pomonella* granulovirus (CpGV): ability to control susceptible and resistant pest populations. Viruses 8, 147.

Graystock, P., Goulson, D., Hughes, W.O., 2015, Parasites in bloom: flowers aid dispersal and transmission of pollinator parasites within and between bee species. Proc. R. Soc. B 282, 20151371.

Greenberg, B., 1970. Sterilizing procedures and agents, antibiotics and inhibitors in mass rearing of insects. Bull. Entomol. Soc. Am. 16, 31–36.

Grosholz, E.D., 1992. Interactions of intraspecific, interspecific, and apparent competition with host-pathogen population dynamics. Ecology 73, 507–514.

Grosholz, E.D., 1993. The influence of habitat heterogeneity on host-pathogen population dynamics. Oecologia 96, 347–353.

Gundersen-Rindal, D., Dupuy, C., Huguet, E., Drezen, J.M., 2013. Parasitoid polydnaviruses: evolution, pathology and applications. Biocontr. Sci. Technol. 23, 1–61.

Guo, H.F., Fang, J.C., Zhong, W.F., Liu, B.S., 2013. Interactions between *Meteorus pulchricornis* and *Spodoptera exigua* multiple nucleopolyhedrovirus. J. Ins. Sci. 13, 12.

Hackett, K.J., Boore, A., Deming, C., Buckley, E., Camp, M., Shapiro, M., 2000. *Helicoverpa armigera* granulovirus interference with progression of *H. zea* nucleopolyhedrovirus disease in *H. zea* larvae. J. Invertebr. Pathol. 75, 99–106.

Hails, R.S., Hernandez-Crespo, P., Sait, S.M., Donnelly, C.A., Green, B.M., Cory, J.S., 2002. Transmission patterns of natural and recombinant baculoviruses. Ecology 83, 906–916.

Hajek, A.E., Tobin, P.C., 2011. Introduced pathogens follow the invasion front of a spreading alien host. J. Anim. Ecol. 80, 1217–1226.

Hajek, A.E., van Nouhuys, S., 2016. Fatal diseases and parasitoids: from competition to facilitation in a shared host. Proc. R. Soc. Lond. B Biol. Sci. 283, 20160154.

Hamm, J.J., Young, J.R., 1974. Mode of transmission of nuclear-polyhedrosis virus to progeny of adult *Heliothis zea*. J. Invertebr. Pathol. 24, 70–81.

Hamm, J.J., Styer, E.L., Federici, B.A., 1998. Comparison of field-collected ascovirus isolates by DNA hybridization, host range, and histopathology. J. Invertebr. Pathol. 72, 138–146.

Han, Y., van Houte, S., Drees, G.F., van Oers, M.M., Ros, V.I., 2015a. Parasitic manipulation of host behaviour: baculovirus SeMNPV EGT facilitates tree-top disease in *Spodoptera exigua* larvae by extending the time to death. Insects 6, 716–731.

Han, Y., van Oers, M.M., van Houte, S., Ros, V.I., 2015b. Virus-induced behavioural changes in insects, in: Mehlhorn, H., (ed.) Host Manipulations by Parasites and Viruses. Springer, Berlin, pp. 149–174.

Harper, J.D., 1986. Interactions between baculoviruses and other entomopathogens, chemical pesticides and parasitoids, in: Granados, R.R., Federici, B.A. (eds.), The Biology of Baculoviruses, Vol. 2. CRC Press, Boca Raton, FL, pp. 133–155.

Harr, J., Guggenheim, R., Boller, T., 1984. High pH values and secretion of ions on leaf surfaces: a characteristic of the phylloplane of Malvaceae. Experientia 40, 935–937.

Harrison, R.L., 2013. Concentration-and time-response characteristics of plaque isolates of *Agrotis ipsilon* multiple nucleopolyhedrovirus derived from a field isolate. J. Invertebr. Pathol. 112, 159–161.

Harrison, R.L., Puttler, B., Popham, H.J., 2008. Genomic sequence analysis of a fast-killing isolate of *Spodoptera frugiperda* multiple nucleopolyhedrovirus. J. Gen. Virol. 89, 775–790.

Hedges, L.M., Brownlie, J.C., O'Neill, S.L., Johnson, K.N., 2008. Wolbachia and virus protection in insects. Science 322, 702–702.

Helms, T.J., Raun, E.S., 1971. Perennial laboratory culture of disease-free insects, in: Burges, H.D., Hussey, N.W. (eds.), Microbial Control of Insects and Mites. Academic Press, New York, pp. 639–634.

Hernández, A., Marina, C.F., Valle, J., Williams, T., 2005. Persistence of *Invertebrate iridescent virus 6* in artificial tropical aquatic environments. Arch. Virol. 150, 2357–2363.

Hewson, I., Brown, J.M., Gitlin, S.A., Doud, D.F., 2011. Nucleopolyhedrovirus detection and distribution in terrestrial, freshwater, and marine habitats of Appledore Island, Gulf of Maine. Microb. Ecol. 62, 48–57.

Hick, P., Evans, O., Looi, R., English, C., Whittington, R.J., 2016. Stability of Ostreid herpesvirus-1 (OsHV-1) and assessment of disinfection of seawater and oyster tissues using a bioassay. Aquaculture 450, 412–421.

Hitchman, R.B., Hodgson, D.J., King, L.A., Hails, R.S., Cory, J.S., Possee, R.D., 2007. Host mediated selection of pathogen genotypes as a mechanism for the maintenance of baculovirus diversity in the field. J. Invertebr. Pathol. 94, 153–162.

Hoa, T.T.T., Zwart, M.P., Phuong, N.T., Oanh, D.T., de Jong, M.C., Vlak, J.M., 2011. Mixed-genotype white spot syndrome virus infections of shrimp are inversely correlated with disease outbreaks in ponds. J. Gen. Virol. 92, 675–680.

Hochberg, M.E., 1989. The potential role of pathogens in biological control. Nature 337, 262–265.

Hochberg, M.E., 1991a. Viruses as costs to gregarious feeding behaviour in the Lepidoptera. Oikos 61, 291–296.

Hochberg, M.E., 1991b. Extra-host interactions between a braconid endoparasitoid, *Apanteles glomeratus*, and a baculovirus for larvae of *Pieris brassicae*. J. Anim. Ecol. 60, 65–77.

Hochberg, M.E., Waage, J.K., 1991. A model for the biological control of *Oryctes rhinoceros* (Coleoptera: Scarabaeidae) by means of pathogens. J. Appl. Ecol. 28, 514–531.

Hodgson, D.J., Vanbergen, A.J., Hartley, S.E., Hails, R.S., Cory, J.S., 2002. Differential selection of baculovirus genotypes mediated by different species of host food plant. Ecol. Lett. 5, 512–518.

Hodgson, D.J., Hitchman, R.B., Vanbergen, A.J., Hails, R.S., Hartley, S.E., Possee, R.D., et al., 2003. The existence and persistence of genotypic variation in nucleopolyhedrovirus populations, in: Hails, R.S., Beringer, J.E., Godfray, H.C.J. (eds.), Genes in the Environment. Blackwell, Oxford, pp. 258–280.

Holmes, S.B., Fick, W.E., Kreutzweiser, D.P., Ebling, P.M., England, L.S., Trevors, J.T., 2008. Persistence of naturally occurring and genetically modified *Choristoneura fumiferana* nucleopolyhedroviruses in outdoor aquatic microcosms. Pest Manag. Sci. 64, 1015–1023.

Hoover, K., Washburn, J.O., Volkman, L.E., 2000. Midgut-based resistance of *Heliothis virescens* to baculovirus infection mediated by phytochemicals in cotton. J. Ins. Physiol. 46, 999–1007.

Hoover, K., Humphries, M.A., Gendron, A.R., Slavicek, J.M., 2010. Impact of viral enhancin genes on potency of *Lymantria dispar* multiple nucleopolyhedrovirus in *L. dispar* following disruption of the peritrophic matrix. J. Invertebr. Pathol. 104, 150–152.

Hoover, K., Grove, M., Gardner, M., Hughes, D.P., McNeil, J., Slavicek, J., 2011. A gene for an extended phenotype. Science 333, 1401–1401.

Hostetter, D.L., Bell, M.R., 1985. Natural dispersal of baculoviruses in the environment, in: Maramorosch, K., Shreman, K.E. (eds.), Viral insecticides for Biological Control. Academic Press, Orlando, FL, pp. 249–284.

Hudson, A.I., Fleming-Davies, A.E., Páez, D.J., Dwyer, G., 2016. Genotype-by-genotype interactions between an insect and its pathogen. J. Evol. Biol. 29, 2480–2490.

Hughes, D.S., Possee, R.D., King, L.A., 1997. Evidence for the presence of a low-level, persistent baculovirus infection of *Mamestra brassicae* insects. J. Gen. Virol. 78, 1801–1805.

Hukuhara, T., 1973. Further studies on the distribution of a nuclear-polyhedrosis virus of the fall webworm, *Hyphantria cunea*, in soil. J. Invertebr. Pathol. 22, 345–350.

Hunter-Fujita, F.R., Entwistle, P.F., Evans, H.F., Crook, N.E., 1998. Insect Viruses and Pest Management. John Wiley & Sons, Chichester.

Hurpin, B., Robert, P.H., 1976. Conservation dans le sol de trois germes pathogènes pour les larves de *Melolontha melolontha* [Col.: Scarabaeidae]. Entomophaga 21, 73–80.

Hussain, M., Asgari, S., 2014. MicroRNAs as mediators of insect host–pathogen interactions and immunity. J. Ins. Physiol. 70, 151–158.

Ignoffo, C.M., 1992. Environmental factors affecting persistence of entomopathogens. Fla. Entomol. 75, 516–525

Ignoffo, C.M., Garcia, C., 1992. Combinations of environmental factors and simulated sunlight affecting activity of inclusion bodies of the *Heliothis* (Lepidoptera: Noctuidae) nucleopolyhedrosis virus. Env. Entomol. 21, 210–213.

Ignoffo, C.M., Hostetter, D.L., Sikorowski, P.P., Sutter, G., Brooks, W.M., 1977. Inactivation of representative species of entomopathogenic viruses, a bacterium, fungus, and protozoan by an ultraviolet light source. Env. Entomol. 6, 411–415.

Ignoffo, C.M., Garcia, C., Saathoff, S.G., 1997. Sunlight stability and rain-fastness of formulations of *Baculovirus heliothis*. Env. Entomol. 26, 1470–1474.

Infante-Rodríguez, D.A., Berber, J.J., Mercado, G., Valenzuela-González, J., Muñoz, D., Williams, T., 2016. Earthworm mediated dispersal of baculovirus occlusion bodies: experimental evidence from a model system. Biol. Control 100, 18–24.

Ishii, T., Takatsuka, J., Nakai, M., Kunimi, Y., 2002. Growth characteristics and competitive abilities of a nucleopolyhedrovirus and an entomopoxvirus in larvae of the smaller tea tortrix, *Adoxophyes honmai* (Lepidoptera: Tortricidae). Biol. Control 23, 96–105.

Ishimwe, E., Hodgson, J.J., Clem, R.J., Passarelli, A.L., 2015. Reaching the melting point: degradative enzymes and protease inhibitors involved in baculovirus infection and dissemination. Virology 479, 637–649.

Jackson, T.A., 2009. The use of *Oryctes* virus for control of rhinoceros beetle in the Pacific islands, in: Hajek, A.E., Glare, T., O'Callaghan, M. (eds.), Use of Microbes for Control and Eradication of Invasive Arthropods. Springer, Dordrecht, pp. 133–140.

Jakubowska, A.K., Vogel, H., Herrero, S., 2013. Increase in gut microbiota after immune suppression in baculovirus-infected larvae. PLoS Pathog. 9, e1003379.

Jakubowska, A.K., D'Angiolo, M., González-Martínez, R.M., Millán-Leiva, A., Carballo, A., Murillo, R., et al., 2014. Simultaneous occurrence of covert infections with small RNA viruses in the lepidopteran *Spodoptera exigua*. J. Invertebr. Pathol. 121, 56–63.

Jakubowska, A.K., Nalcacioglu, R., Millán-Leiva, A., Sanz-Carbonell, A., Muratoglu, H., Herrero, S., Demirbag, Z., 2015. In search of pathogens: transcriptome-based identification of viral sequences from the pine processionary moth (*Thaumetopoea pityocampa*). Viruses 7, 456–479.

Jakubowska, A.K., Murillo, R., Carballo, A., Williams, T., van Lent, J.W., Caballero, P., Herrero, S., 2016. Iflavirus increases its infectivity and physical stability in association with baculovirus. PeerJ 4, e1687.

Jaques, R.P., 1972. Inactivation of foliar deposits of virus of *Trichoplusia ni* (Lepidoptera: Noctuidae) and *Pieris rapae* (Lepidoptera: Pieridae) and tests on protectant additives. Can. Entomol. 104, 1985–1994.

Jaques, R.P., 1974. Occurrence and accumulation of viruses of *Trichoplusia ni* in treated field plots. J. Invertebr. Pathol. 23, 140–152.

Jaques, R.P., 1985. Stability of insect viruses in the environment, in: Maramorosch, K., Sherman, K.E. (eds.), Viral Insecticides for Biological Control. Academic Press, New York, pp. 285–360.

Jehle, J.A., 2010. Nudiviruses: their biology and genetics, in: Assgari, S., Johnson, K. (eds.), Insect Virology. Caister Academic Press, Norfolk, pp. 153–170.

Jiang, J.X., Bao, Y.B., Ji, X.Y., Wan, N.F., 2014. Effect of nucleopolyhedrovirus infection of *Spodoptera litura* larvae on host discrimination by *Microplitis pallidipes*. Biocontr. Sci. Technol. 24, 561–573.

Jura, W.G.Z.O., Otieno, L.H., Chimtawi, M.M.B., 1989. Ultrastructural evidence for trans-ovum transmission of the DNA virus of tsetse, *Glossina pallidipes* (Diptera: Glossinidae). Curr. Microbiol. 18, 1–4.

Kalmakoff, J., Crawford, A.M., 1982. Enzootic virus control of *Wiseana* spp. in the pasture environment. in: Kurstak, E. (ed.), Microbial and Viral Pesticides. Marcel Dekker, New York, pp. 435–448.

Kamita, S.G., Maeda, S., Hammock, B.D., 2003. High-frequency homologous recombination between baculoviruses involves DNA replication. J. Virol. 77, 13053–13061.

Kamita, S.G., Nagasaka, K., Chua, J.W., Shimada, T., Mita, K., Kobayashi, M., et al., 2005. A baculovirus-encoded protein tyrosine phosphatase gene induces enhanced locomotory activity in a lepidopteran host. Proc. Nat. Acad. Sci. U.S.A. 102, 2584–2589.

Kariithi, H.M., van Lent, J., van Oers, M.M., Abd-Alla, A.M., Vlak, J.M., 2013. Proteomic footprints of a member of *Glossinavirus* (Hytrosaviridae): an expeditious approach to virus control strategies in tsetse factories. J. Invertebr. Pathol. 112, S26–S31.

Katsuma, S., 2015a. Baculovirus controls host catapillars (*sic.*) by manipulating host physiology and behavior. Agri. Biosci. Monogr. 5, 1–27.

Katsuma, S., 2015b. Phosphatase activity of *Bombyx mori nucleopolyhedrovirus* PTP is dispensable for enhanced locomotory activity in *B. mori* larvae. J. Invertebr. Pathol. 132, 228–232.

Kaupp, W.J., 1983. Persistence of *Neodiprion sertifer* (Hymenoptera: Diprionidae) nuclear polyhedrosis virus on *Pinus contorta* foliage. Can. Entomol. 115, 869–873.

Kawase, S., Kurstak, E., 1991. Parvoviridae of invertebrates: densonucleosis viruses, in: Kurstak, E. (ed.), Viruses of Invertebrates. Marcel Dekker, New York, pp. 315–344.

Keathley, C.P., Harrison, R.L., Potter, D.A., 2012. Baculovirus infection of the armyworm (Lepidoptera: Noctuidae) feeding on spiny-or smooth-edged grass (*Festuca* spp.) leaf blades. Biol. Control 61, 147–154.

Kemp, E.M., Woodward, D.T., Cory, J.S., 2011. Detection of single and mixed covert baculovirus infections in eastern spruce budworm, *Choristoneura fumiferana* populations. J. Invertebr. Pathol. 107, 202–205.

Killick, H.J., Warden, S.J., 1991. Ultraviolet penetration of pine trees and insect virus survival. Entomophaga 36, 87–94.

Kirkpatrick, B.A., Washburn, J.O., Volkman, L.E., 1998. AcMNPV pathogenesis and developmental resistance in fifth instar *Heliothis virescens*. J. Invertebr. Pathol. 72, 63–72.

Kitajima, E.W., Kim, K.S., Scott, H.A., Gergerich, R.C., 1985. Reovirus-like particles and their vertical transmission in the Mexican bean beetle, *Epilachna varivestis* (Coleoptera: Coccinellidae). J. Invertebr. Pathol. 46, 83–97.

Klemola, N., Klemola, T., Rantala, M.J., Ruuhola, T., 2007. Natural host-plant quality affects immune defence of an insect herbivore. Entomol. Exp. Appl. 123, 167–176.

Knell, R., Begon, M., Thompson, D.J., 1998. Transmission of *Plodia interpunctella* granulosis virus does not conform to the mass action model. J. Anim. Ecol. 67, 592–599.

Knell, R.J., Webberley, K.M., 2004. Sexually transmitted diseases of insects: distribution, evolution, ecology and host behaviour. Biol. Rev. 79, 557–581.

Kolliopoulou, A., Van Nieuwerburgh, F., Stravopodis, D.J., Deforce, D., Swevers, L., Smagghe, G., 2015. Transcriptome analysis of *Bombyx mori* larval midgut during persistent and pathogenic cytoplasmic polyhedrosis virus infection. PloS ONE 10, e0121447.

Kondo, A., Maeda, S., 1991. Host range expansion by recombination of the baculoviruses *Bombyx mori* nuclear polyhedrosis virus and *Autographa californica* nuclear polyhedrosis virus. J. Virol. 65, 3625–3632.

Krokene, P., Heldal, I., Fossdal, C.G., 2013. Quantifying *Neodiprion sertifer* nucleopolyhedrovirus DNA from insects, foliage and forest litter using the quantitative real-time polymerase chain reaction. Agric. Forest Entomol. 15, 120–125.

Kuhn, J.H., Jahrling, P.B., 2010. Clarification and guidance on the proper usage of virus and virus species names. Arch. Virol. 155, 445–453.

Kukan, B., 1999. Vertical transmission of nucleopolyhedrovirus in insects. J. Invertebr. Pathol. 74, 103–111.

Kukan, B., Myers, J.H., 1995. DNA hybridization assay for detection of nuclear polyhedrosis virus in tent caterpillars. J. Invertebr. Pathol. 66, 231–236.

Kunimi, Y., Yamada, E., 1990. Relationship of larval phase and susceptibility of the armyworm, *Pseudaletia separata* Walker (Lepidoptera: Noctuidae) to a nuclear polyhedrosis virus and a granulosis virus. Appl. Entomol. Zool. 25, 289–297.

Kurstak, E., Vago, C., 1967. Transmission of the densonucleosis virus by parasitism of a hymenopteron. Rev. Can. Biol. 26, 311–316.

Kyei-Poku, G.K., Kunimi, Y., 1997. Effect of entomopoxvirus infection of *Pseudaletia separata* larvae on the oviposition behavior of *Cotesia kariyai*. Entomol. Exp. Appl. 83, 93–97.

Lacey, L.A., Brooks, W.M., 1997. Initial handling and diagnosis of diseased insects, in: Lacey, L.A. (ed.), Manual of Techniques in Insect Pathology. Academic Press, New York, pp. 1–16.

Lambrechts, L., Ferguson, N.M., Harris, E., Holmes, E.C., McGraw, E.A., O'Neill, S.L., et al., 2015. Assessing the epidemiological effect of *Wolbachia* for dengue control. Lancet Infect. Dis. 15, 862–866.

Lasa, R., Ruiz-Portero, C., Alcázar, M.D., Belda, J.E., Caballero, P., Williams, T., 2007. Efficacy of optical brightener formulations of *Spodoptera exigua* multiple nucleopolyhedrovirus (SeMNPV) as a biological insecticide in greenhouses in southern Spain. Biol. Control 40, 89–96.

Lawrence, P.O., 2002. Purification and partial characterization of an entomopoxvirus (DlEPV) from a parasitic wasp of tephritid fruit flies. J. Ins. Sci. 2, 1–12.

Lawrence, P.O., Akin, D., 1990. Virus-like particles in the accessory glands of *Biosteres longicaudatus*. Can. J. Zool. 68, 539–546.

Lawrence, P.O., Matos, L.F., 2005. Transmission of the *Diachasmimorpha longicaudata* rhabdovirus (DlRhV) to wasp offspring: an ultrastructural analysis. J. Ins. Physiol. 51, 235–241.

LeBlanc, B.D., Overstreet, R.M., 1991. Effect of desiccation, pH, heat, and ultraviolet irradiation on viability of *Baculovirus penaei*. J. Invertebr. Pathol. 57, 277–286.

Lee, Y., Fuxa, J.R., 2000a. Ingestion and defecation of recombinant and wild-type nucleopolyhedroviruses by scavenging and predatory arthropods. Env. Entomol. 29, 950–957.

Lee, Y., Fuxa, J.R., 2000b. Transport of wild-type and recombinant nucleopolyhedroviruses by scavenging and predatory arthropods. Microb. Ecol. 39, 301–313.

Lee, K.P., Cory, J.S., Wilson, K., Raubenheimer, D., Simpson, S.J., 2006. Flexible diet choice offsets protein costs of pathogen resistance in a caterpillar. Proc. R. Soc. Lond. B Biol. Sci. 273, 823–829.

Levy, S.M., Falleiros, Â.M., Moscardi, F., Gregório, E.A., 2012; The role of peritrophic membrane in the resistance of *Anticarsia gemmatalis* larvae (Lepidoptera: Noctuidae) during the infection by its nucleopolyhedrovirus (AgMNPV). Arthrop. Struct. Dev. 40, 429–434.

Lewis, L.C., Lynch, R.E., Jackson, J.J., 1977. Pathology of a baculovirus of the alfalfa looper, *Autographa californica*, in the European corn borer, *Ostrinia nubilalis*. Env. Entomol. 6, 535–538.

Liebhold, A.M., Plymale, R., Elkinton, J.S., Hajek, A.E., 2013. Emergent fungal entomopathogen does not alter density dependence in a viral competitor. Ecology 94, 1217–1222.

Lin, C.L., Lee, J.C., Chen, S.S., Wood, H.A., Li, M.L., Li, C.F., Chao, Y.C., 1999. Persistent Hz-1 virus infection in insect cells: evidence for insertion of viral DNA into host chromosomes and viral infection in a latent status. J. Virol. 73, 128–139.

Linley, J.R., Nielsen, H.T., 1968. Transmission of a mosquito iridescent virus in *Aedes taeniorhynchus*: II. Experiments related to transmission in nature. J. Invertebr. Pathol. 12, 17–24.

Liu, S., Vijayendran D., Bonning, B.C., 2011 Next generation sequencing technologies for insect virus discovery. Viruses 3, 1849–1869.

Lombarkia, N., Derridj, S., Ioriatti, C., Bourguet, E., 2013. Effect of a granulovirus larvicide, Madex*, on egg-laying of *Cydia pomonella* L. (Lepidoptera: Tortricidae) due to changes in chemical signalization on the apple leaf surface. Afr. Entomol. 21, 196–208.

Lomer, C.J., 1986. Release of *Baculovirus oryctes* into *Oryctes monoceros* populations in the Seychelles J. Invertebr. Pathol. 47, 237–246.

Longdon, B., Jiggins, F.M., 2012. Vertically transmitted viral endosymbionts of insects: do sigma viruses walk alone? Proc. R. Soc. Lond. B Biol. Sci. 279, 3889–3898.

Longdon, B., Wilfert, L., Obbard, D.J., Jiggins, F.M., 2011. Rhabdoviruses in two species of *Drosophila*: vertical transmission and a recent sweep. Genetics 188, 141–150.

López, M., Rojas, J.C., Vandame, R., Williams, T., 2002. Parasitoid-mediated transmission of an iridescent virus. J. Invertebr. Pathol. 80, 160–170.

López-Bueno, A., Rastrojo, A., Peiró, R., Arenas, M., Alcamí, A., 2015. Ecological connectivity shapes quasispecies structure of RNA viruses in an Antarctic lake. Mol. Ecol. 24, 4812–4825.

López-Ferber, M., Simón, O., Williams, T., Caballero, P., 2003. Defective or effective? Mutualistic interactions between virus genotypes. Proc. R. Soc. Lond. B Biol. Sci. 270, 2249–2255.

Lupiani, B., Raina, A.K., Huber, C., 1999. Development and use of a PCR assay for detection of the reproductive virus in wild populations of *Helicoverpa zea* (Lepidoptera: Noctuidae). J. Invertebr. Pathol. 73, 107–112.

Ma, M., Huang, Y., Gong, Z., Zhuang, L., Li, C., Yang, H., et al., 2011. Discovery of DNA viruses in wild-caught mosquitoes using small RNA high throughput sequencing. PLoS ONE 6, e24758.

Magnoler, A., 1968. Laboratory and field experiments on the effectiveness of purified and non-purified nuclear polyhedral virus of *Lymantria dispar* L. Entomophaga 13, 335–344.

Malakar, R., Elkinton, J.S., Carroll, S.D., D'Amico, V., 1999a. Interactions between two gypsy moth (Lepidoptera: Lymantriidae) pathogens: nucleopolyhedrovirus and *Entomophaga maimaiga* (Zygomycetes: Entomophthorales): field studies and a simulation model. Biol. Control 16, 189–198.

Malakar, R., Elkinton, J.S., Hajek, A.E., Burand, J.P., 1999b. Within-host interactions of *Lymantria dispar* (Lepidoptera: Lymantriidae) nucleopolyhedrosis virus and *Entomophaga maimaiga* (Zygomycetes: Entomophthorales). J. Invertebr. Pathol. 73, 91–100.

Marina, C.F., Arredondo-Jiménez, J.I., Castillo, A., Williams, T., 1999. Sublethal effects of iridovirus disease in a mosquito. Oecologia 119, 383–388.

Marina, C., Feliciano, J.M., Valle, J., Williams, T., 2000. Effect of temperature, pH, ion concentration and chloroform treatment on the stability of *Invertebrate iridescent virus* 6. J. Invertebr. Pathol. 75, 91–94.

Marina, C.F., Ibarra, J.E., Arredondo-Jiménez, J.I., Fernández-Salas, I., Liedo, P., Williams, T. 2003. Adverse effects of covert iridovirus infection on life history and demographic parameters of *Aedes aegypti*. Entomol. Exp. Appl. 106, 53–61.

Marina, C.F., Fernández-Salas, I., Ibarra, J.E., Arredondo-Jiménez, J.I., Valle, J., Williams, T., 2005. Transmission dynamics of an iridescent virus in an experimental mosquito population: the role of host density. Ecol. Entomol. 30, 376–382.

Martemyanov, V.V., Dubovskiy, I.M., Rantala, M.J., Salminen, J.P., Belousova, I.A., Pavlushin, S.V., et al., 2012. The effects of defoliation-induced delayed changes in silver birch foliar chemistry on gypsy moth fitness, immune response, and resistance to baculovirus infection. J. Chem. Ecol. 38, 295–305.

Martemyanov, V.V., Kabilov, M.R., Tupikin, A.E., Baturina, O.A., Belousova, I.A., Podgwaite, J.D., et al., 2015a, The enhancin gene: one of the genetic determinants of population variation in baculoviral virulence. Doklady Biochem. Biophys. 465, 351–353.

Martemyanov, V.V., Pavlushin, S.V., Dubovskiy, I.M., Yushkova, Y.V., Morosov, S.V., Chernyak, E.I., et al., 2015b. Asynchrony between host plant and insects-defoliator within a tritrophic system: the role of herbivore innate immunity. PLoS ONE 10, e0130988.

Martin, S.J., Highfield, A.C., Brettell, L., Villalobos, E.M., Budge, G.E., Powell, M., et al., 2012. Global honey bee viral landscape altered by a parasitic mite. Science 336, 1304–1306.

Martínez, G., Christian, P., Marina C.F., Williams, T., 2003. Sensitivity of *Invertebrate iridescent virus* 6 to organic solvents, detergents, enzymes and temperature treatment. Virus Res. 91, 249–254.

Martínez, A.M., Williams, T., López-Ferber, M., Caballero, P., 2005. Optical brighteners do not influence covert baculovirus infection of *Spodoptera frugiperda*. Appl. Env. Microbiol. 71, 1668–1670.

Matthews, H.J., Smith, I., Bell, H.A., Edwards, J.P., 2004. Interactions between the parasitoid *Meteorus gyrator* (Hymenoptera: Braconidae) and a granulovirus in *Lacanobia oleracea* (Lepidoptera: Noctuidae). Env. Entomol. 33, 949–957.

McArt, S.H., Koch, H., Irwin, R.E., Adler, L.S., 2014. Arranging the bouquet of disease: floral traits and the transmission of plant and animal pathogens. Ecol. Lett. 17, 624–636.

McCallum, H., Barlow, N., Hone, J., 2001. How should pathogen transmission be modelled? Trends Ecol. Evol. 16, 295–300.

McMahon, D.P., Fürst, M.A., Caspar, J., Theodorou, P., Brown, M.J., Paxton, R.J., 2015. A sting in the spit: widespread cross-infection of multiple RNA viruses across wild and managed bees. J. Anim. Ecol. 84, 615–624.

McNeil, J., Cox-Foster, D., Gardner, M., Slavicek, J., Thiem, S., Hoover, K., 2010. Pathogenesis of *Lymantria dispar* multiple nucleopolyhedrovirus in *L. dispar* and mechanisms of developmental resistance. J. Gen. Virol. 91, 1590–1600.

Michalsky, R., Pfromm, P.H., CzermaK, P., Sorensen, C.M., Passarelli, A.L., 2008. Effects of temperature and shear force on infectivity of the baculovirus *Autographa californica* M nucleopolyhedrovirus, J. Virol. Meth. 153, 90–96.

Mitsuhashi, W., Miyamoto, K., 2003. Disintegration of the peritrophic membrane of silkworm larvae due to spindles of an entomopoxvirus. J. Invertebr. Pathol. 82, 34–40.

Mitsuhashi, W., Furuta, Y., Sato, M., 1998. The spindles of an entomopoxvirus of Coleoptera (*Anomala cuprea*) strongly enhance the infectivity of a nucleopolyhedrovirus in Lepidoptera (*Bombyx mori*). J. Invertebr. Pathol. 71, 186–188.

Moore, S.D., Hendry, D.A., Richards, G.I., 2011. Virulence of a South African isolate of the *Cryptophlebia leucotreta* granulovirus to *Thaumatotibia leucotreta* neonate larvae. BioControl 56, 341–352.

Moore, S.D., Kirkman, W., Richards, G.I., Stephen, P.R., 2015. The *Cryptophlebia leucotreta* granulovirus – 10 years of commercial field use. Viruses 7, 1284–1312.

Morales-Ramos, J.A., Rojas, M.G., Shapiro-Ilan, D.I. (eds.), 2013. Mass Production of Beneficial Organisms: Invertebrates and Entomopathogens. Academic Press, San Diego, CA.

Morel, G., Fouillaud, M., 1994. Persistence of occluded viruses in the nests of the paper wasp *Polistes hebraeus* (Hym.: Vespidae). Entomophaga 39, 137–147.

Mori, H., Metcalf, P., 2010. Cypoviruses, in: Assgari, S., Johnson, K. (eds.), Insect Virology. Caister Academic Press, Norfolk, pp. 307–324.

Moscardi, F., 1989. Use of viruses for pest control in Brazil: the case of the nuclear polyhedrosis virus of the soybean caterpillar, *Anticarsia gemmatalis*. Mem. Inst. Oswaldo Cruz 84, 51–56.

Moscardi, F., 1999. Assessment of the application of baculoviruses for control of Lepidoptera. Annu. Rev. Entomol. 44, 257–289.

Murillo, R., Muñoz, D., Ruíz-Portero, M.C., Alcázar, M.D., Belda, J.E., Williams, T., Caballero, P., 2007. Abundance and genetic structure of nucleopolyhedrovirus populations in greenhouse substrate reservoirs. Biol. Control 42, 216–225.

Murillo, R., Hussey, M.S., Possee, R.D., 2011. Evidence for covert baculovirus infections in a *Spodoptera exigua* laboratory culture. J. Gen. Virol. 92, 1061–1070.

Murray, J.D., Elkinton, J.S., 1989. Environmental contamination of egg masses as a major component of transgenerational transmission of gypsy-moth nuclear polyhedrosis virus (LdMNPV). J. Invertebr. Pathol. 53, 324–334.

Murray, K.D., Elkinton, J.S., 1992. Vertical distribution of nuclear polyhedrosis virus-infected gypsy moth (Lepidoptera: Lymantriidae) larvae and effects on sampling for estimation of disease prevalence. J. Econ. Entomol. 85, 1865–1872.

Musser, R.O., Kwon, H.S., Williams, S.A., White, C.J., Romano, M.A., Holt, S.M., et al., 2005. Evidence that caterpillar labial saliva suppresses infectivity of potential bacterial pathogens. Arch. Ins. Biochem. Physiol. 58, 138–144.

Mutinelli, F., 2011. The spread of pathogens through trade in honey bees and their products (including queen bees and semen): overview and recent developments. Rev. Sci. Tech. Off. Int. Epizoot. 30, 257–271.

Myers, J.H., Cory, J.S., 2016. Ecology and evolution of pathogens in natural populations of Lepidoptera. Evol. Appl. 9, 231–247.

Myers, J.H., Kuken, B., 1995. Changes in the fecundity of tent caterpillars: a correlated character of disease resistance or sublethal effect of disease? Oecologia 103, 475–480.

Nakai, M., Sakai, T., Kunimi, Y., 1997. Effect of entomopoxvirus infection of the smaller tea tortrix, *Adoxophyes* sp. on the development of the endoparasitoid, *Ascogaster reticulatus*. Entomol. Exp. Appl. 84, 27–32.

Neuvonen, S., Saikkonen, K., Haukioja, E., 1990. Simulated acid rain reduces the susceptibility of the European pine sawfly (*Neodiprion sertifer*) to its nuclear polyhedrosis virus. Oecologia 83, 209–212.

Nordin, G.L., Brown, G.C., Jackson, D.M., 1990. Transovum transmission of two nuclear polyhedrosis viruses (Baculoviridae) by adult tobacco budworm and viral persistence on tobacco foliage. Trans. Kentucky Acad. Sci. 52, 33–39.

Ohba, M., Aizawa, K., 1979. Multiplication of *Chilo* iridescent virus in noninsect arthropods. J. Invertebr. Pathol. 33, 278–283.

Ohbayashi, T., Iwabuchi, K., 1991. Abnormal behavior of the common armyworm *Pseudaletia separata* (Walker) (Lepidoptera: Noctuidae) larvae infected with an entomogenous fungus, *Entomophaga aulicae*, and a nuclear polyhedrosis virus. Appl. Entomol. Zool. 26, 579–585.

Okamura, B., 2016. Hidden infections and changing environments. Integr. Comp. Biol. 56, 620–629.

Okuno, S., Nakai, M., Hiraoka, T., Kunimi, Y., 2002. Isolation of a protein lethal to the endoparasitoid *Cotesia kariyai* from entomopoxvirus-infected larvae of *Mythimna separata*. Ins. Biochem. Mol. Biol. 32, 559–566.

Olofsson, E., 1988a. Dispersal of the nuclear polyhedrosis virus of *Neodiprion sertifer* from soil to pine foliage with dust. Entomol. Exp. Appl. 46, 181–186.

Olofsson, E., 1988b. Environmental persistence of the nuclear polyhedrosis virus of the European pine sawfly in relation to epizootics in Swedish Scots pine forests. J. Invertebr. Pathol. 52, 119–129.

Olofsson, E., 1989. Transmission agents of the nuclear polyhedrosis virus of *Neodiprion sertifer* (Hym.: Diprionidae). Entomophaga 34, 373–380.

O'Reilly, D.R., Miller, L.K., 1991. Improvement of a baculovirus pesticide by deletion of the *egt* gene. Bio/Technol. 9, 1086–1089.

Ouellette, G.D., Buckley, P.E., O'Connell, K.P., 2010. Environmental influences on the relative stability of baculoviruses and vaccinia virus: a review, in: O'Connell, K.P., Skowronski, E.W., Bakanidze, L., Sulakvelidze, A. (eds.), Emerging and Endemic Pathogens: Advances in Surveillance, Detection and Identification. Springer, Dordrecht, pp. 125–149.

Panichareon, B., Khawsak, P., Deesukon, W., Sukhumsirichart, W., 2011. Multiplex real-time PCR and high-resolution melting analysis for detection of white spot syndrome virus, yellow-head virus, and *Penaeus monodon* densovirus in penaeid shrimp. J. Virol. Meth. 178, 16–21.

Parker, B.J., Elderd, B.D., Dwyer, G., 2010. Host behaviour and exposure risk in an insect-pathogen interaction. J. Anim. Ecol. 79, 863–870.

Pascual, L., Jakubowska, A.K., Blanca, J.M., Cañizares, J., Ferré, J., Gloeckner, G., et al., 2012. The transcriptome of *Spodoptera exigua* larvae exposed to different types of microbes. Ins. Biochem. Mol. Biol. 42, 557–570.

Pechan, T., Cohen, A., Williams, W.P., Luthe, D.S., 2002. Insect feeding mobilizes a unique plant defense protease that disrupts the peritrophic matrix of caterpillars. Proc. Nat. Acad. Sci. U.S.A. 99, 13319–13323.

Peng, F., Fuxa, J.R., Richter, A.R., Johnson, S.J., 1999a. Effects of heat-sensitive agents, soil type, moisture, and leaf surface on persistence of *Anticarsia gemmatalis* (Lepidoptera: Noctuidae) nucleopolyhedrovirus. Env. Entomol. 28, 330–338.

Peng, J., Zhong, J., Granados, R.R., 1999b. A baculovirus enhancin alters the permeability of a mucosal midgut peritrophic matrix from lepidopteran larvae. J. Ins. Physiol. 45, 159–166.

Pereira, E.P., Conte, H., Ribeiro, L.D.F.C., Zanatta, D.B., Bravo, J.P., Fernandez, M.A., Brancalhão, R.M.C., 2008. Cytopathological process by multiple nucleopolyhedrovirus in the testis of *Bombyx mori* L., 1758 (Lepidoptera: Bombycidae). J. Invertebr. Pathol. 99, 1–7.

Persson, B., 1981. Population fluctuations of the African armyworm, *Spodoptera exempta* (Walker) (Lepidoptera: Noctuidae), in outdoor cages in Kenya. Bull. Entomol. Res. 71, 289–297.

Pessoa, V., Cunha, F., de Freitas Bueno, A., Bortolotto, O.C., Monteiro, T.S.A., Ramos, V.M., 2014. Persistência do baculovírus anticarsia após diferentes regimes pluviométricos. Ciência Rural 44, 5–10.

Plymale, R.C., Felton, G.W., Hoover, K., 2007. Induction of systemic acquired resistance in cotton foliage does not adversely affect the performance of an entomopathogen. J. Chem. Ecol. 33, 1570–1581.

Plymale, R., Grove, M.J., Cox-Foster, D., Ostiguy, N., Hoover, K., 2008. Plant-mediated alteration of the peritrophic matrix and baculovirus infection in lepidopteran larvae. J. Ins. Physiol. 54, 737–749.

Podgwaite, J.D., Shields, K.S., Zerillo, R.T., Bruen, R.B., 1979. Environmental persistence of the nucleopolyhedrosis virus of the gypsy moth, *Lymantria dispar*. Env. Entomol. 8, 528–536.

Popham, H.J., Nusawardani, T., Bonning, B.C., 2016. Introduction to the use of baculoviruses as biological insecticides, in: Murhammer, D.W. (ed.), Baculovirus and Insect Cell Expression Protocols, 3rd edn. Humana Press, New York, pp. 383–392.

Povey, S., Cotter, S.C., Simpson, S.J., Wilson, K., 2014. Dynamics of macronutrient self-medication and illness-induced anorexia in virally infected insects. J. Anim. Ecol. 83, 245–255.

Prater, C.A., Redmond, C.T., Barney, W., Bonning, B.C., Potter, D.A., 2006. Microbial control of black cutworm (Lepidoptera: Noctuidae) in turfgrass using *Agrotis ipsilon* multiple nucleopolyhedrovirus. J. Econ. Entomol. 99, 1129–1137.

Rabalski, L., Krejmer-Rabalska, M., Skrzecz, I., Wasag, B., Szewczyk, B., 2016. An alphabaculovirus isolated from dead *Lymantria dispar* larvae shows high genetic similarity to baculovirus previously isolated from *Lymantria monacha* – an example of adaptation to a new host. J. Invertebr. Pathol. 139, 56–66.

Raina, A.K., Adams, J.R., Lupiani, B., Lynn, D.E., Kim, W., Burand, J.P., Dougherty, E.M., 2000. Further characterization of the gonad-specific virus of corn earworm, *Helicoverpa zea*. J. Invertebr. Pathol. 76, 6–12.

Rallis, C.P., Burand, J.P., 2002. Pathology and ultrastructure of Hz-2V infection in the agonadal female corn earworm, *Helicoverpa zea*. J. Invertebr. Pathol. 81, 33–44.

Raymond, B., Hails, R.S., 2007. Variation in plant resource quality and the transmission and fitness of the winter moth, *Operophtera brumata* nucleopolyhedrovirus. Biol. Control 41, 237–245.

Raymond, B., Vanbergen, A., Pearce, I., Hartley, S., Cory, J., Hails, R., 2002. Host plant species can influence the fitness of herbivore pathogens: the winter moth and its nucleopolyhedrovirus. Oecologia 131, 533–541.

Raymond, B., Hartley, S.E., Cory, J.S., Hails, R.S., 2005. The role of food plant and pathogen-induced behaviour in the persistence of a nucleopolyhedrovirus. J. Invertebr. Pathol. 88, 49–57.

Raymond, B., Sayyed, A.H., Wright, D.J., 2006. The compatibility of a nucleopolyhedrosis virus control with resistance management for *Bacillus thuringiensis*: co-infection and cross-resistance studies with the diamondback moth, *Plutella xylostella*. J. Invertebr. Pathol. 93, 114–120.

Rebolledo, D., Lasa, R., Guevara, R., Murillo, R., Williams, T., 2015. Baculovirus-induced climbing behavior favors intraspecific necrophagy and efficient disease transmission in *Spodoptera exigua*. PloS ONE 10, e0136742.

Redman, E.M., Wilson, K., Grzywacz, D., Cory, J.S., 2010. High levels of genetic diversity in *Spodoptera exempta* NPV from Tanzania. J. Invertebr. Pathol. 105, 190–193.

Redman, E.M., Wilson, K., Cory, J.S., 2016. Trade-offs and mixed infections in an obligate-killing insect pathogen. J. Anim. Ecol. 85, 1200–1209.

Reeson, A.F., Wilson, K., Cory, J.S., Hankard, P., Weeks, J.M., Goulson, D., Hails, R.S., 2000. Effects of phenotypic plasticity on pathogen transmission in the field in a Lepidoptera-NPV system. Oecologia 124, 373–380.

Reilly, J.R., Hajek, A.E., 2008. Density-dependent resistance of the gypsy moth *Lymantria dispar* to its nucleopolyhedrovirus, and the consequences for population dynamics. Oecologia 154, 691–701.

Reilly, J.R., Hajek, A.E., 2012. Prey-processing by avian predators enhances virus transmission in the gypsy moth. Oikos 121, 1311–1316.

Reineke, A., Asgari, S., 2005. Presence of a novel small RNA-containing virus in a laboratory culture of the endoparasitic wasp *Venturia canescens* (Hymenoptera: Ichneumonidae). J. Ins. Physiol. 51, 127–135.

Ren, W., Renault, T., Cai, Y., Wang, C., 2010. Development of a loop-mediated isothermal amplification assay for rapid and sensitive detection of ostreid herpesvirus 1 DNA. J. Virol. Meth. 170, 30–36.

Renault, S., Petit, A., Benedet, F., Bigot, S., Bigot, Y., 2002. Effects of the *Diadromus pulchellus* ascovirus, DpAV-4, on the hemocytic encapsulation response and capsule melanization of the leek-moth pupa, *Acrolepiopsis assectella*. J. Ins. Physiol. 48, 297–302.

Reyes, A., Christian, P., Valle, J., Williams, T., 2004. Persistence of *Invertebrate iridescent virus* 6 in soil. BioControl 49, 433–440.

Ribeiro, H.C., Pavan, O.H.O., 1994. Baculovirus thermal stability. J. Therm. Biol. 19, 21–24.

Richards, A.R., Christian, P.D., 1999. A rapid bioassay screen for quantifying nucleopolyhedroviruses (Baculoviridae) in the environment. J. Virol. Meth. 82, 63–75.

Richards, A., Cory, J., Speight, M., Williams, T., 1999. Foraging in a pathogen reservoir can lead to local host population extinction: a case study of a Lepidoptera-virus interaction. Oecologia 118, 29–38.

Richardson, M.L., Mitchell, R.F., Reagel, P.F., Hanks, L.M., 2010. Causes and consequences of cannibalism in noncarnivorous insects. Annu. Rev. Entomol. 55, 39–53.

Richter, A.R., Fuxa, J.R., Abdel-Fattah, M., 1987. Effect of host plant on the susceptibility of *Spodoptera frugiperda* (Lepidoptera: Noctuidae) to a nuclear polyhedrosis virus. Env. Entomol. 16, 1004–1006.

Rios-Velasco, C., Gallegos-Morales, G., Rincón-Castro, M.C.D., Cerna-Chávez, E., Sánchez-Peña, S.R., Siller, M.C., 2011. Insecticidal activity of native isolates of *Spodoptera frugiperda* multiple nucleopolyhedrovirus from soil samples in Mexico. Fl. Entomol. 94, 716–718.

Rivkin, H., Kroemer, J.A., Bronshtein, A., Belausov, E., Webb, B.A., Chejanovsky, N., 2006. Response of immunocompetent and immunosuppressed *Spodoptera littoralis* larvae to baculovirus infection. J. Gen. Virol. 87, 2217–2225.

Rohrmann, G.F., 2013. Baculovirus Molecular Biology, 3rd edn. National Center for Biotechnology Information, Bethesda, MD.

Ros, V.I., Houte, S., Hemerik, L., Oers, M.M., 2015. Baculovirus-induced tree-top disease: how extended is the role of egt as a gene for the extended phenotype? Mol. Ecol. 24, 249–258.

Rosario, K., Marinov, M., Stainton, D., Kraberger, S., Wiltshire, E.J., Collings, D.A., et al., 2011. Dragonfly cyclovirus, a novel single-stranded DNA virus discovered in dragonflies (Odonata: Anisoptera). J. Gen. Virol. 92, 1302–1308.

Rothman, L.D., Myers, J.H., 1996. Debilitating effects of viral diseases on host Lepidoptera. J. Invertebr. Pathol. 67, 1–10.

Roy, H.E., Hails, R.S., Hesketh, H., Roy, D.B., Pell, J.K., 2009. Beyond biological control: non-pest insects and their pathogens in a changing world. Ins. Cons. Divers. 2, 65–72.

Ryabov, E.V., 2016. Invertebrate RNA virus diversity from a taxonomic point of view. J. Invertebr. Pathol. doi:10.1016/j.jip.2016.10.002.

Sait, S.M., Gage, M.J.G., Cook, P.A., 1998. Effects of a fertility-reducing baculovirus on sperm numbers and sizes in the Indian Meal Moth, *Plodia interpunctella*. Func. Ecol. 12, 56–62.

Sajap, A.S., Bakir, M.A., Kadir, H.A., Samad, N.A., 2007. Effect of pH, rearing temperature and sunlight on infectivity of Malaysian isolate of nucleopolyhedrovirus to larvae of *Spodoptera litura* (Lepidoptera: Noctuidae). Int. J. Trop. Ins. Sci. 27, 108–113.

Santiago-Alvarez, C., Caballero, P., 1990. Susceptibility of parasitized *Agrotis segetum* larvae to a granulosis virus. J. Invertebr. Pathol. 56, 128–131.

Sarfraz, R.M., Cory, J.S., Myers, J.H., 2013. Life-history consequences and disease resistance of western tent caterpillars in response to localised, herbivore-induced changes in alder leaf quality. Ecol. Entomol. 38, 61–67.

Scotti, P.D., Dearing, S., Mossop, D.W., 1983. Flock house virus: a nodavirus isolated from *Costelytra zealandica* (White) (Coleoptera: Scarabaeida). Arch. Virol. 75, 181–189.

Seki, H., 1986. Effects of physicochemical treatments on a silkworm densonucleosis virus (Yamanashi isolate) of the silkworm, *Bombyx mori*. Appl. Entomol. Zool. 21, 515–518.

Shapiro, M., 1986. In vivo production of baculoviruses, in: Granados, R.R., Federici, B.A. (eds.), The Biology of Baculoviruses, Vol. 2. CRC Press, Boca Raton, FL, pp. 31–62.

Shapiro, M., Domek, J., 2002. Relative effects of ultraviolet and visible light on the activities of corn earworm and beet armyworm (Lepidoptera: Noctuidae) nucleopolyhedroviruses. J. Econ. Entomol. 95, 261–268.

Shapiro, D.I., Fuxa, J.R., Braymer, H.D., Pashley, D.P., 1991. DNA restriction polymorphism in wild isolates of *Spodoptera frugiperda* nuclear polyhedrosis virus. J. Invertebr. Pathol. 58, 96–105.

Shapiro-Ilan, D.I., Fuxa, J.R., Lacey, L.A., Onstad, D.W., Kaya, H.K., 2005. Definitions of pathogenicity and virulence in invertebrate pathology. J. Invertebr. Pathol. 88, 1–7.

Shikano, I., Cory, J.S., 2015. Impact of environmental variation on host performance differs with pathogen identity: implications for host-pathogen interactions in a changing climate. Sci. Rep. 5, 15351.

Shikano, I., Cory, J.S., 2016. Altered nutrient intake by baculovirus-challenged insects: self-medication or compensatory feeding? J. Invertebr. Pathol. 139, 25–33.

Shikano, I., Ericsson, J.D., Cory, J.S., Myers, J.H., 2010. Indirect plant-mediated effects on insect immunity and disease resistance in a tritrophic system. Basic Appl. Ecol. 11, 15–22.

Shikano, I., Olson, G.L., Cory, J.S., 2015. Impact of non-pathogenic bacteria on insect disease resistance: importance of ecological context. Ecol. Entomol. 40, 620–628.

Silva, F.W., Elliot, S.L., 2016. Temperature and population density: interactional effects of environmental factors on phenotypic plasticity, immune defenses, and disease resistance in an insect pest. Ecol. Evol. 6, 3672–3683.

Silva, F.W., Viol, D.L., Faria, S.V., Lima, E., Valicente, F.H., Elliot, S.L., 2013. Two's a crowd: phenotypic adjustments and prophylaxis in *Anticarsia gemmatalis* larvae are triggered by the presence of conspecifics. PLoS ONE 8, e61582.

Silva, F.W., Serrão, J.E., Elliot, S.L., 2016. Density-dependent prophylaxis in primary anti-parasite barriers in the velvetbean caterpillar. Ecol. Entomol. 41, 451–458.

Simón, O., Williams, T., López-Ferber, M., Caballero, P., 2004a. Genetic structure of a *Spodoptera frugiperda* nucleopolyhedrovirus population: high prevalence of deletion genotypes. Appl. Env. Microbiol. 70, 5579–5588.

Simón, O., Williams, T., López-Ferber, M., Caballero, P., 2004b. Virus entry or the primary infection cycle are not the principal determinants of host specificity of *Spodoptera* spp. nucleopolyhedroviruses. J. Gen. Virol. 85, 2845–2855.

Simón, O., Palma, L., Beperet, I., Muñoz, D., López-Ferber, M., Caballero, P., Williams, T., 2011. Sequence comparison between three geographically distinct *Spodoptera frugiperda* multiple nucleopolyhedrovirus isolates: detecting positively selected genes. J. Invertebr. Pathol. 107, 33–42.

Simón, O., Williams, T., López-Ferber, M., Caballero, P., 2012. Deletion of *egt* is responsible for the fast-killing phenotype of natural deletion genotypes in a *Spodoptera frugiperda* multiple nucleopolyhedrovirus population. J. Invertebr. Pathol. 111, 260–263.

Simón, O., Williams, T., Cerutti, M., Caballero, P., López-Ferber, M., 2013. Expression of a peroral infection factor determines pathogenicity and population structure in an insect virus. PLoS ONE 8, e78834.

Singh, R., Levitt, A.L., Rajotte, E.G., Holmes, E.C., Ostiguy, N., Lipkin, W.I., et al., 2010. RNA viruses in hymenopteran pollinators: evidence of inter-taxa virus transmission via pollen and potential impact on non-*Apis* hymenopteran species. PLoS ONE 5, e14357.

Slavicek, J.M., Popham, H.J., 2005. The *Lymantria dispar* nucleopolyhedrovirus enhancins are components of occlusion-derived virus. J. Virol. 79, 10578–10588.

Small, D.A., Moore, N.F., Entwistle, P.F., 1986. Hydrophobic interactions involved in attachment of a baculovirus to hydrophobic surfaces. Appl. Env. Microbiol. 52, 220–223.

Smith, K.M., Loh, E.H., Rostal, M.K., Zambrana-Torrelio, C.M., Mendiola, L., Daszak, P., 2013. Pathogens, pests, and economics: drivers of honey bee colony declines and losses. EcoHealth 10, 434–445.

Smith-Johannsen, H., Witkiewicz, H., Iatrou, K., 1986. Infection of silkmoth follicular cells with *Bombyx mori* nuclear polyhedrosis virus. J. Invertebr. Pathol. 48, 74–84.

Sood, P., Mehta, P.K., Bhandari, K., Prabhakar, C.S., 2010. Transmission and effect of sublethal infection of granulosis virus (PbGV) on *Pieris brassicae* Linn. (Pieridae: Lepidoptera). J. Appl. Entomol. 134, 774–780.

Stasiak, K., Renault, S., Federici, B.A., Bigot, Y., 2005. Characteristics of pathogenic and mutualistic relationships of ascoviruses in field populations of parasitoid wasps. J. Ins. Physiol. 51, 103–115.

Stevenson, P.C., D'Cunha, R.F., Grzywacz, D., 2010. Inactivation of baculovirus by isoflavonoids on chickpea (*Cicer arietinum*) leaf surfaces reduces the efficacy of nucleopolyhedrovirus against *Helicoverpa armigera*. J. Chem. Ecol. 36, 227–235.

Stöhr, A.C., Papp, T., Marschang, R.E., 2016. Repeated detection of an invertebrate iridovirus in amphibians. J. Herpetol. Med. Surg. 26, 54–58.

Stoianova, E., Williams, T., Cisneros, J., Muñoz, D., Murillo, R., Tasheva, E., Caballero, P., 2012. Interactions between an ectoparasitoid and a nucleopolyhedrovirus when simultaneously attacking *Spodoptera exigua* (Lepidoptera: Noctuidae). J. Appl. Entomol. 136, 596–604.

Stoltz, D., Makkay, A., 2003. Overt viral diseases induced from apparent latency following parasitization by the ichneumonid wasp, *Hyposoter exiguae*. J. Ins. Physiol. 49, 483–489.

Strand, M.R., 2010. Polydnaviruses, in: Asgari, S., Johnson, K. (eds.), Insect Virology. Caister Academic Press, Norfolk, pp. 171–197.

Strand, M.R., 2012. Polydnavirus gene products that interact with the host immune system, in: Beckage, N.E., Drezen, J.M. (eds.), Parasitoid Viruses: Symbionts and Pathogens. Elsevier, Oxford, pp. 149–161.

Strand, M.R., Burke, G.R., 2015. Polydnaviruses: from discovery to current insights. Virology 479, 393–402.

Streett, D.A., McGuire, M.R., 1990. Pathogenic diseases of grasshoppers, in: Chapman, R.F., Joern, A. (eds.), Biology of Grasshoppers. John Wiley & Sons, New York, pp. 483–516.

Sun, X., 2015. History and current status of development and use of viral insecticides in China. Viruses 7, 306–319.

Sun, X., Sun, X., Van Der Werf, W., Vlak, J.M., Hu, Z., 2004. Field inactivation of wild-type and genetically modified *Helicoverpa armigera* single nucleocapsid nucleopolyhedrovirus in cotton. Biocontr. Sci. Technol. 14, 185–192.

Tanada, Y., Kaya, H.K., 1993. Insect Pathology. Academic Press, San Diego, CA.

Tanada, Y., Omi, E.M., 1974. Persistence of insect viruses in field populations of alfalfa insects J. Invertebr. Pathol. 23, 360–365

Teixeira, L., Ferreira, A., Ashburner, M., 2008. The bacterial symbiont *Wolbachia* induces resistance to RNA viral infections in *Drosophila melanogaster*. PLoS Biol. 6, e1000002.

Thézé, J., Takatsuka, J., Li, Z., Gallais, J., Doucet, D., Arif, B., et al., 2013. New insights into the evolution of Entomopoxvirinae from the complete genome sequences of four entomopoxviruses infecting *Adoxophyes honmai*, *Choristoneura biennis*, *Choristoneura rosaceana*, and *Mythimna separata*. J. Virol. 87, 7992–8003.

Thézé, J., Cabodevilla, O., Palma, L., Williams, T., Caballero, P., Herniou, E.A., 2014. Genomic diversity in European *Spodoptera exigua* multiple nucleopolyhedrovirus isolates. J. Gen. Virol. 95, 2297–2309.

Thomas, S.R., Elkinton, J.S., 2004. Pathogenicity and virulence. J. Invertebr. Pathol. 85, 146–151.

Thompson, A.C., Sikorowski, P.P., 1980. Fatty acid and glycogen requirement of *Heliothis virescens* infected with cytoplasmic polyhedrosis virus. Comp. Biochem. Physiol. B. 66, 93–97.

Thompson, C.G., Scott, D.W., 1979. Production and persistence of the nuclear polyhedrosis virus of the Douglas-fir tussock moth, *Orgyia pseudotsugata* (Lepidoptera: Lymantriidae), in the forest ecosystem. J. Invertebr. Pathol. 33, 57–65.

Thompson, C.G., Scott, D.W., Wickman, B.E., 1981. Long-term persistence of the nuclear polyhedrosis virus of the Douglas-fir tussock moth, *Orgyia pseudotsugata* (Lepidoptera: Lymantriidae), in forest soil. Env. Entomol. 10, 254–255.

Vail, P.V., Hostetter, D.L., Hoffmann, D.F., 1999. Development of the multi-nucleocapsid nucleopolyhedroviruses (MNPVs) infectious to loopers (Lepidoptera: Noctuidae: Plusiinae) as microbial control agents. Integr. Pest Man. Rev. 4, 231–257.

Valles, S.M., Oi, D.H., Yu, F., Tan, X.X., Buss, E.A., 2012. Metatranscriptomics and pyrosequencing facilitate discovery of potential viral natural enemies of the invasive Caribbean crazy ant, *Nylanderia pubens*. PLoS ONE 7, e31828.

van der Werf, W., Hemerik, L., Vlak, J.M. and Zwart, M.P., 2011. Heterogeneous host susceptibility enhances prevalence of mixed-genotype micro-parasite infections. PLoS Comput. Biol. 7, e1002097.

van Houte, S., Ros, V.I., Oers, M.M., 2013. Walking with insects: molecular mechanisms behind parasitic manipulation of host behaviour. Mol. Ecol. 22, 3458–3475.

van Houte, S., Ros, V.I., van Oers, M.M., 2014a. Hyperactivity and tree-top disease induced by the baculovirus AcMNPV in *Spodoptera exigua* larvae are governed by independent mechanisms. Naturwissenschaften 101, 347–350.

van Houte, S., van Oers, M.M., Han, Y., Vlak, J.M., Ros, V.I., 2014b. Baculovirus infection triggers a positive phototactic response in caterpillars to induce "tree-top" disease. Biol. Lett. 10, 20140680.

van Oers, M.M., 2010. Genomics and biology of Iflaviruses, in: Assgari, S., Johnson, K. (eds.), Insect Virology. Caister Academic Press, Norfolk, pp. 231–250.

van Oers, M.M., Lampen, M.H., Bajek, M.I., Vlak, J.M., Eker, A.P., 2008. Active DNA photolyase encoded by a baculovirus from the insect *Chrysodeixis chalcites*. DNA Repair 7, 1309–1318.

Vasconcelos, S.D., Williams, T., Hails, R.S., Cory, J.S., 1996a. Prey selection and baculovirus dissemination by carabid predators of Lepidoptera. Ecol. Entomol. 21, 98–104.

Vasconcelos, S.D., Cory, J.S., Wilson, K.R., Sait, S.M., Hails, R.S., 1996b. The effect of baculovirus infection on the mobility of *Mamestra brassicae* L. (Lepidoptera: Noctuidae) larvae at different developmental stages. Biol. Control 7, 299–306.

Vieira, C.M., Tuelher, E.S., Valicente, F.H., Wolff, J.L.C., 2012. Characterization of a *Spodoptera frugiperda* multiple nucleopolyhedrovirus isolate that does not liquefy the integument of infected larvae. J. Invertebr. Pathol. 111, 189–192.

Vilaplana, L., Wilson, K., Redman, E.M., Cory, J.S., 2010. Pathogen persistence in migratory insects: high levels of vertically-transmitted virus infection in field populations of the African armyworm. Evol. Ecol. 24, 147–160.

Virto, C., Zárate, CA., López-Ferber, M., Murillo, R., Caballero, P., Williams, T., 2013. Gender-mediated differences in vertical transmission of a nucleopolyhedrovirus. PLoS ONE 8, e70932.

Virto, C., Navarro, D., Tellez, M.M., Herrero, S., Williams, T., Murillo, R., Caballero, P., 2014. Natural populations of *Spodoptera exigua* are infected by multiple viruses that are transmitted to their offspring. J. Invertebr. Pathol. 122, 22–27.

Wagner, G.W., Webb, S.R., Paschke, J.D., Campbell, W.R., 1974. A picornavirus isolated from *Aedes taeniorhynchus* and its interaction with mosquito iridescent virus. J. Invertebr. Pathol. 24, 380–382.

Wan, N.F., Jiang, J.X., Li, B., 2016. Effect of host plants on the infectivity of nucleopolyhedrovirus to *Spodoptera exigua* larvae. J. Appl. Entomol. 140, 636–644.

Wang, G., Zhang, J., Shen, Y., Zheng, Q., Feng, M., Xiang, X., Wu, X., 2015. Transcriptome analysis of the brain of the silkworm *Bombyx mori* infected with *Bombyx mori* nucleopolyhedrovirus: a new insight into the molecular mechanism of enhanced locomotor activity induced by viral infection. J. Invertebr. Pathol. 128, 37–43.

Wang, P., Granados, R.R., 1997. An intestinal mucin is the target substrate for a baculovirus enhancin. Proc. Nat. Acad. Sci. U.S.A. 94, 6977–6982.

Wang, P., Granados, R.R., 2000. Calcofluor disrupts the midgut defense system in insects. Ins. Biochem. Mol. Biol. 30, 135–143.

Washburn, J.O., Haas-Stapleton, E.J., Tan, F.F., Beckage, N.E., Volkman, L.E., 2000. Co-infection of *Manduca sexta* larvae with polydnavirus from *Cotesia congregata* increases susceptibility to fatal infection by *Autographa californica* M nucleopolyhedrovirus. J. Ins. Physiol. 46, 179–190.

Watanabe, H., Shimizu, T., 1980. Epizootiological studies on the occurrence of densonucleosis in the silk-worm, *Bombyx mori*, reared at sericultural farms. J. Sericult. Sci. Japan 49, 485–492.

Webb, R.E., Peiffer, R.A., Fuester, R.W., Valenti, M.A., Thorpe, K.W., White, G.B., Shapiro, M., 1999. Effects of Blankophor BBH, a virus-enhancing adjuvant, on mortality of gypsy moth (Lepidoptera: Lymantriidae). J. Entomol. Sci. 34, 391–403.

Webb, R.E., Shapiro, M., Thorpe, K.W., Peiffer, R.A., Fuester, R.W., Valenti, M.A., et al., 2001. Potentiation by a granulosis virus of Gypchek, the gypsy moth (Lepidoptera: Lymantriidae) nuclear polyhedrosis virus product. J. Entomol. Sci. 36, 169–176.

Webster, C.L., Waldron, F.M., Robertson, S., Crowson, D., Ferrari, G., Quintana, J.F., et al., 2015. The discovery, distribution, and evolution of viruses associated with *Drosophila melanogaster*. PLoS Biol. 13, e1002210.

Weissman, D.B., Gray, D.A., Pham, H.T., Tijssen, P., 2012. Billions and billions sold: pet-feeder crickets (Orthoptera: Gryllidae), commercial cricket farms, an epizootic densovirus, and government regulations make for a potential disaster. Zootaxa 3504, 67–88.

Wennmann, J.T., Köhler, T., Alletti, G.G., Jehle, J.A., 2015. Mortality of cutworm larvae is not enhanced by Agrotis segetum granulovirus and Agrotis segetum nucleopolyhedrovirus B coinfection relative to single infection by either virus. Appl. Env. Microbiol. 81, 2893–2899.

White, A., Bowers, R.G., Begon, M., 1999. The spread of infection in seasonal insect-pathogen systems. Oikos 85, 487–498.

White, A., Watt, A.D., Hails, R.S., Hartley, S.E., 2000. Patterns of spread in insect-pathogen systems: the importance of pathogen dispersal. Oikos 89, 137–145.

Williams, T., 1993. Covert iridovirus infection of blackfly larvae. Proc. R. Soc. Lond. B Biol. Sci. 251, 225–230.

Williams, T., 2008. Natural invertebrate hosts to iridoviruses (Iridoviridae). Neotrop. Entomol. 37, 615–632.

Williams, T., Bergoin, M., van Oers, M.M., 2016. Diversity of large DNA viruses of invertebrates. J. Invertebr. Pathol. doi:10.1016/j.jip.2016.08.001.

Williams, T., Ward, V., 2010. Iridoviruses, in: Assgari, S., Johnson, K. (eds.), Insect Virology. Caister Academic Press, Norfolk, pp. 123–152.

Wilson, K., Cotter, S.C., 2009. Density-dependent prophylaxis in insects, in: Whitman, D.W., Ananthakrishnan, T.N. (eds.), Phenotypic Plasticity of Insects: Mechanisms and Consequences. Science Publishers, Plymouth, pp. 381–420.

Woods, S.A., Elkinton, J.S., Podgwaite, J.D., 1989. Acquistion of nuclear polyhedrosis virus from tree stems by newly emerged gypsy moth (Lepidoptera: Lymantriidae) larvae. Environ. Entomol. 18, 298–301.

Wu, Y.L., Wu, C.P., Liu, C.Y.Y., Hsu, P.W.C., Wu, E.C., Chao, Y.C., 2011. A non-coding RNA of insect HzNV–1 virus establishes latent viral infection through microRNA. Sci. Rep. 1, 60.

Xia, X., Yu, Y., Weidmann, M., Pan, Y., Yan, S., Wang, Y., 2014. Rapid detection of shrimp white spot syndrome virus by real time, isothermal recombinase polymerase amplification assay. PloS ONE 9, e104667.

Xia, X., Yu, Y., Hu, L., Weidmann, M., Pan, Y., Yan, S., Wang, Y., 2015. Rapid detection of infectious hypodermal and hematopoietic necrosis virus (IHHNV) by real-time, isothermal recombinase polymerase amplification assay. Arch. Virol. 160, 987–994.

Xu, P., Liu, Y., Graham, R.I., Wilson, K., Wu, K., 2014. Densovirus is a mutualistic symbiont of a global crop pest (*Helicoverpa armigera*) and protects against a baculovirus and Bt biopesticide. PLoS Pathog, 10, e1004490.

Young, S.Y., 1990. Influence of sprinkler irrigation on dispersal of nuclear polyhedrosis virus from host cadavers on soybean. Env. Entomol. 19, 717–720.

Young, S.Y., Kring, T.J., 1991. Selection of healthy and nuclear polyhedrosis virus infected *Anticarsia gemmatalis* (Lep.: Noctuidae) as prey by nymphal *Nabis roseipennis* (Hemiptera: Nabidae) in laboratory and on soybean. Entomophaga 36, 265–273.

Young, S.Y., Yearian, W.C., 1974. Persistence of *Heliothis* NPV on foliage of cotton, soybean, and tomato. Env. Entomol. 3, 253–255.

Young, S.Y., Yearian, W.C., 1986. Movement of a nuclear polyhedrosis virus from soil to soybean and transmission in *Anticarsia gemmatalis* (Hübner) (Lepidoptera: Noctuidae) populations on soybean. Env. Entomol. 15, 573–580.

Young, S.Y., Yearian, W.C., 1987. *Nabis roseipennis* adults (Hemiptera: Nabidae) as disseminators of nuclear polyhedrosis virus to *Anticarsia gemmatalis* (Lepidoptera: Noctuidae) larvae. Env. Entomol. 16, 1330–1333.

Young, S.Y., Yearian, W.C., 1989. Persistence and movement of nuclear polyhedrosis virus on soybean plants after death of infected *Anticarsia gemmatalis* (Lepidoptera: Noctuidae). Env. Entomol. 18, 811–815.

Young, S.Y., Yearian, W.C., Kim, K.S., 1977. Effect of dew from cotton and soybean foliage on activity of *Heliothis* nuclear polyhedrosis virus. J. Invertebr. Pathol. 29, 105–111.

Yue, C., Schröder, M., Gisder, S., Genersch, E., 2007. Vertical-transmission routes for deformed wing virus of honeybees (*Apis mellifera*). J. Gen. Virol. 88, 2329–2336.

Zelazny, B., 1972. Studies on *Rhabdionvirus oryctes*: I. Effect on larvae of *Oryctes rhinoceros* and inactivation of the virus. J. Invertebr. Pathol. 20, 235–241.

Zelazny, B., 1976. Transmission of a baculovirus in populations of *Oryctes rhinoceros*. J. Invertebr. Pathol. 27, 221–227.

Zhou, Y., Wu, J., Lin, F., Chen, N., Yuan, S., Ding, L., et al., 2015. Rapid detection of *Bombyx mori* nucleopolyhedrovirus (BmNPV) by loop-mediated isothermal amplification assay combined with a lateral flow dipstick method. Mol. Cell. Probes, 29, 389–395.

Zwart, M.P., Elena, S.F., 2015. Matters of size: genetic bottlenecks in virus infection and their potential impact on evolution. Annu. Rev. Virol. 2, 161–179.

Zwart, M.P., van Oers, M.M., Cory, J.S., van Lent, J.W., van der Werf, W., Vlak, J.M., 2008. Development of a quantitative real-time PCR for determination of genotype frequencies for studies in baculovirus population biology. J. Virol. Meth. 148, 146–154.

Zwart, M.P., Hemerik, L., Cory, J.S., de Visser, J.A.G., Bianchi, F.J., Van Oers, M.M., et al., 2009. An experimental test of the independent action hypothesis in virus-insect pathosystems. Proc. R. Soc. Lond. B Biol. Sci. 276, 2233–2242.

8

Bacteria

Trevor A. Jackson[1], Colin Berry[2] and Maureen O'Callaghan[1]

[1] AgResearch Ltd, Lincoln Research Centre, Christchurch, New Zealand
[2] Cardiff School of Biosciences, Cardiff University, Cardiff, UK

8.1 Introduction

Bacteria are one of the most successful and diverse groups of living organisms, with members occupying every conceivable habitat from hot-water springs and deep oceans to rainforests and deserts. They are closely associated with all lifeforms, and insects are no exception; bacteria range from common environmental contaminants of the insect body to obligate symbionts. Given the abundance of bacteria in the environment, it is no surprise that they can always be isolated from dead and diseased insects.

The biology and genetics of bacterial entomopathogens have recently been reviewed by Jurat-Fuentes and Jackson (2012) and Glare et al. (2017), and their role as pathogens was described by Boucias and Pendland (1998). Entomopathogenic species are found among the Gram-positive Firmicutes and the Gram-negative Proteobacteria, with even the Tenericutes (Spiroplasmas) including entomopathogenic species. As with other pathogens, the dynamics of bacterial entomopathogens will be determined by various factors, including host and pathogen density, transmission, virulence, and persistence (Chapter 1). The environment and the condition of the host can also play a major role in the development of insect diseases. The role of stress (resulting from crowding, poor nutrition, and adverse environmental conditions) in the development of epizootics was raised by Steinhaus (1960) and Tanada (1964) in their early works on insect pathology.

It is not surprising that the first records of bacterial pathogens come from dead insects closely associated with human activities, including culture of silkworm, beekeeping, and grain storage (Kreig, 1987). In these conditions, insects are found at high densities and are sometimes stressed, and transmission from diseased to susceptible insects can be aided by human management and the inadvertent contamination of healthy insect colonies. Observations of disease in managed insect populations raised awareness of the potential for diseases to impact natural populations. One of the earliest recognized insect epizootics was a bacterial disease of grasshoppers observed by d'Herelle (1911). However, early researchers were often frustrated by variability in the host response to exposure to putative bacterial entomopathogens. For example, Stephenson (1959) produced high mortality among desert locusts (*Schistocerca gregaria*) in laboratory cages

Ecology of Invertebrate Diseases, First Edition. Edited by Ann E. Hajek and David I. Shapiro-Ilan.
© 2018 John Wiley & Sons Ltd. Published 2018 by John Wiley & Sons Ltd.

with treatments of a laboratory-isolated strain of *Serratia marcescens*, but field results were less convincing. He attributed the difference in results to high density and canni- balism among the caged insects. The importance of such cofactors has led some groups of bacteria to be viewed as opportunistic pathogens (Boucias and Pendland, 1998), with infection occurring only when a certain combination of environmental and biotic con- ditions holds.

Bacteria often carry extrachromosomal DNA as plasmids, allowing exchange of genetic material. The role of horizontal gene transfer in the evolution of insect pathogens is well established in *Bacillus thuringiensis* (Bt) (Box 8.1) (Jurat-Fuentes and Jackson, 2012). Genetic interchange may lead to common "toxin complexes" being found among different genera (Waterfield et al., 2001), or in the loss of plasmids, leading to loss of pathogenicity (Arronson et al., 1982; Gonzalez and Carlton, 1984).

The ecology of bacterial entomopathogens has been reviewed previously (Tanada, 1964; Krieg, 1987), but more recent reviews in insect pathology have focused on molecular genetics, bioassay techniques, and field use of microbial insecticides. This chapter updates current knowledge of the ecology of entomopathogenic bacteria, the use of new tools to examine their interactions with the environment, and factors affect- ing their distribution and abundance. It includes case studies on specific bacteria and looks at how bacterial ecology affects the use and impact of bacterial entomopathogens in microbial control.

8.2 Bacterial Pathogens and Associations with Insects

As already noted, bacteria will always be associated with insects, and dead insects often undergo septicemia as saprophytic bacteria take over the cadaver. In a list of 1900 diag- nostic accessions of dead and diseased insects diagnosed at the German "Laboratory for Diagnosis, Histo- and Cytopathology of Arthropod Diseases," a high proportion of the 450 examined species were colonized by bacteria, but, as noted by the authors, bacteria are generally found in dead insects and are not necessarily the cause of death (Kleespies et al., 2008). Thus, identification of an entomopathogen requires a system and some rules. Entomopathogenic bacteria have usually been discovered by isolation from dead insects or by screening bacteria isolated from the environment using bioassays. To determine whether a bacterial isolate is pathogenic, Koch's postulates must be tested. These state that a microbe cannot be described as a pathogen unless it can be isolated from a diseased host, purified, multiplied, and used to produce the same disease symp- toms in a healthy host (Kaya and Vega, 2012). Koch's postulates provide the foundation for determination of pathogenicity, but it is important that dose and conditions of test- ing are also noted.

A list of some of the most well-known bacterial entomopathogens is presented in Table 8.1. Bt was first recognized from silkworms by Ishiwata, and later from flour moths by Berliner (as detailed in Beegle and Yamamoto, 1992). Once the association of pathogens with dead insects had been made, and interest was raised, insect pathology became a discipline associated with laboratories in the United States, Germany, France, and other countries, leading to the discovery of a wide range of bacterial entomo- pathogens, including a large number of variants of Bt (Jurat-Fuentes and Jackson, 2012).

Table 8.1 Sources of iconic species/strains of entomopathogenic bacteria by date of discovery.

Bacterium	Classification: family	Source	References
Bacillus thuringiensis	Bacillaceae	Diseased silkworms (*Bombyx mori*)	Ishiwata (1901/02) (in Beegle and Yamamoto, 1992)
Paenibacillus (Bacillus) *larvae*	Paenibacillaceae	Honeybee (*Apis mellifera*)	White (1906) (in Bailey, 1983)
Melissococcus pluton	Enterococcaceae	Honeybee (*A. mellifera*)	White (1908) (in Bailey, 1983)
Bacillus thuringiensis var. *thuringiensis*	Bacillaceae	Flour moth *Ephestia* (*Anagasta*) *kühniella*	Berliner (1911) (in Beegle and Yamamoto, 1992)
Bacillus thuringiensis var. *kurstaki*	Bacillaceae	Flour moth *E. kühniella*	Kurstak (1962)
Paenibacillus (Bacillus) *popilliae*	Paenibacillaceae	Hemolymph of milky diseased Japanese beetle (*Popillia japonica*) larva	Dutky (1940) (in Dutky, 1963)
Paenibacillus lentimorbus	Paenibacillaceae	Hemolymph of milky diseased Japanese beetle (*P. japonica*) larva	Dutky (1940) (in Dutky, 1963)
Rickettsiella popilliae	Coxiellaceae	Japanese beetle (*P. japonica*)	Dutky and Gooden (1952)
Rickettsiella melolonthae	Coxiellaceae	May beetle (*Melolontha* sp.)	Willie and Martignoni (1952)
Bacillus thuringiensis var. *kurstaki*	Bacillaceae	Flour moth (*E. kühniella*)	Kurstak (1962)
Bacillus thuringiensis var. *kurstaki* (HD1)	Bacillaceae	Diseased pink bollworm (*Pectinophora gossipiella*)	Dulmage (1970)
Serratia marcescens	Enterobacteriaceae	*Schistocerca gregaria* (lab colony)	Stephenson (1959)
Lysinibacillus (Bacillus) *sphaericus*	Bacillaceae	Cool weather mosquito (*Culiseta incidens*)	Kellen et al. (1965)
Paenibacillus (Bacillus) *alvei*	Bacillaceae	Mosquitoes	Singer (1973)
Bacillus circulans	Bacillaceae	Mosquitoes	Singer (1973)
Rickettsiella costelytrae	Coxiellaceae	New Zealand grass grub (*Costelytra zealandica*)	Moore et al. (1974)
B. thuringiensis var. *israelensis*	Bacillaceae	Mosquito larvae and mud	Goldberg and Margalit (1977)
Xenorhabdus nematophila	Enterobacteriaceae	Nematodes and dead insects	Thomas and Poinar (1979)
B. thuringiensis var. *tenebrionis*	Bacillaceae	Dead mealworm (*Tenebrio molitor*) larva	Krieg et al. (1983)

(Continued)

Table 8.1 (Continued)

Bacterium	Classification: family	Source	References
Serratia entomophila	Enterobacteriaceae	Diseased *Costelytra zealandica* larvae	Grimont et al. (1988)
Streptomyces avermitilis	Actinobacteraceae	Soil	Turner and Schaeffer (1989)
Saccharopolyspora spinosa	Actinobacteraceae	Soil	Mertz and Yao (1990)
Photorhabdus luminescens	Enterobacteriaceae	Nematodes and dead insects	Poinar et al. (1980); Boemare et al. (1993)
Brevibacillus (Bacillus) *laterosporus*	Bacillaceae	Dead insects	Orlova et al. (1998)
Pseudomonas entomophila	Pseudomonadaceae	Fruit fly (*Drosophila melanogaster*)	Vodovar et al. (2005)
Chromobacterium subtsugae	Niesseriaceae	Forest soil	Martin et al. (2007)
Yersinia entomophaga	Enterobacteriaceae	Dead *Costelytra zealandica*	Hurst et al. (2011a)
Chromobacterium Csp-P	Niesseriaceae	Yellow fever mosquito (*Aedes aegypti*)	Ramirez et al. (2014)
Serratia nematodiphila	Enterobacteriaceae	Pond sediments	Patil et al. (2012)

The history of the discovery of Bt as an insect pathogen has been summarized previously (Beegle and Yamamoto, 1992; Côté, 2007). In brief, a disease-causing bacterium killing silkworms (*Bombyx mori*) was first noted by Ishiwata in Japan in 1901 and a decade later a similar bacterium was isolated by Ernst Berliner in Thuringia, Germany, from diseased populations of the Mediterranean flour moth (*Ephestia kühniella*) in a grain store. Berliner designated this species "*Bacillus thuringiensis*." Grain store residues have been a good source of isolates of Bt. In 1962, Edouard Kurstak recovered a new isolate from *E. khüniella*, and in the United States Howard Dulmage obtained an isolate from diseased *Pectinophora gossipiella*. Both isolates were found to belong to the same serotype and became the iconic *Bacillus thuringiensis kurstaki* (Btk). The abundance of lepidoptera-active toxins in Bt isolated from lepidopteran hosts suggested a pathotypic association, which indicated that dead insects from specific target groups could be a good source of new selective toxins. An International Reference Centre was established by the World Health Organization (WHO), where dead mosquito larvae collected by scientists from around the world could be examined. This initiative yielded mosquito-active isolates from the *Bacillus* (*Lysinibacillus*) *sphaericus* and *Bacillus alvei-circulans* groups (Singer, 1973, 1981). This approach was continued, and eventually the highly dipteran-active *Bacillus thuringiensis israelensis* (Bti) was isolated from mosquito-infested pools in Israel (Goldberg and Margalit, 1977) and found to be the predominant biotype from these environments (Brownbridge and Margalit, 1987). Soil insects, in particular the Coleoptera, are generally less susceptible to Bt toxins than their foliar-feeding counterparts. However, awareness of the potential

---

Box 8.1 *Bacillus thuringiensis* – opportunist toxin producer or true pathogen?

Bacillus thuringiensis (Bt) is a ubiquitous bacterium that has been isolated from soil and other environmental samples from nearly all parts of the world, including Antarctica. The species is characterized by a resting sporangium containing a spore and a distinctive parasporal crystal and, with the advance of science, by an increasing number of phenotypic and genotypic markers. Bt came to the attention of scientists, and in particular insect pathologists, with the discovery of insecticidal isolates that have been developed as bioinsecticides for a range of pests. Epizootics of disease caused by Bt among insects are rare, and frequently no susceptible hosts can be found for novel isolates. This has led to a view that Bt is not truly an insect pathogen, but rather an environmental bacterium that can produce toxins of utility for insect pest control. However, Bt is indistinguishable from *Bacillus cereus*, except that its plasmids encode insecticidal toxins. The synthesis of toxins poses an energetic burden on the cell, and studies have shown that Bt is a poor competitor with other microbes in the environment. Pathogenicity provides toxin-bearing strains with a comparative advantage over their nonpathogenic counterparts in the presence of insects. Increasingly, associations between isolate, pathotype, and a susceptible host are emerging, suggesting that at least some strains or biotypes of Bt are true pathogens.

---

of Bt and the continued search for new isolates have identified isolates active against coleopteran species. A novel strain, *Bacillus thuringiensis tenebrionis* (Btt), was isolated from dead larvae of *Tenebrio molitor* (Huger et al., 1986). Sharpe (1976) showed that larvae of *P. japonica* were susceptible to a strain of *Bacillus thuringiensis galleriae* (Btg). Ohba et al. (1992) discovered a strain of *Bacillus thuringiensis japonensis* (Btj) with activity against scarabs, and Asano et al. (2003) reported high activity of another strain of Btg (strain SDS-502) from Japanese soil against the ruteline scarab *Anomala cuprea*. Further coleopteran activity was shown in a Bt strain isolated from soil in China (Yu et al., 2006).

Following its accidental introduction into the United States early in the 20th century, Japanese beetle (*Popillia japonica*) spread across the eastern seaboard, where the larvae caused extensive damage to turf and lawns. "Milky disease" was discovered in the invasive populations and found to be caused by two species of bacteria: *Paenibacillus* (*Bacillus*) *popilliae* (Box 8.2) and *Paenibacillus* (*Bacillus*) *lentimorbus* (Dutky, 1963). Milky disease caused by *P. popilliae* or *P. lentimorbus* has subsequently been found infecting scarab larvae from a wide range of species distributed around the world (Klein and Jackson, 1992). The original differentiation of the two species was based on the presence of a parasporal crystal in *P. popilliae* that was absent in *P. lentimorbus*. Molecular typing confirmed the two species groupings, but indicated that differentiation on the basis of the parasporal body was inadequate as it can be present or absent in both species (Harrison et al., 2000). Molecular analysis also showed intraspecific variation which may be related to geographical isolation and host specificity (Rippere et al., 1999; Dingman et al., 2009). Another species from the same genus, *Paenibacillus larvae*, is notorious as the cause of American foul brood of honeybees, which has been distributed with domesticated honeybees to many parts of the world (James and Li, 2012).

Box 8.2 *Paenibacillus popilliae* – a true survivor

Paenibacillus popilliae (Bacillaceae) is the cause of milky disease among larvae of the beetle family Scarabaeidae. Spores are ingested from the soil by scarab larvae feeding on plant roots or other organic material. The spores germinate in the insect gut and release vegetative cells which enter the midgut epithelial cells and multiply before crossing the epithelial basal membrane and entering the hemocoel. Unrecognized by the insect humoral defenses, bacteria multiply until they have filled the body cavity, producing an ivory-white appearance in the infected larva. The disease has a long chronic phase in which the larvae continue feeding until the spore load per larva is maximized. As food resources are consumed, the bacteria begin to sporulate, with a spore and crystal produced in a resistant sporangium. *Paenibacillus popilliae* is considered an obligate pathogen as it is hard to germinate and grow *in vitro*. On death of the infected larva, resistant sporangia are released into the soil, where they will survive for many years. Milky disease, caused by *P. popilliae* and *P. lentimorbus*, has been recognized in scarab larvae throughout the world although isolated bacterial strains appear to show some level of host specificity. While milky-diseased larvae are not uncommon, epizootics are rare and seem to be associated with environmental conditions.

Soil samples (Monnerat et al., 1992, 2004), eutrophic ponds (Yousten et al., 1985), and insects (Kellen et al., 1965; Singer, 1973; Nishiwaki et al., 2007) have been the sources of entomopathogenic isolates of *Lysinibacillus sphaericus*. *L. sphaericus* is probably composed of a number of related species (Gomez-Garzon, 2016), which can be grouped by DNA homology (*L. sphaericus* DNA homology group IIA) but are not easily distinguished morphologically (Krych et al., 1980; Massie et al., 1985). A survey of 35 strains indicated that the toxic strains were almost clonal (Ge et al., 2011). Toxicity is most notable against mosquitoes, with highest activity against *Culex*, *Anopheles*, and *Mansonia* strains and variable toxicity to strains in the genera *Aedes* and *Ochlerotatus* (Berry, 2012). Activity against sandflies (*Phlebotomus* sp.) (Robert et al., 1997; Wahba 2000; Wermelinger et al., 2000) and other invertebrates has also been reported (Bone and Tinelli, 1987; Key and Scott, 1992). *L. sphaericus* group IIA strains may produce a range of toxins, including the Mtx toxins expressed during vegetative growth and, in the high-toxicity strains, spore-associated crystals of the BinA/BinB binary toxin and/or the Cry48/Cry49 toxin pair (Berry, 2012). The latter has a very limited target range and is only known to kill mosquitoes in the genus *Culex* (Jones et al., 2008).

Bacteria of the Enterobacteriaceae are frequently associated with dead insects, but their role in mortality is often questioned. Grimont et al. (1979) characterized *Serratia* spp. isolates from insects, but it was not until 1981 that a bacterium from the genus causing a consistent pathology was identified. This came from the soil-dwelling New Zealand grass grub, *Costelytra zealandica*. First tentatively identified as *Hafnia alvei* (Trought et al., 1982) and later defined as *Serratia marinorubra* (Stuki et al., 1984), the pathogenic strain that caused amber disease in *C. zealandica* larvae was finally identified as a new species, *Serratia entomophila* (Box 8.3) (Grimont et al., 1988). Pathogenic strains of *Serratia proteamaculans* were also isolated from diseased *C. zealandica*. More recently, another pathogenic species, *Yersinia entomophaga*, was isolated from a field-collected *C. zealandica* larva that had died in the laboratory (Hurst et al., 2011a).

Box 8.3 *Serratia entomophila* – a facultative specialist

"Facultative specialist" seems to be a contradiction in terms, but *Serratia entomophila*, the causative agent of amber disease in a single species of a soil-dwelling scarab beetle, provides an excellent example of this type of entomopathogen. *Serratia* spp. (Enterobacteriaceae) are ubiquitous soil bacteria capable of saprophytic growth in soil and cultivation on artificial media. Isolates of both *S. entomophila* and *S. proteamaculans* in New Zealand have been found to contain the pADAP plasmid, which confers pathogenicity against the endemic New Zealand grass grub (*Costelytra zealandica*). Free-living bacteria in the soil are ingested by feeding *C. zealandica* larvae and enter the midgut, where the Sep toxins and antifeeding prophages are released, leading to cessation of feeding and gut clearance by the infected larva, which is referred to as amber disease. Following a long period of chronic infection and multiplication of bacteria in the alimentary tract, bacteria break through the gut lining into the hemocoel, where they multiply, causing septicemia and death, before being released back into the soil. Amber disease builds up in *C. zealandica* populations in a delayed density-dependent manner and is commonly associated with *C. zealandica* populations in older pastures. The pathogenic strains can persist in the soil but decline over time in the absence of susceptible insect hosts. The disease can be dispersed from infected populations by flying adult beetles. *Costelytra zealandica* is the only insect species that has been found, to date, to be susceptible to the pADAP-bearing *Serratia* isolates.

The consistent cream-yellow coloration of *Galleria mellonella* larvae after infection by the entomopathogenic nematode *Steinernema* (*Neoaplectana*) *carpocapsae* led Thomas and Poinar (1965) to suspect bacterial infection, which was confirmed with the isolation of *Achromobacter nematophilus* (later *Xenorhabdus nematophila*) from infected larvae. The bacterium was a potential pathogen with little or no oral toxicity but which produced high levels of mortality after injection of just a few cells (Poinar and Thomas, 1965). The subsequent discovery and description of heterorhabditid nematodes (Poinar, 1976), which caused their insect hosts to turn red after infection, led to the isolation of the pigmented, bioluminescent bacterium described as *Xenorhabdus luminescens* (Thomas and Poinar, 1979) (later *Photorhabdus luminescens*; Boemare et al., 1993). More recent studies using molecular identification techniques have demonstrated the diversity among entomopathogenic nematodes of the genera *Steinernema* and *Heterorhabditis*, with the characterization of many distinct nematode species, each paired with a specific bacterium from the genera *Xenorhabdus* (for Steinernematidae) or *Photorhabdus* (for Heterorhabditidae) (Lewis and Clarke, 2012). The potential of other bacterial species to act as nematode symbionts has been debated, but the slug parasite *Phasmarhabditis hermaphrodita* could be cultured on a range of bacterial species (Wilson, 1995), some of which are known to include facultative pathogens with the ability to overcome insect host defenses. Recently, *Serratia nematodiphila* was found to be symbiotically associated with the nematode *Heterorhabditidoides chongmingensis* (Zang et al., 2009), and *Serratia* sp. SCBI has been identified as a symbiont of the free-living nematode *Caenorhabditis briggsae* (Abebe et al., 2011).

As already discussed, dead or diseased insects have been the key source of novel pathogens over the past century, but screening from soil and environmental sources has become an important strategy for the isolation of new strains and species. For example, scarab-active Asian strains of Bt were isolated from soil samples (Asano et al., 2003; Yu et al., 2006), and *Chromobacterium subtsugae* was isolated from forest soil by Martin et al. (2007). More recently, a mosquito active isolate, *Chromobacterium* (Csp_P), was isolated from the midgut of *Aedes aegypti* (Ramirez et al., 2014). However, screening of environmental samples does not always yield many new isolates of interest. When Meadow et al. (1992) sampled a British flour mill, they found Bt could be isolated from 72% of the samples, but that 55% of the nearly 500 Bt isolates collected in the study were not toxic to the lepidopteran and dipteran insects tested. Similarly, when Martin and Travers (1989) screened over 1000 soil samples collected from different environments worldwide, they recovered Bt from 70% of the samples, but this had no particular relationship to the presence of insects and 40% of the isolates showed no activity to either lepidopteran or dipteran insects. In contrast, mosquito active isolates of Bti and *L. sphaericus* have been frequently isolated from water and mud (Brownbridge and Margalit, 1986; Singer, 1981), and a new species with mosquito larvicidal activity, *Serratia nematodiphila*, was isolated from pond sediments (Patil et al., 2012). Soil has also been the source of *Streptomyces* spp. (Family Streptomycetaceae) that produce insecticidal toxins, including the avermectins. A number of secondary metabolites have been discovered that are active against a range of insects and nematodes (Ruiu, 2015). During a screening program for antibiotics, *Saccharopolyspora spinosa*, the source of spinosads, was isolated from soil in sugar mill waste (Mertz and Yao, 1990).

The frequent isolation of Bt from dead and diseased insects has led to the assumption that Bt is an insecticidal species, but this assumption may be an artifact of research directed toward diseased and dead insects (Côté, 2007). The widespread occurrence of Bt from samples without insects and the wide variety of toxins carried by the bacterium led Côté to conclude that Bt should be considered an environmental bacterium with some insecticidal properties. This issue will be discussed further later in the chapter. The variability in pathogenicity, both between isolates within a species and even within specific isolates, underlines the need for rigorous tests to prove Koch's postulates before an isolate can be declared pathogenic to a particular insect species. Given the genetic fluidity of microbes, pathogenicity is a property of strains, not species; this highlights the need for well-preserved reference isolates when claims of pathogenicity are made.

8.3   Pathogenicity and Virulence

Pathogenicity has been defined as the ability to cause disease, while virulence has been considered the disease-producing power of an organism (Chapter 1) (Kaya and Vega, 2012). Opportunistic pathogens (Boucias and Pendland, 1998) are those that are found in an insect's environment and that, given particular conditions, cause death of an insect host. Common environmental species, such as *Pseudomonas aeruginosa* and *Serratia marcescens*, have been frequently found causing death in insect colonies but are rarely encountered in natural populations (Bucher and Stevens, 1957; Bucher, 1963). Bulla et al. (1975) considered such opportunists "dubious" pathogens, which is reflected in the lack of attention they have received in intervening years. Bucher (1960) coined the term "potential pathogens" for those bacteria with limited invasive ability but strong capability to grow within an insect's hemocoel. An example is mortality of the Argentine stem weevil (Jackson and McNeill, 1998; Jackson et al., 2004) caused by septicemia after contamination of parasitoid wasp ovipositors with *S. marcescens*. The "potential pathogen" concept has been further demonstrated by the entomopathogenic nematode symbionts *Xenorhabdus* spp. and *Photorhabdus* spp., both of which proliferate in the hemocoel once delivered by the nematode host (Forst et al., 1997). Bucher (1960) considered that a potential pathogen should be able to produce mortality with a dose of $<10^4$ bacteria delivered to the hemocoel.

8.3.1 Pathogenicity

Bacterial pathogens must be ingested into the alimentary tract or vectored into the insect hemocoel to have any pathogenic effect on the insect host. Once inside the insect, bacteria are exposed to particular conditions of the insect gut or hemocoel that may activate the cell. This might be as dramatic as the lysis of the sporangium and spore germination in *P. popilliae*, the solubilization of the crystals of Bt to release protoxins (Jurat-Fuentes and Jackson, 2012), or the induction and release of Sep toxins by *S. entomophila* (Hurst et al., 2007a). The liberated cells or toxins degrade the insect gut cells, eventually allowing vegetative bacteria to enter the insect hemocoel and multiply in the nutrient-rich body cavity, causing septicemia and death of the host (Fig. 8.1).

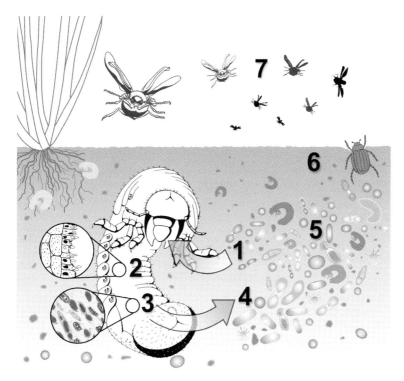

Fig 8.1 Disease cycle for a soil scarabaeid larva/bacterial infection. Bacteria are ingested from the soil while the insect feeds on roots and organic matter (1). Bacteria initiate pathogenesis in the gut (2) and invade the hemocoel, causing septicemia (3). Bacteria produced in the cadaver are liberated into the soil, where facultative organisms may multiply (4). Bacteria remain in stasis in the soil profile until ingested by a susceptible host (1) or dispersed by mobile host stages (6). Bacteria are introduced to new populations by adults seeking mates (7).

Box 8.4 *Rickettsiella* – the silent assassin

Rickettsiella spp. (Coxiellaceae) bacteria are known as pathogens of Coleoptera and Diptera and have been isolated from insects in many parts of the world. These minute bacteria must be ingested from the soil, but mechanisms of infection are not well known. The bacterium infects the hemocytes and fat bodies of the host larvae, eventually filling these organs with bacteria and associated crystals and causing death of the host. Diagnosis is difficult, but the occurrence in a wide number of species suggests that this may be an unrecognized "silent assassin" affecting populations of many insects.

Another important distinction for our discussion of the ecology of bacterial entomopathogens is that between facultative and obligate pathogen forms. Facultative pathogens are easily cultured in the laboratory and can be mass produced (Jackson, 2017). They can infect and multiply within the host insect, but can also grow readily in the environment, given adequate nutrients. Most entomopathogenic bacteria are facultative pathogens, including all strains of Bt and *L. sphaericus*, as well as members of the Enterobacteriaceae, such as *Serratia* spp. Their ease of growth on artificial media raises the intriguing question of their dynamics in the environment beyond the host and the relative importance of pathogenic and saprophytic growth phases. Relatively few of the bacterial entomopathogens are considered obligate pathogens of insects; these multiply only within the host insect, and as such will not readily grow on artificial media (e.g., *P. popilliae*). The Coxiellaceae (Genus *Rickettsiella*; Box 8.4) are also obligate pathogens. Failure to utilize environmental nutrients seems at first to be a disadvantage, but in combination with survival and transmission strategies, IT avoids the waste of pathogenic propagules in the absence of a susceptible host.

Pathogenesis has been described in detail for some host–pathogen interactions. Milky disease of the Scarabaeidae is initiated after spores of *P. popilliae* are ingested from the soil by feeding larvae. The spores germinate and the sporangium bursts in the high-pH and enzyme-rich conditions of the midgut to release vegetative rods which enter the midgut cells by phagocytosis (Splittstoesser et al., 1978). Multiplication occurs in the regenerative nidi of the midgut before the bacterial cells penetrate the basal membrane of the gut and enter the hemocoel where they continue to multiply (Kawanishi et al., 1978). Once in the hemolymph of a susceptible scarab larva, vegetative cells of *P. popilliae* establish a chronic infection, which may last for several weeks, replicating through several cycles without triggering a hemocytic reaction. By the end of the infection, the insect takes on an ivory-white appearance due to the high density of sporulated cells within the hemocoel and body tissues of the larvae, before eventual death of the infected insect and release of about 10^9 cells to the soil (Klein and Jackson, 1992). Milky disease is also produced in scarab larvae by *P. lentimorbus*, which often lacks the parasporal crystal found in *P. popilliae*. Infection by *P. larvae*, the cause of American foul brood, will likewise fill infected honeybee larvae with up to 7.5×10^8 spores per larva (James, 2011; James and Li, 2012). Similarly, *Rickettsiella popilliae* multiplies through a chronic infection to produce enormous numbers of minute cells within the fat body lobes and hemocytes of an infected insect, filling it to such an extent that the insect takes on a white or blue appearance (Krieg, 1963). *Serratia entomophila* also establishes a chronic infection after ingestion by its host, *C. zealandica*, but this is confined for several weeks to the alimentary tract (Jackson et al., 2001). Once in the gut, the ingested bacteria release Sep proteins and antifeeding prophages (Hurst et al., 2007a,b) which cause the infected insect to cease feeding and void its midgut of organic material (Jackson et al., 1993, 2001). Some of the ingested bacteria adhere to the cuticular lining of the gut wall (Wilson et al., 1992; Jackson et al., 1993), but most grow through the alimentary tract, adhering to food particles (Hurst and Jackson, 2002), where the population reaches 10^6–10^7 bacteria per larva (Jackson et al., 2001). The Sep toxins interfere with enzyme regulation and degrade the midgut cytoskeletal network (Gatehouse et al., 2008; Marshall et al., 2012) and after a significant period without feeding, the larva is weakened and the hemocoel is breached. The bacteria continue to multiply in the cadaver, where populations can reach 10^8–10^9 cells per larva.

Disease progression in insects infected with Bt is more rapid; this has been summarized by Jurat-Fuentes and Jackson (2012), Glare et al. (2017), and others. Briefly, spores and crystals are ingested by feeding larvae and enter the alimentary tract where the spores germinate and the crystal toxin proteins are released, causing cessation of feeding and paralysis, at least of the midgut. The breakdown of the midgut epithelial barrier allows invasion of the hemocoel by midgut bacteria, leading to a fatal septicemia in the infected insect. Cessation of feeding occurs shortly after ingestion of a pathogenic dose and death follows within 2–3 days. Angus and Heimpel (1959) defined three host types based on the characteristics of infection. Host types I and II experience paralysis and death after ingestion of only the crystal toxin, but for type III hosts both spores and crystals are required to produce mortality. Exposure to pure toxin or toxins expressed through plant tissue will obviously not lead to multiplication of the Bt cell, but for crystal/spore products it is a different story. Spores in the midgut germinate to release vegetative cells which multiply in the midgut and take advantage of the deterioration of the midgut epithelial barrier to enter the hemocoel and multiply in the hemolymph and body tissues. On consumption of all the suitable nutrients provided by the cadaver, the vegetative cells sporulate, and spores are liberated to the environment on decay of the cadaver. Unlike *P. popilliae*, which has the ability to pack the host with a consistently high density of spores prior to death and release from the cadaver, Bt is less successful in reproduction within the host insect. For example, Takatsuka and Kunimi (1998) found that Bt serovar *aizawai* grew rapidly in laboratory-treated *E. kühniella* larvae, producing a relatively low yield of 1×10^5 colony-forming units (CFUs) per larva. Raymond et al. (2008a) reported low reproduction of cells in cadavers of *Plutella xylostella* killed by Bt. Furthermore, Takatsuka and Kunimi (2000) showed that poor growth and sporulation of Bt in insect cadavers resulted from inhibition by competitive gut bacterial flora. The effect of bacterial competition on Bt multiplication was confirmed by Raymond et al. (2007), who reported that the Bt yield from *P. xylostella* larvae killed after ingestion of Btk was reduced when a competitive strain of *B. cereus* was also introduced. In a further study, Raymond et al. (2008b) found that co-infection with antibiotic-producing strains of *B. cereus* increased the production of Bt in the cadaver by reducing the abundance of the commensal gut flora. Contrary to the negative impact of competitive gut flora on Bt multiplication, it was suggested that the insecticidal activity of Bt might be dependent on the presence of other gut flora (Broderick et al., 2006), but with closer scrutiny it has been concluded that other bacterial species are not essential (Johnston and Crickmore, 2009; Raymond et al., 2009b; van Frankenhuyzen et al., 2010), although they may contribute to larval mortality (Broderick et al., 2009).

The spore-forming bacterium *L. sphaericus* is primarily of interest as a pathogen of mosquitoes (Fam. Culicidae). This species has a worldwide distribution and can be frequently isolated from soil and soil-aquatic systems. Highly toxic strains contain a parasporal crystal, which is released after ingestion of the spores by feeding mosquito larvae. The *L. sphaericus* crystals may contain the Cry48 and Cry49 two-component toxin (Jones et al., 2008) and/or, more commonly, the binary toxin composed of BinA and BinB proteins (Berry, 2012). The Bin toxin causes swelling of the midgut, paralysis, and degradation of tissues until the vegetative cells penetrate the hemolymph, causing septicemia and death. *Lycinibacillus sphaericus* Group IIA strains produce vegetative toxins (Mtx1, Mtx2, Mtx3, Mtx4) that may play a role in enhancing toxicity once spores have germinated. Interestingly, all *L. sphaericus* genomes reported from all DNA homology groups contain a gene encoding sphaericolysin, a cholesterol-dependent cytolysin first reported from strain A3-2 (Nishiwaki et al., 2007) with injection toxicity against *Blatella germanica* and *Spodoptera litura*. This toxin is part of a highly conserved toxin family that is present in many Gram-positive entomopathogens, including Bt and *P. alvei* (Silva-Filha et al., 2014).

Yersinia entomophaga (Hurst et al., 2011b) also has an acute impact on infected larvae. As little as 4 hours after ingestion of a toxic dose, treated larvae began to convulse, vomit, and expel frass (Hurst et al., 2014), taking on a temporary amber appearance before lapsing into a moribund state as bacteria break through into the hemocoel, which results in death within 48 hours. Bacterial quantification indicated 9×10^9 CFU of *Y. entomophaga* per cadaver of *C. zealandica* larvae within 72 hours of infection. Similar rapid death of the host follows nematode-aided inoculation of *Photorhabdus* or *Xenorhabdus* cells, which release a range of cytolysins/hemolycins and toxins to overcome the host's defenses, allowing the bacteria to proliferate and transform the cadaver into nutrients for the invading nematodes (Nielsen-LeRoux et al., 2012).

As originally noted by Bucher (1963), entomopathogenic bacteria can be divided into pathotypes – groups of organisms, usually within a microbial species, which will produce the same symptoms in the host. The concept of the pathotype was applied to Bt by Krieg (1983) to differentiate between lepidopteran, dipteran, and coleopteran active isolates. Underpinning our definition of pathotype is the specificity of the bacterium to a species or group of insects. *Serratia entomophila* has only been shown to be infectious to a single species of insect, *C. zealandica* (Jackson, 2007). But, interestingly, *Yersinia entomophaga*, another bacterium containing similar Tc toxins that is isolated from the same host in New Zealand, has a much wider host range (Hurst et al., 2011a). The coleopteran-active Bt strains also show a high degree of specificity, but this is not necessarily associated with the host of origin. *Bacillus thuringiensis tenebrionis* (Btt), originally isolated from the stored product pest *T. molitor*, was found to be highly effective against chrysomelid beetles but showed little or no activity against lepidopteran and dipteran targets (Krieg, 1983). Asano et al. (2003) showed high activity of Btg (strain SDS-502) from Japanese soil against ruteline scarabs from Japan (Asano et al., 2003), but, conversely, Bt strain 185 from China showed a high activity against the native melolonthine scarab *Holotrichia parallela* but no toxicity to the ruteline *Anomala cuprea* or a range of other test insects (Yu et al., 2006).

As demonstrated by Bt, pathogenicity will be determined by the genes or combination of genes carried by a particular isolate (Jurat-Fuentes and Jackson, 2012). In bacteria, many pathogenicity genes are carried on plasmids, which can be transferred by conjugation between bacterial cells. In the laboratory, Glare et al. (1996) were able to transfer the pADAP plasmid, which encodes the genetic determinants of amber disease (Hurst et al., 2000), between isolates of *Serratia* strains and species, and with it they transferred the ability to cause amber disease in *C. zealandica*. Orthologues of the *sep* genes have also been found on plasmids within strains of *Serratia liquefaciens* and *Yersinia fredericksenii*, also isolated from but nonpathogenic to *C. zealandica* larvae (Dodd et al., 2006), suggesting the *sep* genes are part of a horizontally mobile region. In addition, preliminary experiments have demonstrated transfer of pADAP by conjugation within the *C. zealandica* larvae (M. O'Callaghan, unpublished). Studies of Bt strains isolated from leaves of the same plant showed a wide variation in genomic markers (Collier et al., 2005; Bizzarri et al., 2008) and the pattern of plasmids in these isolates suggests that conjugal transfer is a common event in this species (Bizzarri and Bishop, 2008).

8.3.2 Virulence

Virulence is considered a measurable characteristic of the ability of the microbe to cause disease and is intended for within-group or within-species pathogen comparisons (Chapter 1) (Kaya and Vega, 2012). For a particular bacterium/host combination, virulence is often determined by standardized bioassay (O'Callaghan et al., 2012). Two common measures of virulence are the dose response (LD_{50}) and the time to kill (LT_{50}) (Shapiro-Ilan et al., 2012). The dose response will have the greatest effect on the ecology of entomopathogens, as a low infective dose will provide the opportunity to maximize multiplication of the invasive microbe. The LD_{50} for the enterobacterial pathogens *Serratia* and *Yersinia* against scarab larvae is about 10^4 ingested bacteria (Jackson et al., 2001; Hurst et al., 2014), which can allow subsequent $10\,000\times$ multiplication in the cadaver of a diseased insect. A low LD_{50} toward a particular host will provide an advantage to pathogenic strains; Kreig (1987) reported as much as a 4 log difference in toxicity to mosquito larvae among strains of *L. sphaericus*. However, strains of pathogenic *Serratia* spp. showed little variation in virulence toward *C. zealandica* larvae (O'Callaghan et al., 1992), suggesting that other factors such as environmental persistence may have been more important in the evolution of the relationship. Milner (1981) reported a high LD_{50} of 10^7 cells for infection of scarab larvae by *P. popilliae*, which minimizes the opportunity for multiplication in the host, but it is possible that insect stress or environmental factors can increase the apparent virulence of these bacteria. Low field efficacy of applied *P. popilliae* led Redmond and Potter (1995) to suggest that the bacterium had lost virulence, but the continued occurrence of field infections indicated that other environmental factors were at play. In addition to the antagonistic competition from gut bacteria, which can moderate virulence as discussed earlier, Raymond et al. (2008a) found that maximum multiplication in the cadaver occurred after dosing with intermediate densities of Bt.

The time taken to cause death in the host insect (LT_{50}) will have a lesser effect on bacterial multiplication than the absolute dose, but will be important in allowing the infective agent to recycle between hosts. Bacteria released rapidly from a decaying cadaver will be available to infect healthy, susceptible insects in the population, while pathogens causing chronic diseases may be held with the infected insect for a long period and thus be unavailable to cause fresh infections.

8.4   Disease Transmission

Transmission, the journey of the infective propagule to encounter a new susceptible host, is one of the key factors determining successful continuance of a bacterial entomopathogen. Most bacteria infect their hosts through horizontal transmission. Entomopathogenic cells ingested during insect feeding cause infection, the pathogen multiples within the host, and new propagules are released to the environment to infect more hosts. There is little evidence to suggest transovarian (within the egg) vertical transmission of bacterial pathogens, although transovum passage (the contamination of egg batches, e.g., on the surfaces of eggs) by infected females cannot be totally discounted. Transmission will be favored at high insect densities, and it is not surprising that bacterial pathogens are frequently found in laboratory colonies of insects (e.g. Konecka et al., 2007) which must be kept scrupulously clean or treated with antibiotics to avoid bacterial-induced colony collapse.

In the natural habitat, where insect hosts are more dispersed, transmission is more problematic, and entomopathogenic bacteria have developed a number of strategies to meet the challenge. The non-spore-forming bacterial nematode symbionts, *Xenorhabdus* and *Photorhabdus* spp., are fastidious and are considered poor survivors in the environment (Poinar and Thomas, 1967), but they have solved the problem of transmission by engaging in transport between hosts in the intestine of the free-living stage of the host nematode where they are protected from environmental challenges (Chapter 11). On death of a diseased soil-dwelling scarab host, non-spore-forming *Serratia* spp. are released into the soil to be ingested by feeding larvae, allowing a further cycle of disease. Disease is more common in high-density *C. zealandica* populations, as the proximity of hosts favors transmission. However, pathogenic strains do not persist in the soil in the absence of hosts and, to establish new foci of infection, it appears that *S. entomophila* is dispersed from infected sites by adult beetles during the flight season (O'Callaghan and Jackson, 1993a), which explains the dispersal of bacteria and the colonization of untreated areas by pathogenic bacteria after the flight season (Zydenbos et al., 2016). Adult beetles are also a possible mechanism for dispersal of *Rickettsiella melolonthae* (Hurpin and Robert, 1977).

While Bt is considered a common environmental bacterium as well as an insect pathogen, evidence of horizontal transmission in field infections is sparse. Takatsuka and Kunimi (1998) concluded that Bt serovar *aizawai* produced in *E. kühniella* larvae could not be transmitted horizontally in laboratory tests, but Raymond et al. (2010a) demonstrated transmission from dead to live *P. xylostella* larvae on leaf surfaces. However, spores and toxin crystals are only transient on the leaf surfaces, and initiation of a fresh cycle of infections among insects feeding on the leaves will require a further input of spores and crystals as inocula. Jensen et al. (2003) speculated that Bt and other members of the *B. cereus* group persist in a reservoir through nonpathogenic colonization of invertebrate guts. Raymond et al. (2010b) considered bacterial recolonization of plant material through contamination or endophytic transport to the leaves, but raised the problem that both crystals and spores are needed to achieve toxicity and replication. While infection of hosts by naturally occurring Bt is obviously complex and a rare event, the ubiquitous distribution of Bt in soil could be explained by the persistence of the spores for very long periods. A rare combination of events – susceptible insects, optimal temperature and nutrients – could provide conditions for germination and

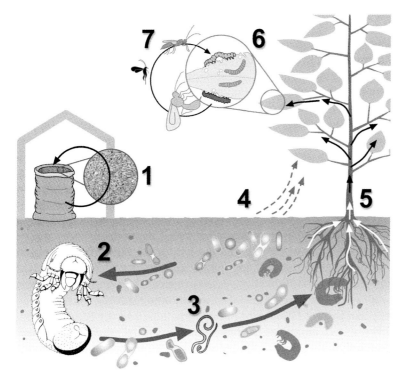

Fig 8.2 Range of different methods by which bacterial entomopathogens can be transmitted. (1) Bacterial cycling within grain stores. (2) Horizontal transmission between insects in soil. (3) Transport by nematodes between hosts. (4) Bacteria from the soil reservoir deposited on fresh foliage. (5) Transmission through plants as endophytes (putative). (6) Horizontal transmission on the phylloplane. (7) Transmission by parasitoids.

multiplication leading to a pathogenic spore/crystal mix in the environment and a cycle of disease. The scarab pathogen *P. popilliae* appears to have carried the "resistant spore strategy" to the extreme; its ultradormant spore (Milner, 1977) enables the bacterium to persist in favorable scarab habitats for many years between cycles of the host. Mechanisms of entomopathogenic bacterial transmission are summarized in Fig. 8.2.

8.5 Survival in the Environment

The ability of bacterial entomopathogens to persist in the environment in the absence of their host contributes to the successful transmission of disease, especially when hosts are transient. The abiotic factors generally considered most important in persistence of bacterial entomopathogens include sunlight, temperature, humidity/water, and chemicals, influenced by the substrate-foliage, soil or water. In addition to abiotic stresses, persistence of bacterial entomopathogens in the environment will also be impacted by a range of biotic factors, including competition with other microbes for nutrients and niches, predation by protozoa and nematodes, and adverse specific interactions with bacteriophages.

8.5.1 Soil

The occurrence of Bt in most of the 350 soil samples examined by Martin and Travers (1989) and its status as a facultative pathogen capable of growing on a wide range of nutrients have suggested to some that it is primarily an environmental bacterium (Côté, 2007). But others (Raymond et al., 2010a) have suggested that the soil is a "sump" for spores, not an active growth environment. While spores can survive for extended periods in the soil (Glare and O'Callaghan, 2000), West et al. (1985a) found that Bt would not grow under most natural soil conditions, and, while growth could be stimulated by autoclaving the soil or adding nutrients, it was not suited to multiplication in natural soil. Furthermore, West (1984) found that the Bt parasporal protein crystals were rapidly broken down after application to arable soils, with a concurrent loss in insecticidal activity. Successful horizontal transmission with infection via spores ingested from the soil is challenging as Bt spores will have to germinate, undergo a cycle of vegetative growth, and then sporulate to produce the parasporal crystal with Cry toxins pathogenic to the insect host (Raymond et al., 2010a). The difficulties posed by horizontal transmission from the soil have led to proposals for plant transfer (see later). Transmission from the soil pool seems less problematic for Bt Type III hosts, where the spore and crystal are needed for infection, and interestingly, for the scarab beetles described earlier, these relationships between the host and the Bt strain seem more specific. Transmission from the soil microbial pool is certainly the mode of infection for scarab larvae and milky disease. Larvae feeding on plant roots and soil organic matter ingest *P. popilliae* spores which only germinate in the host insect gut after a combination of activating conditions. In the soil, the spore is dormant and appears to persist for many years (Klein, 1992). Stahley et al. (1992) found that only about 1% of *P. popilliae* spores would germinate on a selective medium. The difficulty of breaking dormancy was demonstrated by Milner (1977) who used a series of seven heat shocks (70 °C for 20 minutes) to break the dormancy of *P. popilliae* var. *rhopaea* before isolating vegetative cells on selective media. However, peroral injection of spores into the alimentary tract (Dingman, 1996) and direct injection into the hemocoel (Koppenhöfer et al., 2012) both produce infections in the larvae, indicating the influence of host factors in spore germination. Long dormancy broken only by the specific conditions in the insect gut seems to benefit *P. popilliae* in its association with a transient host that may pass many years between occupations of specific sites. Perhaps the lifecycle of Bt is similar, and for these sporeformers transmission is just a waiting game. A genetic link between activation and pathogenicity has been demonstrated in Bti, where genes encoding an increased activation to germinate in alkaline conditions have been found encoded on the same plasmid as the mosquitocidal toxins (Abdoarrahem, 2009). Since the midgut of mosquito larvae is alkaline (Dadd, 1975), this may represent an adaptation to end dormancy on encountering a potential host environment.

The ubiquitous soil bacteria *Serratia* spp. (Grimont and Grimont, 1978) are easily cultured so that the ecology of insect pathogenic forms can be readily studied. A range of microbiological, biochemical, and molecular methods (O'Callaghan and Jackson, 1993b; Jackson et al., 1997; Monk et al., 2010) have been used to study the ecology of these facultative pathogens, *S. entomophila* and *S. proteamaculans* (the causative agents of amber disease of *C. zealandica*), in soil. Both pathogenic and nonpathogenic strains of each species can be readily isolated from New Zealand pasture soils. Surveys of pasture soils indicated that *C. zealandica* pathogenic *Serratia* were generally absent from young pastures and sites where the host was absent (O'Callaghan et al., 1999). *Serratia proteamaculans* appears to be more widespread, with the presence of populations of *S. entomophila* tending to be more closely linked with populations of the host. Where present, *Serratia* populations averaged 5×10^4 CFU/g soil, but the distribution within pastures was uneven and populations as high as 10^7 CFU/g soil were recorded in locations with high *C. zealandica* density. While not studied intensively, population levels vary with season, with low to nondetectable levels found in months when the host species is in pupal, adult, and egg stages (August–March), increasing as larvae become infected with the disease (May–July) (M. O'Callaghan, pers. obs.). As already discussed, a diseased larva may contain as many as 10^8 *Serratia* cells (M. O'Callaghan, unpublished data), which are released into soil at the time of death, and a strong correlation was found between numbers of pathogenic *Serratia* in soil during June–July and prevalence of amber disease in *C. zealandica* (O'Callaghan et al., 1999). The density of diseased larvae will, to a large extent, determine the number of bacteria available to cause disease in the subsequent season (Barlow, 1999). As the disease causes the host population to decline, the production of pathogenic bacteria will be reduced, allowing the host population to recover in a delayed density-dependent oscillation until density is such that transmission is favored and a further epizootic occurs (Jackson, 1993). A disease cycle can be disrupted by the application of an insecticide. In a study comparing the efficacy of a chemical insecticide (diazinon) with *S. entomophila* for long-term management of *C. zealandica*, insecticide treatment caused a rapid decline in larval numbers (Zydenbos et al., 2016). However, the insect population resurged to high, damaging levels 2 years after treatment in the absence of naturally occurring amber disease and with low levels of bacteria in the soil. In contrast, application of bacteria established a pathogenic *S. entomophila* population which was maintained over several years in the residual grass grub population and which also spread to control plots such that larval populations were maintained below damaging levels.

The seasonal increase in soil populations and the maintenance of *Serratia* populations across years may also be the result of saprophytic growth. Nonpathogenic modes of reproduction have been reported in other entomopathogens (in particular, fungi) and are common for many disease-causing bacteria in mammals (e.g., *E. coli*). *Serratia* spp. are capable of rapid growth in soil in the absence of competition with other microorganisms (e.g., in sterile soil) and in response to the addition of nutrients to soil (O'Callaghan et al., 1988). Godfray et al. (1999) illustrated the benefit of a saprophytic phase in *Serratia–C. zealandica* population dynamics through modeling. However, both nonpathogenic and pathogenic strains of *Serratia* are found in New Zealand pasture soils where both can coexist. Nonpathogenic strains are widespread, while pathogenic strains are most commonly found in localized areas where grass grub populations are present. *In vitro* growth experiments show that nonpathogenic strains have a more rapid growth rate (M. O'Callaghan, unpublished data), which is consistent with other studies demonstrating the metabolic burden or cost of maintaining a plasmid within a bacterial population (e.g., Dahlberg and Chao 2003). However, the presence of the host insect provides an advantage to the pathogenic strains, which have often been detected from larvae in soil when none can be recovered from the surrounding soil. The larva provides an ideal environment for multiplication and persistence as it provides nutrition and refuge from environmental stresses prevalent in soil (e.g., drying–rewetting cycles, temperature fluctuations, etc.). Within the larva, the pathogenic strain faces competition only from a limited gut microflora, whereas in soil, it must compete with the more numerous, diverse, and highly competitive soil microflora. Once released from the dead larva, populations of pathogenic *Serratia* strains decline to low numbers in the face of environmental stresses and biotic factors such as competition.

As non-spore-forming bacteria, *Serratia* strains are highly susceptible to environmental stresses; in laboratory experiments, populations declined rapidly in dry soils, with the decline becoming more rapid as the soil temperature increased from 10 to 20 °C (O'Callaghan et al., 2001). The rates of decline observed in these experiments mirrored those seen in the field following application of the bacterium for grass grub control. The impact of desiccation stress on the *Serratia* population (and on other grass grub pathogens) in soil can also be inferred from field observations, where grass grub outbreaks in pasture commonly occur after dry summers (Barlow and Jackson, 1998).

8.5.2 Aqueous Environments

The bacteria *L. sphaericus* and Bti are associated with mosquitoes and blackflies in aqueous environments. The persistence of *L. sphaericus* in aquatic environments appears to be higher than that of Bti, possibly due to its ability to recycle through insect hosts (Nicolas et al., 1987; Correa and Yousten, 1995), but also due to its lower tendency to adhere to particulates that are lost through sedimentation (Yousten et al., 1992). Lower adherence to particulates may also be the reason that *L. sphaericus* is better able to persist than Bti in polluted water (Silapanuntakul et al., 1983; Nicolas et al., 1987). The environments within which *L. sphaericus* can grow may be limited as the bacterium is unable to ferment most sugars (White and Lotay 1980; Massie et al., 1985) because it lacks sugar transporters and the metabolic enzymes required for fermentation (Han et al., 2007; Hu et al., 2008). Ingestion of spores by non-target species showed no evidence that the bacteria had actively reproduced in midge larvae, snails, or oysters, but viable spores could be recovered from these organisms, indicating their persistence (Yousten et al., 1991).

8.5.3 On the Phylloplane and *In Planta*

Survival of Bt on the phylloplane is much shorter than in the soil, as indicated by the need for regular sprays of Bt to protect plant foliage from insect pest feeding. Glare and O'Callaghan (2000) reviewed application data and found that Bt survived on the foliage for a few weeks, at best. Maduell et al. (2008) found that Bt grows poorly on leaves, although it is more resistant to desiccation than more common epiphytic bacteria. Lack of growth in the phylloplane and poor persistence pose a problem for horizontal transmission as spores (and parasporal crystals) must contaminate the leaf surface for a sufficient period to allow an infective dose to be ingested by feeding larvae. Raymond et al. (2010a) showed that horizontal transmission could occur when a decaying cadaver was added to a leaf with healthy feeding larvae. Presumably, transmission can also occur by recycling of spores after application of sprays, but initiation of infection from environmental sources remains problematic. One possible solution is that Bt is transmitted *in planta* as Bt strains have been identified as endophytes within plant tissues (McInroy and Kloepper, 1995). Further studies indicated that Bt placed on the soil might be re-isolated from leaf surfaces (Maduell et al., 2007; Bizzarri and Bishop, 2008; Monnerat et al., 2009; Raymond et al., 2010b; Botelho et al., 2012) and this appears to be mediated via endophytic migration through the xylem (Monnerat et al., 2009). Vidal-Quist et al. (2013) showed that the ability to colonize *Arabidopsis* roots was related to the phylogeny of the Bt strains used. In experiments examining relationships between bacteria, insects, and plants, Raymond et al. (2010b) found that isolates from a single pathogenic genotype (Bt ST8) were better than nonpathogenic genotypes in the endophytic and epiphytic colonization of seedlings from soil. Furthermore, the presence of insects nearly doubled the density of bacterial spores (Bt/*B. cereus*) in the soil and the proportion of strains (isolates) expressing insecticidal toxins, indicating that the applied Bt was acting as a true pathogen and multiplying in the cadavers of infected insects and subsequently accumulating in the soil (Raymond et al., 2010b). However, given the rarity of cadavers and the transient low numbers of Bt on the plant surfaces, the authors concluded that leaves were colonized from the soil reservoir rather than through host-to-host transmission on the leaves. Thus, while the potential for Bt to be endophytic has been established, the ability of strains to colonize plants is variable (Vidal-Quist et al., 2013) and the impact of endophytic Bt on feeding insects is yet to be proven.

The putative benefits to plants from an association with Bt may be derived from the uptake of insecticidal toxins, but Bt may also afford plants a degree of protection from other pathogens. Many plant pathogens can increase their virulence through a quorum-sensing mechanism mediated by the level of N-acyl homoserine lactone. Several Bt strains, however, have been shown to produce AiiA, a quorum-quenching enzyme that degrades this signal (Lee et al., 2002). The role of this protein in attenuating the virulence of *Erwinia carotovora* and maintaining the rhizosphere population of Bt has been demonstrated (Park et al., 2008) and incorporation of the gene encoding AiiA into the plant *Amorphophallus konjac* increases its resistance to soft rot caused by *E. carotovora* (Ban, 2009). Like *B. cereus* strains, Bt may produce the antibiotic Zwittermicin A, which has activity against plant pathogenic fungi and some activity against Gram-negative and Gram-positive bacteria (Silo-Suh, 1998; Kevany et al., 2009). Another Bt in the rhizosphere has been shown to produce bacteriocins and to act as a soya root-nodulating agent and growth promoter when co-inoculated with *Bradyrhizobium japonicum* (Bai et al., 2002), indicating yet another way in which Bt strains might be beneficial to plants and maintain their presence in the environment.

8.6 Population Dynamics: Epizootics and Enzootics

Despite the frequent isolation of bacteria from dead and diseased insects, reports of epizootics of bacteria in insect populations are rare (some recorded epizootics are listed in Table 8.2). Fuxa and Tanada (1987) discussed the terminology in relation to medical epidemiology and defined an epizootic as "an unusually large number of cases of disease" (see also Chapter 1). However, as has been frequently noted, the term is not well defined, although an epizootic is generally considered to occur when a significant proportion of the population is infected concurrently with a disease-causing agent. Epizootics, like epidemics, are characterized by changes in disease prevalence over time and often occur in insect populations in a delayed density-dependent manner

Table 8.2 Some reported epizootics of bacterial diseases.

Bacterium	Host	References
Coccobacillus acridiorum	*Schistocerca pallens*	d'Herelle (1911)
Paenibacillus popilliae	*Popillia japonica*	White (1941)
Rickettsiella melolonthae	*Melolontha* spp.	Niklas (1957–69) (in Jackson and Glare, 1992)
Bacillus thuringiensis	*Ephestia* spp.	Vankova and Purrini (1979)
Paenibacillus popilliae	*Costelytra zealandica*	Wigley and East (1985)
Paenibacillus popilliae	*Cyclocephala parallela*	Boucias et al. (1986)
Rickettsiella popilliae	*Tipula paludosa*	Krieg (1987)
Bacillus thuringiensis var. *israelensis*	*Culex pipiens*	Margalit (1990)
Paenibacillus popilliae	*Amphimallon solstitialis*	Glare et al. (1993)
Serratia entomophila	*Costelytra zealandica*	Jackson (1990)
Paenibacillus popilliae	*Cyclocephala hirta*	Kaya et al. (1993)
Serratia entomophila	*Costelytra zealandica*	O'Callaghan et al. (1999)
Bacillus thuringiensis	*Mythimna loreyi*	Porcar and Caballero (2000)
Bacillus thuringiensis var. *aizawai*	*Plodia interpunctella* (laboratory cultures)	Shojaaddini et al. (2012)
Bacillus thuringiensis	*Cydia pomonella* (laboratory cultures)	Konecka et al. (2007)

(Barlow, 1999) causing collapses in the host population (Fig. 8.3). A disease may move from an epizootic to an enzootic phase with a low disease prevalence that persists over time.

Epizootics of bacterial disease were first recorded by Ishiwata and Berliner (see Beegle and Yamamoto, 1992) in artificial conditions of silkworm rearing and grain stores, as noted earlier. The first recorded field epizootic was encountered by d'Herelle (1911) when he found a bacterial disease among swarms of the locust *Schistocerca pallens* arriving in the Yucutan from Guatemala. He noted dysentery and death among locusts and found he could isolate the causative bacterium (which he identified as *Coccobacillus acridiorum*) and induce epizootics in healthy populations by applying the bacterium to foliage in advance of the progressing swarm. He reported positive results and carried out similar studies in Argentina and Tunisia (d'Herelle, 1912, 1914).

As the invasion of the Japanese beetle spread in the eastern United States, White (1941) described natural epizootics of milky disease of Japanese beetle caused by *P. popilliae* at three sites in New Jersey. Milky disease levels reached 67% of larvae infected and White noted the importance of disease in early-instar larvae and the rise in disease prevalence as temperatures rose from spring to summer. An epizootic of milky disease was reported by Boucias et al. (1986) among larvae of the scarab *Cyclocephala parallela* in Florida sugarcane fields, with prevalence rising from 16 to

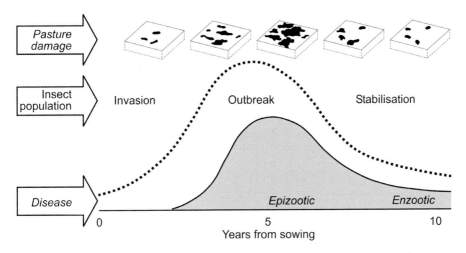

Fig 8.3 Phases of amber disease during an outbreak of grass grub (*Costelytra zealandica*) in newly sown pastures in New Zealand. The grass grub outbreak grows to a peak after 4–5 years, with densities rising rapidly in the absence of disease. Disease enters the high-density insect population, causing an epizootic and population crash. Disease next enters an enzootic phase, where it persists within the population, causing death and a lower population density. Pasture damage from root feeding increases as the population grows but is reduced to acceptable levels with the crash in the population.

30% as the season progressed. East and Wigley (1985) concluded that an epizootic of milky disease was responsible for the collapse of a population of *C. zealandica* in New Zealand pasture during summer. Milky disease more commonly occurs at enzootic levels in established scarab populations. In a survey of 115 grassland pastures infested with *C. zealandica* in Canterbury, New Zealand, Jackson (1990) reported that milky disease was present in 16%, with a mean prevalence of 4% infection. In Europe, Hurpin and Robert (1972) reported a low level of milky disease in *Melolontha melolontha*, which peaked in the summer of the second year of the insect's 3-year lifecycle; these authors, however, considered the disease to be of minor importance in suppression of the pest. Also in summer, Glare et al. (1993) reported 4.6% milky disease in a population of *Amphimallon solstitialis* in Darmstadt, Germany. Kaya et al. (1993) encountered a greater impact of milky disease in *Cyclocephala hirta* in California where they reported infection levels of up to 10%.

Despite its frequent isolation from dead and diseased insects, Bt has rarely been associated with epizootics of disease beyond stored products. Burges and Hurst (1976) found that spot applications of Bt to the surface of grain could induce epizootics of disease in larvae of *Plodia interpunctella* but concluded that natural disease could only curb infestations in grain debris. Vankova and Purrini (1979) isolated Bt associated with natural epizootics of disease in Lepidoptera in old watermills in Yugoslavia. Krieg (1987) considered that epizootics of Bt-induced disease in grain stores resulted from the behavior of the infected larvae, which accumulate at the surface of the grain substrate prior to death. This is exactly where the moths lay their eggs, and the emergent larvae are susceptible and provide a fresh substrate for growth of the bacterium. Detection of Bt-causing epizootics in field crop pests is extremely rare. A natural epizootic of Bt infecting the lepidopteran *Mythimna loreyi* was reported by Porcar and Caballero

(2000), who isolated Bt serovar *aizawai* from the dead larvae. However, the presence of Bt on plant foliage appears to be quite common (Damgaard et al., 1997), although presumably not in sufficient quantities to cause insect infections.

The association of Bti and dipteran pests may be more direct. Bti was isolated from a stagnant pond where "a high concentration of dead and dying *Culex pipiens* larvae in an epizootic situation was found as a thick grayish white carpet on the surface of the water body" (Margalit, 1990). Bti has subsequently been frequently recovered from mosquito- and blackfly-inhabited pools, suggesting the bacterium is produced in these habitats through epizootics (Brownbridge and Margalit, 1987). Infections of *Rickettsiella* can be recognized in advanced stages by the white or blue appearance of the infected larvae. Niklas (in Jackson and Glare, 1992) found large numbers of *Melolontha* spp. larvae on the soil surface in German forests following epizootics of "blue disease" caused by *Rickettsiella melolonthae*. *Rickettsiella pyronotae* has also been found in New Zealand soils, causing epizootics in larvae of the pasture scarab *Pyronota setosa*. Amber disease caused by strains of *Serratia* spp. was discovered while scientists were looking for the cause of population collapses in *C. zealandica* (Trought et al., 1982). Infected larvae were clearly recognizable from their clear amber appearance, and diseased larvae were commonly found in older pastures where infection levels could reach more than 50% (Jackson, 1990).

When reporting epizootics, disease levels are usually reported as percent infection recorded at the time of sampling. However, this will underestimate the impact of disease if the disease recycles through the population (see Chapter 1). The underestimation will be greatest where the infected, identifiable stage is short and transient and least where there is a long chronic period before death and the pathogenic propagules remain contained within the insect.

8.7 Evolution

While it is well known that bacteria have a huge potential for genetic change and evolution, evidence suggests that many important relationships evolved a long time ago. *Paenibacillus popilliae* and *P. lentimorbus* infections are known only from the coleopteran family Scarabaeidae. The precursors of the current scarab beetle species are known from the fossil record in the Mesozoic period before the breakup of the Pangaea supercontinent about 175 million years ago (Moron, 1984; Ahrens et al., 2014). Scarab beetles can be found on all habitable continents, and with them infections of *Paenibacillus* spp. that appear to show host specificity (Klein, 1992), suggesting that they coevolved with the beetles and emerged from ancient origins. The evolved strategy of *P. popilliae* in causing milky disease has been highly successful and the presence of apparently specific strains of the bacterium in scarab populations on all continents suggests that the bacterium coevolved with this family. The suggestion of ancient origins is supported by molecular evidence from Zeigler (2013) who showed that 16S rRNA sequences of *P. popilliae* and *P. lentimorbus* from different origins were closely related, but that they had only a distant relationship to *P. larvae*, the cause of American foul brood in bees. Despite the apparent success of this model for 200 million years, milky disease pathogenesis remains restricted to *P. popilliae* and *P. lentimorbus* infections of the Scarabaeidae, with no evolution to other insect hosts.

There is similar evidence that *Rickettsiella* infections of Scarabaeidae form another coevolved group with ancient origins. Leclerque et al. (2012) found distinct but similar genotypes of *Rickettsiella* spp. from different species of Scarabaeidae in Europe and Oceania. The unique associations of *Photorhabdus* spp. bacteria with *Heterorhabditis* spp. nematodes and of *Xenorhabdus* spp. bacteria with *Steinernema* spp. nematodes from all continents also suggested to Poinar and Grewal (2012) evidence of coevolution before the Pangaea separation. Poinar (2011) further suggested that the entomopathogenic nematode species and their associated bacteria would have evolved twice, from free-living, bacteria-feeding nematodes in the Mesozoic Period.

Not all entomopathogenic bacteria–host relationships suggest such a long coevolution. The presence of variants of the Tc gene endowing insect pathogenicity in a wide range of bacterial species indicates genetic mobility. The plasmid-borne *sep* genes of *S. entomophila* and *S. proteamaculans* and the monospecific nature of the disease interaction with *C. zealandica* larvae suggest that this is a more recent association, but the link between the Tc genes and pathogenic relationships of *Xenorhabdus*, *Photorhabdus*, *Yersinia*, and *Serratia* genera of the Enterobacteriaceae indicates a deep underpinning relationship that could also have its origins in the Mezozoic. However, some relationships are specific and isolated geographically, which could indicate a more recent evolution. The *sep* genes of *S. entomophila* are contained on a 153 kb plasmid, which, in the laboratory, could be transferred to other *Serratia* species, including nonpathogenic strains of *S. proteamaculans, S. marcescens, and S. liquefaciens*, as well as some other members of the Enterobacteriaceae (*Klebsiella* sp., *Enterobacter* sp., and *E. coli*) (Glare et al., 1996). In all cases, receipt of the plasmid caused amber disease symptoms in the grass grub host. Despite the potential for the plasmid to be transferred between bacterial hosts, pathogenicity to *C. zealandica* in nature is confined to isolates of *S. entomophila* and *S. proteamaculans* found in the insect's soil environment, and the pADAP plasmid has only been found in isolates from New Zealand.

In Bt, strain evolution may be driven quite rapidly with exchange of toxin-coding plasmids in the environment (Bizzarri et al., 2008) and the spread of invertebrate toxicity within the range of the *B. cereus sensu lato* group (Soufiane and Côté, 2010). The wide range of closely related Cry toxins, particularly from the 3-domain toxin class, shows clear evidence of evolution of different domains at different rates (Bravo, 1997), with potential hotspots for adaptation (Wu et al., 2007). This may involve an evolutionary "arms race" with target insects that develop resistance to individual toxins (Bravo et al., 2013). The ability of plasmid-borne Bt toxin genes to evolve by mutation and homologous recombination (evidenced by remnant fragments of toxin genes) and reassort exploiting transposase sites that flank many toxin genes (Kronstad and Whiteley, 1984; Mahillon et al., 1994; Berry et al., 2002), coupled with the ability of the plasmids to transfer between host strains, provides a fluidity to the evolution of Bt.

8.8 Ecology Guiding Use of Bacterial Entomopathogens in Microbial Control

Bacteria have been commercialized for insect pest control and are often referred to as "bacterial insecticides," but this term often overlooks their ecology which should be considered in order to guide the use of bacterial entomopathogens in pest control.

Bt has been the most successful organism to be used in microbial control and its success has been attributed to both its toxic ability and its ease of production through large-scale fermentation. While delivery of the toxin has parallels with chemical pesticide application and it is applied in **inundative** applications (for definitions, see Chapter 13), the benefits of selectivity should also be considered. Selectivity has allowed Bt to become the cornerstone of many IPM programs, and the lack of residue tolerances for Bt products means that they may be used for pest control close to harvesting. Bti has been widely used in many public health programs (Becker and Lüthy, 2017) and in eradication campaigns for lepidopterous pests, such as the campaign against tussock moth in Auckland, New Zealand, where the pest was successfully eradicated, but only after more than 40 spray applications of Btk to some areas (Hajek and Tobin, 2010). Following widespread application, accumulation of Bt spores in the environment has been noted, but the consequences of this are not yet understood.

In contrast, strategies for the application of *P. popilliae* are **inoculative** and have been based on the cell's robustness, persistence in the environment, and ability to recycle. In the major campaign to control the invasive wave of Japanese beetle in the eastern United States in the 1940s, spots of spore powder were applied to the turf surface in a gridlike pattern, relying on rain and organic material turnover to introduce the bacterium into the root zone where the damaging larvae feed (Klein et al., 2007). During the campaign, 111 tons of "milky spore" powder were applied to more than 160 000 sites. This was successful in establishing milky disease in Japanese beetle populations and was reported to suppress the pest population. The campaign was terminated when milky disease became widespread through the infested zone, suggesting either that it was being widely redistributed from the treated populations or that there had been multiple introgressions into the invasive pest population from the original native source. Over a period of more than 100 years, the pest has spread westward and the invasive front has now reached the Rocky Mountains. However, introduction of *P. popilliae* has not been demonstrably successful in slowing the spread and it is not being used in current eradication of new outbreaks on the west coast of the United States. Klein (1992) has suggested that infectivity of *P. popilliae* has declined in recent years. Redmond and Potter (1995) were unsuccessful in attempts to introduce the bacterium into field populations of Japanese beetle in Kentucky and Koppenhöfer et al. (2000) were also unsuccessful in their attempts to introduce *P. popilliae* into populations of the susceptible *C. hirta* in California. The latter authors concluded that, without understanding of the relationship between the spore and the soil environment, the bacterium should be regarded as a general suppressive factor rather than a microbial insecticide. Attempts to introduce it into the invasive population on the Azores were not successful (M. Klein, pers. comm.) and the beetle has spread farther and is now established on the European mainland.

Bacterial products based on *Serratia entomophila* are **inoculative** and have been applied in order to initiate a cycle of disease in the target *C. zealandica* population, but at a density and distribution that ensure uptake and establishment. Although the bacterium is present in the wider environment and will eventually enter the pest population through natural transmission, direct application allows the disease epizootic to be initiated earlier in the pest invasive cycle in new pastures and thus avoids the damage peak. In contrast to *P. popilliae*, *S. entomophila* is a non-spore-forming bacterium and its application to the soil surface is ineffective as ultraviolet (UV) light and desiccation quickly kill any surface-applied bacteria. To overcome this problem, *S. entomophila* has

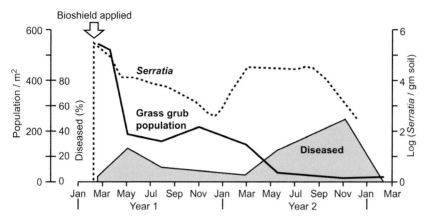

Fig 8.4 Amber disease dynamics in New Zealand following commercial application of Bioshield, a product containing the bacterium *Serratia entomophila*, to control the grass grub *Costelytra zealandica*. Bacteria were applied in February to actively feeding grass grub larvae. Disease levels rise to about 25% and contribute to maintaining a level of 10^3–10^4 pathogenic bacteria per gram of soil. The grass grubs pupate, emerge as adults, and lay eggs in November–December, but the second generation of larvae encounters pathogenic bacteria and drops to low levels suffering from high proportions of disease.

been delivered into the soil profile using modified seed drills to establish bacteria in rows through the pasture. Densities of bacteria in the rows are sufficient to infect larvae feeding in the root zone, and a level of 20% infection within 6 weeks of application is usually sufficient to establish a recycling epizootic of bacteria (Townsend, 2004). Once established in the population, bacteria will recycle, causing fresh waves of infection with each successive year in which *C. zealandica* persists in the pasture (Fig. 8.4). The objective of application is to reduce the population below the damage threshold and limit economic loss. The bacterium was applied to more than 15 000 ha of pastureland from 1992 to 2002 as the liquid product Invade (Jackson, 2003) and has continued to be used as the granule Bioshield (Jackson, 2017). Market penetration has been limited by the high cost of specialist application.

Thus, the strategies for use of entomopathogenic bacteria in microbial control will depend on the intrinsic properties of the product, live cells, or toxin and the ecology of the interaction, whether the objective is control or inoculation. Both factors should be considered in the development of new organisms into microbial products. It is important to note that one of the greatest potential benefits of bacterial entomopathogens will be persistence, which can be achieved by recycling, cell stability, dormant spore stages, or a combination of all three.

8.9 Conclusion

Bacteria are always associated with insects and, through evolution, have developed relationships either for their mutual benefit (symbiosis) or to their own advantage by multiplying on the resource provided by the abundance of insects (pathogenesis). Given conditions of stress, both in the laboratory and after field collection, many insects will succumb to bacterial infection and perhaps die. But these opportunistic pathogens probably have little effect on natural populations in the field. Discovery of entomopathogens with potential for development as microbial control agents, or as natural regulators of insect populations in the field, has required a combination of thorough observation of pest populations in the field and careful pathology studies in the laboratory. Targeted surveys of post-outbreak pest populations have yielded useful candidate strains. New entomopathogenic bacterial species and strains have been discovered by taking samples from the insect's environment, isolating pure cultures and screening by bioassay. This process is greatly simplified by having a clear target organism, or group of organisms, and a selective medium for culture. Molecular tools are being increasingly used to identify species and strains of bacteria, but attribution of pathogenesis to a species or defined genetic grouping is complicated by gene flow among bacteria. Bacteria often have mobile genetic elements and the transfer of genetic material between isolates, species, and genera is becoming increasingly recognized, especially through the comparison of gene sequences. However, even before gene sequencing, common pathologies suggested transposition of Bt *cry* genes or of genes controlling the Tc toxins. The fluidity of pathogenicity genes means that bacterial strains or isolates are pathogenic, not the species as a whole.

Given the enormous variety of bacteria, their genetic fluidity, their close association with insects, and the relatively common isolation of opportunistic bacterial entomopathogens, it is surprising that so few bacterial species have been found as true insect pathogens and that so few insect active genes have been isolated. The apparently limited number of bacterial entomopathogens may be related to host factors (known or unknown) that result in pathogenesis that is specific (*Serratia entomophila* and *C. zealandica*), group-related (*P. popilliae* and scarab larvae), or part of a looser range of pathotypes compared to that existing within the collections of Bt. The complexities of pathogenesis are illustrated by *P. popilliae*, which remains an enigma: why a bacterium that was successfully used 80 years ago and occurs in natural epizootics cannot be harnessed again for invasive scarab control is unknown (Klein, 1992).

Pathogenesis is obviously a successful strategy in the ecology of entomopathogenic strains of bacteria, as they can be readily found in nature. To achieve this success, bacterial entomopathogens have used different ecological strategies ranging from assisted transmission by nematodes and parasitoids, through facultative growth within the soil environment, to extreme persistence of spores. Understanding the ecology of bacterial entomopathogens will lead to a better comprehension of insect population cycles and improved prediction of pest outbreaks or population regulation. Improved ecological understanding of bacterial entomopathogens will assist in the design of application strategies for better microbial control of insect pests.

References

Abebe, E., Abebe-Akele, F., Morrison, J., Cooper, V., Thomas, W.K., 2011. An insect pathogenic symbiosis between a *Caenorhabditis* and *Serratia*. Virulence 2, 158–161.

Abdoarrahem, M.M., Gammon, K., Dancer, B.N., Berry, C., 2009. A genetic basis for the alkaline-activation of germination in *Bacillus thuringiensis* subsp. *israelensis*. Appl. Environ. Microb. 75, 6410–6413.

Ahrens, D., Schwarzer, J., Vogler, A.P., 2014. The evolution of scarab beetles tracks the sequential rise of angiosperms and mammals. Philos. Trans. R. Soc. Lond. B Biol. Sci. 281, 20141470

Angus, T.A., Heimpel, A.M., 1959. Inhibition of feeding, and blood pH changes, in lepidopterous larvae infected with crystal-forming bacteria. Can. Entomol. 91, 352–358.

Aronson, A.I., Tyrell, D.J., Fitz-James, P.C., Bulla, L.A. Jr., 1982. Relationship of the syntheses of spore coat protein and parasporal crystal protein in *Bacillus thuringiensis*. J. Bacteriol. 151, 399–410

Asano, S., Yamashita, C., Iizuka, T., Takeuchi, K., Yamanaka, S., Cerf, D., Yamamoto, T., 2003. A strain of *Bacillus thuringiensis* subsp. *galleriae* containing a novel *cry8* gene highly toxic to *Anomala cuprea* (Coleoptera: Scarabaeidae). Biol. Contr. 28, 191–196.

Bai, Y., D'Aoust, F., Smith, D.L., Driscoll, B.T., 2002. Isolation of plant-growth-promoting *Bacillus* strains from soybean root nodules. Can. J. Microbiol. 48, 230–238.

Bailey, L., 1983. *Melissococcus pluton*, the cause of European foulbrood of honey bees (*Apis* spp.). J. Appl. Bacteriol. 55, 65–69.

Ban, H., Chai, X., Lin, Y., Zhou, Y., Peng, D., Zhou, Y., et al., 2009. Transgenic *Amorphophallus konjac* expressing synthesized acyl-homoserine lactonase (aiiA) gene exhibit enhanced resistance to soft rot disease. Plant Cell Rep., 28: 1847–1855.

Barlow, N.D., 1999. Models in biological control: a field guide, in: Hawkins,B.A., Cornell, H.V. (eds.), Theoretical Approaches to Biological Control. Cambridge University Press, Cambridge, pp. 43–70.

Barlow, N.D., Jackson, T.A. (1998). Predicting and managing pasture pest outbreaks in New Zealand, in: Zalucki, M.P., Drew, R.A.I., White, G.G. (eds.). Proc. Sixth Australasian Applied Entomological Research Conference, Brisbane, pp. 73–81.

Barlow, N.D., Jackson, T.A., Townsend, R.J., 1996. Predicting Canterbury grass grub outbreaks: the role of temperature. Proc. NZ Plant Prot. Conf. 49, 262–265.

Becker, N., Lüthy P., 2017. Mosquito control with entomopathogenic bacteria in Europe, in: Lacey, L.A. (ed.), Microbial Control of Insect and Mite Pests. Academic Press, New York, pp. 379–392.

Beegle, C.C., Yamamoto, T., 1992. History of *Bacillus thuringiensis* Berliner research and development. Can. Entomol. 124, 587–616.

Berry, C., 2012. The bacterium, *Lysinibacillus sphaericus*, as an insect pathogen. J. Invertebr. Pathol. 109, 1–10.

Berry, C., O'Neil, S., Ben-Dov, E., Jones, A.F., Murphy, L., Quail, M.A., et al., 2002. Complete sequence and organization of pBtoxis, the toxin-coding plasmid of *Bacillus thuringiensis* subsp. *israelensis*. Appl. Environ. Microb. 68, 5082–5095

Bizzarri, M.F., Bishop, A.H., 2008. The ecology of *Bacillus thuringiensis* on the phylloplane: colonization from soil, plasmid transfer, and interaction with larvae of *Pieris brassicae*. Microb. Ecol., 56, 133–139.

Bizzarri, M.F., Prabhakar, A., Bishop, A.H., 2008. Multiple-locus sequence typing analysis of *Bacillus thuringiensis* recovered from the phylloplane of clover (*Trifolium hybridum*) in vegetative form. Microb. Ecol. 55, 619–625.

Boemare, N.E., Akhurst, A.J, Mournat, R.G. 1993. DNA relatedness between *Xenorhabdus* spp. (Enterobacteriaceae), symbiotic bacteria of entomopathogenic nematodes and a proposal to transfer to a new genus, *Photorhabdus* gen. nov. Int. J. Syst. Bacteriol. 43, 249–255.

Botelho Praça, L., Menezes Mendes Gomes, A.C., Cabral, G., Soares Martins, E., Ryoiti Sujii, E., Gomes Monnerat, R., 2012. Endophytic colonization by Brazilian strains of *Bacillus thuringiensis* on cabbage seedlings grown *in vitro*. Bt Research 3, 11–19.

Bone, L.W., Tinelli, R., 1987. *Trichostrongylus colubriformis*: larvicidal activity of toxic extracts from *Bacillus sphaericus* (strain 1593) spores. Exp. Parasitol. 64, 514–516.

Boucias, D., Pendland, J.C., 1998. Principles of insect pathology. Springer, New York.

Boucias, D. G., Cherry, R.H., Anderson, D.L., 1986. Incidence of *Bacillus popilliae* in *Ligyrus subtropicus* and *Cyclocephala parallela* (Coleoptera: Scarabaeidae) in Florida sugarcane fields. Environ. Entomol. 15, 703–705

Bravo, A., 1997. Phylogenetic relationships of *Bacillus thuringiensis* delta-endotoxin family proteins and their functional domains. J. Bacteriol. 179, 2793–2801

Bravo, A., Gomez, I., Porta, H., Garcia-Gomez, B. I., Rodriguez-Almazan, C., Pardo, L., Soberon, M., 2013. Evolution of *Bacillus thuringiensis* Cry toxins insecticidal activity. Microb. Biotechnol. 6, 17–26

Broderick, N.A., Raffa, K.F., Handelsman, J., 2006. Midgut bacteria required for *Bacillus thuringiensis* insecticidal activity. Proc. Nat. Acad. Sci. U.S.A. 103, 15 196–15 199.

Broderick, N.A., Robinson, C.J., McMahon, M.D., Holt, J., Handelsman, J., Raffa, K.F., 2009. Contributions of gut bacteria to *Bacillus thuringiensis*-induced mortality vary across a range of Lepidoptera. Bmc Biology, 7.

Brownbridge, M., Margalit, J., 1986. New *Bacillus thuringiensis* strains isolated in Israel are highly toxic to mosquito larvae. J. Invertebr. Pathol. 48, 216–222.

Brownbridge, M., Margalit, J., 1987. Mosquito active strains of *Bacillus sphaericus* isolated from soil and mud samples collected in Israel. J. Invertebr. Pathol. 50, 106–112.

Bucher, G.E., 1960. Potential bacterial pathogens of insects and their characteristics. J. Insect Pathol. 2, 172–195

Bucher, G.E., 1963. Nonsporulating bacterial pathogens, in: Steinhaus, E.A. (ed.), Insect Pathology: An Advanced Treatise. Academic Press, New York, pp. 117–147.

Bucher, G.E., Stephens, J.M., 1957. A disease of grasshoppers caused by the bacterium *Pseudomonas aeruginosa* (Schroeter) Miguela. Can. J. Microbiol. 3, 611–625.

Bulla B.A., 1975. Bacteria as insect pathogens. Annu. Rev. Microbiol. 29, 163–90.

Burges, H.D., Hurst, J.A., 1977. Ecology of *Bacillus thuringiensis* in storage moths. J. Invertebr. Pathol. 30, 131–139.

Collier, F.A., Elliot, S.L., Ellis, R.J., 2005. Spatial variation in *Bacillus thuringiensis cereus* populations within the phyllosphere of broad-leaved dock (*Rumex obtusifolius*) and surrounding habitats. FEMS Microbiol. Ecol. 54, 417–425.

Correa, M., Yousten, A.A., 1995. *Bacillus sphaericus* spore germination and recycling in mosquito larval cadavers. J. Invertebr. Pathol. 66, 76–81.

Côté, J.C., 2007. How early discoveries about *Bacillus thuringiensis* prejudiced subsequent research and use, in: Vincent, C., Goettel, M.S., Lazarovits, G. (eds.), Biological Control: A Global Perspective. CABI, Wallingford, 169–180.

Dadd, R.H., 1975. Alkalinity within the midgut of mosquito larvae with alkaline-active digestive enzymes. J. Insect Physiol. 21, 1847–1853.

Dahlberg C., Chao L., 2003. Amelioration of the cost of conjugative plasmid carriage in *Eschericha coli* K12. Genetics 165, 1641–1649

Damgaard, P.H., Hansen, B.M., Pedersen, J.C., Eilenberg, J., 1997. Natural occurrence of *Bacillus thuringiensis* on cabbage foliage and insects associated with cabbage crops. J. Appl. Microbiol. 82, 253–258.

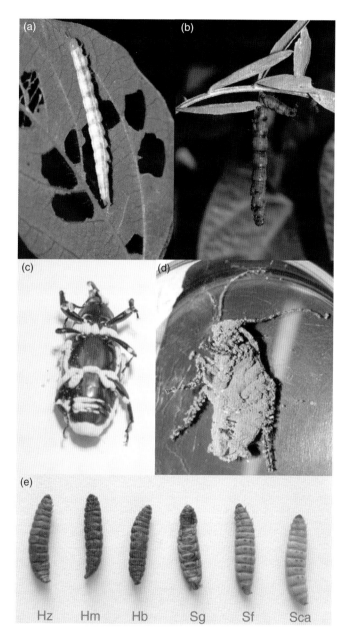

Fig. 2.1 Larva of the green clover worm moth, *Hypenas cabra*, infected with granulovirus (GV), with change of color to become increasingly whiter (a), before dying (b). Infection caused by the fungus *Beauveria bassiana* in an adult red weevil (*Rhynchophorus ferrugineus*) (c) and by *Metarhizium anisopliae* in an adult locust (d), with the distinct colorations of each genus. Larvae of the wax worm *Galleria mellonella* killed by entomopathogenic nematodes display different colors depending on the species: dark green/gray for *Heterorhabditis zealandica* (Hz), dark red for *H. megidis* (Hm) and *H. bacteriophora* (Hb), dark gray for *Steinernema glaseri* (Sg), brown for *Steinernema feltiae* (Sf), and creamy for *Steinernema carpocapsae* (Sca). *Source:* Courtesy of Gerry Carner, Surendra Dara, Stefan Jaronski, and Rubén Blanco-Pérez.

Ecology of Invertebrate Diseases, First Edition. Edited by Ann E. Hajek and David I. Shapiro-Ilan
© 2018 John Wiley & Sons Ltd. Published 2018 by John Wiley & Sons Ltd.

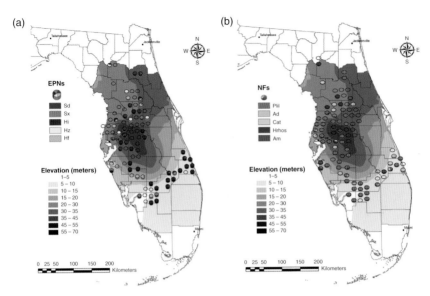

Fig. 2.3 Spatial patterns of the abundance of the most commonly encountered species of entomopathogenic nematodes (EPNs) and nematophagous fungi (NF) in natural areas across the Florida peninsula (*n* = 91 georeferenced localities), measured by species-specific primers and probes in qPCR assays. (a) EPNs: *Heterorhabditis floridensis* (Hf), *H. indica* (Hi), *H.zealandica* (Hz), *Steinernema diaprepesi* (Sd), and *S.* sp. *glaseri* group (Sx). (b) NF: *Purpureocillium lilacinus* (Plil), *Arthrobotrys dactyloides* (Ad), *A. musiformis* (Am), *Catenaria* sp. (Cat), and *Hirsutella rhossiliensis* (Hrhos). *Source:* Campos-Herrera et al. (2016a). Reproduced with permission of Elsevier.

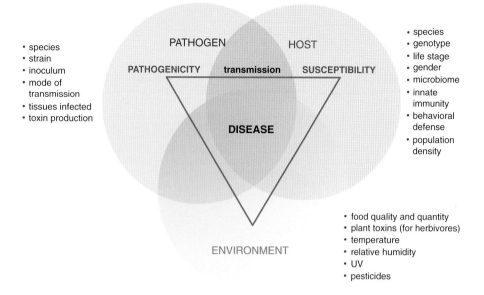

• species
• strain
• inoculum
• mode of
  transmission
• tissues infected
• toxin production

PATHOGEN   HOST

PATHOGENICITY   **transmission**   SUSCEPTIBILITY

DISEASE

ENVIRONMENT

• species
• genotype
• life stage
• gender
• microbiome
• innate
  immunity
• behavioral
  defense
• population
  density

• food quality and quantity
• plant toxins (for herbivores)
• temperature
• relative humidity
• UV
• pesticides

Fig. 4.1 Some of the interacting host, pathogen, and environmental factors determining the development of insect disease.

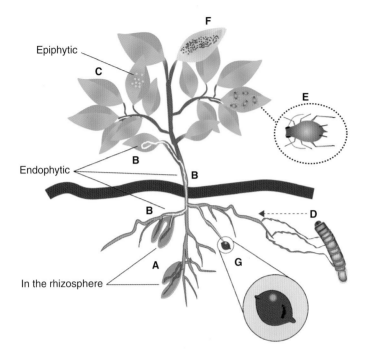

Epiphytic

Endophytic

In the rhizosphere

Fig. 9.1 Plant–fungus–invertebrate associations and interactions. Some hypocrealean invertebrate pathogenic fungi can associate with plants by (A) proliferating as saprophytes in the areas around roots (the rhizosphere), (B) growing inside plant tissues as endophytes below- and aboveground, or (C) being present on plant surfaces as epiphytes. (D) Hyphae from fungal-infected insect cadavers in the soil can connect to roots and thereby transfer nutrients from an insect cadaver to a plant. Endophytic fungi can have trophic effects by (E) reducing population growth or fitness of insect herbivores and (F) reducing disease symptoms caused by plant pathogens. In addition to growing endophytically, (G) some fungi can infect nematode eggs. Figure prepared by Dr. Helen Hesketh, Centre for Ecology & Hydrology, UK © NERC (CEH).

Fig. 14.1 Different types of closed, semi-open, and outdoor facilities for large- or medium-scale production of insects. (a) Closed production facilities for lesser mealworm (*Alphitobius diaperinus*). Larvae are reared in plastic boxes on separate shelves. The Netherlands. (b) Automatic separation of *Tenebrio molitor* larvae from substrate. The Netherlands. (c) Closed production facilities for black soldier fly (BSF; *Hermetia illucens*). Flies are reared and eggs laid in a climate-controlled room. South Africa. (d,e) Semi-open production facilities for house cricket (*Acheta domesticus*) and black cricket (*Gryllus bimaculatus*). New cohorts are started in the different sections at different time intervals. Thailand. (f) Semi-open production facilities for silkworms (*Bombyx mori*). Pupae are kept on open shelves. South Africa. (g) *Bombyx mori* larvae on mulberry leaves on open shelves. Fresh leaves are added regularly. South Africa. (h) Alfalfa leafcutter bee (*Megachile rotundata*). Loose cocoons during overwintering in trays before release in spring. United States. (i) Outdoor nest for mason bees (*Osmia rufa*). Bee pupae can be artificially added in the cardboard cylinders, but naturally occurring *O. rufa* will also use the nest. Denmark. (j) Outdoor nest at a site where both naturally occurring *Osmia rufa* and leafcutter bee (*Megachile* sp., holes covered with green leaves) are nesting. Denmark. (k) Outdoor hives for honeybees (*Apis mellifera*) near tropical forest with high biological diversity. Tanzania. (l) Outdoor hives for honeybees in an agricultural landscape with medium biological diversity. Denmark. (m) Outdoor hives for honeybees in an almond plantation with low biological diversity. California, USA. *Source*: (a,b) Courtesy of Proti-Farm, The Netherlands; (c) Courtesy of Agri-Protein, South Africa; (d,e) Courtesy of Afton Halloran, University of Copenhagen, Denmark; (f,g,i,j) Courtesy of Jørgen Eilenberg; (h) Courtesy of Rosalind James; (k–m) Courtesy of Annette Bruun Jensen, Univ. Copenhagen, Denmark.

Fig. 14.1 (*Continued*)

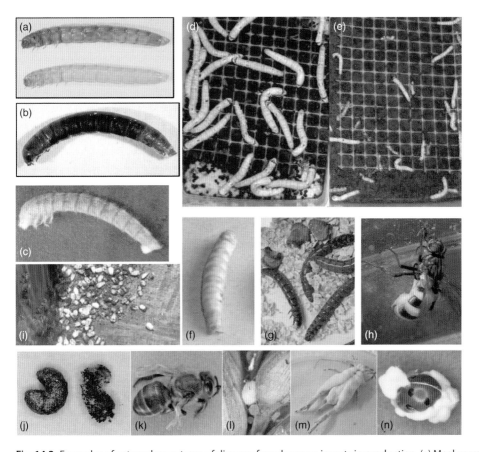

Fig. 14.2 Examples of external symptoms of diseases found among insects in production. (a) Mealworm (*Tenebrio molitor*). Upper: Uninfected larvae. Lower: Larva infected with nematode *Steinernema feltiae*. (b) *Tenebrio molitor* larva infected with unknown bacterium. (c) *Tenebrio molitor* larva infected with the fungus *Beauveria bassiana*. (d) Silkworm (*Bombyx mori*), fifth-instar larvae, uninfected. (e) *Bombyx mori*, fifth-instar larvae, infected with bacterium *Enterococcus mundtii*. (f) *Bombyx mori*, fifth-instar larvae, infected with virus Bombyx mori nucleopolyhedrovirus (BmNPV). (g) Giant mealworm (*Zophobas morio*) larvae infected with the bacterium *Pseudomonas* sp. (h) House fly (*Musca domestica*) adult, infected with the fungus *Entomophthora muscae*. (i) Honeybee (*Apis mellifera*). Specimens infected with fungus *Ascosphaera apis* outside hive. (j) Pupae of alfalfa leafcutter bees (*Megachile rotundata*) infected with *Ascosphaera aggregata* (left) and *Ascosphaera proliperda* (right). (k) *A. mellifera* adult infected with deformed wing virus (DWV). (l) Cereal aphids (*Sitobion avenae*). Central specimen infected with the fungus *Pandora neoaphidis*. (m) House cricket (*Acheta domesticus*), adult, infected with the fungus *Beauveria bassiana*. (n) Seven-spotted ladybird (*Coccinella septempunctata*) infected with the fungus *Beauveria bassiana*. *Source*: (a–f,h,k) Eilenberg et al. (2015). Reproduced with permission of Wageningen Academic Publishers; (g) Courtesy of Gabriela Maciel-Vergara; (i) Courtesy of Irfan Kandemir; (l) Courtesy of Jørgen Eilenberg; (m) Courtesy of Sunday Ekesi, ICIPE, Kenya; (n) Courtesy of Jørgen Eilenberg.

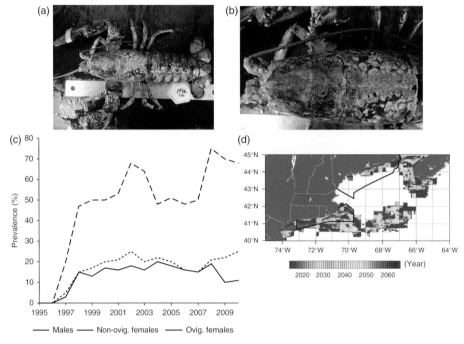

Fig. 15.4 Aspects of epizootic shell disease (ESD) of the American lobster: (a) heavy infection of the dorsal carapace of a lobster from southern Massachusetts; (b) close-up showing the necrotic appearance of the shell, the loss of the rostrum in the animal, and the dark melanotic response in the heavily damaged central portion of the lesion; (c) prevalence of ESD on sub-legal lobsters from inshore waters of Rhode Island – note the predilection for ovigerous (egg-bearing) females that cannot molt out of the disease; (d) forecast of when summer bottom temperatures will be over 12 °C in the Gulf of Maine and Long Island Sound, color-coded by year. *Source*: (a,b) Photos from author's collection; (c) Castro and Somers (2012). Reproduced with permission of National Shellfisheries Association; (d) Maynard, http://rstb.royalsocietypublishing.org/content/371/1689/20150208. Used under CC BY https://creativecommons.org/licenses/by/4.0/.

d'Herelle, F., 1911. Sur une épizootie de nature bactérienne sévissant sur les sauterelles au Mexique. C. R. Acad. Sci. Paris Ser. D. 152, 1413–1415.

d'Herelle, F., 1912. Sur la propagation, dans la Républic Argentine, de l'épizootie des sauterelles du Mexique. C. R. Acad. Sci. Paris Ser. D. 154, 623–625.

d'Herelle, F., 1914. Le coccobacille des sauterelles. Ann. Inst. Pasteur (Paris) 28, 280–328.

Dingman, D.W., 1996. Description and use of a peroral injection technique for studying milky disease. J. Invertebr. Pathol. 67, 102–104.

Dingman, D.W., 2008. Geographical distribution of milky disease bacteria in the eastern United States based on phylogeny. J. Invertebr. Pathol. 97, 171–181.

Dingman, D.W., 2009. DNA fingerprinting of *Paenibacillus popilliae* and *Paenibacillus lentimorbus* using PCR-amplified 16S–23S rDNA intergenic transcribed spacer (ITS) regions. J. Invertebr. Pathol. 100, 16–21.

Dodd, S.J., Hurst, M.R., Glare, T.R., O'Callaghan, M., Ronson, C.W., 2006. Occurrence of sep insecticidal toxin complex genes in *Serratia* spp. and *Yersinia frederiksenii*. Appl. Environ. Microb. 72, 6584–6592.

Dulmage, H.T., 1970. Insecticidal activity of HD-1, a new isolate of *Bacillus thuringiensis* var. *alesti*. J. Invertebr. Pathol. 15, 232–239.

Dutky, S.R. 1963. The milky diseases, in: Steinhaus, E.A. (ed.), Insect Pathology, Vol. 2. Academic Press, New York, pp. 75–115.

Dutky, S.R., Gooden, E.L. 1952. *Coxiella popilliae*, n. sp. a *Rickettsia* causing blue disease of Japanese beetle larvae. J. Bacteriol. 63, 743–750.

East, R. Wigley, P.J. 1985. Causes of grass grub (*Costelytra zealandica* White) population collapse in the northern North Island of New Zealand. Proceedings of the 4th Australasian Conference on Grassland Invertebrate Ecology, pp. 191–200.

Forst, S., Dowds, B., Boemare, N., Stackebrandt, E. 1997. Bugs that kill bugs. Ann. Rev. Microbiol. 51, 47–72.

Fuxa, J.R., Tanada Y., 1987. Epidemiological concepts applied to insect epizootiology, in: Fuxa, J.R., Tanada, Y. (eds.). Epizootiology of Insect Diseases. John Wiley & Sons, New York.

Gatehouse, H.S., Marshall, S.D.G., Simpson, R.M., Gatehouse, L.N., Jackson, T.A., Christeller, J.T., 2008. *Serratia entomophila* inoculation causes a defect in exocytosis in *Costelytra zealandica* larvae. Insect Mol. Biol. 17, 375–385.

Ge, Y., Hu, X., Zheng, D., Wu, Y., Yuan, Z., 2011. Allelic diversity and population structure of *Bacillus sphaericus* revealed by multilocus sequence typing. Appl. Environ. Microb. 77, 5553–5556.

Glare, T.R., Hurst, M.R.H., Grkovic, S. 1996. Plasmid transfer among several members of the family Enterobacteriaceae increases the number of species capable of causing experimental amber disease in grass grub. FEMS Microbiol. Lett. 139, 117–120

Glare, T., O'Callaghan, M., 2000. *Bacillus thuringiensis*: Biology, Ecology and Safety. John Wiley & Sons, New York.

Glare, T.R., Jackson, T.A., Zimmermann, G., 1993. Occurrence of *Bacillus popilliae* and two nematode pathogens in populations of *Amphimallon solstitialis* (Col. Scarabaeidae) near Darmstadt, Germany. Entomophaga 38, 441–450.

Glare, T.R., Jurat-Fuentes, J.L., O'Callaghan, M., 2017. Basic and applied research: entomopathogenic bacteria, in: Lacey, L.A. (ed.), Microbial Control of Insect and Mite Pests. Academic Press, New York, pp. 47–67.

Godfray, H.C.J., Briggs, C.J., Barlow, N.D., O'Callaghan, M., Glare, T.R., Jackson, T.A., 1999. A model of insect pathogen dynamics in which a pathogenic bacterium can also reproduce saprophytically. Philos. Trans. R. Soc. Lond. B Biol. Sci. 266, 233–240.

Goldberg, L.J., Margalit, J., 1997. A bacterial spore demonstrating rapid larvicidal activity against *Anopheles sergentii, Uranotaenia unguiculata, Culex univitattus, Aedes aegypti* and *Culex pipiens*. Mosq. News. 37, 355–358.

Gomez-Garzon, C., Hernandez-Santana, A., Dussan, J., 2016. Comparative genomics reveals *Lysinibacillus sphaericus* group comprises a novel species. BMC Genom. 17, 709.

Gonzalez, J.M., Jr., Carlton, B.C., 1984. A large transmissible plasmid is required for crystal toxin production in *Bacillus thuringiensis* variety *israelensis*. Plasmid 11, 28–38

Grimont, P.A., Grimont, F., 1978. The genus *Serratia*. Annu. Rev. Microbiol. 32, 221–248.

Grimont, P.A., Grimont, F., Lysenko, O., 1979. Species and biotype identification of *Serratia* strains associated with insects. Curr. Microbiol. 2, 139–142.

Grimont, P.A.D., Jackson, T.A., Ageron, E., Noonan, M.J., 1988. *Serratia entomophila* sp. nov. associated with amber disease in the New Zealand grass grub *Costelytra zealandica*. Int. J. Syst. Bacteriol. 38, 1–6.

Guidi, V., Lehner, A., Luthy, P., Tonolla, M., 2013. Dynamics of *Bacillus thuringiensis* var. *israelensis* and *Lysinibacillus sphaericus* spores in urban catch basins after simultaneous application against mosquito larvae. PLoS ONE 8(2), e55658.

Hajek, A.E., Tobin, P.C., 2010. Micro-managing arthropod invasions: eradication and control of invasive arthropods with microbes. Biol. Invasions 12, 2895–2912.

Han, B., Liu, H., Hu, X., Cai, Y., Zheng, D., Yuan, Z., 2007. Molecular characterization of a glucokinase with broad hexose specificity from *Bacillus sphaericus* strain C3-41. Appl. Environ. Microb. 73, 3581–3586.

Harrison, H., Robin, P., Yousten, A.A., 2000. *Paenibacillus* associated with milky disease in central and south American scarabs. J. Invertebr. Pathol. 76, 169–175.

Hu, X., Fan, W., Han, B., Liu, H., Zheng, D., Li, Q., et al., 2008. Complete genome sequences of the mosquitocidal bacterium *Bacillus sphaericus* C3-41 and comparisons with closely related *Bacillus* species. J. Bacteriol. 190, 2892–2902.

Huger, A.M., Krieg, A., Langenbruch, G.A., Schnetter, W., 1986. Discovery of a new strain of *Bacillus thuringiensis* effective against Coleoptera. Mitteilungen aus der Biologischen Bundesanstalt fur Land und Forstwirtschaft 233, 83–96.

Hurpin, B., 1971. The capacities for survival and persistence of the virulence of *Rickettsiella melolonthae*, agent of the rickettsiosis of larvae of *Melolontha* sp. (Coleoptera Scarabaeidae). Annales de Zoologie Ecologie Animal 3, 151–181.

Hurpin, B., Robert, P.H., 1972. Comparison of the activity of certain pathogens of the cockchafer *Melolontha melolontha* in plots of natural meadowland. J. Invertebr. Pathol. 19, 291–298.

Hurpin, B., Robert, P.H., 1976. Conservation in the soil of three microorganisms pathogenic to the larvae of *Melolontha melolontha* (Col.: Scarabaeidae). Entomophaga 21, 73–80.

Hurpin, B., Robert, P.H., 1977. Effects on a natural population of *Melolontha melolontha* (Col.: Scarabaeidae) of the introduction of *Rickettsiella melolonthae* and *Entomopoxvirus melolonthae*. Entomophaga 22, 85–91.

Hurst, M.R., Jackson, T.A., 2002. Use of the green fluorescent protein to monitor the fate of *Serratia entomophila* causing amber disease in the New Zealand grass grub, *Costelytra zealandica*. J. Microbiol. Meth. 50, 1–8.

Hurst, M.R.H., Glare, T.R., Jackson, T.A., Ronson, C.W., 2000. Plasmid-located pathogenicity determinants of *Serratia entomophila*, the causal agent of amber disease of grass grub, show similarity to the insecticidal toxins of *Photorhabdus luminescens*. J. Bacteriol. 182, 5127–5138.

Hurst, M.R., Beard, S.S., Jackson, T.A., Jones, S.M., 2007a. Isolation and characterization of the *Serratia entomophila* antifeeding prophage. FEMS Microbiol. Lett. 270, 42–48.

Hurst, M.R., Jones, S.M., Tan, B., Jackson, T.A., 2007b. Induced expression of the *Serratia entomophila* Sep proteins shows activity towards the larvae of the New Zealand grass grub *Costelytra zealandica*. FEMS Microbiol. Lett. 275, 160–167.

Hurst, M.R., Becher, S.A., Young, S.D., Nelson, T.L., Glare, T.R., 2011a. *Yersinia entomophaga* sp. nov., isolated from the New Zealand grass grub *Costelytra zealandica*. Int. J. Syst. Evol. Micro. 61, 844–849.

Hurst, M.R.H., Jones, S.A., Binglin, T., Harper, L.A., Jackson, T.A., Glare, T.R., 2011b. The main virulence determinant of *Yersinia entomophaga* MH96 is a broad-host-range toxin complex active against insects. J. Bacteriol. 193, 1966–1980.

Hurst, M.R., van Koten, C., Jackson, T.A., 2014. Pathology of *Yersinia entomophaga* MH96 towards *Costelytra zealandica* (Coleoptera; Scarabaeidae) larvae. J. Invertebr. Pathol. 115, 102–107.

Jackson, T.A., 1990. Biological control of grass grub in Canterbury. Proceedings of the New Zealand Grassland Association 52, 217–220.

Jackson, T.A., 1993. Advances in the microbial control of pasture pests in New Zealand. Proceedings of the Sixth Australasian Grassland Invertebrate Ecology Conference, pp. 304–311.

Jackson, T.A., 2003. Environmental safety of inundative application of a naturally occurring biocontrol agent, *Serratia entomophila*, in: Hajek, H.A. (ed.), Environmental Impacts of Microbial Insecticides: Need and Methods for Risk Assessment. Kluwer Academic Publishers, Amsterdam, pp. 169–176.

Jackson, T.A., 2007. A novel bacterium for control of grass grub, in: Vincent, C., Goettel, M.S., Lazarovits, G. (eds.), Biological Control: A Global Perspective. CABI, Wallingford, pp. 160–168.

Jackson, T.A., 2017. Entomopathogenic Bacteria: Mass Production, Formulation, and Quality Control, in: Lacey, L.A. (ed.), Microbial Control of Insect and Mite Pests. Academic Press, New York, pp. 125–139.

Jackson, T.A., Glare, T.R., 1992. Rickettsial diseases of scarabs, in: Jackson, T.A., Glare T.R. (eds.), Use of Pathogens in Scarab Pest Management. Intercept, Andover, pp. 33–42.

Jackson, T.A., McNeill, M.R., 1998. Premature death in parasitized *Listronotus bonariensis* adults can be caused by bacteria transmitted by the parasitoid *Microctonus hyperodae*. Biocontrol Sci. Tech. 8, 389–396.

Jackson, T.A., Huger, A.M., Glare, T.R., 1993. Pathology of amber disease in the New Zealand grass grub *Costelytra zealandica* (Coleoptera: Scarabaeidae). J. Invertebr. Pathol. 61, 123–130.

Jackson, T.A., Townsend, T.R., Nelson, T.L., Richards, N.K., Glare, T.R., 1997. Estimating amber disease in grass grub populations by visual assessment and DNA colony blot analysis. Proceedings of the 50th New Zealand Plant Protection Conference, pp. 165–168.

Jackson, T.A., Boucias, D.G., Thaler, J.O., 2001. Pathobiology of amber disease, caused by *Serratia* spp., in the New Zealand grass grub, *Costelytra zealandica*. J. Invertebr. Pathol. 78, 232–243.

Jackson, T.A., McNeill, M.R., Madurappulige, D., 2004. Use of *Serratia* spp. bacteria for monitoring behaviour of parasitoids. Int. J. Pest Manage. 50, 173–176.

James, R.R., 2011. Potential of ozone as a fumigant to control pests in honey bee (Hymenoptera: Apidae) hives. J. Econ. Entomol. 104, 353–359.

James, R.R., Li, Z., 2012. From silkworms to bees: diseases of beneficial insects, in: Vega, F.E., Kaya, H.K. (eds.), Insect Pathology, 2nd edn. Academic Press, San Diego, CA, pp. 425–459.

Jensen, G.B., Hansen, B.M., Eilenberg, J., Mahillon, J., 2003. The hidden lifestyles of *Bacillus cereus* and relatives. Environ. Micro. 5, 631–640.

Johnston, P.R., Crickmore, N., 2009. Gut bacteria are not required for the insecticidal activity of *Bacillus thuringiensis* toward the tobacco hornworm, *Manduca sexta*. Appl. Environ. Microb. 75, 5094–5099.

Jones, G.W., Wirth, M.C., Monnerat, R.G., Berry, C., 2008. The Cry48Aa-Cry49Aa binary toxin from *Bacillus sphaericus* exhibits highly-restricted target specificity. Environ. Microb. 10, 2418–2424.

Jurat-Fuentes, J.L., Jackson, T.A., 2012. Bacterial entomopathogens, in: Vega, F.E., Kaya, H.K. (eds.), Insect Pathology, 2nd edn. Academic Press, San Diego, CA, pp. 265–349.

Kawanishi, C.Y., Splittstoesser, C.M., Tashiro, H., 1978. Infection of the european chafer, *Amphimallon majalis*, by *Bacillus popilliae*: ultrastructure. J. Invertebr. Pathol. 31, 91–102.

Kaya, H.K., Klein, M.G., Burlando, T.M., 1993. Impact of *Bacillus popilliae*, *Rickettsiella popilliae* and entomopathogenic nematodes on a population of the scarabaeid, Cyclocephala hirta. Biocontrol Sci. Tech. 3, 443–453.

Kaya, H.K., Klein, M.G., Burlando, T.M., Harrison, R.E., Lacey, L.A., 1992. Prevalence of two *Bacillus popilliae* Dutky morphotypes and blue disease in *Cyclocephala hirta* LeConte (Coleoptera: Scarabaeidae) populations in California. Pan-Pacific Entomol. 68, 38–45.

Kaya, H.K., Vega, F.E., 2012. Scope and basic principles of insect pathology, in: Vega, F.E., Kaya, H.K. (eds.), Insect Pathology, 2nd edn. Academic Press, San Diego, CA, pp. 1–12.

Kellen, W.R., Clark, T.B., Lindegren, J.E., Ho, B.C., Rogoff, M.H., Singer, S. 1965. *Bacillus sphaericus* Neide as a pathogen of mosquitoes. J. Invertebr. Pathol. 7, 442–448.

Kevany, B.M., Rasko, D.A., Thomas, M.G., 2009. Characterization of the complete zwittermicin A biosynthesis gene cluster from *Bacillus cereus*. Appl. Environ. Microb. 75, 1144–1155.

Key, P.B., Scott, G.I. 1992. Acute toxicity of the mosquito larvicide, *Bacillus sphaericus*, to the grass shrimp, *Palaemonetes pugio*, and mummichog, *Fundulus heteroclitus*. Bull. Environ. Contam. Tox. 49, 425–430.

Kleespies, R.G., Huger, A.M., Zimmermann, G., 2008. Diseases of insects and other arthropods: results of diagnostic research over 55 years. Biocontrol Sci. Tech. 18, 439–482.

Kleespies, R.G., Marshall, S.D.G., Schuster, C., Townsend, R.J., Jackson, T.A., Leclerque, A., 2011. Genetic and electron-microscopic characterization of *Rickettsiella* bacteria from the manuka beetle, *Pyronota setosa* (Coleoptera: Scarabaeidae). J. Invertebr. Pathol. 107, 206–211.

Klein, M.G. 1992. Use of *Bacillus popilliae* in Japanese beetle control, in: Jackson, T.A., Glare, T.R. (eds.), Use of Pathogens in Scarab Pest Management. Intercept, Andover, pp. 179–189.

Klein, M.G., Jackson, T.A., 1992. Bacterial diseases of scarabs, in: Jackson, T.A., Glare, T.R. (eds.), Use of Pathogens in Scarab Pest Management. Intercept, Andover, pp. 43–61.

Klein, M.G., Grewal, P.S., Jackson, T.A., Koppenhöfer, A.M., 2007. Lawn, turf and grassland pests, in: Lacey, L.A., Kaya, H.K. (eds.), Field Manual of Techniques for the Application and Evaluation of Entomopathogens, 2nd edn. Kluwer, Dordtrecht, pp. 655–675.

Knell, R.J., Begon, M., Thompson, D.J., 1996. Transmission dynamics of *Bacillus thuringiensis* infecting *Plodia interpunctella*: a test of the mass action assumption with an insect pathogen. Philos. Trans. R. Soc. Lond. B Biol. Sci. 263, 75–81.

Konecka, E., Kazanowski, A., Ziemnicka, J., Ziemnicki, K. 2007. Molecular and phenotypic charaterisation of *Bacillus thuringiensis* isolated during epizootics in *Cydia pomonella* L. J. Invertebr. Pathol. 94, 56–63.

Konecka, E., Baranek, J., Kaznowski, A., 2014. Crystalline protein profiling and cry gene detection in *Bacillus thuringiensis* strains isolated during epizootics in *Cydia pomonella* L. Biol. Lett. 51, 83–92.

Koppenhöfer, A.M., Wilson, M.G., Brown, I., Kaya, H.K., Gaugler, R. 2000. Biological control agents for white grubs (Coleoptera: Scarabaeidae) in anticipation of establishment of Japanese beetle in California. J. Econ. Entomol. 93, 71–80.

Koppenhöfer, A.M., Jackson, T.A., Klein, M.G., 2012. Bacteria for use against soil-inhabiting insects, in: Lacey, L.E. (ed.), Manual of Techniques in Invertebrate Pathology. Elsevier, Amsterdam, pp. 129–149.

Krieg A. 1963. Rickettsiae and Rickettsioses, in: Steinhaus, E.A. (ed.), Insect Pathology: An Advanced Treatise, Vol. 2. Academic Press, New York, pp. 577–617.

Krieg, A. 1987. Diseases caused by bacteria and other prokaryotes, in: Fuxa, J.R., Tanada, Y. (eds.), Epizootiology of Insect Diseases. John Wiley & Sons, New York, pp. 323–355.

Krieg, A., 1986. The discovery of *Bacillus thuringiensis* by Dr. Ernst Berliner: a milestone in insect pathology and microbial control of pest insects – a retrospective and prospective view. Mitteilungen aus der Biologischen Bundesanstalt fur Land und Forstwirtschaft 233, 11–24.

Krieg, A., Huger, A.M., Langenbruch, G.A., Schnetter, W., 1983. *Bacillus thuringiensis* var. *tenebrionis*: a new pathotype effective against larvae of Coleoptera. Zeitschrift fur Angewandte Entomologie 96, 500–508.

Kronstad, J.W., Whiteley, H.R., 1984. Inverted repeat sequences flank a *Bacillus thuringiensis* crystal protein gene. J. Bacteriol. 160, 95–102

Krych, V.K., Johnson, J.L., Yousten, A.A., 1980. Deoxyribonucleic acid homologies among strains of *Bacillus sphaericus*. Int. J. Sys. Bacteriol. 30, 476–484.

Kurstack, E., 1962. Donnees sur L'epizootic bacterienne naturalle provoquee par un *Bacillus* du type *Bacillus thuringiensis* sur *Epiphestia kuhniella* Zeller. Entomophaga Mem. Hors Ser. 2, 245–247.

Leclerque, A., Kleespies, R.G., Schuster, C., Richards, N.K., Marshall, S.D.G., Jackson, T.A., 2012. Multilocus sequence analysis (MLSA) of "*Rickettsiella costelytrae*" and "*Rickettsiella pyronotae*," intracellular bacterial entomopathogens from New Zealand. J. Appl. Microbiol. 113, 1228–1237.

Lee, S.J., Park, S.Y., Lee, J.J., Yum, D.Y., Koo, B.T., Lee, J.K., 2002. Genes encoding the N-acyl homoserine lactone-degrading enzyme are widespread in many subspecies of *Bacillus thuringiensis*. Appl. Environ. Microb. 68, 3919–3924.

Lekakarn, H., Promdonkoy, B., Boonserm, P., 2015. Interaction of *Lysinibacillus sphaericus* binary toxin with mosquito larval gut cells: Binding and internalization. J. Invertebr. Pathol. 132, 125–131.

Maduell, P., Armengol, G., Llagostera, M., Lindow, S., Orduz, S., 2007. Immigration of *Bacillus thuringiensis* to bean leaves from soil inoculum or distal plant parts. J. Appl. Micro. 103, 2593–2600.

Maduell, P., Armengol, G., Llagostera, M., Orduz, S., Lindow, S., 2008. *B. thuringiensis* is a poor colonist of leaf surfaces. Microb. Ecol. 55, 212–219.

Mahillon, J., Rezsöhazy, R., Hallet, B., Delcour, J., 1994. IS*231* and other *Bacillus thuringiensis* transposable elements: a review. Genetica 93, 13–26.

Margalit, J., 1990. Discovery of *Bacillus thuringiensis israelensis*, in: de Barjac, H., Sutherland, D. (eds.), Bacterial Control of Mosquitos and Blackflies. Springer, Amsterdam, pp. 3–9.

Marshall, S.D.G., Hares, M.C., Jones, S.A., Harper, L.A., Vernon, J.R., Harland, D.P., et al., 2012. Histopathological effects of the Yen-Tc toxin complex from *Yersinia entomophaga* MH96 (Enterobacteriaceae) on the *Costelytra zealandica* (Coleoptera: Scarabaeidae) larval midgut. Appl. Environ. Microb. 78, 4835–4847.

Martin, P.A.W., Travers, R.S., 1989. Worldwide abundance and distribution of *Bacillus thuringiensis* isolates. Appl. Environ. Microb. 55, 2437–2442.

Martin, P.A.W., Gundersen-Rindal, D., Blackburn, M., Buyer, J., 2007. *Chromobacterium subtsugae* sp. nov., a betaproteobacterium toxic to Colorado potato beetle and other insect pests. J. Syst. Evol. Microbiol. 57, 993–999.

Massie, J., Roberts, G., White, P.J., 1985. Selective isolation of *Bacillus sphaericus* from soil by use of acetate as the only major source of carbon. Appl. Environ. Microb. 49, 1478–1481.

McInroy, J., Kloepper, J., 1995. Survey of indigenous bacterial endophytes from cotton and sweetcorn. Plant Soil 173, 337–342.

Meadows, M.P., Ellis, D.J., Butt, J., Jarrett, P., Burges, H.D., 1992. Distribution, frequency, and diversity of *Bacillus thuringiensis* in an animal feed mill. Appl. Environ. Microb. 58, 1344–1350.

Mertz, F.P., Yao, R.C., 1990. *Saccharopolyspora spinosa* sp. nov. isolated from soil collected in a sugar mill rum still. nt. J. Syst. Evol. Microb. 40, 34–39.

Milner, R.J., 1977. A method for isolating milky disease, *Bacillus popilliae* var. *rhopaea* spores from the soil. J. Invertebr. Pathol. 30, 283–287.

Milner, R.J., 1981. A novel milky disease organism from Australian scarabaeids: field occurrence, isolation, and infectivity. J. Invertebr. Pathol. 37, 304–309.

Monk, J., Young, S.D., Vink, C.J., Winder, L.M., Hurst, M.R.H., O'Callaghan, M., 2010. Q-PCR and high-resolution DNA melting analysis for simple and efficient detection of biocontrol agents, in: Ridgway, H.J., Glare, T.R., Wakelin, S.A., O'Callaghan, M. (eds.), Paddock to PCR: Demystifying Molecular Technologies for Practical Plant Protection. New Zealand Plant Protection, Christchurch, pp. 117–124.

Monnerat, R., de Silva, S.F., Dias, D.S., Martins, E.S., Praca, L.B., Jones, G.W., et al., 2004. Screening of Brazilian *Bacillus sphaericus* strains for high toxicity against *Culex quinquefasciatus* and *Aedes aegypti*. J. Appl. Entomol. 128, 469–473.

Monnerat, R.G., Batista, A.C., de Medeiros, P.T., Martins, É.S., Melatti, V.M., Praça, L.B., et al., 2007. Screening of Brazilian *Bacillus thuringiensis* isolates active against

Spodoptera frugiperda, Plutella xylostella and *Anticarsia gemmatalis*. Biol. Control 41, 291–295.

Monnerat, R.G., Soares, C.M., Capdeville, G., Jones, G., Soares Martins, E., Praça, L., et al., 2009. Translocation and insecticidal activity of *Bacillus thuringiensis* bacteria living inside of plants. Microb. Biotechnol. 2, 512–520.

Monnerat Schenkel, R.G., Nicolas, L., Frachon, E., Hamon, S., 1992. Characterization and toxicity to mosquito larvae of four *Bacillus sphaericus* strains isolated from Brazilian soils. J. Invertebr. Pathol. 60, 10–14.

Moore, S.G., Kalmakoff, J., Miles, J.A.R., 1974. An iridescent virus and a rickettsia from the New Zealand grass grub *Costelytra zealandica* (Coleoptera: Scarabaeidae). N.Z. J. Zool. 1, 205–210.

Morón, M.A., 2004. Escarabajos: 200 millones de años de Evolución, 2nd edn. Instituto de Ecología, Saragossa.

Nicolas, L., Dossou-Yovo, J., Hougard, J.-M., 1987. Persistence and recycling of *Bacillus sphaericus* 2362 spores in *Culex quinquefasciatus* breeding sites in West Africa. Appl. Microbiol. Biotechnol. 25, 341–345.

Nielsen-LeRoux, C., Gaudriault, S., Ramarao, N., Lereclus, D., Givaudan, A., 2012. How the insect pathogen bacteria *Bacillus thuringiensis* and *Xenorhabdus/Photorhabdus* occupy their hosts. Curr. Opin. Microbiol. 15, 220–231.

Nishiwaki, H., Nakashima, K., Ishida, C., Kawamura, T., Matsuda, K., 2007. Cloning, functional characterization, and mode of action of a novel insecticidal pore-forming toxin, Sphaericolysin, produced by *Bacillus sphaericus*. Appl. Environ. Microb. 73, 3404–3411.

O'Callaghan, M., Jackson, T.A., 1993a. Adult grass grub dispersal of *Serratia entomophila*. Proceedings of the 46th New Zealand Plant Protection Conference, pp. 235–236.

O'Callaghan, M., Jackson, T.A., 1993b. Bacteriophage can inhibit the development of amber disease in the New Zealand grass grub, *Costelytra zealandica* (White). J. Invertebr. Pathol. 62, 319–320.

O'Callaghan, M., Jackson, T.A., 1993c. Isolation and enumeration of *Serratia entomophila* – a bacterial pathogen of the New Zealand grass grub, *Costelytra zealandica*. J. Appl. Bacteriol. 75, 307–314.

O'Callaghan, M., Jackson, T.A., Glare, T.R., 1997. *Serratia entomophila* bacteriophages: host range determination and preliminary characterization. Can. J. Microbiol. 43, 1069–1073.

O'Callaghan, M., Jackson, T.A., Noonan, M.J., 1988. Ecology of *Serratia entomophila* in soil. Proceedings of the 5th Australasian Grassland Conference on Grassland Invertebrate Ecology, pp. 69–75.

O'Callaghan, M., Jackson, T.A., Mahanty, H.K., 1992. Selection, development and testing of phage-resistant strains of *Serratia entomophila* for grass grub control. Biocontrol Sci. Tech. 2, 297–305.

O'Callaghan, M., Young, S.D., Barlow, N.D., Jackson, T.A., 1999. The ecology of grass grub pathogenic *Serratia* spp. in New Zealand pastures. Proceedings of the 7th Australasian Grassland Invertebrate Ecology Conference, pp. 85–91.

O'Callaghan, M., Gerard, E.M., Johnson, V.W., 2001. Effect of soil moisture and temperature on survival of microbial control agents in soil. N.Z. Plant Protect. 54, 128–135.

O'Callaghan, M., Glare, T.R., Lacey, L.A., 2012. Bioassay of bacterial entomopathogens against insect larvae, in: Lacey, L.E. (ed.), Manual of Techniques in Invertebrate Pathology. Elsevier, Amsterdam, pp. 101–127.

Ohba, M., Iwahana, H., Asnao, S., Suzuki, N., Sato, R., Hori, H., 1992. A unique isolate of *Bacillus thuringiensis*serovar *japonensis* with a high larvicidal activity specific for scarabaeid beetles. Lett. Appl. Microbiol. 14, pp. 54–57.

Ogier, J.C., Pages, S., Tailliez, P., 2011. Molecular phylogeny of *Heterorhabditis* and *Steinernema* and their symbiotic bacteria. What is true and what is wrong: impact on the evolutionary history of these organisms? IOBC/WPRS Bull. 66, 309–310.

Orlova, M.V., Smirnova, T.A., Ganushkina, L.A., Yacubovich, V.Y., Azizbekyan, R.R., 1998. Insecticidal activity of *Bacillus laterosporus*. Appl. Environ. Microb. 64, 2723–2725.

Park, S.J., Park, S.Y., Ryu, C.M., Park, S.H., Lee, J.K., 2008. The role of AiiA, a quorum-quenching enzyme from *Bacillus thuringiensis*, on the rhizosphere competence. J. Microbiol. Biotechnol. 18, 1518–1521.

Patil, C.D., Patil, S.V., Salunke, B.K., Salunkhe, R.B., 2012. Insecticidal potency of bacterial species *Bacillus thuringiensis* SV2 and *Serratia nematodiphila* SV6 against larvae of mosquito species *Aedes aegypti, Anopheles stephensi,* and *Culex quinquefasciatus.* Parasitol. Res. 110, 1841–1847.

Pettersson, B., Rippere, K.E., Yousten, A.A., Priest, F.G., 1999. Transfer of *Bacillus lentimorbus* and *Bacillus popilliae* to the genus *Paenibacillus* with emended descriptions of *Paenibacillus lentimorbus* comb. nov. and *Paenibacillus popilliae* comb. nov. Int. J. Syst. Bacteriol. 49, 531–540.

Poinar, G.O. Jr. 1976. Description and biology of a new insect parasitic rhabditoid, *Heterorhabditis bacteriophora* n. gen., n. sp. (Heterorhabditidae, n. fam.) (Rhabditida (Oerley)). Nematologica 21, 463–470.

Poinar, G.O. Jr., 1979. Nematodes for Biological Control of Insects. CRC Press, Boca Raton, FL.

Poinar, G.O. Jr. 2011. The Evolutionary History of Nematodes. Leiden, Brill.

Poinar, G.O. Jr., Grewal, P.S., 2012. History of entomopathogenic nematology. J. Nematol. 44, 153–161.

Poinar, G.O. Jr., Thomas, G.M. 1965. A new bacterium, *Achromobacter nematophilus* sp. nov. (Achromobactericeae, Eubacteriales) associated with a nematode. Int. Bull. Bact. Nomen. Tax. 15, 249–252.

Poinar, G.O. Jr., Thomas, G., Haygood, M., Nealson, K.H., 1980. Growth and luminescence of the symbiotic bacteria associated with the terrestrial nematode, *Heterorhabditis bacteriophora*. Soil Biol. Biochem. 12, 5–10.

Porcar, M., Caballero, P., 2000. Molecular and insecticidal characterization of a *Bacillus thuringiensis* strain isolated during a natural epizootic. J. Appl. Microbiol. 89, 309–316.

Ramirez, J.L., Short, S.M., Bahia, A.C., Saraiva, R.G., Dong, Y., Kang, S., et al. 2014. *Chromobacterium* Csp_P reduces malaria and dengue infection in vector mosquitoes and has entomopathogenic and in vitro anti-pathogen activities. PLoS Pathog. 10, e1004398.

Raymond, B., Bonsall, M., 2009. How do toxin producing *Bacillus thuringiensis* strains persist in the field? An evolutionary-ecology perspective. IOBC/WPRS Bull. 45, 211–214.

Raymond, B., Davis, D., Bonsall, M.B., 2007. Competition and reproduction in mixed infections of pathogenic and non-pathogenic *Bacillus* spp. J. Invertebr. Pathol. 96, 151–155.

Raymond, B., Elliot, S.L., Ellis, R.J., 2008a. Quantifying the reproduction of *Bacillus thuringiensis* HD1 in cadavers and live larvae of *Plutella xylostella*. J. Invertebr. Pathol. 98, 307–313.

Raymond, B., Lijek, R.S., Griffiths, R.I., Bonsall, M.B., 2008b. Ecological consequences of ingestion of *Bacillus cereus* on *Bacillus thuringiensis* infections and on the gut flora of a lepidopteran host. J. Invertebr. Pathol. 99, 103–111.

Raymond, B., Ellis, R.J., Bonsall, M.B., 2009a. Moderation of pathogen-induced mortality: the role of density in *Bacillus thuringiensis* virulence. Biol. Lett. 5, 218–220.

Raymond, B., Johnston, P.R., Nielsen-LeRoux, C., Lereclus, D., Crickmore, N., 2010a. *Bacillus thuringiensis*: an impotent pathogen? Trends Microbiol. 18, 189–194.

Raymond, B., Johnston, P.R., Wright, D.J., Ellis, R.J., Crickmore, N., Bonsall, M.B., 2009b. A mid-gut microbiota is not required for the pathogenicity of *Bacillus thuringiensis* to diamondback moth larvae. Environ. Microb. 11, 2556–2563.

Raymond, B., Wyres, K.L., Sheppard, S.K., Ellis, R.J., Bonsall, M.B., 2010b. Environmental factors determining the epidemiology and population genetic structure of the *Bacillus* cereus group in the field. PLoS Pathog. 6, e1000905.

Redmond, C.T., Potter, D.A. 1995. Lack of efficacy of in vivo- and putatively in vitro-produced *Bacillus popilliae* against field populations of Japanese beetle (Coleoptera: Scarabaeidae) grubs in Kentucky. J. Econ. Entomol. 88: 846–854.

Rippere, K.E., Tran, M.T., Yousten, A.A., Hilu, K.H., Klein, M.G., 1998. *Bacillus popilliae* and *Bacillus lentimorbus*, bacteria causing milky disease in Japanese beetles and related scarab larvae. Int. J. Syst. Bacteriol. 48, 395–402.

Robert, L.L., Perich, M.J., Schlein, Y., Jacobson, R.L.,Wirtz, R.A., Lawyer, P.G., Githure, P.I., 1997. Phlebotomine sand fly control using bait-fed adults to carry the larvicide *Bacillus sphaericus* to the larval habitat. J. Am. Mosq. Control Assoc. 13, 140–144.

Ruiu, L., 2015. Insect pathogenic bacteria in integrated pest management. Insects 6, 352–367.

Shapiro-Ilan, D.I., Bruck, D.J., Lacey, L.A., 2012. Principles of epizootiology and microbial control, in: Vega, F.E., Kaya, H.K. (eds.), Insect Pathology, 2nd edn. Academic Press, San Diego, CA, pp. 29–72.

Sharpe, E.S. 1976. Toxicity of the parasporal crystal of *Bacillus popilliae* to Japanese beetle larvae. J. Invertebr. Pathol. 27, 421–422.

Shojaaddini M., López M.J., Moharramipour S., Khodabandeh M., Talebi A.A., Vilanova C., et al., 2012. A *Bacillus thuringiensis* strain producing epizootics on *Plodia interpunctella*: a case study. J. Stored Prod. Res. 48, 52–60.

Silapanuntakul, S., Pantuwatana, S., Bhumiratana, A., Chaaroensiri, K., 1983. The comparative persistence of toxicity of *Bacillus sphaericus* strain 1593 and *Bacillus thuringiensis* serotype H–14 against mosquito larvae in different kinds of environments. J. Invertebr. Pathol. 42, 387–392.

Silva-Filha, M.H.N.L., Berry, C., Regis, L., 2014. *Lysinibacillus sphaericus*: toxins and mode of action, applications for mosquito control and resistance management, in: Dhadialla, T.S., Gill, S.S. (eds.), Advances in Insect Physiology: Insect Midgut and Insecticidal Proteins. Elsevier, Oxford, pp. 89–176.

de Silva, O.S., Prado, G.R., de Silva, J.L.R., Silva, C.E., de Costa, M., Heermann, R., 2013. Oral toxicity of *Photorhabdus luminescens* and *Xenorhabdus nematophila* (Enterobacteriaceae) against *Aedes aegypti* (Diptera: Culicidae). Parasitol. Res. 112, 2891–2896.

Singer, S., 1973. Insecticidal activity of recent bacterial isolates and their toxins against mosquito larvae. Nature 244, 110–111.

Singer, S. 1981. Potential of *Bacillus sphaericus* and related spore-forming bacteria for pest control, in: Burges, H.D. (ed.), Microbial Control of Pests and Plant Diseases 1970–1980. Academic Press, New York, pp. 283–298.

Soufiane, B., Cote, J.C. (2010) *Bacillus thuringiensis* serovars *bolivia, vazensis* and *navarrensis* meet the description of *Bacillus weihenstephanensis*. Curr. Microbiol. 60, 343–34.

Splittoesser, C.M., Kawanashi, C.Y., Tashiro, H. 1978. Infection of the European chafer by *Bacillus popilliae*; light and electron microscope observations. J. Invertebr. Pathol. 31, 84–90.

Stahly, D.P., Klein, M.G., 1992. Problems with in vitro production of spores of *Bacillus popilliae* for use in biological control of the Japanese beetle. J. Invertebr. Pathol., 60, 283–291.

Stahly, D.P., Takefman, D.M., Livasy, C.A., Dingman, D.W. 1992. Selective medium for quantitation of *Bacillus popilliae* in soil and in commercial spore powders. Appl. Environ. Microb. 58, 740–743.

Steinhaus, E.A. 1960. Importance of environmental factors in the insect-microbe ecosystem. Bacteriol. Rev. 24, 365–373.

Steinhaus, E.A. (ed.), 1963. Insect Pathology: An Advanced Treatise, Vols. 1 & 2. Academic Press, New York.

Stephenson, J.P. 1959. Epizootiology of a disease of the desert locust, *Schistocerca gregaria* (Forskal) caused by nonchromogenic strains of *Serratia marcescens* Bizo. J. Insect Pathol. 1, 232–244.

Stucki, G., Jackson, T.A., Noonan, M.J., 1984. Isolation and characterisation of *Serratia* strains pathogenic for larvae of the New Zealand grass grub *Costelytra zealandica*. N.Z. J. Sci. 27, 255–260.

Takatsuka, J., Kunimi Y. 1998. Replication of *Bacillus thuringiensis* in larvae of the Mediterranean flour moth, *Ephestia kuehniella* (Lepidoptera; Pyralidae): growth, sporulation and insecticidal activity of parasporal crystals. Appl. Entomol. Zool. 33, 479–486.

Takatsuka, J., Kunimi, Y. 2000. Intestinal bacteria affect growth of *Bacillus thuringiensis* in larvae of the oriental tea tortrix, *Homona agnanima* Diakonoff (Lepidoptera: Tortricidae). J. Invertebr. Pathol. 76, 222–226.

Tanada, Y. 1964. Epizootiology of insect diseases, in: de Bach, P. (ed.), Biological Control of Insect Pests and Weeds. Chapman and Hall, London, pp. 548–578.

Thomas, G.M., Poinar G.O. Jr., 1979. *Xenorhabdus* gen. nov., a genus of entomopathogenic, nematophilic bacteria of the family Enterobacteriaceae. Int. J. Syst. Evol. Microbiol. 29, 352–360.

Townsend, R.J., Ferguson, C.M., Proffitt, J.R., Slay, M.W.A., Swaminathan, J., Day, S., et al., 2004. Establishment of *Serratia entomophila* after application of a new formulation for grass grub control. N.Z. Plant Prot. 57, 10–12.

Trought, T.E.T., Jackson, T.A., French, R.A., 1982. Incidence and transmission of a disease of grass grub (*Costelytra zealandica*) in Canterbury. N.Z. J. Exper. Agric. 10, 79–82.

Turner, M.J., Schaeffer, J.M. 1989. Mode of action of ivermectin, in: Campbell, W.C. (ed.), Ivermectin and Amermectin. Springer Verlag, New York, pp. 73–88.

van Frankenhuyzen, K., Liu, Y., Tonon, A., 2010. Interactions between *Bacillus thuringiensis* subsp. *kurstaki* HD-1 and midgut bacteria in larvae of gypsy moth and spruce budworm. J. Invertebr. Pathol. 103, 124–131.

Vankova, J., Purrini, K., 1979. Natural epizooties caused by bacilli of the species *Bacillus thuringiensis* and *Bacillus cereus*. Z. Angew. Entomol. 88, 216–221.

Vidal-Quist, J.C., Rogers, H.J., Mahenthiralingam, E., Berry, C., 2013. *Bacillus thuringiensis* colonises plant roots in a phylogeny-dependent manner. FEMS Microbiol. Ecol. 86, 474–489.

Vodovar, N., Vinals, M., Liehl, P., Basset, A., Degrouard, J., Spellman, P., et al., 2005. *Drosophila* host defense after oral infection by an entomopathogenic *Pseudomonas* species. Proc. Nat. Acad. Sci. U.S.A. 102, 11 414–11 419.

Wahba, M.M., 2000. The influence of *Bacillus sphaericus* on the biology and histology of *Phlebotomus papatasi*. J. Egypt Soc. Parasitol. 30, 315–323.

Waterfield, N.R., Ciche, T., Clarke, D., 2009. *Photorhabdus* and a host of hosts. Ann. Rev. Microbiol. 63, 557–574.

Waterfield, N.R., Bowden, D.J., Fetherston, J.D. Perry, R.D., French-Constant, R.H. 2001. The *tc* genes of *Photorhabdus*: a growing family. Trends Microbiol. 9, 185–191.

Wermelinger, E. D., Zanuncio, J. C., Rangel, E. F., Cecon, P. R., Rabinovitch, L., 2000. Toxicity of *Bacillus* species to larvae of *Lutzomyia longipalpis* (L. & N.) (Diptera: Psychodidae: Phlebotominae). An. Soc. Entomol. Brasil. 29, 609–614.

West, A.W., Burges, H.D., White, R.J., Wyborn, C.H., 1984. Persistence of *Bacillus thuringiensis* parasporal crystal insecticidal activity in soil. J. Invertebr. Pathol. 44, 128–133.

West, A.W., Burges, H.D., Dixon, T.J., Wyborn, C.H., 1985a. Effect of incubation in non-sterilised and autoclaved arable soil on survival of *Bacillus thuringiensis* and *Bacillus cereus* spore inocula. N.Z. J. Agric. Res. 28, 559–566.

West, A.W., Burges, H.D., Dixon, T.J., Wyborn, C.H., 1985b. Survival of *Bacillus thuringiensis* and *Bacillus cereus* spore inocula in soil: effects of pH, moisture, nutrient availability and indigenous microorganisms. Soil Biol. Biochem. 17, 657–665.

White, R.T., 1941. Development of milky disease on Japanese beetle larvae under field conditions. J. Econ. Entomol. 34, 213–215.

White, P.J., Lotay, H.K., 1980. Minimal nutritional requirements of *Bacillus sphaericus* NCTC9602 and 26 other strains of this species: the majority grow and sporulate with acetate as sole major source of carbon. J. Gen. Microbiol. 118, 13–19.

Willie, H., Martignioni, M.E., 1952. Vorläufige Mitteilung über einen neuen Krankheitstyphus beim Engerling von *Melolontha vulgaris* F. Schweiz Z. Allg. Pathol. Bakteriol., 15, 470–475.

Wilson, C.J., Mahanty, H.K., Jackson, T.A., 1992. Adhesion of bacteria (*Serratia* spp.) to the foregut of grass grub (*Costelytra zealandica* (White)) larvae and its relationship to the development of amber disease. Biocontrol Sci. Tech. 2, 59–64.

Wilson, C.R., Jackson, T.A., Mahanty, H.K., 1993. Preliminary characterization of bacteriophages of *Serratia entomophila*. J. Appl. Bacteriol. 744, 484–489.

Wilson, M.J., Glen, D.M., George, S.K., Pearce, J.D., 1995. Selection of a bacterium for the mass production of *Phasmarhabditis hermaphrodita* (Nematoda: Rhabditidae) as a biocontrol agent for slugs. Fund. Appl. Nematol. 18, 419–425.

Wu, J.Y., Zhao, F.Q., Bai, J., Deng, G., Qin, S., Bao, Q.Y., 2007 Adaptive evolution of *cry* genes in *Bacillus thuringiensis*: implications for their specificity determination. Geno. Prot. Bioinfo. 5, 102–110.

Yamaguchi, T., Sahara, K., Bando, H., Asano, S., 2008. Discovery of a novel *Bacillus thuringiensis* Cry8D protein and the unique toxicity of the Cry8D-class proteins against scarab beetles. J. Invertebr. Pathol. 99, 257–262.

Yousten, A.A., Benfield, E.F., Campbell, R.P., Foss, S.S., Genthner, F.J., 1991. Fate of *Bacillus sphaericus* 2362 spores following ingestion by nontarget invertebrates. J. Invertebr. Pathol. 58, 427–435.

Yousten, A.A., Fretz, S.B., Jelley, S.A., 1985. Selective medium for mosquito-pathogenic strains of *Bacillus sphaericus*. Appl. Environ. Microb. 49, 1532–1533.

Yousten, A.A., Genthner, F.J., Benfield, E.F., 1992. Fate of *Bacillus sphaericus* and *Bacillus thuringiensis* serovar *israelensis* in the aquatic environment. J. Am. Mosq. Cont. Assoc. 8, 143–148.

Yu, H., Zhang, J., Huang, D., Gao, J., Song, F., 2006. Characterization of *Bacillus thuringiensis* strain Bt185 toxic to the asian cockchafer: *Holotrichia parallela*. Curr. Microbiol. 53, 13–17.

Zhang, C., Yang, S., Xu, M., Sun, J., Liu, H., Liu, J., et al., 2009. *Serratia nematodiphila* sp. nov., associated symbiotically with the entomopathogenic nematode *Heterorhabditidoides chongmingensis* (Rhabditida: Rhabditidae). J. Syst. Evol. Microbiol. 59, 1603–1608.

Zydenbos, S.M., Townsend, R.J., Lane, P.M.S., Mansfield, S., O'Callaghan, M., Koten, C.V., Jackson, T.A., 2016. Effect of *Serratia entomophila* and diazinon applied with seed against grass grub populations on the North Island volcanic plateau. N.Z. Plant Protection 69, 86–93.

9

Fungi

Ann E. Hajek[1] and Nicolai V. Meyling[2]

[1] Department of Entomology, Cornell University, Ithaca, NY, USA
[2] Department of Plant and Environmental Sciences, University of Copenhagen, Frederiksberg, Denmark

9.1   Introduction

Fungal pathogens are well known as important natural enemies of invertebrate hosts. The earliest examples that have been found are insects in amber with visible fungal infections, dated to 100–110 million years ago (Vega et al., 2012). The germ theory of disease was first proven close to 200 years ago, when Agostino Bassi documented that the fungal pathogen *Beauveria bassiana* infected and killed silkworms (*Bombyx mori*) in silk production facilities (Tanada and Kaya, 1993). Perhaps this discovery that micro-organisms can kill animals was first achieved with a fungal entomopathogen rather than another type of pathogen because fungal structures resulting from infections are often visually obvious without using a microscope. In addition, insects are ubiquitous in most terrestrial environments, where they can compete with humans for resources, vector human diseases, or are industrially propagated (e.g., for silk production). The fact that insects compete with humans for resources has largely provided the impetus for studies of insect pathogenic fungi. Epizootics caused by these entomopathogenic fungi, or fungal entomopathogens, can be extremely obvious, especially when high levels of infection control outbreak host populations (Carruthers and Soper, 1987; Pell et al., 2001; Roy et al., 2010). Therefore, suggestions that fungi be used to control pestiferous invertebrates followed after the germ theory of disease, as early scientists learned more about fungal pathogens. By 1879, in Russia, Elie Metchnikoff began working toward the use of *Metarhizium anisopliae* for control of soil-borne agricultural pests. Development of entomopathogenic fungi for control of invertebrate pests has continued as a main focus of research today, and 171 products based on entomopathogenic fungi were listed in 2007 (Faria and Wraight, 2007). The potential for using fungi for pest control, as well as for understanding methods for protecting beneficial arthropods from fungal infection, drives the direction of both basic and applied research today. In addition, fungal pathogens are also gaining notoriety and interest as causes of emerging diseases (those new to regions; see Chapter 16) of invertebrate and vertebrate wildlife and plants around the world (Fisher et al., 2012).

There is abundant diversity among fungal pathogens of invertebrates, with estimates of approximately 1000 species in >100 genera based on pathogens of insects alone (Vega et al., 2012). However, fungi infecting invertebrates frequently have few morphological characteristics for differentiating species, complicating species delineation. Thus, the addition of molecular-based characters allowed important progress to be made toward differentiating species and their evolutionary relationships, and the numbers of species will therefore only increase. For example, molecular analysis of one of the most common and well-studied genera, *Metarhizium*, demonstrated that what previously might have been identified as one species, *M. anisopliae*, is instead nine cryptic species (Bischoff et al., 2009). As we learn more about the identity of individual fungal species, we will also need to understand differences in their biology and ecology.

Fungal entomopathogens are unique among some invertebrate pathogens because they usually infect by penetrating through the host integument, as opposed to bacteria and viruses, which usually must be consumed and subsequently infect through the gut wall. Microsporidia, previously considered protists but now known to be fungi, are an exception as they also infect through the gut wall (see Section 9.1.1 and Chapter 10), and entomopathogenic nematodes infect through body openings and the cuticle (Chapter 11). Regardless, fungi have long been considered especially important pathogens for hosts that do not ingest food by chewing, such as hemipteran insects with piercing sucking mouthparts.

Fungal infection via penetration through the cuticle is hypothesized as potentially having a negative side, because spores are exposed externally to the environment during the infection process. This exposure is considered one basis for the environmental sensitivity of entomopathogenic fungi. Therefore, studies of ecological relations between fungi, hosts, and the environment have been paramount toward understanding disease ecology and the potential for exploiting fungal infections to control invertebrates.

Based on the importance of fungal pathogens infecting invertebrates and our increasing knowledge about them, the biology has been reviewed relatively recently (Butt et al., 2001; Vega et al., 2012), as has the previously neglected evolutionary ecology (Boomsma et al., 2014). Reviews specifically about the ecology and epizootiology of fungal pathogens infecting insect hosts have been published (Carruthers and Soper, 1987; Glare and Milner, 1991; Hajek, 1997; Meyling and Eilenberg, 2007; Roy et al., 2010). In this chapter, we will investigate the recent results of studies of the ecology of fungal pathogens, including ecological insights obtained by implementation of molecular tools. This chapter will span a spectrum of invertebrates as hosts, although emphasis will be on pathogens of terrestrial insects, which have been the focus of most ecological research.

9.1.1 Fungal Systematics and Taxonomy

The Fungi comprise an individual kingdom in the tree of life, which is divided into several phyla containing pathogens of invertebrates. The advances of molecular-based phylogenies have significantly altered the traditional systematics of fungi, and recent revisions have included the microsporidia in this kingdom as a separate phylum (James et al., 2006; McLaughlin and Spatafora, 2015). However, the microsporidia are traditionally studied using methods and concepts different than those appropriate for most other fungi and are in taxonomic terms classified like animals by the International Code of Zoological Nomenclature (Minnis, 2015), while "true" fungi follow the International Code of Nomenclature for algae, fungi and plants (ICN); consequently, the microsporidia are covered separately, in Chapter 10.

The ability to infect invertebrates (predominantly insects and arachnids) has arisen several times in divergent clades of fungi (Vega et al., 2012; Boomsma et al., 2014), but the main important phyla of fungal pathogens infecting invertebrates are Entomophthoromycotina and Ascomycota. The subphylum Entomophthoromycotina was recently erected due to the abandonment of the paraphyletic Zygomycota and creation of the phylum Zoopagomycota (Spatafora et al., 2016). This subphylum contains two classes; in one class, two orders (Entomophthorales and Neozygitales) include pathogens of arthropods, nematodes, and tardigrades (Boomsma et al., 2014). The otherwise widespread fungal phylum Basidiomycota does not include many invertebrate pathogens, although a few species utilize soil nematodes (Nordbring-Hertz et al., 2011) or scale insects (Henk and Vilgalys, 2007) as hosts. The phylum Ascomycota contains the vast majority of all described fungal species (Blackwell, 2011), including the majority of the most well-known invertebrate pathogens, which are placed in the subphylum Pezizomycotina. Within this subphylum, the class Orbiliomycetes contains several species that have evolved specialized structures to capture nematodes in soil (Nordbring-Hertz et al., 2011; Pfister, 2015), but many of these can be considered predators rather than pathogens. The class Eurotiomycetes, order Onygenales, comprises a range of bee-specialist pathogens within the genus *Ascosphaera*, while the class Laboulbeniomycetes contains species that are obligate ectoparasites of many species of arthropods. However, the most studied fungal pathogens are placed in the class Sordariomycetes, order Hypocreales (Vega et al., 2012), which will be referred to as hypocrealean fungi.

Within the Hypocreales, multiple ways to obtain nutrition from different substrates, including bio- and necrotrophism, have evolved with the ability to infect and kill arthropods found in taxa of several hypocrealean families (Spatafora et al., 2007). In particular, the family Clavicipitaceae was recently split into three main clades (Sung et al., 2007), each redefined as individual families. Hence, the most commonly studied genera of invertebrate pathogens are distributed among the three families Clavicipitaceae, Cordycipitaceae, and Ophiocordycipitaceae.

Another significant realization based on recent molecular phylogenies was that some important genera of hypocrealean insect pathogens were polyphyletic and thus needed redefined generic names. It is beyond the scope of this chapter to describe all of the taxonomic changes that have been made in recent years. Examples would be that some insect and nematode pathogens in the genus *Paecilomyces* are now renamed in the genera *Isaria* and *Purpureocillium* (Luangsa-Ard et al., 2005, 2011), and that *Lecanicillium* and *Simplicillium* are the correct genera for insect pathogenic species previously identified as *Verticillium* (Zare and Gams, 2001).

The mode of reproduction for the Ascomycota has caused some taxonomic challenges, because they have a pleomorphic lifecycle; that is, the species undergo two distinct morphological phases: an asexual (mitotic) life stage (identifiable as one or more anamorph taxa) and a sexual (meiotic) life stage, the teleomorph (Vega et al., 2012). Traditionally, each of these morphological entities was assigned a different scientific name, leading to a dual nomenclature for the same organism. Since 2012, based on molecular characterization, the dual nomenclature for pleomorphic fungi has been formally abandoned, and now only one generic name is allowed, under the rule One

Fungus = One Name (1F = 1N), so for an organism with dual nomenclature, one of the two names must be cancelled and the other retained. Based on complex decisions, either the teleomorphic or the anamorphic name is retained to represent both states. Of interest to most invertebrate pathologists, it is likely that the genus *Metarhizium* will be retained in the family Clavicipitaceae, while several familiar anamorphic generic names in the Cordycipitaceae such as *Beauveria*, *Lecanicillium*, and *Isaria* will potentially be formally submerged into *Cordyceps* (R.A. Humber, pers. comm.). It is currently advised to keep updated regarding changes in the valid nomenclature for invertebrate pathogenic fungi.

The 1F = 1N principle does not directly affect the nomenclature of Entomophthoromycotina, but continued molecular phylogenetic and genomic studies might still alter systematic affiliation of species within this phylum. For example, associations will be clarified as molecular methods are used to work out the names of species in the form-genus *Tarichium*, for which only azygospores and zygospores are known.

9.1.2 Relevance of Fungal Systematics and Taxonomy in Ecology

Assigning scientific names to organisms is a discipline for taxonomists, but in order to communicate explicitly and transparently, scientists of other research disciplines must pay attention to the development of systematic and taxonomic revisions within their fields. Two of the most frequently studied genera of hypocrealean entomopathogenic fungi, *Metarhizium* and *Beauveria*, were recently shown by phylogenetic analyses of multiple genes to each be made up of numerous cryptic species (Rehner and Buckley, 2005; Bischoff et al., 2009; Rehner et al., 2011). Consequently, the species *Metarhizium anisopliae sensu lato* (*s.l.*; in the broad sense) was divided into several species by Bischoff et al. (2009), and the genus was further expanded to also include *Nomuraea* and a few other genera by Kepler et al. (2014). Fungal isolates that were traditionally identified as *M. anisopliae s.l.* by morphological characteristics in fact frequently belong to *Metarhizium brunneum* or *Metarhizium robertsii* (Gao et al., 2011; Klingen et al., 2015). For *Beauveria*, morphological characters are unsuitable for distinguishing some widespread species, including the most commonly recovered *Beauveria bassiana* and *Beauveria pseudobassiana* (Rehner et al., 2011).

The morphological characters traditionally used for species identification are few and ambiguous for species identification of *M. anisopliae s.l.* and *B. bassiana s.l.*, so reliable assignment of isolates to species can only be achieved by DNA sequencing. Ultimately, a single "barcode" region to sequence will be advantageous. While the nuclear ribosomal internal transcribed spacer (ITS) region of rDNA is proposed as a universal barcode for fungi (Schoch et al., 2012), this region provides insufficient resolution for species identification within *Metarhizium* and *Beauveria* (Rehner and Buckley, 2005; Bischoff et al., 2009). A useful marker to reliably identify species of *Metarhizium* is the 5′ primed end of the gene translation elongation factor 1-alpha (5′-TEF) (Bischoff et al., 2009; Kepler et al., 2014), while *Beauveria* species can be identified by sequencing the intergenic region Bloc (Rehner et al., 2011). In general, TEF provides a useful marker for Ascomycete fungi (Stielow et al., 2015), and it is also convenient for identification of *Beauveria* spp. (Meyling et al., 2012). However, identification should not simply be based on noncritical comparisons with sequences using the Basic Local Alignment Search Tool (BLAST) in the National Center for Biotechnology

Information (NCBI) databases, but should include sequence alignment with verified reference sequences from the original authoritative publications, followed by phylogenetic analysis. Full identification and reference to isolates studied should further include a combination of deposition of the isolate in a culture collection and submission of the sequence to a public repository such as GenBank at NCBI. Such approaches will greatly improve future transparency of experimental research reported in the literature. If it is not possible to molecularly identify the researched isolates, and identifications are based solely on morphological characters, then the isolates must be identified "in the broad sense" (e.g., *M. anisopliae s.l.* or *B. bassiana s.l.*). However, if the same isolates are used repeatedly for research purposes, they should be molecularly identified.

9.2   Fungal Biology and Pathology

Fungal pathogens infecting invertebrates have diverse life strategies, yet there are some commonalities. Most species are obligate killers and are semelparous, only reproducing by forming vast amounts of infective units during a single episode after the host is dead. In these cases, optimal fungal fitness is dependent on an infected host remaining alive until the fungus is prepared to produce spores, or on the cadaver being preserved so that it is only or principally used by the fungus through continued growth and sporulation. However, some fungal pathogens can be called iteroparous, as they release spores over prolonged periods of time. This includes spore release by *Ophiocordyceps camponoti-rufipedis*, a pathogen of carpenter ants with prolonged spore production and release; it has been hypothesized that this process is prolonged because there is no specific persistent stage for this fungus (Andersen et al., 2012) while other fungi that are semelparous produce specific persistent stages (Section 9.2.1.1). Conidial release has also been documented as cycling over shorter intervals, in association with environmental hydration and subsequent drying (Hajek and Soper, 1992; Sawyer et al., 1997).

Fungal pathogens are unique among invertebrate pathogens because, for most species, spores attach to the host cuticle and use enzymatic and mechanical means to grow through the cuticle and enter the host hemocoel; however, in the cases of bee pathogens in the genus *Ascosphaera* and mosquito pathogens in the genus *Culicinomyces*, infection is *per os*. Studies have shown that the strategy of penetration of fungi through the insect cuticle can backfire when a fungal pathogen is attempting to infect a molting insect, in which case the growing hypha can be shed along with the molt skin, thereby preventing penetration into the host (Vey and Fargues, 1977; Luz et al., 2003).

9.2.1   Biology and Pathology of Major Groups of Fungal Pathogens

As the biology and pathology of the two major groups of fungi, the Entomophthoromycotina, order Entomophthorales, and the Ascomycota, order Hypocreales, are extremely different, we will discuss the two groups separately.

9.2.1.1   Entomophthoromycotina, Entomophthorales
Species of Entomophthorales are obligate bio- or necrotrophic pathogens and do not grow in nature when outside of a host. This is consistent with the fact that many of these

species are difficult to cultivate *in vitro* and some have never been cultured. It is also in agreement with the fact that many pathogens in this group are quite host-specific and restricted to only one host genus or family.

Conidia of entomophthoralean fungi are usually coated with preformed mucous so that when they land on a host, they adhere to the integument before penetrating the cuticle (Pell et al., 2001). Conidia generally do not require nutrients for growth. Sometimes, appressoria (specialized attachment cells below which insect pathogenic fungi penetrate the host cuticle) are produced when the fungus is going to penetrate the cuticle; cuticular penetration is thought to occur via a combination of mechanical pressure and enzymatic degradation of the cuticle (Pell et al., 2001). After penetrating the host cuticle, cells grow in the hemolymph. Some species grow as wall-less protoplasts initially, before changing and growing as hyphal bodies with cell walls, while other species grow only as either protoplasts or hyphal bodies while within the host. It is thought that growth as wall-less protoplasts helps the fungus to avoid the invertebrate immune response, through which sugars present in fungal cell walls may be detected. Initially during infections, the fungal cells increase in number in the hemolymph and then begin to attack tissues, so that by the time the host dies, the body cavity is filled with fungal cells. Entomophthoralean species variously produce several kinds of spores, but of two main types: conidia and resting spores. Whether conidia and resting spores are produced from the same host cadaver or not depends on the species.

After host death, the fungus grows out of the host cadaver, often through nonsclerotized sections of the cuticle, such as between segments, and produces asexual primary conidia externally, which are often actively ejected from the cadaver. If a primary conidium does not land on a new host, it can germinate to produce a secondary conidium, which often, but not always, resembles the primary conidium but is slightly smaller. Secondary conidia are actively ejected by the primary, and when not landing on a host, this process can also continue with production of tertiary conidia. Thus, for some species, each propagule produced from a host cadaver has three active chances to reach a new host. For other species, primary conidia are not infective but secondary conidia, which have different forms, are infective (Pell et al., 2001). Conidia that are infective are ready to germinate and infect immediately and are considered relatively short-lived after being ejected from conidiophores.

In contrast to short-lived and active conidia, most entomophthoralean species also produce persistent zygospores or azygospores; both of these spore types are called resting spores. Resting spores are produced within cadavers after host death and are dormant after production. Entomophthoralean species often switch from conidia to resting spore production under conditions when hosts will no longer be present (e.g., in later instar larvae, at lower temperatures, or when photoperiod decreases) (Hajek, 1997). Resting spores have thick walls and, once cadavers disintegrate, are deposited in the soil, where they can be present in large numbers, creating a reservoir, especially in the surface layers of the soil (Hajek et al., 1998). Resting spore germination over time has been studied in few systems, but for the gypsy moth (*Lymantria dispar*) pathogen *Entomophaga maimaiga* only a percentage of the resting spores in the reservoir germinate each year (Hajek et al., 2004), so the reservoir persists even in years with few hosts and/or little fungal activity. When resting spores germinate, they produce a germ conidium, which is also actively ejected and can infect when it lands on a host (Pell et al., 2001). Resting spores of *E. maimaiga* have been shown experimentally to retain the capacity to germinate for 6 years (Weseloh and Andreadis, 1997) and empirically soil-borne resting spores germinated after 11–12 years in the soil (Hajek et al., 2000). The infection of 13- or 17-year periodical cicadas (*Magicicada* spp.) via *Massospora cicadina* resting spores was suggested in the past as evidence that resting spores of this species are very long-lived. However, Duke et al. (2002) demonstrated that resting spores of *M. cicadina* can infect cicadas 10 months after production, so these resting spores do not have to remain dormant for 13 or 17 years.

9.2.1.2 Ascomycota, Hypocreales

Anamorphs of species in the Order Hypocreales are some of the most well known and most abundant entomopathogenic fungi worldwide, although the associated teleomorphs often have restricted ranges and can be difficult or impossible to find. Some of these species are facultative pathogens that are able to grow to some extent in the environment, including within plants (see Section 9.6.1), generally as saprotrophs. Thus, these species are generally easier to grow on artificial nutrient media *in vitro*. The most well-known species in the genera *Beauveria* and *Metarhizium* can possess broader host ranges, although some species in these genera can also have narrower host ranges (Hu et al., 2014). Interestingly, the teleomorphs of Hypocreales may have narrower host ranges than their corresponding anamorphs (Boomsma et al., 2014).

The teleomorphs of the hypocrealean fungi produce ascospores, which are actively discharged, while the anamorphs produce spores, named conidia, which rely on passive transmission (Boomsma et al., 2014). Initial conidial adhesion to hosts is passive and nonspecific. Conidia of some species are covered with a proteinaceous rodlet layer and are hydrophobic, while those of other species have smooth surfaces and are hydrophilic (Vega et al., 2012). Adhesion by conidia of *M. anisopliae* is also mediated by adhesin-like proteins on the conidial surfaces. Conidia require carbon and nitrogen on the host cuticle to germinate and penetrate the cuticle (Vega et al., 2012). Species of *Metarhizium* have been studied extensively and found to produce proteases, chitinases, and lipases for cuticular penetration. Hypocrealean anamorphs often grow within the host hemocoel as blastospores or hyphal bodies. In contrast to fungi of Entomophthoromycotina, species in Hypocreales are well known for producing secondary metabolites when growing within hosts. These are chemicals that are not directly needed by the fungus for growth or reproduction. Many studies have been conducted demonstrating that a multitude of secondary metabolites are produced by different fungal species (Molnár et al., 2010; Rohlfs and Churchill, 2011). While the role of most of these chemicals in pathogenesis remains in question, it seems clear that secondary metabolites produced by both *Beauveria* and *Metarhizium* impair cellular defenses that are part of the host immune response. However, in at least one trial with *Metarhizium anisopliae* against three species of hosts, isolates that produced more of the cyclic depsipeptides called destruxins *in vitro* killed hosts more quickly *in vivo* (Kershaw et al., 1999). Kershaw et al. (1999) hypothesized that this trial demonstrated diversity in the strategies of different isolates in *Manduca sexta* larvae: (i) one isolate employed a "toxin strategy," in which the fungus grew little in the hemolymph and produced enough destruxins to assist in host death, while (ii) other isolates displayed a "growth strategy," which relied on profuse fungal growth in the hemolymph, eventually resulting in host starvation leading to host death. This latter strategy would be more similar to entomophthoralean growth within infected hosts.

9.2.2 Distribution Patterns and Habitat Associations of Invertebrate Pathogenic Fungi

Most fungal pathogens of invertebrates are ubiquitous in many terrestrial ecosystems of most climatic regions and propagule reservoirs are predominantly harbored in the soil environment when occurring outside the host (Hesketh et al., 2010). Many hypocrealean entomopathogenic fungi are readily extracted from soil samples using standard methods such as the use of insects as baits (where susceptible insects – mostly larval stages – are placed in soil and thus exposed to potential pathogenic inoculum in the sample) (Zimmermann, 1986). Insect baits have been used to isolate hypocrealean fungi from soils in the Arctic (Meyling et al., 2012) and tropical regions (Rezende et al., 2015), although most reports are from temperate or subtropical areas (see Vega et al., 2012 for details). However, selective isolation of certain fungal taxa depends on choice of bait insect species (Klingen et al., 2002; Goble et al., 2010) and incubation temperature (Mietkiewski and Tkaczuk, 1998), so these aspects must be considered when designing surveys of soil samples and during interpretation of data. Based on data from the limited species of bait insects that are typically used for distributional studies, it is likely that these studies underestimate the diversity of soilborne entomopathogenic fungi. Aphids have been used as insect baits for entomophthoralean fungi (Nielsen et al., 2003), but careful attention must be paid to the potential hosts because these fungi are frequently very host-specific (Hajek et al., 2012). Species of Entomophthoromycotina with broad host ranges such as *Conidiobolus coronatus* are occasionally reported from surveys of soil samples using the wax moth (*Galleria mellonella*) as bait insect (Chandler et al., 1997; Keller et al., 2003; Meyling and Eilenberg, 2006a).

Host cadavers filled with fungal biomass are expected to form a refuge for fungi; this resource is defended from invasion by opportunistic microorganisms by *B. bassiana* through upregulation of the production of secondary metabolites such as tenellin and beauvericin in the final stages of the infection and in the cadavers, particularly under humid conditions (Boucias and Pendland, 1998; Lobo et al., 2015). Fungal pathogens in the Entomophthoromycotina and Hypocreales have generally evolved different strategies for survival outside of the host environment (Boomsma et al., 2014). Entomophthoralean fungi produce thick-walled resting spores in or on the host cadaver that are resilient to ambient conditions in the environment (see Section 9.2.2.1). In contrast, anamorphic hypocrealean fungi do not create specifically long-lived spores in nature, but instead produce vast amounts of conidia on the outsides of host cadavers. The number of viable conidia on cadavers is expected to decline relatively rapidly, but the investment in large numbers of infective units provides sufficiently plentiful environmental inocula to ensure transmission. In Finland, conidia of *M. anisopliae s.l.* were observed to persist for 3 years after application to agricultural field soil, while those of *B. bassiana s.l.* persisted for at least 1 year (Vänninen et al., 2000). Although conidia of hypocrealean entomopathogenic fungi have been considered to be dormant infective structures responsible for fungal survival in the soil environment, recent evidence has demonstrated that some fungal taxa, particularly *Metarhizium* spp., can proliferate in the soil in association with plant roots, not only persisting but building up fungal biomass, probably so that they can survive prolonged periods without arthropod hosts (see Section 9.6.1) (Hu and St. Leger, 2002; Wang et al., 2011; Klingen et al., 2015).

Inoculum of hypocrealean entomopathogenic fungi can also be isolated from aboveground habitats such as phylloplanes of herbs and shrubs (Meyling and Eilenberg,

2006b), which may constitute an additional ecologically important reservoir for fungal transmission. Propagules of *Beauveria* spp. have been reported to occur on phylloplanes of various common plant species, including trees, with increasing densities through the season (Meyling and Eilenberg, 2006b; Ormond et al., 2010; Garrido-Jurado et al., 2015). Recently, Howe et al. (2016) demonstrated that propagules of *Lecanicillium* spp. and *Isaria* spp. were most abundant on phylloplanes of lime trees (*Tilia* x *europaea*) 13 m high in the canopy, close to the trunk rather than in the outer part of the crown. The occurrences of these fungal taxa correlated with occurrences of infections of the invasive ladybird *Harmonia axyridis* collected in the same trees, indicating that the insects may become infected in their aboveground habitat (Howe et al., 2016). Therefore, insects may acquire fungal inoculum not only in the soil environment but also in arboreal habitats if conditions are conducive for infection. To date, the occasional aboveground natural occurrence of *Metarhizium* spp. has been reported from phylloplanes of diverse agroforestry ecosystems in Spain, although *B. bassiana* was predominantly isolated in the areas investigated (Garrido-Jurado et al., 2015). While *Beauveria* spp. are reported from both aboveground habitats and the soil, *Metarhizium* spp. are generally collected from belowground habitats (Meyling and Eilenberg, 2007; Meyling et al., 2011; Behie et al., 2015).

9.2.2.1 Patterns of Fungal Abundance and Distribution: Insights from the Use of Molecular Markers

It was recently realized that because important and widespread hypocrealean entomopathogenic fungal genera, including *Metarhizium* and *Beauveria*, are composed of cryptic species, fungal diversity within ecosystems cannot be assessed by traditional morphological identifications alone (Rehner and Buckley, 2005; Bischoff et al., 2009; Rehner et al., 2011; Kepler et al., 2014). An implication of this is that surveys of the natural occurrence of entomopathogenic fungi using only morphological criteria for identification will most likely underestimate diversity, which may affect conclusions about factors governing abundance and distribution in the ecosystems investigated. For example, a fungal population of *M. anisopliae s.l.* identified by morphology (Meyling et al., 2011) potentially represents a community of several sympatric populations of *Metarhizium* spp. (Steinwender et al., 2014). There is still much to be investigated regarding fungal diversity using new molecular methods, especially for species identification and quantification (e.g., Creer et al., 2016). In the temperate regions studied to date, the *Metarhizium* communities in agricultural soils are made up of several species, but predominantly *M. robertsii* and *M. brunneum*. In Ontario, Canada, Bidochka et al. (1998) conducted a survey of entomopathogenic fungi in agricultural and forest soils and concluded that *M. anisopliae* (*s.l.*) was most abundant in the former and *B. bassiana* (*s.l.*) in the latter, although both taxa were recovered from both habitat types. Molecular characterizations of the same fungal isolates with general markers later revealed that *Metarhizium* isolates from agricultural and forest soils, respectively, clustered in two separate clades (see Section 9.5) (Bidochka et al., 2001). In fact, the clade recovered most frequently in agricultural soil samples was *M. robertsii*, while that in forest samples was *M. brunneum* (Bischoff et al., 2009). The relatively high frequency of *M. robertsii* in the soil environment of open, exposed habitats such as meadows in Ontario and agricultural fields in the mid-Atlantic United States was confirmed by Wyrebek et al. (2011) and Kepler et al. (2015) by use of specific markers.

In contrast, investigations from northern Europe (Denmark) have revealed *M. brunneum* to be most frequently isolated from agricultural soils. A detailed survey in an experimental farming system was initially characterized by Meyling et al. (2011) using morphological characters for identification, and *M. anisopliae s.l.* constituted almost 80% of the entomopathogenic fungal isolates obtained using insect baits. However, repeated isolations from the same site, which included DNA sequencing for identification, revealed that the *Metarhizium* community was predominantly composed of *M. brunneum*, but also included *M. robertsii* and *M. majus* (Steinwender et al., 2014, 2015). Furthermore, similar frequency distributions of fungal species were recovered at the site in consecutive years using insect bait (Steinwender et al., 2014) and selective agar media (Steinwender et al., 2015), indicating that these methods can both isolate the same genotypes and be used for characterization of the *Metarhizium* community in soils. However, in other agricultural field sites in Denmark, *Metarhizium flavoviride* was the most frequently recovered *Metarhizium* species in soil samples by both insect baits and selective media (Meyling and Eilenberg, 2006a; Keyser et al., 2015), suggesting that communities, at least in Denmark, are locality-specific and likely not defined by management practice.

There is currently less knowledge available regarding the diversity, natural occurrence, and distribution of *Beauveria* spp. than for *Metarhizium* spp. in agroecosystems, but co-occurring cryptic populations also exist in *Beauveria*. Meyling and Eilenberg (2006a) reported *B. bassiana s.l.* to be the most frequently isolated entomopathogenic fungus from both agricultural and hedgerow soil samples using the insect bait method within an agroecosystem in Denmark. However, DNA sequencing of a broad selection of the isolates showed that while only a single *B. bassiana* clade could be documented in the agricultural soil, several clades of *B. bassiana*, as well as *B. brongniartii* and *B. pseudobassiana*, coexisted in the soil of the hedgerow bordering the field (Meyling et al., 2009).

It therefore appears that entomopathogenic fungal communities in the soil environment cannot be universally predicted by overall habitat criteria but that community compositions are quite variable among localities, although some patterns have been reported for habitats defined by certain biotic and abiotic factors (Vega et al., 2012; see below). However, conclusions on the relationships between fungal occurrence and various (expected) predictive factors will be incomplete if only morphological characters are used for identification in studies, such as those reported by Quesada-Moraga et al. (2007), Jabbour and Barbercheck (2009), and Clifton et al. (2015). Without knowledge of whether the *Metarhizium* species pool is composed of *M. robertsii*, *M. brunneum*, or other species, and which is most abundant, it is not possible to predict important factors contributing to the occurrence of this species, as was seen in the Canadian study by Bidochka et al. (1998, 2001). Although tedious laboratory work is necessary, future studies focusing on natural fungal communities should include careful consideration of the sampling plan, isolation procedures, and molecular characterization in order to provide in-depth coverage and understanding of the diversity and community structure of hypocrealean invertebrate fungal pathogens.

Few investigations of *Metarhizium* distribution and diversity in the tropics have been conducted. It was recently shown that sugarcane agroecosystems in Brazil harbored diverse *Metarhizium* communities, comprising predominantly divergent clades within *M. anisopliae sensu stricto* (indicating further cryptic speciation within this species), as well as two taxonomically undescribed clades (Rezende et al., 2015). Further characterizations of Brazilian isolates have revealed the occurrence of several additional species, as well as the yet undescribed *Metarhizium* clades, to be widespread in both natural and agricultural habitats of Brazil (Rocha et al., 2013; de Castro, 2016).

9.2.3   Factors Governing Diversity Patterns of Fungal Pathogens

Hosts and pathogens are expected to coevolve in ways that counteract infection and ways that overcome host defenses, respectively, in a coevolutionary arms race (Pedrini et al., 2015). Such patterns are likely to be observed in obligate pathogens that are relying exclusively on hosts for survival and reproduction, such as the entomophthoralean fungi. Coevolution processes can lead to host specialization, which is often seen for this fungal group, with some species/clades specifically infecting a single host species (Jensen et al., 2001). With insects being the most diverse animal group on earth, significant fungal diversity could be expected among entomopathogens although we still have very limited knowledge on this subject (Blackwell, 2011).

In order for virulence of pathogens and resistance or tolerance of hosts to evolve, genetic variation must be present in both the host and the pathogen populations. To date, evaluations of such variation have principally been undertaken using entomophthoralean fungi. Studies of aphid pathogenic fungi and pea aphids (*Acyrthosiphon pisum*) (which did not harbor facultative endosymbionts; see Section 9.6.2) have found that aphid lines challenged with different strains of *Pandora neoaphidis* vary in susceptibility (Parker et al., 2014), so genotypic diversity in susceptibility among hosts has been demonstrated. In order to investigate genetic variability within and among populations of fungal pathogens, numerous studies of community structure and host specialization at different geographic scales have been conducted with *Entomophthora* spp. and *P. neoaphidis* (see De Fine Licht et al., 2016). One study evaluated epizootics caused by the fly pathogen *Entomophthora muscae* in five house fly (*Musca domestica*) populations located within 33 km of each other in Denmark. Results suggested a panmictic organization of several lineages of *E. muscae*, with a high level of reciprocal migration within the area (Lihme et al., 2009). Another study of *E. muscae* revealed sympatric occurrence of two different clades, each with subpopulations, based on multilocus genotyping (Gryganskyi et al., 2013). One of these closely related clades specialized in a phytophagous fly, *Delia radicum*, earlier in the spring, while the other specialized in a predatory fly, *Coenosia tigrina*, later in spring, although host exploitation did not differ absolutely for the different fungal clades. This unique study demonstrated recombination but limited gene exchange between the closely related populations of *E. muscae*, which were partially separated based on host use and seasonality. Thus, at least for the most host-specialized entomophthoralean fungi, development of new species can perhaps occur as strains of the fungus become isolated in their use of different host species.

Many entomopathogenic species of hypocrealean fungi have broad host ranges and are therefore less likely to coevolve with their hosts, although there are exceptions (e.g., *Metarhizium acridum* exhibits a much narrower host spectrum than most other species of *Metarhizium*; Hu et al., 2014). Most *Beauveria* and *Metarhizium* isolates tested in laboratory bioassays are able to infect host species of several insect orders (i.e., they have broad physiological host range), and isolates of the same clades of *Beauveria* spp.

have also been found to cause infections in diverse host species within the same habitat (Meyling et al., 2009). For these fungal species, the host may not be the main factor governing diversity. As an alternative, Bidochka et al. (2001) posed the hypothesis that for the widespread generalist fungal pathogens such as *Metarhizium* and *Beauveria*, the particular abiotic factors in the environment select for the fungal diversity patterns observed in different habitats. This habitat-selection hypothesis, as opposed to the traditional host-selection hypothesis of pathogenicity, is based on the assumption that the hypocrealean fungi spend major timespans outside of hosts, where they are exposed to environmental abiotic conditions. This idea is supported by observations that tolerance of fungal isolates to abiotic stresses such as ultraviolet (UV) radiation and temperature extremes is correlated with the particular habitats of origin (see Section 9.5) (Bidochka et al., 2001, 2002). Fungal abundance patterns were unrelated to virulence, so the most frequent fungal group would be expected to be the one best adapted for survival in the present abiotic conditions until a new host can become infected. However, in an individual agroecosystem, several *Metarhizium* species representing many genotypes were found to coexist in the soil environment over consecutive years (Steinwender et al., 2014, 2015), and isolates of these genotypes showed variation in UV tolerance and temperature growth optima (Steinwender, 2013). Among the isolates tested, the most abundant genotype, belonging to *M. brunneum*, exhibited relatively high tolerance to UV and high conidial production from host cadavers, while growth rates *in vitro* were intermediate (Steinwender, 2013). These two traits could be hypothesized as being important for enhanced conidial survival outside hosts, as well as for increasing the population once a host is successfully colonized. Alternatively, Wyrebeck et al. (2011) proposed that the species/clades of *Metarhizium* most abundant in the soil environment are adapted to the root systems of the particular plants characteristic of the given habitat (see Section 9.6). However, plant specificity of certain *Metarhizium* genotypes has not been tested experimentally. Whatever the mechanism, it is evident that local communities of *Metarhizium* and *Beauveria* are highly diverse (e.g. Meyling et al., 2009; Steinwender et al., 2014) and that this diversity is likely maintained by the habitat providing sufficient diversity in both biotic and abiotic niches to allow for coexistence of closely related fungal genotypes.

9.3 Dynamics of Fungal Pathogens

9.3.1 Disease Transmission

Fungal pathogens predominantly rely on horizontal means of transmission, where infective spores produced from a cadaver will infect other susceptible individuals after some period of time in the environment. However, variability in mechanisms of transmission abound; sometimes (but rarely), spores are produced before hosts are dead. Spores can be transmitted directly from an infected individual or cadaver to a healthy individual or indirectly from a host to the environment and then to another host. For entomophthoralean species, instead of or along with conidia, resting spores are produced, which are not ready to germinate immediately but require prolonged periods of time and conducive environmental conditions before they will produce infective germ conidia, resulting in a significant delay in transmission from a reservoir.

For infection to occur, the amount of inoculum that is successfully transmitted to a new host must be greater than a threshold pathogen load, below which infection does not occur. This relationship is considered highly different for hypocrealean and entomophthoralean fungal species, in particular. Among Hypocreales, dose–response bioassays demonstrate sigmoidal relationships for *Beauveria* and *Metarhizium* spp. between dose and mortality due to infection, demonstrating that an Allee effect occurs at lower pathogen densities (Hughes et al., 2004; Uma Devi and Uma Maheswara Rao, 2006). Above the threshold for infection, the mass action principle is thought to drive interactions, determining that the probability of a host becoming infected is directly positively related to the number of conidia encountered (McCallum et al., 2001).

In contrast, transmission of Entomophthorales is not considered as dependent on large doses for successful infection, and single conidia are thought to be able to initiate infections (Bellini et al., 1992; Steinkraus, 2006). Conidia of Entomophthorales are larger than those of Hypocreales and carry more of their own resources for the initiation of growth. It has been hypothesized that the larger size of entomophthoralean conidia could be associated with the lower doses of conidia needed for successful infection to occur (Steinkraus, 2006).

Successful transmission is also strongly influenced by climate (see Section 9.5) and host density. Models of *B. bassiana* infection of Colorado potato beetles (*Leptinotarsa decemlineata*) indicated that horizontal transmission was more sensitive to the proportion of cadavers that were sporulating but less sensitive to the time between host death and sporulation (Long et al., 2000). A model was developed based on the effects of temperature on time to 90% mortality after *M. acridum* was applied as a biopesticide against four species of grasshoppers in Africa and Europe (Klass et al., 2007). Ambient temperature data adjusted by delays to fungal growth due to thermoregulation (see Section 9.4.1) resulted in good estimates of pathogen virulence for five of six comparisons. An important aspect of studies of ecology for infectious diseases is the quantification of transmission (see Chapter 1). Coefficients measuring transmission of *M. acridum* in grasshopper populations have been derived (Thomas et al., 1995; Arthurs et al., 2001), as has an estimate of the transmission efficiency of *E. maimaiga* in gypsy moth populations (i.e., one in 4 million conidia that are produced from a cadaver reach a host and infect) (Hajek et al., 1993).

Transmission can be influenced strongly by host behavior and is more successful when hosts are gregarious, so that the chance that a spore will reach a healthy host individual is increased with decreasing distance between hosts. This is true for colonial insects like aphids; when the entomophthoralean pathogen *P. neoaphidis* actively ejects conidia from a cadaver, chances are great that a conidium will successfully land on an aphid and infect, because within colonies, aphids are located near one another (Steinkraus, 2006). Many studies have been conducted investigating the potential for horizontal transmission of conidia from contaminated to healthy hosts during mating (e.g., Reyes-Villanueva et al., 2011; Ugine et al., 2014). The chances of transfer of doses high enough for infection are increased when potential hosts are actively attracted to sporulating cadavers, as for the males of the spider mite *Tetranychus urticae*, which are attracted to female cadavers producing conidia of *Neozygites floridana* (Trandem et al., 2015). Chances of successful transmission can also be greater for insects living in groups, where transmission from parent directly to offspring and horizontal transmission can both occur more readily. For example, female alfalfa leafcutting bees (*Megachile rotundata*) can become contaminated with spores of chalkbrood (*Ascosphaera aggregata*) when they must chew through cells of dead, diseased siblings in order to leave their natal nests (James, 2011). When the bees subsequently create their own nests, lay eggs, and deposit pollen provisions, the pollen can become contaminated with these fungal spores, leading to infection of the offspring.

Social insects employ methods that have been termed "social immunity" to compensate for the potential for high pathogen transmission rates due to high levels of interactions within high-density groups of closely related individuals. Social immunity includes collectively performed sanitary actions, ranging from self- and allogrooming to removal of infected colony members or nest relocation (Cremer et al., 2007). Also acting to decrease transmission rates in groups of social insects can be the fact that as conidia are transferred from a contaminated individual to a healthy individual via normal activities in dense colonies, the amount of inoculum being passed effectively decreases with succeeding passages (Konrad et al., 2012). One model has shown that even with high contact rates, in a social insect colony the dose of fungal spores becomes diluted with repeated horizontal transmission, which can reduce pathogen loads to below the threshold levels needed for infection (Novak and Cremer, 2015).

Behavior of hosts, social or not, can influence transmission of fungal pathogens. Many studies have reported that potential insect hosts are known to either avoid fungal pathogens or not (see Baverstock et al., 2010). Studies have also shown that susceptible hosts are sometimes attracted to fungal pathogens; for example, black vine weevils (*Otiorhynchus sulcatus*) can be attracted to *M. brunneum* (referred to as *M. anisopliae* in Kepler and Bruck, 2006) and collembolans (Dromph and Vestergaard, 2002) and female mosquitoes (George et al., 2013) can be attracted to *B. bassiana*, although these hosts are susceptible to the pathogens. These examples are not thought to be caused by parasitic manipulation of hosts, because these fungi have broad host ranges. However, in more coevolved host–pathogen systems, fungal manipulation of host behavior acts to increase transmission (Roy et al., 2006). The most obvious and well-known example is pre-death climbing, which acts to locate cadavers at higher positions so that spore dispersal will be enhanced; such behaviors are known for numerous specialized host–pathogen associations. In some instances, the fungus can grow out of the cadaver, producing rhizoids to attach the cadaver in the elevated location (Roy et al., 2006), or the host can bite on to the substrate so that the cadaver remains in position after host death; it is assumed that the biting is due to manipulation by the fungus, which acts to increase transmission (e.g., Hughes et al., 2011).

Fungal transmission to hosts can also be influenced by members of the biotic community. For example, as leaves of crop plants grew and increased in area, the effective dose of *M. anisopliae* conidia that would be contacted by herbivorous larvae of mustard beetles (*Phaedon cochleariae*) often decreased (Inyang et al., 1998). In a direct interaction, coccinellids have been reported eating aphids infected with *P. neoaphidis*, as well as sporulating cadavers (Roy et al., 1998), and desert locusts (*Schistocerca gregaria*) infected with *M. acridum* have been shown to have reduced ability to escape from predators (Arthurs and Thomas, 2001). However, the presence or activity of predators and parasitoids can result indirectly in enhanced movement by aphids, resulting in increased contact with infective spores and therefore increased infection. On a local scale, some ants and coccinellids can vector spores of the aphid pathogen *P. neoaphidis* from one location to another (Baverstock et al., 2010).

9.3.2 Fungal Dispersal

Fungi differ among groups in their means for dispersal, although spore movement through the environment is usually considered the major pathway. Active spore dispersal ranges from conidia being forcibly ejected from host cadavers, as with most species in the Entomophthoromycotina and hypocrealean teleomorphs, to swimming flagellated spores in species of the Phylum Blastocladiomycota, such as the mosquito pathogen *Coelomomyces* (Gleason et al., 2010). For terrestrial Entomophthoromycotina, chances for landing on a host or not are random, but the probability of reaching a host is increased via production of supernumerary conidia (see Section 9.2.1.1). Airborne conidial dispersal by species of Entomophthoromycotina can occur with diel periodicity, which maximizes discharge at night, thereby minimizing conidial exposure to UV radiation while increasing exposure to higher humidity and lower temperatures. Discharge of primary conidia of *Neozygites fresenii* from cadavers of cotton aphids (*Aphis gossypii*) resulted in 24% of conidia settling on substrates around aphids and 76% entering the air (Steinkraus, 2006). Numerous studies have shown that entomophthoralean conidia become airborne in episodes, and densities up to 90 000 primary conidia/m^3 of air have been reported.

Models of dispersal of the gypsy moth pathogen *E. maimaiga* strongly suggest that some conidia are dispersed locally while others escape the tree canopy and disperse over longer distances; such two-stage dispersal helps to account for the rapid spread by *E. maimaiga* in North America between 1989 and 1992 (Dwyer et al., 1998; Weseloh, 2003). The concept of spore clouds moving longer distances is consistent with findings that *E. maimaiga* can spread closely behind the low-density populations of gypsy moth at the edge of the ever-expanding gypsy moth populations in North America (Hajek and Tobin, 2011); the cloud can disperse across bodies of water (Tobin and Hajek, 2012), and it has been found infecting sentinel gypsy moth larvae ahead of the spreading populations of this host (A.E. Hajek, unpublished data).

Another means of active dispersal by entomopathogenic fungi is the movement of infected hosts. Aphids with wings, which can disperse long distances as air-plankton, have been shown to be more susceptible to infections by the entomophthoralean *P. neoaphidis*. When aphids were trapped at >30 m above the ground, 36.6% of >7000 winged aphids were infected with pathogens, including at least seven species of fungal pathogens (Feng et al., 2004).

Hypocrealean species do not actively discharge conidia and appear to depend in part on abiotic means such as wind and rain for dispersal (Meyling and Eilenberg, 2007). For example, studies with *B. bassiana* have shown that rainfall is associated with movement of soilborne conidia onto phylloplanes (Bruck and Lewis, 2002). Another means of movement has been shown via association of *Metarhizium* with plant roots, which can result in dispersal of *Metarhizium* within the soil environment (Keyser et al., 2014). In addition, hosts can provide dispersal of fungal spores on their surfaces; for example, aphid predators were associated with movement of *B. bassiana* conidia from soil or from sporulating cadavers to phylloplanes of nettle plants (Meyling et al., 2006).

9.3.3 Fungal Environmental Survival and Persistence

It is generally considered that fungi infecting terrestrial invertebrates predominantly need to persist in the soil either during periods when environmental conditions are not agreeable (e.g., winter, droughts) or when hosts are not present or active. However, beyond this, hypocrealean and entomophthoralean fungi differ significantly in their means of environmental persistence. Hypocrealean fungi are thought not to produce specialized structures for persistence in the field, although microsclerotia (persistent melanized hyphal bundles, typical of many plant pathogens) of *M. brunneum* can be produced *in vitro* (Jackson and Jaronski, 2009); however, these structures are not known from the field for any hypocrealean entomopathogens.

Hypocrealean species are generally thought to persist for from a few days to >1 year. This has been observed in the soil, with abiotic and biotic conditions influencing persistence (Vänninen et al., 2000; Jaronski, 2007). The relatively rapid decline in conidial viability is counterbalanced by very high production of conidia. Moreover, the saprophytic phase of hypocrealean fungi can ensure persistence without the presence of hosts (e.g., *Metarhizium* has been shown to survive well near plant roots, where it can grow as a saprophyte using resources released from roots; Hu and St. Leger, 2002; Wang et al., 2005, 2011), and this dual lifestyle enhances persistence (see Section 9.6.1).

Resting spores of the majority of entomophthoralean fungi are found at or near the soil surface when they germinate; on germination, they forcibly eject an infective germ conidium, which can become airborne. At least for the gypsy moth larval pathogen *E. maimaiga*, infections initiated by germ conidia produce only conidia, while later cycles of infection leading to death and spore production during the approximately 2-month larval season will gradually switch to producing only resting spores in later instars, which ensures persistence (Hajek, 2001).

9.3.4 Impacts on Host Population Densities over Space and Time

Occurrences of epizootics in invertebrate populations caused by fungal pathogens have been reported from a diversity of invertebrate hosts (Carruthers and Soper, 1987; Pell et al., 2001; Klingen and Haukeland, 2006; Baverstock et al., 2010). However, summaries across systems or over longer periods of time or larger areas are infrequent. Usually, only fungal pathogens impacting invertebrates that are pests in agricultural fields or in managed to semimanaged forests have been the subjects of studies (Hesketh et al., 2010), and in some cases, these are emergent pathogens (i.e., pathogens new to science or to the particular location). In addition, although the term "epizootic" is defined as an unusual level of infection (see Chapter 1), the diseases that are usually studied are those causing high levels of infection during epizootics (versus an uncommon pathogen occasionally causing low levels of infection during epizootics). Disease prevalence and quantification of the pathogen in the environment are among the data that are usually required to study epizootics.

For epizootics to develop, hosts, pathogens, and environmental conditions (often referred to as the "disease triangle") must allow ample pathogen survival and infection. Pathogens can increase exponentially by infecting, developing within, and killing hosts (the infection cycle), then following this pattern again and again. A model of the gypsy moth–*E. maimaiga* system suggested that from four to nine infection cycles can occur within a season of approximately two spring months for this pathogen that only infects larval stages of this univoltine host (Hajek et al., 1993). The number of cycles in this system per season is strongly influenced by temperature and ambient moisture levels.

Models that have been used to study epizootics of diseases caused by fungal entomo-pathogens are based on an assumption that if there are more hosts and more inoculum, there will be more infection (= density dependence) (Hesketh et al., 2010). While this has been documented regardless of high- or low-density host populations in some instances (e.g., Kamata, 2000), in some systems density independence has been reported (Monzón et al., 2008). For the gypsy moth pathogen *E. maimaiga*, while density depend-ence has often been documented (Hajek et al., 2015; Hajek and van Nouhuys, 2016), this is not always the case (Liebhold et al., 2013). In the case of Liebhold et al. (2013), where infection was reported as density independent, it was hypothesized that the relation-ship between host population and infection level could be density independent under conditions where heterogeneity in abiotic conditions plays a larger part than host density.

The emphasis in studies of epizootics has usually been on temporal variability in impacts of fungal pathogens on hosts, as opposed to spatial variation. However, studies of the metapopulation dynamics of *E. maimaiga* and the gypsy moth found that this pathogen, which actively disperses via airborne conidia, was present at virtually all sites studied throughout four US states, but that levels of infection varied at different sites (Hajek et al., 2015). This heterogeneity could have been in part due to the differ-ent sample sizes from different sites, but it was also due, once again, to the strong impact that variability in abiotic conditions at different sites can play in the develop-ment of epizootics (Hajek et al., 2015). Spatial simulations of epizootics caused by *B. bassiana*, which is not actively dispersed, in Russian wheat aphid (*Diuraphis noxia*) populations reported pockets of infected versus healthy hosts in wheat, the preferred host plant (Knudsen and Schotzko, 1999). This pattern occurred because the aphids did not move while on wheat, resulting in local extinction of infected aggregations of aphids, but also in the escape of non-infected aggregations. For a nonpreferred host where some aphids departed to locate preferred food, this spatial pattern of aggrega-tion of infections did not occur.

A major focus of studies of pests is whether a fungal pathogen regulates pest popula-tions. The optimal method for such a study is to use life tables, following mortality caused by different agents throughout different life stages of the host. However, for invertebrates, life-table studies are generally only conducted focusing on the stages that cause damage (often neglecting eggs, pupae, and adults). These types of studies require that researchers look beyond the pathogen of interest to include predators, parasitoids, and other pathogens. In the gypsy moth–*E. maimaiga* system, the fungal pathogen, acting with density dependence, caused greater levels of host mortality in both high- and low-density populations compared with the Lymantria dispar multinu-cleopolyhedrovirus (LdMNPV) and four species of parasitoids (Hajek et al., 2015; Hajek and van Nouhuys, 2016). In these studies, outbreak gypsy moth populations crashed to low levels and low-density gypsy moth populations remained at low densi-ties. For Entomophaga aulicae nucleopolyhedrovirus (EaNPV) and Orgyia leucostigma nucleopolyhedrovirus (OlNPV) infecting outbreak populations of the white-marked tussock moth (*Orgyia leucostigma*) in Nova Scotia, infections by both pathogens were abundant but pathogen-induced mortality of hosts was not high enough to completely control the outbreaks (van Frankenhuyzen et al., 2002).

Pathogens that remain more at enzootic levels are rarely studied at the population level, even though chronic infections in populations certainly occur. Colonies of the carpenter ant (*Camponotus rufipes*) are chronically infected at lower levels by *Ophio-cordyceps camponoti-rufipedis* (Loreto et al., 2014), but, spatially, infection does not occur within the colonies due to social immunity. Infected worker ants leave the nests and die above ant trails, and conidia released from their cadavers subsequently fall on to the trails and infect further workers leaving the colonies to forage. Infection levels in this system remain consistently low because only the worker caste – the older ants that are dispersing and spending less time within the colony – is infected by the pathogen.

9.4 Interactions between Fungal Pathogens and Host Individuals

9.4.1 Host Responses to Fungal Pathogens to Prevent or Cure Infections

Some taxa of entomopathogenic fungi are able to induce behavioral responses in their infected hosts to enhance chances of further pathogen transmission, as already mentioned. However, the hosts can also mount responses toward pathogens in order to mitigate the negative consequences of infection, either by avoiding exposure, reducing the infection risk, or limiting the costs of being infected by tolerating the pathogen, or potentially even by clearing the infection altogether (Schmid-Hempel, 2011). Much knowledge of how insects avoid and defend themselves against pathogens has been collected through study systems regarding the evolutionary ecology of insects that have implemented fungi as model pathogens. For invertebrates, it would be beneficial if they could avoid exposure to pathogens altogether. Schmid-Hempel (2011) divided the host defense sequence into pre- and post-infection defenses.

Pre-infection defenses are responses that reduce exposure and infection risk, meaning that they are mounted before or during acquisition of pathogen propagules. Avoiding encounter with high densities of fungal spores through evaluating whether a particular habitat contains the pathogen and avoiding it can be advantageous to the host, particularly for mobile invertebrates, which inevitably will be exposed to pathogens in their environments. While foraging on leaf arenas, the predatory bug *Anthocoris nemorum* avoided leaf areas that were inoculated with *B. bassiana*, and females preferred to oviposit in leaf tissues without *B. bassiana* conidia (Meyling and Pell, 2006). Similarly, the ladybird *Coccinella septempunctata* was able to detect and avoid *B. bassiana* when the fungus was present as free conidia and as sporulating conspecific cadavers on both leaves and soil surfaces (Ormond et al., 2010), with soil representing the overwintering site for the adult ladybirds. Since hypocrealean entomopathogenic fungi are ubiquitous in the soil environment, subterranean insects will become exposed, but avoiding fungal hotspots would likely decrease infection risk. Consequently, soil-dwelling coleopteran larvae and mole crickets avoid soil patches experimentally inoculated with *Metarhizium* sp. and *B. bassiana*, respectively (Villani et al., 1994; Thompson and Brandenburg, 2005).

Some soil arthropods, such as collembolans, are generally resistant to infections by *Metarhizium* spp., *Beauveria* spp., and other entomopathogens (Broza et al., 2001), and collembolan species can be attracted to fungi (Dromph and Vestergaard, 2002). Intriguingly, pharaoh ant (*Monomorium pharaonis*) workers scouting for new nest sites were attracted to potential nests containing conspecific cadavers sporulating with *M. brunneum* (Pontieri et al., 2014), and the authors hypothesized that this behavior could be adaptively explained by immunizing the colony for future exposure to the fungus. Limited knowledge is available on pre-infection host responses toward the specialist fungal pathogens of the Entomophthorales, but pea aphids have been shown to be unable to avoid host plants containing sporulating cadavers with *P. neoaphidis* (Baverstock et al., 2005), so that fungal transmission is not hampered by host behavioral responses.

Parasitoids respond to a range of cues in their habitat in order to evaluate patch quality, predominantly to locate host patches and avoid intraguild predation. A particular case of intraguild predation is when the intraguild predator is a generalist pathogen that infects both the other natural enemy and its prey, such as a parasitoid and its host. In this case, the female parasitoid should avoid hosts and/or host patches containing the pathogen in order to improve survival for itself and its offspring and thereby increase overall fitness. Indeed, the belowground hymenopteran parasitoid *Trybliographa rapae* of the cabbage maggot (*Delia radicum*) laid more eggs in uninfected hosts than in hosts infected by *M. brunneum* or *B. bassiana* (Rännbäck et al., 2015). In choice experiments conducted at a larger spatial scale, *T. rapae* preferentially selected cabbage plants without *M. brunneum* root inoculation rather than plants inoculated with relatively high densities of *M. brunneum* conidia (Cotes et al., 2015). This response correlated with the emission of a specific fungus-derived volatile cue from the *M. brunneum*-inoculated plants. However, there is little evidence of exactly which fungus-related volatiles cause behavioral responses in insects, although volatile profiles characterized for *B. bassiana* and *M. anisopliae* isolates have been shown to be repellent to the termite *Macrotermes michaelseni* (Mburu et al., 2011, 2013), indicating that avoidance behavior can be elicited by these cues. Similarly, chemical compounds of an isolate of *Isaria fumosorosea* were characterized by Yanagawa et al. (2015), who showed that *Coptotermes formosanus* termites avoided these cues if freely present but were attracted to them if fungal spores were present on nestmates. The latter behavior was associated with allogrooming (Yanagawa et al., 2015), a particular hygienic response of social insects toward pathogen exposure (Cremer et al., 2007).

If contact with the pathogen cannot be avoided, behaviors other than grooming can be deployed by insects as pre-infection defensive tools. Prophylactically treating substrates with resins collected from conifer trees protects workers and offspring of wood ants (*Formica paralugubris*) from infections by various microorganisms, including *Metarhizium* and *Beauveria* (Chapuisat et al., 2007; Brütsch and Chapuisat, 2014); the resin acts as an efficient disinfectant. Also, compounds derived from the insects themselves have antimicrobial activity, which can be used to sanitize against fungal conidia (e.g., garden ants, *Lasius neglectus*, use formic acid to disable conidia removed by grooming brood inoculated with *M. brunneum*; Tragust et al., 2013). The production and application of antimicrobial compounds used for the sanitation of cuticular surfaces and surroundings is collectively referred to as the external immune defense (see Otti et al., 2014 for examples). External sanitation is also found in nonsocial insects, but should principally be expected in species with a restricted habitat, such as a localized food source or a nest where offspring are raised; otherwise, this strategy would become too costly (Otti et al., 2014). Interestingly, Pedrini et al. (2015) showed that *B. bassiana* has evolved a countermeasure against benzoquinones, which are pre-infection defense secretions produced by flour beetles (*Tribolium castaneum*). These compounds inhibit spore germination and fungal growth, but the enzyme benzoquinone oxidoreductase produced by *B. bassiana* reduces the deleterious effect on the fungus (Pedrini et al.,

2015). However, the inactivation of the insect defense mechanism is incomplete, indicating that the fungus is lagging behind in the arms race, as would also be expected for a generalist pathogen.

Another pre-infection response to increased risk of pathogen infection would be to invest in an efficient protective barrier that prevents pathogen entry (Schmid-Hempel, 2011), which is of particular relevance to protection against fungal pathogens. In insects, melanization of the cuticle constitutes an investment in barrier protection against infections (Wilson et al., 2001), but the risk of acquiring infectious diseases is increased under high-density conditions. Accordingly, both *S. gregaria* (Wilson et al., 2002) and the flour beetle *Tenebrio molitor* (Barnes and Sita-Jothy, 2000) showed density-dependent cuticular melanization that made individuals from dense populations more resistant to *M. acridum* and *Metarhizium* sp., respectively. The melanization response observed in the cuticle is confounded with investment in the internal immune response (Otti et al., 2014) through upregulation of the phenoloxidase (PO) cascade (Barnes and Siva-Jothy, 2000), an important component of the innate immune response that is active once a pathogen has entered the host (Boughton et al., 2011).

Inside the host, the presence of pathogens will raise a post-infection defense response (Schmid-Hempel, 2011). It is beyond the scope of this chapter to cover the internal immune defense of invertebrates (this topic is reviewed elsewhere, e.g., Boughton et al., 2011), but some specific responses by insects as a consequence of fungal infections have been demonstrated. House flies (*Musca domestica*) exhibit "behavioral fever" (a type of thermoregulation) whereby individuals infected by *Entomophthora* spp. will actively search for warm microhabitats in which to raise their body temperature above levels suitable for fungal growth (Roy et al., 2006). Early in infection, the choice of higher temperatures by *M. domestica* infected with *E. muscae* killed the fungal cells within them, curing the infection (Watson et al., 1993). Later in infection, flies fevered but were unable to clear the infection, although the fever suppressed fungal development in the host (tolerance of infection), allowing females that had fevered to oviposit more eggs than infected flies that did not have access to high-temperature microhabitats (Roy et al., 2006). Interestingly, behavioral fever may be more related to the host biology than fungal taxonomy, since house flies also exhibit behavioral fever when infected by *B. bassiana*. In this case, fevering suppresses development of *B. bassiana*, resulting in more oviposition time for females, but at the cost of reduced egg viability (Anderson et al., 2013). Migratory locusts (*Locusta migratoria*) choose to elevate their body temperatures when infected with *M. acridum* (and, once again, the impact of higher temperatures is greater early in infections), and even short bouts of higher temperatures resulted in tolerance to the fungus until body temperatures became lower (Ouedraogo et al., 2004). When *S. gregaria* fever, locusts successfully tolerate *M. acridum* infections and produce viable offspring (Elliot et al., 2002). When fruit flies (*Drosophila melanogaster*) became infected by *M. robertsii*, they demonstrated a contrasting behavioral response by selecting microhabitats that were cooler than those preferred by its uninfected conspecifics (Hunt et al., 2016). In this case, low temperatures prevented further fungal development inside the host, resulting in tolerance of the pathogen by the host. Reproduction was delayed until later in life compared with uninfected control flies; hence, the major cost was in early-age reproductive output (Hunt et al., 2016). Rännbäck et al. (2015) observed increased oviposition by fungus-exposed *T. rapae* parasitoids that later succumbed to infection, and hypothesized that the elevated egg laying was a response to early-anticipated death, which resulted in investment in rapid reproductive output.

Far-ranging species, such as ground beetles, are expected to invest principally in internal immune defenses (Otti et al., 2014). Indeed, adult ground beetles and staphylinid beetles have highly melanized cuticles, indicating that their investment strategy, at least in part, constitutes a pre-infection defense. These beetles, with both internal and external defenses, have been reported as being highly resistant to hypocrealean entomopathogenic fungi (e.g., Riedel and Steenberg, 1998). Similarly, soil-dwelling herbivores may be expected to invest in internal immune defenses if they disperse widely within the soil environment. In contrast, Otti et al. (2014) argue that species confined to a limited spatial habitat should invest in external immune responses, such as sanitation of body surfaces; this could apply to soil-dwelling beetle larvae, which feed on roots and have a relatively thin cuticle (i.e., little barrier defense) but nonetheless can live for years in the soil. Indeed, prior to emerging as adults and infesting the tree canopy, pecan weevils (*Curculio caryae*) typically spend 2–3 years in the soil in pupal chambers or cells. Shapiro-Ilan and Mizel (2015) demonstrated that soil from *C. caryae* pupal chambers suppressed activity of *B. bassiana* conidia compared to soil from outside pupal chambers. Such localized sanitation would be expected to prevent infection during the soil-dwelling stages.

9.5 Impact of Abiotic Factors on Infected Hosts and Pathogen Inocula

Entomopathogenic fungi are generally sensitive to abiotic environmental factors (Jaronski, 2007, 2010; Vidal and Fargues, 2007); survival and activity of fungi can be strongly impacted by temperature, moisture, light, aeration, and surrounding pH (Deacon, 2006). These environmental factors can impact fungal pathogens at any stage (e.g., while growing inside of hosts, when infective stages are being produced, during dispersal and infection, during dormancy). The scientific literature is replete with studies investigating how environmental conditions, but especially temperature, moisture, and light, influence entomopathogenic fungi (e.g., Jaronski, 2010; Rangel et al., 2015). A particular emphasis has been on laboratory studies comparing the activities of different isolates of fungal species under controlled environmental conditions in order to evaluate fitness and environmental stress tolerances. In particular, studies have shown that physical, chemical, and nutritional conditions during mycelial growth can impact tolerance to heat or UV-B radiation in subsequently produced conidia, although at times increased tolerances were accompanied by decreased conidial production (Rangel et al., 2015).

Often, fungal isolates from different geoclimatic origins display differential environmental stress tolerances in response to differing temperatures and moisture regimes (see Section 9.2.2.1) (Vidal and Fargues, 2007). Comparing fungal isolates within the same species (all drawn from the northern hemisphere), those originating from more southern areas are often tolerant to higher temperatures than those from northern sites (e.g., Bidochka et al., 2002). A study comparing fungal isolates for control of black vine weevil larvae during early spring, demonstrated that Norwegian isolates survived better under cold conditions than an isolate of *M. brunneum* originating in Austria (Klingen et al., 2015).

Moisture levels, often quantified as ambient relative humidity (RH), dew, or rainfall, are generally considered the most environmentally limiting condition for fungi. High humidities (e.g., >90%) are generally thought to be required for sporulation, germination, and infection by hypocrealean species (e.g., Inglis et al., 2001). However, in the semi-arid African Sahel, production of spores by *M. acridum* from grasshopper cadavers required high humidity associated with rainfall, although subsequent infection did not appear to have this requirement (Arthurs et al., 2001). Ambient moisture levels are generally positively associated with the development of epizootics (Hajek et al., 2001). In agreement with this, moisture levels were the most important environmental factors associated with levels of gypsy moth larval infection by *E. maimaiga* over 3 years, at the end of which time outbreak host populations crashed in association with high levels of infection (Reilly et al., 2014). In addition, frequency of rainfall was positively associated with levels of infection by *E. maimaiga* during a long-term study of low-density host populations (A.E. Hajek, unpublished data). Rainfall can also have devastating impacts on entomopathogenic fungi following application to foliage for biological control, as conidia are washed into the soil (Jaronski, 2010). However, high levels of ambient moisture are not always necessary for infection, because entomopathogenic fungi can utilize micro-environmental moisture. Thus, stored products pests, which live in fairly dry environments, can become infected by *Isaria fumosorosea* (syn. *Paecilomyces fumosoroseus*) (Michalaki et al., 2007). Species within the entomophthoralean genus *Neozygites* create specialized spores called capilliconidia for infection of smaller, walking insects; this type of spore is better able to survive and retain activity in areas where low RH is common.

Along with moisture levels, temperature is critically important, both in relation to temperature extremes that restrict fungal activity and in terms of impacting the range of fungal activity between high and low extremes. As thermotolerances occur across a range of temperatures, temperature tolerance is less exacting than moisture tolerance, as generally the levels of moisture required are very high. Within moderate temperature ranges, temperature impacts the rapidity with which a fungus can function. Comparisons of *Metarhizium* isolates (described in Section 9.2.2.1) demonstrated that *M. brunneum* isolates from forest soils were relatively cryophilic (e.g., active to 8 °C), while *M. robertsii* isolates from agricultural soils were relatively thermophilic (e.g., active at 37 °C) and more tolerant to UV light (Bidochka et al., 2001; M.J. Bidochka, pers. comm.). Comparable temperature-dependent growth patterns were observed for separate clades of *B. bassiana* of agricultural and forest ecosystem origin (Bidochka et al., 2002), suggesting clade/species-specific adaptations to particular abiotic factors. However, a contrasting pattern was seen in an agroecosystem in Denmark, where isolates of *M. robertsii* had high relative growth activity at 12.5 °C compared to co-occurring isolates of *M. brunneum*, which was the most abundant species in the agricultural soils (Steinwender, 2013). These results support the hypothesis by Bidochka et al. (2001, 2002) that the clades/species most abundant in agricultural soils have limited activity at lower temperatures and these traits are likely not species-related, but rather represent ecological traits of locally adapted fungal clades.

Of great importance is how quickly after infection hosts are killed, and this can be strongly influenced by temperature. For pest-control purposes, the goal would be to prevent damage by pests that impacts humans before the pests die from infections. Empirical studies and models of the activity of *B. bassiana* (developed to target mosquitoes) showed that from 10 to 34 °C, 90% of two key malaria vectors died within their extrinsic incubation period, so that transmission of the parasite *Plasmodium*, which causes malaria, did not occur (Heinig et al., 2015).

Of course, in the field, conditions are not constant, and results from laboratory studies of variable conditions or interactions are more informative relative to ecology. On testing fluctuating moisture and temperature regimes to simulate temperate, subtropical, and arid regions, quiescent conidia of *I. fumosorosea* persisted optimally under the temperate regime (Bouamama et al., 2010a). When quiescent conidia were exposed to various wetting–drying cycles under variable temperature and humidity regimes, conidial persistence was higher on wet foliage and on foliage exposed to the cycles than on dry foliage (Bouamama et al., 2010b). In addition, temperature and moisture treatments were less limiting under dark conditions than under daytime light conditions (Vidal and Fargues, 2007).

Light can impact pathogenic fungi in numerous specific ways. First, a major emphasis of studies has been on the impact of UV on fungal spores, especially when spores are deposited on the surfaces of leaves of agricultural crops to provide pest control, as these spores are usually directly exposed to sunlight. UV-B is considered more damaging than UV-A, and therefore has been the focus of many studies (Jaronski, 2010). Most entomopathogenic fungi are sensitive to UV radiation, and spores often cannot survive more than a few hours of direct UV exposure (Fernandes et al., 2015). However, once again, this is often dependent on the fungal isolate and is not necessarily species-specific (Bidochka et al., 2001; Jaronski, 2010). As one mechanism for protection against UV damage, darker pigmentation of *M. anisopliae* conidia has been associated with greater tolerance of UV (Braga et al., 2006). Furthermore, fungal inoculum was found at higher frequencies on leaf surfaces inside lime tree canopies than at the canopy edges (Howe et al., 2016), indicating protection of inoculum from adverse conditions, such as UV light, by the shade of the canopy. Fungi can also be influenced by diurnal cycles of light and dark. Numerous entomophthoralean species cause host mortality in diurnal cycles, with hosts often dying before scotophase (Pell et al., 2001; Nielsen and Hajek, 2006). Because most entomophthoralean species must kill a host before new spores are produced, this allows the fungus to take advantage of cooler, moister conditions at night for sporulation. Light intensity did not influence the production of primary conidia or infective capilliconidia by *Neozygites floridana*, although the numbers of spores produced decreased under constant light conditions compared with constant dark or 12 hours light/12 hours dark (de Castro et al., 2013).

Finally, abiotic features such as soil type, soil pH, presence of nutrients (for hypocrealean fungi), and presence of agrochemicals can impact the survival and growth of pathogenic fungi. Soil texture impacts the infection of insects, but the effect is tempered by levels of soil moisture, which display differential interactions with different soil types (Jaronski, 2007).

9.6 Impact of Biotic Factors on Pathogenic Fungi

9.6.1 Endophytic and Rhizosphere Associations of Invertebrate Fungal Pathogens

Some taxa of invertebrate pathogenic fungi have evolved adaptations for utilizing living plants as substrates, and these lifestyles have recently received increased attention from researchers following the initial documentations of such plant associations by

Beauveria (Bing and Lewis, 1991) and *Metarhizium* (Hu and St. Leger, 2002). This topic has recently been reviewed (Behie and Bidochka, 2014; Quesada-Moraga et al., 2014; Vidal and Jaber, 2015; Moonjely et al., 2016), and here we will mainly focus on aspects of ecological relevance, including trophic interactions. Curiously, research on *B. bassiana* endophytes has mostly focused on applied ecological aspects (e.g., experimental establishment and trophic effects), while research on plant associations for *Metarhizium* spp. has, to a large extent, focused on the mechanisms behind the fungus–plant interactions.

In particular, *B. bassiana* and *Metarhizium* spp. have been shown to associate with many different plant species, with *B. bassiana* being able to grow endophytically in both below- and aboveground plant organs and *Metarhizium* spp. mostly occuring only in relation to the root system (Behie et al., 2015). *Beauveria bassiana* remains internalized within host plant tissues without causing damage to the plant, so it can be regarded as a true endophyte, being a microorganism that is asymptomatically harbored within plant tissue (Petrini, 1991), while *Metarhizium* spp. appear to have most of their relative biomass in the rhizosphere, with limited intimate hyphal growth inside root tissues (Sasan and Bidochka, 2012; Behie and Bidochka, 2014), defining them more appropriately as rhizosphere-competent (Hu and St. Leger, 2002). Nevertheless, the discovery of the close associations of these fungi with plants has spurred a wealth of recent research focused toward potentially utilizing the fungus–plant interaction for plant protection and crop production (e.g., Moonjely et al., 2016).

9.6.1.1 Natural Occurrence and Distribution of Invertebrate Pathogenic Fungi as Plant Associates

Naturally occurring fungal endophytes are ubiquitous in diverse tissues of most, if not all, plant taxa (Rodriguez et al., 2009). Among these, *B. bassiana* has occasionally been reported when plant tissues have been surveyed for culturable fungi (Vega et al., 2008, 2010), suggesting that this species is merely a member of diverse endophytic communities and that any effects due to occupation of a plant by endophytes should not be attributed to *B. bassiana* in isolation. Behie et al. (2015) isolated *B. bassiana* from root, stem, and leaf tissues of field-collected plants, indicating that colonization is possible throughout the plant, which was confirmed by experimental laboratory inoculations. Interestingly, isolation of *Metarhizium* spp. was reported to be localized to the root systems of investigated plants in both field-collected and experimental plant material (Behie et al., 2015). Although several species of *Metarhizium* can be isolated from the same field-collected plants (Behie et al., 2015; Steinwender et al., 2015), some level of plant specificity was reported by Wyrebek et al. (2011), who isolated *M. robertsii* more frequently from roots of grasses and herbaceous plants in open habitats, and *M. brunneum* and *M. guizhouense* mostly from roots of trees. Using an *in vivo* isolation approach involving exposing bait insects to field-collected roots with adhering rhizosphere soil, Fisher et al. (2011) also reported a high frequency of *M. guizhouense* from tree roots (spruce and fir species), with *M. brunneum* most frequently isolated from roots of blueberry and strawberry.

The repeated isolation of several similar multilocus genotypes of *M. brunneum* and *M. robertsii* in the same agroecosystem in consecutive years (Steinwender et al., 2014, 2015), using first insect baits for soil sampling and then the *in vitro* method for isolation from both exterior and interior of roots (Wyrebek et al., 2011), indicated that both entomopathogenic and rhizosphere association traits of *Metarhizium* spp. are widespread. It was hypothesized that the nonspecific association of *Metarhizium* spp. with different crops could be caused by the occasional recruitment of fungal associates by plants from the surrounding soil environment (Steinwender et al., 2015), reflecting the entomopathogenic fungal community of the soil of the particular habitat rather than a coevolved symbiosis. However, it remains to be resolved to what degree plant communities govern the distribution of *Metarhizium* spp. in the soil of different habitats, or whether the patterns observed are dictated by abiotic factors (as suggested by Bidochka et al., 2001: see Section 9.5), and to what extent species occurrences are caused by historical colonization events. In a long-term field experiment, a transformed isolate of *M. robertsii* became established in the soil environment of a turf grass field within the naturally occurring *Metarhizium* community for at least 4 years (Wang et al., 2011), indicating that establishment in a newly colonized habitat will be possible if rhizosphere competence is a trait of the immigrating genotype.

9.6.1.2 Experimental Inoculations of Plants with Entomopathogenic Fungi

The application potential of *B. bassiana* as an endophyte has recently received a lot of attention, and researchers have reported the successful experimental inoculation of many plant species from several families (reviewed by Quesada-Moraga et al., 2014; Vidal and Jaber, 2015). Fewer studies have included investigations of the trophic effects of the endophytic association, and often negative effects on herbivore population growth have been reported with only occasional evidence of fungal infection (McKinnon et al., 2017). Plant pathogens are also reported to be suppressed when plants are inoculated with *B. bassiana* and *Lecanicillium* spp. (Ownley et al., 2010). Thus, it appears that the ecological effects of the fungus–plant association are pronounced, but that mechanisms involving *B. bassiana* are likely related not to its pathogenicity but rather to responses produced either by the fungus or the plant that affect population growth of herbivores or establishment of plant pathogens. It has been mentioned that the mechanisms involved could be related to fungal production of secondary metabolites by *B. bassiana* (Quesada-Moraga et al., 2014), but plants also respond to the presence of microbes by induction of systemically produced compounds (Ownley et al., 2010; Pineda et al., 2010), which could also play an important role. Future studies of *B. bassiana* endophytes should consider the mechanistic aspect of plant responses to fungal inoculation as an important part of the fungus–plant–insect interaction.

In contrast to plant association studies for *B. bassiana*, research into the rhizosphere associations of several *Metarhizium* species has mostly focused on particular fungal-derived mechanisms involved in fungus–plant interactions, while addressing ecological questions to a much lesser extent. However, several insights have been made, including that: (i) *Metarhizium* spp. can build up biomass in the rhizosphere over extended periods in the absence of insect hosts (Hu and St. Leger, 2002; Wang et al., 2011; Klingen et al., 2015), presumably by utilizing carbohydrates in root exudates (Fang and St. Leger, 2012); (ii) *Metarhizium* spp. can transport nitrogen from infected insect cadavers in the soil to plants via hyphal connections with the roots (Behie et al., 2012; Behie and Bidochka 2014); and (iii) the generalist entomopathogenic *Metarhizium* species have a diverse genetic toolbox, with genes for both insect and plant associations (Gao et al., 2011; St. Leger et al., 2011). There remain different views as to whether entomopathogenicity in *Metarhizium* evolved earlier than plant association (Spatafora et al., 2007) or whether the ability to infect insects is a derived trait originating from a plant-associated ancestor (Gao et al., 2011; Moonjely et al., 2016). Nonetheless, the dual ability to use both plants and insects as resources opens new ecologically relevant questions (Boomsma et al., 2014). While *Metarhizium* spp. are able to persist and propagate hyphal biomass alongside and in competition with a large microbial community in the rhizosphere, the insect host constitutes a unique nutritional resource for producing a burst of new infective units after infection. Furthermore, the association with roots could be viewed as a means of obtaining mobility in the soil by an otherwise immobile organism, in order to locate new insect hosts for infection (Keyser et al., 2014), while the saprotrophic lifestyle ensures survival without insect hosts (Wang et al., 2011). It is also important to note that so far, very little focus has been placed on responses in the plant as a consequence of these fungal associations, although overall plant growth promotion has been reported (Moonjely et al., 2016). Unravelling such responses – both ecological effects and the mechanisms behind induced molecular pathways – is important to fully understanding and appreciating the application potential of the fungus–plant association.

Interestingly, the nematophagous fungus *Pochonia chlamydosporia* is also an endophyte of various plants (Larriba et al., 2014), colonizing different tissues (Behie et al., 2015), and *P. chlamydosporia* has been reported to elicit growth promotion and defense pathways in barley (Larriba et al., 2015). Since *Pochonia* is the sister group of *Metarhizium* (Kepler et al., 2014), similar responses should be studied in plants inoculated with rhizosphere-associating *Metarhizium* spp. isolates. It should also be noted that fungal taxa from genera not usually considered invertebrate-pathogenic may be superior to *B. bassiana* in insect suppression when established as endophytes (Akutse et al., 2013). Therefore, invertebrate pathogenic fungi occurring as endophytes and/or root associates may not necessarily be the ultimate plant mutualists or "silver bullets" capable of both enhancing plant growth and providing protection against pests in agriculture. However, realization of the multifaceted lifestyles of these fungi should spur unparadigmatic research into plant–fungus–insect interactions that will potentially establish novel platforms for plant protection, which will most likely succeed by integrating other microorganisms in order to gain the full benefit of their differential adaptations.

9.6.1.3 Direct and Indirect Fungal Interactions with Insects and Plants

The various interactions and resulting effects that invertebrate pathogenic fungi have with other organisms can be defined as direct or indirect depending on whether they are between an actor and a receiver/responder (direct) or are mediated by a third party (indirect). We have emphasized these aspects in relation to fungal entomopathogens elsewhere (Meyling and Hajek, 2010), and here we will just briefly highlight some central aspects of significance. In the traditional view of host–pathogen interactions, fungal pathogens have a direct interaction with insects via infection; in these cases, the direct effect is negative to the insects and positive to the fungi. If the particular insect is a herbivore and the infection causes a decline in insect population size resulting in reduced tissue damage due to herbivory, then the fungus has an indirect interaction with the plant, mediated by the insect, causing a positive density-mediated indirect effect through the reduction of insect density. However, since several fungal taxa are now known to be associated with plants in the rhizosphere and/or as endophytes, multiple interactions can be predicted, both direct and indirect (Fig. 9.1). Negative effects

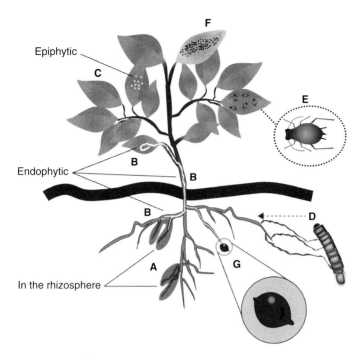

Fig. 9.1 Plant–fungus–invertebrate associations and interactions. Some hypocrealean invertebrate pathogenic fungi can associate with plants by (A) proliferating as saprophytes in the areas around roots (the rhizosphere), (B) growing inside plant tissues as endophytes below- and aboveground, or (C) being present on plant surfaces as epiphytes. (D) Hyphae from fungal-infected insect cadavers in the soil can connect to roots and thereby transfer nutrients from an insect cadaver to a plant. Endophytic fungi can have trophic effects by (E) reducing population growth or fitness of insect herbivores and (F) reducing disease symptoms caused by plant pathogens. In addition to growing endophytically, (G) some fungi can infect nematode eggs. Figure prepared by Dr. Helen Hesketh, Centre for Ecology & Hydrology, UK © NERC (CEH). (*See color plate section for the color representation of this figure.*)

other than infections observed in insects feeding on plants endophytically colonized by *B. bassiana* (e.g. Gurulingappa et al., 2010) will then be considered as indirect if they are mediated by an elicited plant response or as direct if they are caused by compounds produced by the fungus. For example, in numerous systems, tritrophic interaction studies have shown that insect or mite susceptibility to a fungal pathogen, as well as pathogen development, can differ based on the plant eaten by the herbivorous host (see Wekesa et al., 2011). Complex interaction networks can thus be predicted depending on the physical location on the fungus and its mode of action. From an applied perspective, such speculations might be of purely academic interest as long as the insect population is suppressed due to fungal activity, but clarifying the mode of action is actually part of the registration process for biocontrol agents within the European Union (EU Commission, 2013; R. Gwynn, pers. comm.).

9.6.2 Interactions between Host Symbionts and Fungal Pathogens

Knowledge of symbionts associated with insects has grown – including knowledge about defenses provided by symbionts to counter fungal entomopathogens. In particular, aphids have reduced immune systems and a diverse community of facultative endosymbionts (see Barribeau et al., 2014). Scarborough et al. (2005) were the first to demonstrate that pea aphids hosting a bacterial endosymbiont were protected by it against a fungal pathogen. In this study, the fungal pathogen investigated was *P. neoaphidis* and the endosymbiont was a facultative secondary symbiont, *Regiella insecticola*, which can be removed using antibiotics. Pea aphids hosting *R. insecticola* were more resistant to *P. neoaphidis* infection, and among those that became infected, sporulation was lower than for pea aphids not harboring the endosymbiont, although the levels of the latter response differed by aphid clone. Subsequent studies demonstrated protection against *P. neoaphidis* when pea aphids were hosting the distantly related facultative symbionts *Rickettsia*, *Rickettsiella*, and *Spiroplasma* (Łukasik et al., 2013). Presence of *R. insecticola* also protected against the aphid-specific entomophthoralean fungus *Zoophthora occidentalis*, but not the less host-specific *B. bassiana* (Parker et al., 2013). Pea aphids hosting both *Spiroplasma* and X-type (an endosymbiont belonging to the Enterobacteriaceae) endosymbionts were protected against *P. neoaphidis*, but presence of X-type in non-infected insects had a cost as fecundity declined compared with pea aphids with *Spiroplasma* but without X-type (Heyworth and Ferrari, 2015).

Now that protection from fungal pathogens by aphid endosymbionts is well established, at least for the pea aphid as a model system, researchers have begun investigating the evolutionary and ecological implications of communities of symbionts as part of host microbiomes. Although the majority of studies to date have been conducted in laboratories, in one field study the presence of the eight heritable facultative endosymbionts associated with pea aphids was variable in field populations. At least in clover fields, higher levels of *R. insecticola* were associated with lower aphid mortality from fungal infections (Smith et al., 2015). We expect that studies investigating endosymbionts, insect hosts, and fungal pathogens will continue to explore interactions under ecologically realistic conditions, as well as expand to additional host systems.

Ants constitute a very different type of symbiont when tending aphids; these mutualists gain honeydew from aphids and, in return, provide protection against predators and parasitoids. Studies have now shown that ants tending aphids also detect when aphids have been killed by *P. neoaphidis* and remove cadavers from aphid colonies (Nielsen et al., 2010). Ants can recognize *P. neoaphidis* conidia on cuticles of healthy aphids, which they sanitize by grooming.

9.6.3 Interactions between Fungal Pathogens and Other Natural Enemies

Research is often focused on one host infected with one parasite or pathogen, but Thomas et al. (2003) make a good case that mixed infections by different pathogens are probably common in nature. When sharing a host, entomopathogenic fungi interact with other natural enemies (i.e., other pathogens, parasitoids, and predators), and the resulting effects can be either synergistic or antagonistic if the two enemies have an interaction in the host, or additive if no interaction occurs (Roy and Pell, 2000; Cedergreen et al., 2013). However, it is important to use well-established definitions of these concepts, such as "independent action" or "response addition," as emphasized by Cedergreen et al. (2013) and Backhaus and Faust (2012), to compare observed and expected effects. Although these authors based their models within an ecotoxicological framework, the concepts are readily applicable to invertebrate pathology, where mortality is usually studied as the response variable. Based on the diversity and abundance of insect natural enemies, an equivalent diversity of interactions is possible, and an increasing number of studies have been exploring this subject.

9.6.3.1 Interactions among Co-infecting Pathogens

A majority of studies have been conducted on interactions between two co-infecting pathogens and have not extended to interactions between three. Speed of kill (virulence) and sporulation (fitness) of *M. acridum* were influenced by co-infection with *B. bassiana*, although the *B. bassiana* strains being tested were largely avirulent against *S. gregaria* (Thomas et al., 2003). Co-infection in this system of *M. acridum* and one strain of *B. bassiana* resulted in a synergistic effect in relation to virulence, although sporulation declined. When two strains of *Metarhizium* spp. competed within the same *G. mellonella* larval host, the more virulent strain (in terms of both killing and speed of sporulation) was the successful competitor and was able to reproduce (Staves and Knell, 2010). On infection of *Plutella xylostella* by two strains of the entomophthoralean *Zoophthora radicans*, whichever strain was present in higher conidial concentration in experimental arenas caused the infection; co-infection, in which both strains reproduced, only occurred when hosts were exposed to high doses of both pathogens (Morales-Vidal et al., 2013). On comparison between two different species of entomophthoralean fungi, the species with the higher concentration of conidia in the environment was more successful at infecting *P. xylostella*, regardless of virulence (Guzmán-Franco et al., 2009). Further studies in this same system demonstrated that the order of infection of *P. xylostella* by *Z. radicans* and *Pandora blunckii* determined which species would successfully reproduce when the host was dead (Zamora-Macorra et al., 2012). When larvae were co-inoculated with both fungal species, the prior residency times of the pathogens and the order of inoculation affected which species reproduced, and it was rare that both fungal species reproduced from one cadaver (Sandoval-Aguilar et al., 2015).

Fewer studies have been conducted evaluating interactions between fungal and other types of pathogens. One study investigating co-infections of the gypsy moth fungal pathogen *E. maimaiga* with LdMNPV reported that the order of infection strongly influenced reproduction by the two pathogens (Malakar et al., 1999). In addition, *E. maimaiga* generally killed hosts faster than LdMNPV, and the prevalence of infections in which the virus reproduced increased only when larvae were infected with the virus 10 days before fungal infection. This is consistent with the theory that during co-infections (also called superinfections), the most virulent strain infecting a host will be the winner (Nowak and May, 1994), and therefore the one that is transmitted.

Studies of co-infection between entomopathogenic nematodes and fungal pathogens have demonstrated results ranging from antagonism to additive effects to synergy. Barbercheck and Kaya (1991) showed that when *Spodoptera exigua* larvae were first infected with *B. bassiana*, they were more susceptible to *Heterorhabditis bacteriophora*. Ansari et al. (2008) reported synergistic interactions when larvae of black vine weevils were exposed to *M. anisopliae* conidia and either simultaneously or subsequently to any of three species of entomopathogenic nematodes. In contrast, Shapiro-Ilan et al. (2004) reported antagonism as a result of some pairs of fungi and nematodes that were combined for testing against the pecan weevil. Other studies have shown additive effects of fungi and entomopathogenic nematodes (Ansari et al., 2006; Neumann and Shields,

2008). In contrast, when different strains and species of *Metarhizium* spp. co-infected *G. mellonella* larvae along with the nematode *Steinernema feltiae*, the less virulent strains of *M. anisopliae* were more successful at reproduction (Staves and Knell, 2010). Community-level studies with *B. bassiana* and three species of entomopathogenic nematodes against two insect hosts showed that with greater biodiversity of nematodes, but the same pathogen density, host mortality increased (Jabbour et al., 2011).

9.6.3.2 Interactions of Fungal Pathogens with Parasitoids and Predators

Fungal pathogens can directly infect some predators and parasitoids, and sometimes larval parasitoids within hosts. Alternatively, fungi can indirectly affect parasitoids and predators by infecting the host needed for food (for the predator) or development (for the parasitoid). Research has emphasized the latter interaction where a shared resource is utilized.

Interactions between parasitoids in the Hymenoptera and Diptera and entomopathogenic fungi have often been studied in an effort to understand how these agents can be used on the same crops with additive or even synergistic effects (Roy and Pell, 2000; Furlong and Pell, 2005). Usually, fungi develop and kill hosts faster than parasitoids, and the parasitoids then die because they are not ready to pupate; therefore, in many systems, the prior residence time for the parasitoid is critically important to determining the outcome. Thus, parasitoids can usually only successfully develop in shared hosts if the fungus infects significantly later than them. In contrast, the aphelinid parasitoid of the Russian wheat aphid and the fungus *Isaria fumosorosea* have similar development times, so the fungus does not have a complete advantage over the parasitoid, which can sometimes successfully develop in aphids that have been infected by the fungus first (Mesquita and Lacey, 2001). If parasitoids successfully develop after sharing a host with a fungal pathogen, adults may have reduced fecundity and longevity and the fungus may have reduced spore production (Tamayo-Mejía et al., 2015).

As a positive interaction between fungi and predators or parasitoids, fungal transmission can be increased by the activity of the predators/parasitoids (Section 9.3.1) (Baverstock et al., 2010). This can occur when the predators or parasitoids act as vectors or the hosts/prey increase activity to avoid these natural enemies. Predators and parasitoids can also either avoid utilizing fungal-infected insects or can feed on them preferentially (Furlong and Pell, 2005). For example, predatory mites feeding on cassava green mites (*Mononychellus tanajoae*) feed on fungal-infected hosts more frequently than healthy hosts, but mites eating fungal-infected hosts have reduced survival and oviposition (Agboton et al., 2013).

Relatively few studies have been conducted investigating interactions between fungal pathogens and other natural enemies at the population level in order to evaluate the impact of the species of entomopathogenic fungus on the densities of hosts, parasitoids, predators, and other pathogens. Kistner et al. (2015) found that mechanical transmission of the pathogen *Entomophaga grylli* by ants significantly aided transmission, and therefore reduced grasshopper (*Camnula pellucida*) populations. Baverstock et al. (2009) conducted experiments in polytunnels with the entomophthoralean *P. neoaphidis*, the parasitoid *Aphidius ervi*, and the predator *C. septempunctata*, all of which attack pea aphids (Baverstock et al., 2009). They found that while the predator ate some infected aphids (and thus acted as an intraguild predator), this was compensated by the additional transmission due to increased aphid movement in the presence of the predator, and the parasitoid was not negatively impacted by the fungal pathogen. In contrast, a study of alfalfa weevil populations in eastern North America showed that the fungal pathogen *Zoophthora phytonomi* could displace two species of parasitoids introduced to provide classical biological control (see Furlong and Pell, 2005). However, the parasitoid species that prefered later larval instars, in which less fungal infection was seen, was the superior competitor of the two. Regardless, when rainfall was abundant and epizootics occurred, parasitoid populations declined, leading to instability in the system during dry years (Harcourt, 1990). Another study of multiple natural enemies involved the gypsy moth larval pathogen *E. maimaiga*, LdMNPV, and four species of parasitoids, with comparisons between low-density host populations and a range of other host densities (Hajek and van Nouhuys, 2016). In both contexts, the activity of the fungus was density-dependent (infection levels increased at higher host densities), and the fungus was the most abundant natural enemy. The fungus developed faster than any of the other natural enemies, and few parasitoids usually developed to adulthood in the hosts. In contrast, levels of virus infection were low, but, because the virus reproduced to some extent before a co-infected host larva died, LdMNPV could successfully co-infect with *E. maimaiga*.

9.6.4 Mycoparasitism of Fungal Pathogens

Most invertebrate pathogenic fungi produce many spores to compensate for the fact that a large proportion of spores will not reach a host and successfully infect. As already discussed, beyond the high chance of each spore not reaching a host, abiotic conditions must be amenable for fungal spore survival and infection. However, many studies conducted with soil have found more infections to occur in sterile soil than in nonsterilized soil, suggesting that fungal spores can also be inactivated by members of the soil biota (Shimuza et al., 2002) through antibiosis by excretion of secondary metabolites (Vos et al., 2015) or because fungi are consumed by mycoparasites. For example, a mycoparasite has been reported decreasing *B. bassiana* sporulation from the cadaver of a Colorado potato beetle (*Leptinotarsa decemlineata*) (Posada et al., 2004). In ant graveyards where cadavers of the carpenter ants *C. rufipes* killed by *O. camponoti-rufipedis* were collected, 55.4% had been hyperparasitized by mycoparasites and therefore would not sporulate (Andersen et al., 2012).

Resting spores of the entomophthoralean pathogen of gypsy moth (*E. maimaiga*) were shown to decrease significantly in number over time after an epizootic (Hajek et al., 2004), which initiated studies searching for potential mycoparasites. A mycoparasitic chytrid was found parasitizing *E. maimaiga* resting spores exposed to soil from hardwood forests in southern Ohio, USA (Hajek et al., 2013). Further investigations of soils with histories of epizootics caused by *E. maimaiga* from New York State and Pennsylvania, USA yielded ascomycete and oomycete mycoparasites (Castrillo and Hajek, 2015). Results from these few studies strongly suggest the need for further research on the impact of mycoparasitism on environmental persistence of invertebrate-pathogenic fungi.

Fungi with adaptations for mycoparasitism are used as biological control agents against plant pathogens (e.g., *Trichoderma* spp.) (Vos et al., 2015), and such traits could potentially compromise the compatibility of other fungal biological control agents (Keyser et al., 2016). Laboratory *in vitro* tests indicated that a *Metarhizium* isolate was susceptible to several isolates of mycoparasitic fungi, but that the virulence was largely unaffected when *Metarhizium* was applied to an insect together with a mycoparasite

(Krauss et al., 2004). Combinations of *Metarhizium* spp. and the mycoparasitic fungus *Clonostachys rosea* in seed treatments to control soil insects and a seedborne pathogen suggested no interaction effect, so that insects feeding on roots became infected by *Metarhizium* while the pathogen's symptoms were controlled by *C. rosea* (Keyser et al., 2016). The exact mode of action of fungi such as *Trichoderma* spp. or *C. rosea* is not always clear; whether mycoparasitism, antibiosis, or competition is occurring, these antagonists are often just broadly referred to as "antagonists."

9.7   Use of Pathogenic Fungi for Biological Control of Invertebrates

Fungal pathogens are used to provide biological control in numerous ways. In fact, external penetration through the cuticles of hosts means that this is the group of pathogens that is considered first for the control of herbivorous hemipteran species, because these hosts do not chew the surfaces of leaves, which largely excludes infection by viruses, bacteria, and microsporidia, which infect through the digestive tract.

The primary type of biological control emphasized for fungal pathogens is inundative augmentation, in which biological control agents are applied for immediate impact based on the propagules released (Eilenberg et al., 2001). In many instances, the fungus persists for some period of time and successive waves of infection occur, but the primary goal of this approach is to create temporary epizootics that manage pest populations. Over 150 products have been developed for inundative use around the world (Faria and Wraight, 2007). Almost all of these products are species in the Hypocreales, including fungal strains with somewhat broad host ranges, so that companies producing them have larger markets. Fungi of the other major group of arthropod-pathogens, the Entomophthorales, have narrower host ranges, which has deterred development by industry and, due to the more complex nutritional requirements of many of these species, has meant that large-scale methods for mass production have rarely been developed.

The hypocrealean fungi available for inundation are often chosen for use as replacements for synthetic chemical pesticides in agriculture, but care must be taken so that these living organisms provide optimal control. The correct doses of fungal propagules must be applied to locations where the targeted pests are present; historically, entire areas were blanketed with spores, but more recently researchers have been trying to concentrate infective spores over narrower areas where hosts are present. Different concerns must be addressed if the pest is present on foliage versus in the soil. For foliar applications, temperature, moisture, and UV radiation are major factors impacting efficacy, while temperature, moisture, and soil characteristics, including biotic factors, impact fungal pathogens applied to the soil (Jaronski, 2010). Fungal pathogens are best used inundatively to target pests at lower density during population buildup so that applications are preventative and not reacting to high densities of pests.

As already mentioned, some species of hypocrealean fungi utilize nematodes as resources. This group has mostly been ignored by invertebrate pathologists but is traditionally studied by plant pathologists, with the aim of controlling plant parasitic nematodes. In particular, the species *P. lilacinum* (syn. *Paecilomyces lilacinus*) and *P. chlamydosporia* (syn. *Verticillium chlamydosporium*), which infect nematode eggs, have been investigated, and a few commercial products have been developed for inundative biological control (Kerry, 2000), but a range of other microbes acting against nematodes are also recognized for their biological control potential (Li et al., 2015). Recently, these fungi were considered as biological control agents against the egg stages of animal parasitic nematodes (eggs deposited in the soil environment contaminate pastures, which constitutes a major problem for free-ranging livestock and poultry; Thapa et al., 2015), while fungi from other orders infecting the adult worms also have been investigated (Braga and Araújo, 2014). Although the hypocrealean nematophagous fungi have shown the ability to inactivate the thick-shelled eggs of animal parasitic nematodes *in vitro* (reviewed by Braga and Araújo, 2014), their biological control potential remains to be evaluated under more realistic settings in nonsterile soil. The nematophagous and entomopathogenic fungi of *Pochonia* and *Metarhizium* within Hypocreales are phylogenetically very closely related (Kepler et al., 2014) and are likely to produce comparable enzymes, such as chitinases and proteases, to break down the eggshell and cuticle of their hosts. Both genera are demonstrated to be plant-associated, whether as endophytes and/or as rhizosphere colonizers (Behie et al., 2014; Larriba et al., 2014; Quesada-Moraga et al., 2014), indicating overlapping adaptive strategies and ecologies. Novel approaches to biological control should consider exploring these adaptations of the fungi for more than one function applicable to plant protection.

The spectrum of biotic and abiotic factors affecting the efficacy of fungal pathogens in biological control highlights the importance of considering them as living organisms and not just applying spores using the pesticide paradigm (i.e., spray-and-kill). An ecological approach to biological control, taking into consideration the many environmental factors affecting fungi, may be a more sustainable route for the implementation of fungi in pest management. Although potentially more complex than spray-and-kill, such an approach is likely also to emphasize to users that fungi have different modes of action than chemical pesticides and that control by fungi is achieved over a longer timespan (days) than for chemicals, which can have immediate effects. Ecological approaches are incorporated in two biological control strategies: classical and conservation biological control.

Classical biological control is the introduction of natural enemies to a new location, with the goal of permanent establishment and control (Eilenberg et al., 2001); this method has most often been used against invasive species, which often have few natural enemies in the area being invaded. While the use of fungal pathogens for classical biological control is in no way as extensive as the use of insect parasitoids and predators for this purpose, among introductions of all arthropod pathogens, 49.3% of classical biological control programs introduced fungi and microsporidia as opposed to viruses, bacteria, or nematodes (Hajek and Delalibera, 2010). Once again, a major type of host targeted has been within the Hemiptera. Classical biological control with fungal pathogens began being used long ago, with the first releases recorded in 1894–95 against white grubs in sugarcane in Australia (Hajek et al., 2005). Across all introductions, approximately one-third of the agents that have been released have become established, but for many programs, especially earlier ones, results have not always been reported. Complete success in pest control has rarely been achieved for programs using this strategy, although some releases have provided significant control (e.g., *Neozygites tanajoae* released against the cassava green mite (*Mononychellus tanajoa*) in Africa (Hajek and Delalibera, 2010) and *E. maimaiga* released against the gypsy moth in Bulgaria (Georgiev et al., 2013)).

Conservation biological control is aimed at altering the environment to provide resources for natural enemies to build up their populations and, in this way, increase pest control (Eilenberg et al., 2001). For conservation biological control strategies to be successful, a thorough ecological understanding of community interactions and effects is necessary (Pell et al., 2010). Studies of conservation biological control with fungal pathogens are limited, but have focused on fungal populations present in the crop environment and in habitats surrounding crops (Meyling and Eilenberg, 2007). One fundamental idea is to sustain high populations of fungi in crop margins on alternative hosts with the aim of spilling over fungal inoculum to pests within the crops, thereby suppressing pest populations through apparent competition (Meyling and Hajek, 2010). Among the methods investigated in an attempt to increase fungal populations is the increase of humidity at appropriate times, whether by application of water via irrigation, by decreasing crop spacing to create more humid microenvironments, or by changing harvesting practices (Pell et al., 2010). To improve the persistence of fungal pathogens in the crop environments, tillage practices can be altered to preserve the reservoir of entomopathogenic fungi, either at the top of the soil or at the locations within the soil where pests occur. Consideration can also be given to whether crop residues need to be burned, as this destroys persistent fungal structures. The presence and abundance of arthropod fungal pathogens in areas adjacent to crops, in the context of the persistence and provision of inoculum reservoirs, has mostly been studied for aphid pathogenic entomophthoralean fungi. The most successful example is a program developed to monitor densities of cotton aphids so that pesticides are not applied when an epizootic by *Neozygites fresenii* is certain to occur (Pell et al., 2010). Growing host plants to support populations of non-pest aphids in order to sustain *P. neoaphidis* has also been investigated (Ekesi et al., 2005).

An area that has only been explored in a limited fashion is how responses of insects toward fungal pathogen challenge (as emphasized in Section 9.4.1) can support or compromise biological control strategies depending on which party shows the response. Pre-infection defenses such as avoidance of fungal-inoculated habitats would be in support of biological control if the responses were exhibited by other natural enemies of a target pest. For instance, predatory insects avoiding patches with fungal spores could direct their prey search to patches where the fungus is not present at high densities, potentially reaching complementary effects against the pest population (Straub et al., 2008). Similar regulation effects could be achieved by combining fungal pathogens and parasitoids, if the parasitoids could avoid patches with high fungal densities (Cotes et al., 2015; Rännbäck, 2015; Rännbäck et al., 2015). However, pre-infection defenses displayed by the target pest could be detrimental to biological control with fungal pathogens if the pest population avoided or reduced infection risks. In particular, social insects such as ants, termites, and wasps represent important invasive species that are difficult to control. Application of fungal inoculum to their colonies can elicit various defensive responses, such as allogrooming (Ugelvig and Cremer, 2007), burial of sporulating cadavers (Myles 2002), and seclusion behavior of infected individuals (Bos et al., 2012), all of which limit the effects of pathogens. Hence, novel strategies for inoculation and dispersal must be considered to overcome species-specific defenses of social insects in order to achieve successful biological control (Harris et al., 2000). These considerations emphasize that knowledge of community interactions and behavioral ecology is necessary to exploit fungal pathogens for efficient biological control, as well as to identify pitfalls and constraints.

9.8   Conclusion

Our understanding of the ecology of fungal pathogens infecting invertebrates is increasing, but there is still much to learn. Over the past several decades, the names of many fungal species have been (and will likely continue to be) changed, both based on the ever-increasing use of molecular techniques providing much-needed phylogenetic data and due to the 1F = 1N campaign across the field of mycology. It is important that invertebrate pathologists keep themselves updated on taxonomic developments. Furthermore, species identifications of fungal pathogens must be based on molecular markers in future studies, in order to enable clear and explicit communication about which taxa are involved. With molecular methods for identification, it is also now evident that many responses reported in the literature on the generalist pathogens within the Hypocreales are not species-specific, but rather isolate- or strain-specific. Thus, we must now use heightened specificity for studies and for any generalizations about responses by fungal pathogens. Molecular methods will also assist in allowing a better understanding of the host specialization of fungal pathogens, as well as of host–pathogen interactions. In addition, interest in and information about interactions between fungal pathogens and other members of the biotic community will certainly continue to increase as we learn that community impacts are ubiquitous. As a few examples, endosymbionts can protect against or assist fungal pathogens, predatory insects can increase transmission by increasing host movement or can decrease transmission by consuming sporulating cadavers, and mycoparasites can decrease inoculum in environmental reservoirs. Much recent knowledge has been based on the genetic toolbox for *Metarhizium* spp. in particular, especially focusing on interactions of invertebrate pathogenic fungi with plants, whether as epiphytes, endophytes, or rhizosphere inhabitants. Future studies of fungus–plant interactions should consider the mechanistic aspects of plant responses to fungal inoculation as an important part of associations, and these investigations should be conducted in collaboration between insect pathologists and plant biologists. Realizing the multifaceted lifestyles of these fungi and their many interactions with organisms other than invertebrate hosts should spur new avenues of both basic and applied research that will potentially establish novel platforms for pest control and plant protection. Finally, ecosystem-level studies of the development of epizootics across space and time, and of associated biotic and abiotic factors favoring transmission and infection, will continue to provide valuable information toward understanding and predicting when, where, and why epizootics caused by fungal pathogens will occur.

Acknowledgments

We thank Helen Hesketh, Centre for Ecology and Hydrology, for her assistance with the manuscript and for her excellent artwork, and Richard Humber for his comments on the taxonomy and systematics of fungi.

References

Agboton, B.V., Hanna, R., Onzo, A., Vidal, S., von Tiedemann, A., 2013. Interactions between the predatory mite *Typhlodromalus aripo* and the entomopathogenic fungus *Neozygites tanajoae* and consequences for the suppression of their shared prey/host *Mononychellus tanajoa*. Exp. Appl. Acarol. 60, 205–217.

Akutse, K.S., Maniania, N.K., Fiaboe, K.K.M., van den Berg, J., Ekesi, S., 2013. Endophytic colonization of *Vicia faba* and *Phaseolus vulgaris* (Fabaceae) by fungal pathogens and their effects on the life-history parameters of *Liriomyza huidobrensis* (Diptera: Agromyzidae). Fung. Ecol. 6, 293–301.

Andersen, S.B., Ferrari, M., Evans, H.C., Elliot, S.L., Boomsma, J.J., Hughes, D.P., 2012. Disease dynamics in a specialized parasite of ant societies. PLoS ONE 7(5), e36352.

Anderson, R.D., Blanford, S., Thomas, M.B., 2013. House flies delay fungal infection by fevering: at a cost. Ecol. Entomol. 38, 1–10.

Ansari, M.A., Shah, F.A., Tirry, L., Moens, M., 2006. Field trials against *Hoplia philanthus* (Coleoptera: Scarabaeidae) with a combination of an entomopathogenic nematode and the fungus *Metarhizium anisopliae* CLO 53. Biol. Control 39, 453–459.

Ansari, M.A., Shah, F.A., Butt, T.M., 2008. Combined use of entomopathogenic nematodes and *Metarhizium anisopliae* as a new approach for black vine weevil, *Otiorhynchus sulcatus*, control. Entomol. Exp. Appl. 129, 340–347.

Arthurs, S., Thomas, M.B., 2001. Behavioural changes in *Schistocerca gregaria* following infection with fungal pathogens: implications for susceptibility to predation. Entomol. Exp. Appl. 26, 227–234.

Arthurs, S.P., Thomas, M.B., Lawton, J.L., 2001. Seasonal patterns of persistence and infectivity of *Metarhizium anisopliae* var. *acridum* in grasshopper cadavers in the Sahel. Entomol. Exp. Appl. 100, 69–76.

Backhaus, T., Faust, M., 2012. Predictive environmental risk assessment of chemical mixtures: a conceptual framework. Environ. Sci. Technol. 46, 2564–2573.

Barbercheck, M.E., Kaya, H.K., 1991. Competitive interactions between entomopathogenic nematodes and *Beauveria bassiana* (Deuteromycotina, Hyphomycetes) in soilborne larvae of *Spodoptera exigua* (Lepidoptera, Noctuidae). Environ. Entomol. 20, 707–712.

Barnes, A.I., Siva-Jothy, M.T., 2000. Density-dependent prophylaxis in the mealworm beetle *Tenebrio molitor* L. (Coleoptera: Tenebrionidae): cuticular melanization is an indicator of investment in immunity. Proc. R Soc. B 267, 177–182.

Barribeau, S.M., Parker, B.J., Gerardo, N.M., 2014. Exposure to natural pathogens reveals costly aphid response to fungi but not bacteria. Ecol. Evol. 4, 488–493.

Baverstock, J., Alderson, P.G., Pell, J.K., 2005. *Pandora neoaphidis* transmission and aphid foraging behavior. J. Invertebr. Pathol. 90, 73–76.

Baverstock, J., Clark, S.J., Alderson, P.G., Pell, J.K., 2009. Intraguild interactions between the entomopathogenic fungus *Pandora neoaphidis* and an aphid predator and parasitoid at the population scale. J. Invertebr. Pathol. 102, 167–172.

Baverstock, J., Roy, H.E., Pell, J.K., 2010. Entomopathogenic fungi and insect behavior: from unsuspecting hosts to targeted vectors. BioControl 55, 89–102.

Behie, S.W., Bidochka, M.J., 2014. Ubiquity of insect-derived nitrogen transfer to plants by endophytic insect-pathogenic fungi: an additional branch of the soil nitrogen cycle. Appl. Environ. Microbiol. 80, 1553–1560.

Behie, S.W., Jones, S.J., Bidochka, M.J., 2015. Plant tissue localization of the endophytic insect pathogenic fungi *Metarhizium* and *Beauveria*. Fung. Ecol. 13, 112–119.

Behie, S.W., Zelisko, P.M., Bidochka, M.J., 2012. Endophytic insect-parasitic fungi translocate nitrogen directly from insects to plants. Science 336, 1576–1577.

Bellini, R., Mullens, B.A., Jespersen, J.B., 1992. Infectivity of two members of the *Entomophthora muscae* complex (Zygomycetes: Entomophthorales) for *Musca domestica* (Dipt.: Muscidae). Entomophaga 37, 11–19.

Bidochka, M.J., Kasperski, J.E., Wild, G.A.M., 1998. Occurrence of the entomopathogenic fungi *Metarhizium anisopliae* and *Beauveria bassiana* in soils from temperate and near-northern habitats. Can. J. Bot. 76, 1198–1204.

Bidochka, M.J., Kamp, A.M., Lavender, T.M., Dekoning, J., De Croos, J.N.A., 2001. Habitat association in two genetic groups of the insect-pathogenic fungus *Metarhizium anisopliae*: uncovering cryptic species? Appl. Environ. Microbiol. 67, 1335–1342.

Bing, L.A., Lewis, L.C., 1991. Suppression of *Ostrinia nubilalis* (Hubner) (Lepidoptera, Pyralidae) by endophytic *Beauveria bassiana* (Balsamo) Vuillemin. Environ. Entomol. 20, 1207–1211

Bischoff, J. F., Rehner, S. A., Humber, R. A., 2009. A multilocus phylogeny of the *Metarhizium anisopliae* lineage. Mycologia 101, 512–530.

Blackwell, M., 2011. The fungi: 1, 2, 3, … 5.1 million species? Am. J. Bot. 98, 426–438.

Boughton, R.K., Joop, G., Armitage, S.A.O., 2011. Outdoor immunology: methodological considerations for ecologists. Funct. Ecol. 25, 81–100.

Boomsma, J.J., Jensen, A.B., Meyling, N.V., Eilenberg, J., 2014. Evolutionary interaction networks of insect pathogenic fungi. Annu. Rev. Entomol. 59, 467–485.

Bos, N., Lefèvre, T., Jensen, A.B., d'Ettorre, P., 2012. Sick ants become insociable. J. Evol. Biol. 25, 342–351.

Bouamama, N., Vidal, C., Fargues, J., 2010a. Effects of fluctuating moisture and temperature regimes on the persistence of quiescent conidia of *Isaria fumosorosea*. J. Invertebr. Pathol. 105, 139–144.

Bouamama, N., Vidal, C., Fargues, J., 2010b. Effects of wetting on the persistence of quiescent conidia of *Isaria fumosorosea*. Biol. Control 55, 203–210.

Boucias, D.G., Pendland, J.C., 1998. Principles of Insect Pathology. Kluwer, Boston, MD.

Braga, F.R., Araújo, J.V., 2014. Nematophagous fungi for biological control of gastrointestinal nematodes in domestic animals. Appl. Microbiol. Biotechnol. 98, 71–82.

Braga, G.U., Rangel, D.E., Flint, S.D., Anderson, A.J., Roberts, D.W., 2006. Conidial pigmentation is important to tolerance against solar-simulated radiation in the entomopathogenic fungus *Metarhizium anisopliae*. Photochem. Photobiol. 82, 418–422.

Broza, M., Pereira, R.M., Stimac, J.L., 2001. The nonsusceptibility of soil Collembola to insect pathogens and their potential as scavengers of microbial pesticides. Pedobiologia 45, 523–534.

Bruck, D.J., Lewis, L.C., 2002. Rainfall and crop residue effects on soil dispersion and *Beauveria bassiana* spread to corn. Appl. Soil Ecol. 20, 183–190.

Brütsch, T., Chapuisat, M., 2014. Wood ants protect their brood with tree resin. Anim. Behav. 93, 157–161.

Butt, T.M., Jackson, C.W., Magan, N., (eds.), 2001. Fungi as Biocontrol Agents: Progress, Problems and Potential. CABI, Wallingford.

Carruthers, R.I., Soper, R.S., 1987. Fungal diseases, in: Fuxa, J.R., Tanada, Y. (eds.), Epizootiology of Insect Diseases. John Wiley & Sons, New York, pp. 357–416.

Castrillo, L.A., Hajek, A.E., 2015. Detection of presumptive mycoparasites associated with *Entomophaga maimaiga* resting spores in forest soils. J. Invertebr. Pathol. 124, 87–89.

de Castro, T.R., 2016. Abundance, genetic diversity and persistence of *Metarhizium* spp. fungi from soil of strawberry crops and their potential as biological control agents against the two-spotted spider mite *Tetranychus urticae*. PhD thesis, University of Sao Paulo and University of Copenhagen.

de Castro, T.R., Wekesa, V.W., de Andrade Moral, R., Demétrio, C.G.B., Delalibera Jr., I., Klingen, I., 2013. The effects of photoperiod and light intensity on the sporulation of Brazilian and Norwegian isolates of *Neozygites floridana*. J. Invertebr. Pathol. 114, 230–233.

Cedergreen, N., Svendsen, C., Backhaus, T., 2013. Chemical mixtures: concepts for predicting toxicity, in: Jorgensen, S.E. (ed.), Encyclopedia of Environmental Management. CRC Press, Boca Raton, FL, pp. 2572–2581.

Chandler, D., Hay, D., Reid, A.P., 1997. Sampling and occurrence of entomopathogenic fungi and nematodes in UK soils. Appl. Soil Ecol. 5, 133–141.

Chapuisat, M., Oppliger, A., Magliano, P., Christe, P., 2007. Wood ants use resin to protect themselves against pathogens. Proc. R. Soc. B 274, 2013–2017.

Clifton, E.H., Jaronski, S.T., Hodgson, E.W., Gassmann, A.J., 2015. Abundance of soil-borne entomopathogenic fungi in organic and conventional fields in the Midwestern USA with an emphasis on the effect of herbicides and fungicides on fungal persistence. PLoS ONE 10(7), e0133613.

Cotes, B., Rännbäck, L.-M., Björkman, M., Norli, H.R., Meyling, N.V., Rämert, B., Anderson, P., 2015. Habitat selection of a parasitoid mediated by volatile cues informing on host and intraguild predator densities. Oecologia 179, 151–162.

Creer, S., Deiner, K., Frey, S., Porazinska, D., Taberlet, P., Thomas, W.K., Potter, C., et al., 2016. The ecologist's field guide to sequence-based identification of biodiversity. Meth. Ecol. Evol. 7, 1008–1018.

Cremer, S., Armitage, S.A.O., Schmid-Hempel, P., 2007. Social immunity. Curr. Biol. 17, R693-R702.

Deacon, J.W., 2006. Fungal Biology, 4th edn. Blackwell, Malden, MA.

De Fine Licht, H., Hajek, A.E., Jensen, A.B., Eilenberg, J., 2016. Utilizing genomics to study entomopathogenicity in the fungal phylum *Entomophthoromycota*: a review of current genetic resources. Adv. Gen. 94, 41–65.

Dromph, K.M., Vestergaard, S., 2002. Pathogenicity and attractiveness of entomopathogenic hyphomycete fungi to collembolans. Appl. Soil Ecol. 21, 197–210.

Duke, L., Steinkraus, D.C., English, J.E., Smith, K.G., 2002. Infectivity of resting spores of *Massospora cicadina* (Entomophthorales: Entomophthoraceae), an entomopathogenic fungus of periodical cicadas (*Magicicada* spp.) (Homoptera: Cicadidae). J. Invertebr. Pathol. 80, 1–6.

Dwyer, G., Elkinton, J.S., Hajek, A.E., 1998. Spatial scale and the spread of a fungal pathogen of gypsy moth. Amer. Nat. 152, 485–494.

Eilenberg, J., Hajek, A., Lomer, C., 2001. Suggestions for unifying the terminology in biological control. BioControl 46, 387–400.

Ekesi, S., Shah, P.A., Clark, S.J., Pell, J.K., 2005. Conservation biological control with the fungal pathogen *Pandora neoaphidis*: implications of aphid species, host plant and predator foraging. Agric. For. Entomol. 7, 21–30.

Elliot, S.L., Blanford, S., Thomas, M.B., 2002. Host-pathogen interactions in a varying environment: temperature, behavioural fever and fitness. Proc. R. Soc. Lond. Ser. B: Biol. Sci. 269, 1599–1607.

EU Commission 2013. Commission Regulation (EU) No 283/2013 of 1 March 2013 setting out the data requirements for active substances, in accordance with Regulation (EC) No. 1107/2009 of the European Parliament and of the Council concerning the placing of plant protection products on the market. Official Journal of the European Union, L93.

Fang, W., St. Leger, R.J., 2012. Enhanced UV resistance and improved killing of malaria mosquitoes by photolyase transgenic entomopathogenic fungi. PLoS ONE 7(3), e43069.

Faria, M.R. de, Wraight, S.P., 2007. Mycoinsecticides and mycoacaricides: a comprehensive list with worldwide coverage and international classification of formulation types. Biol. Control 43, 237–257.

Feng, M.G., Chen, C., Chen, B., 2004. Wide dispersal of aphid-pathogenic Entomophthorales among aphids relies upon migratory alates. Environ. Microbiol. 6, 510–516.

Fernandes, E.K.K., Rangel, D.E.N., Braga, G.U.L., Roberts, D.W., 2015. Tolerance of entomopathogenic fungi to ultraviolet radiation: a review on screening of strains and their formulation. Curr. Genet. 61, 427–440.

Fisher, J.J., Rehner, S.A., Bruck, D.J., 2011. Diversity of rhizosphere associated entomopathogenic fungi of perennial herbs, shrubs and coniferous trees. J. Invertebr. Pathol. 106, 289–295.

Fisher, M.C., Henk, D.A., Briggs, C.J., Brownstein, J.S., Madoff, L.C., McCraw, S.L., Gurr, S.J., 2012. Emerging fungal threats to animal, plant and ecosystem health. Nature 484, 186–194.

Furlong, M.J., Pell, J.K., 2005. Interactions between entomopathogenic fungi and arthropod natural enemies, in: Vega, F.E., Blackwell, M. (eds.), Insect-Fungal Associations: Ecology and Evolution. Oxford University Press, Oxford, pp. 51–73.

Gao, Q., Jin, K., Ying, S.H., Zhang, Y., Xiao, G., Shang, Y., Duan, Z., et al., 2011. Genome sequencing and comparative transcriptomics of the model entomopathogenic fungi *Metarhizium anisopliae* and *M. acridum*. PLOS Genet. 7(1), e1001264.

Garrido-Jurado, I., Fernandez-Bravo, M., Campos, C., Quesada-Moraga, E., 2015. Diversity of entomopathogenic Hypocreales in soil and phylloplanes of five Mediterranean cropping systems. J. Invertebr. Pathol. 130, 97–106.

George, J., Jenkins, N.E., Blanford, S., Thomas, M.B., Baker, T.C., 2013. Malaria mosquitoes attracted by fatal fungus. PLoS ONE 8(5), e62632.

Georgiev, G., Mirchev, P., Rossnev, B., Petkov, P., Georgieva, M., Pilarska, D., et al., 2013. Potential of *Entomophaga maimaiga* Humber, Shimazu and Soper (Entomophthorales) for suppressing *Lymantria dispar* (Linnaeus) outbreaks in Bulgaria. Compt. Rend. Acad. Bulg. Sci. 66, 1025–1032.

Glare, T.R., Milner, R.J., 1991. Ecology of entomopathogenic fungi, in: Arora, D.K., Ajello, L., Mukerji, K.G. (eds.), Handbook of Applied Mycology: Humans, Animals and Insects, Vol. 2. Marcel Dekker, New York, pp. 547–612.

Gleason, F.H., Marano, A.V., Johnson, P., Martin, W.W., 2010. Blastocladian parasites of invertebrates. Fung. Biol. Rev. 24, 56067.

Goble, T.A., Dames, J.F., Hill, M.P., Moore, S.D., 2010. The effects of farming system, habitat type and bait type on the isolation of entomopathogenic fungi from citrus soils in the eastern Cape Province, South Africa. BioControl 55, 399–412.

Gurulingappa, P., Sword, G.A., Murdoch, G., McGee, P.A., 2010. Colonization of crop plants by fungal entomopathogens and their effects on two insect pests when in planta. Biol. Control 55, 34–41.

Guzmán-Franco, A.W., Clark, S.J., Alderson, P.G., Pell, J.K., 2009. Competition and co-existence of *Zoophthora radicans* and *Pandora blunckii*, two co-occurring fungal pathogens of the diamondback moth, *Plutella xylostella*. Mycol. Res. 113, 1312–1321.

Gryganskyi, A.P., Humber, R.A., Stajich, J.E., Mullens, B., Anishchenko, I.M., Vilgalys, R., 2013. Sequential utilization of hosts from different fly families by genetically distinct, sympatric populations within the *Entomophthora muscae* species complex. PLoS ONE 8(8), e71168.

Hajek, A.E., 1997. Ecology of terrestrial fungal entomopathogens. Adv. Micro. Ecol. 15, 193–249.

Hajek, A.E., 2001. Larval behavior in *Lymantria dispar* increases risk of fungal infection. Oecologia 126, 285–291.

Hajek, A.E., Delalibera, Jr., I., 2010. Fungal pathogens as classical biological control agents against arthropods. BioControl 55, 147–158.

Hajek, A.E., Soper, R.S., 1992. Temporal dynamics of *Entomophaga maimaiga* after death of gypsy moth (Lepidoptera: Lymantriidae) larval hosts. Environ. Entomol. 21, 129–135.

Hajek, A.E., Tobin, P.C., 2011. Introduced pathogens follow the invasion front of a spreading alien host. J. Anim. Ecol. 80, 1217–1226.

Hajek, A.E., van Nouhuys, S., 2016. Interactions among fatal diseases and parasitoids driven by density of a shared host. Proc. R. Soc. B 283, 20160154.

Hajek, A.E., Larkin, T.S., Carruthers, R.I., Soper, R.S., 1993. Modeling the dynamics of *Entomophaga maimaiga* (Zygomycetes: Entomophthorales) epizootics in gypsy moth (Lepidoptera: Lymantriidae) populations. Environ. Entomol. 22, 1172–1187.

Hajek, A.E., Bauer, L., McManus, M.L., Wheeler, M.M., 1998. Distribution of resting spores of the *Lymantria dispar* pathogen *Entomophaga maimaiga* in soil and on bark. BioControl 43, 189–200.

Hajek, A.E., Shimazu, M., Knoblauch, B., 2000. Isolating a species of Entomophthorales using resting spore-bearing soil. J. Invertebr. Pathol. 75, 298–300.

Hajek, A.E., Wraight, S.P., Vandenberg, J.D., 2001. Control of arthropods using pathogenic fungi, in: Pointing, S.B., Hyde, K.D. (eds.), Bio-Exploitation of Fungi. Fungal Diversity Press, Hong Kong, pp. 309–347.

Hajek, A.E., Strazanac, J.S., Wheeler, M.M., Vermeylen, R., Butler, L., 2004. Persistence of the fungal pathogen *Entomophaga maimaiga* and its impact on native Lymantriidae. Biol. Control 30, 466–471.

Hajek, A. E., McManus, M.L., Delalibera Jr., I., 2005. Catalogue of introductions of pathogens and nematodes for classical biological control of insects and mites. USDA, Forest Service. FHTET–2005-05. Available from: http://www.fs.fed.us/foresthealth/technology/pdfs/catalogue.pdf (accessed May 8, 2017).

Hajek, A.E., Papierok, B., Eilenberg, J., 2012. Methods for study of the Entomophthorales, in: Lacey, L.A. (ed.), Manual of Techniques in Invertebrate Pathology. Elsevier, New York, pp. 285–316.

Hajek, A.E., Longcore, J.E., Simmons, D.R., Peters, K., Humber, R.A., 2013. Chytrid mycoparasitism of entomophthoralean azygospores. J. Invertebr. Pathol. 114, 333–336.

Hajek, A.E., Tobin, P.C., Haynes, K.J., 2015. Replacement of a dominant viral pathogen by a fungal pathogen does not alter the synchronous collapse of a forest insect outbreak. Oecologia 177, 785–797.

Harcourt, D.G., 1990. Displacement of *Bathyplectes curculionis* (Thoms.) (Hymenoptera: Ichneumonidae) by *B. anurus* (Thoms.) in eastern Ontario populations of the alfalfa weevil, *Hypera postica* (Gyll.) (Coleoptera: Curculionidae). Can. Entomol. 122, 641–645.

Harris, R.J., Harcourt, S.J., Glare, T.R., Rose, E.A.F., Nelson, T.J., 2000. Susceptibility of *Vespula vulgaris* (Hymenoptera: Vespidae) to generalist entomopathogenic fungi and their potential for wasp control. J Invertebr Pathol, 75, 251–258.

Heinig, R.L., Paaijmans, K.P., Hancock, P.A., Thomas, M.B., 2015. The potential for fungal biopesticides to reduce malaria transmission under diverse environmental conditions. J. Appl. Ecol. 52, 1558–1566.

Henk, D.A., Vilgalys, R., 2007. Molecular phylogeny suggests a single origin of insect symbiosis in the Pucciniomycetes with support for some relationships within the genus *Septobasidium*. Am. J. Bot. 94, 1515–1526.

Hesketh, H., Roy, H.E., Eilenberg, J., Pell, J.K. Hails, R.S., 2010. Challenges in modelling complexity of fungal entomopathogens in semi-natural populations of insects. BioControl 55, 55–73.

Heyworth, E.R., Ferrari, J., 2015. A facultative endosymbiont in aphids can provide diverse ecological benefits. J. Evol. Biol. 28, 1753–1780.

Howe, A.G., Ravn, H.P., Jensen, A.B., Meyling, N.V., 2016. Spatial and taxonomical overlap of fungi on phylloplanes and invasive alien ladybirds with fungal infections in tree crowns of urban green spaces. FEMS Microbiol. Ecol. 92, article no. fiw143.

Hu, G., St. Leger, J., 2002. Field studies using a recombinant mycoinsecticide (*Metarhizium anisopliae*) reveal that it is rhizosphere competent. Appl. Environ. Microbiol. 68, 6383–6387.

Hu, X., Xiao, G., Zheng, P., Shang, Y., Su, Y., Zhang, X., et al., 2014. Trajectory and genomic determinants of fungal-pathogen speciation and host adaptation. Proc. Natl. Acad. Sci. U.S.A. 111, 16796–16801.

Hughes, D.P., Andersen, S.B., Hywel-Jones, N.L., Himaman, W., Billen, J., Boomsma, J.J., 2011. Behavioral mechanisms and morphological symptoms of zombie ants dying from fungal infection. BMC Ecol. 11, 13.

Hughes, W.H.O., Petersen, K.S., Ugelvig, L.V., Pedersen, D., Thomsen, L., Poulsen, M., Boomsma J.J., 2004. Density-dependence and within-host competition in a semelparous parasite of leaf-cutting ants. BMC Evol. Biol. 4, 45.

Hunt, V.L., Zhong, W., McClure, C.D., Mlynski, D.T., Duxbury, E.M.L., Charnley, A.K., Priest, N.K., 2016. Cold-seeking behaviour mitigates reproductive losses from fungal infection in *Drosophila*. J. Anim. Ecol. 85, 178–186.

Inglis, G.D., Goettel, M.S., Butt, T.M., Strasser, H., 2001, Use of hyphomycetous fungi for managing insect pests, in: Butt, T.M., Jackson, C., Magan, N. (eds.), Fungi as Biocontrol Agents: Progress, Problems and Potential. CABI, Wallingford, pp. 23–69.

Inyang, E.N., Butt, T.M., Ibrahim, L., Clark, S.J., Pye, B.T., Beckett, A., Archer, S., 1998. The effect of plant growth and topography on the acquisition of conidia of the insect pathogen *Metarhizium anisopliae* by larvae of *Phaedon cochleariae*. Mycol. Res. 102, 1365–1374.

Jabbour, R. Barbercheck, M.E., 2009. Soil management effects on entomopathogenic fungi during the transition to organic agriculture in a feed grain rotation. Biol. Control 51, 435–443.

Jabbour, R., Crowder, D.W., Aultman, E.A., Snyder, W.E., 2011. Entomopathogen biodiversity increases host mortality. Biol. Control 59, 277–283.

Jackson, M.A., Jaronski, S.T., 2009. Production of microsclerotia of the fungal entomopathogen *Metarhizium anisopliae* and their potential for use as a biocontrol agent for soil-inhabiting insects. Mycol. Res. 113, 842–850.

James, T.Y., Kauff, F., Schoch, C.L., Matheny, P.B., Hofstetter, V., Cox, C.J., et al., 2006. Reconstructing the early evolution of Fungi using a six-gene phylogeny. Nature 443, 818–822.

James, R.R., 2011. Chalkbrood transmission in the alfalfa leafcutting bee: the impact of disinfecting bee cocoons in loose cell management systems. Environ. Entomol. 40, 782–787.

Jaronski, S.T., 2007. Soil ecology of the entomopathogenic Ascomycetes: a critical examination of what we (think) we know, in: Ekesi, S., Maniania, N.K. (eds.), Use of Entomopathogenic Fungi in Biological Pest Management. Research Signpost, Kerala, pp. 91–144.

Jaronski, S.T., 2010. Ecological factors in the inundative use of fungal entomopathogens. BioControl 55, 159–185.

Jensen, A.B., Thomsen, L., Eilenberg, J., 2001. Intraspecific variation and host specificity of *Entomophthora muscae sensu stricto* isolates revealed by random amplified polymorphic DNA, universal primed PCR, PCR-restriction fragment length polymorphism, and conidial morphology. J. Invertebr. Pathol. 78, 251–259.

Kamata, N., 2000. Population dynamics of the beech caterpillar, *Syntypistis punctatella*, and biotic and abiotic factors. Popul. Ecol. 42, 267–278.

Keller, S., Kessler, P., Schweizer, C., 2003. Distribution of insect pathogenic soil fungi in Switzerland with special reference to *Beauveria brongniartii* and *Metarhizium anisopliae*. BioControl 48, 307–319.

Kepler, R.M., Bruck, D.J., 2006. Examination of the interaction between the black vine weevil (Coleoptera: Curculionidae) and an entomopathogenic fungus reveals a new tritrophic interaction. Environ. Entomol. 35, 1021–1029.

Kepler, R.M., Humber, R.A, Bischoff, J.F., Rehner, S.A., 2014. Clarification of generic and species boundaries for *Metarhizium* and related fungi through multigene phylogenetics. Mycologia 106, 811–829.

Kepler, R.M., Ugine, T.A., Maul, J.E., Cavigelli, M.A., Rehner, S.A., 2015. Community composition and population genetics of insect pathogenic fungi in the genus *Metarhizium* from soils of a long-term agricultural research system. Environ. Microbiol. 17, 2791–2804.

Kerry, B.R., 2000. Rhizosphere interactions and the exploitation of microbial agents for the biological control of plant-parasitic nematodes. Annu. Rev. Phytopathol. 38, 423–441.

Kershaw, M.J., Moorhouse, E.R., Bateman, R., Reynolds, S.E., Charnley, A.K., 1999. The role of destruxins in the pathogenicity of *Metarhizium anisopliae* for three species of insects. J. Invertebr. Pathol. 74, 213–223.

Keyser, C.A., Thorup-Kristensen, K., Meyling, N.V., 2014. *Metarhizium* seed treatment mediates fungal dispersal *via* roots and induces infections in insects. Fung. Ecol. 11, 122–131.

Keyser, C.A., de Fine Licht, H.H., Steinwender, B.M., Meyling, N.V., 2015. Diversity within the entomopathogenic fungal species *Metarhizium flavoviride* associated with agricultural crops in Denmark. BMC Microbiol. 15, art. 249.

Keyser, C.A., Jensen, B., Meyling, N.V., 2016. Dual effects of *Metarhizium* spp. and *Clonostachys rosea* against an insect and a seed borne pathogen in wheat. Pest Mgmt. Sci. 72, 517–526.

Kistner, E.J., Saums, M., Belovsky, G.E., 2015. Mechanical vectors enhance fungal entomopathogen reduction of the grasshopper pest *Camnula pellucida* (Orthoptera: Acrididae). Environ. Entomol. 44, 144–152.

Klass, J.I., Blanford, S., Thomas, M.B., 2007. Use of a geographic information system to explore spatial variation in pathogen virulence and the implications for biological control of locusts and grasshoppers. Agric. For. Entomol. 9, 201–208.

Klingen, I., Westrum, K., Meyling, N.V., 2015. Effect of Norwegian entomopathogenic fungal isolates against *Otiorhynchus sulcatus* larvae at low temperatures and their persistence in strawberry rhizospheres. Biol, Control 81, 1–7.

Klingen, I., Haukeland, S., 2006. The soil as a reservoir for natural enemies of pest insects and mites with emphasis on fungi and nematodes, in: Eilenberg, J., Hokkanen. H.M.T. (eds.), An Ecological and Societal Approach to Biological Control. Springer, Heidelberg, pp. 145–211.

Klingen, I., Eilenberg, J., Meadow, R., 2002. Effects of farming system, field margins and bait insect on the occurrence of insect pathogenic fungi in soils. Agric. Ecosyst. Environ. 91, 191–198.

Klingen, I., Westrum, K., Meyling, N.V., 2015. Effect of Norwegian entomopathogenic fungal isolates against *Otiorhynchus sulcatus* larvae at low temperatures and persistence in strawberry rhizospheres. Biol. Control 81, 1–7.

Knudsen, G.R., Schotzko, D.J., 1999. Spatial simulation of epizootics caused by *Beauveria bassiana* in Russian wheat aphid populations. Biol. Control 16, 318–326.

Konrad, M., Vyleta, M.L., Theis, F.J., Stock, M., Tragust, S., Klatt, M., Drescher, V., et al., 2012. Social transfer of pathogenic fungus promotes active immunization in ant colonies. PLOS Biol. 10(4), e1001300.

Krauss, U., Hidalgo, E., Arroyo, C., Piper, S.R., 2004. Interaction between the entomopathogens *Beauveria bassiana*, *Metarhizium anisopliae* and *Paecilomyces fumosoroseus* and the mycoparasites *Clonostachys* spp., *Trichoderma harzianum* and *Lecanicillium lecanii*. Biocontr. Sci. Technol. 14, 331–346.

Larriba, E., Jaime, M.D.L.A, Carbonell-Caballero, J., Conesa, A., Dopazo, J., Nislow, C., et al., 2014. Sequencing and functional analysis of the genome of a nematode egg-parasitic fungus, *Pochonia chlamydosporia*. Fung. Gen. Biol. 65, 69–80.

Larriba, E., Jaime, M.D.L.A, Nislow, C., Martin-Nieto, J., Lopez-Llorca, L.V., 2015. Endophytic colonization of barley (*Hordeum vulgare*) roots by the nematophagous fungus *Pochonia chlamydosporia* reveals plant growth promotion and a general defense and stress transcriptomic response. J. Plant Res. 128, 665–678.

Li, J., Zou, C., Xu, J., Ji, X., Niu, X., Yang, J., et al., 2015. Molecular mechanisms of nematode-nematophagous microbe interactions: basis for biological control of plant-parasitic nematodes. Annu. Rev. Phytopathol. 53, 67–95.

Liebhold, A.M., Plymale, R.C., Elkinton, J.S., Hajek, A.E., 2013. Emergent fungal entomopathogen does not alter density dependence in a viral competitor. Ecology 94, 1217–1222.

Lihme, M., Jensen, A.B., Rosendahl, S., 2009. Local scale genetic structure of *Entomophthora muscae* epidemics. Fung. Ecol. 2, 81–86.

Lobo, L.S., Luz, C., Fernandes, E.K.K., Juárez, M.P., Pedrini, N., 2015. Assessing gene expression during pathogenesis: Use of qRT-PCR to follow toxin production in the entomopathogenic fungus *Beauveria bassiana* during infection and immune response of the insect host *Triatoma infestans*. J. Invertebr. Pathol. 128, 14–21.

Long, D.W., Drummond, F.A., Groden, E., Donahue, D.W., 2000. Modelling *Beauveria bassiana* horizontal transmission. Agric. For. Entomol. 2, 19–32.

Loreto, R.G., Elliot, S.L., Freitas, M.L.R., Pereira, T.M., Hughes, D.P., 2014. Long-term disease dynamics for a specialized parasite of ant societies: a field study. PLoS ONE 9(8), e103516.

Luangsa-Ard, J.J., Hywel-Jones, N.L., Manoch, L., Samson, R.A., 2005. On the relationships of *Paecilomyces* sect. Isarioidea species. Mycol. Res. 109, 581–589.

Luangsa-Ard, J., Houbraken, J., van Doorn, T., Hong, S.-B., Borman, A.M., Hywel-Jones, N.L., Samson, R.A., 2011. *Purpureocillium*, a new genus for the medically important *Paecilomyces lilacinus*. FEMS Microbiol. Lett. 321, 141–149.

Łukasik, P, van Asch, M., Guo, H., Ferrari, J., Godfray, H.C.J., 2013. Unrelated facultative endosymbionts protect aphids against a fungal pathogen. Ecol. Lett. 16, 214–218.

Luz, C., Fargues, J., Romaña, C., 2003. Influence of starvation and blood meal-induced moult on the susceptibility of nymphs of *Rhodnius prolixus* Stål (Hem., Triatominae) to *Beauveria bassiana* (Bals.) Vuill. infection. J. Appl. Entomol. 12, 153–156.

Malakar, R., Elkinton, J.S., Hajek, A.E., Burand, J.P., 1999. Within-host interactions of *Lymantria dispar* L. (Lepidoptera: Lymantriidae) nucleopolyhedrosis virus (LdNPV) and *Entomophaga maimaiga* (Zygomycetes: Entomophthorales). J. Invertebr. Pathol. 73, 91–100.

Mburu, D.M., Ndung'u, M.W., Maniania, N.K., Hassanali, A., 2011. Comparison of volatile blends and gene sequences of two isolates of *Metarhizium anisopliae* of different virulence and repellency toward the termite *Macrotermes michaelseni*. J. Exp. Biol. 214, 956–962.

Mburu, D.M., Maniania, N.K., Hassanali, A., 2013. Comparison of volatile blends and nucleotide sequences of two *Beauveria bassiana* isolates of different virulence and repellency towards the termite *Macrotermes michaelseni*. J. Chem. Ecol. 39, 101–108.

McCallum, H., Barlow, N., Hone, J., 2001. How should pathogen transmission be modelled? Tr. Ecol. Evol. 16, 295–300.

McKinnon, A., Saari, S., Raad, M., Moran-Diez, M., Meyling, N.V., Glare, T., 2017. *Beauveria bassiana* as an endophyte: a critical review on associated methodology and biocontrol potential. BioControl 62, 1–17.

McLaughlin, D., Spatafora, J.W. (eds.), 2015. The Mycota, VII, Systematics and Evolution, Part B, 2nd edn. Springer, Heidelberg.

Ment, D., Gindin, G., Rot, A., Soroker, V., Glazer, I., Barel, S., Samish, M., 2010. Novel technique for quantifying adhesion of *Metarhizium anisopliae* conidia to the tick cuticle. Appl. Environ. Microbiol. 76, 3521–3528.

Mesquita, A.L.M., Lacey, L.A., 2001. Interactions among the entomopathogenic fungus *Paecilomyces fumosoroseus* (Deuteromycotina: Hyphomycetes), the parasitoid *Aphelinus asychis* (Hymenoptera: Aphelinidae) and their aphid host. Biol. Control 22, 51–59.

Meyling, N.V., Eilenberg, J., 2006a. Occurrence and distribution of soil borne entomopathogenic fungi within a single organic agro-ecosystem. Agric. Ecosyst. Environ. 113, 336–341

Meyling, N.V., Eilenberg, J., 2006b. Isolation and characterisation of *Beauveria bassiana* isolates from phylloplanes of hedgerow vegetation. Mycol. Res. 110, 188–195

Meyling, N.V., Eilenberg, J., 2007. Ecology of the entomopathogenic fungi *Beauveria bassiana* and *Metarhizium anisopliae* in temperate agroecosystems: potential for conservation biological control. Biol. Control 43, 145–155.

Meyling, N.V., Hajek A.E., 2010. Principles from community and metapopulation ecology: application to fungal entomopathogens. BioControl 55, 39–54.

Meyling, N.V., Pell, J.K., 2006. Detection and avoidance of an entomopathogenic fungus by a generalist insect predator. Ecol. Entomol. 31, 162–171.

Meyling, N.V., Pell, J.K., Eilenberg, J., 2006. Dispersal of *Beauveria bassiana* by the activity of nettle insects. J. Invertebr. Pathol. 93, 121–126.

Meyling, N.V., Lübeck, M., Buckley, E.P., Eilenberg, J., Rehner, S.A., 2009. Community composition, host-range and genetic structure of the fungal entomopathogen *Beauveria* in adjoining agricultural and semi-natural habitats. Molec. Ecol. 18, 1282–1293.

Meyling, N.V., Thorup-Kristensen, K., Eilenberg, J., 2011. Below- and aboveground abundance and distribution of fungal entomopathogens in experimental conventional and organic cropping systems. Biol. Control 59, 180–186.

Meyling, N.V., Pilz, C., Keller, S., Widmer, F., Enkerli, J., 2012. Diversity of *Beauveria* spp. isolates from pollen beetles *Meligethes aeneus* in Switzerland. J. Invertebr. Pathol. 109, 76–82.

Michalaki, M.P., Athanassiou, C.G., Steenberg, T., Buchelos, C.T., 2007. Effect of *Paecilomyces fumosoroseus* (Wise) Brown and Smith (Ascomycota: Hypocreales) alone or in combination with diatomaceous earth against *Tribolium confusum* Jacquelin du Val (Coleoptera: Tenebrionidae) and *Ephestia kuehniella* Zeller (Lepidoptera: Pyralidae). Biol. Control 40, 280–286.

Mietkiewski, R., Tkaczuk, C., 1998. The spectrum and frequency of entomopathogenic fungi in litter, forest soil and arable soil. IOBC/WPRS Bull. 21(4), 41–44.

Minnis, A.M., 2015. The shifting sands of fungal naming under the ICN and the one name era for Fungi, in: McLaughlin, D.J., Spatafora, J.W. (eds.), The Mycota, VII, Systematics and Evolution, Part B, 2nd edn. Springer, Heidelberg, pp. 179–203.

Molnár, I., Gibson, D.M., Krasnoff, S.B., 2010. Secondary metabolites from entomopathogenic Hypocrealean fungi. Nat. Prod. Rep. 27, 1241–1275.

Monzón, A.J., Guharay, F., Klingen, I., 2008. Natural occurrence of *Beauveria bassiana* in *Hypothenemus hampei* (Coleoptera: Curculionidae) poplations in unsprayed coffee fields. J. Invertebr. Pathol. 97, 134–141.

Moonjely, S., Barelli, L., Bidochka, M.J., 2016. Insect pathogenic fungi as endophytes. Adv. Gen. 94, 107–135.

Morales,-Vidal, S., Alatorre-Rosas, R., Clark, S.J., Pell, J.K., Guzmán-Franco, A.W., 2013. Competition between isolates of *Zoophthora radicans* co-infecting *Plutella xylostella* popluations. J. Invertebr. Pathol. 113, 137–145.

Myles, T.G., 2002. Alarm, aggregation, and defense by *Reticulitermes flavipes* in response to a naturally occurring isolate of *Metarhizium anisopliae*. Sociobiology 40, 243–255.

Neumann, G., Shields, E.J., 2008. Multiple-species natural enemy approach for biological control of alfalfa snout beetle (Coleoptera: Curculionidae) using entomopathogenic nematodes. J. Econ. Entomol. 101, 1533–1539.

Nielsen, C., Agrawal, A.A., Hajek, A.E., 2010. Ants defend aphids against lethal disease. Biol. Lett. 6, 205–208.

Nielsen, C., Hajek, A.E., Humber, R.A., Bresciani, J., Eilenberg, J., 2003. Soil as an environment for winter survival of aphid-pathogenic Entomophthorales. Biol. Control 28, 92–100.

Nielsen, C., Hajek, A.E., 2006. Diurnal pattern of death and sporulation of *Entomophaga maimaiga*-infected *Lymantria dispar*. Entomol. Exp. Appl. 118, 237–243.

Nordbring-Hertz, B., Jansson, H.-B., Tunlid, A., 2011. Nematophagous fungi, in: Encyclopedia of Life Sciences. John Wiley & Sons, Chichester.

Novak, S., Cremer, S., 2015. Fungal disease dynamics in insect societies: optimal killing rates and the ambivalent effect of high social interaction rates. J. Theor. Biol. 372, 54–64.

Nowak, M.A., May, R.M., 1994. Superinfection and the evolution of parasite virulence. Proc. R. Soc. B 255, 81–89.

Ormond, E.L., Thomas, A.P.M., Pell, J.K., Freeman, S.N., Roy, H.E., 2010. Avoidance of a generalist entomopathogenic fungus by the ladybird, *Coccinella septempunctata*. FEMS Microbiol. Ecol. 77, 229–237.

Otti, O., Tragust, S., Feldhaar, H., 2014. Unifying external and internal immune defences. Tr. Ecol. Evol. 29, 625–634.

Ouedraogo, R.M., Goettel, M.S., Brodeur, J., 2004. Behavioral thermoregulation in the migratory locust; a therapy to overcome fungal infection. Oecologia, 138, 312–319.

Ownley, B.H., Gwinn, K.D., Vega, F.E., 2010. Endophytic fungal entomopathogens with activity against plant pathogens: ecology and evolution. BioControl 55, 113–128.

Parker, B.J., Spragg, C.J., Altincicek, B., Gerardo, N.M., 2013. Symbiont-mediated protection against fungal pathogens in pea aphids: a role for pathogen specificity? Appl. Environ. Microbiol. 79, 2455–2458.

Parker, B.J., Garcia, J.R., Gerardo, N.M., 2014. Genetic variation in resistance and fecundity tolerance in a natural host-pathogen interaction. Evolution 68, 2421–2429.

Pell, J., Eilenberg, J., Hajek, A.E., Steinkraus, D.C., 2001. Biology, ecology and pest management potential of Entomophthorales, in: Butt, T.M., Jackson, C.W., Magan, N. (eds.), Fungi as Biocontrol Agents: Progress, Problems and Potential. CABI, Wallingford, pp. 71–153.

Pell, J.K., Hannam, J.J., Steinkraus, D.C., 2010. Conservation biological control using fungal entomopathogens. BioControl 55, 187–198.

Pedrini, N., Ortiz-Urquiza, A., Huarte-Bonnet, C., Fan, Y., Juárez, M.P., Keyhani, N.O., 2015. Tenebrionid secretions and a fungal benzoquinone oxidoreductase form competing components of an arms race between a host and pathogen. Proc. Natl. Acad. Sci. U.S.A. 112, E3651–E3660.

Petrini, O., 1991. Fungal endophytes of tree leaves, in: Andrews, J., Hirano, S. (eds.), Microbial Ecology of Leaves. Springer, New York, pp. 179–197.

Pfister, D.H., 2015. Pezizomycotina: Pezizomycetes, Orbiliomycetes, in: McLaughlin, D.J., Spatafora, J.W. (eds.), The Mycota, VII, Systematics and Evolution, Part B, 2nd edn. Springer, Heidelberg, pp. 35–56.

Pineda, A., Zheng, S.-J., van Loon, J.J.A., Pieterse, C.M.J., Dicke, M., 2010. Helping plants to deal with insects: the role of beneficial soil-borne microbes Tr. Plant Sci. 15, 507–514

Pontieri, L., Vojvodic, S., Graham, R., Pedersen, J.S., Linksvayer, T.A., 2014. Ant colonies prefer infected over uninfected nest sites. PLoS ONE, 9, e111961.

Posada, F., Vega, F.E., Rehner, S.A., Blackwell, M., Weber, D., Suh, S.O., Humber, R.A., 2004. *Syspastospora parasitica*, a mycoparasite of the fungus *Beauveria bassiana* attacking the Colorado potato beetle, *Leptinotarsa decemlineata*: a tritrophic association. J. Insect Sci. 4, 24.

Quesada-Moraga, E., Navas-Cortés, J.A., Maranhao, E.A.A., Ortiz-Urquiza, A., Santiago-Alvarez, C., 2007. Factors affecting the occurrence and distribution of entomopathogenic fungi in natural and cultivated soils. Mycol. Res. 111, 947–966.

Quesada-Moraga, E., Herrero, N., Zabalgogeazcoa, I., 2014. Entomopathogenic and nematophagous fungal endophytes, in: Verma, V.C., Gange, A.C. (eds.), Advances in Endophytic Research. Springer, India, pp. 85–99.

Rännbäck, L.-M., Cotes, B., Anderson, P., Rämert, B., Meyling, N.V., 2015. Mortality risk from entomopathogenic fungi affects oviposition behavior in the parasitoid wasp *Trybliographa rapae*. J. Invertebr. Pathol. 124, 78–86.

Rännbäck, L.-M., 2015. Biological control strategies against the cabbage root fly *Delia radicum*: Effect of predators, parasitoids and pathogens. Doctoral thesis no. 2015:53, Faculty of Landscape Architecture, Horticulture and Crop Production Science, Swedish University of Agricultural Sciences.

Rangel, D.E.N., Braga, G.U.L., Fernandes, E.K.K., Keyser, C.A., Hallsworth, J.E., Roberts, D.W., 2015. Stress tolerance and virulence of insect-pathogenic fungi are determined by environmental conditions during conidial formation. Curr. Genet. 61, 383–404.

Rehner, S.A., Buckley, E.P., 2005. A *Beauveria* phylogeny inferred from nuclear ITS and EF1-alpha sequences: evidence for cryptic diversification and links to *Cordyceps* teleomorphs. Mycologia 97, 84–98

Rehner, S.A., Minnis, A.M., Sung, J.-M., Luangsa-ard, J.J., DeVotto, L., Humber, R.A., 2011. Phylogenetic systematics of the anamorphic, entomopathogenic *Beauveria* (Hypocreales, Ascomycota). Mycologia 103, 1055–1073.

Reilly, J.R., Hajek, A.E., Liebhold, A.M., Plymale, R.S., 2014. The impact of *Entomophaga maimaiga* on outbreak gypsy moth population: the role of weather. Environ. Entomol. 43, 632–641.

Reyes-Villanueva, F., Garza-Hernandez, J.A., Garcia-Munguia, A.M., Tamez-Guerra, P., Howard, A.F.V., Rodriguez-Perez, M.A., 2011. Dissemination of *Metarhizium anisopliae* of low and high virulence by mating behavior in *Aedes aegypti*. Para. Vectors 4, 171.

Rezende, J.M., Zanardo, A.B.R., Lopes, M.S., Delalibera, I., Rehner, S.A., 2015. Phylogenetic diversity of Brazilian *Metarhizium* associated with sugarcane agriculture. BioControl 60, 495–505.

Riedel, W., Steenberg, T., 1998. Adult polyphagous coleopterans overwintering in cereal boundaries: winter mortality and susceptibility to the entomopathogenic fungus *Beauveria bassiana*. BioControl 43, 175–188.

Rocha, L.F.N., Inglis, P.W., Humber, R.A., Kipnis, A., Luz, C., 2013. Occurrence of *Metarhizium* spp. in Central Brazilian soils. J. Basic Microbiol. 53, 251–259.

Rodriguez, R.J., White, J.F., Arnold, A.E., Redman, R.S., 2009. Fungal endophytes: diversity and functional roles. New Phytol. 182, 314–330.

Rohlfs, M., Churchill, A.C.L., 2011. Fungal secondary metabolites as modulators of interactions with insects and other arthropods. Fung. Gen. Biol. 48, 23–34.

Roy, H.E., Pell, J.K., 2000. Interactions between entomopathogenic fungi and other natural enemies: Implications for biological control. Biocontr. Sci. Technol. 10, 737–752.

Roy, H.E., Pell, J.K., Clark, S.J., Alderson, P.T., 1998. Implications of predator foraging on aphid pathogen dynamics. J. Invertebr. Pathol. 71, 236–247.

Roy, H.E., Steinkraus, D.C., Eilenberg, J., Hajek, A.E., Pell, J.K., 2006. Bizarre interactions and endgames: Entomopathogenic fungi and their arthropod hosts. Annu. Rev. Entomol. 51, 331–357.

Roy, H., Vega, F., Chandler, D., Goettel, M., Pell, J., Wajnberg, E. (eds.), 2010. The Ecology of Fungal Entomopathogens. Springer, Dordrecht.

Sandoval-Aguilar, J.A., Guzmán-Franco, A.W., Pell, J.K., Clark, S.J., Alatorre-Rosas, R., Santillán-Galicia, M.T., et al., 2015. Dynamics of competition and co-infection between *Zoophthora radicans* and *Pandora blunckii* in *Plutella xylostella* larvae. Fung. Biol. 17, 1–9.

Sasan, R.K., Bidochka, M.J., 2012. The insect-pathogenic fungus *Metarhizium robertsii* (Clavicipitaceae) is also an endophyte that stimulates plant root development. Am. J. Bot. 99, 101–107.

Sawyer, A.J., Ramos, M.E., Poprawski, T.J., Soper. R.S., Carruthers, R.I., 1997. Seasonal patterns of cadaver persistence and sporulation by the fungal pathogen *Entomophaga grylli* (Fresenius) Batko (Entomophthorales: Entomophthoraceae) infecting *Camnula pellucida* (Scudder) (Orthoptera: Acrididae). Mem. Entomol. Soc. Can. 129(S171), 355–374.

Scarborough, C.L., Ferrari, J., Godfray, H.C.J., 2005. Aphid protected from pathogen by endosymbiont. Science 310, 1781.

Schmid-Hempel, P., 2011. Evolutionary Parasitology. The integrated study of infections, immunology, ecology and genetics. Oxford University Press, New York.

Schoch, C.L., Seifert, K.A., Huhndorf, S., Robert, V., Spouge, J.L., Levesque, C.A., Chen, W., 2012. Nuclear ribosomal internal transcribed spacer (ITS) region as a universal DNA barcode marker for Fungi. Proc. Natl. Acad. Sci. U.S.A. 109, 6241–6246.

Shapiro-Ilan, D.I., Jackson, M., Reilly, C.C., Hotchkiss, M.W., 2004. Effects of combining an entomopathogenic fungi or bacterium with entomopathogenic nematodes on mortality of *Curculio caryae* (Coleoptera: Curculionidae). Biol. Control 30, 119–126.

Shapiro-Ilan, D.I., Mizel III, R.F., 2015. An insect pupal cell with antimicrobial properties that suppress an entomopathogenic fungus. J. Invertebr. Pathol. 124, 114–116.

Shimuza, M., Maehara, N., Sato, H., 2002. Density dynamics of the entomopathogenic fungus *Beauveria bassiana* Vuillemin (Deuteromycotina: Hyphomycetes) introduced into forest soil, and its influence on other soil microorganisms. Appl. Entomol. Zool. 37, 263–269.

Smith, A.H., Łukasik, P., O'Connor, M.P., Lee, A., Mayo, G., Drott, M.T., et al., 2015. Patterns, causes and consequences of defensive microbiome dynamics across multiple scales. Molec. Biol. 24, 1135–1149.

Spatafora, J.W., Sung, G.H., Sung, J.M., Hywel-Jones, N.L., White, J.F., 2007. Phylogenetic evidence for an animal pathogen origin of ergot and the grass endophytes. Molec. Ecol. 16, 1701–1711.

Spatafora, J.W., Chang, Y., Benny, G.L., Lazarus, K., Smith, M.E., Berbee, M.L., et al., 2016. A phylum-level phylogenetic classification of zygomycete fungi based on genome-scale data. Mycologia 108, 1028–1046.

St. Leger, R.J., Wang, C., Fang, W., 2011. New perspectives on insect pathogens. Fung. Biol. Rev. 25, 84–88.

Steinkraus, D.C., 2006. Factors affecting transmission of fungal pathogens of aphids. J. Invertebr. Pathol. 92, 125–131.

Steinwender, B.M., 2013. Entomopathogenic fungi *Metarhizium* spp. in the soil environment of an agroecosystem: molecular diversity, root association and ecological characteristics. PhD thesis, University of Copenhagen, Faculty of Science, Department of Plant and Environmental Sciences.

Steinwender, B.M., Enkerli, J., Widmer, F., Eilenberg, J., Thorup-Kristensen, K., Meyling, N.V., 2014. Molecular diversity of the entomopathogenic fungal *Metarhizium* community within an agroecosystem. J. Invertebr. Pathol. 123, 6–12.

Steinwender, B.M., Enkerli, J., Widmer, F., Eilenberg, J., Kristensen, H.L., Bidochka, M.J., Meyling, N.V. 2015. Root isolations of *Metarhizium* spp. from crops reflect diversity in the soil and indicate no plant specificity. J. Invertebr. Pathol. 132, 142–148.

Stielow, J.B., Lévesque, C.A., Seifert, K.A., Meyer, W., Irinyi, L., Smits, D., et al., 2015. One fungus, which genes? Development and assessment of universal primers for potential secondary fungal DNA barcodes. Persoonia 35, 242–263.

Straub, C.S., Finke, D.L., Snyder, W.E., 2008. Are the conservation of natural enemy biodiversity and biological control compatible goals? Biol. Control 45, 225–237.

Sung, G.-H., Hywel-Jones, N.L., Sung, J.-M., Luangsa-ard, J.J., Srestha, B., Spatafora, J.W., 2007. Phylogenetic classification of *Cordyceps* and the clavicipitaceous fungi. Stud. Mycol. 57, 5–59.

Tamayo-Mejía, F., Tamez-Guerra, P., Guzmán-Franco, A.W., Gomez-Flores, R., 2015. Can *Beauveria bassiana* Bals. (Vuill.) (Ascomycetes: Hypocreales) and *Tamarixia triozae* (Bruks) (Hymenoptera: Eulophidae) be used together for improved biological control of *Bactericera cockerelli* (Hemiptera: Triozidae)? Biol. Control 90, 42–48.

Tanada, Y., Kaya, H.K., 1993. Insect Pathology, Academic Press, San Diego, CA.

Thapa, S., Meyling, N.V., Katakam, K.K., Thamsborg, S.M., Mejer, H., 2015. A method to evaluate relative ovicidal effects of soil microfungi on thick-shelled eggs of animal-parasitic nematodes. Biocontr. Sci. Technol. 25, 756–767.

Thomas, M.B., Wood, S.N., Lomer, C.J., 1995. Biological control of locusts and grasshoppers using a fungal pathogen: the importance of secondary cycling. Proc. R. Soc. Lond. B 259, 265–270.

Thomas, M.B., Watson, E.L., Valverde-Garcia, P., 2003. Mixed infections and insect-pathogen interactions. Ecol. Lett. 6, 183–188.

Thompson, S.R., Brandenburg, R.L., 2005. Tunneling responses of mole crickets (Orthoptera: Gryllotalpidae) to the entomopathogenic fungus, *Beauveria bassiana*. Environ. Entomol. 34, 140–147.

Tobin, P.C., Hajek, A.E., 2012. Release, establishment, and initial spread of the fungal pathogen *Entomophaga maimaiga* in island populations of *Lymantria dispar*. Biol. Control 63, 31–39.

Tragust, S., Mitteregger, B., Barone, V., Konrad, M., Ugelvig, L.V., Cremer, S., 2013. Ants disinfect fungus-exposed brood by oral uptake and spread of their poison. Curr. Biol. 23, 76–82.

Trandem, N., Bhattarai, U.R., Westrum, K., Kjøberg Knudsen, G., Klingen, I., 2015. Fatal attraction: male spider mites prefer females killed by the mite-pathogenic fungus *Neozygites floridana*. J. Invertebr. Pathol. 128, 6–13.

Ugelvig, L.V., Cremer, S., 2007. Social prophylaxis: group interaction promotes collective immunity in ant colonies. Curr. Biol. 17, 1967–1971.

Ugine, T.A., Peters, K.E., Gardescu, S., Hajek, A.E., 2014. The effect of time post-exposure and gender on horizontal transmission of *Metarhizium brunneum* conidia between mating pairs of Asian longhorned beetles (Coleoptera: Cerambycidae). Environ. Entomol. 43, 1552–1560.

Uma Devi, K., Uma Maheswara Rao, C., 2006. Allee effect in the infection dynamics of the entomopathogenic fungus *Beauveria bassiana* (Bals.) Vuill. on the beetle, *Mylabris pustulata*. Mycopathologia 161, 385–394.

van Frankenhuyzen, K., Ebling, P., Thurston, G., Lucarotti, C., Royama, T., Guscott, R., et al., 2002. Incidence and impact of *Entomophaga aulicae* (Zygomycetes:

Entomophthorales) and a nucleopolyhedrovirus in an outbreak of the whitemarked tussock moth (Lepidoptera: Lymantriidae). Can. Entomol. 134, 825–845.

Vänninen, I., Tyni-Juslin, J., Hokkanen, H., 2000. Persistence of augmented *Metarhizium anisopliae* and *Beauveria bassiana* in Finnish agricultural soils. BioControl 45, 201–222.

Vega, F.E., Posada, F., Aime, A.M., Pava-Ripoll, M., Infante, F., Rehner, S.A., 2008. Entomopathogenic fungal endophytes. Biol. Control, 46, 72–82.

Vega, F.E., Simpkins, A., Aime, M.A., Posada, F., Peterson, S.W., Rehner, S.A., et al., 2010. Fungal endophyte diversity in coffee plants from Columbia, Hawai'i, Mexico and Puerto Rico. Fung. Ecol. 3, 122–138.

Vega, F.E., Meyling, N.V., Luangsa-Ard, J.J., Blackwell, M., 2012. Fungal entomopathogens, in: Vega, F.E., Kaya, H.K. (eds.), Insect Pathology, 2nd edn. Academic Press, Amsterdam, pp. 171–220.

Vey, A., Fargues, J., 1977. Histological and ultrastructural studies of *Beauveria bassiana* infection in *Leptinotarsa decemlineata* larvae during ecdysis. J. Invertebr. Pathol. 30, 207–215.

Vidal, C., Fargues, J., 2007. Climatic constraints for fungal biopesticides, in: Ekesi, S., Maniania, N.K. (eds.), Use of Entomopathogenic Fungi in Biological Pest Management. Research Signpost, Kerala, pp. 39–55.

Vidal, S., Jaber, L.R., 2015. Entomopathogenic fungi as endophytes: plant-endophyte-herbivore interactions and prospects for use in biological control. Curr. Sci. 109, 46–54.

Villani, M.G., Krueger, S.R., Schroeder, P.C., Consolie, F., Consolie, N.H., Preston-Wilsey, L.M., Roberts, D.W., 1994. Soil application effects of *Metarhizium anisopliae* on Japanese beetle (Coleoptera: Scarabaeidae) behavior and survival in turfgrass microcosms. Environ. Entomol. 23, 502–513.

Vos, C.M.F., de Cremer, K., Cammue, B.P.A., de Connick, B., 2015. The toolbox of *Trichoderma* spp. in the biocontrol of *Botrytis cinerea* disease. Mol. Plant Pathol. 16, 400–412.

Wang, C.S., Hu, G., St. Leger, R.J., 2005. Differential gene expression by *Metarhizium anisopliae* growing in root exudate and host (*Manduca sexta*) cuticle or hemolymph reveals mechanisms of physiological adaptation. Fung. Gen. Biol. 42, 704–718.

Wang, S., O'Brien, T.R., Pava-Ripoll, M., St. Leger, R.J., 2011. Local adaptation of an introduced transgenic insect fungal pathogen due to new beneficial mutations. Proc. Natl. Acad. Sci. U.S.A. 108, 20449–20454.

Watson, D.W., Mullens, B.A., Petersen, J.J., 1993. Behavioral fever response of *Musca domestica* (Diptera: Muscidae) to infection by *Entomophthora muscae* (Zygomycetes: Entomophthorales). J. Invertebr. Pathol. 61, 10–16.

Wekesa, V.W., Vital, S., Silva, R.A., Ortega, E.M.M., Klingen, I., Delalibera Jr., I., 2011. The effect of host plants on *Tetranychus evansi* and *Tetranychus urticae* (Acari: Tetranychidae) and on their fungal pathogen *Neozygites floridana* (Entomophthorales: Neozygitaceae). J. Invertebr. Pathol. 107, 139–145.

Weseloh, R.M., 2003. Short and long range dispersal in the gypsy moth (Lepidoptera: Lymantriidae) fungal pathogen, *Entomophaga maimaiga* (Zygomycetes: Entomophthorales). Environ. Entomol. 32, 111–122.

Weseloh, R.M., Andreadis, T.G., 1997. Persistence of resting spores of *Entomophaga maimaiga*, a fungal pathogen of the gypsy moth, *Lymantria dispar*. J. Invertebr. Pathol. 69, 195–196.

Wilson, K., Cotter, S.C., Reeson, A.F., Pell, J.K., 2001. Melanism and disease resistance in insects. Ecol. Lett. 4, 637–649.

Wilson, K., Thomas, M.B., Blanford, S., Doggett, M., Simpson, S.J., Moore, S.L., 2002. Coping with crowds: density-dependent disease resistance in desert locusts. Proc. Natl. Acad. Sci. U.S.A. 99, 5471–5475.

Wyrebek, M., Huber, C., Sasan, R.K., Bidochka, M.J., 2011. Three sympatrically occurring species of *Metarhizium* show plant rhizosphere specificity. Microbiology 157, 2904–2911.

Yanagawa, A., Imai, T., Akino, T., Toh, Y., Yoshimura, T., 2015. Olfactory cues from pathogenic fungus affect the direction of motion of termites, *Coptotermes formosanus*. J. Chem. Ecol. 41, 1118–1126.

Zamora-Macorra, E.J., Guzmán-Franco, A.W., Pell, J.K., Alatorre-Rosas, R., Suarez-Espinoza, J., 2012. Order of inoculation affects the success of co-invading entomopathogenic fungi. Neotrop. Entomol. 41, 521–523.

Zare, R., Gams, W., 2001. A revision of *Verticillium* section *Prostrata*. IV. The genera *Lecanicillium* and *Simplicillium* gen. nov. Nova Hedwigia 3, 1–50.

Zimmermann, G., 1986. The *Galleria* bait method for detection of entomopathogenic fungi in soil. J. Appl. Entomol. 102, 213–215.

10

Microsporidia

Gernot Hoch[1] and Leellen F. Solter[2]

[1] *Department of Forest Protection, BFW Austrian Research Centre for Forests, Vienna, Austria*
[2] *Illinois Natural History Survey, Prairie Research Institute, University of Illinois, Champaign, IL, USA*

10.1   Introduction

Microsporidia are obligate intracellular parasites that have refined "the art of living together" with their hosts by minimizing damage in order to achieve optimum reproduction and transmission to new hosts (Vavra and Lukes, 2013). Microsporidia have two core stages in their lifecycle: the meront and the spore (Smith, 2009). The meront is the proliferative stage that is adapted for intracellular parasitism and usually develops in the host cell cytoplasm. The spore is the infective stage and can persist for a limited time outside the host cell. Microsporidian spores are equipped with a unique infection apparatus: a polar filament coiled within the spore that everts to "inject" the contents of the spore into the host cell. The meront and spore stages are found in all microsporidian species; species-specific lifecycles, however, can range from relatively simple developmental cycles to considerably more complex cycles with a variety of developmental stages and obligatory intermediate hosts. Between 1300 and 1500 microsporidian species from approximately 187 genera have been described from various animal orders, including all vertebrate orders and many invertebrate taxa; a few species have been reported to infect certain protists (Keeling and Fast, 2002). Most described species have been isolated from arthropods and fish (Vavra and Lukes, 2013), with more than 700 species known from insects (Becnel and Andreadis, 1999). The assumption that the number of described microsporidia species is a minute portion of those actually existing was corroborated by a study in which 22 novel species were isolated from soil, sand, and compost (Ardila-Garcia et al., 2013). The taxonomy and biology of the Microsporidia are extensively covered in the text *Microsporidia: Pathogens of Opportunity* (Weiss and Becnel, 2014).

The discovery of microsporidia was a landmark in early insect pathology studies. Pasteur discovered the first disease to be attributed to microsporidia – the pebrine disease that threatened the silkworm industry in Europe – and developed measures to control it (Pasteur, 1870). Nägeli (1857) described the causative agent of the silkworm pebrine as *Nosema bombycis*. Since the early days of discovery, microsporidiosis in insects has been intensively studied; consequently, nearly half of the described microsporidian

Ecology of Invertebrate Diseases, First Edition. Edited by Ann E. Hajek and David I. Shapiro-Ilan.
© 2018 John Wiley & Sons Ltd. Published 2018 by John Wiley & Sons Ltd.

genera have an insect as the type host (Becnel and Andreadis, 1999). Several genera have been studied for their potential in biological control (Solter and Becnel, 2000), and one species, *Paranosema* (*Nosema, Antonospora*) *locustae*, was developed and registered as a microbial insecticide against grasshoppers (Henry and Oma, 1981). Additionally, as indicated by the circumstances of their initial discovery, microsporidia cause significant problems in managed colonies of insects such as silkworms (*Bombyx mori*) and honeybees (*Apis mellifera*), as well as insects being reared for biological control and research purposes. Moreover, they pose potential serious problems in mass rearing of insects for food and feed (Stentiford et al., 2016) (see also Chapters 14–16).

After the initial mention of a relationship to fungi (Nägeli, 1857), microsporidia were classified within the Protozoa, then as Phylum Microspora erected by Sprague (1977). The phylum was renamed Microsporidia by Sprague and Becnel (1998) based on precedence. Microsporidia and other protists lacking mitochondria were placed in the kingdom Archezoa by Cavalier-Smith (1983), but this was never generally accepted (Vavra and Larsson, 1999). There is currently agreement that microsporidia should be placed in or near the kingdom Fungi (James et al., 2006; Hibbet et al., 2007). The exact relationship to the Fungi remains unresolved at the genetic level, and morphological characteristics and lifecycle strategies are very different from those of other fungal Opisthokonta (Vavra and Lukes, 2013).

Microsporidia are highly adapted obligate, intracellular parasites and are primarily characterized by a reduction in morphology, metabolism, and genome (Keeling and Fast, 2002). One stage, the environmentally resistant, infective spore, is adapted to survive outside the host cell. The spore wall consists of a proteinaceous, electron-dense exospore, a thicker chitinous, electron-transparent endospore layer, and the plasma membrane. The most unique feature of microsporidian spores is the polar filament, which, upon germination, unwinds and is everted, penetrating the membrane of the host cell. The contents of the spore are injected through this filament into the host cell. The polar filament is an autapomorphic character that strongly identifies microsporidia as a monophyletic taxon (Vavra and Lukes, 2013).

10.1.1 Mechanisms of Infection

Microsporidia typically gain access to new hosts by oral ingestion of spores. Spores germinate in the gut lumen of the host and infect cells of the midgut. Environmental stimuli, such as alterations in pH, dehydration followed by rehydration, or the presence of certain ions, trigger spore germination, but specific triggers are known for very few species. Germination occurs when the osmotic pressure in the spore increases and the polar filament is discharged (Undeen, 1990; Keohane and Weiss, 1999). When the filament strikes and pierces a host cell, the sporoplasm (membrane-bound content of the spore) travels through the filament and emerges into the host cell cytoplasm, generally escaping immune system detection as a foreign invader (Keeling and Fast, 2002).

From the midgut, where the first vegetative reproduction takes place, the pathogen may spread to other organs, depending on the tissue tropism of the species. Some species produce a spore type that germinates within the host cells (described later) and appears to be responsible for dissemination to other tissues. However, the mechanisms of pathogen spread within the host are still not fully understood (Vavra and Lukes, 2013). Another important pathway for infection is vertical transmission from infected female host to the offspring via the embryo (transovarial transmission), or egg-surface (transo-vum transmission) (Tanada and Kaya, 1993; Becnel and Andreadis, 1999). Additionally, microsporidia can potentially gain access through the host cuticle by ovipositing parasitic wasps (Brooks, 1993).

10.1.2 Microsporidian Life Cycles

After entering the new host cell through the polar filament, the sporoplasm develops into a vegetative stage, the meront, followed by a proliferation phase, merogony (Fig. 10.1a). Triggered by an as yet unknown signal, vegetative stages enter sporogony, terminating in the formation of spores. Reproductive stages are strictly intracellular and often cause no visible damage to host cells until sporulation (Vavra and Lukes, 2013). Two different sporulation sequences have been described from some microsporidian taxa. One spore type is produced at an early stage of infection and spontaneously germinates inside the host cells. These "primary spores" (Iwano and Kurtii, 1995) are hypothesized to be a mechanism for cell-to-cell transmission (Avery and Anthony, 1983; Iwano and Ishihara, 1991; Iwano and Kurtti, 1995; Sagers et al., 1996). The spores produced in the second sporulation sequence later in development do not germinate inside the host cells. These "environmental spores" (Maddox et al., 1999) can survive outside host cells and enable transmission to new hosts (Fig. 10.1b). Overall, the developmental cycles vary widely among microsporidian taxa and can reach considerable levels of complexity, involving production of various spore types. Some groups require intermediate hosts, for example, the genus *Amblospora*, which infects mosquitoes and copepods (Andreadis, 1985; Micieli et al., 2000). For a review of representative lifecycles of entomopathogenic microsporidia, see Becnel and Andreadis (1999) and Solter et al. (2012a).

10.1.3 Pathology

In arthropods, microsporidian infection can range from chronic to acute depending on such variables as initial dosage, condition of the host, and life stage exposed, but also the

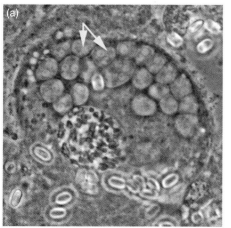 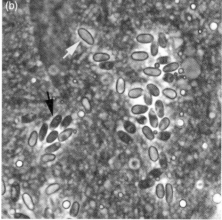

Fig. 10.1 (a) Vegetative microsporidia (meronts and sporonts) within the cytoplasm of an *Otiorhynchus sulcatus* host cell. (b) Immature spores (black arrow) and mature spores (white arrow) in *O. sulcatus* midgut tissues. *Source*: Courtesy of L. Solter.

inherent interaction of a particular microsporidian species with its host. Virulent microsporidia can lead to premature death of hosts, and sublethal effects can become manifest in increased susceptibility to environmental stress, reduced pupal weight, reduced adult lifespan, and/or reduced fecundity (Tanada and Kaya, 1993; Becnel and Andreadis, 1999; Solter and Becnel, 2000). The specific organs and tissues that support infection differ depending on the host–pathogen species association. This has important implications for the nature of infection in the host, including pathogen virulence, host response, and the mechanisms of transmission, as outlined in more detail later. Most commonly, fat body tissues and the midgut epithelial tissues are sites of microsporidian infection (Becnel and Andreadis, 1999), but many species produce systemic infections. Abundant sporulation and cell hypertrophy in infected tissues can produce a puffy whitish appearance, and well-developed microsporidiosis can be detected macroscopically in aquatic hosts with transparent cuticles (Weiser, 1961; Becnel and Andreadis, 1999).

Despite the fact that microsporidia typically do not cause detectable harm to the cells of their hosts until sporulation begins, and that they may resemble an "intracytoplasmic symbiont rather than a typical parasite" (Vavra and Lukes, 2013), there are measurable pathological effects prior to mortality. Because microsporidia are completely dependent on their host cells during development, they absorb nutrients as well as energy in the form of adenosine triphosphate (ATP) directly from the host cells (Weidner and Trager, 1973; Weidner et al., 1999; Katinka et al., 2001; Williams et al., 2008). With the onset of sporulation, the integrity of host cells is eventually damaged and the function of organs may be severely impaired. Reduced growth and development of infected hosts is a frequently reported effect of microsporidian infections. Weight gain and growth rate are typically reduced and larval development prolonged in both hemimetabolous and holometabolous insects (e.g., Hoch and Schopf, 2001; Rath et al., 2003; Down et al., 2004a; Tounou et al., 2011). Microsporidia-infected amphipods also show reduced weight (Terry et al., 1998). Reduced food consumption has been reported for various lepidopteran larvae infected with *Nosema* or *Vairimorpha* species (Haque et al., 1999; Rath et al., 2003; Pollan et al., 2009). Moreover, the host insect may be depleted of nutrients such as carbohydrates and lipids, particularly when the metabolically important fat body is infected (Hoch et al., 2002; Rivero et al., 2007; Mayack and Naug, 2010). This could be caused by uptake of nutrients by the microsporidia, disturbance of synthetic processes in the infected fat body, or increased energy metabolism in host cells induced by the pathogen. The effects on host physiology may also cause disturbance of hormone metabolism. Elevated titers of juvenile hormone (JH) have been reported from lepidopteran hosts as a consequence of microsporidian infection (e.g., the bright-line brown-eye (tomato moth; *Lacanobia oleracea*) infected with *Vairimorpha necatrix*; Down et al., 2008). Larvae of the gypsy moth (*Lymantria dispar*) infected with *Vairimorpha disparis* exhibit significantly reduced activity of the juvenile hormone esterase (JHE), a JH-degrading enzyme (Karlhofer et al., 2012), suggesting that hormonal disturbance could be an effect of impaired function of the fat body tissue where JHE is produced.

If microsporidian infection does not prevent development of the host into the adult stage, the number of host progeny is often reduced. Examples are moth species (Siegel et al., 1986; Bauer and Nordin, 1989; Goertz et al. 2008), lady beetles (Joudrey and Bjornson, 2007), and grasshoppers (Maniania et al., 2008). A reduced number of eggs in the amphipod *Gammarus duebeni* was shown to be correlated to lower weight of females infected with a *Nosema* sp. (Terry et al., 1998).

A remarkable effect of microsporidia in amphipods is the feminization of infected individuals. This can lead to profound alterations in the sex ratio, such as reported for *G. duebeni*, where feminization caused by *Nosema granulosis* occurred in 66% of infected young hosts (Terry et al., 1998). Vertical transmission is highly efficient, with values around 90% in this host–pathogen system (Terry et al., 1998; Ironside et al., 2003b). Virulence is low, which facilitates vertical transmission. Feminization and other sex-altering or sex-killing effects due to microsporidian infection have not been reported in insects.

10.2   Host Population

10.2.1   Susceptibility to Microsporidiosis

Early immature stages of host insects are typically more susceptible to microsporidian infections than later instars, and the effects are more profound (Solter, 2014). Of the many studies of infections in lepidopteran larvae that have been reported, examples include a detailed dataset from *L. oleracea* infected with *V. necatrix* that nicely illustrates an increase in LD_{50} values (spore dosages required to kill 50% of challenged individuals) with advancing larval stage (Down et al., 2004a). Third-instar larvae of the sugarcane borer (*Diatraea saccharalis*) had higher LT_{50} values (median lethal times) and significantly lower mortality than first instars infected with a *Nosema* sp. (Simoes et al., 2015). Infectivity of *Nosema whitei* is much higher in young larvae of the tenebrionid beetle *Tribolium castaneum*; a logistic regression model describes this age dependence of susceptibility (Blaser and Schmid-Hempel, 2005). Additionally, susceptibility of the locust *Schistocerca gregaria* to *P. locustae* is lower at late nymphal stages; percentage of individuals succumbing to infection decreases and survival time increases with increasing age of initial infection (Tounou et al., 2011). Infection in embryos as a consequence of transovarial transmission often leads to high mortality of neonates, as reported for European corn borers (*Ostrinia nubilalis*) infected with *Nosema pyrausta* (Andreadis, 1986).

Not all insects are susceptible to microsporidian pathogens during their immature stages. Larvae of *A. mellifera* were shown to be physiologically susceptible to *Nosema ceranae* in laboratory tests (Eiri et al., 2015), but newly emerged adults in infected hives were not shown to be infected (Smart and Sheppard, 2011), nor did newly emerged adults subsequently develop infections when isolated from infected workers (Huang and Solter, 2013).

10.2.2   Immune Response

Due to the unique mechanism of infecting host cells via "injection" directly into the host cell and intracellular development, it appears that microsporidia often do not elicit a host immune response. When pathogen stages do not occur free in the hemolymph of the host and the integrity of host tissue is not damaged, the infection, particularly in the early stages, may not be recognized. However, there are reports of immune responses to infection. Nodulation by hemocytes and melanization can be particularly strong in nonhabitual hosts (Brooks, 1970a,b; Solter et al., 1997; Solter and Maddox, 1998). Nevertheless, immune responses occur in habitual hosts as well when damage of tissue occurs. *Locusta migratoria* and *Gryllus bimaculatus* respond to fat body infection by *P. locustae* and *Paranosema grylli*, respectively, by forming nodules with melanin deposits around heavily infected cells. The production of enlarged, malformed spores was interpreted as a result of detrimental effects of melanine metabolites in this system (Tokarev et al., 2007). Comparing responses to infection by different microsporidian species in *L. dispar* larvae, Hoch et al. (2004) reported that the strongest increases in hemocyte counts and melanization were elicited by the two most virulent species tested, both of which are naturally occurring in the host and reproduce at massive levels in the host fat body. The response likely stemmed from tissue damage that began at the time of intensive sporulation. A strong hemocytic response was reported when silk glands of *L. dispar* larvae were physically damaged by a *Nosema* infection (David and Pilarska, 1988). Likewise, the fruit fly *Drosophila melanogaster* responds to infection with *Tubulinosema kingi* with a marked increase in hemocyte counts, which is assumed to be induced by damage to tissue integrity (Vijendravarma et al., 2008).

Immune defense responses observed in most studied host–microsporidia systems have not been reported to prevent the progression of infections (Texier et al., 2010). Experimental immune suppression of *L. dispar* larvae by the polydnavirus and venom of a parasitic braconid did not allow successful infection by microsporidian species for which untreated *L. dispar* are nonpermissive. However, this treatment did increase reproduction of naturally occurring microsporidia (Hoch et al., 2009a).

The role of host insect hemocytes in microsporidian infections is not fully understood. As illustrated earlier, they have been reported to participate in the immune response to infection by, for example, nodulation. Phagocytosis of microsporidian spores by hemocytes has been observed *in vitro* (Kurtz et al., 2000). However, some authors have concluded that microsporidia can also utilize hemocytes as host cells. Infected hemocytes may function as a means of transport to various organs and tissues (Weiser, 1961, 1978; David and Weiser, 1994; Johnson et al., 1997). A conclusive experimental clarification of the role of hemocytes in the development of microsporidian infections is still lacking, but it may differ depending on such variables as the particular host–pathogen species interactions, tissue tropism, and virulence.

10.2.3 Behavioral Response

Microsporidia do not appear to cause the negative geotropism observed for some viral or fungal infections; studies by van Frankenhuyzen et al. (2007) and Chapman (2010) gave no indication of altered location of microsporidia-infected hosts. However, other effects on host behavior can occur. The energy stress due to carbohydrate depletion in *N. ceranae*-infected *A. mellifera* (Mayack and Naug, 2010) was connected with higher hunger levels and resulting modified behavior in infected insects. Infected forager bees showed increased gustatory response to sucrose solution and reduced willingness to pass food (trophallaxis) to other bees (Naug and Gibbs, 2009). In addition, infected workers began foraging earlier in their lifecycles and died sooner, guarded more frequently, and fed the queen less frequently than uninfected workers (Wang and Moeller, 1970 (*Nosema apis*); Goblirsch et al., 2013 (*N. ceranae*)). Infected bees may also take longer to return to the hive when foraging (Kralj and Fuchs, 2010).

Microsporidia can also affect dispersal behavior. The flight duration and distance of *O. nubilalis* adults are significantly reduced when they are infected with *N. pyrausta*, most likely due to the energetic stress imposed by the pathogen (Dorhout et al., 2011). Female *Choristoneura fumiferana* adults infected with *N. fumiferanae* are less likely to emigrate from the tree where they developed, perhaps due to smaller size of infected females (Everleigh et al., 2007). A long-term study of the bark beetle *Ips typographus* caught in pheromone traps suggested that infection with *Chytridiopsis typographi* might negatively affect flight ability or interfere with pheromone perception (Wegensteiner et al., 2010).

A dysfunctional hormone metabolism (see Section 10.1.3.) may produce additional consequences deleterious to the behavior of infected hosts. Importantly, behavioral changes may reduce the damage produced by pest insects. *Locusta migratoria* infected with *P. locustae* changed from the gregarious form to the less damaging solitary form; a change that became apparent after an experimental application of microsporidia but before infections led to a reduction in host density (Fu et al., 2010).

10.3   Pathogen Population

10.3.1   Virulence of Microsporidian Pathogens

As already illustrated, microsporidia are well adapted for living inside the host cell, harnessing energy and nutrients from the host and optimizing their reproduction. Therefore, most microsporidian species are, at least in the early stages of infections, pathogens of relatively low virulence. They do not release toxins into the host, and often the disease is of chronic nature (Solter et al., 2012a). The level of virulence typically depends on tissue tropism of the microsporidian pathogen, with species infecting tissues of high metabolic activity, such as the insect fat body, having the most detrimental effects and causing up to 100% host mortality in the immature stages. This has been demonstrated for a range of *Vairimorpha* species infecting lepidopteran larvae (e.g., Hague et al., 1999; Down et al., 2004a; Goertz and Hoch, 2008a). Likewise, grasshoppers infected by *P. locustae*, which also infects the fat body, suffer high mortality (Canning, 1962; Henry and Oma, 1981). Prior to death of the host, a severe depletion of nutrients, as well as a disturbance of hormone metabolism, becomes evident. Although hosts are killed in the larval or nymphal stage by these microsporidian species, unless the initial spore dosage is atypically high, host death does not occur before the infected tissues have been fully utilized by the parasite and billions of microsporidian spores have been produced in the host cells. Other microsporidia infecting insects, including species in the genera *Nosema* and *Endoreticulatus* infecting Lepidoptera, are typically less virulent (Andreadis, 1981; Bauer and Nordin, 1989; Solter et al., 2005). Virulence is also usually lower for microsporidian species that infect only the digestive tract of their hosts (Maddox et al., 1998); for example, mortality of *L. dispar* larvae infected with *Endoreticulatus schubergi* is relatively low (Goertz and Hoch, 2008a). Moreover, *E. schubergi* infection has astonishingly little effect on the level of nutrients in the host (Hoch et al., 2009b), unlike infections with species that target the fat body tissues. Microsporidia that partially or completely depend on vertical transmission need to be of relatively low virulence in order to be successfully transmitted to the next host generation (see Section 10.4.2).

10.3.2 Host Specificity

A high level of host specificity is assumed for parasitic organisms that are intimately associated with the host – especially the host cell – because most species likely will have evolved and co-speciated with their original hosts (Vavra and Lukes, 2013). For most microsporidian species, the host range is limited, but not necessarily restricted to one host species. Many reports that host ranges are generally broad were based on mid-20th century laboratory host range studies using unusually high dosages of purified spores; later field studies suggested that putative alternate hosts, even if adventitiously infected in the field, would probably not transmit the pathogen to the conspecific population (Solter et al., 1998b). A few species, however – including those infecting mammalian hosts – have relatively broad natural host ranges (Solter et al., 2010; Hinney et al., 2016).

10.3.2.1 Physiological vs. Ecological Host Specificity

Laboratory host range testing usually represents a "worst case" infectivity assessment of a microsporidian species or isolate. Dosages of purified infective spores that produce 100% infection rates in a natural host are often in the range of 10^3–10^4 spores per microliter, a single bolus that is probably unusual in field situations. Insect species that are susceptible in the laboratory (physiologically susceptible) may seldom or never become infected in the field because (i) they do not encounter the infective spores due to differences in microhabitat or timing of activity between potential and natural hosts; (ii) the number of spores typically encountered is not sufficient to infect a marginally susceptible host; or (iii) light early infections are successfully resisted by the host immune response to a pathogen with which it has not evolved. Evidence for these resistance factors was shown in a series of laboratory and field host range studies for several species of microsporidia infecting *L. dispar* (Solter et al., 1997, 2000, 2010; Solter and Maddox, 1998).

10.3.2.2 Host Range

Rarely, species in several insect families within a taxonomic order are infected by one microsporidian species. One example of a microsporidium with a naturally broad host range is *P. locustae*, which is known to infect 20 orthopteran species representing 14 genera in two families in Argentina, where the pathogen became established after release as a biological control agent (Lange, 2010). Another species with a broad host range, *Cystosporogenes* sp. (likely *Cystosporogenes operophterae*), was recovered from field-collected larvae representing eight families of Lepidoptera in a Central European broadleaved forest (Solter et al., 2010). An experimental, inundative release of *V. disparis* (originally isolated from *L. dispar*) in Slovakia produced infections in native lepidopteran larvae from nine species in four families. However, no infections were recovered from the same species the year following the release (Solter et al., 2010), nor were the susceptible species infected in areas where *V. disparis* was naturally occurring in *L. dispar* (Solter et al., 2000).

The host range of *Nosema bombi* appears to be broad; it was reported in 22 species of bumblebees in the United States (Cordes et al., 2012) and at least eight species in Europe (Tay et al., 2005). However, all susceptible host species are members of the genus *Bombus*. *Nosema ceranae* appears to infect *Apis* and some *Bombus* species (Pilschuk et al., 2009; Li et al., 2012; Gamboa et al., 2015), but *N. apis* has only been recovered from *Apis* species. Microsporidia in the genus *Nosema* that are closely related to or conspecific with the type species *N. bombycis* and isolates in the genus *Endoreticulatus* that are genetically homologous with *E. schubergi* are found globally in different lepidopteran species. It has not been determined if these isolates represent holarctic species utilizing a broad host range or if they are separate, closely related species within a species complex, but evidence suggests that the *Nosema* clade has broadly radiated within the Lepidoptera (Kyei-Poku et al., 2008).

Another aberration to the typically narrow host range occurs when microsporidia infect parasitoids that co-occur with the pathogen in a common host insect. Examples are the hymenopteran endoparasitoids *Macrocentrus grandii* and *Cotesia flavipes*, which acquire infection by *N. pyrausta* and *Nosema* sp., respectively, from their lepidopteran hosts (Andreadis, 1980; Siegel et al., 1986; Simoes et al., 2012), and *Pediobius foveolatus*, which becomes infected by a *Nosema* sp. from its coleopteran host, *Epilachna varivestis* (Own and Brooks, 1986). However, in many other host–parasitoid–pathogen systems, endoparasitoids do not acquire microsporidian infections from their hosts (reviewed in Brooks, 1993; Moawed et al., 1997; Hoch et al., 2000).

10.3.2.3 Alternate Hosts

Some microsporidian species require alternate hosts for completion of their lifecycle; in some cases, the hosts are from different invertebrate classes. For example, *Amblyospora* sp. from the mosquito *Culex annulirostris* also infects the copepod *Mesocyclops albicans* (Sweeney et al., 1985). Meiospores of *Amblyospora connecticus* are produced in larval *Aedes cantator* but are not orally infectious to these mosquitoes; they are directly infectious to the copepod *Acanthocyclops vernalis*, which is the intermediate host required to complete the lifecycle of the microsporidium (Andreadis, 1985, 1988). Although hosts from different classes are infected, the intimate relationship with each of their hosts confers a high level of specificity of the microsporidium to the two host species. *Amblyospora connecticus* was experimentally transmitted to 4 of 19 tested mosquito species, but mature spores were produced only in *Aedes epactius*, and vertical transmission did not occur in any of the susceptible hosts. The microsporidium was able to complete its lifecycle only in its original host (Andreadis, 1989).

10.3.2.4 Microsporidia Crossing the Invertebrate–Vertebrate Barrier

Microsporidia have been recognized as emerging pathogens in humans since *Enterozytozoon biennusi* was described from an AIDS patient in 1985 (Wittner, 1999). Since then, more than 10 microsporidian species, all of them parasites of vertebrate hosts, have been found to infect humans. Most often, the patients were immune-compromised because of HIV infection (Kotler and Orenstein, 1999) or were users of immunosuppressant drugs (Didier and Weiss, 2011). The report of a fatal skeletal muscle infection with *Anncaliia* (*Nosema*, *Brachiola*) *algerae*, a mosquito pathogen, in a patient undergoing immunosuppressive treatment (Coyle et al., 2004) sheds new light on the potential of some microsporidia to cross host phyla and is a remarkable example of the potential host range of some microsporidia taxa. Interestingly, *Encephalitzoon romaleae*, a species isolated from lubber grasshopppers (*Romalea microptera*) (Lange et al., 2009), is the closest known relative of *Encephalitozoon hellem* and *Encephalitozoon cuniculi*, species that infect humans. *E. hellem*, a bird/human pathogen, possesses an insect gene insertion that is also found in *E. romaleae* (Selman et al. 2011).

10.3.3   Persistence in the Environment

A crucial question concerning microsporidia that infect host species that have distinct generations is how they can persist during periods when no susceptible host stages are present or when the host population density is very low. In the native range of *L. dispar* in southeastern Europe, microsporidia were regularly recovered over a 15-year study, even in low-density host populations (Pilarska et al., 1998). *Paranosema locustae* persisted following an application in grasshopper populations throughout a 9-year study, and prevalence remained above 35% (Miao et al., 2012). Persistence during periods of very low host density must rely on either vertical transmission or tolerance of adverse conditions, which can be highly variable and often harsh in terrestrial environments.

The infective microsporidian spore is protected by a dense spore wall consisting of a proteinaceous, electron-dense exospore and thick, chitinous, electron-transparent endospore (Vavra and Larsson, 1999). Aquatic environments usually provide less changable conditions, and species infecting aquatic invertebrates typically have a thinner endospore and are intolerant of freezing or drying (Solter et al., 2012a). Nevertheless, microsporidia infecting hosts in ephemeral aquatic habitats must be adapted to persist. *Amblyospora albifasciati*, for example, was shown to persist in copepod and mosquito host populations after transient pools were dry for a month (Micieli et al., 2001); spores may be able to tolerate limited desiccation or may be protected within the bodies of the hosts during these periods. The discovery of a new microsporidian genus infecting nematodes that inhabit deep-sea methane seeps on the Pacific Ocean floor demonstrates that these pathogens are able to persist in extreme environments. No information exists regarding whether spores can survive outside the host in these conditions; based on histology, a sexual route of transmission is assumed (Sapir et al., 2014).

Spores of microsporidia are killed at high temperatures with varying thresholds for different species (Maddox, 1973). Microsporidia within their insect hosts are generally negatively affected by high temperatures; 37 °C appears to be a developmental threshold (Becnel and Andreadis, 1999). *Nosema apis* spore development was not possible in infected *A. mellifera* hosts at 37 °C, while the optimum temperature for microsporidian reproduction was 33 °C (Martin-Hernandez et al., 2009). Temperatures of 40 °C and above reduce the spore viability of *N. apis* (Malone et al., 2001). Heat treatment has been used to cure insect colonies of microsporidian infection; for example, incubation of *O. nubilalis* eggs at 43.3 °C for 30 minutes produced larvae that were free of *N. pyrausta* infection (Raun, 1961). Exposure of microsporidia-infected phytoseid mites to 33 °C for 7 days reduced vertical transmission and consequently reduced disease prevalence in treated mite colonies (Olsen and Hoy, 2002).

Spores of some terrestrial microsporidia were reported to survive dessication for up to 270 days at room temperature (Maddox, 1977). Conditions that protect spores from detrimental microbial degradation can be beneficial for some species; for example, spores of *N. whitei* infecting tenebrionid beetles can survive extended periods in dry host cadavers (Undeen and Vavra, 1997).

Cooler temperatures increase the survival of microsporidian spores, but conditions under which spores can survive vary for different species. Refrigeration of highly purified spores in deionized or distilled water enabled survival of some species for several years (up to 10 years was observed for *A. algerae*) (Undeen and Vavra, 1997), but another mosquito pathogen, *Edhazardia aedis*, does not survive refrigerator temperatures (approximately 4 °C). Even when applied to soil, spores of *V. necatrix* were infective to *Trichoplusia ni* after 1 year when stored at 4 °C in the laboratory. On the other hand, spores were inactivated within 60 days when applied to the soil in the field (Chu and Jaques, 1981).

Spores of many terrestrial microsporidia survive freezing and can therefore be stored, at least for short periods of time, at −20 to −30 °C. Frequently, addition of glycerol to the spore suspension is required (Undeen and Vavra, 1997). Storage of spores in liquid nitrogen (−196 °C) is an excellent option for long-term storage of terrestrial microsporidia; in one study, spores survived for up to the maximum tested period of 25 years (Maddox and Solter, 1996). *Nosema disstriae* spores suspended in water retained some infectivity after three cycles of freezing and thawing in liquid nitrogen, and addition of glycerol allowed for 13 cycles (Undeen and Solter, 1996). However, repeated freezing and thawing generally degrades spores. Aquatic species do not survive storage in liquid nitrogen.

Conditions encountered by spores are more variable and detrimental under field conditions. Even the spores of species infecting terrestrial insects are likely to survive only in suitable habitats and microclimates. Nevertheless, survival time of approximately 1 year allows for sufficient viability and transmission to the next host generation. The *L. dispar* pathogen *V. disparis*, a species that is dependent on horizontal transmission, survives in the spore stage and retains infectivity throughout the winter when inside host cadavers. However, this has been demonstrated only when cadavers are placed in leaf litter and soil, and when there are periods of continuous snow cover. When cadavers were exposed on tree trunks, spores lost infectivity (Goertz and Hoch, 2008b). Spores of *E. schubergi*, on the other hand, were infective after 8 months' storage in cadavers on the ground, or after 4 months on tree trunks. This pathogen is also transmitted vertically by contamination of the egg surface during oviposition (Goertz and Hoch, 2008b). Another means of persisting in a host population is by infection of hibernating host stages. *Nosema pyrausta* overwinters in fifth-instar *O. nubilalis* larvae (Andreadis, 1986) and *N. fumiferanae* overwinters in second-instar *C. fumiferanae* larvae (van Frankenhuyzen et al., 2007).

As already mentioned, aquatic microsporidia do not survive freezing in most cases. Spores of the mosquito pathogen *E. aedis* are damaged at temperatures slightly above freezing and do not survive at 5 °C for more than 2 days (Undeen and Becnel, 1992). Other aquatic microsporidia must survive harsher and more variable environments, so there appear to be exceptions to the lethal effects of drying and cold temperatures in aquatic microsporidia. Spores of *Hamiltosporidium tvärminnensis* (formerly misidentified as *Octosporea bayeri*), a parasite of the cladoceran *Daphnia magna*, retained infectivity after 5 weeks of dessication outside of the host (Vizoso et al., 2005). Such tolerance to dry conditions is an important adaptation for aquatic microsporidia inhabiting unstable environments such as rock pools, which can become dry for some periods. This environment is further challenging for microsporidia, since freezing occurs in winter (Lass et al., 2011).

Microsporidia appear to be universally intolerant of ultraviolet (UV) radiation. Early studies of various entomopathogenic species, mostly pathogens of leaf-feeding Lepidoptera, showed that spore viability was destroyed within a few hours of exposure to sunlight (reviewed in Benz, 1987). Consequently, UV light can be used for disinfection. Dry spores of *Octosporea muscaedomesticae* were killed after 15 minutes' exposure to UV light from a germicidal lamp, or after 30 minutes when they were in aqueous suspension (Teetor and Kramer, 1977). Low- and medium-intensity UV light was demonstrated to be suitable for inactivation of spores of the human pathogen *Encephalitozoon intestinalis* in drinking water (Huffman et al., 2002). In the environment, UV radiation is probably a major factor in spore degradation.

10.4 Transmission

Transmission from host to host is central to host–parasite interactions and the key to the persistence of obligate intracellular parasites in host populations. Microsporidia can be transmitted horizontally from one host individual to another, or vertically from parent to offspring (definitions following Onstad et al., 2006). Horizontal transmission of microsporidia is typically achieved by ingestion of food contaminated with spores. Other, less common possibilities (shown only in laboratory studies) are intrahemocelic injection of spores by insect parasitoids and venereal transmission (Becnel and Andreadis, 1999). Tissue tropism of a microsporidian species determines the potential transmission pathways. When the alimentary tract is the target tissue for development, mature infective spores are usually released along with contaminated feces while infected hosts are still alive. On the other hand, when spore production is restricted to the fat body, spores are only released after host death. In both cases, success of transmission depends on the release of high numbers of infective spores into the environment. Pathways of horizontal transmission often have been assumed based on histological identification of infected tissues; quantitative experimental data are surprisingly scarce. Infection of the reproductive organs is prerequisite for transovarial, vertical transmission; however, if oviposited eggs are externally contaminated with spores from the infected parent, vertical transmission is also possible.

Transmission of a microsporidian species may occur by more than one pathway, and both horizontal and vertical transmission are possible for a single species. Dunn and Smith (2001) hypothesized that vertical transmission exerts strong selective pressure for low virulence in the sex that transmits the microsporidian parasite. Horizontally transmitted microsporidia should be selected to maximize reproduction, and, thus, parasite burden in the host is high. This is often associated with higher virulence.

10.4.1 Horizontal Transmission

10.4.1.1 Transmission from Living Hosts

Infection of the midgut epithelial cells and production of mature infective spores leads to contamination of the environment with spore-containing feces throughout the lifetime of the infected host. The honeybee parasites *N. apis* and *N. ceranae* are restricted to the midgut tissues, and spores are released in the feces. Hygienic behaviors lead to contamination of the mouthparts when the bees remove fecal material from the hive (Huang and Solter, 2013), and spores may be passed from bee to bee via trophallaxis, the passing of pollen or nectar from worker bees to nurse bees (Smith, 2012). Many microsporidia infecting lepidopteran hosts are transmitted orally via feces-contaminated food. *N. pyrausta* is horizontally transmitted among *O. nubilalis* larvae by fecal contamination of corn stalk tunnels (Andreadis, 1987), particularly in the second generation, while infection in first-generation larvae results mainly from vertical transmission (Siegel et al., 1988). *E. schubergi* sporulates in the midgut epithelium of *L. dispar* larvae (Zwölfer, 1927), and high numbers of spores, up to 10^8 per larva, are then continuously released until the host pupates or dies (Goertz and Hoch, 2008a). *Nosema lymantriae* does not produce infective spores in the midgut epithelium of *L. dispar* larvae but is also transmitted when host larvae are still alive. In this case, infected Malpighian tubules are apparently the source of spores that are released with feces (Weiser, 1957; Weiser and Linde, 1998). Spores of *Thelohania solenopsae*, a microsporidian parasite of red fire ants (*Solenopsis* spp.) are released by adult workers with fecal fluids and (in low numbers) crop fluids. The feces are probably the major route of infection to other colony members, with crop fluids being of minor importance (Williams et al., 1998; Oi et al., 2001; Chen et al., 2004).

Efficacy of transmission via feces depends on the behavior of infectious and susceptible hosts and on environmental factors. As already illustrated, any behavior that involves manipulating fecal material (including coprophagy) or contaminated food with mouthparts will increase the likelihood for transmission. Fecal contamination of areas where gregarious hosts aggregate may increase prevalence of the disease, and microsporidian infections often become problematic in laboratory colonies or mass rearings of insects. Meridic diet offered as a food source may further support transmission, as the quality of feces is often softer or stickier compared to that produced by eating natural foods, which maximizes contamination of the food. When lepidopteran larvae are feeding on foliage in the field, moisture or precipitation may reduce transmission by dilution of spores or may increase transmission by spreading of spores. Transmission of *N. lymantriae* on feces-contaminated dry foliage was significantly lower than transmission of the microsporidium under conditions of simulated light rain (Steyer, 2010). Apparently, moisture contributed to sticking of fecal pellets to the foliage, or more spores were released from the fecal pellets onto leaf surfaces (Fig. 10.2a). The impact of different host feeding behaviors on microsporidian transmission is illustrated by the dynamics of *Nosema* spp. infections in young larvae of *C. fumiferana* and *Choristoneura pinus*. While early instars of *C. fumiferana* mine host tree needles, early instars of *C. pinus* feed gregariously in cones for several weeks. *N. fumiferanae* builds up slowly in *C. fumiferanae* populations, while the gregarious behavior of *C. pinus* enhances *Nosema* sp.

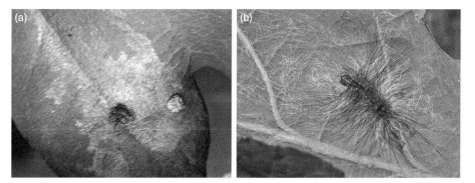

Fig. 10.2 (a) *Lymantria dispar* frass dissolving on a host plant leaf (oak) in rain and dew. (b) Silk and exuvia of a *L. dispar* larva on host plant leaf. *Source*: Courtesy of L. Solter.

(99.4% homology to *N. fumiferanae*; Kyei-Poku et al., 2008) transmission at an early stage and leads to rapid buildup of disease prevalence when *C. pinus* population density increases (van Frankenhuyzen et al., 2011).

Silk (Fig. 10.2b) or regurgitate from infected hosts may be another source of spores, enabling transmission when hosts are still alive. Spores of *Nosema portugal* develop in silk glands and are found in silk strands produced by infected *L. dispar* larvae, a potential pathway for horizontal transmission (Jeffords et al., 1987; Maddox et al., 1999). The larvae regularly follow silk trails during their daily movements between feeding and resting places on trees (McManus and Smith, 1972). On the other hand, spore release with silk does not appear to be important for the closely related *N. lymantriae*, a microsporidium that likewise causes massive infection and sporulation in silk glands of *L. dispar* larvae; rather, *Nosema lymantriae* infection impairs the function of the glands so that infected larvae cease production of silk (Goertz and Hoch, 2008a). However, regurgitation of spores cannot be ruled out as one possible route of spore release.

10.4.1.2 Transmission after Host Death

Horizontal transmission can occur after death of the infected host when spores are produced in the fat body tissues. In this case, the infected tissue is typically hypertrophied and completely filled with spores when the host succumbs to the infection. Locusts acquire *P. locustae* infections when feeding on cadavers of infected conspecific individuals (Henry, 1972; Lockwood, 1989). Feeding on cadavers and predation on infected hosts, in addition to ingestion of spores in water bodies, have been proposed as horizontal transmission pathways among amphipods, as shown for *Gammarus* and *Niphargus* species (Weigand et al., 2016). Coccinellid beetle larvae become infected with *Tubulinosema hippodamiae* when feeding on infected coccinellid eggs; 100% transmission occurred between conspecifics, but also between coccinellid species (Saito and Bjornson, 2006). This transmission route may occur frequently in the coccinellid family, because cannibalistic behavior is common.

For hosts that feed on cadavers less commonly, the success of horizontal transmission of fat-body parasites depends on release of spores into the environment when cadavers decompose. Microsporidia are not motile, and therefore cannot actively disperse from their host; nor do the cadavers liquefy like those infected with baculoviruses. It is thus unlikely that microsporidian spores are evenly dispersed in the environment. Considering that an *L. dispar* nucleopolyhedrovirus strain lacking the gene for host liquefaction is transmitted with lower success in the field than the strain possessing this gene (D'Amico et al., 2013), we expect a similar handicap for microsporidia. The exact mode of microsporidia transmission in foliage feeders or mining insects via cadavers is still not completely resolved. A simple mathematical model simulating transmission of *N. lymantriae* in *L. dispar* larvae sufficiently predicted transmission by contamination of foliage with spore-laden feces, but transmission with host cadavers as the spore source was not well predicted (Goertz and Hoch, 2011). The behavior of infected hosts, localization of cadavers, and, consequently, unknown distribution of spores were believed to be the reasons.

Cadavers of *D. magna* killed by *H. tvärminnensis* do not rupture and release spores into the water immediately after death. Vizoso et al. (2005) assumed that bacterial or detritivore activity might accelerate this process. Moreover, predators can play an important role in spreading spores of microsporidia developing from the fat body of the infected hosts. Movement of undigested spores through the gut of a predator may lead to a better dispersal of spores in the environment, as well as a reduction of the latent period; this was demonstrated for the predatory bug *Podisus maculiventris* feeding on larvae of *L. oleracea* infected by *V. necatrix*. By feeding on infected hosts and defecating undigested spores on the host food plant, the predator caused increased transmission of the microsporidium *V. necatrix* (Down et al., 2004b). The carabid *Calosoma syco-phanta* likewise increased transmission of the fat-body pathogen *V. disparis* among *L. dispar* larvae by spreading spores at a time when infected hosts otherwise would still be alive and not disseminating spores (Goertz and Hoch, 2013). Since sporulation of both of these *Vairimorpha* spp. is restricted to the fat body, predation resulted in an earlier release of infectious stages. Changes in the duration of the latent period, as with dispersal by a predator, had a very strong impact on transmission according to a mathematical simulation model for *V. disparis* (Goertz and Hoch, 2011).

10.4.1.3 Transmission by Parasitoid Vectors

Intrahemocelic injection is one possible route for transmission of microsporidia and can be achieved by oviposition of parasitoids that have developed in infected hosts. Transmission is most likely to occur when the parasitoids become infected with a microsporidian pathogen of the host – as is the case for the braconid endoparasitoid *M. grandii* developing in *O. nubilalis* larvae infected with *N. pyrausta* (Siegel et al., 1986) and for *C. flavipes* in *D. saccharalis* larvae infected with an undescribed *Nosema* species (Simoes et al., 2012). When an infected parasitoid oviposits into an uninfected host, the infection may be transmitted to the parasitoid offspring. However, an infection of the parasitoid acquired when developing in infected hosts does not necessarily lead to transmission to new hosts (Futerman et al., 2006). Another mechanism for transmitting microsporidia from infected to uninfected hosts is by sequential stinging of infected and uninfected hosts (e.g., Laigo and Tamashiro, 1967; Hamm et al., 1983; Own and Brooks, 1986). In such cases, ovipositors of the parasitoids become contaminated with spores. While the potential for parasitoid vectored transmission of microsporidia has been shown in laboratory studies, it has not been documented in field situations.

10.4.1.4 Effects of Host Development and Host–Microsporidia Interactions on Transmission

Temperature affects development of both host and microsporidium, and consequences for interactions between temperature and transmission can be expected. In a comprehensive model for interaction between *N. pyrausta* and its host, *O. nubilalis*, temperature and the timing of processes were variables that strongly influenced the outcome of the simulation (Onstad and Maddox, 1989). The effect of temperature was experimentally demonstrated for *N. lymantriae* infecting *L. dispar* larvae. Spores were released from larvae in feces 7 days earlier when infected larvae were reared at 24 °C versus 18 °C. Moreover, at 24 °C, higher total numbers of spores were released although infected larvae also died earlier at the higher temperature (Pollan et al., 2009).

Microsporidia that depend on horizontal transmission should optimize the production of infectious stages in their hosts. This is achieved by a complete invasion of the infected tissue by the microsporidium and is often accompanied by hypertrophy of the organ (Weiser, 1961; Becnel and Andreadis, 1999). Optimal utilization of the host as a resource is also reflected by good correlations between host body mass and spore load (Blaser and Schmid-Hempel, 2005; Hoch et al., 2009a). Extension of larval lifespan, allowing hosts to grow larger and consequently to produce more microsporidian spores, may also be advantageous for the pathogen, as was concluded for *N. whitei* infecting *T. castaneum* (Blaser and Schmid-Hempel, 2005) and *V. necatrix* infecting *L. oleracea* (Down et al., 2008). For the final outcome of the transmission process in both cases, it may be important that the benefit of higher pathogen reproduction is not outweighed by an extended latent period for these fat-body pathogens that are not released prior to death of the host.

10.4.2 Vertical Transmission

Vertical transmission, the direct transfer of pathogens from parent to progeny, is common among the microsporidia. Vertical transmission occurs in all microsporidian lineages and is therefore considered to be an ancestral trait of the phylum (Terry et al., 2004). Transmission can be transovarial; in this case, the passage of the pathogen from female host to egg occurs within the ovary (Onstad et al., 2006). Transovum transmission is the more general term that covers transmission via the embryo or contaminated egg surface, but often it is used when referring to passage of the pathogen by contamination of the egg surface (Solter, 2014). Either way, vertical transmission is important in the persistence of a microsporidian species in host populations during periods of low density. In an evolutionary perspective, vertical and horizontal transmission are strategies that exert selective pressures in different directions. While a pathogen that is horizontally transmitted should maximize reproduction in the host and is often associated with high virulence, selective pressure should favor low virulence of vertically transmitted pathogens in order to maximize survival and reproduction of the female host (Dunn and Smith, 2001). *Nosema* species in Lepidoptera are frequently transmitted both horizontally and vertically, and virulence in this clade is relatively lower than that of the related *Vairimorpha* species (Maddox et al., 1998), which are not vertically transmitted (Goertz and Hoch, 2008b). Similarly, vertically transmitted *Nosema empoasce* did not appear to cause deleterious effects in the potato leafhopper (*Empoasca fabae*) host (Ni, 1993; Ni et al., 1996). Such limitation of detrimental effects on the host is a prerequisite for persistence of a vertically transmitted pathogen in a host population. Reduced reproduction of a microsporidium maintains a lower pathogen burden, and infection may even be restricted to the gonads.

Feminization of the host, observed for some microsporidia infecting crustacean hosts, can further support vertical transmission (Terry et al., 1998; Dunn and Smith, 2001; Ironside et al., 2003b). Screening of microsporidian infections in amphipod crustaceans revealed vertical transmission of microsporidia in all tested host species. None of the pathogens caused patent pathogenesis, and a majority led to female bias (Terry et al., 2004), apparently due to suppression of androgenic gland differentiation (Jahnke et al., 2013). *Nosema granulosis* and *Dictyocoela duebenum*, for example, can lead to extremely female-biased offspring in *G. duebeni*. This increase of infected female offspring is hypothesized to enhance transovarial transmission; horizontal transmission becomes less important and is rare or even lacking in these species (Ironside et al., 2003a; Ironside and Alexander, 2015). A vertically transmitted microsporidium that causes a female bias in the host sex ratio may even lead to a competitive advantage, as was concluded for the invasive amphipod *Crangonyx pseudogracilis*. *Fibrillonosema crangonycis*, a feminizing microsporidian pathogen retained by this North American host during invasion into Europe, may have increased the invasive potential of the host by increasing its population growth rate while exerting no negative pressure because of its low virulence (Slothouber Galbreath et al., 2010). On the other hand, offspring of *D. magna* infected with *H. tvärminnensis* were shown to be slightly male-biased. This microsporidium is transmitted both vertically and horizontally. Males produced significantly higher spore loads than females, so more male offspring may enhance horizontal transmission and allow rapid spread of the pathogen (Roth et al., 2008). *Daphnia magna* populations in rock pools are highly unstable; nevertheless, *H. tvärminnensis* prevalence reached a remarkable 50% in infected populations (Ebert et al., 2001).

Overall, the majority of microsporidia that are vertically transmitted are also horizontally transmitted (Becnel and Andreadis, 1999; Dunn and Smith, 2001) and therefore depend on a balance of traits required for both transmission mechanisms. Microsporidia infecting lepidopteran hosts frequently use both routes. *Nosema bombycis*, the causal agent of pebrine disease in silk worms, is a systemic pathogen that is both horizontally and vertically transmitted (Kawarabata, 2003), and the control of vertical transmission has been key for controlling pebrine disease in the silk industry (Smith, 2009). A combination of vertical and horizontal infection is also important for other *Nosema* species, such as *N. pyrausta* infecting *O. nubilalis* (Lewis et al., 2009), *N. lymantriae* and *N. portugal* infecting *L. dispar* (Novotny and Weiser, 1993; Maddox et al., 1999; Goertz and Hoch, 2008b), and *N. fumiferanae* infecting *C. fumiferana* (van Frankenhuyzen et al., 2007). In all cases, the gonads of the females are infected and the microsporidium is transferred within the embryo. Typically, the pathogen burden in the embryo is sufficiently low to allow neonate larva to emerge. This type of transovarial transmission can be highly efficient, with, for example, close to 100% of progeny from infected *C. fumiferana* females being infected with *N. fumiferanae*, a pathogen of rather low virulence. The percentage of infected larvae can be lower when the spore load in the mother is lower (van Frankenhuyzen et al., 2007). Vertical transmission of *N. lymantriae* (notably a pathogen of higher virulence) in *L. dispar* occurs at lower levels: between 35 and 72% of offspring of infected mothers are infected (Novotny and Weiser, 1993; Goertz and Hoch, 2008b).

Infection of the gonads is not necessarily linked to or required for vertical transmission. *V. disparis* primarily infects the fat body tissues of *L. dispar* larvae but also infects the gonads (Vavra et al., 2006). However, the pathogen appears to be too virulent for vertical transmission to occur at significant levels, since infected larvae do not survive to adulthood even when inoculated with a very low dosage (Goertz and Hoch, 2008b). On the other hand, infection by *E. schubergi* is restricted to the midgut tissue of the host, yet 8–29% of the offspring of infected mothers can be infected. Transmission occurs by contamination of the egg surface (Goertz and Hoch, 2008b). Egg surface contamination was also used as a means to inoculatively release *N. portugal* into *L. dispar* populations. Hatching larvae consume about 50% of the chorion and thereby ingest infective dosages of the microsporidium (Jeffords et al., 1988).

For an insect like *L. dispar* that spends almost 10 months of the year in the egg stage, susceptible larvae are only available for a short period of time in the spring. A combination of vertical transmission and a variety of horizontal transmission pathways

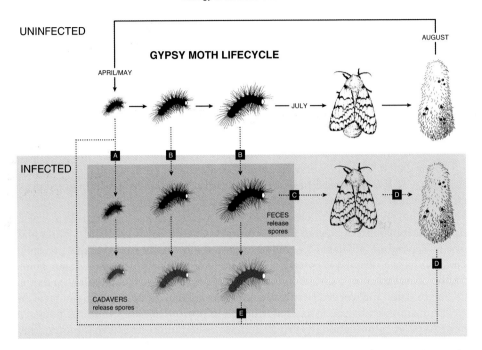

Fig. 10.3 Horizontal and vertical transmission of *Nosema lymantriae* in Central European *Lymantria dispar* populations. (A) Infected egg masses and spore-laden carcasses from the previous season are sources of infection for young larvae hatching in spring. (B) Newly infected larvae begin to release spores with feces after a latent period of about 2 weeks. After death, carcasses are an additional source of spores. These spores infect older larvae. (C) Older larvae that acquire the infection but do not die, due to lower susceptibility or ingesting a small number of spores, may develop into infected adults. (D) Infected females deposit eggs; the microsporidium is transmitted vertically. This guarantees persistence of *N. lymantriae* in the host population over the long period when gypsy moth is only present in the egg stage. (E) Spores also can survive the winter months inside cadavers in protected places. *Source*: Courtesy of D. Ruffatto and G. Hoch.

allows persistence and spread in the host population (Fig. 10.3.). Vertical transmission should be a particularly important trait supporting the persistence of a microsporidium in the host population, especially during the periods of low host density that are known to occur with this univoltine outbreak species. Indeed, microsporidia were recovered from sampled *L. dispar* populations in Austria, Slovakia, and Bulgaria over consecutive years and regardless of host population density (Pilarska et al., 1998; Hoch et al., 2001). It was concluded that vertical transmission was the key for year-to-year persistence of *N. portugal* following an experimental introduction into a North American *L. dispar* population. Vertically infected larvae may have served as a source for horizontal transmission to susceptible larvae in the year following release (Jeffords et al., 1989).

Cryptic feeding and solitary habits of a host may decrease the likelihood of horizontal transmission, and therefore vertical transmission may be of greater importance for microsporidia infecting such hosts. An example is infection of the bark beetle *Tomicus piniperda* by *Canningia tomici*. The pathogen infects midgut tissues, fat body, Malpighian tubules, and gonads. Gonadotropism and high prevalence of infection in filial generations suggest that vertical transmission occurs (Kohlmayr et al.,

2003). Opportunies to contact spores in the environment are limited because larvae feed individually in their galleries and newly eclosed beetles are isolated during maturation feeding in pine twigs.

10.5 Epizootiology

Microsporidia express species-specific variations in infectivity, virulence, tissue tropism, transmission, lifecycle complexity, environmental persistence, and host taxa preferences, among other characteristics, which combine to define a wide range of specific pathogen–host interactions, epizootic potential, and impacts on the host populations. In general, microsporidia are considered to be chronic pathogens that are host density-dependent. However, while examples of "textbook" epizootic responses to host density have been recorded for some microsporidia, examples are also known of species that persist at low prevalence in relatively low-density host populations but do not increase appreciably as host density increases or persist at high enzootic levels, apparently with little impact on the host population. Examples of the wide variation in pathogen–host population dynamics are provided in this section.

10.5.1 Microsporidian Prevalence in Invertebrate Populations and Impact on Host Populations

Nosema pyrausta infecting *O. nubilalis* is a well-known example of a density-dependent pathogen–host interaction (Hill and Gary, 1979; Andreadis, 1986). *Nosema pyrausta* causes systemic infections in the host and is both horizontally and vertically transmitted in the second (summer) host generation; it persists in infected overwintering *O. nubilalis* larvae and is transovarially transmitted to the spring generation by infected females (Siegel et al., 1986). *Nosema pyrausta* can persist at relatively high enzootic levels (>5%) in the host population when *O. nubilalis* populations are low, increasing to over 60% prevalence when outbreaks of the host occur (Lewis et al., 2006). Host population declines usually follow such epizootics (Andreadis, 1984). *Nosema pyrausta* impacts include high mortality in transovarially infected larvae, slow larval growth and development, reduced adult fecundity, and reduced adult lifespan (Lewis et al., 2009).

Although *N. bombi* infecting *Bombus* spp. has a broad range within the host genus, enzootic and epizootic prevalence levels vary widely among host species. In a United States-wide survey, Cordes et al. (2012) reported high *N. bombi* prevalence in several *Bombus* species, but much lower prevalence in large collections of other, often sympatric, *Bombus* species. Similar varying infection levels among *Bombus* spp. were reported from Alaska (Koch and Strange, 2012), western Europe (Paxton et al., 2008), and West Siberia (Vavilova et al., 2015). *Nosema bombi* has deleterious effects on the host (Otti and Schmid-Hempel, 2007), but whether it seriously impacts populations of host species that have high enzootic prevalence is not known. Several of the US *Bombus* species with high *N. bombi* prevalence appear to be declining, notably the western species *Bombus occidentalis* (Cameron et al., 2011). However, Koch and Strange (2012) reported high *N. bombi* prevalence in *B. occidentalis* in Alaska, where the host does not appear to be in decline.

An example of a microsporidium that is maintained at very low prevalence levels in the host is the highly virulent *V. disparis*. Prevalence of the pathogen, which appears to depend on horizontal transmission in the *L. dispar* host population (Goertz and Hoch, 2008b), hovered between 1 and 5% over a 13-year period in one Bulgarian study site, with reports of 15–30% prevalence in rare epizootics (Pilarska et al., 1998). This epizootic peak is considerably lower than the enzootic level in several other microsporidia–host systems, such as *N. fumiferanae* in *C. fumiferana* (>50%) (Thomson, 1960; Wilson, 1973, 1977), *Larsoniella duplicati* in the bark beetle *Ips duplicatus* (detected in 80% of samples, 10–57% prevalence) (Holusa et al., 2009), or *Nematocenator marisprofundi* in *Desmodora marci*, a nematode inhabiting deep-sea methane seeps (50–60% prevalence) (Sapir et al., 2014). *Nosema carpocapsae* was regularly detected at high prevalence in *Cydia pomonella* in various regions in Germany over 2 decades. Prevalence in adult moths fluctuated between 22 and 57% and between 34 and 60% in two populations monitored for more than 15 consecutive years. No correlation to population dynamics of the host was reported. However, because infection reduces fecundity and fertility of the host, an impact on host populations was assumed (Zimmermann et al., 2013).

Prevalence of some microsporidian species may also fluctuate seasonally in hosts that are present year-round. *Amblyospora* spp. prevalence in the blackfly *Simulium pertinax* was highest in two host populations during the summer months in Brazil (Araújo-Coutinho et al., 2004). *Nosema apis* prevalence was reported to be highly seasonal, peaking in spring, when it was the only microsporidium known to infect *A. mellifera* in the United States (Bailey and Ball, 1991). It was also apparently seasonal in the African subspecies *A. mellifera scutellata* where *N. ceranae* was not present (Fries et al., 2003). Chen et al. (2012) reported that *N. ceranae* prevalence in *A. mellifera* was correlated with temperature, noting that peak prevalence occurred in winter months in Taiwan, when average temperatures were 15 °C. Prevalence peaks at moderate temperatures were also recorded during the spring months in the United States, 12–17 °C in Virginia (Traver and Fell, 2011), and 15 °C in California (Oliver, 2011). Chen et al. (2012) suggested that the correlation with moderate temperatures may be related to buildup of *A. mellifera* colonies to higher densities in the spring months.

Caution should be used when comparing reported prevalence data for microsporidia because detection sensitivity varies between microscopic and molecular screening methods. Microscopic surveys of microsporidia in host populations typically only identify sporulating infections and may not adequately detect the actual prevalence of infection. Blaker et al. (2014) reported that nonsporulating infections were detected in bumble bees by polymerase chain reaction (PCR). PCR detected far more *N. bombi* infections in individual *Bombus* spp. than did microscopy (Klee et al., 2006; Rutrecht et al., 2007; Blaker et al., 2014), however, PCR is highly sensitive and may detect small numbers of spores in the gut contents of an insect that do not result in infections. Moreover, it is questionable whether nonsporulating infections can be transmitted, or even if they deleteriously impact the host (Blaker et al., 2014).

10.5.2 Microsporidia in Cultured Insects

Microsporidia can decimate insect colonies that are mass-reared for human use. The classic example of silkworm colony loss documented in France in the 19th century due to *N. bombycis* infection remains a serious problem in sericulture today (Chakrabarty et al., 2013; Fu, 2016). Other examples include colonies of pest insects such as *C. fumiferana* (van Frankenhuyzen et al., 2004), *O. nubilalis* (L.F.S., personal observation), and *D. saccharalis* (Simoes et al., 2012), reared to produce biological control agents, as well as managed *A. mellifera*.

Analyses of *N. ceranae* impact in *A. mellifera* are controversial, with some reports suggesting that *N. ceranae* has a prominent role in colony decline (Higes et al., 2008, 2009; Bromenshenk et al., 2010) and others demonstrating heavily infected but otherwise strong colonies (Martin et al., 2013; Pohorecka et al., 2014; Budge et al., 2015). *Apis mellifera* may be a relatively new host for *N. ceranae* (Maside et al., 2015), but in the United States and other southern temperate climates it currently appears to be the dominant microsporidian species and may have outcompeted *N. apis* (Chen et al., 2009; Martínez et al., 2012). The eusocial behavior of *A. mellifera*, as well as management practices that crowd large numbers of colonies into relatively small areas, may serve to increase the prevalence of *N. ceranae* (Roberts and Hughes, 2014), resulting in a high enzootic presence of the pathogen (Kielmanowicz et al., 2015). Whether consistently high prevalence is strongly detrimental to *A. mellifera* colonies has been difficult to analyze, partly due to the presence of other pathogens and parasites in the same colonies. Loss of colonies may be caused by one overwhelming pathogen or parasite, but it is more likely to result from multiple stress factors (Evans and Schwartz, 2011; Cornman et al., 2012; Kielmanowicz et al., 2015).

Microsporidian infections in arthropods that are consumed as human food are important both because of the potential damage to a food source and potential zoonoses (Stentiford et al., 2016). *Enterocytozoon hepatopenaei* is an emerging pathogen in two farmed shrimp species in Asia: the Asian tiger shrimp, *Penaeus modon* (Tourtip et al. 2009) and the Pacific white shrimp, *Penaeus vannamei*, in which overall pathogen prevalence has reached >65% (Rajendran, et al. 2016). This species is a congener of *Enterocytozoon bieneusi*, a human pathogen. It causes degeneration of the hepatopancreas and has been strongly associated with white feces syndrome in shrimps, causing significant mortality (Flegel, 2012). Another microsporidium infecting penaid shrimps, *Agmasoma penaei*, was managed by removal of a fish "carrier" from water supply canals (Flegel, 2012).

10.5.3 Microsporidia as Potential Biological Control Agents

The general consensus among researchers studying microsporidia as potential biological control agents is that they are not suitable for development as microbial pesticides. They are probably better suited as classical biological control agents, inoculatively released into populations of their natural hosts where prevalence is low or the pathogen is not present. Because invertebrate microsporidia can only be produced in living cells and, with a few exceptions, do not reproduce well in tissue culture (reviewed by Solter et al. 2012b; Molestina et al. 2014), labor-intensive mass rearing of a susceptible host is the only option for production. Spore concentrations need to be so high for inundative release that sufficient inoculum cannot be produced efficiently or cost-effectively. Another limitation for the use of microsporidia as microbial pesticides is that these pathogens generally lack the necessary virulence for "quick kill" (Lacey et al., 2001; Solter et al., 2012a).

Paranosema locustae, a pathogen of grasshoppers and locusts, is the only microsporidium that is commercially produced, partly because its effectiveness lies in application strategies other than repeated inundative releases. A one-time release of *P. locustae* into selected grasshopper populations in the Argentine Pampas and in China successfully established in the target species and several other pest orthopterans. *Paranosema locustae* continues to suppress grasshopper populations years after introduction (Lange and Azzaro, 2008; Shi et al., 2009). Bjornson and Oi (2014) have provided an excellent general overview of the use of microsporidia in biological control.

References

Andreadis, T.G., 1980. *Nosema pyrausta* infection in *Macrocentrus grandii*, a braconid parasite of the European corn borer, *Ostrinia nubilalis*. J. Invertebr. Pathol. 35, 229–233.

Andreadis, T.G., 1981. Impact of *Nosema pyrausta* on field populations of *Macrocentrus grandii*, an introduced parasite of the European corn borer, *Ostrinia nubilalis*. J. Invertebr. Pathol. 39, 298–302.

Andreadis, T.G., 1984. Epizootiology of *Nosema pyrausta* in field populations of the European corn borer (Lepidoptera: Pyralidae). Environ. Entomol. 13, 882–887.

Andreadis, T.G., 1985. Experimental transmission of a microsporidian pathogen from mosquitoes to an alternate copepod host. Proc. Natl. Acad. Sci. USA 82, 5574–5577.

Andreadis, T.G., 1986. Dissemination of *Nosema pyrausta* in feral populations of the European corn borer, *Ostrinia nubilalis*. J. Invertebr. Pathol. 48, 335–343.

Andreadis, T.G., 1987. Horizontal transmission of *Nosema pyrausta* (Microsporida: Nosematidae) in the European corn borer, *Ostrinia nubilalis* (Lepidoptera: Pyralidae). Environ. Entomol. 16, 1124–1129.

Andreadis, T.G., 1988. *Amblyospora connecticus* sp. nov. (Microsporida: Amblyosporidae): Horizontal transmission studies in the mosquito *Aedes cantator* and formal description. J. Invertebr. Pathol. 52, 90–101.

Andreadis, T.G., 1989. Host specificity of *Amblyospora connecticus* (Microsporida: Amblyosporidae), a polymorphic microsporidian parasite of *Aedes cantator* (Diptera: Culicidae). J. Medical Entomol. 26, 140–145.

Araújo-Coutinho, C.J.P.C., Nascimento, E.S., Figueiró, R., Becnel, J.J., 2004. Seasonality and prevalence rates of microsporidia in *Simulium pertinax* (Diptera: Simuliidae) larvae in the region of Serra dos Órgãos, Rio de Janeiro, Brasil. J. Invertebr. Pathol. 85, 188–191.

Ardila-Garcia, A.M., Raghuram, N., Sihota, P., Fast, N.M., 2013. Microsporidian diversity in soil, sand, and compost of the Pacific Northwest. J. Eukaryot. Microbiol. 60, 601–608.

Avery, S.W., Anthony, D.W., 1983. Ultrastructural study of early development of *Nosema algerae* in *Anopheles albimanus*. J. Invertebr. Pathol. 42, 87–95.

Bailey L., Ball, B.V., 1991. Honey Bee Pathology. Academic Press, San Diego, CA.

Bauer, L.S., Nordin, G.L., 1989. Effect of *Nosema fumiferanae* (Microsporida) on fecundity, fertility, and progeny performance of *Choristoneura fumiferana* (Lepidoptera: Tortricidae). Environ. Entomol. 18, 261–265.

Becnel, J.J., Andreadis, T.G., 1999. Microsporidia in insects, in: Wittner, M., Weiss, L. M. (eds.), The Microsporidia and Microsporidiosis. ASM Press, Washington, DC, pp. 447–501.

Benz, G., 1987. Environment, in: Fuxa, J.R., Tanada, Y. (eds.), Epizootiology of Insect Diseases. John Wiley & Sons, New York, pp. 177–214.

Bjornson, S., Oi, D., 2014. Microsporidia biological control agents and pathogens of beneficial insects, in: Weiss, L.M., Becnel, J.J. (eds.), Microsporidia, Pathogens of Opportunity. John Wiley & Sons, Oxford, pp. 635–670.

Blaker, E.A., Strange, J.P., James, R.R., Monroy, F.P., Cobbs, N.S., 2014: PCR reveals high prevalence of non/low sporulating *Nosema bombi* (microsporidia) infections in bumble bees (*Bombus*) in Northern Arizona. J. Invertebr. Pathol. 123, 25–33.

Blaser, M., Schmid-Hempel, P., 2005. Determinants of virulence for the parasite *Nosema whitei* in its host *Tribolium castaneum*. J. Invertebr. Pathol. 89, 251–257.

Bromenshenk, J.J., Henderson, C.B., Wick, C.H., Stanford, M.F., Zulich, A.W., Jabbour, R.E., Deshpande, S.V. et al. 2010. Iridovirus and microsporidian linked to honey bee colony decline. PLoS ONE 5(10), e13181.

Brooks, W.M., 1970a. The inflammatory response of the tobacco hornworm, *Manduca sexta*, to infection by the microsporidian, *Nosema sphingidis*. J. Invertebr. Pathol. 17, 87–93.

Brooks, W.M., 1970b. Protozoan infections of insects with emphasis on inflammation, in: Proceedings of the IV International Colloquium on Insect Pathology. College Park, MD, pp. 11–27

Brooks, W.M., 1993. Host-parasitoid-pathogen interactions, in: Beckage, N.E., Thompson, S.N., Federici, B.A. (eds.), Parasites and Pathogens of Insects, Vol. 2: Pathogens. Academic Press, San Diego, CA, pp. 231–272.

Budge, G.E., Pietravalle, S., Brown, M., Laurenson, L., Jones, B., Tomkies, V., Delaplane, K.S., 2015. Pathogens as predictors of honey bee colony strength in England and Wales. PLOS ONE 10(7), e0133228.

Cameron, S.A., Lozier, J.D., Strange, J.P., Koch, J.B., Cordes, N., Solter, L.F., Griswold, T.L., 2011. Recent widespread decline of some North American bumble bees: current status and causal factors. Proc. Natl. Acad. Sci. 108, 662–667.

Canning, E.U., 1962. The pathogenicity of *Nosema locustae* Canning. J. Insect Pathol. 4, 248–256.

Cavalier-Smith, T., 1983. A 6-kingdom classification and a unified phylogeny, in: Schenk, H.E.A., Schwemmler, W.S. (eds.), Endocytobiology II: Intracellular Space as Oligogenetic Ecosystem, Walter de Gruyter, Berlin, New York, pp. 1027–1034.

Chapman, D., 2010. Feeding habits and behaviour of *Lymantria dispar* (Lep.: Lymantriidae) when infected with the microsporidian pathogen *Vairimorpha disparis*. Master's thesis, Universität für Bodenkultur, Vienna.

Chakrabarty, S., Saha, A.K., Bindroo, B.B., Manna, B., Kumar. S.N., 2013. An improved method for the detection of pebrine (*Nosema bombycis* N.) spores in silkworm. Appl. Biol. Res. 15, 91–96.

Chen, J.S.C., Snowden, K., Mitchell, F., Sokolova, J., Fuxa, J., Vinson, S.B., 2004. Sources of spores for the possible horizontal transmission of *Thelohania solenopsae* (Microspora: Thelohaniidae) in the red imported fire ants, *Solenopsis invicta*. J. Invertebr. Pathol. 85, 139–145.

Chen, Y-P., Evans, J.P., Zhou, L., Boncristiani, H., Kimura, K., Xiao, T., Litkowski, A.M., et al., 2009. Asymmetrical coexistence of *Nosema ceranae* and *Nosema apis* in honey bees. J. Invertebr. Pathol. 101, 204–209.

Chen, Y-W., Chung, W-P., Wang, C-H., Solter, L.F., Huang, W-F., 2012. *Nosema ceranae* infection intensity highly correlates with temperature. J. Invertebr. Pathol. 111, 264–267.

Chu, W.H., Jaques, R.P., 1981. Factors affecting infectivity of *Vairimorpha necatrix* (Microsporidia: Nosematidae) in *Trichoplusia ni* (Lepidoptera: Noctuidae). Can. Entomol. 113, 93–102.

Cordes, N., Huang, W-F., Strange, J.P., Cameron, S.A., Griswold, T.L., Lozier, J.D., Solter, L.F., 2012. Interspecific geographic distribution and variation of the pathogens *Nosema bombi* and *Crithidia* species in United States bumble bee populations. J. Invertebr. Pathol. 109, 209–216.

Cornman, R.S., Tarpy, D.R., Chen, Y., Jeffreys, L., Lopez, D., Pettis, J.S., et al., 2012. Pathogen webs in collapsing honey bee colonies. PLOS One 7, e43562.

Coyle, C.M., Weiss, L.M., Rhodes, L.V., Cali, A., Takvorian, P.M., Brown, D.F., Visvesvara, G.S., et al., 2004. Fatal myositis due to the microsporidian *Brachiola algerae*, a mosquito pathogen. N. Eng. J. Med. 351, 42–47.

D'Amico, V., Slavicek, J., Podgwaite, J.D., Webb, R., Fuester, R., Peiffer, R.A., 2013. Deletion of *v-chiA* from a baculovirus reduces horizontal transmission in the field. Appl. Environ. Microbiol. 79, 4056–4064.

David, L., Pilarska, D., 1988. Pathological changes in the silk glands and mortality of caterpillars of *Lymantria dispar* L. caused by a *Nosema* microsporidian. Acta Entomol. Bohemoslov. 85, 257–261.

David, L., Weiser, J., 1994. Role of hemocytes in the propagation of a microsporidian infection in larvae of *Galleria mellonella*. J. Invertebr. Pathol. 63, 212–213.

Didier, E.S., Weiss, L.M., 2011. Microsporidiosis: not just in AIDS patients. Curr. Opin. Infect. Dis. 24, 490–495.

Dorhout, D.L., Sappington, T.W., Lewis, L.C., Rice, M.E., 2011. Flight behaviour of European corn borer infected with *Nosema pyrausta*. J. Appl. Entomol. 135, 25–37.

Down, R.E., Bell, H.A., Kirkbride-Smith, A.E., Edwards, J.P., 2004a: The pathogenicity of *Vairimorpha necatrix* (Microspora: Microsporidia) against the tomato moth, *Lacanobia oleracea* (Lepidoptera: Noctuidae) and its potential use for the control of lepidopteran glasshouse pests. Pest Manag. Sci. 60, 755–764.

Down, R.E., Bell, H.A., Matthews, H.J., Kirkbride-Smith, A.E., Edwards, J.P., 2004b. Dissemination of the biocontrol agent *Vairimorpha necatrix* by the spined soldier bug, *Podisus maculiventris*. Entomol. Exp. Appl. 110, 103–114.

Down, R.E., Bell, H.A., Bryning, G., Kirkbride-Smith, A.E., Edwards, J.P., Weaver, R.J., 2008. Infection by the microsporidium *Vairimorpha necatrix* (Microspora: Microsporidia) elevates juvenile hormone titres in larvae of the tomato moth, *Lacanobia oleracea* (Lepidoptera: Noctuidae). J. Invertebr. Pathol. 97, 223–229.

Dunn, A.M., Smith, J.E., 2001. Microsporidian life cycles and diversity: the relationship between virulence and transmission. Microb. Inf. 3, 381–388.

Ebert, D., Hottinger, J.W., Pajunen, V.I., 2001. Temporal and spatial dynamics of parasite richness in a *Daphnia* metapopulation. Ecology 82, 417–3434.

Eiri, D.M., Suwannapong, G., Endler, M., Nieh, J.C., 2015. *Nosema ceranae* can infect honey bee larvae and reduces subsequent adult longevity. PLOS ONE 10(5), e0126330.

Evans J.D., Schwarz, R.S., 2011. Bees brought to their knees: microbes affecting honey bee health. Trends Microbiol. 19, 614–620.

Everleigh, E.S., Lucarotti, C.J., McCarthy, P.C., Morin, B., Royama, T., Thomas, A.W., 2007. Occurrence and effects of *Nosema fumiferanae* infections on adult spruce budworm caught above and within the forest canopy. Agric. For. Entomol. 9, 247–258.

Flegel, T.W., 2012. Historic emergence, impact and current status of shrimp pathogens in Asia. J. Invertebr. Pathol. 110, 166–173.

Fries, I., Slemenda, S.B., da Silva, A., Pieniazek, N.J., 2003. African honey bees (*Apis mellifera scutellata*) and nosema (*Nosema apis*) infections. J. Apicultur. Res. 42, 13–15.

Fu, X.J., Hunter, D.M., Shi, W.P., 2010. Effect of *Paranosema (Nosema) locustae* (Microsporidia) on morphological phase transformation of *Locusta migratoria manilensis* (Orthoptera: Acrididae). Biocontr. Sci. Technol. 20, 683–693.

Fu, Z., He, X., Cai, S., Liu, H., He, X., Li, M., Lu, X., 2016. Quantitative PCR for detection of *Nosema bombycis* in single silkworm eggs and newly hatched larvae. J. Microbiol. Methods 120, 72–78.

Futerman, P.H., Layen, S.J., Kotzen, M.L., Franzen, C., Kraaijeveld, A.R., Godfray, H.C.J., 2006. Fitness effects and transmission routes of a microsporidian parasite infecting *Drosophila* and its parasitoids. Parasitology 132, 479–492.

Gamboa, V., Ravoet, J., Brunain, M., Smagghe, G., Meeus, I., Figueroa, J., Riano, D., et al., 2015. Bee pathogens found in *Bombus atratus* from Colombia: a case study. J. Invertebr. Pathol. 129, 36–39.

Goblirsch, M., Huang, Z.Y., Spivak, M., 2013. Physiological and behavioral changes in honey bees (*Apis mellifera*) induced by *Nosema ceranae* infection. PLOS ONE 8(3), e58165.

Goertz, D., Golldack, J., Linde, A., 2008. Two different and sublethal isolates of *Nosema lymantriae* (Microsporidia) reduce the reproductive success of their host, *Lymantria dispar*. Biocontr. Sci. Technol. 18, 419–430.

Goertz, D., Hoch, G., 2008a. Horizontal transmission pathways of terrestrial microsporidia: A quantitative comparison of three pathogens infecting different organs in *Lymantria dispar* L. (Lep.: Lymantriidae) larvae. Biol. Control 44, 196–206.

Goertz, D., Hoch, G., 2008b. Vertical transmission and overwintering of microsporidia in the gypsy moth, *Lymantria dispar*. J. Invertebr. Pathol. 99, 43–48.

Goertz, D., Hoch, G., 2011. Modeling horizontal transmission of microsporidia infecting gypsy moth, *Lymantria dispar* (L.), larvae. Biol. Control 56, 263–270.

Goertz, D., Hoch, G., 2013. Influence of the forest caterpillar hunter *Calosoma sycophanta* on the transmission of microsporidia in larvae of the gypsy moth *Lymantria dispar*. Agric. For. Entomol. 15, 178–186.

Hamm, J.J., Nordlund, D.A., Mullinix, B.G., 1983. Interaction of the microsporidium *Vairimorpha* sp. with *Microplitis croceipes* (Cresson) and *Cotesia marginiventris* (Cresson) (Hymenoptera: Braconidae), two parasitoids of *Heliothis zea* (Boddie) (Lepidoptera: Noctuidae). Environ. Entomol. 12, 1547–1550.

Haque, Md.A., Canning, E.U., Wright, D.J., 1999. Entomopathogenicity of *Vairimorpha* sp. (Microsporidia) in the diamondback moth, *Plutella xylostella* (Lepidoptera: Yponomeutidae). Bull. Entomol. Res. 89, 147–152.

Henry, J.E., 1972. Epizootiology of infections by *Nosema locustae* Canning (Microsporida: Nosematidae) in grasshoppers. Acrida 1, 111–120.

Henry, J.E., Oma, E.A., 1981. Pest control by *Nosema locustae*, a pathogen of grasshoppers and crickets, in: Burges, H.D. (ed.), Microbial Control of Pests and Plant Diseases 1979–1980. Academic Press, New York, pp. 573–568.

Hibbett, D.S., Binder, M., Bischoff, J.F., Blackwell, M., Cannon, P.F., Eriksson, O.E., et al., 2007. A higher-level phylogenetic classification of the Fungi. Mycol. Res. 111, 509–547.

Higes, M., Martín-Hernández, R., Botías, C., Bailón, E.G., González-Porto, A.V., Barrios, L., et al., 2008. How natural infection by *Nosema eranae* causes honeybee colony collapse. Environ. Microbiol. 10, 2659–2669.

Higes, M., Martín-Hernández, R., Garrido-Bailón, E., González-Porto, A.V., García-Palencia, P., Meana, A., et al., 2009. Honeybee colony collapse due to *Nosema ceranae* in professional apiaries. Environ. Microbiol. Rep. 1, 110–113.

Hill, R.E., Gary, W.J., 1979. Effects of the microsporidian, *Nosema pyrausta*, on field populations of European corn borers in Nebraska. Environ. Entomol. 8, 91–95.

Hinney, B., Sak, B., Joachim, A., Kváč, M., 2016. More than a rabbit's tail – *Encephalitozoon* spp. in wild mammals and birds. Int. J. Parasitol.: Parasites Wildl. 5, 76–87.

Hoch, G., Schopf, A., Maddox, J.V., 2000. Interactions between an entomopathogenic microsporidium and the endoparasitoid *Glyptapanteles liparidis* within their host, the gypsy moth larva. J. Invertebr. Pathol. 75, 59–68.

Hoch, G., Schopf, A., 2001. Effects of *Glyptapanteles liparidis* (Hym.: Braconidae) parasitism, polydnavirus, and venom on development of microsporidia infected and uninfected *Lymantria dispar* (Lep.: Lymantriidae) larvae. J. Invertebr. Pathol. 77, 37–43.

Hoch, G., Zubrik, M., Novotny, J., Schopf, A., 2001. The natural enemy complex of the gypsy moth, *Lymantria dispar* (Lep., Lymantriidae) in different phases of its population dynamics in eastern Austria and Slovakia – a comparative study. J. Appl. Entomol. 125, 217–227.

Hoch, G., Schafellner, C., Henn, M.W., Schopf, A., 2002. Alterations in carbohydrate and fatty acid levels of *Lymantria dispar* larvae caused by a microsporidian infection and potential adverse effects on a co-occurring endoparasitoid, *Glyptapanteles liparidis*. Arch. Insect Biochem. Physiol. 50, 109–120.

Hoch, G., Solter, L.F., Schopf, A., 2004. Hemolymph melanization and alterations in hemocyte numbers in *Lymantria dispar* larvae following infections with different entomopathogenic microsporidia. Entomol. Exp. Appl. 113, 77–86.

Hoch, G., Solter, L.F., Schopf, A., 2009a. Treatment of *Lymantria dispar* (Lepidoptera, Lymantriidae) host larvae with polydnavirus/venom of a braconid parasitoid increases spore production of entomopathogenic microsporidia. Biocontr. Sci. Technol. 19, S1, 35–42.

Hoch, G., Pilarska, D.K., Dobart, N., 2009b. Effect of midgut infection with the microsporidium *Endoreticulatus schubergi* on carbohydrate and lipid levels in *Lymantria dispar* larvae. J. Pest Sci. 82, 351–356.

Holusa, J., Weiser, J., Zizka, Z., 2009. Pathogens of the spruce bark beetles *Ips typographus* and *Ips duplicatus*. Centr. Eur. J. Biol. 4, 567–573.

Huang, W-F., Solter, L.F., 2013. Comparative development and tissue tropism of *Nosema apis* and *Nosema ceranae*. J. Invertebr. Pathol. 113, 35–41.

Huffman, D.E., Gennaccaro, A., Rose, J.B., Dussert, B.W., 2002. Low- and medium-pressure UV inactivation of microsporidia *Encephalitozoon intestinalis*. Water Res. 36, 3161–3164.

Ironside, J.E., Alexander, J., 2015. Microsporidian parasites feminise hosts without paramyxean co-infection: support for convergent evolution of parasitic feminization. Int. J. Parasitol. 45, 427–433.

Ironside, J.E., Dunn, A.M., Rollinson, D., Smith, J.E., 2003a. Association with host mitochondrial haplotypes suggests that feminizing microsporidia lack horizontal transmission. J. Evol. Biol. 16, 1077–1083.

Ironside, J.E., Smith, J.E., Hatcher, M.J., Sharpe, R.G., Rollinson, D., Dunn, A.M., 2003b. Two species of feminizing microsporidian parasite coexist in populations of *Gammarus duebeni*. J. Evol. Biol. 16, 467–473.

Iwano, H., Ishihara, R., 1991. Dimorphism of spores of *Nosema* spp. in cultured cell. J. Invertebr. Pathol. 57, 211–219.

Iwano, H., Kurtti, T.J., 1995. Identification and isolation of dimorphic spores from *Nosema furnacalis* (Microspora: Nosematidae). J. Invertebr. Pathol. 65, 230–236.

Jahnke, M., Smith, J.E., Dubuffet, A., Dunn, A.M. 2013. Effects of feminizing microsporidia on the masculinizing function of the androgenic gland in *Gammarus duebeni*. J. Invertebr. Pathol. 112, 146–151.

James, T.Y., Kauff, F., Schoch, C.L., Matheny, P.B., Hofstetter, V., Cox, C.J., et al., 2006. Reconstructing the early evolution of Fungi using a six-gene phylogeny. Nature 443, 818–822.

Jeffords, M.R., Maddox, J.V., O'Hayer, K.W., 1987. Microsporidian spores in gypsy moth larval silk: a possible route of horizontal transmission. J. Invertebr. Pathol. 49, 332–333.

Jeffords, M.R., Maddox, J.V., McManus, M.L., Webb, R.E., Wieber, A., 1988. Egg contamination as a method for the inoculative release of exotic microsporidia of the gypsy moth. J. Invertebr. Pathol. 51, 190–196.

Jeffords, M.R., Maddox, J.V., McManus, M.L., Webb, R.E., Wieber, A., 1989. Evaluation of the overwintering success of two European microsporidia inoculatively released into gypsy moth populations in Maryland. J. Invertebr. Pathol. 53, 253–240.

Johnson, M.A., Becnel, J.J., Undeen, A.H., 1997. A new sporulation sequence in *Edhazardia aedis* (Microsporidia: Culicosporidae), a parasite of the mosquito *Aedes aegypti* (Diptera: Culicidae). J. Invertebr. Pathol. 70, 69–75.

Joudrey, P., Bjornson, S., 2007. Effects of an unidentified microsporidium on the convergent lady beetle, *Hippodamia convergens* Guérin-Méneville (Coleoptera: Coccinellidae), used for biological control. J. Invertebr. Pathol. 94, 140–143.

Karlhofer, J., Schafellner, C., Hoch, G., 2012. Reduced activity of juvenile hormone esterase in microsporidia-infected *Lymantria dispar* larvae. J. Invertebr. Pathol. 110, 126–128.

Katinka, M.D., Duprat, S., Cornillot, E., Metenier, G., Thomarat, F., Prensier, G., et al., 2001. Genome sequence and gene compaction of the eukaryote parasite *Encephalitozoon cuniculi*. Nature 414, 450–453.

Kawarabata, T., 2003. Biology of microsporidians infecting the silkworm, *Bombyx mori*, in Japan. J. Insect Biotechnol. Sericology 72, 1–32.

Keeling, P.J., Fast, N.M., 2002. Microsporidia: Biology and evolution of highly reduced intracellular parasites. Ann. Rev. Microbiol. 56, 93–116.

Keohane, E.M., Weiss, L.M., 1999. The structure, function, and composition of the microsporidian polar tube, in: Wittner, M., Weiss, L.M. (eds.), The Microsporidia and Microsporidiosis. ASM Press, Washington, DC, pp. 196–224.

Kielmanowicz, M.G., Inberg, A., Lerner, I.M., Golani, Y., Brown, N., Turner, C.L., et al., 2015. Prospective large-scale field study generates predictive model identifying major contributors to colony losses. PLOS Pathogens, e1004816.

Klee, J., Tay, W.T., Paxton, R.J., 2006. Specific and sensitive detection of *Nosema bombi* (Microsporidia: Nosematidae) in bumble bees (*Bombus spp: Hymenoptera: Apidae) by PCR of partial rRNA gene sequences. J. Invertebr. Pathol. 91, 98–104.

Koch, J.B., Strange, J.P., 2012. The status of *Bombus occidentalis* and *B. moderatus* in Alaska with special focus on *Nosema bombi* incidence. Northwest Sci. 86, 212–220.

Kohlmayr, B., Weiser, J., Wegensteiner, R., Händel, U., Zizka, Z., 2003. Infection of *Tomicus piniperda* (Col., Scolytidae) with *Canningia tomici* sp. n. (Microsporidia, Unikaryonidae). J. Pest Sci. 76, 65–73.

Kotler, D.P., Orenstein, J.M., 1999. Clinical syndromes associated with microsporidiosis, in: Wittner, M., Weiss, L.M. (eds.), The Microsporidia and Microsporidiosis. ASM Press, Washington, DC, pp. 258–292.

Kralj, J., Fuchs, S., 2010. *Nosema* sp. influences flight behavior of infected honey bees (*Apis mellifera*) foragers. Apidology 41, 21–28.

Kurtz, J., Nahif, A.A., Sauer, K.P., 2000. Phagocytosis of *Vairimorpha* sp. (Microsporida, Nosematidae) spores by *Plutella xylostella* and *Panorpa vulgaris* hemocytes. J. Invertebr. Pathol. 75, 237–239.

Kyei-Poku, G., Gauthier, D., van Frankenhuyzen, K., 2008. Molecular data and phylogeny of *Nosema* infecting lepidopteran forest defoliators in the genera *Choristoneura* and *Malacosoma*. J. Eukaryot. Microbiol. 55, 51–58.

Lacey, L.A., Frutos, R., Kaya, H.K., Vail, P., 2001. Insect pathogens as biological control agents: do they have a future? Biol. Control 21, 230–248.

Laigo, F.M., Tamashiro, M., 1967. Interactions between a microsporidian pathogen of the lawn-armyworm and the hymenopterous parasite *Apanteles marginiventris*. J. Invertebr. Pathol. 9, 546–554.

Lange, C.E., 2010. *Paranosema locustae* (Microsporidia) in grasshoppers (Orthoptera: Acridoidea) of Argentina: field host range expanded. Biocontr. Sci. Technol. 20, 1047–1054.

Lange, C.E., Azzaro, F.G., 2008. New case of long-term persistence of *Paranosema locustae* (Microsporidia) in melanopline grasshoppers (Orthoptera: Acrididae: Melanoplinae) of Argentina. J. Invertebr. Pathol. 99, 357–359.

Lange, C.E., Johny, S., Baker, M.D., Whitman, D.W., Solter, L.F., 2009. A new *Encephalitozoon* species (Microsporidia) isolated from the lubber grasshopper, *Romalea microptera* (Beauvois) (Orthoptera: Romaleidae). J. Parasitol. 95, 976–986.

Lass, S., Hottinger, J.W., Fabbro, T., Ebert, D., 2011. Converging seasonal prevalence dynamics in experimental epidemics. BMC Ecology 11, 14.

Lewis, L.C., Bruck, D.J., Prasifka, J.R., Raun, E.S., 2009. *Nosema pyrausta*: its biology, history, and potential role in a landscape of transgenic insecticidal crops. Biol. Control 48, 223–231.

Lewis, L.C., Sumerford, D.V., Bing, L.A., Gunnarson, R.D., 2006. Dynamics of *Nosema pyrausta* in natural populations of the European corn borer, *Ostrinia nubilalis*: A six-year study. BioControl 51, 627–642.

Li, J., Chen, W., Wu, J., Peng, W., An, J., Schmid-Hempel, P., Schmid-Hempel, R., 2012. Diversity of *Nosema* associated with bumblebees (*Bombus* spp.) from China. Int. J. Parasitol. 42, 49–61.

Lockwood, J.A., 1989. Ontogeny of cannibalism in rangeland grasshoppers (Orthoptera: Acrididae). J. Kan. Entomol. Soc. 62, 534–541.

Maddox, J.V., 1973. The persistence of the microsporidia in the environment. Misc. Publ. Entomol. Soc. Am. 9, 99–104.

Maddox, J.V., 1977. Stability of entomopathogenic protozoa. Misc. Publ. Entomol. Soc. Am. 10, 3–18.

Maddox, J.V., Solter, L.F., 1996. Long-term storage of viable microsporidian spores in liquid nitrogen. J. Eukaryot. Microbiol. 43, 211–225.

Maddox, J.M., McManus, M.L., Solter, L.F., 1998. Microsporidia affecting forest lepidoptera, in: McManus, M.L., Liebhold, A.M. (eds.), Proceedings: Population Dynamics, Impacts, and Integrated Management of Forest Defoliating Insects. Gen. Techn. Rep. NE–247. USDA Forest Service, Radnor, PA, pp. 198–205.

Maddox, J.V., Baker, M.D., Jeffords, M.R., Kuras, M., Linde, A., Solter, L.F., McManus, M.L., et al., 1999. *Nosema portugal*, n. sp., isolated from gypsy moths (*Lymantria dispar* L.) collected in Portugal. J. Inverbr. Pathol. 73, 1–14.

Malone, L.A., Gatehouse, H.S., Tregidga, E.L., 2001. Effects of time, temperature, and honey on *Nosema apis* (Microsporidia: Nosematidae), a parasite of the honeybee, *Apis mellifera* (Hymenoptera: Apidae). J. Invertebr. Pathol. 77, 258–268.

Maniania, N.K., Vaughan, L.J., Ouna, E., 2008. Susceptibility of immature stages of the locusts *Schistocerca gregaria* and *Locusta migratoria migratorioides* to the microsporidium *Johenrea locustae* and effects of infection on feeding and fertility in the laboratory. Biocontr. Sci. Technol. 18, 913–920.

Martin, S.J., Hardy, J., Villalobos, E., Martín-Hernández, R., Nikaido, S., Higes, M. 2013. Do the honeybee pathogens *Nosema ceranae* and deformed wing virus act synergistically? Environ. Microbiol. Rep. 5, 506–510.

Martín-Hernández, R., Meana, A., García-Palencia, P., Marín, P., Botías, C., Garrido-Bailón, E., et al., 2009. Effect of temperature on the biotic potential of honeybee microsporidia. Appl. Environm. Microbiol. 75, 2554–2557.

Martínez, J., Leal, G., Conget, P., 2012. *Nosema ceranae*, an emergent pathogen of *Apis mellifera* in Chile. Parasitol. Res. 111, 601–607.

Maside, X., Gómez-Moracho, T., Jara, L., Martín-Hernández, R., De la Rúa, P., Higes, M., Bartolomé, C., 2015. Population genetics of *Nosema apis* and *Nosema ceranae*: One host (*Apis mellifera*) and two different histories. PLOS ONE 10(12), e0145609.

Mayack, C., Naug, D., 2010. Parasitic infection leads to decline in hemolymph sugar levels in honeybee foragers. J. Insect Physiol. 56, 1572–1575.

McManus, M.L., Smith, H.R., 1972. Importance of the silk trail in the diel behaviour of late instars of the gypsy moth. Environ. Entomol. 1, 793–795.

Miao, J., Guo, Y., Shi, W., 2012. The persistence of *Paranosema locustae* after application in Qinghai Plateau, China. Biocontr. Sci. Technol. 22, 733–735.

Micieli, M.V., Garcia, J.J., Becnel, J.J., 2000. Horizontal transmission of *Amblyospora albifasciati* Garcia and Becnel, 1994 (Microsporidia: Amblyosporidae), to a copepod intermediate host and the neotropical mosquito, *Aedes albifasciatus* (Macquart, 1837). J. Invertebr. Pathol. 75, 76–83.

Micieli, M.V., Garcia, J.J., Andreadis, T.G., 2001. Epizootiological studies of *Amblyospora albifasciati* (Microsporidiida: Amblyosporidae) in natural populations of *Aedes albifasciatus* (Diptera: Culicidae) and *Mesocyclops annulatus* (Copepoda: Cyclopidae) in a transient floodwater habitat. J. Invertebr. Pathol. 77, 68–74.

Moawed, S.M., Marei, S.S., Saleh, M.R., Matter, M.M., 1997. Impact of *Vairimorpha ephestiae* (Microsporidia: Nosematidae) on *Bracon hebetor* (Hymenoptera: Braconidae), an external parasite of the American bollworm, *Heliothis armigera* (Lepidoptera: Noctuidae). Eur. J. Entomol. 94, 561–565.

Molestina, R., Becnel, J.J., Weiss, L.M., 2014. Culture and propagation of microsporidia, in: Weiss, L.M., Becnel, J.J. (eds.), Microsporidia, Pathogens of Opportunity. John Wiley & Sons, Oxford, pp. 457–467.

Nägeli, K.W., 1857. Über die neue Krankheit der Seidenraupe und verwandte Organismen. Bot. Ztg. 15, 760–761.

Naug, D., Gibbs, A., 2009. Behavioral changes mediated by hunger in honeybees infected with *Nosema ceranae*. Apidiologie 40, 595–599.

Ni, X., 1993. A new microsporidium associated with the potato leafhopper, *Empoasca fabae* (Harris) (Homoptera: Cicadellidae). PhD dissertation, University of Missouri.

Ni, X., Backus, E.A., Maddox, J.V., 1996. Transmission mechanisms of *Nosema empoascae* (Microspora: Nosematidae) in *Empoasca fabae* (Homoptera: Cicadellidae). J. Invertebr. Pathol. 69, 269–275.

Novotny, J., Weiser, J., 1993. Transovarial transmission of *Nosema lymantriae* (*Protozoa, Microsporidia*) in the gypsy moth *Lymantria dispar* L. Biologia 48, 125–129.

Oi, D.H., Becnel, J.J., Williams, D.F., 2001. Evidence of intracolony transmission of *Thelohania solenopsae* (Microsporida: Thelohaniidae) in red imported fire ants (Hymenoptera: Formicidae) and the first report of spores from pupae. J. Invertebr. Pathol. 78, 128–134.

Oliver, R., 2011. An update on the "nosema cousins." Am. Bee J. 151, 1153–1158.

Olsen, L.E., Hoy, M.A., 2002. Heat curing *Metaseiulus occidentalis* (Nesbitt) (Acari, Phytoseiidae) of a fitness-reducing microsporidium. J. Invertebr. Pathol. 79, 173–178.

Onstad, D.W., Maddox, J.V., 1989. Modelling the effects of the microsporidium, *Nosema pyrausta*, on the population dynamics of the insect, *Ostrinia nubilalis*. J. Invertebr. Pathol. 53, 410–421.

Onstad, D.W., Fuxa, J.R., Humber, R.A., Oestergaard, J., Shapiro-Ilan, D.I., Gouli, V.V., et al., 2006. Abridged glossary of terms used in invertebrate pathology, 3rd edn. Society for Invertebrate Pathology. Available from: http://www.sipweb.org/resources/glossary.html (accessed May 8, 2017).

Otti, O., Schmid-Hempel, P., 2007. *Nosema bombi*: a pollinator parasite with detrimental fitness effects. J. Invertebr. Pathol. 96, 118–124.

Own, O.S., Brooks, W.M., 1986. Interactions of the parasite *Pediobius foveolatus* (Hymenoptera: Eulophidae) with two *Nosema* spp. (Microsporida: Nosematidae) of the Mexican bean beetle (Coleoptera: Coccinellidae). Environ. Entomol. 15, 32–39.

Pasteur, L. 1870. Études sur la maladie des vers à soie. Gauthier-Villars, Paris.

Paxton, R.J., Klee, J., Korpela, S., Fries, I., 2008. *Nosema ceranae* has infected *Apis mellifera* in Europe since at least 1998 and may be more virulent than *Nosema apis*. Apidiologie 38, 558–565.

Pilarska, D.K., Solter, L.F., Maddox, J.V., McManus, M.L., 1998. Microsporidia from gypsy moth (*Lymantria dispar* L.) populations in Central and Western Bulgaria. Acta Zool. Bulg. 50, 109–113.

Plischuk, S., Martín-Hernández, R., Prieto, L., Lucia, M., Botías, C., Meana, A., Abrahamovich, A.H., et al., 2009. South American native bumblebees (Hymenoptera: Apidae) infected by *Nosema ceranae* (Microsporidia), an emerging pathogen of honeybees (*Apis mellifera*). Environ. Microbiol. Rep. 1, 131–135.

Pohorecka, K., Bober, A., Skubida, M., Zdanska, D., Toroj, K., 2014. A comparative study of environmental conditions, bee management and the epidemiological situation in apiaries varying in the level of colony losses. J. Apic. Sci. 58, 107–132.

Pollan, S., Goertz, D., Hoch, G., 2009. Effect of temperature on development of the microsporidium *Nosema lymantriae* and disease progress in the host *Lymantria dispar* (L. 1758). Mitt. dtsch. Ges. allg. angew. Entomol. 17, 173–178.

Rajendran, K.V., Shivam, S., Praveena, P.E., Rajan, J.J.S, Kumar, S., Avunje, S., et al., 2016. Emergence of *Enterocytozoon hepatopenaei* (EHP) in farmed *Penaeus* (*Lipopenaeus*) *vannamei* in India. Aquaculture 454, 272–280.

Rath, S.S., Prasad, B.C., Sinha, B.R.R.P., 2003. Food utilization efficiency in fifth instar larvae of *Antheraea mylitta* (Lepidoptera: Saturniidae) infected with *Nosema* sp. and its effect on reproductive potential and silk production. J. Invertebr. Pathol. 83, 1–9.

Raun, E.S., 1961. Elimination of microsporidiosis in laboratory-reared European corn borers by the use of heat. J. Invertrebr. Pathol. 3, 446–448.

Rivero, A., Agnew, P., Bedhomme, S., Sidobre, C., Michalakis, Y., 2007. Resource depletion in *Aedes aegypti* mosquitoes infected by the microsporidia *Vavraia culicis*. Parasitol. 134, 1355–1362.

Roberts, K.E., Hughes, W.O.H., 2014. Horizontal transmission of a parasite is influenced by infected host phenotype and density. Parasitol. 142, 395–405.

Roth, O., Ebert, D., Vizoso, D.B., Bieger, A., Lass, S., 2008. Male-biased sex-ratio distortion caused by *Octosporea bayeri*, a vertically and horizontally-transmitted parasite of *Daphnia magna*. Int. J. Parasitol. 38, 969–979.

Rutrecht, S.T., Klee, J., Brown, J.F., 2007. Horizontal transmission of *Nosema bombi* to its adult bumble bee hosts: effects of dosage, spore source and host age. Parasitol. 134, 1719–1726.

Sagers, J.B., Munderloh, U.G., Kurtti, T.J., 1996. Early events in the infection of a *Helicovrpa zea* cell line by *Nosema furnacalis* and *Nosema pyrausta* (Microspora: Nosematidae). J. Invertebr. Pathol. 67, 28–34.

Saito, T., Bjornson, S., 2006. Horizontal transmission of a microsporidium from the convergent lady beetle, *Hippodamia convergens* Guérin-Méneville (Coleoptera: Coccinellidae), to three coccinellid species of Nova Scotia. Biol. Contr. 39, 427–433.

Sapir, A., Dillman, A.R., Connon, S.A., Grupe, B.M., Ingels, J., Mundo-Ocampo, M., et al., 2014. Microsporidia-nematode associations in methane seeps reveal basal fungal parasitism in the deep sea. Front. Microbiol. 5, 43.

Selman, M., Pombert, J-F., Solter, L., Farinelli, L., Weiss, L.M., Keeling, P., Corradi, N. 2011. Acquisition of an animal gene by microsporidian intracellular parasites. Curr. Biol. 21, R576–R577.

Shi, W-P., Wang, Y-Y., Lv, F., Guo, C., Cheng, X., 2009. Persistence of *Paranosema* (*Nosema*) *locustae* (Microsporidia: Nosematidae) among grasshopper (Orthoptera: Acrididae) populations in the Inner Mongolia rangeland, China. BioControl 54, 77–84.

Siegel, J.P., Maddox, J.V., Ruesink, W.G., 1986. Impact of *Nosema pyrausta* on a braconid, *Macrocentrus grandii*, in central Illinois. J. Invertebr. Pathol. 47, 271–276.

Siegel, J.P., Maddox, J.V., Ruesink, W.G., 1988. Seasonal progress of *Nosema pyrausta* in the European corn borer, *Ostrinia nubilalis*. J. Invertebr. Pathol. 52, 130–136.

Simões, R.A., Reis, L.G., Bento, J.M.S., Solter, L.F., Delalibera Jr., I., 2012. Biological and behavioral parameters of the parasitoid *Cotesia flavipes* (Hymenoptera: Braconidae) are altered by the pathogen *Nosema* sp. (Microsporidia: Nosematidae). Biol. Contr. 63, 164–171.

Simões, R.A., Feliciano, J.R., Solter, L.F., Delalibera Jr., I., 2015. Impacts of *Nosema* sp. (Microsporidia: Nosematidae) on the sugarcane borer, *Diatraea saccharalis* (Lepidoptera: Crambidae). J. Invertebr. Pathol. 129, 7–12.

Slothouber Galbreath, J.G.M., Smith, J.E., Becnel, J.J., Butlin, R.K., Dunn, A.M., 2010. Reduction in post-invasion genetic diversity in *Crangonyx pseudogracilis* (Amphipoda: Crustacea): a genetic bottleneck or the work of hitchhiking vertically transmitted microparasites? Biol. Invasions 12, 191–209.

Smart, M.D., Sheppard, W.S., 2011. *Nosema ceranae* in age cohorts of the western honey bee (*Apis mellifera*). J. Invertebr. Pathol. 109, 148–151.

Smith, J.E., 2009. The ecology and evolution of microsporidian parasites. Parasitol. 136, 1901–1914.

Smith, M.L., 2012. The honey bee parasite *Nosema ceranae*: transmissible via food exchange? PLOS ONE 7, e43319.

Solter, L.F., 2014. Epizootiology of microsporidiosis in invertebrate hosts, in: Weiss, L.M., Becnel, J.J. (eds.), Microsporidia: Pathogens of Opportunity. John Wiley & Sons, Oxford, pp. 165–194.

Solter, L.F, Becnel, J.J., 2000. Entomopathogenic microsporidia, in: Lacey, L.A., Kaya, H.K. (eds.), Field Manual of Techniques in Invertebrate Pathology. Kluwer, Dordrecht, pp. 231–254.

Solter, L.F., Maddox, J.V., 1998. Physiological host specificity of microsporidia as an indicator of ecological host specificity. J. Invertebr. Pathol. 71, 207–216.

Solter, L.F., Maddox, J.V., McManus, M.L., 1997. Host specificity of microsporidia (Protista: Microspora) from European populations of *Lymantria dispar* (Lepidoptera: Lymantriidae) to indigenous North American Lepidoptera. J. Invertebr. Pathol. 69, 135–150.

Solter, L.F., Pilarska, D.K., Vossbrinck, C.F., 2000. Host specificity of microsporidia pathogenic to forest Lepidoptera. Biol. Contr. 19, 48–56.

Solter, L.F., Maddox, J.V., Vossbrinck, C.R., 2005. Physiological host specificity: a model using the European corn borer, *Ostrinia nubilalis* (Hübner) (Lepidoptera: Crambidae) and microsporidia of row crop and other stalk-boring hosts. J. Invertebr. Pathol. 90, 127–130.

Solter, L.F., Pilarska, D.K., McManus, M.L., Zubrik, M., Patocka, J., Huang, W-F., Novotny, J., 2010. Host specificity of microsporidia pathogenic to the gypsy moth, *Lymantria dispar* (L.): field studies in Slovakia. J. Invertebr. Pathol. 105, 1–10.

Solter, L.F., Becnel, J.J., Oi, D.H., 2012a. Microsporidian entomopathogens, in: Vega, F.E., Kaya, H.K. (eds.), Insect Pathology. Academic Press, London, Waltham, San Diego, CA, pp. 221–263.

Solter, L.F., Becnel, J.J., Vavra, J., 2012b. Research methods for entomopathogenic microsporidia and other protists, in: Lacey, L.A. (ed.) Manual of Techniques in Invertebrate Pathology. Academic Press, San Diego, CA, pp. 329–371.

Sprague, V., 1977. Systematics of the microsporidia, in: Bulla Jr., L.A., Cheng, T.C. (eds.), Comparative Pathobiology, vol. 2. Plenum Press, New York.

Sprague, V., Becnel, J.J., 1998. Note on the name-author-date combination for the taxon MICROSPORIDIES Balbiani, 1882, when ranked as a phylum. J. Invertebr. Pathol. 71, 91–94.

Stentiford, G.D., Becnel, J., Weiss, L., Keeling, P., Didier, E., Williams, B., Bjornson, S., et al., 2016. Microsporidia – emergent pathogens in the global food chain. Trends Parasitol. 32, 336–348.

Steyer, C., 2010. Die Rolle von Faeces und simuliertem Regen bei der Übertragung der Mikrosporidien *Nosema lymantriae* und *Endoreticulatus schubergi* bei *Lymantria dispar*. Master's thesis, Universität für Bodenkultur Wien.

Sweeney, A.W., Hazard, E.I., Graham, M.F., 1985. Intermediate host for an *Amblyospora* sp. (Microspora) infecting the mosquito, *Culex annulirostris*. J. Invertebr. Pathol. 46, 98–102.

Tanada, Y., Kaya, H.K., 1993. Insect Pathology. Academic Press, San Diego, CA.

Tay, W.T., O'Mahony, E.M., Paxton, R.J., 2005. Complete rRNA gene sequences reveal that the microsporidium *Nosema bombi* infects diverse bumblebee (*Bombus* spp.) hosts and contains multiple polymorphic sites. J. Eukaryot. Microbiol. 52, 505–513.

Teetor, G.E., Kramer, J.P., 1977: Effect of ultraviolet radiation on the microsporidian *Octosporea muscaedomesticae* with reference to protectants provided by the host *Phormia regina*. J. Invertebr. Pathol. 30, 348–353.

Terry, R.S., Smith, J.E., Dunn, A.M., 1998. Impact of a novel, feminising microsporidium on its crustacean host. J. Eukaryot. Microbiol. 45, 497–501.

Terry, R.S., Smith, J.E., Sharpe, R.G., Rigaud, T., Littlewood, D.T.J., Ironside, J.E., Rollinson, D., et al., 2004. Widespread vertical transmission and associated host sex-ratio distortion within the eukaryotic phylum Microspora. Proc. R. Soc. Lond. B 271, 1783–1789.

Texier, C., Vidau, C., Viguès, B., El Alaoui, H., Delbac, F., 2010. Microsporidia: a model for minimal parasite–host interactions. Curr. Opin. Microbiol. 13, 443–449.

Thomson, H.M., 1960. The possible control of a budworm infestation by a microsporidian disease. Can. Dept. Agric. Bi-Mon. Prog. Rept. 16, 1.

Tokarev, Y.S., Sokolova, Y.Y., Entzeroth, R., 2007. Microsporidia–insect host interactions: Teratoid sporogony at the sites of host tissue melanization. J. Invertebr. Pathol. 94, 70–73.

Tourtip, S., Wongtripop, S., Stentiford, G.D., Bateman, K.S., Sriurairatana, S., Chavadej, J. Sritunyalucksana, K., et al., 2009. *Enterocytozoon hepatopenaei* sp. Nov. (Microsporida: Enterocytozoonidae), a parasite of the black tiger shrimp *Penaeus monodon* (Decapoda: Penaeidae): fine structure and phylogenetic relationships. J. Invertebr. Pathol. 102, 21–29.

Tounou, A.K., Kooyman, C., Douro-Kpindou, O.K., Gumedzoe, Y.M., Poehlingn, H.M., 2011. Laboratory assessment of the potential of *Paranosema locustae* to control immature stages of *Schistocerca gregaria* and *Oedaleus senegalensis* and vertical transmission of the pathogen in host populations. Biocontr. Sci. Technol. 21, 605–617.

Traver, B.E., Fell, R.D., 2011. Prevalence and infection intensity of *Nosema* in honey bee (*Apis mellifera* L.) colonies in Virginia. J. Invertebr. Pathol. 107, 43–49.

Undeen, A.H., 1990. A proposed mechanism for the germination of microsporidian (Protozoa: Microspora) spores. J. Theor. Biol. 142, 223–235.

Undeen, A.H., Becnel, J.J., 1992. Longevity and germination of *Edhazardia aedis* (Microspora: Amblyosporidae) spores. Biocontr. Sci. Technol. 2, 247–256.

Undeen, A.H., Solter, L.F., 1996. The sugar content and density of living and dead microsporidian (Protozoa: Microspora) spores. J. Invertebr. Pathol. 67, 80–91.

Undeen, A.H., Vavra, J., 1997. Research methods for entomopathogenic protozoa. In: Lacey, L.A. (ed.), Manual of Techniques in Insect Pathology. Academic Press, San Diego, CA, pp. 117–151.

van Frankenhuyzen, K., Ebling, P., McCron, B., Ladd, T., Gauthier, D., Vossbrinck, C., 2004. Occurrence of *Cystosporogenes* sp. (Protozoa, Microsporidia) in a multi-species insect production facility and its elimination from a colony of the eastern spruce budworm, *Choristoneura fumiferanae* (Clem.) (Lepidoptera: Tortricidae). J. Invertebr. Pathol. 87, 16–28.

van Frankenhuyzen, K., Nystrom, C., Liu, Y., 2007. Vertical transmission of *Nosema fumiferanae* (Protozoa, Microsporidia) and consequences for distribution, post-diapause emergence and dispersal of second-instar larvae of the spruce budworm, *Choristoneura fumiferana* (Clem.) (Lepidoptera: Tortricidae). J. Invertebr. Pathol. 96, 173–182.

van Frankenhuyzen, K., Ryall, K., Liu, Y., Meating, J., Bolan, P., Scarr, T., 2011. Prevalence of *Nosema* sp. (Microsporidia: Nosematidae) during an outbreak of the jack pine budworm in Ontario. J. Invertebr. Pathol. 108, 201–208.

Vavilova, V., Sormacheva, I., Woyciechowski, M, Eremeeva, N., Fet, V., Strachecka, A., Bayborodin, S.I., et al., 2015. Distribution and diversity of *Nosema bombi* (Microsporidia: Nosematidae) in the natural populations of bumblebees (*Bombus* spp.) from West Siberia. Parasitol. Res. 114, 3373–3383.

Vavra, J., Hylis, M., Vossbrinck, C.R., Pilarska, D.K., Linde, A., Weiser, J., McManus, M.L., et al., 2006. *Vairimorpha disparis* n. comb. (Microsporidia: Burenellidae): a redescription and taxonomic revision of *Thelohania disparis* Timofejeva 1956, a microsporidian parasite of the gypsy moth *Lymantria dispar* (L.) (Lepidoptera: Lymantriidae). J. Eukaryot. Microbiol. 53, 292–304.

Vavra, J., Larsson, J.I.R., 1999. Structure of the microsporidia, in: Wittner, M., Weiss, L.M. (eds.), The Microsporidia and Microsporidiosis. ASM Press, Washington, DC, 7–84.

Vavra, J., Lukes, J., 2013. Microsporidia and "The Art of Living Together," in: Rollinson, D. (ed.), Advances in Parasitology. Academic Press, San Diego, CA, pp. 253–320.

Vijendravarma, R.K, Godfray, H.C.J., Kraaijeveld, A.R., 2008. Infection of *Drosophila melanogaster* by *Tubulinosema kingi*: stage-specific susceptibility and within-host proliferation. J. Invertebr. Pathol. 99, 239–241.

Vizoso, D.B., Lass, S., Ebert, D., 2005. Different mechanisms of transmission of the microsporidium *Octosporea bayeri*: a cocktail of solutions for the problem of parasite permanence. Parasitol. 130, 501–509.

Wang D.I., Moeller, F.E., 1970. The division of labor and queen attendance behavior of *Nosema* infected worker honeybees. J. Econ. Entomol. 63: 1539–1541.

Wegensteiner, R., Dedryver, C.-A., Pierre, J.-S., 2010. The comparative prevalence and demographic impact of two pathogens in swarming *Ips typographus* adults: a quantitative analysis of long term trapping data. Agric. For. Entomol. 12, 49–57.

Weidner, E., Trager, W., 1973. ATP in the extracellular survival of an intracellular parasite *Nosema michaelis* microsporidia. J. Cell Biol. 57, 586–591.

Weidner, E., Findley, A.M., Dolgikh, V., Sokolova, J., 1999. Microsporidian biochemistry and physiology, in: Wittner, M., Weiss, L.M. (eds.), The Microsporidia and Microsporidiosis. ASM Press, Washington, DC, pp. 172–195.

Weigand, A.M., Kremers, J., Grabner, D.S., 2016. Shared microsporidian profiles between an obligate (*Niphargus*) and facultative subterranean amphipod population (*Gammarus*) at sympatry provide indications for underground transmission pathways. Limnologica 58, 7–10.

Weiser, J., 1957: Mikrosporidien des Schwammspinners und Goldafters. Z. Angew. Entomol. 40, 509–527.

Weiser, J., 1961. Die Mikrosporidien als Parasiten der Insekten. Monografien zur angewandten Entomologie. Beihefte Z. Angew. Entomol. 17, 1–149.

Weiser, J., 1978. Transmission of microsporidia to insects via injection. Vest. Cesk. Zool. Spol. 42, 311–317.

Weiser, J., Linde, A., 1998. Microsporidian early spores as confusing factor in taxonomy: *Nosema muscularis* vs. *N. lymantriae*. Abstracts of the VIth European Congress of Entomology. Ceské Budéjovice, p. 571.

Weiss, L.M., Becnel, J.J., (eds.) 2014. Microsporidia: Pathogens of Opportunity. John Wiley & Sons, Oxford.

Williams, D.F., Knue, G.J., Becnel, J.J., 1998 Discovery of *Thelohania solenopsae* from the imported fire ant, *Solenopsis invicta*, in the United States. J. Invertebr. Pathol. 71, 175–176.

Williams, B.A.P., Haferkamp, I., Keeling, P.J., 2008. An ADP/ATP-specific mitochondrial carrier protein in the microsporidian *Antonospora locustae*. J. Mol. Biol. 375, 1249–1257.

Wilson, G.G., 1973. Incidence of microsporidia in a field population of spruce budworm. Can. For. Serv. Bi-mon. Res. Notes 29, 35–36.

Wilson, G.G., 1977. The effects of feeding microsporidian (*Nosema fumiferana*) spores to naturally infected spruce budworm (*Choristoneura fumiferana*). Can. J. Zool. 55, 249–250.

Wittner, M., 1999. Historic perspective on the microsporidia: expanding horizons, in: Wittner, M., Weiss, L.M. (eds.), The Microsporidia and Microsporidiosis. ASM Press, Washington, DC, pp. 1–6.

Zimmermann, G., Huger, A.M., Kleespies, R.G., 2013. Occurrence and prevalence of insect pathogens in populations of the codling moth, *Cydia pomonella* L.: a long-term diagnostic survey. Insects 4, 425–446.

Zwölfer, W. 1927. Die Pebrine des Schwammspinners und Goldafters, eine neue wirtschaftlich bedeutungsvolle Infektionskrankheit. Z. Angew. Entomol. 12, 498–500.

11

Nematodes

David I. Shapiro-Ilan[1], Ivan Hiltpold[2] and Edwin E. Lewis[3]

[1] USDA-ARS, SEA, SE Fruit and Tree Nut Research Unit, Byron, GA, USA
[2] Department of Entomology and Wildlife Ecology, University of Delaware, Newark, DE, USA
[3] Department of Entomology and Nematology, University of California – Davis, Davis, CA, USA

11.1   Introduction

11.1.1   Diversity and Life Histories

The types of relationships between nematodes and arthropods are quite diverse. Thirteen different suborders of the Nematoda are associated in some way with insects (Blaxter, 2011). Note that taxonomic authorities are included in this chapter to provide clarification after past uncertainties in the literature for certain organisms discussed herein. The various relationships with insects can be categorized based on the amount of damage suffered by the host. These categories range from benign phoretic nematodes that are vectored by insects, having little or no effect on the host, to direct and lethal parasites. This chapter focuses on a group in the Family Rhabditidae that are called entomopathogenic nematodes (EPNs), which occupy the lethal extreme of this continuum. However, we include short descriptions of nematodes in the families Mermithidae and Neotylenchidae. These three groups of nematodes have very different life-history characteristics, and stand to illustrate the diversity of parasitic relationships between nematodes and insects, although some characteristics are common to all species. All nematode species have six life stages: the egg, four immature stages, and the adult. For parasitic nematodes, there is usually at least one stage that is tolerant to environmental extremes and serves as the infective stage, or transmission stage. Reproduction may occur inside or outside the parasite's host, but most of the nematode's nutrition is provided by the host, both directly and indirectly.

Mermithids are obligate lethal parasites of many arthropods, leeches, and other invertebrate species. Perhaps the best known is *Romanomermis culicivorax* Ross and Smith, which is a parasite of aquatic mosquito larvae that has been used in biological control programs. Hosts are invaded by the preparasitic stage (second-stage juvenile), which hatches from eggs found in the substrate. The preparasites do not feed and must find a host quickly, as they live for only about 3 days. *Romanomeris culicivorax* preparasites swim rapidly through the water to find a host and penetrate through the cuticle. After penetration, the nematodes develop within the host for 7–10 days.

Ecology of Invertebrate Diseases, First Edition. Edited by Ann E. Hajek and David I. Shapiro-Ilan.
© 2018 John Wiley & Sons Ltd. Published 2018 by John Wiley & Sons Ltd.

The postparasite nematodes leave the host, sink to the bottom of the body of water, and continue to develop into adults in the substrate. The host dies immediately after nematode emergence. Male and female nematodes mate in the substrate once they reach adulthood, and females oviposit there. Eggs hatch within about 3 weeks at 27 °C (Platzer, 1981).

There is substantial variation among species of mermithids, since the habitats they exploit are diverse. For example, the nematode *Mermis nigrescens* Dujardin produces eggs that are affixed to leaf surfaces, awaiting ingestion by their grasshopper hosts (Platzer et al., 2005). Once inside the host gut, the eggs hatch within an hour, and the juveniles penetrate into the hemocoel. Mature juveniles exit the grasshopper, descend to the soil, and develop into adults. Adults mate in spring or summer and remain in the soil for about 1 year, after which time females emerge to deposit eggs.

The neotylenchid *Deladenus* (= *Beddingia*) *siricidicola* Bedding is a facultative parasite with a very complicated lifecycle (summarized by Lewis and Clarke, 2012). This species of nematode parasitizes the Sirex woodwasp (*Sirex noctilio* F.) and has been used in several successful biological control efforts (Bedding, 2009). In the absence of hosts, the nematodes are free-living, and complete their lifecycle by feeding on the fungus *Amylostereum areolatum* (Bedding and Akhurst, 1978). This fungus is inoculated into pines along with the nematodes by the woodwasp. The free-living nematodes can pass through many generations without a host, but must parasitize a wasp periodically or they lose their ability to infect a host (Bedding, 2009). The parasitic lifecycle occurs when these nematodes are in the presence of woodwasp larvae, which are detected by the presence of high levels of CO_2 and an acidic pH. Preparasitic mated female nematodes penetrate through the larval host's cuticle using a stylet and take in nourishment through their cuticle. The female grows quickly, sometimes up to 1000 times the size of the preparasite; it can stay in this state for months. At the time the woodwasp ecloses in order to leave the tree, the female nematode produces juveniles that can invade the wasp's ovaries and eggs, rendering the wasp sterile. The wasp then "nemaposits," instead of ovipositing, to inoculate the new tree with the nematodes. The free-living lifecycle then resumes until woodwasp larvae are encountered in the new tree.

All species of EPNs share many commonalities: they are obligate parasites, they are all associated with symbiotic bacteria, and they have a single transmission stage, known as the infective juvenile (IJ) stage, which occurs in the third juvenile stage (Fig. 11.1). Throughout most scientific literature, and in the remainder of this chapter, when the term "nematode" is used in reference to an EPN, the nematode–bacterium complex is the actual topic. The EPNs that are currently available commercially for biological control of insects are classified into two families, the Heterorhabditidae and Steinernematidae, which are not phylogenetically closely related but share many characteristics via convergent evolution. The Heterorhabditidae comprise a single genus, *Heterorhabditis*, which includes 21 species (Table 11.1). Steinernematidae has two genera, *Steinernema* (96 described species) (Table 11.2) and *Neosteinernema* (1 species). The list of EPN species has been increasing steadily over time (Poinar, 1990, Stock and Hunt, 2005; Lewis and Clarke, 2012; Shapiro-Ilan et al., 2016a), and therefore our lists (Tables 11.1 and 11.2) are not meant to be exhaustive, but only to represent the current diversity at the time this volume is published. The Heterorhabditidae are associated with bacteria in the genus *Photorhabdus* and the Steinernematidae with *Xenorhabdus* spp. bacteria. A third group of nematodes that meets the criteria to be categorized as EPNs has recently been

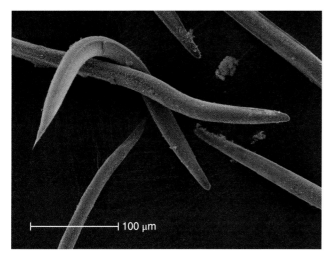

Fig. 11.1 Scanning electron micrograph (SEM) of infective juveniles of the entomopathogenic nematode *Heterorhabditis bacteriophora*. *Source*: Coutesy of Ivan Hiltpold.

Table 11.1 List of *Heterorhabditis* species.

1) *H. amazonensis* Andaló, Nguyen, & Moino	11) *H. indica* Poinar, Karunakar, & David
2) *H. atacamensis* Egington, Buddie Moore, France, Merino, & Hunt	12) *H. marelata* Liu & Berry
	13) *H. megidis* Poinar, Jackson, & Klein
3) *H. bacteriophora* Poinar	14) *H. mexicana* Nguyen, Shapiro-Ilan, Stuart, McCoy, James, & Adams
4) *H. baujardi* Phan, Subbotin, Nguyen, and Moens	
5) *H. beicherriana* Li, Liu, Nermut, Puza, & Mracek	15) *H. noenieputensis,* Malan, Knoetze, & Tiedt
	16) *S. pakistanense* Shahina, Anis, Reid, Rowe, & Maqbool
6) *H. brevicaudis* Liu	
7) *H. downesi* Stock, Burnell, and Griffin	17) *H. poinari* Kakulia and Mikaia
8) *H. floridensis* Nguyen, Gozel, Koppenhöfer, & Adams	18) *H. safricana* Malan, Nguyen, De Waal, & Tiedt
9) *H. georgiana* Nguyen, Shapiro-Ilan, & Mbata	19) *H. sonorensis* Stock, Rivera-Orduno, & Flores-Lara
10) *H. gerrardi* Plitchta, Joyce, Clarke, Waterfield, & Stock	20) *H. taysearae* Shamseldean, El-Sooud, Abd-Elgawad, & Saleh
	21) *H. zealandica* Poinar

described in the family Rhabditidae, genus *Oscheius*. It was first recognized by Zhang et al. (2008). This group, initially described as the genus *Heterorhabditoides*, and later combined into the genus *Oscheius*, carries bacterial symbionts, one of which has been identified as *Serratia nematodiphila* (Zhang et al., 2009), but other members of the genus are associated with multiple bacterial species (Torres-Barragan et al., 2011). This chapter will focus on the Heterorhabditidae and Steinernematidae, because the other rhabditids that are considered EPNs are not as well characterized ecologically and are not used in biological control applications. Another nematode species with a similar life history is the mollusk parasite *Phasmarhabditis hermaphrodita* (Schneider), which kills

Table 11.2 List of *Steinernema* species.

1) *S. abbasi* Elawad, Ahmad, & Reid
2) *S. aciari* Qiu, Hu, Zhou, Mei, Nguyen, & Pang
3) *S. affine* (Bovien)
4) *S. arenarium* (Artyukhovsky)
5) *S. anatoliense* Hazir, Stock, & Keskin
6) *S. ashiuense* Phan, Takemoto, & Futai
7) *S. asiaticum* Anis, Shahina, Reid, & Rowe
8) *S. apuliae* Triggiani, Mracek, & Reid
9) *S. australe* Edgington, Buddie, Tymo, Hunt, Nguyen, France, Merino, & Moore
10) *S. backanense* Phan, Spiridonov, Subbotin, & Moens
11) *S. balochiense*, Fayyaz, Khanum, Ali, Solangi, Gulsher, & Javed
12) *S. bifurcatum* Fayyaz, Yan, Qiu, Han, Gulsher, Khanum, & Javed
13) *S. boemarei*, Lee, Sicard, Skeie, & Stock,
14) *S. brazilense* Nguyen, Ginarte, Leite, Santos, & Haracava
15) *S. bicornutum* Tallosi, Peters, and Ehlers
16) *S. cameroonense* Kanga, Trinh, Waeyenberge, Spiridonov, Hauser, & Moens
17) *S. carpocapsae* (Weiser)
18) *S. caudatum* Xu, Wang, & Li
19) *S. ceratophorum* Jian, Reid, & Hunt
20) *S. changbaiense* Ma, Chen, De Clercq, Han, & Moens
21) *S. cholashanense* Nguyen, Puza, & Mracek
22) *S. citrae* Stokwe, Malan, Nguyen, Knoetze, & Tiedt
23) *S. costaricense* Uribe-Lorio, Mora, & Stock
24) *S. colombiense* Lopez-Nunez, Plichta, Gongora-Botero, & Stock
25) *S. cubanum* Mracek, Hernandez and Boemare
26) *S. cumgarense* Phan, Spiridonov, Subbotin, & Moens
27) *S. diaprepesi* Nguyen and Duncan
28) *S. eapokense* Phan, Spiridonov, Subbotin, & Moens
29) *S. ethiopiense* Tamiru, Waeyenberge, Hailu, Ehlers, Půža, & Mráček
30) *S. everestense* Khatri-Chhetri, Waeyenberge, Spiridonov, Manandhar, & Moens
31) *Steinernema fabii* Abate, Malan, Tiedt, Wingfield, Slippers, & Hurley
32) *S. feltiae* (Filipjev)
33) *S. glaseri* (Steiner)
34) *S. goweni* San-Blas, Morales-Montero, Portillo, Nermu, & Puza
35) *S. guangdongense* Qiu, Fan, Zhou, Pang, & Nguyen
36) *S. hebeiense* Chen, Li, Yan, Spiridonov, & Moens
37) *S. hermaphroditum* Stock, Griffin, & Chaerani
38) *S. ichnusae* Tarasco, Mráček, Nguyen, & Triggiani
39) *S. innovationi*, Çimen, Lee, Hatting, Hazir, & Stock
40) *S. intermedium* (Poinar, 1985)
41) *S. jeffreyense*, Malan, Knoetze, & Tiedt
42) *S. jollieti* Spiridonov, Krasomil-Osterfeld, & Moens
43) *S. karii* Waturu, Hunt & Reid
44) *S. huense* Phan, Mráček, Půža, Nermů, Jarošová
45) *S. khoisanae* Nguyen, Malan, & Gozel
46) *S. kraussei* (Steiner)
47) *S. kushidai* Mamiya
48) *S. lamjungense* Khatri-Chhetri, Waeyenberge, Spiridonov, Manandhar, & Moens
49) *S. litorale* Yoshida
50) *S. loci* Phan, Nguyen & Moens
51) *S. leizhouense* Nguyen, Qiu, Zhou, & Pang
52) *S. longicaudum* Shen & Wang
53) *S. masoodi* Ali, Shaheen, Pervez, & Hussain
54) *S. minutum* Maneesakorn, Grewal, & Chandrapatya
55) *S. monticolum* Stock, Choo, & Kaya
56) *S. neocurtillae* Nguyen and Smart
57) *S. nepalense* Khatri-Chhetri, Waeyenberge, Spiridonov, Manandhar, & Moens
58) *S. nyetense* Kanga, Trinh, Waeyenberge, Spiridonov, Hauser, & Moens
59) *S. oregonense* Liu and Berry
60) *S. pakistanense* Shahina, Anis, Reid, Rowe, & Maqbool
61) *S. papillatum* San-Blas' Portillo, Nermut, Půža, Morales-Montero
62) *S. phyllophagae* Nyguyen & Buss
63) *S. poinari*, Mrácek, Puza, & Nermut
64) *S. puertoricense* Roman & Figueroa

Table 11.2 (Continued)

65) *S. pui* Qiu, Zhao, Wu, Lv, & Pang	82) *S. silvaticum* Sturhan, Spiridonov, & Mracek
66) *S. puntauvense* Uribe-Lorio, Mora, & Stock	83) *Steinemema surkhetense* Khatri-Chhetri, Waeyenberge, Spiridonov, Manandhar, & Moens
67) *S. pwaniensis* Puza, Nermut, Mracek, Gengler, & Haukeland	84) *S. tami* Pham, Nguyen, Reid, & Spiridonov
68) *S. rarum* (de Doucet)	85) *S. texanum* Nguyen, Stuart, Andalo, Gozel, & Rogers
69) *S. riobrave* Cabanillas, Poinar, & Raulston	86) *S. thanhi* Phan, Nguyen, & Moens
70) *S. ritteri* de Doucet and Doucet	87) *S. thermophilum* Ganguly and Singh
71) *S. robustispiculum* Phan, Subbotin, Waeyenberge, & Moens	88) *S. tielingense* Ma, Chen, Li, Han, Khatri-Chhetri, & De Clercq
72) *S. sacchari,* Nthenga, Knoetze, Berry, Tiedt, & Malan	89) *S. tophus,* Çimen, Lee, Hatting, Hazir, & Stock
73) *S. sangi* Phan, Nguyen, & Moens	90) *S. unicornum* Egington, Buddie Tymo, France, Merino, & Hunt
74) *S. sasonense* Phan, Spiridonov, Subbotin, & Moens	91) *S. vulcanicum* Clausi, Longo, Rappazzo, Tarasco, Vinciguerra
75) *S. scapterisci* Nguyen & Smart	92) *S. websteri* Cutler and Stock
76) *S. scarabaei* Stock and Koppenhöfer	93) *S. weiseri* Mracek, Sturhan, & Reid
77) *S. schliemanni* Spiridonov, Wayenberge, & Moens	94) *S. xinbinense* Ma, Chen, De Clercq, Wayenberge, Han, & Moens
78) *S. seemae* Ali, Shaheen, Pervez, & Hussain	95) *S. xueshanense* Mracek, Qi-Zhi, & Nguyen
79) *S. serratum* Liu	96) *S. yirgalemense* Nguyen, Tesfamariam, Gozel, Gaugler, & Adams
80) *S. siamkayai* Stock, Somsook, & Kaya	
81) *S. sichuanense* Mracek, Nguyen, Tailliez, Boemare, & Chen	

slugs. This species carries more than a single species of bacteria in the gut, and these bacteria provide nutrition to the nematodes, but their other functions are unknown (Rae et al., 2010) and so they are not included under the term "EPN."

EPNs share many aspects of their lifecycles and development, and a generalized description of the lifecycle follows. However, it should be noted that there are species- and genus-specific differences in the specifics of the lifecycle and reproductive biology. The IJ stage is the only one of the six life stages that occurs outside the host, and it serves as the transmission stage. This is a specialized third-stage juvenile, analogous to the dauer stage in *Caenorhabditis elegans* (Maupas), and is more resistant to environmental extremes than any of the parasitic stages. The IJ is sealed within the second-stage juvenile cuticle, referred to as the sheath, which likely affords extra protection to environmental extremes (Rickert-Campbell and Gaugler, 1991). The sole function of the IJ is to locate, recognize, and infect a host; during this stage, there is no development, feeding, or reproductive behavior. The lifespan of IJs varies by species and temperature, but is generally measured in weeks or months, and depends on metabolizing lipid stores deposited during development inside the host (Lewis et al., 1995). The host is invaded through either natural openings (like the mouth, anus, or spiracles) or sometimes via direct penetration. Once inside the hemocoel, the IJ molts and releases the symbiotic bacteria within a few hours, by regurgitation or defecation (Ciche and Ensign, 2003). The bacteria either avoid or suppress the insect immune system (Goodrich-Blair and Clarke, 2007) and kill the host with a combination of toxins and septicemia within a few days. At this point, there are some differences in development between nematodes in the Heterorhabditidae and in the Steinernematidae. *Heterorhabditis* spp. IJs develop into hermaphrodites for the first generation, but subsequent generations include males, true females, and hermaphrodites. *Steinernema* spp. IJs develop into males and females and are always amphimictic (i.e., reproduction in which sperm and eggs come from separate individuals), except for the species *Steinernema hermaphroditum* Stock, Griffin & Chaerani (Griffin et al., 2000). Inside the host, between one and three generations of nematodes can occur (depending on the size of the host) over the course of 10–30 days (depending on temperature and the species of EPN). The next generation of IJs develops in response to a buildup of waste products of the nematodes and the declining nutritional quality inside the host. Once the IJs exit the host, they begin to search for a new host.

EPNs are characterized by their association with symbiotic bacteria, which they vector from host to host during transmission. The relationship between the nematode and the bacteria is mutualistic. The nematode relies on the bacteria to act as the primary killing agents (although the nematode also contributes to killing the host), to serve as food, and to maintain a suitable environment inside the host for nematode reproduction via the production of antibiotics (which inhibit secondary invaders). The bacteria rely on the nematode to vector them from host to host, as they have no invasive capabilities and cannot persist in the environment without the nematode; the nematode also contributes to suppressing the host's immune system. Each IJ carries between 10 and 300 individual bacterial cells, varying among species and individuals. Steinernematids carry the bacteria attached to the interior of a specialized vesicle in the anterior part of the gut (Martens and Goodrich-Blair, 2005), whereas in heterorhabditids the bacteria are attached to the pre-intestinal valve (Ciche et al., 2008). Interestingly, the degree of fidelity between the nematode and its symbiotic bacteria varies from pair to pair and has been characterized multiple times (Lewis and Clarke, 2012). Briefly, each nematode species is associated with a single bacterial species, but each bacterial species may be associated with multiple nematode species. For example, *S. karii* Waturu, Hunt & Reid is only associated with *X. hominickii* Tailliez et al., but this bacterial species is also found in two other species of *Steinernema*. The taxonomy and systematics of the nematode species are moving at a faster pace than that of the bacteria, so there are many EPN species that are associated with unknown species of bacteria.

11.1.2   EPN Distribution

EPNs have been isolated from all continents except Antarctica, and from practically every region where anyone has made a concerted effort to search. They are exceedingly common soil organisms. Their use as biological control agents may add to the wide distribution of commercially available species, since they are commonly applied at a rate of 2.5 billion IJs per hectare and there are records of applied populations becoming established after application (e.g., Parkman et al., 1994). EPNs have been isolated from numerous different locations and habitat types, but almost always in soil that is moist, is less than 50% clay, and has limited salinity (Kaspi et al., 2010). Natural populations of EPNs are usually found in clumped distributions (Stuart et al., 2006), even when the habitat is relatively homogeneous, such as turfgrass (Campbell et al, 1996). In fact, when efforts have been made to apply EPNs for biological control, an evenly applied population of nematodes has been found to display a clumped structure within a matter of days (Wilson et al., 2003).

11.2 Transmission

Transmission of EPNs is horizontal. There are no reports of vertical transmission of EPNs, although there are reports of EPNs infecting insect eggs (Kalia et al., 2014), and therefore it may be conceivable (although unlikely) that vertical transmission occurs. Horizontal transmission occurs primarily from soil to insect, although transfer may also occur via other substrates or from infected insect to live host.

The infection process of EPN IJs is very well studied due to the importance this has to EPNs' use as biological control agents (Lewis, 2002). The IJ is most often the life stage that is formulated and applied to the soil to reduce pest populations. Because there are no competing goals for foraging IJs (e.g., they do not feed, mate, or reproduce during this stage), everything an individual does should improve the chances that it will encounter an acceptable host.

11.3 Host Population

Information on the host ranges of EPN species is available for only a few of the species. Essentially, the host ranges of species of economic importance or perceived economic potential, such as *S. carpocapsae* (Weiser), *S. glaseri* (Steiner), *S. feltiae* (Filipjev), *S. kraussei* (Steiner), *S. scapterisci* Nguyen & Smart, *S. scarabaei* Stock and Koppenhöfer, *S. riobrave* Cabanillas, Poinar, & Raulston, *H. bacteriophora* Poinar, *H. indica* Poinar, Karunakar, & David, and *H. megidis* Poinar, Jackson, & Klein, have been studied extensively because they are used or studied to target certain pest species for biological control. However, for most EPN species, host ranges remain enigmatic for two main reasons. First, most EPNs are isolated from soil using *Galleria mellonella* (L.) as a bait insect. This insect is extremely susceptible to infection, and so provides little useful information about natural host affiliations. Second, natural infections are not often found in the field, but see Peters (1996) for a partial list of natural host associations based on the observance of natural infections. Additionally, the vast majority of EPN species have simply not been characterized for their biological traits, such as host range.

Potential hosts can defend against EPN infection using physical barriers, avoidance behaviors, and their immune system. Scarab larvae (Coleoptera: Scarabaeidae) have two examples of physical barriers to infection by pathogens. First, they have an extremely thick peritrophic membrane, through which IJs must penetrate to access the hemocoel via the gut. Second, structures called "sieve plates" cover the spiracles of scarab larvae, effectively acting as screens that block the entry of EPNs through these openings. Wireworms – larvae of beetles in the family Elateridae – have similar physical barriers to infection, in that their spiracles are composed of thin slits through which IJs cannot fit, and their pre-oral cavity contains forward-projecting hairs that are thought to exclude IJs (Eidt and Thurston, 1995) and potentially other pathogens. Larvae of the

Japanese beetle (*Popillia japonica* Newman) have been shown to exhibit two defensive behaviors in response to contact with EPN IJs (Gaugler et al., 1994). When even a single IJ was placed on the cuticle of a larva, the insect engaged in a series of grooming behaviors that would remove the IJ; when the larva was restrained inside a small cage and restricted from grooming, EPN infection rates increased. Japanese beetle larvae were also recorded moving away from areas of soil where large numbers of IJs were introduced (Schroeder et al., 1993).

The immune system of insects also offers protection from EPN infection. The effectiveness of the immune response varies greatly among combinations of hosts and parasites, and is a major factor in shaping the host range of EPN species. Overcoming the immune response is achieved in a number of ways, and both the nematodes and the bacteria play a role. *Steinernema* spp. can kill hosts without their bacteria, partly due to the production of a number of proteases that suppress the reaction of the immune system soon after invasion (summarized in Lewis and Clarke, 2012). *Heterorhabditis* spp. are only virulent in conjunction with their bacterial partners, but they elicit a reduced immune response in hosts when invading without their bacteria, perhaps by reducing the phagocytic capability of the hemocytes (Eleftherianos et al., 2010), although this is not well understood. The symbiotic bacteria play a complementary role in overcoming the host immune system. *Photorhabdus* spp. bacteria are recognized by the host immune system, whereas *Xenorhabdus* spp. actively suppress it. Both *Xenorhabdus* and *Photorhabdus* also produce several toxins that are lethal when injected into hosts.

11.4 Pathogen Population

11.4.1 Pathogenicity and Virulence

Most EPN species are pathogenic (able to infect and cause disease) in a diverse array of hosts spanning several insect orders (Grewal et al., 2005; Shapiro-Ilan et al., 2016a). For example, *S. carpocapsae* has an experimental host range of >200 insects across 10 orders when challenging potential hosts in the laboratory (Poinar, 1979); the nematode's ecological host range (i.e., those species infected in nature), however, may be more limited. There are some nematode species that have narrower host ranges, such as *S. scarabaei*, which is specific to Scarabaeidae (Koppenhöfer and Fuzy, 2003), and *S. scapterisci*, which is specific to Orthoptera (Nguyen and Smart, 1991). Among EPNs that are pathogenic to a particular host, virulence (disease-causing power) can vary greatly among nematode species and strains within species (Grewal et al., 2002; Shapiro-Ilan et al., 2002a, 2011). Due to environmental factors or other influences, the most virulent EPN strains or species determined under laboratory screening may not necessarily translate into the most efficacious under field conditions (Shapiro-Ilan et al., 2012a). For example, *S. feltiae* and *S. riobrave* were both observed to possess high virulence to plum curculio (*Conotrachelus nenuphar* (Herbst)) in the laboratory (Shapiro-Ilan et al., 2002b, 2011), but under field conditions in Georgia, USA only *S. riobrave* applications caused high levels of control because the temperatures were too high for optimal activity by *S. feltiae* (Shapiro-Ilan et al., 2004a). Nonetheless, to maximize biocontrol efficacy, virulence is a critical factor in choosing the most appropriate nematode for the target pest (Shapiro-Ilan et al., 2002a, 2006a).

11.4.2 Persistence and Recycling

There are two mechanisms that will enable EPNs to persist in the environment: longevity and recycling. Longevity refers to the length of time nematodes can persist as IJs outside the host and recycling refers to reproductive capacity as a mechanism to sustain the population in the environment. The innate longevity of EPN species and strains varies (Shapiro-Ilan et al., 2006b). Barring adverse conditions, IJs of species in both EPN genera can live 8 months or more (Kaya and Stock, 1997; Hass et al., 2001), yet in nature the longevity of EPNs will generally be shortened due to biotic or abiotic factors, as described later. Generally, following soil applications, EPN populations remain high enough to provide effective control for 2–8 weeks (Shapiro-Ilan et al., 2016a). Some level of recycling commonly occurs after nematode application, but it is usually not sufficient to achieve multiseason control. However, some exceptions exist; when environmental conditions are conducive (e.g., soil moisture), hosts are abundant, and depending on the nematode species or strain, multiseason or multiyear suppression has been observed (Klein and Georgis, 1992; Shields et al., 1999). Factors that affect the level of recycling include soil type, ground cover, insect host species, and host density (Kaya, 1990; Klein and Georgis, 1992).

11.4.3 Dispersal and Foraging Behavior

Dispersal and foraging behaviors are key components in EPN ecology, as they impact the individual nematode's ability to complete its lifecycle, as well as influencing the spatial distribution of nematodes on a population level. EPN dispersal can be passive or active. Passive dispersal occurs through wind, rain, and human activity. It may also occur through phoresy, such as via mites (Epsky et al., 1988) or earthworms (Shapiro-Ilan and Brown, 2013). EPNs can also be transported by host species; this differs from phoresy in that one of the organisms is harmed. For example, prior to succumbing to an infection, adult *P. japonica* and adult large pine weevils (*Hylobius abietis* L.) (Kruitbos et al., 2009) were reported to transport EPNs. Active EPN dispersal is affected by various factors, including nematode species, moisture (a water film is required for movement), and soil parameters (e.g., sandier soils are generally more conducive to movement) (Kaya, 1990). EPNs navigate in soil by responding to various cues, such as temperature gradients, CO_2, nitrogenous compounds, vibration, and electromagnetic fields (Burman and Pye, 1980; Torr, et al., 2004; Dillman et al., 2012; Ilan et al., 2013).

EPN foraging strategies have been classified as a continuum between ambusher and cruiser types (Lewis et al., 1992; Lewis, 2002). Cruisers, such as *S. glaseri* and *H. bacteriophora*, are highly mobile and actively seek hosts throughout the soil profile. In contrast, ambushers, such as *S. carpocapsae* and *S. scapterisci*, are characterized by relative low motility and a tendency to stay near the soil surface, where they may stand on their tails waiting for a host to pass (Campbell and Gaugler, 1993). An interesting behavior exhibited by standing IJs is jumping, which has been linked to host-finding behavior, since host cues stimulate directional jumping (Campbell and Kaya, 1999). Most nematode species, however, are positioned between these two extremes and are intermediate foragers (e.g., *S. riobrave* and *S. feltiae*) (Lewis, 2002). It has been suggested that cruiser nematodes are more effective at controlling sedentary insect pests below the soil surface and ambushers are more effective against mobile pests at the soil surface. However, there are many instances in which cruisers and ambushers have effectively controlled various insects regardless of host mobility or location in the soil profile (Shapiro-Ilan et al., 2014a). Furthermore, some reports suggest the host-seeking strategy may simply be a factor influenced by the environment (Kruitbos et al, 2010). Nonetheless, vast evidence exists that is consistent with the ambusher/cruiser model (e.g., in comparing dispersal of *S. glaseri* or other cruiser types with *S. carpocapsae*) (Schroeder and Beavers, 1987; Lewis et al., 1992; Grewal et al., 1994a; Lewis, 2002; Wennemann et al., 2004; Lewis and Clarke, 2012), and thus the paradigm remains a useful baseline for understanding EPN foraging.

Foraging IJs experience strong selection pressure to make good infection decisions, because once the host is penetrated and the nematode molts, there is no chance for reversal. Host-finding is followed by host recognition, and EPNs have been shown to respond most strongly behaviorally to insects in which their reproduction is supported (Lewis et al., 1996). Once a host is located and recognized, a mass attack strategy is employed, where several to hundreds of conspecific IJs invade a single host. Fushing et al. (2008) described the structure of host invasion mathematically. The premise was that an uninfected host presents two types of risk to IJs that invade. First, the host immune system is intact and may kill off the first few nematodes that invade the hemocoel. Second, for steinernematids, where at least one member of each sex must enter a host, there is a chance that even if the first nematode invading is successful, they may ultimately fail to reproduce. A number of hypotheses about how risk-sensitivity could shape EPN invasion strategies have been developed based on the idea that a recently invaded host will be a much more valuable resource than a host that is healthy. Indeed, IJs respond more strongly to cues emitted from hosts already infected by conspecific nematodes than to those of uninfected hosts (Grewal et al., 1996), and they opt to invade infected hosts over healthy hosts when presented with a choice between the two (Fushing et al., 2008).

Relatively recent advances in the study of EPN foraging behavior have been made with reference to specialization and group dynamics. Bal et al. (2014) discovered that, although most *S. carpocapsae* display ambusher foraging behavior, a small proportion are specialized "sprinters" that disperse faster and farther than certain cruiser species (e.g., *H. bacteriophora*). The sprinter behavior may be an adaptive strategy evolved to reduce intraspecific competition and facilitate foraging success when proximal hosts are scarce (Bal et al., 2014). Another recently discovered driver of EPN foraging and infection behavior at the population level stems from a tendency of IJs to move and infect in aggregate (i.e., displaying group behavior) (Fushing et al., 2008; Shapiro-Ilan et al., 2014b). This result may also explain why EPN populations so quickly become aggregated in natural settings: conceivably, chemical pheromones induce aggregative behavior akin to what is observed in *C. elegans* (Choe et al., 2012).

11.5 Abiotic Environmental Factors

11.5.1 Soil Moisture

In order to move in the ground, EPNs require soil aggregates to be covered with a film of water thick enough to allow nematode movement. Anoxic conditions and low surface tension in saturated soils are not favorable to EPN movement (Wallace, 1971). Therefore, moisture is a critical factor affecting EPN survival and behavior (e.g., Grant et al., 2003).

In Florida, a redundancy analysis based on 53 citrus groves showed that factors having an impact on soil moisture significantly affected EPN community richness, altering the control of Diaprepes root weevil (*Diaprepes abbreviates* L.) (Stuart et al., 2008; Campos-Herrera et al., 2013).

EPNs have evolved several physiological and behavioral strategies to overcome water stress. Salame et al. (2015) showed that the desiccation-intolerant EPN *H. bacteriophora* moved vertically toward layers of soil where moisture was more suitable, whereas *S. carpocapsae* IJs, which are more tolerant of desiccation, were recovered from dryer layers of soil. In the context of pest management, desiccation tolerance certainly is an asset, especially because several application techniques expose EPNs to dry conditions (e.g., Shapiro-Ilan et al., 2015a). Screening for natural desiccation tolerance (Seenivasan et al., 2014), hybridization (between desiccation tolerant and less tolerant strains), or selective breeding for desiccation tolerance (e.g., Shapiro-Ilan et al., 2005; Nimkingrat et al., 2013) has shown promising results with several species.

11.5.2 Soil Temperature

Temperature is an important factor for the survival, infection, and optimal development of EPNs (Grewal et al., 1994b). Temperature profoundly impacts EPN development (e.g., Kaya, 1977), respiration (e.g., Lindegren et al., 1986), survival and persistence (e.g., Salem et al., 2008; Susurluk, 2013), host-finding (e.g., Byers et al., 1982), and infectivity (e.g., Gouge et al., 1999). In northwestern Europe, the optimal temperature for EPN activity is 20–25 °C (Griffin, 1993). Yet, heat-tolerant nematodes were isolated from arid environments in the Middle East (e.g., Shamseldean et al., 1994; Shapiro-Ilan et al., 1996a) and are useful for biocontrol applications in warmer climates. On the other end of the spectrum, Khatri-Chhetri et al. (2010) found various steinernematids in freezing Himalayan regions in Nepal.

In the context of EPN mass production and biological control, storage temperature impacts EPN effectiveness (e.g., Koppenhöfer et al., 2013). Selective breeding or hybridization can again provide temperature stress-tolerant strains (e.g., Ehlers et al., 2005; Shapiro-Ilan et al., 2005).

11.5.3 Soil Characteristics and Chemistry

Heterorhabditids preferentially occur in sandy soil, whereas steinernematids are generally dominant in soils rich in organic matter (Stuart et al., 2015 and references therein). Soil texture also affects EPNs, particularly the porosity, as it can restrict the movement of several EPN species (Portillo-Aguilar et al., 1999). EPNs tolerate a wide range of soil pH, although strong acidic or alkaline pH reduces their efficacy (Kung et al., 1990; Jaworska, 1993). Salt concentrations also affect EPN behavior (Nielsen et al., 2011); foraging of *S. glaseri*, for instance, is inhibited by high concentrations of NaCl, KCl, and $CaCl_2$ (Thurston et al., 1994a). Certain soil nutrients are associated with EPN abundance (Hoy et al., 2008) and can be manipulated to favor EPN-mediated pest control. EPNs generally show little response to the application of synthetic fertilizers (e.g., Bednarek et al., 1997). Composted manure is beneficial to EPNs (Duncan et al., 2007), whereas fresh manure or high rates of urea adversely impact their persistence and pathogenicity (Shapiro et al., 1996b). Heavy metal buildup from intense fertilizer usage can also negatively impact EPN populations (Sun et al., 2016). Interactions between

EPNs and chemical pesticides may be synergistic, additive, or antagonistic depending on the nematode and host species, specific chemical components, and concentrations (Koppenhöfer and Grewal, 2005); for example, the neonicotinoid insecticide imidacloprid has a synergistic interaction with EPNs when targeting white grub scarab larvae (Koppenhöfer et al., 2000).

11.5.4 Ultraviolet Light

In the context of pest control, ultraviolet (UV) radiation is a major concern when applying nematodes below- and aboveground. UV radiation and sunlight rapidly reduce EPN pathogenicity (e.g., Gaugler et al., 1992). The impact of UV varies across EPN strains and species (Shapiro-Ilan et al., 2015b). Application aboveground is particularly problematic, but UV-absorbent gels can protect EPNs from detrimental effects while providing them with appropriate levels of moisture (e.g., Shapiro-Ilan et al., 2016b). Developments in protection against UV radiation and desiccation will open new opportunities for the application of these biological control agents (Hiltpold, 2015; Shapiro-Ilan et al., 2015a).

11.6 Biotic Interactions

11.6.1 Interactions with Predators and Pathogens, Including Intraguild Competition

In soil, EPNs compete with fungi, bacteria, and other nematode species, including other EPNs, for the same resources. In these competitive interactions, the symbiotic bacteria *Photorhabdus* spp. and *Xenorhabdus* spp. play protective roles against several microorganisms. These bacteria secrete secondary metabolites, which can act as antifungal, antibacterial, nematicidal, insecticidal, and/or repellent compounds (Lewis et al., 2015 and references therein). Symbiotic bacteria also play a role in preventing other EPNs from attacking previously infected hosts (e.g., Christen et al., 2007), thereby reducing competition for food resources. As with interactions with chemical insecticides, various interactions (e.g., antagonism, additivity, synergy) can occur between EPNs and other entomopathogens or nonmicrobial biocontrol agents (Shapiro-Ilan et al., 2016a). Examples of entomopathogens having antagonistic interactions with EPNs include *Beauveria bassiana* (Barbercheck and Kaya, 1990; Brinkman and Gardner, 2000) and *Isaria fumosorosea* (Shapiro-Ilan et al., 2004b). In contrast, synergistic interactions have been reported with *Paenibacillus popilliae* (Thurston et al., 1994b), *Bacillus thuringiensis* (Koppenhöfer and Kaya, 1997), and the entomopathogenic fungus *Metarhizium anisopliae* (Ansari et al., 2010).

At higher trophic levels, parasitoids and the ubiquity of predatory invertebrates in soil certainly impact EPN occurrence and persistence. Noncompetitive and competitive interactions between parasitic Hymenoptera and EPNs have been observed (Lacey et al., 2003; Everard et al., 2009). Interestingly, Mbata and Shapiro-Ilan (2010) reported that IJs preferentially infected parasitized host larvae of the Indian meal-moth (*Plodia interpunctella* (Hübner)) over nonparasitized larvae, yet parasitoids did not choose differently between EPN-infected and non-infected hosts. Predators have a negative impact on EPNs; for instance, under laboratory conditions, adult female mites (*Sancassania polyphyllae* Zachvatkin) can consume more than 80% of *S. feltiae*

IJs within 24 hours (Karagoz et al., 2007), potentially interfering with pest control (Ekmen et al., 2010). Within the predatory guild, ants scavenge on EPN-infected cadavers within a few days of EPN infection (Gulcu et al., 2012; Ulug et al., 2014). Inter- and intraspecific competitive interactions between EPNs also occur inside the host; male *Steinernema* spp. coil around one another in the host hemolymph, often causing nematode death (O'Callaghan et al., 2014; Zenner et al., 2014).

11.6.2 Cues Used in Host-Finding and Navigation

In the soil darkness, EPNs use various chemical and physical cues to locate and navigate toward their hosts. EPNs respond to insect-emitted volatiles that trigger various foraging behaviors (e.g., Dillman et al., 2012). EPNs also rely on physical cues to locate their host. Torr et al. (2004) demonstrated that EPNs move toward artificial vibrations in the ground, potentially showing that they respond to vibration generated by feeding or moving insect hosts. EPNs also navigate along electrical fields similar to those emitted by living insects (e.g., Shapiro-Ilan et al., 2012b; Ilan et al., 2013).

11.6.3 Tri-trophic Interactions (Plant, Insect, Nematode)

Not only do nematodes respond to host cues, but they also react to plant signals. Insect damage on roots alters the typical volatile blends emitted by healthy roots, resulting in attraction of EPNs toward their hosts feeding on roots (Boff et al., 2001; Van Tol et al., 2001; Rasmann et al., 2005; Ali et al., 2010; Hiltpold et al., 2010b; Rasmann et al., 2011; Turlings et al., 2012; Tonelli et al., 2016). These volatiles diffuse well in the ground (Hiltpold et al., 2008) and are detected as far as 10 m from the source (Ali et al., 2012). Certain root exudates improve EPN infectiousness while hampering plant parasitic nematodes (Hubbard et al., 2005; Hiltpold et al., 2015). In addition to insect-induced root volatiles impacting EPN behavior, the roots themselves influence EPN foraging, as the nematodes exploit roots to facilitate their movement in the ground (Demarta et al., 2014).

Selective breeding of EPNs (Hiltpold et al., 2010a) or genetically engineered volatile emissions (Degenhardt et al., 2009) improves pest control efficacy, which can result in pest-control efficacy similar to that of chemical insecticides (e.g., Hiltpold et al., 2010b).

11.7 Applied Ecology and Aspects in Microbial Control

11.7.1 Production, Formulation, and Application

EPNs are cultured using *in vivo* or *in vitro* methods (solid or liquid fermentation) (Shapiro-Ilan et al., 2012c, 2014a). *In vivo* production is used in small-scale rearing and in commercial production (e.g., by several companies in North America). The process has low start-up costs, but is limited by costs of labor and insect hosts, as well as a low economy of scale (Shapiro-Ilan et al., 2012c, 2014a). Hence, *in vivo* production efficiency can be enhanced by automating the process from insect rearing to nematode harvest (Shapiro-Ilan et al., 2014a). The majority of commercial EPN production is accomplished via *in vitro* liquid fermentation, which is generally deemed to be the most efficient approach (Shapiro-Ilan et al., 2014a). *In vitro* solid fermentation is considered intermediate between *in vivo* and liquid culture in terms of start-up costs, labor requirements, and efficiency of production. Some studies have indicated lower quality or efficacy in EPNs produced in liquid culture relative to *in vivo*-produced products, whereas in other cases no differences were detected as a result of culture method. Yields and quality in fermentation are affected by bioreactor parameters and media components.

Various EPN formulations may be used for aqueous application, including activated charcoal, alginate and polyacrylamide gels, clay, diatomaceous earth, paste, peat, polyurethane sponge, vermiculite, and water-dispersible granules (WDGs) (Shapiro-Ilan et al., 2012c). Most formulations are geared toward providing stability and ease of use. Formulations or adjuvants can also provide protection from UV or desiccation (e.g., the recent use of fire gels and polymers) (Schroer and Ehlers, 2005; Shapiro-Ilan and Dolinski, 2015; Shapiro-Ilan et al., 2015a, 2016a,b).

EPNs can be applied for biocontrol purposes using most standard agricultural equipment, including various sprayers and irrigation systems (Shapiro-Ilan et al., 2012c; Shapiro-Ilan and Dolinski, 2015). The choice of equipment can impact EPN efficacy and may vary according to the needs of the cropping system (Fife et al., 2005; Lacey et al., 2006a; Shapiro-Ilan et al. 2006a, 2012c). Nonetheless, a number of studies did not detect differences in efficacy when comparing application equipment (Nilsson and Gripwall, 1999; Shapiro-Ilan et al., 2015a). Although EPNs are usually applied in aqueous suspension, alternative approaches exist, such as baits, capsules, and infected host cadavers. EPNs applied directly to the target site in their infected hosts (the "cadaver method") mimic conditions of natural emergence in soil. The cadaver approach is advantageous to the *in vivo* EPN producer because process costs are reduced; additionally, EPNs applied in the cadaver approach have been reported to be more dispersive, infective, and efficacious than those applied in aqueous suspension (Shapiro-Ilan et al., 2014a; Dolinski et al., 2015). Hiltpold et al. (2012) developed another way to apply EPNs directly to the target site: they encapsulated the EPN in capsules applied in the ground. This method offers the advantage of luring the insect target to the capsule by coating it with chemical attractants and feeding stimulants (Hiltpold et al., 2012, 2016), although it is yet to be refined (Hiltpold, 2015).

11.7.2 Approaches to Microbial Control

The reader is referred to Chapter 13 for full definitions and descriptions of the four approaches to microbial control: inundative, classical, inoculative, and conservation/environmental manipulation. As a general rule, EPNs are used in microbial control in an inundative fashion; that is, with little or no expectation of recycling and with an understanding that reapplication will be required annually or seasonally (Shapiro-Ilan et al. 2012a; Shapiro-Ilan and Dolinski, 2015). There are numerous examples of EPNs being used effectively in inundative applications, such as against black vine weevil (*Otiorhynchus sulcatus* (F.)), borers (*Synanthedon* spp.), Diaprepes root weevil (*D. abbreviatus*), fungus gnats (Diptera: Sciaridae), western flower thrips (*Frankliniella occidentalis* (Pergande)), white grubs (Coleoptera: Scarabaeidae), and other insects (Shapiro-Ilan et al., 2014a, 2016a) (see Table 11.3).

There are some exceptions (albeit relative few) in which EPNs have been utilized in other microbial control approaches, including classical, inoculative, and conservation methods. A prime example of classical biocontrol (introduction) using EPNs is the

Table 11.3 Major pests targeted commercially with entomopathogenic nematodes (EPNs) in an inundative approach.

Common name[a]	Scientific name	Order: family	Key crop(s) targeted	Primary EPNs used
Armyworms	various	Lepidoptera: Noctuidae	Vegetables	Sc, Sf, Sr
Artichoke plume moth	*Platyptilia carduidactyla* (Riley)	Lepidoptera: Pterophoridae	Artichoke	Sc
Banana moth	*Opogona sacchari* Bojer	Lepidoptera: Tineidae	Ornamentals	Hb, Sc
Banana root borer	*Cosmopolites sordidus* (Gemar)	Coleoptera: Curculionidae	Banana	Sc, Sf, Sg
Billbug	*Sphenophorus* spp.	Coleoptera: Curculionidae	Turf	Hb, Sc
Black cutworm	*Agrotis ipsilon* (Hufnagel)	Lepidoptera: Noctuidae	Turf, vegetables	Sc
Black vine weevil	*Otiorhynchus sulcatus* (F.)	Coleoptera: Curculionidae	Berries, ornamentals	Hb, Hd, Hm, Hmeg, Sc, Sg
Borers (e.g., peachtree borer)	*Synanthedon* spp. and other sesiids	Lepidoptera: Sesiidae	Fruit trees and ornamentals	Hb, Sc, Sf
Citrus root weevil	*Pachnaeus* spp.	Coleoptera: Curculionidae	Citrus, ornamentals	Sr, Hb
Codling moth	*Cydia pomonella* (L.)	Lepidoptera: Tortricidae	Pome fruit	Sc, Sf
Corn earworm	*Helicoverpa zea* Boddie	Lepidoptera: Noctuidae	Vegetables	Sc, Sf, Sr
Corn rootworm	*Diabrotica* spp.	Coleoptera: Chrysomelidae	Vegetables	Hb, Sc
Cranberry girdler	*Chrysoteuchia topiaria* (Zeller)	Lepidoptera: Crambidae	Cranberries	Sc
Crane fly	various	Diptera: Tipulidae	Turf	Sc
Diaprepes root weevil	*Diaprepes abbreviatus* (L.)	Coleoptera: Curculionidae	Citrus, ornamentals	Hb, Sr
Fungus gnats	various	Diptera: Sciaridae	Mushrooms, greenhouse	Sf, Hb
Iris borer	*Macronoctua onusta* Grote	Lepidoptera: Noctuidae	Iris	Hb, Sc
Large pine weevil	*Hylobius abietis* (L.)	Coleoptera: Curculionidae	Forest plantings	Hd, Sc
Leafminers	*Liriomyza* spp.	Diptera: Agromyzidae	Vegetables, ornamentals	Sc, Sf
Navel orangeworm	*Amyelois transitella* (Walker)	Lepidoptera: Pyralidae	Nut and fruit trees	Sc

(Continued)

Table 11.3 (Continued)

Common name[a]	Scientific name	Order: family	Key crop(s) targeted	Primary EPNs used
Plum curculio	*Conotrachelus nenuphar* (Herbst)	Coleoptera: Curculionidae	Fruit trees	Sr, Sf
Scarab grubs	Various	Coleoptera: Scarabaeidae	Turf, ornamentals	Hb, Sc, Sg, Hz
Shore flies	*Scatella* spp.	Diptera: Ephydridae	Ornamentals	Sc, Sf
Strawberry root weevil	*Otiorhynchus ovatus* (L.)	Coleoptera: Curculionidae	Berries	Hm, Sc
Sweetpotato weevil	*Cylas formicarius* (Summers)	Coleoptera: Curculionidae	Sweet potato	Hb, Sc, Sf
Western flower thrips	*Frankliniella occidentalis* (Pergande)	Thysanoptera: Thripidae	Greenhouse, flowers	Sc, Sf

a) At least one scientific paper reported ≥75% suppression of these pests in the field; the table is not meant to be an exhaustive list. Hb, *Heterorhabditis bacteriophora*; Hd, *H. downesi*; Hm, *H. marelata*; Hmeg, *H. megidis*; Hz, *H. zealandica*; Sc, *Steinernema carpocapsae*; Sf, *S. feltiae*; Sg, *S. glaseri*; Sk, *S. kushidai*; Sr, *S. riobrave*.

successful release and establishment of *S. scapterisci*, which was imported from Uruguay for control of exotic mole crickets (*Scapteriscus* spp.) (Parkman et al., 1994; Frank 2009). Klein and Georgis (1992) reported an example of inoculative control against *P. japonica*, in which *H. bacteriophora* provided 99% mortality 1 year after application; the exceptional level of persistence was attributed to a high host density and conducive environmental conditions. Conservation of EPNs can be enhanced through manipulation of the soil environment, such as by the addition of mulches or crop residues that provide protection from UV or desiccation (Shapiro et al., 1999; Lacey et al., 2006b). In an innovative example of conservation, citrus trees were planted in islands of imported soil that was more conducive to EPN control than endemic soils; as a result, 68% more adult *D. abbreviatus* weevils were captured in traps in native soil plots relative to the imported sandy soil islands, which favored EPN persistence and efficacy (Duncan et al., 2013).

11.8 Conclusion

EPNs are best known as commercially successful biocontrol agents used to control a wide variety of economically important pests. Their success in biocontrol is based on various factors, including major advances in production technology, the discovery of efficacious EPN strains and species, and huge efforts to characterize field efficacy against numerous pests. Nonetheless, the market share that EPNs represent relative to the overall pesticide industry remains small, primarily due to cost competitiveness. Biocontrol using EPNs can be expanded through further advances in strain enhancement and discovery, formulation and application technology, and environmental manipulation (Shapiro-Ilan et al., 2012a,c).

In addition to their utility in biocontrol, EPNs can play an important role in natural population regulation, and their life histories and population dynamics may serve as model systems for ecological studies. Because they are simply and inexpensively maintained in laboratory settings, EPNs can also be valuable tools for demonstrating ecological and behavioral principles in classroom settings. EPNs are unique among entomopathogen groups in terms of their dispersal capability and propensity to distribute themselves in the environment. New realms of ecological study that have been recently pioneered and are poised to be expanded include tri-trophic interactions (plant–insect–nematode) and group behavior. Advances in understanding EPN ecology, such as infection and foraging behavior, will also contribute to improving biocontrol efforts. Further, EPNs are an especially interesting group to research due to their relatedness to the highly studied but nonpathogenic free-living nematode *C. elegans*. Their relatedness allows for a multitude of comparisons on molecular and behavioral levels, addressing the interrelationships of pathogenic and nonpathogenic species in an ecosystem. In addition, available EPN species are (i) representative of a broader group of organisms, (ii) amenable to experimental manipulation, and (iii) available at reasonable cost for research purposes. All of these aspects make EPNs ideal model organisms to disentangle belowground interactions at various trophic levels and to provide a better understanding of the complexity of soil-dwelling invertebrate ecology.

References

Ali, J. G., Alborn, H.T., Stelinski, L.L., 2010. Subterranean herbivore-induced volatiles released by citrus roots upon feeding by *Diaprepes abbreviatus* recruit entomopathogenic nematodes. J. Chem. Ecol. 36, 361–368.

Ali, J.G., Alborn, H.T., Campos-Herrera, R., Kaplan, F., Duncan, L.W., Rodriguez-Saona, C., Koppenhöfer, A.M., Stelinski, L.L., 2012. Subterranean, herbivore-induced plant volatile increases biological control activity of multiple beneficial nematode species in distinct habitats. PLoS ONE 7, e38146.

Ansari, M.A., Shah, F.A., Butt, T.M., 2010. The entomopathogenic nematode *Steinernema kraussei* and *Metarhizium anisopliae* work synergistically in controlling overwintering larvae of the black vine weevil, *Otiorhynchus sulcatus*, in strawberry growbags. Biocontrol Sci. Technol. 20, 99–105.

Bal, H.K., Taylor, R.A., Grewal, P.S., 2014. Ambush foraging entomopathogenic nematodes employ 'sprinters' for long-distance dispersal in the absence of hosts. J. Parasitol. 100, 422–432.

Barbercheck, M.E., Kaya, H.K., 1990. Interactions between *Beauveria bassiana* and the entomogenous nematodes *Steinernema feltiae* and *Heterorhabditis heliothidis*. J. Invertebr. Pathol. 55, 225–234.

Bedding, R.A. 2009. Controlling the pine-killing woodwasp, *Sirex noctilio*, with nematodes, in: Hajek, A.E., Glare, T.R., O'Callaghan, M. (eds.) Use of Microbes for Control and Eradication of Invasive Arthropods. Springer, Dordrecht, pp. 213–235.

Bedding, R. A., Akhurst, R.J., 1978. Geographical distribution and host preferences of *Deladenus* species (Nematoda: Neotylenchidae) parasitic in siricid woodwasps and associated hymenopterous parasitoids. Nematologica 24, 286–294.

Bednarek, A., Gaugler, R., 1997. Compatibility of soil amendments with entomopathogenic nematodes. J Nematol. 29, 220–227.

Blaxter, M.L., 2011. Nematodes: the worm and its relatives. PLoS Biol. 9, e1001050.

Boff, M.I.C., Zoon, F.C., Smits, P.H., 2001. Orientation of *Heterorhabditis megidis* to insect hosts and plant roots in a Y-tube sand olfactometer. Entomol. Experimental. Applic. 98, 329–337.

Brinkman, M.A., Gardner, W.A., 2000. Possible antagonistic activity of two entomopathogens infecting workers of the red imported fire ant (Hymenoptera: Formicidae). J. Entomol. Sci. 35, 205–207.

Burman, M., Pye, A.E., 1980. *Neoaplectana carpocapsae*: movements of nematode populations on a thermal gradient. Exp. Parasitol. 49, 258–265.

Byers, J.A., Poinar, G.O. Jr., 1982. Location of insect hosts by the nematode, *Neoaplectana carpocapsae*, in response to temperature. Behaviour 79, 1–10.

Campbell, J.F., Gaugler, R., 1993. Nictation behavior and its ecological implications in the host search strategies of entomopathogenic nematodes (Heterorhabditidae and Steinernematidae). Behaviour 126, 155–170.

Campbell, J.F., Kaya, H.K., 1999. How and why a parasitic nematode jumps. Nature 397, 485–486.

Campbell, J.F., Lewis, E.E., Yoder, F., Gaugler, R., 1996. Spatial and temporal distribution of entomopathogenic nematodes in turf. Parasitol. 113, 473–482.

Campos-Herrera, R., Pathak, E., El-Borai, F.E., Stuart, R.J., Gutiérrez, C., Rodríguez-Martín, J.A., et al., 2013. Geospatial patterns of soil properties and the biological control potential of entomopathogenic nematodes in Florida citrus groves. Soil Biol.Biochem. 66, 163–174.

Ciche, T.A., Ensign, J.C., 2003. For the insect pathogen *Photorhabdus luminescens*, which end of the nematode is out? App. Environ. Microbiol. 69. 1890–1897.

Ciche, T.A., Kim, K., Kaufmann-Daszczuk, B., Nguyen, K.C.Q., Hall, D.H., 2008. Cell invasion and matricide during *Photorhabdus luminescens* transmission by *Heterorhabditis bacteriophora* nematodes. App. Environ. Microbiol. 74, 2275–2287.

Christen, J.M., Campbell, J.F., Lewis, E.E., Shapiro-Ilan, D.I., Ramaswamy, S.B., 2007. Responses of the entomopathogenic nematode, *Steinernema riobrave* to its insect hosts, *Galleria mellonella* and *Tenebrio molitor*. Parasitol. 134, 889–898.

Choe, A., Von Reuss, S.H., Kogan, D., Gasser, R.B., Platzer, E.G., Schroeder, F.C., Sternberg, P.W., 2012. Ascaroside signaling is widely conserved among nematodes. Curr. Biol. 22, 772–780.

Degenhardt, J., Hiltpold, I., Köllner, T.G., Frey, M., Gierl, A., Gershenzon, J., et al., 2009. Restoring a maize root signal that attracts insect-killing nematodes to control a major pest. Proc Natl. Acad. Sci. U.S.A. 106, 13 213–13 218.

Demarta, L., Hibbard, B.E., Bohn, M.O., Hiltpold, I., 2014. The role of root architecture in foraging behavior of entomopathogenic nematodes. J. Invertebr. Pathol. 122, 32–39.

Dolinski, C., Shapiro-Ilan, D.I., Lewis, E.E., 2015. Insect cadaver applications: pros and cons, in: Campos-Herrera, R. (ed.), Nematode Pathogenesis of Insects and Other Pests: Ecology and Applied Technologies for Sustainable Plant and Crop Protection. Springer, Cham, pp. 207–230.

Dillman, A.R., Guillermin, M.L., Lee, J.H., Kim, B., Sternberg, P.W., Hallem, E.A., 2012. Olfaction shapes host–parasite interactions in parasitic nematodes. Proc Natl. Acad. Sci. U.S.A. 109(35), E2324–2333.

Duncan, L.W., Graham, J.H., Zellers, J., Bright, D., Dunn, D.C., El-Borai, F.E., Porazinska, D.L., 2007. Food web responses to augmenting the entomopathogenic nematodes in bare and animal manure-mulched soil. J. Nematol. 39, 176–189.

Duncan, L.W., Stuart, R.J., El-Borai, F.E., Campos-Herrera, R., Pathak, E., Giurcanu, M., Graham, J.H., 2013. Modifying orchard planting sites conserves entomopathogenic nematodes, reduces weevil herbivory and increases citrus tree growth, survival and fruit yield. Biol. Control 64, 26–36.

Ehlers, R. U., Oestergaard, J., Hollmer, S., Wingen, M., Strauch, O., 2005. Genetic selection for heat tolerance and low temperature activity of the entomopathogenic nematode-bacterium complex *Heterorhabditis bacteriophora-Photorhabdus luminescens*. Biocontrol. 50, 699–716.

Eidt, D.C, Thurston, G.S., 1995. Physical deterrents to infection by entomopathogenic nematodes in wireworms (Coleoptera: Elateridae) and other soil insects. Canadian Entomologist. 127, 423–429.

Ekmen, Z. I., Hazir, S., Cakmak, I., Ozer, N., Karagoz, M., Kaya, H. K., 2010. Potential negative effects on biological control by *Sancassania polyphyllae* (Acari: Acaridae) on an entomopathogenic nematode species. Biol. Control. 54, 166–171.

Eleftherianos, I., Joyce, S., Ffrench-Constant, R. H., Clarke, D. J., Reynolds, S. E., 2010. Probing the tri-trophic interaction between insects, nematodes and *Photorhabdus*. Parasitology. 137, 1695–1706.

Epsky, N.D., Walter, D.E., Capinera, J.L., 1988. Potential role of nematophagous microarthropods as biotic mortality factors of entomogenous nematodes (Rhabditida: Steinernematidae, Heterorhabditidae). J. Econ. Entomol. 81, 821–825.

Everard, A., Griffin, C.T., Dillon, A.B., 2009. Competition and intraguild predation between the braconid parasitoid *Bracon hylobii* and the entomopathogenic nematode *Heterorhabditis downesi*, natural enemies of the large pine weevil, *Hylobius abietis*. Bull. Entomol. Res. 99, 151–161.

Fife, J.P., Ozkan, H.E., Derksen, R.C., Grewal, P.S., Krause, C.R., 2005. Viability of a biological pest control agent through hydraulic nozzles. Transactions ASAE 48, 45–54.

Frank, H.J., 2009. *Steinernema scapterisci* as a biological control agent of *Scapteriscus* mole crickets, in: Hajek, A.E, Glare, T.R., O'Callaghan, M. (eds.), Use of Microbes for Control and Eradication of Invasive. Springer, Cham, pp. 115–132.

Fushing, H., Zhu, L., Shapiro-Ilan, D.I., Campbell, J.F., Lewis, E.E., 2008. State-space based mass event-history model. I: Many decision making agents with one target. Ann. Appl. Stat. 2, 1503–1522.

Gaugler, R., Wang, Y., Campbell, J., 1994. Aggressive and evasive behaviors in *Popillia japonica* (Coleoptera: Scarabaeidae) larvae: Defenses against entomopathogenic nematode attack. J. Invertebr. Pathol. 64, 193–199.

Gaugler, R., Bednarek, A., Campbell, J. F., 1992. Ultraviolet inactivation of Heterorhabditid and Steinernematid nematodes. J. Invertebr. Pathol. 59, 155–160.

Goodrich-Blair, H., Clarke, D. J. 2007. Mutualism and pathogenesis in *Xenorhabdus* and *Photorhabdus*: two roads to the same destination. Mol. Microbiol. 64. 260–268.

Gouge, D.H., Lee, L.L., Henneberry, T.J., 1999. Effect of temperature and Lepidopteran host species on entomopathogenic nematode (Nematoda: Steinernematidae, Heterorhabditidae) infection. Environ. Entomol. 28, 876–883.

Grant, J.A., Villani, M.G., 2003. Soil moisture effects on entomopathogenic nematodes. Environ. Entomol. 32, 80–87.

Grewal, P.S., Lewis, E.E., Gaugler, R., Campbell, J.F., 1994a. Host finding behaviour as a predictor of foraging strategy in entomopathogenic nematodes. Parasitol. 108, 207–215.

Grewal, P.S., Selvan, S., Gaugler, R., 1994b. Thermal adaptation of entomopathogenic nematodes-niche breadth for infection, establishment and reproduction. J. Therm. Biol. 19, 245–253.

Grewal, P., Lewis, E.E., Gaugler, R., 1996. Response of infective stage parasites (Rhabditida: Steinernematidae) to volatile cues from infected hosts. J. Chem. Ecol. 23, 503–515.

Grewal, P.S., Grewal, S.K., Malik, V.S., Klein, M.G., 2002. Differences in susceptibility of introduced and native white grub species to entomopathogenic nematodes from various geographic localities. Biol. Control 24, 230–237.

Grewal, P.S., Ehlers, R-U, Shapiro-Ilan, D.I. (eds.), 2005. Nematodes as Biological Control Agents, CABI, Wallingford.

Griffin, C.T., Chaerani, R., Fallon, D., Reid, A. P., Downes, M.J., 2000. Occurrence and distribution of entomopathogenic nematodes *Steinernema* spp. and *Heterorhabditis indica* in Indonesia. J. Helminthol. 74, 143–150.

Griffin, C.T., 1993. Temperature responses of entomopathogenic nematodes: implications for the success of biological control programmes, in: Bedding, R.A., Akhurst, R., Kaya, H.K. (eds.), Nematodes and the Biological Control of Insect Pests. CSIRO Publishing, Clayton, pp. 115–126.

Gulcu, B., Hazir, S., Kaya, H. K., 2012. Scavenger deterrent factor (SDF) from symbiotic bacteria of entomopathogenic nematodes. J. Invertebr. Pathol. 110, 326–333.

Hass, B., Downes, M.J., Griffin, C.T., 2001. Correlation between survival in water and persistence of infectivity in soil of *Heterorhabditis* spp. isolates. Nematol. 3, 573–579.

Hiltpold, I., 2015. Prospects in the application technology and formulation of entomopathogenic nematodes for biological control of insect pests, in: Campos-Herrera, R. (ed.), Nematode Pathogenesis of Insects and Other Pests: Ecology and Applied Technologies for Sustainable Plant and Crop Protection. Springer, Cham, pp. 187–205.

Hiltpold, I., Hibbard, B.E., 2016. Neonate larvae of the specialist herbivore *Diabrotica virgifera virgifera* do not exploit the defensive volatile (*E*)-β-caryophyllene in locating maize roots. J. Pest Sci. 80(4), 853–858.

Hiltpold, I., Turlings, T.C.J., 2008. Belowground chemical signalling in maize: when simplicity rhymes with efficiency. J. Chem. Ecol. 34, 628–635.

Hiltpold, I., Baroni, M., Toepfer, S., Kuhlmann, U., Turlings, T.C.J., 2010a. Selection of entomopathogenic nematodes for enhanced responsiveness to a volatile root signal helps to control a major root pest. J. Exp. Biol. 213, 2417–2423.

Hiltpold, I., Toepfer, S., Kuhlmann, U., Turlings, T.C.J., 2010b. How maize root volatiles influence the efficacy of entomopathogenic nematodes against the western corn rootworm? Chemoecology. 20, 155–162.

Hiltpold, I., Hibbard, B.E., French, B.W., Turlings, T.C.J., 2012. Capsules containing entomopathogenic nematodes as a Trojan horse approach to control the western corn rootworm. Plant and Soil. 358, 11–25.

Hiltpold, I., Jaffuel, G., Turlings, T.C.J., 2015. The dual effects of root-cap exudates on nematodes: from quiescence in plant-parasitic nematodes to frenzy in entomopathogenic nematodes. J. Exp. Bot. 66, 603–611.

Hoy, C.W., Grewal, P.S., Lawrence, J.L., Jagdale, G., Acosta, N., 2008. Canonical correspondence analysis demonstrates unique soil conditions for entomopathogenic nematode species compared with other free-living nematode species. Biol. Control. 46, 371–379.

Hubbard, J.E., Flores-Lara, Y., Schmitt, M., McClure, M.A., Stock, S.P., Hawes, M.C., 2005. Increased penetration of host roots by nematodes after recovery from quiescence induced by root cap exudate. Nematol. 7, 321–331.

Ilan, T., Kim-Shapiro, D.B., Bock, C., Shapiro-Ilan, D.I., 2013. The impact of magnetic fields, electric fields and current on the directional movement of *Steinernema carpocapsae*. Int. J. Parasitol. 43, 781–784.

Kalia, V., Sharma, G., Shapiro-Ilan, D.I., Ganguly, S. 2014. Biocontrol potential of *Steinernema thermophilum* and its symbiont *Xenorhabdus indica* against lepidopteran pests: virulence to egg and larval stages. J. Nematol. 46, 18–26.

Jaworska, M., 1993. Investigations on the possibility of using entomophilic nematodes in reduction of *Cephalcia abietis* (L.) (Hym., Pamhillidae) population. Polish J. Entomol. 62, 201–212.

Karagoz, M., Gulcu, B., Cakmak, I., Kaya, H.K., Hazir, S., 2007. Predation of entomopathogenic nematodes by *Sancassania* sp. (Acari: Acaridae). Exp. Appl. Acarol. 43, 85–95.

Kaspi, R., Ross, A., Hodson, A., Stevens, G., Kaya, H., Lewis, E.E. 2010. Foraging efficacy of the entomopathogenic nematode *Steinernema riobrave* in different soil types from California citrus groves. Appl. Soil Ecol. 45, 243–253.

Kaya, H.K., 1977. Development of the DD-136 strain of *Neoaplectana carpocapsae* at constant temperatures. J. Nematol. 9, 346–349.

Kaya, HK., 1990. Soil ecology, in: Gaugler, R., Kaya, H.K. (eds.), Entomopathogenic Nematodes in Biological Control. CRC Press. Boca Raton, FL, pp. 93–116.

Kaya, H.K., Stock, S.P., 1997. Techniques in insect nematology, in: Lacey, L.A. (ed.), Manual of Techniques in Insect Pathology. Academic Press, San Diego, CA, pp. 281–324.

Khatri-Chhetri, H.B., Waeyenberge, L., Manandhar, H.K., Moens, M., 2010. Natural occurrence and distribution of entomopathogenic nematodes (Steinernematidae and Heterorhabditidae) in Nepal. J. Invertebr. Pathol. 103, 74–78.

Klein, M.G., Georgis, R., 1992. Persistence of control of Japanese beetle (Coleoptera: Scarabaeidae) larvae with steinernematid and heterorhabditid nematodes. J. Econ. Entomol. 85, 727–730.

Koppenhöfer, A.M., Fuzy, E.M., 2003. Ecological characterization of *Steinernema scarabaei*, a scarab-adapted entomopathogenic nematode from New Jersey. J. Invertebr. Pathol. 83, 139–148.

Koppenhöfer, A.M., Grewal, P.S., 2005. Compatibility and interactions with agrochemicals and other biocontrol agents, in: Grewal, P.S., Ehlers, R.-U., Shapiro-Ilan, D.I. (eds.), Nematodes as Biological Control Agents. CABI, Wallingford, pp. 363–381.

Koppenhöfer, A.M., Kaya, H.K., 1997. Additive and synergistic interactions between entomopathogenic nematodes and *Bacillus thuringiensis* for scarab grub control. Biol. Control 8, 131–137.

Koppenhöfer, A.M., Grewal, P.S., Kaya, H.K., 2000. Synergism of imidacloprid and entomopathogenic nematodes against white grubs: the mechanism. Entomol. Exp. Appl. 94, 283–293.

Koppenhöfer, A.M., Ebssa, L., Fuzy, E.M., 2013. Storage temperature and duration affect *Steinernema scarabaei* dispersal and attraction, virulence, and infectivity to a white grub host. J. Invertebr. Pathol. 112, 129–137.

Kruitbos, L.M., Heritage, S., Wilson, M.J., 2009. Phoretic dispersal of entomopathogenic nematodes by *Hylobius abietis*. Nematology 11, 419–427.

Kruitbos, L.M., Heritage, S., Hapca, S., Wilson, M.J., 2010. The influence of habitat quality on the foraging strategies of the entomopathogenic nematodes *Steinernema carpocapsae* and *Heterorhabditis megidis*. Parasitology 137, 303–309.

Kung, S.-P., Gaugler, R., Kaya, H. K., 1990. Influence of soil pH and oxygen on persistence of *Steinernema* spp. J. Nematol. 22, 440–445.

Lacey, L.A., Unruh, T.R., Headrick, H.L., 2003. Interactions of two parasitoids (Hymenoptera: Ichneumonidae) of codling moth (Lepidoptera: Tortricidae) with the entomopathogenic nematode *Steinernema carpocapsae* (Rhabditida: Steinernematidae). J. Invertebr. Pathol. 83, 230–239.

Lacey, L.A., Arthurs, S.P., Unruh, T.R., Headrick, H., Fritts, R. Jr., 2006a. Entomopathogenic nematodes for control of codling moth (Lepidoptera: Tortricidae) in apple and pear orchards: effect of nematode species and seasonal temperatures, adjuvants, application equipment and post-application irrigation. Biol. Control 37, 214–223.

Lacey, L.A., Arthurs, S.P., Granatstein, D., Headrick, H., Fritts, R. Jr., 2006b. Use of entomopathogenic nematodes (Steinernematidae) in conjunction with mulches for control of codling moth (Lepidoptera: Tortricidae). J. Entomol. Sci. 41, 107–119.

Lewis, E.E., 2002. Behavioral ecology, in: Gaugler, R. (ed.), Entomopathogenic Nematology. CABI, Wallingford, pp. 205–224.

Lewis, E.E., Clarke, D.J., 2012. Nematode parasites and entomopathogens, in: Vega, F.E., Kaya, H.K. (eds.), Insect Pathology, 2nd edn. Elsevier, Amsterdam, pp. 395–424.

Lewis, E.E., Gaugler, R., Harrison, R., 1992. Entomopathogenic nematode host finding: response to host contact cues by cruise and ambush foragers. Parasitol. 105, 309–319.

Lewis, E. E., Selvan, S., Campbell, J.F., Gaugler, R., 1995. Changes in foraging behaviour during the infective stage of entomopathogenic nematodes. Parasitol. 110, 583–590.

Lewis, E. E., Ricci, M., Gaugler, R., 1996. Host recognition behavior reflects host suitability for the entomopathogenic nematode, *Steinernema carpocapsae*. Parasitol. 113, 573–579.

Lewis, E.E., Hazir, S., Hodson, A., Gulcu, B., 2015. Trophic relationships of entomopathogenic nematodes in agricultural habitats, in: Campos-Herrera, R. (ed.), Nematode Pathogenesis of Insects and Other Pests: Ecology and Applied Technologies for Sustainable Plant and Crop Protection. Springer, Cham, pp. 139–163.

Lindegren, J.E., Rij, R.E., Ross, S.R., Fouse, D.C., 1986. Respiration rate of *Steinernema feltiae* infective juveniles at several constant temperatures. J. Nematol. 18, 221–224.

Martens, E. C., Goodrich-Blair, H., 2005. The *Steinernema carpocapsae* intestinal vesicle contains a subcellular structure with which *Xenorhabdus nematophila* associates during colonization initiation. Cell. Microbiol. 7, 1723–1735.

Mbata, G.N., Shapiro-Ilan, D.I., 2010. Compatibility of *Heterorhabditis indica* (Rhabditida: Heterorhabditidae) and *Habrobracon hebetor* (Hymenoptera: Braconidae) for biological control of *Plodia interpunctella* (Lepidoptera: Pyralidae). Biol. Control 54, 75–82.

Nielsen, A.L., Spence, K.O., Nakatani, J., Lewis, E.E., 2011. Effect of soil salinity on entomopathogenic nematode survival and behaviour. Nematol. 13, 859–867.

Nilsson, U., Gripwall, E., 1999. Influence of application technique on the viability of the biological control agents V*erticillium lecanii* and *Steinernema feltiae*. Crop Prot. 18, 53–59.

Nguyen, K.B., Smart, G.C. Jr., 1991. Pathogenicity of *Steinernema scapterisci* to selected invertebrates. J. Nematol. 23, 7–11.

Nimkingrat, P., Strauch, O., Ehlers, R. U., 2013. Hybridisation and genetic selection for improving desiccation tolerance of the entomopathogenic nematode *Steinernema feltiae*. Biocontrol Sci. Technol. 23, 348–361.

O'Callaghan, K.M., Zenner, A.N.R.L., Hartley, C.J., Griffin, C.T., 2014. Interference competition in entomopathogenic nematodes: male *Steinernema* kill members of their own and other species. Int. J. Parasitol. 44, 1009–1017.

Parkman, J.P., Frank, J.H., Nguyen, K.B., Smart, G.C. Jr., 1994. Inoculative release of *Steinernema scapterisci* (Rhabditida: Steinernematidae) to suppress pest mole crickets (Orthoptera: Gryllotapidae) on golf courses. Environ. Entomol. 23, 1331–1337.

Peters, A., 1996. The natural host range of *Steinernema* and *Heterorhabditis* spp. and their impact on insect populations. Biocontrol Sci. Technol. 6, 389–402.

Platzer, E.G., 1981. Biological control of mosquitoes with mermithids. J. Nematol. 13, 257–262.

Platzer, E.G., Mullens, B.A., Shamseldean, M.M., 2005. Mermithid nematodes, in: Grewal, P.S., Ehlers, R.-U., Shapiro-Ilan, D.I. (eds.), Nematodes as Biological Control Agents. CABI, Wallingford, pp. 411–418.

Poinar, G.O. Jr., 1979. Nematodes for Biological Control of Insects. CRC Press, Boca Raton, FL.

Poinar, G. O. Jr., 1990. Biology and taxonomy of *Steinernematidae* and *Heterorhabditidae*, in: Gaugler, R., Kaya, H. K. (eds.), Entomopathogenic Nematodes in Biological Control. CRC Press, Boca Raton, FL, pp. 23–62.

Portillo-Aguilar, C., Villani, M.G., Tauber, M.J., Tauber, C. A., Nyrop, J.P., 1999. Entomopathogenic nematode (Rhabditida: Heterorhabditidae and steinernematidae) response to soil texture and bulk density. Environ. Entomol. 28, 1021–1035.

Rae, R.G., Tourna, M., Wilson, M.J., 2010. The slug parasitic nematode *Phasmarhabditis hermaphrodita* associates with complex and variable bacterial assemblages that do not affect its virulence. J. Invertebr. Pathol. 104, 222–226.

Rasmann, S., Köllner, T.G., Degenhardt, J., Hiltpold, I., Toepfer, S., Kuhlmann, U., et al., 2005. Recruitment of entomopathogenic nematodes by insect-damaged maize roots. Nature 434, 732–737.

Rasmann, S., Erwin, A.C., Halitschke, R., Agrawal, A.A., 2011. Direct and indirect root defences of milkweed (*Asclepias syriaca*): trophic cascades, trade-offs and novel methods for studying subterranean herbivory. J. Ecol. 99, 16–25.

Rickert-Campbell, L., Gaugler, R., 1991. Role of the sheath in desiccation tolerance of two entomopathogenic nematodes. Nematologica 37, 324–332.

Salame, L., Glazer, I., 2015. Stress avoidance: vertical movement of entomopathogenic nematodes in response to soil moisture gradient. Phytoparasitica 43, 647–655.

Salem, S.A., Abdel-Rahman, H.A., Zebitz, C.P.W., Saleh, M.M.E., Ali, F.I., El-Kholy, M.Y., 2008. Survival, pathogenicity and propagation of entomopathogenic nematodes under different temperatures. Egypt J. Biol. Pest Control 18, 91–98.

Schroeder, P.C., Villani, M.G., Ferguson, C.S., Nyrop, J.P., Shields, E.J., 1993. Behavioral interactions between Japanese beetle (Coleoptera: Scarabaeidae) grubs and an entomopathogenic nematode (Nematoda: Heterorhabditidae) within turf microcosms. Environ. Entomol. 22, 595–600.

Schroeder, W.J., Beavers, J.B., 1987. Movement of the entomogenous nematodes of the families Heterorhabditidae and Steinernematidae in soil. J. Nematol. 19, 257–259.

Schroer, S., Ehlers, R.-U., 2005. Foliar application of the entomopathogenic nematode *Steinernema carpocapsae* for biological control of diamondback moth larvae (*Plutella xylostella*). Biol. Control 33, 81–86.

Seenivasan, N., Sivakumar, M., 2014. Screening for environmental stress-tolerant entomopathogenic nematodes virulent against cotton bollworms. Phytoparasitica 42, 165–177.

Shamseldean, M.M., Abd-Elgawad, M.M., 1994. Natural occurrence of insect pathogenic nematodes (Rhabditida: Heterorhabditidae) in Egyptian soils. Afro–Asian J. Nematol. 4, 151–154.

Shapiro-Ilan, D.I., Brown, I., 2013. Earthworms as phoretic hosts for *Steinernema carpocapsae* and *Beauveria bassiana*: implications for enhanced biological control. Biol. Control. 66, 41–48.

Shapiro-Ilan, D.I., Dolinski, C., 2015. Entomopathogenic nematode applications technology, in: Campos-Herrera, R. (ed.), Nematode Pathogenesis of Insects and Other Pests: Ecology and Applied Technologies for Sustainable Plant and Crop Protection. Springer, Cham, pp. 231–254.

Shapiro, D.I., Glazer, I., Segal, D., 1996a. Trait stability in and fitness of the heat tolerant entomopathogenic nematode *Heterorhabditis bacteriophora* IS5 strain. Biol. Control 6, 238–244.

Shapiro, D.I., Tylka, G.L., Lewis, L.C., 1996b. Effects of fertilizers on virulence of *Steinernema carpocapsae*. Appl. Soil Ecol. 3, 27–34.

Shapiro, D.I., Obrycki, J.J., Lewis, L.C., Jackson, J.J., 1999. Effects of crop residue on the persistence of *Steinernema carpocapsae*. J. Nematol. 31, 517–519.

Shapiro-Ilan, D.I., Gouge, D.H., Koppenhöfer, A.M., 2002a. Factors affecting commercial success: case studies in cotton, turf and citrus, in: Gaugler, R. (ed.), Entomopathogenic Nematology. CABI, Wallingford, pp. 333–356.

Shapiro-Ilan, D.I., Mizell, R.F., Campbell, J.F., 2002b. Susceptibility of the plum curculio, *Conotrachelus nenuphar*, to entomopathogenic nematodes. J. Nematol. 34, 246–249.

Shapiro-Ilan, D.I, Mizell, R.F., Cottrell, T.E., Horton, D.L., 2004a. Measuring field efficacy of *Steinernema feltiae* and *Steinernema riobrave* for suppression of plum curculio, *Conotrachelus nenuphar*, larvae. Biol. Control 30, 496–503.

Shapiro-Ilan, D.I., Jackson, M., Reilly, C.C., Hotchkiss, M.W., 2004b. Effects of combining an entomopathogenic fungi or bacterium with entomopathogenic nematodes on mortality of *Curculio caryae* (Coleoptera: Curculionidae). Biol. Control 30, 119–126.

Shapiro-Ilan, D.I., Stuart, R.J. McCoy, C.W., 2005. Targeted improvement of *Steinernema carpocapsae* for control of the pecan weevil, *Curculio caryae* (Horn) (Coleoptera: Curculionidae) through hybridization and bacterial transfer. Biol. Control 34, 215–221.

Shapiro-Ilan, D.I., Gouge, D.H., Piggott, S.J., Patterson Fife, J., 2006a. Application technology and environmental considerations for use of entomopathogenic nematodes in biological control. Biol. Control 38, 124–133.

Shapiro-Ilan, D.I., Stuart, R.J., McCoy, C.W., 2006b. A comparison of entomopathogenic nematode longevity in soil under laboratory conditions. J. Nematol. 38, 119–129.

Shapiro-Ilan, D.I., Leskey, T.C., Wright, S.E., 2011. Virulence of entomopathogenic nematodes to plum curculio, *Conotrachelus nenuphar*: effects of strain, temperature, and soil type. J. Nematol. 43, 187–195.

Shapiro-Ilan, D.I., Bruck, D.J., Lacey, LA., 2012a. Principles of epizootiology and microbial control, in: Vega, F.E., Kaya, H.K. (eds.), Insect Pathology, 2nd edn. Elsevier, Amsterdam, pp. 29–72.

Shapiro-Ilan, D.I., Lewis, E.E., Campbell, J.F., Kim-Shapiro, D.B., 2012b. Directional movement of entomopathogenic nematodes in response to electrical field: effects of species, magnitude of voltage, and infective juvenile age. J. Invertebr. Pathol. 109, 34–40.

Shapiro-Ilan, D.I., Han, R., Dolinski, C., 2012c. Entomopathogenic nematode production and application technology. J. Nematol. 44, 206–217.

Shapiro-Ilan, D.I., Han, R., Qiu, X., 2014a. Production of entomopathogenic nematodes, in: Morales-Ramos, J., Rojas, G., Shapiro-Ilan, D.I (eds.), Mass Production of Beneficial Organisms: Invertebrates and Entomopathogens. Academic Press, Amsterdam, pp., 321–356.

Shapiro-Ilan, D.I., Lewis, E.E., Schliekelman, P., 2014b. Aggregative group behavior in insect parasitic nematode dispersal. Int. J. Parasitol. 44, 49–54.

Shapiro-Ilan, D.I., Cottrell, T.E., Mizell, R.F. III., Horton, D.L., and Abdo, Z., 2015a. Field suppression of the peachtree borer, *Synanthedon exitiosa*, using *Steinernema carpocapsae*: effects of irrigation, a sprayable gel and application method. Biol. Control 82, 7–12.

Shapiro-Ilan, D.I. Hazir, S., Leite, L., 2015b. Viability and virulence of entomopathogenic nematodes exposed to ultraviolet radiation. J. Nematol. 47, 184–189.

Shapiro-Ilan, D.I., Hazir, S., and Glazer, I., 2016a. Basic and applied research: entomopathogenic nematodes, in: Lacey, L.A. (ed.), Microbial Agents for Control of Insect Pests: From Discovery to Commercial Development and Use. Academic Press, Amsterdam, pp. 253–267.

Shapiro-Ilan, D.I., Cottrell, T.E., Mizell, R.F. III, Horton, D.L., 2016b. Efficacy of *Steinernema carpocapsae* plus fire gel applied as a single spray for control of the lesser peachtree borer, Synanthedon pictipes. Biol. Control 94, 33–36.

Shields, E.J., Testa, A., Miller, J.M., Flanders, K.L., 1999. Field efficacy and persistence of the entomopathogenic nematodes *Heterorhabditis bacteriophora* "Oswego" and *H. bacteriophora* "NC" on Alfalfa snout beetle larvae (Coleoptera: Curculionidae). Environ. Entomol. 28, 128–136.

Stock, S.P., Hunt, D.J., 2005. Morphology and systematics of nematodes used in biocontrol, in: Grewal, P.S., Ehlers, R-U, Shapiro-Ilan, D.I. (eds.), Nematodes as Biocontrol Agents. CABI, Wallingford, pp. 3–43.

Stuart, R.J., Barbercheck, M.E., Grewal, P. S., Taylor, R.A.J., Hoy, C.W., 2006. Population biology of entomopathogenic nematodes: concepts, issues, and models. Biol. Control 38, 80–102.

Stuart, R.J., El Borai, F.E., Duncan, L.W., 2008. From augmentation to conservation of entomopathogenic nematodes: trophic cascades, habitat manipulation and enhanced biological control of *Diaprepes abbreviatus* root weevils in Florida citrus groves. J. Nematol. 40, 73–84.

Stuart, R.J., Barbercheck, M.E., Grewal, P.S., 2015. Entomopathogenic nematodes in the soil environment: Distributions, interactions and the influence of biotic and abiotic factors, in: Campos-Herrera, R. (ed.), Nematode Pathogenesis of Insects and Other

Pests: Ecology and Applied Technologies for Sustainable Plant and Crop Protection. Springer, Cham, pp. 97–137.

Sun, Y., Bai, G., Wang, Y., Zhang, Y., Pan, J., Cheng, W., et al., 2016. The impact of Cu, Zn and Cr salts on the relationship between insect and plant parasitic nematodes: a reduction in biocontrol efficacy. Appl. Soil Ecol. 107, 108–115.

Susurluk, I.A., 2013. Persistence and foraging behaviour of heat tolerant heterorhabditis bacteriophora strain in soil. Russian J. Nematol. 21, 51–57.

Thurston, G.S., Ni, Y., Kaya, H.K., 1994a. Influence of salinity on survival and infectivity of entomopathogenic nematodes. J. Nematol. 26, 345–351.

Thurston, G.S., Kaya, H.K., Gaugler, R., 1994b. Characterizing the enhanced susceptibility of milky disease-infected scarabaeid grubs to entomopathogenic nematodes. Biol. Control 4, 67–73.

Tonelli, M., Peñaflor, M.F.G.V., Leite, L.G., Silva, W.D., Martins, F., Bento, J.M.S., 2016. Attraction of entomopathogenic nematodes to sugarcane root volatiles under herbivory by a sap-sucking insect. Chemoecol. 26, 59–66.

Torr, P., Heritage, S., Wilson, M.J., 2004. Vibrations as a novel signal for host location by parasitic nematodes. Int. J. Parasitol. 34, 997–999.

Torres-Barragan, A., Suazo A., Buhler, W.G., Cardoza, Y.J., 2011. Studies on the entomopathogenicity and bacterial associates of the nematode *Oscheius carolinensis*. Biol. Control 59, 123–129.

Turlings, T.C.J., Hiltpold, I., Rasmann, S., 2012. The importance of root-produced volatiles as foraging cues for entomopathogenic nematodes. Plant and Soil 359, 51–60.

Ulug, D., Hazir, S., Kaya, H.K., Lewis, E., 2014. Natural enemies of natural enemies: the potential top-down impact of predators on entomopathogenic nematode populations. Ecol. Entomol. 39, 462–469.

Van Tol, R.W.H.M., van der Sommen, A.T.C., Boff, M.I.C., Van Bezooijen, J., Sabelis, M. W., Smits, P. H., 2001. Plants protect their roots by alerting the enemies of grubs. Ecol. Lett. 4, 292–294.

Wallace, H.R., 1971. Abiotic influence in the soil environment, in: Zuckerman, B.M., Mai, W.F., Rohde, R.A. (eds.), Plant Parasitic Nematodes. Academic Press, New York, pp. 257–280.

Wennemann, L., Shanks, C.H., Smith, K.A., 2004. Movement of entomopathogenic nematodes in soils of *Fragaria* spp. Comm. Agric. Appl. Biol. Sci. 69, 347–357.

Wilson, M.J., Lewis, E.E., Yoder, F., Gaugler, R., 2003. Application pattern and persistence of the entomopathogenic nematode *Heterorhabditis bacteriophora*. Biol. Control. 26, 180–188.

Zenner, A.N.R.L., O'Callaghan, K.M., Griffin, C.T., 2014. Lethal fighting in nematodes is dependent on developmental pathway: male-male fighting in the entomopathogenic nematode *Steinernema longicaudum*. PLOS ONE 9, e89385.

Zhang, C.X., Yang, S.Y., Xu, M.X., Sun, J., Liu, H., Liu, J.R., et al., 2009. A novel species of *Serratia*, family Enterobacteriaceae: *Serratia nematodiphila* sp.nov., symbiotically associated with entomopathogenic nematode *Heterorhabditidoides chongmingensis* (Rhabditida: Rhabditidae). Int. J. Syst. Evol. Microbiol. 59, 1603–1608.

Section IV

Applied Ecology of Invertebrate Pathogens

12

Modeling Insect Epizootics and their Population-Level Consequences
Bret D. Elderd

Department of Biological Sciences, Louisiana State University, Baton Rouge, LA, USA

12.1   Introduction

Mechanistic models of species interactions and disease transmission (e.g., Kermack and McKendrick, 1927; Rohani et al., 2002; Wootton, 2005) play an important role in helping us understand the patterns we see in nature and the processes responsible (Kendall, 2015). The importance of grounding ecological experiments in a testable theoretical framework is often understated (Scheiner, 2013), but can prove fruitful when done correctly. With regard to insect epizootics, interest in insect pathogens can be traced back to ancient China, where silkworm populations started to succumb to a then unknown pathogen. While the implications for infected individuals were obvious, the culprits were not discovered for centuries (Cory and Myers, 2003). In fact, it was only a few decades ago that insect pathogens were linked, via mechanistic models, to the boom-and-bust dynamics often seen in naturally occurring insect populations (Anderson and May, 1980). Early on, Brown (1987) described the importance of considering the theoretical perspective and using simulation models to gain a better understanding of epizootic dynamics; this field continues to grow (e.g., Onstad and Carruthers, 1990; Dwyer et al., 2000; Stuart et al., 2006; Hesketh et al., 2010; Shapiro-Ilan et al., 2012). The ideas and concepts associated with model formulation and testing have also continued to develop. These newly developed tools, along with standard analytical tools, will prove to be incredibly useful to any individual interested in insect epizootics, regardless of their training or research focus.

In general, epizootic models tend to focus on either the short-term dynamics associated with a single epizootic (e.g., Dwyer et al., 1997) or the long-term population-level consequences of multiple epizootic events (e.g., Dwyer et al., 2004). By combining models with data regardless of the time scale considered, the study of insect epizootics has led to a better understanding of the processes responsible for driving the population dynamics of outbreaking insects. From a management/application perspective, the construction of mechanistic models helps to highlight the potential benefits or unforeseen consequences of using insect pathogens to control population outbreaks (Hochberg, 1989; Reilly and Elderd, 2014). From a broader perspective, insect epizootics have helped to develop both models and ideas that are central to a great deal of ecology and

Ecology of Invertebrate Diseases, First Edition. Edited by Ann E. Hajek and David I. Shapiro-Ilan.
© 2018 John Wiley & Sons Ltd. Published 2018 by John Wiley & Sons Ltd.

to address fundamental ecological questions such as what causes populations to cycle (Barraquand et al., 2017). These insights continue to develop as new data are collected and models are refined.

Regardless of the approach one takes toward research, everyone uses models. There are three basic categories of models, which are not necessarily mutually exclusive: empirical, theoretical, and simulation-based (Hobbs and Hooten, 2015). The vast majority of science is conducted using empirical approaches. The empirical approach describes relationships and patterns between variables that have been measured and is exemplified by familiar statistical tools like regression and analysis of variance. Empirical approaches summarize relationships from a phenomenological perspective and not necessarily a mechanistic one. For example, in an empirical setting, a regression conducted on disease incidence and some independent variable (e.g., host density) may show that disease incidence increases as the independent variable increases. Here, the slope simply describes the rate of change or the pattern observed in the data, but, in this example, the model does not explicitly imply a mechanism. This measured relationship advances our understanding of the epizootic process but does not describe the mechanism driving the process.

On the other hand, theoretical models are mechanistic in nature. Historically, theoretical models have been considered overly simplistic, since they often focus on only a single interaction, or on a limited number of parameters. While theoretical models have provided a great deal of insight into ecological dynamics (e.g., the Lotka–Volterra equations), strictly theoretical approaches alone can easily become focused on the elegance of the method and lose sight of the ecology involved (Levin, 2012). However, when these models are confronted with data, the range of possible behaviors (e.g., population cycles or no cycles) quickly narrows and the models become quite powerful. Simulation-based models also take advantage of data that has been collected, but these models contain multiple interactions and can easily become relatively complex. Thus, simulation models are parameter-rich or high-dimensional. They often fit the data well, but it can be difficult to determine which of the many parameters drives the observed patterns (Elderd et al., 2006). Models constructed to understand epizootic dynamics or other ecological processes do not have to fall into a single category and may draw from multiple categories (Hobbs and Hooten, 2015). Thus, the three categories serve as useful heuristics for thinking about model development and linking models to data.

These three modeling perspectives are often seen as strictly distinct from other. However, this viewpoint draws on historical precedence and does not take into account how prevalent it has become to combine mechanistic models of increasing complexity with data. In the world of modeling insect epizootics, when Anderson and May (1980) compared theoretical model outcomes to empirical data, they opened up a new window from which to see the world. In their frequently cited works, Anderson and May (1979, 1980) combined ideas from classic predator–prey models and susceptible–infected–recovered (SIR) models, which essentially combined long-term dynamic models with short-term epizootic models, to understand natural observations of insect populations. Using this approach, Anderson and May (1980) demonstrated that pathogens could be responsible for the boom-and-bust cycles associated with the long-term dynamics of insect populations. They continued to expand upon the mechanisms responsible for short-term epizootic events and long-term population dynamics for species controlled by pathogen outbreaks (Anderson and May, 1981). Thus, they started a rich literature in which the idea of combining models with data serves as a cornerstone for understanding insect epizootics.

In their original work, Anderson and May (1980) showed that insect population models that invoked host–pathogen interactions qualitatively displayed the same dynamics as observational data collected in the field. Following this, more rigorous methods of analyzing observational time series and field data began to take hold. These methods often advocated a likelihood-based approach that simply asked how likely were the data that had been collected to be correct, given the predicated dynamics from the model. If the model did not do well in terms of predicting the actual data, it was either set aside or refined, and the investigative process continued. The model, in this instance, can be considered equivalent to a hypothesis that is tested with data. The framework for this likelihood-based approach culminated in the influential book, *The Ecological Detective* (Hilborn and Mangel, 1997), which advocated not only confronting models with data but also testing multiple models at once. One method for doing so, which comes from the field of information theory, is the Akaike information criterion (AIC) (Burnham and Anderson, 2002). The AIC operates on the principle of parsimony to choose the best model. The most complicated model (i.e., the model with the most parameters) will always fit the data better (i.e., the model will be more likely) than less complicated ones. However, fitting the more complicated model comes at a cost, because a more complicated model will do a poor job of predicting the next set of data to be collected. Thus, the AIC balances between model fit and model complexity (Section 12.4.2). Just as the AIC draws upon likelihood-based methods, so Bayesian methods are based on likelihood approaches. Bayesian methods, particularly hierarchical Bayesian models, have become increasingly popular and provide the flexibility to analyze data and compare multiple models (Hobbs and Hooten, 2015). For a worked example from the Bayesian perspective, see Section 12.5.1. Overall, a lot of progress has been made since the initial explorations of purely theoretical models and the fitting of empirical models to data, and this progress will continue in the future. While likelihood-based frameworks work well for short-term epizootic events, different tactics are often required when examining long-term boom-and-bust cycles (Section 12.6.2) (e.g., Kendall et al., 1999; Turchin, 2003; Dwyer et al., 2004). Together, each of the methods mentioned here that focuses on combining models with data represents a suite of tools for understanding what drives epizootics over both the short and the long term.

12.2   The Pathogen and its Hosts

While there are a number of diseases described throughout this book that have a great deal of impact on invertebrate populations and epizootic dynamics, I will focus on diseases caused by baculoviruses (see also Chapter 7). Although the focus here is on baculoviruses, models have been developed to describe epizootic dynamics for a variety of insect pathogens. These include, but are certainly not limited to, fungal pathogens (e.g., Edelstein et al., 2005; Scholte et al., 2005; Hesketh et al., 2010) and nematodes (e.g., Stuart et al., 2006). For example, Scholte et al. (2005) used an enthomopathogenic fungal model to highlight the efficacy of a vector control method to decrease the population of a malaria vector, *Anopheles gambiae*. Thus, even though baculoviruses provide the biological motivation for model development here, the methods outlined can be applied to many other biological systems. However, since all good mechanistic models need to be motivated by the biology of the system, baculoviruses represent a good place to start, given their importance in driving epizootic dynamics (Cory and Myers, 2003) and the use of mechanistic models in describing these dynamics (e.g., Dwyer et al., 1997; Elderd et al., 2013). To reiterate, while the biology and the associated models throughout draw on baculoviruses as examples, the methodologies discussed have quite a broad use in enhancing our understanding of epizootic dynamics as a whole.

Baculovirus infections begin when a susceptible individual consumes occlusion bodies (OBs), often containing multiple copies of the virus. If enough OBs are consumed, the individual becomes fatally infected. Sublethal or covert infections also occur (Roy et al., 2009), but at relatively low levels (Myers et al., 2000). Covert infections may contribute to the persistence of pathogens at low host densities (Roy et al., 2009) and function in a manner similar to vertical transmission between mothers and their offspring, which also allows pathogens to persist at relatively low host densities (Anderson and May, 1981). However, covert infections likely do not drive the boom-and-bust cycles associated with epizootic dynamics. If a lethal rather than a sublethal infection occurs, the infection process moves through a number of stages before the death of the host, which can release millions of OBs into the environment; transmission resulting from a sublethal infection would be minimal in comparison unless that infection became lethal.

In Lepidoptera, the infection process begins when a larva consumes foliage on which OBs reside. Upon entering the midgut, the outer protein coat of the OB dissolves, releasing the virions. The virions then infect the host's midgut cells and the infection eventually becomes systemic. After a period of time, the virus essentially liquefies the internal structure of the host as a result of producing more virus. Upon the host's death, the outer larval integument splits open and releases virus into the environment to begin the infection process again (Cory and Hoover, 2006; Elderd, 2013). Over time, the virus degrades due to ultraviolet (UV) light exposure.

An epizootic can begin when first-instar larvae become infected (Dwyer et al., 1997). However, there is a delay between infection of the first instars and release of the next round of virus. During this delay, which varies depending upon the virus and the species infected, the healthy larvae molt to second, third, or fourth instars. Due to the baculovirus infection, the infected individuals do not molt (Miller, 1997). After the infected larvae die and OBs are released on to the surrounding leaf tissue, the larger healthy larvae that have developed into later instars consume the virus, since it now resides on the leaf tissue on which they are feeding, and disease prevalence increases. The epizootic ends due to either larval pupation or "epizootic burnout," which occurs when there is a lack of infected individuals in the population to continue the spread of the disease (Dwyer et al., 2000; Fuller et al., 2012). The delay between infection and host death has important consequences for both the short- and the long-term dynamics of pathogen-driven host populations, such as lepidopteran species prone to epizootics.

While observational data in these systems can be readily collected, baculovirus systems also lend themselves to experimental approaches that test hypotheses surrounding transmission dynamics. To initiate an experiment that follows the natural progression of a baculovirus-driven epizootic, first instars can be lethally infected with a dose of OBs. Once infected, the first instars are confined on an experimental plant or branch using mesh bags. Mesh bags stop larvae from escaping and prevent the virus from degrading due to UV light exposure. The infected individuals then die and, after death, release OBs on to the leaf tissue. Once the first instars have died, healthy third or fourth instars are placed into the mesh bag. These individuals are then allowed to feed for a period of time. Afterwards, the larvae are collected and reared in individual cups until death or pupation (Dwyer et al., 1997; Elderd et al., 2008; Elderd and Reilly, 2014). Infection can be easily diagnosed visually given the drastic manner in which the infection process slowly consumes the larva. Additionally, since the OBs are quite large and can be seen under a light microscope (Elderd, 2013), any potential infections can be readily confirmed. In the simplest approach, one can manipulate the amount of pathogen in the system (the independent variable) and record the fraction of insects surviving (the dependent variable). Thus, experiments that manipulate multiple factors such as temperature and the amount of pathogen in the system (Elderd and Reilly, 2014) can be readily performed. The data produced can then be combined with any suite of models to test the associated hypothesis.

12.3   Modeling Disease Transmission: A Single Epizootic

The models used to understand short-term epizootic dynamics associated with a single event can be traced back to Kermack and McKendrick (1927), who developed the SIR model to describe epidemic dynamics. Instead of SIR dynamics, baculovirus systems consist of susceptible individuals, infected individuals, and pathogen, since there is little evidence that infected individuals recover. If the simplifying assumption is made that all baculovirus infections are lethal, we need only consider the number of susceptibles and the amount of pathogen in the system, since all infected individuals eventually become pathogen (Dwyer et al., 2000). This assumption is met by the experimental methods described earlier. Mathematically, the equation for the susceptible larvae S takes the form of the following differential equation:

$$\frac{dS}{dt} = -\beta SV. \tag{12.1}$$

Here, the change in susceptible larvae over time is simply a product of the disease transmission coefficient β times the number of susceptibles S and the amount of virus in the system V. The transmission parameter β encompasses the whole of the infection process and can be thought of as the fraction of encounters between the virus and a susceptible larva that leads to an instantaneous infection. As with all models (empirical, mechanistic, or simulation-based), it is important to consider all the assumptions. The main ones for the model of susceptible populations given here are that per capita transmission $\left(\text{i.e., } \dfrac{1}{S}\dfrac{dS}{dt} \right)$ is linear and that all individuals are equally susceptible to becoming infected. Relaxing this assumption, or changing the model structure to better fit the biology of the system and the data, leads to new insights into the transmission process. Equation 12.1, however, serves as a useful starting point.

   By integrating equation 12.1 and using experimental data, estimates of the transmission rate β can be easily calculated. In an experiment, the amount of virus or the number of cadavers in the system at the beginning of the experiment $V(0)$ is known, as is the initial number of susceptibles in the experimental treatment $S(0)$. Here, 0 refers to the start of the experiment. After conducting the experiment until time T, the number of susceptible individuals (i.e., the number of individuals that pupate rather than die from an infection) is also known, $S(T)$. These data can be easily plugged into the integral of equation 12.1, which is integrated from time 0 to T. The integral of equation 12.1 is simply:

$$-\ln\left[\frac{S(T)}{S(0)}\right] = \beta V(0)T. \tag{12.2}$$

Thus, by regressing the cadaver density against the negative natural log of the fraction uninfected $(-\ln[S(T)/S(0)])$ with no intercept term, an estimate of the transmission rate can be calculated directly from the data (Elderd et al., 2008); this is simply the slope of the line (Fig. 12.1, dashed line). But unlike standard regression models, which calculate the slope (and the intercept) as a phenomenological relationship, here, the slope is linked directly to the disease transmission rate β. Thus, the mechanism can be explicitly inferred from the mathematical model of the process.

12.3.1 Phenomenological and Mechanistic Models

While equation 12.1 represents a simple model of transmission dynamics, it is a reasonable representation of epizootic dynamics, and it and similar forms have been used extensively (e.g., Hochberg, 1989; Hochberg and Waage, 1991; Boots, 2004). However, the linear model does not always fit the data collected. Some baculovirus epizootic data show a decidedly nonlinear or curvilinear fit (Dwyer et al., 1997; Elderd and Reilly, 2014), such that infection rates at higher pathogen levels are less than expected if the

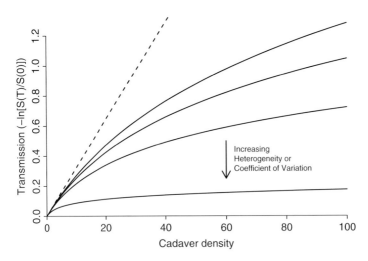

Fig. 12.1 Effect on transmission of increasing the transmission rate's coefficient of variation C in equation 12.5. The solid lines represent populations in which risk varies across individuals. The dashed line represents a population in which all individuals are equally at risk (equation 12.2). Each line uses the same value for the transmission rate of the virus.

linear model held true (equation 12.1). A simple solution to this problem would be to raise the number of susceptibles or the amount of virus by a power (Hochberg, 1991), which would result in a nonlinear model that could better fit the data. This phenomenological model then takes the form:

$$\frac{dS}{dt} = -\beta S^g V^h. \tag{12.3}$$

Here, g and h are the nonlinear effects on transmission of susceptible and infected population densities, respectively (Hochberg, 1991). However, while this power model will fit the nonlinear data better, the biological mechanism or mechanisms driving the nonlinear fit remain unknown. In this instance, what exactly does g or h mean from a biological standpoint?

A potential mechanism that may drive the nonlinearity in infection rates goes back to one of the main of assumptions of the linear model: that all individuals have the same transmission rate β. In Dwyer et al. (1997), the authors assumed that individuals differ in their susceptibility to virus. Essentially, some individuals are more susceptible than average and others are less susceptible than average. Thus, there was not a single transmission rate, but a mean transmission rate with some variability about the mean rate. Therefore, the transmission rate became a distribution rather than a single point estimate. The modified equation accounting for differences in susceptibility (i.e., heterogeneity in the transmission rate) thus becomes:

$$\frac{dS}{dt} = -\bar{\beta} \left[\frac{S(t)}{S(0)} \right]^{C^2} SV. \tag{12.4}$$

Here, $\bar{\beta}$ is the mean transmission rate. The transmission rate is scaled by the ratio of the number of susceptibles currently in the population $S(t)$ divided by the number of susceptibles at the start of the epizootic $S(0)$. The ratio is raised to the square of the coefficient of variation C associated with the transmission rate. Integrating equation 12.4 results in:

$$-\ln \left[\frac{S(T)}{S(0)} \right] = \frac{1}{C^2} \ln \left(1 + \bar{\beta} C^2 V(0) T \right). \tag{12.5}$$

Here, T is once again the time that the experiment ran. For equation 12.5, instead of estimating just β from the data, two parameters need to be estimated, $\bar{\beta}$ and C. For any single level of heterogeneity, at low pathogen levels, highly susceptible individuals become infected and transmission rises quickly (Fig. 12.1, solid lines). However, as pathogen levels increase, transmission tapers off, since only highly resistant individuals remain in the population. As the heterogeneity in the population increases, the coefficient of variation C increases, which results in fewer individuals becoming infected at the end of the epizootic as pathogen levels increase (Fig. 12.1). If, instead, C decreases and goes to zero (i.e., little variability in C), the dynamics become similar to the linear equation (equation 12.2). While equation 12.5 was developed with epizootics in mind, it borrows from work by Anderson and May (1991) on HIV spread and how varying contact rates influence HIV transmission. Thus, equation 12.4 represents another example of the give and take between epizootiological and epidemiological research.

Once a model is developed, it is important to test it. Dwyer et al. (1997) exemplified this approach by showing the stepwise process of confronting models with data. In a series of experiments on the invasive gypsy moth (*Lymantria dispar*) and its species-specific baculovirus, Lymantria dispar multinucleopolyhedrovirus (LdMNPV), the authors tested whether the linear (equation 12.1) or the nonlinear (equation 12.4) model explained the data better, using a series of experimental epizootics. However, it should be noted that baculoviruses do not represent the only pathogen in the system. *Entomophaga maimaiga*, a fungal pathogen, also infects gypsy moth larva (Hajek, 1999), but infection rates can be either density-independent (Liebhold et al., 2013) or density-dependent (Hajek et al., 2015) according to the weather conditions (Hajek and van Nouhuys, 2016). LdMNPV, unlike *E. maimaiga*, is always strongly density-dependent (Liebhold et al., 2013). Thus, the linear and nonlinear models focused on baculovirus transmission along with host and pathogen densities (Dwyer et al., 1997) are appropriate given the biology of the baculovirus system.

For short-term epizootic events, the experimental data clearly support the nonlinear model for the gypsy moth (see Dwyer et al., 1997, fig. 3). Interestingly, in a comparison of lab-reared larvae with feral larvae, the degree of heterogeneity in transmission was much less in the former. Since the lab-reared larvae were not exposed to virus, the authors hypothesized that the level of heterogeneity should be less than for gypsy moth larvae reared from eggs collected from the wild. As new models have been developed and tested with empirical data, new mechanistic insights into what drives gypsy moth baculovirus epizootics continue to be gained. Other models have shown that rapid host evolution (Elderd et al., 2008), interactions between plant defensive chemicals and virus (Elderd et al., 2013), and heterogeneity in the pathogen (Fleming-Davies et al., 2015) are all important factors driving gypsy moth epizootic dynamics. All of the preceding models rest upon the mechanistic backbone of equation 12.4, which has continued to move the field forward. They also highlight that by combining mechanistic model development with experimental data, new insights can be continually gained.

12.4   Fitting Models to Data

To determine whether heterogeneity or other factors can be invoked as drivers of epizootic dynamics, a model must be compared to either experimental or observational data. Standard approaches fit the model using the well-trodden path of frequentist statistics. These methods determine fit either through the amount of variation explained or through the significance of a term in the model (e.g., the slope in a regression), as dictated by a null-hypothesis test and its subsequently generated p-value. However, the use of p-values continues to fall out of favor, as evidenced by a policy statement from the American Statistical Association (ASA) (Wasserstein and Lazar, 2016). This statement cautions against overreliance on the p-value as a valid statistic to indicate whether an effect drives the patterns seen in data. Additional problems arise when using this approach to compare multiple non-nested models. Often, when there are multiple mechanisms suspected of driving the dynamics in a variety of nonlinear or linear ways, non-nested models quickly accumulate. The question of how best to compare multiple models to the data remained a problem until information theory began to gain a foothold in the wildlife literature (Anderson et al., 2000) and was highlighted in two influential books (Hilborn and Mangel, 1997; Burnham and Anderson, 2002).

12.4.1 Akaike Information Criterion

An information-theoretic approach to data analysis became widely used after the publication of Burnham and Anderson (2002). This approach allows a researcher to compare multiple models (i.e., alternative hypotheses) and determine which best fit the data. This is in direct contrast to classical statistics, which focuses on either accepting or rejecting a null hypothesis. The rejection of the null hypothesis simply means that the null model does not fit the data and is not an acceptance of the alternative model. Thus, all of our inference is based on the null model. In comparing multiple models, Burnham and Anderson (2002) focused on the use of the AIC, which operates according to the principle of parsimony or Occam's razor. That is, in deciding which is the best model, the researcher must shave away all that is unnecessary. Thus, in constructing a mechanistic model, one wants to construct a model with the smallest number of parameters that best explains the data. Essentially, there needs to be a balance between underfitting (i.e., too few parameters) and overfitting (i.e., too many parameters) models.

Hirotu Akaike, a Japanese statistician, developed a simple formula that corrects for constructing models that are too simple or too complex. The formula states:

$$\text{AIC} = -2L(\text{Data} \mid \Theta) + 2K. \tag{12.6}$$

Here, $L(\text{Data} \mid \Theta)$ is the log likelihood of the data given the model parameters Θ, and K is the number of parameters in the model (Burnham and Anderson, 2002). In a standard regression model, the log likelihood of the slope and the intercept is often calculated using the sum of squares of the difference or the error between the data and the model's predictions, assuming that the error is normally distributed. To calculate the log likelihood for data associated with infections, the error in the model often follows a binomial distribution, since the data are counts of infected and non-infected individuals (e.g., Elderd and Reilly, 2014). By using the equation 12.6 and, if necessary, the associated correction for small sample sizes AIC_c (see Burnham and Anderson, 2002), the raw AIC score can be calculated. The model with the lowest score is the best-fit model. Models with too few parameters are less likely and have a low log likelihood. Models with too many parameters (i.e., larger values of K), while fitting the data better, are penalized by adding to their AIC score. The best model thus represents a balance between model fit and complexity.

To fully gauge the degree of support for a model or for one of the alternative hypotheses, the differences between the best-fit model and the other models – the ΔAIC – should first be calculated. The formula for calculating the ΔAIC is:

$$\Delta \text{AIC}_i = \text{AIC}_i - \min(\text{AIC}), \tag{12.7}$$

where i is the model being considered and min(AIC) is the minimum of all AIC model scores. Thus, the best-fit model, which is the model with the lowest AIC score, has a ΔAIC of 0. Models that have $\Delta \text{AIC} > 10$ are considered poor fits to the data, those with values between 4 and 7 have little support, and those with values greater than 0 but less than 2 have substantial support (Burnham and Anderson, 2002).

ΔAIC scores can, in turn, be used to calculate AIC weights, which are the weights of evidence for the relative likelihood of particular models given the models considered. AIC weights are calculated using:

$$w_i = \frac{\exp(-0.5\Delta\text{AIC}_i)}{\sum_{r=1}^{R} \exp(-0.5\Delta\text{AIC}_r)}, \tag{12.8}$$

where w_i is the weight of evidence for model i given all R models (i.e., better models are reflected by higher weights). These weights allow for a direct comparison between alternative hypotheses and can be used to gain further insight via multimodel inference (Burnham and Anderson, 2002).

12.4.2 An Example of the AIC in Action

To give a concrete example of the use of the AIC in analyzing epizootic data, I will draw on a series of experiments examining the effects of global climate change on baculovirus transmission in the fall armyworm (*Spodoptera frugiperda*) (Elderd and Reilly, 2014). The fall armyworm is a multivoltine crop pest that overwinters in Florida and Texas (Pitre and Hogg, 1983). As springtime temperatures increase, the fall armyworm reinvades the entire extent of its range by migrating northward until it reaches Ontario, Canada. As adults, female fall armyworms lay eggs in clusters. After the eggs hatch, there are six larval instars (Pitre and Hogg, 1983). They then pupate for 7–37 days depending upon the temperature (Sparks, 1979), emerge to mate, and continue their lifecycle. The species, like many lepidopterans, exhibits boom-and-bust dynamics. As the population increases during the boom phase, infestations occur, which can be widespread (Fuxa, 1982) and potentially devastating to farmers (Hinds and Dew, 1915).

Spodoptera frugiperda nucleopolyhedrovirus (SfNPV), a species-specific baculovirus, represents an important mortality source for the fall armyworm (Richter et al., 1987). Prior to an epizootic, a viral reservoir in the soil provides the initial inoculation of virus (Fuxa and Geaghan, 1983). After 4–6 days, initially infected larvae die (De Oliveira, 1999), while uninfected larvae grow to third or fourth instars (Pitre and Hogg, 1983). The older instars become infected by consuming the contaminated foliage on which the first instars died. Over time, virus particles degrade due to UV light exposure (Miller, 1997). The epizootic dynamics in the fall armyworm are very similar to those in other baculovirus-driven populations, like the gypsy moth. Thus, we can start with the same base model and modify it to answer whatever questions are posed. In this instance, how will rising temperatures affect transmission dynamics?

To determine how increasing temperatures affect epizootic dynamics, we established a series of control and experimental plots. In the experimental plots, we manipulated temperature using open-top chambers (OTCs) (Marion et al., 1997), which significantly raised temperatures in experimental as compared to control plots (Elderd and Reilly, 2014). In each of the $40.1\,\text{m}^2$ plots, we placed a single soybean plant (*Glycine max*) of the same variety. Each plant was covered in a mesh bag, and a varying number of first-instar baculovirus-infected larvae (0, 15, 30, or 60) were placed on top. Once the first instars died, we placed 20 healthy fourth instars on the plant and allowed them

Table 12.1 AIC and WAIC results for the single-epizootic model. AIC_c scores, ΔAIC_c, AIC_c weights, and WAIC for each model. For the models considered, climate effect is due to OTCs raising temperatures in the experimental plots. The nonlinear model (equation 12.5) assumes heterogeneity in disease risk. For the linear model (equation 12.2), when $C = 0$, no difference in risk is assumed. $\bar{\beta}$ refers to estimates of mean transmission rate. Best-fit model is in bold.

Model	AIC_c	ΔAIC_c	AIC_c wt	WAIC
1. No climate effect, linear model	57.8	10.2	0.00	108.8
2. No climate effect, nonlinear model	52.7	5.1	0.03	106.4
3. Climate effect, linear model	52.5	4.9	0.04	107.4
4. Climate effect, nonlinear model	50.1	2.5	0.12	106.2
5. Climate effect, differences in C only	**47.6**	**0.0**	**0.42**	**103.3**
6. Climate effect, differences in $\bar{\beta}$ only	48.6	1.0	0.25	106.1

Note: AIC_c is based on an information theoretic approach to comparing multiple models. WAIC can be considered the Bayesian equivalent.

to feed. After 2 days, we collected the larvae and reared them until pupation or death. Baculovirus deaths were confirmed and recorded.

For heuristic purposes, in order to demonstrate the utility of the AIC approach, I will only consider a set of simple models (Table 12.1) based on equations 12.2 and 12.5. Additionally, although the experiment was conducted three separate times over the course of a number of years, I will only analyze the data from a single year. A complete analysis of the experiments and data is presented in Elderd and Reilly (2014).

Using the AIC approach for the climate-change experiment, the best-supported model, given the plausible models considered, is the one where $\bar{\beta}$ is the same for both treatments but the treatments have different values of C (Table 12.1). Note that the results are presented in terms of the small sample correction AIC_c, as recommended given the sample size (Burnham and Anderson, 2002). By examining ΔAIC_c, the support for the best-fit model (as compared to the null models) is strong, given that the null linear and nonlinear models have ΔAIC_c values between 5 and 10. Thus, increasing temperature has an effect on disease transmission. This does not hold true for all of the models considered. There is also support for the model where only $\bar{\beta}$ differs. The same pattern can be seen when examining the AIC_c weights (Table 12.1). For all of the data associated with these experiments, the general trend holds that C differs and that as temperatures rise, C exponentially decreases (Elderd and Reilly, 2014). Thus, as the climate warms, the nonlinear dynamics of the host–pathogen interaction become more and more similar to those of interactions governed by the linear model (Fig. 12.1).

12.5 A Bayesian Approach

Bayesian analysis has become another increasingly popular approach to fitting models to data. Like information-theoretic approaches and classic statistical approaches, Bayesian approaches are likelihood-based. That is, the results of the analysis hinge on how likely the data are given the model. However, there is an added component: prior information about the system. The basis for the approach stems from Bayes' theorem, which states:

$$P(\Theta \mid \text{Data}) \propto \pi(\Theta)\mathcal{L}(\text{Data} \mid \Theta) \tag{12.9}$$

where the posterior probability of the model parameters Θ given the data is proportional to ($\propto$), the prior probability of the parameters $\pi(\Theta)$ times the likelihood of the data given the model parameters $\mathcal{L}(\text{Data} \mid \Theta)$. In the past, the implementation of a Bayesian approach was often limited due to the complexity of the computations associated with the analysis. Recently, a proliferation of Bayesian books with ecological perspectives (e.g., Clark, 2007; Kéry, 2010; Hobbs and Hooten, 2015) and the availability of freeware programs (e.g., WinBugs, JAGS, STAN) have made Bayesian approaches much more accessible.

A distinct advantage of Bayesian methods is that they provide a framework for incorporating prior information about a system (e.g., preliminary studies), which is especially valuable when data are sparse. Typically, prior information enters into the classical analysis framework in the discussion when the authors state whether their current findings are similar to or different from those of previous studies (Hille Ris Lambers et al., 2005). In a Bayesian approach, the prior contains quantitative information and becomes a parameter in the analysis ($\pi(\Theta)$ in equation 12.9). If no prior information is available, vague priors can be used, which contain relatively little information. Explicitly stating a prior can be controversial to some, but if individuals are uncomfortable selecting a prior, the easiest way to minimize prior influence is to overwhelm it with data (Hobbs and Hooten, 2015). However, the use of informed priors makes the most of previously hard-won data and represents a powerful approach to developing mechanistic models for understanding epizootic dynamics.

A fundamental difference between a Bayesian approach and more classical approaches stems from the difference in how the parameters are treated. Classic frequentist approaches assume that a parameter's value is fixed and that the exact estimate becomes better resolved as sample size increases (Hobbs and Hooten, 2015). In contrast, Bayesian approaches assume that a parameter is a random variable drawn from a distribution. This is the difference between a single value for quantifying disease transmission rates, which is estimated with increasing precision, and a distribution of uncertainty reflecting the inherent variability of the transmission rate (Ellison, 2004; Hobbs and Hilborn, 2006). A more in-depth examination of Bayesian analysis from a philosophical perspective, as touched upon earlier, can be found elsewhere in the literature (e.g., Dennis, 1996; Ellison, 1996, 2004).

12.5.1 Fitting a Bayesian Model

For the linear model (equation 12.1), and assuming there is no difference in the treatment effects, a simple Bayesian model can be constructed such that:

$$y_i \sim \text{binomial}(p_i, N_i), \tag{12.10}$$

$$\ln(p_i) = -\beta V(0)T, \tag{12.11}$$

$$\beta \sim \text{lognormal}(0,1000). \tag{12.12}$$

The number of survivors or non-infected larvae is distributed ($\sim$) binomially with a probability p_i given an initial number of healthy larvae N_i. Here, p_i is simply the fraction of uninfected larvae ($S(T)/S(0)$). Thus, equation 12.11 is equivalent to equation 12.1. The disease transmission rate β has a prior probability that is log normally distributed with a mean of 0 and a variance of 1000. Thus, the prior is considered vague and contains little information. The resulting posterior for each replicate i becomes:

$$\overbrace{P(\beta \mid y_i)}^{\text{Posterior}} \propto \underbrace{\text{binomial}\left(y_i \mid e^{-\beta V(0)T}, N_i\right)}_{\text{Likelihood}} \overbrace{\text{lognormal}(\beta \mid 0, 1000)}^{\text{Prior}}, \tag{12.13}$$

which, following equation 12.9, explicitly shows the relationship between the posterior and its Bayesian components, the likelihood and the prior.

To obtain posterior estimates of the disease transmission rate, one needs to fit the models (Table 12.1) to the data using Markov chain Monte Carlo (MCMC) methods. This can be done via a freeware package called JAGS (or "Just another Gibbs sampler") (Plummer et al., 2003), which can be run directly in R (Yu-Sung and Masanao, 2015), or else by writing the MCMC code directly in R (R Core Team, 2015). The standard approach to constructing the posterior consists of running multiple chains at various starting points, trimming their beginnings, and combining them. For each model (Table 12.1), five separate chains, each for 50 000 iterations, were run in JAGS, with the initial conditions for each chosen randomly. The first 10 000 iterations were discarded as "burn-in" to eliminate any transients associated with the initial conditions. All other iterations were retained to serve as estimates of the posterior distribution. The chains were not thinned, where thinning entails keeping only every mth iteration of the chain and discarding all others, as can be common practice. Link and Eaton (2012) showed that thinning is inefficient and reduces the precision of the parameter estimates. Using standard metrics, MCMC convergence was assessed by examining both within-chain and between-chain convergence. If the chains do not converge, the model is doing a poor job of estimating the parameters associated with the data. Two common metrics involve calculating the Brooks–Gelman–Rubin and the Heidelberger–Welch diagnostics. The Brooks–Gelman–Rubin statistic compares within- and between-chain variation (Brooks and Gelman, 1998), with values less than 1.1 indicating good between-chain convergence (Gelman and Hill, 2006). The Heidelberger–Welch diagnostic tests for stationarity (Heidelberger and Welch, 1983). Specifically, it tests whether or not a sample chain's mean changes over the entire MCMC sample. If the mean does not change, the chain is stationary. It is always a good idea to visually inspect the chains in addition to making sure that they have converged (i.e., that the draws from the various chains overlap and that the chains are stationary). When the chains converge, all are combined to produce the posterior distribution.

To assess overall model fit to the data, Gelman and Hill (2006) recommend carrying out posterior predictive checks. Posterior predictive checks use a standard discrepancy statistic, such as the sum of squared deviations of observed values from predictions, to examine how well the fitted model can generate new data. The simulated new data based on MCMC draws and the actual data are compared by measuring the lack of fit to model predictions. Large differences in fit between the two data sets indicate that the model misfits the actual data and should be modified. Lack of fit can be examined visually or can be used to compute a Bayesian p-value (p_B), which quantifies the frequency with which the discrepancy for the simulated data is greater than the discrepancy for the actual data. Values between 0.15 and 0.85 indicate that the model fits the data well (Hobbs and Hooten, 2015).

Like AIC, there are equivalent Bayesian methods for comparing multiple non-nested models. The approach used will depend on the data collected and the analysis to be conducted (Hooten and Hobbs, 2015). For the epizootic field experiments previously described, the Watanabe–Akaike information criterion (WAIC) is perfectly suitable. The WAIC is similar to the AIC in that there are two components of the formula to parsimoniously balance model fit and model complexity. The WAIC takes a similar form to the AIC and is calculated by computing:

$$\text{WAIC} = -2\text{lppd} + 2p_w, \tag{12.14}$$

where lppd is the log posterior predictive density and p_w is the associated parameter penalty (Gelman et al., 2014) for overfitting the model. The posterior predictive density is based on how well the posterior estimates of the model (e.g., transmission rate) predict new data. Since there are no new data, we simply ask how well each estimate of the transmission rate from the posterior MCMC sample does in predicting the data at hand. The second term, which determines the effective number of parameters in the Bayesian model, is the sum of the variance associated with the log posterior predictive density (Hobbs and Hooten, 2015). Formally, the WAIC can be written as:

$$\text{WAIC} = -2\sum_{i=1}^{n}\log\left(\frac{\sum_{j=1}^{J}\left(y_i \mid \Theta^j\right)}{J}\right) + 2\sum_{i=1}^{n}\text{Var}\left(\log\left(y_i \mid \Theta\right)\right),$$

where n is the number of observations, J is the number of samples of the posterior, y_i represents each data point, and Θ^j are the parameter estimates from a single sample j of the posterior (Hobbs and Hooten, 2015). The WAIC represents just one metric that can be used to validate Bayesian models. While this formula can appear daunting, numerous resources exist that can help one in either understanding or calculating the WAIC (e.g., Gelman et al., 2014; Hobbs and Hooten, 2015; Hooten and Hobbs, 2015). Another popular method is cross-validation, whereby some data are used to fit the model and others are left out to test how well the model does in fitting them. In fact, the penalty term for the WAIC can be considered an approximation to cross-validation (Gelman et al., 2014). The most appropriate metric will depend upon the model being fit, the data, and the manner in which the data have been collected (Hooten and Hobbs, 2015).

12.5.2 An Example of the WAIC in Action

The same experiment and data from the AIC example can also be analyzed from a Bayesian perspective, and the associated WAIC scores calculated. For the models considered, the rankings are similar to the AIC results (Table 12.1). The model with the lowest WAIC scores is still the model where C differs but the transmission rate $\bar{\beta}$ stays

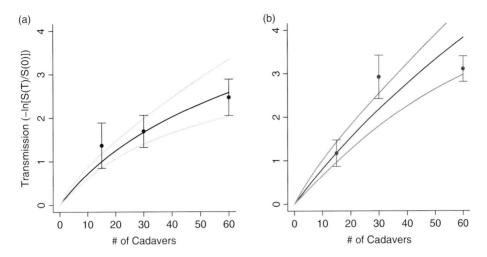

Fig. 12.2 Best-fit model (Table 12.1) for (a) control and (b) warmed plots. The solid dark line represents the predicted fraction infected (equation 12.5) given the median Bayesian estimates of $\bar{\beta}$ and C for each treatment. The lighter shaded lines correspond to the 90% credible intervals (CIs) associated with these estimates. The points represent the mean of the data. The bars are the standard errors.

the same across treatments. Note, for the WAIC, that there are no equivalent metrics associated with model comparisons, such as weights used in the AIC (see earlier). However, the use of the WAIC continues to be developed and refined. When applied to fall armyworm virus data, the model results show that the coefficient of variation C increases as temperatures increases, which results in an increase in overall transmission at higher cadaver densities (Fig. 12.2). Using the WAIC, the same conclusion can be drawn: that when temperatures rise the coefficient of variation associated with transmission declines and the dynamics become more and more similar to linear transmission dynamics (Elderd and Reilly, 2014). At the end of the day, both the AIC and the WAIC result in the same best model. The advantage of using a Bayesian framework becomes more readily apparent as the models considered become increasingly complicated.

12.6 Long-Term Dynamics

The focus, so far, has been on single occurrences of a high prevalence of disease in a population (i.e., a single epizootic). Considerable research also focuses on modeling the long-term dynamics of insect populations driven by semiregular epizootic events. As this research has shown via the use of mechanistic models, epizootics drive or help drive the boom-and-bust population cycles often associated with insects, particularly those of economic concern.

As previously mentioned, Anderson and May's (1980) seminal paper combined ideas from two often disparate fields of research: predator–prey dynamics and epidemiology. Most previous efforts in modeling disease outbreaks focused on single epizootic events. These models are best exemplified in the epidemiological literature as the SIR models (Kermack and McKendrick, 1927), in which a main assumption is that the overall population size does not change over the course of an epidemic or epizootic (Anderson and

May, 1979). This approach works well with questions focused on near-term consequences, such as, "How many individuals will become infected over the course of an
epizootic?" On the other hand, predator–prey models focus on the long-term population dynamics of prey and their predators, which are based on the classic work of Lotka
(1932) and Volterra (1926). Anderson and May used ideas from both fields to construct
a model showing that larch bud moth (*Zeiraphera diniana*) outbreaks could be driven by
host–pathogen interactions (Anderson and May, 1980). Surprisingly, prior to their work,
ecologists generally ignored the ability of pathogens to control the population dynamics
of an insect (Anderson and May, 1981). Interestingly, more recent work on the same
larch bud moth system has shown that parasitoids, not pathogens, drive the boom-and-
bust cycles (Kendall et al., 1999; Turchin, 2003). When expanding the model to include
spatial dynamics, dispersal, along with plant quality, can play an important role (Bjørnstad
et al., 2002). The change in the driver of the cycle from the pathogen to the parasitoid
exemplifies the importance of continually confronting observational data with mechanistic models and modifying a model as new data and new hypotheses emerge.

12.6.1 Long-Term Dynamics: Confronting Models with Data

For the univoltine gypsy moth, the short-term dynamics associated with epizootics during the larval phase and the long-term dynamics associated with adult reproduction can
be considered separately. First, the epizootic occurs (a within-generation process), and
then reproduction occurs (a between-generation process). A number of mechanistic
models have been developed to describe this within- and between-generation process
(e.g., Dwyer et al., 2004; Bjørnstad et al., 2010; Elderd et al., 2013). The general gestalt of
these models is summarized nicely by Fuller et al. (2012).

To start off, consider the short-term or within-generation dynamics, which are governed by a series of differential equations that track the entirety of the epizootic process.
The equations are:

$$\frac{dS}{dt} = -\bar{\beta} \left[\frac{S(t)}{S(0)} \right]^{C^2} SV, \tag{12.15}$$

$$\frac{dE_1}{dt} = \bar{\beta} \left[\frac{S(t)}{S(0)} \right]^{C^2} SV - m\delta E_1, \tag{12.16}$$

$$\frac{dE_i}{dt} = m\delta E_{i-1} - m\delta E_i \left(i = 2, \ \dots, \ m \right), \tag{12.17}$$

$$\frac{dV}{dt} = m\delta E_m - \mu V. \tag{12.18}$$

Here, the equivalent terms have the same meanings as before (see equation 12.4). A
major change from the classic SIR model is reflected in the fact that there is now an
exposed class E, within which there are a number of different stages E_i. The individuals in each i th stage, E_i, have consumed enough virus to become infected but have not
yet succumbed to the virus and become pathogen, V. If there is only a single infected

class in the model, some larvae will instantly become pathogen, as exposed individuals continually move at an exponential rate out of the single exposed class (Keeling and Rohani, 2008). By allowing for m total stages, the infected stages becomes a sum of exponential distributions, which is a gamma distribution with a mean of $1/\delta$, where δ is the average speed of kill, and a variance of $1/m\delta$. The number of stages depends upon both the mean and the variance estimates of the speed of kill. For gypsy moth larvae, the best estimates are $1/\delta$ = 12 days and m = 20 (Fuller et al., 2012). To reiterate, equations 12.15–12.18 only describe the within-season dynamics of the insect host when it is susceptible and succumbs to the baculovirus.

Long-term or between-season dynamics of the host population track host reproduction after the epizootic ends. Recall, the epizootic ends either due to the uninfected individuals pupating or due to epizootic burnout (Dwyer et al., 2000; Fuller et al., 2012). At the end of the epizootic, the equations describing the long-term dynamics are:

$$N_{n+1} = \lambda N_n \left[1 - I\left(N_n, Z_n\right) \right] \left(1 - \frac{abN_n}{b^2 + N_n^2} \right), \tag{12.19}$$

$$Z_{n+1} = fN_n I\left(N_n, Z_n\right) + \gamma Z_n. \tag{12.20}$$

Here, N_n and Z_n are the densities of the hosts and the cadavers before the epizootic in generation n and $I\left(N_n, Z_n\right)$ is the fraction of the larvae that become infected (equations 12.15–12.18). The net reproductive rate is λ. For outbreaking insects, population densities are kept at low levels during inter-outbreak periods by generalist predators or parasitoids (Dwyer et al., 2004). For gypsy moth populations, this can take the form of a Type III functional response. The fraction surviving predation is represented by the term $1 - abN_n / \left(b^2 + N_n^2\right)$, where a is the maximum predation rate and b is the saturation constant. Baculovirus densities depend upon the survival f of virus derived from the current generation and the survival of virus γ from previous generations. While it is likely that sublethal or covert infections play only a small role in the long-term dynamics, the preceding model also adequately describes covert infections. It assumes that some fraction of the virus survives from one generation to the next, which could be derived from covert infections. As long as this fraction is density-independent, the model provides an accurate accounting of covert infections (Elderd et al., 2013). Over the course of multiple generations, the modeling consists in stringing together the short-term (e.g., one season for univoltine gypsy moths) epizootic followed by adult reproduction, which sets the stage for the next epizootic.

12.6.2 Time-Series Diagnostics

While fitting models to data using results from short-term experiments draws directly from the standard statistical literature, long-term data sets represent a different problem from an analytical perspective. They are often observational and constitute a classic example of an "inverse problem" (Kendall et al., 1999), such that the data collected may arise due to many different mechanistic processes (e.g., intraspecific density-dependent regulation vs. host–pathogen interactions). How best to decide which mechanisms may be responsible for the observed data is central to understanding what drives the boom-and-bust cycles associated with long-term epizootic dynamics.

For many of these observational data sets, the data are not directly fitted to the model. For instance, a number of papers exploring gypsy moth long-term dynamics use defoliation data as a proxy for gypsy moth population numbers (e.g., Dwyer et al., 2004; Elderd et al., 2008; Bjørnstad et al., 2010). To compare the model output with the observational data, authors often rely on matching various metrics associated with the time series of the data (e.g., average period between peak outbreaks or defoliation events) with the model output. Directly fitting the model to the data becomes increasingly problematic if the dynamics of the system are chaotic, since the model and the data are sensitive to initial conditions (Dwyer et al., 2004). Thus, instead of directly fitting the data to determine which model drives the observed dynamics, "time-series" probes are advocated (Kendall et al., 1999; Turchin, 2003).

Kendall et al. (1999) were among the first advocates in the ecological literature to push for the use of "time-series" probes by combining time-series statistics with mechanistic population models. Previous to this paper, most time-series analyses consisted of fitting nonmechanistic models that could be considered biologically naïve to observational data. On the other side of the coin were the theoretical population ecologists who constructed biologically explicit models that elicited general patterns seen in the data but often did not use standard goodness–of-fit metrics to see if their models stood up to the data (Kendall et al., 1999). The use of "time-series" probes blends the two historic approaches by combining biologically reasonable models with time-series analytical approaches.

There are three steps to this approach. The first consists in constructing the mathematical model describing the long-term dynamics (e.g., equations 12.19–12.20). In the second, once the model is constructed, it is parameterized using independent data and/ or other time-series data. In the third step, the model predictions are compared to the time series using a suite of statistical probes, such as average period, amplitude, autocorrelation, and spectral density functions. This is done by simulating the model to generate a suite of synthetic time series and comparing the simulated dynamics to the actual time series. In the end, if the model fits the data using the time-series probes, the hypothesis driving the mechanistic models is worth pursuing. While Kendall et al. (1999) used data from Nicholson's classic blowfly populations to demonstrate the utility of this approach, the usefullness of time-series probes in understanding long-term epizootic dynamics has been demonstrated time and time again (e.g., Dwyer et al., 2004; Johnson et al., 2006; Abbott and Dwyer, 2008; Elderd et al., 2008; Bjørnstad et al., 2010).

While others have directly compared model output to data using classical goodness-of-fit measures for non-epizootically driven time series (Ives et al., 2008), the time-series probe approach remains popular. However, methods described by Ives et al. (2008) hold promise from both a standard-likelihood perspective and from a more Bayesian one (Barraquand et al., 2017) for examining cyclic dynamics. Thus, likelihood- and Bayesian-based approaches may also hold promise in the realm of modeling epizootic dynamics.

To understand how and whether a method works, as outlined in Ives et al. (2008), simulating data represents a useful first step. Exploratory analyses consist simply in using a model (e.g., equations 12.19–12.20) to create fake data by adding process or measurement error to the model's deterministic skeleton. The simulated data are analyzed using a method of choice (e.g., a Bayesian approach), and the results are compared to the known simulated truth (Kéry, 2010).

To examine the methods used in Ives et al. (2008) from a Bayesian perspective, data were simulated using the nondimensionalized version of the burnout approximation model in Dwyer et al. (2000). The simulation used three equations to represent the dynamics, as follows:

$$N_{n+1} = \lambda \varepsilon_n N_n \left[1 - I\left(N_n, Z_n\right) \right],$$ (12.21)

$$Z_{n+1} = \phi N_n I\left(N_n, Z_n\right) + \gamma Z_n,$$ (12.22)

$$1 - I\left(N_n, Z_n\right) = \left[1 + C^2 N_n I\left(N_n, Z_n\right) + Z_n \right]^{-1/C^2}.$$ (12.23)

Here, ϕ is the product of pathogen survival and mean susceptibility of newly emerging larvae (Dwyer et al., 2000) and ε_n is a log normally distributed random variable with a median of 1 and a standard deviation of σ. For simplicity of presentation, γ is set to 0 and there are no generalist predators in the model. To understand the boom-and-bust dynamics of the insect host population given the preceding, there are only three parameters in the nondimensionalized model that matter: λ, C, and ϕ. All of the other parameters simply move the population mean up or down, and do not affect the period or amplitude of the population cycles. In terms of the simulated data, the analysis only uses the time series associated with the host population, N, as an input. Overall, the Bayesian approach

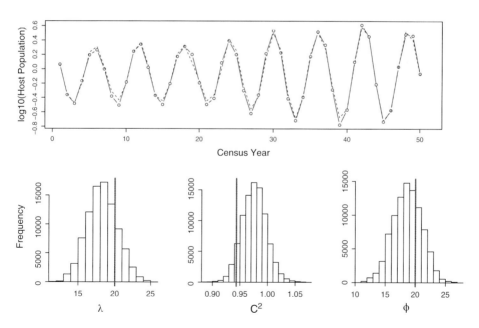

Fig. 12.3 Bayesian estimates of a simulated time series using equations 12.15–12.18, assuming epizootic burnout (Dwyer et al. 2000) without generalist predators. In the top panel, the simulated data are represented by the solid line and the best estimate of the dynamics by the dashed line. The bottom three panels show the associated histograms for each of the parameters estimated from the model. The vertical lines represent the true values of the parameters used to simulate the data.

does a reasonable job of predicting the dynamics and the model's associated true parameter values (Fig. 12.3). The next step is to confront the same modeling framework with data.

12.7   Modifying and Applying the Model

It is important to remember that models, like hypotheses, are not static. The development of epizootic models, whether regarding single epizootic events or the long-term dynamics of the host population, have continued and will continue to evolve as more data are collected and new hypotheses emerge. For the gypsy moth, the development of changes in the basic long-term model still continues. While earlier models focused on host–pathogen dynamics alone (Dwyer et al., 1997, 2000), later models included the importance of generalist predators (equations 12.19–12.20) (Dwyer et al., 2004), the effect of host–density on disease resistance (Reilly and Hajek, 2008), host evolutionary dynamics (Elderd et al., 2008), and host food resources (Bjørnstad et al., 2010; Elderd et al., 2013). The models that incorporated changes in host food source or food quality also incorporated spatial components to characterize outbreak dynamics in different forest types.

While the preceding highlights model development with regard to gypsy moth populations, model development and modification are also important to understanding epizootic dynamics in other insect species (e.g., Hochberg, 1989; Boots and Mealor, 2007; Elderd and Reilly, 2014).

Epizootic models can also be used to examine applied problems, which can be studied by modeling epizootic dynamics before enacting management strategies or policies. As a number of insect species are pests, pathogens and parasites represent potential biocontrol agents. While the easiest way to deal with a short-term infestation may be to release pathogens into the environment, the long-term effects often remain unknown. The main reasons for this stem from the need for long-term observational data collected over multiple epizootic periods (often decades) and the lack of data collected during a series of long-term experimental trials (i.e., controls and experimental plots). A prime example of such a successful combination of experimental trials and model development in a non-insect system is found in Hudson et al. (1998), who focused on controlling population declines in the red grouse (*Lagopus lagopus scoticus*) caused by a parasitic nematode. The field component of this research took almost a decade at multiple sites, which might not be possible in the midst of an insect infestation. However, the use of mechanistic models of insect epizootic dynamics for applied problems within the confines of one's computer allows the researcher to conduct experiments via simulation. This approach has led to insights into the effectiveness of pathogens as biocontrol agents for a number of insect taxa (Hochberg, 1989; Hochberg and Waage, 1991; Reilly and Elderd, 2014). Reilly and Elderd (2014), in their model of the gypsy moth–baculovirus system, showed that spraying focused only on the short-term effects of controlling an insect population may lead to unexpected and undesirable consequences. For certain spraying regimes, the gypsy moth population can be maintained at constant population levels instead of exhibiting boom-and-bust cyclic dynamics, but the average population size is relatively high. This, in turn, may result in constant defoliation of the forests that these biocontrol efforts are trying to protect. Overall, while the mechanistic models provide a deeper understanding of what drives epizootics in insect populations, they also provide a useful tool for asking questions of an applied nature.

12.8   Conclusion

The use of models to understand epizootic dynamics has a long history in the ecological literature. Much of the past debate concerning which methodology is best suited for moving the field forward centered on the historic false dichotomy between empirical and theoretical approaches, while sometimes invoking simulation-based methods. However, the ability to confront models with data has led to new and exciting developments in the field, since models can now be used as hypotheses to drive research questions. While using the preceding techniques and ideas may seem easy to some and daunting to others, they do not necessarily need to be mastered by all. Instead, they represent a framework to begin a conversation about questions that can be answered, how to design empirical studies, and how best to use the data produced. The reason the false dichotomy of empiricism and theory continues to blur stems from more individuals being able to speak in multiple languages. Thus, mastering each technique is not essential, but being able to communicate across the false divide is. As the dialogue advances and individuals speak across their own expertise, the biology of the system becomes better connected to the mechanistic framework, which leads to a better understanding of what drives the epizootic process.

Acknowledgments

The author would like to thank Jennifer Kluse and all the lab and field personnel who helped along the way – a list way too numerous to include given the page limits. He would also like to thank the University of Colorado, Boulder, Department of Applied Mathematics and Vanja Dukic for both physical space and refuge to write this chapter. The author thanks Fred Barraquand for providing him coding advice used in the long-term dynamics model. Lastly, the author would like to thank his mentors, Greg Dwyer and Dan Doak, who got him started on this journey. This work was funded by NSF grant 1316334 as part of the joint NSF-NIH-USDA Ecology and Evolution of Infectious Diseases program.

References

Abbott, K.C., Dwyer, G., 2008. Using mechanistic models to understand synchrony in forest insect populations: the North American Gypsy Moth as a case study. Am. Nat. 172, 613–624.

Anderson, D., Burnham, K., Thompson, W.L., 2000. Null hypothesis testing: problems, prevalence, and an alternative. J. Wildlife Manage. 64, 912–923.

Anderson, R.M., May, R.M., 1979. Population biology of infectious diseases: part I. Nature 280, 361–367.

Anderson, R.M., May, R.M., 1980. Infectious diseases and population cycles of forest insects. Science 210, 658–661.

Anderson, R.M., May, R.M., 1981. The population dynamics of microparasites and their invertebrate hosts. Philos. Trans. R. Soc. B Biol. Sci. 291, 451–524.

Anderson, R.M., May, R.M., 1991. Infectious Disease of Humans: Dynamics and Control. Oxford University Press, Oxford.

Barraquand, F., Louca, S., Abbott, K.C., Cobbold, C.A., Cordoleani, F., DeAngelis, D.L., et al., 2017. Moving forward in circles: challenges and opportunities in modeling population cycles. Ecol. Letters. In Press.

Bjørnstad, O.N., Peltonen, M., Liebhold, A.M., Baltensweiler, W., 2002. Waves of larch budmoth outbreaks in the European Alps. Science 298, 1020–1023.

Bjørnstad, O.N., Robinet, C., Liebhold, A.M., 2010. Geographic variation in North American gypsy moth cycles: subharmonics, generalist predators, and spatial coupling. Ecology 91, 106–118.

Boots, M. 2004. Modelling insect diseases as functional predators. Physiol. Entomol. 29, 237–239.

Boots, M., Mealor, M., 2007. Local interactions select for lower pathogen infectivity. Science 315, 1284–1286.

Brooks, S.P., Gelman A., 1998. General methods for monitoring convergence of iterative simulations. J. Comp. Graph. Stat. 7, 434–455.

Brown, G., 1987. Modeling, in: Fuxa, J., Tanada, Y. (eds.), Epizootiology of Insect Diseases. John Wiley & Sons, Chichester, pp. 43–68.

Burnham, K., Anderson, D., 2002. Model selection and multimodal inference: a practical information-theoretic approach. Springer, New York.

Clark, J.S., 2007. Models for Ecological Data: An Introduction. Princeton University Press, Princeton, NJ.

Cory, J.S., Hoover, K., 2006. Plant-mediated effects in insect-pathogen interactions. Trends Ecol. Evol. 21, 278–286.

Cory, J.S., Myers, J.H., 2003. The ecology and evolution of insect baculoviruses. Annu. Rev. Ecol. Evol. Syst. 34, 239–272.

De Oliveira, M. 1999. South America, in: Hunter-Fujita, F., Entwistle, P., Evans H., Crook, N., (eds.), Insect Viruses and Pest Management. John Wiley & Sons, Chichester, pp. 339–355.

Dennis, B., 1996. Discussion: should ecologists become Bayesians? Ecol. Appl. 6, 1095–1103.

Dwyer, G., Dushoff, J., Elkinton, J.S., Levin. S.A., 2000. Pathogen-driven outbreaks in forest defoliators revisited: building models from experimental data. Am. Nat. 156, 105–120.

Dwyer, G., Dushoff, J., Yee, S.H., 2004. The combined effects of pathogens and predators on insect outbreaks. Nature 430, 341–345.

Dwyer, G., Elkinton, J.S., Buonaccorsi, J.P., 1997. Host heterogeneity in susceptibility and disease dynamics: tests of a mathematical model. Am. Nat. 150, 685–707.

Edelstein, J.D., Trumper, E.V., Lecuona, R.E., 2005. Temperature-dependent development of the entomopathogenic fungus *Nomuraea rileyi* (Farlow) Samson in *Anticarsia gemmatalis* (Hubner) larvae (Lepidoptera : Noctuidae). Neotropical Entomol. 34, 593–599.

Elderd, B.D., 2013. Developing models of disease transmission: insights from the ecology of baculovirus-driven systems. PLoS Pathog. 9, e1003372.

Elderd, B.D., Dukic, V.M., Dwyer, G., 2006. Uncertainty in predictions of disease spread and public health responses to bioterrorism and emerging diseases. Proc. Natl. Acad. Sci. U.S.A. 103, 15693–15697.

Elderd, B.D., Dushoff, J., Dwyer, G., 2008. Host-pathogen dynamics, natural selection for disease resistance, and forest-defoliator outbreaks. Am. Nat. 172, 829–842.

Elderd, B.D., Haynes, K., Rehill, B., Dwyer, G., 2013. Interactions between an induced plant defense and a pathogen drive outbreaks of a forest insect. Proc. Natl. Acad. Sci. U.S.A. 110, 14978–14983.

Elderd, B.D., Reilly, J., 2014. Warmer temperatures increase disease transmission and outbreak intensity in a host-pathogen system. J. Anim. Ecol. 83, 838–849.

Ellison, A.M., 1996. An introduction to Bayesian inference for ecological research and environmental decision-making. Ecol. Appl. 6, 1036–1046.

Ellison, A.M. 2004. Bayesian inference in ecology. Ecol. Letters 7, 509–520.

Fleming-Davies, A.E., Dukic, V., Andreasen, V., Dwyer, G., 2015. Effects of host heterogeniety on pathogen diversity and evolution. Ecol. Letters 18, 1252–1261.

Fuller, E., Elderd, B., Dwyer, G., 2012. Pathogen persistence in the environment and insect-baculovirus interactions: disease-density thresholds, epidemic burnout, and insect outbreaks. Am. Nat. 179, E70–E96.

Fuxa, J.R., 1982. Prevalence of viral infections in populations of fall armyworm, *Spodoptera frugiperda*, in Southeastern Louisiana. Environ. Entomol. 11, 239–242.

Fuxa, J.R., Geaghan, J.P., 1983. Multiple-regression analysis of factors affecting prevalence of nuclear polyhedrosis virus in *Spodoptera frugiperda* (Lepidoptera, Noctuidae) populations. Environ. Entomol. 12, 311–316.

Gelman, A., Carlin, J.B., Stern, H.S., Rubin, D.B., 2014. Bayesian data analysis. Taylor & Francis, Philadelphia, PA.

Gelman, A., Hill, J., 2006. Data analysis using regression and multilevel/hierarchical models. Cambridge University Press, Cambridge.

Hajek, A.E., 1999. Pathology and epizootiology of *Entomophaga maimaiga* infections in forest Lepidoptera. Microbiol. Mol. Biol. Rev. 63, 814–835.

Hajek, A.E., Tobin, P.C., Haynes, K.J., 2015. Replacement of a dominant viral pathogen by a fungal pathogen does not alter the collapse of a regional forest insect outbreak. Oecologia 177, 785–797.

Hajek, A.E., van Nouhuys, S., 2016. Fatal diseases and parasitoids: from competition to facilitation in a shared host. Proc. R. Soc. B Biol. Sci. 283, 20160154.

Heidelberger, P., Welch, P.D., 1983. Simulation run length control in the presence of an initial transient. Oper. Res. 31, 1109–1144.

Hesketh, H., Roy, H.E., Eilenberg, J., Pell, J.K., Hails, R.S., 2010. Challenges in modelling complexity of fungal entomopathogens in semi-natural populations of insects. Biocontrol 55, 55–73.

Hilborn, R., Mangel, M., 1997. The Ecological Detective: Confronting Models with Data. Princeton University Press, Princeton, NJ.

Hille Ris Lambers, J., Clark, J.S., Lavine, M., 2005. Implications of seed banking for recruitment of southern Appalachian woody species. Ecology, 86 85–95.

Hinds, W., Dew, J., 1915. The Grass Worm or Fall Army Worm. Technical Report Bulletin no. 186, Alabama Agricultural Experiment Station.

Hobbs, N.T., Hilborn, R., 2006. Alternatives to statistical hypothesis testing in ecology: a guide to self teaching. Ecol. Appl. 16, 5–19.

Hobbs, N.T., Hooten, M.B., 2015. Bayesian models: a statistical primer for ecologists. Princeton University Press, Princeton, NJ.

Hochberg, M.E., 1989. The potential role of pathogens in biological control. Nature 337, 262–265.

Hochberg, M.E., 1991. Nonlinear transmission rates and the dynamics of infectious disease. J. Theor. Biol. 153, 301–321.

Hochberg, M.E., Waage, J.K., 1991. A model for the biological control of *Oryctes rhinoceros* (Coleoptera, Scarabaeidae) by means of pathogens. J. Appl. Ecol. 28, 514–531.

Hooten, M.B., Hobbs, N.T., 2015. A guide to Bayesian model selection for ecologists. Ecol. Monogr. 85, 3–28.

Hudson, P.J., Dobson, A.P., Newborn, D., 1998. Prevention of population cycles by parasite removal. Science 282, 2256–2258.

Ives, A.R., Einarsson, A., Jansen, V.A.A., Gardarsson, A., 2008. High-amplitude fluctuations and alternative dynamical states of midges in Lake Myvatn. Nature 452, 84–87.

Johnson, D.M., Liebhold, A.M., Tobin, P.C., Bjørnstad, O.N., 2006. Allee effects and pulsed invasion by the gypsy moth. Nature 444, 361–363.

Keeling, M.J., Rohani, P., 2008. Modeling infectious diseases in humans and animals. Princeton University Press, Princeton, NJ.

Kendall, B.E., 2015. Some directions in ecological theory. Ecology 96, 3117–3125.

Kendall, B.E., Briggs, C.J., Murdoch, W.W., Turchin, P., Ellner, S.P., McCauley, E., et al., 1999. Why do populations cycle? A synthesis of statistical and mechanistic modeling approaches. Ecology 80, 1789–1805.

Kermack, W.O., McKendrick, A.G., 1927. A contribution to the mathematical theory of epidemics. Proc. R. Soc. A Math. Phys. Eng. Sci. 115, 700–721.

Kéry, M., 2010. Introduction to WinBUGS for Ecologists. Academic Press, Boston, MA.

Levin, S.A., 2012. Towards the marriage of theory and data. Interface Focus 2, 141–143.

Liebhold, A.M., Plymale, R., Elkinton, J.S., Hajek, A.E., 2013. Emergent fungal entomopathogen does not alter density dependence in a viral competitor. Ecology 94, 1217–1222.

Link, W.A., Eaton, M.J., 2012. On thinning of chains in MCMC. Methods Ecol. Evol. 3, 112–115.

Lotka, A.J., 1932. Contribution to the mathematical theory of capture I. Conditions for capture. Proc. Natl. Acad. Sci. U.S.A. 18, 172–178.

Marion, G.M., Henry, G.H.R., Freckman, D.W., Johnstone, J., Jones, G., Jones, M.H., et al. 1997. Open-top designs for manipulating field temperature in high-latitude ecosystems. Global Change Biol. 3, 20–32.

Miller, L.K. (ed.). 1997. Baculoviruses. Kluwer Academic, Dordrecht.

Myers, J.H., Malakar, R., Cory, J.S., 2000. Sublethal nucleopolyhedrovirus infection effects on female pupal weight, egg mass size, and vertical transmission in gypsy moth (Lepidoptera : Lymantriidae). Environ. Entomol. 29, 1268–1272.

Onstad, D.W., Carruthers, R.I., 1990. Epizootiological models of insect diseases. Annu. Rev. Entomol. 35, 399–419.

Pitre, H.N., Hogg, D.B., 1983. Development of the fall armyworm (Lepidoptera, Noctuidae) on cotton, soybean and corn. J. Georgia Entomol. Soc. 18, 182–187.

Plummer, M. et al., 2003. JAGS: a program for analysis of Bayesian graphical models using Gibbs sampling, in: Proceedings of the 3rd International Workshop on Distributed Statistical Computing, Vol. 124, p. 125. Technische Universit at Wien, Vienna.

R Core Team., 2015. R: a language and environment for statistical computing. R Foundation for Statistical Computing, Vienna. Available from: https://www.R-project. org/ (accessed May 8, 2017).

Reilly, J.R., Elderd, B.D., 2014. Effects of biological control on long-term population dynamics: identifying unexpected outcomes. J. Appl. Ecol. 51, 90–101.

Reilly, J.R., Hajek, A.E., 2008. Density-dependent resistance of the gypsy moth *Lymantria dispar* to its nucleopolyhedrovirus, and the consequences for population dynamics. Oecologia 154, 691–701.

Richter, A.R., Fuxa, J.R., Abdelfattah, M., 1987. Effect of host plant on the susceptibility of *Spodoptera frugiperda* (Lepidoptera, Noctuidae) to a nuclear polyhedrosis virus. Environ. Entomol. 16, 1004–1006.

Rohani, P., Keeling, M.J., Grenfell, B.T., 2002. The interplay between determinism and stochasticity in childhood diseases. Am. Nat., 159 469–481.

Roy, H.E., Hails, R.S., Hesketh, H., Roy, D.B., Pell, J.K., 2009. Beyond biological control: non-pest insects and their pathogens in a changing world. Insect Conserv. Div. 2, 65–72.

Scheiner, S.M., 2013. The ecological literature, an idea-free distribution. Ecol. Letters 16, 1421–1423.

Scholte, E.J., Ng'habi, K., Kihonda, J., Takken, W., Paaijmans, K., Abdulla, S., et al., 2005. An entomopathogenic fungus for control of adult African malaria mosquitoes. Science 308, 1641–1642.

Shapiro-Ilan, D., Bruck, D., Lacey, L., 2012. Principles of epizootiology and microbial control, in: Tanada, Y., Kaya, H.K. (eds.), Insect Pathology. Academic Press, Amsterdam, pp. 29–72.

Sparks, A.N., 1979. Review of the biology of the fall armyworm (Lepidoptera, Noctuidae). Florida Entomol. 62, 82–87.

Stuart, R.J., Barbercheck, M.E., Grewal, P.S., Taylor, R.A.J., Hoy, C.W., 2006. Population biology of entomopathogenic nematodes: concepts, issues, and models. Biol. Control 38, 80–102.

Turchin, P., 2003. Complex Population Dynamics: A Theoretical/Empirical Synthesis. Princeton University Press, Princeton, NJ.

Volterra, V., 1926. Fluctuations in the abundance of a species considered mathematically. Nature 118, 558–560.

Wasserstein, R.L., Lazar, N.A., 2016. The ASA's statement on p-values: context, process, and purpose. Am. Stat. 70, 129–133.

Wootton, J.T., 2005. Field parameterization and experimental test of the neutral theory of biodiversity. Nature 433, 309–312.

Yu-Sung, S., Masanao, Y., 2015. R2jags: using R to Run "JAGS." Available from: https://CRAN.R-project.org/package=R2jags (accessed May 8, 2017).

13

Leveraging the Ecology of Invertebrate Pathogens in Microbial Control

Surendra K. Dara[1], Tarryn A. Goble[2] and David I. Shapiro-Ilan[3]

[1] *University of California Cooperative Extension, Division of Agriculture and Natural Resources, San Luis Obispo, CA, USA*
[2] *Department of Entomology, Cornell University, Ithaca, NY, USA*
[3] *USDA-ARS, SEA, SE Fruit and Tree Nut Research Unit, Byron, GA, USA*

13.1 Basics of Microbial Control and Approaches

Microbial control is the management of pests with beneficial microorganisms, which may include bacteria, fungi, nematodes, protists, and viruses. There are four basic approaches that include the release of microbial control agents (MCAs) through classical, inoculative, or inundative approaches in order to introduce organisms in either concentrated or large areas, as well as environmental manipulation, which helps to conserve the released or existing organisms (Fuxa, 1987; Shapiro-Ilan et al., 2012). Microbial control falls within the realm of biological control, where biocontrol agents such as parasitoids, predators, and MCAs are either released or conserved for pest management (Helyer et al., 2014). Inoculative and inundative approaches are considered for augmentation of MCAs that may already exist, but at low levels. A diverse array of MCAs have been commercialized and each is effective against certain pests, depending on their feeding habits and habitat (Table 13.1).

13.1.1 Classical Microbial Control

Classical microbial control (also referred to as the introduction/establishment approach) typically involves exploration of invertebrate pathogens in the native range of invasive pests, followed by importation, mass production, evaluation of specificity and efficacy, and release of the pathogen in the new range where the invasive pest is a problem. This approach has also been used to introduce exotic MCAs for control of native pests. An example of classical microbial control involves the cassava green mite (*Mononychellus tanajoa*), a pest in the neotropics, which was introduced into sub-Saharan Africa and caused significant yield losses to cassava (Lyon, 1973). The entomopthoralean fungus *Neozygites tanajoae* (=*N. floridana*), which causes natural epizootics in *M. tanajoa* populations in Brazil, was successfully introduced from this country into Benin, West Africa to reduce pest populations (Hountondji et al., 2002). Another successful example is the introduction of entomophthoralean, *Entomophaga maimaiga*, into the United States (Hajek et al., 1998) and Europe (Zúbrik et al., 2016) in order to control the gypsy moth

Ecology of Invertebrate Diseases, First Edition. Edited by Ann E. Hajek and David I. Shapiro-Ilan.
© 2018 John Wiley & Sons Ltd. Published 2018 by John Wiley & Sons Ltd.

Table 13.1 Microbial control agents (MCAs) from different pathogen groups: examples of commercialized organisms and recommended target pests.

MCA	Target pests
Bacteria	
Bacillus sphaericus	Diptera (mosquitoes)
B. thuringiensis subsp. *aizawai*	Lepidoptera
B. thuringiensis subsp. *israelensis*	Diptera
B. thuringiensis subsp. *kurstaki*	Lepidoptera
B. thuringiensis subsp. *tenebrionis*	Coleoptera
Chromobacterium subtsugae	Acarina, Coleoptera, Homoptera, Hemiptera, and
Paenibacillus popilliae and	Lepidoptera
P. lentimorbus	Japanese beetle, *Popillia japonica*
Fungi	
Beauveria bassiana	One or more pests of Acarina, Coleoptera, Diptera,
B. brongniartii	Hemiptera, Hymenoptera, Lepidoptera, Orthoptera,
Hirsutella thompsonii	Thysanoptera, and others
Isaria fumosorosea	Diptera (mosquitoes)
Lagenidium giganteum	Mainly Acarina
Lecanicillium lecanii	Mainly Coccidae
L. longisporum	Mainly Aphididae
L. muscarium	Mainly Thysanoptera and Hempitera
Metarhizium anisopliae	Plant-parasitic nematodes
M. brunneum	
Paecilomyces lilacinus	
Entomopathogenic nematodes	
Heterorhabditis bacteriophora Poinar	Various pests among the Coleoptera, Lepidoptera,
H. indica Poinar, Karunakar & David	Diptera, Thysanoptera, Siphonaptera, and Hemiptera;
H. marelata Liu & Berry	except *S. scapterisci*, which is specific to Orthoptera
H. megidis Poinar, Jackson & Klein	(Shapiro-Ilan et al. 2014)
H. zealandica Poinar	Slugs
Steinernema carpocapsae (Weiser)	
S. feltiae (Filipjev)	
S. glaseri (Steiner)	
S. kraussei (Steiner)	
S. riobrave	
S. scapterisci	
S. longicaudum	
Other Nematodes	
Phasmarhabditis hermaphrodita	
Protozoa	
Nosema locustae	Orthoptera and some Lepidoptera

Table 13.1 (Continued)

MCA	Target pests
Viruses	
Granuloviruses (GVs)	
Cydia pomonella GV	Lepidoptera
Thaumatotibia leucotreta GV	Codling moth, *Cydia pomonella*
Pthorimaea operculella GV	False codling moth, *Thaumatotibia leucotreta*
Pieris rapae GV	Potato tuberworm, *Phthorimaea operculella*
Plutella xylostella GV	Imported cabbageworm, *Pieris rapae*
Nucleopolyhedroviruses (NPVs)	
Agrotis ipsilon multicapsid NPV	Diamondback moth, *Plutella xylostella*
Anticarsia gemmatalis NPV	Black cutworm, *Agrotis ipsilon*
Autographa californica multicapsid NPV	Velvetbean caterpillar, *Anticarsia gemmatalis*
Helicoverpa armigera multicapsid NPV	Alfalfa looper, *Autographa californica*
Helicoverpa zea single nucleocapsid NPV	Cotton bollworm, *Helicoverpa armigera*
Spodoptera exigua multicapsid NPV	Corn earworm, *H. zea*
Spodoptera littoralis multicapsid NPV	Beet armyworm, *Spodoptera exigua*
Lymantria dispar multicapsid NPV	Cotton leafworm, *S. littoralis*
	Gypsy moth, *Lymantria dispar*

(*Lymantria dispar*). The first successful example of classical microbial control with entomopathogenic nematodes (EPNs) occurred in Florida, where *Steinernema scapterisci* was imported from Uruguay, successfully established, and used to provide control of mole crickets (*Scapteriscus* spp.) (Parkman and Smart, 1996), which had previously been introduced from Argentina and Uruguay into turf grass, pastures, and vegetables. Populations of the devastating coconut rhinoceros beetle (*Oryctes rhinoceros*) were effectively controlled by the introduction of *Oryctes* virus from Malaysia into different countries in the Australasia and South Pacific areas (Shyam Prasad et al., 2008). Several successful examples of classical microbial control of insect and mite pests using multiple categories of pathogens are described by Hajek et al. (2005).

13.1.2 Inoculative Release

In this approach, exotic or native MCAs produced in large quantities are introduced, multiply in the habitat, and provide some level of long-term control (multiseason or multiyear). However, periodic re-releases may be required and permanent establishment is not expected. *Lecanicillium lecanii* is used in an inoculative approach to the control of green peach aphid (*Myzus persicae*) in greenhouses (Gullan and Cranston, 2014). In the case of the entomophthoralean fungus *Pandora neoaphidis*, which is difficult to culture *in vitro*, releasing infected live aphids helped with the dispersal of the pathogen (Dara and Semtner, 1996). Released aphids would disperse among their populations, eventually die from pathogen infections, and spread the inoculum to healthy individuals. Hajek et al. (2007) discussed similar approaches, involving the release of insects infected with *Entomophaga praxibuli* against the differential grasshopper

(*Melanoplus differentialis*), *Entomophaga aulicae* against the browntail moth (*Euproctis chryssorhoea*), *Neozygites fresenii* against the cotton aphid (*Aphis gossypii*), and multiple species of microsporidia against *L. dispar*. The bacterium *Paenibacillus popilliae* has been used in inoculative applications for the control of the Japanese beetle (*Popillia japonica*) in lawns (Klein et al., 2007).

13.1.3 Inundative Release

This approach is somewhat similar to the usual mode of application for conventional pesticides. Mass-produced native or exotic MCAs are applied at regular intervals, without the expectation of significant recycling toward multiseason or multiyear control.

13.1.4 Conservation/Environmental Manipulation

Environmental manipulation to conserve native or released MCAs helps their preservation and propagation in the pest environment and increases their success. Providing environmental conditions that are ideal for the survival and multiplication of MCAs and avoiding or minimizing agricultural practices that are detrimental to these agents are important aspects of conservation. Primarily, conservation methods for MCAs are geared toward overcoming their sensitivity to harsh environmental conditions such as exposure to ultraviolet (UV) radiation, high temperatures, or dry conditions (Kung et al., 1991; Jones et al., 1993; Braga et al., 2001; Shapiro-Ilan et al., 2012). Predicting the epizootics of *N. fresenii* in aphids on cotton helped delay pesticide applications (Steinkraus, 2007), and agronomic practices such as conservation tillage and planting ground covers increased the abundance of EPNs and fungi in vegetable crops (Hummel et al., 2002). Specific examples are discussed later in the chapter.

13.2 Ecological Considerations

The success of microbial control depends on the primary goal of eliciting an outbreak of disease in the pest population. Thus, the major factors influencing an epizootic include: the MCA (host specificity, dispersal, virulence, and pathogen density), the target pest population (host density, behavioral and genetic resistance), transmission, and various environmental factors (Shapiro-Ilan et al., 2012). An ideal MCA would be an organism that is able to achieve a successful epizootic in a pest population, is easy to mass-produce, formulate, and apply, and can persist under adverse environmental conditions.

13.2.1 Host Specificity

Both host specificity and a lack of it can be beneficial in microbial control. For classical microbial control, a high degree of specificity is generally required (e.g., to minimize unintended impact on non-target organisms). However, entomopathogens that have a narrow host range may be less likely to be developed by commercial companies as inundative agents due to the limited market size. Specificity can range from within single insect orders or families to single genera or species. For example, *Bacillus thuringiensis* (Bt) or *Bacillus thuringiensis kurstaki* (Btk) is specific to Lepidoptera, *Bacillus thuringiensis israelensis* (Bti) to Diptera, and *Bacillus thuringiensis tenebrionis* (Btt) to

Coleoptera, while *P. popilliae* is specific to *P. japonica* (Federici, 1999). Several viruses are also specific to a particular species of insect and are effective in controlling that species without any impact on other ones. Compared to viruses and certain bacteria, in general, the most commonly used species of entomopathogenic fungi (EPF) (i.e., hypocrealean Ascomycetes) have broader host ranges because infection is caused by cuticular penetration of the propagules and they are not restricted to entry into the host by ingestion (Inglis et al., 2001). EPF such as *B. bassiana*, *Hirsutella thompsonii*, *Isaria fumosorosea*, *Metarhizium anisopliae*, and *Metarhizium brunneum* are especially ideal for controlling insects with sucking mouthparts, but can also infect coleopterans, lepidopterans, and orthopterans that have chewing mouthparts. EPNs also are generally effective against multiple orders of insects.

13.2.2   Dispersal Ability

The ability of MCAs to disperse passively or actively and spread infections impacts microbial control success. Rain or irrigation and wind currents can help with the dispersal of MCAs (Tanada and Kaya, 1990; Inglis et al., 2001; Ekesi et al., 2005; Jaronski, 2010). Although at least some species in each pathogen group (except viruses) are motile, EPNs are the only group that, as a whole, primarily use active dispersal to locate hosts (Tanada and Kaya, 1990). Infective juveniles (IJs) disperse from infected cadavers in search of new hosts. In some cases, EPNs can use earthworms, certain mites, and isopods as phoretic hosts for their dispersal (Shapiro-Ilan and Brown, 2013). Movement of infected live hosts or natural enemies within or between fields is another major source of dispersal for multiple pathogens (Steinkraus et al., 1995; Ekesi et al., 2005). Several taxa of Entomophthoromycota forcibly discharge their conidia, which increases their chances of encountering a host (Humber, 2016). Conidia of *P. neoaphidis* (= *Erynia neoaphidis*) can be discharged to a maximum distance of 9 mm horizontally and 8 mm vertically from the aphid cadavers (Hemmati et al., 2001), while the conidia of *Entomopthora muscae* and *E. schizophorae*, pathogens of the house fly (*Musca domestica*) can be discharged up to 87.5 mm distance (Six and Mullens, 1996). In the case of the entomophthoralean fungus *Strongwellsea castrans*, infected flies, with a hole formed in the abdomen due to fungal infection, fly around for a few days dispersing infective conidia (Nair and McEwen, 1973).

13.2.3   Virulence

Virulence, the disease-causing power of an organism, is one of the most important attributes for an MCA and is primarily related to its genetic makeup. Different strains of MCAs have a wide range of virulence against a particular pest or multiple species of pests. When an MCA is commercialized, factors such as environmental stability, safety to non-target organisms, and ease of industrial-scale production are also considered, instead of selecting a strain based entirely on high virulence (Ravensberg, 2011).

13.2.4   Pathogen Density

A high density and widespread spatial distribution of the MCAs in the field is critical to the success of microbial control as it increases the contact between susceptible pests and inoculum. Except for EPNs, which actively seek out a host, most MCAs rely on host arthropods coming in contact with their inocula to initiate infection. Infection process can occur by penetration through cuticle or natural openings or by ingestion. Inoculative and inundative microbial control attempt to introduce more inoculum into the environment in order to start an epizootic.

13.2.5 Host-Related Factors

Various host-related factors can increase or decrease the likelihood of an epizootic. Certain pathogens influence the behavior of the host insect in a manner that helps the spread of the infection. Caterpillars infected with some baculoviruses and locusts and grasshoppers infected with *Entomophaga grylli* move toward the top of the plant before they die, thereby increasing the spread of infective propagules to healthy hosts on the lower parts of canopy (Pell et al., 2001; Harrison and Hoover, 2012). Alate aphids and flies infected with some entomophthoralean fungi disperse and spread the infection (Pell et al., 2001). On the other hand, grasshoppers and other insects can display what is known as behavioral fever, where infected individuals move toward warmer parts of the canopy to increase body heat and thus reduce the progression of infection (Inglis et al., 2001). Other behaviors, such as grooming or quarantining infected individuals in social insects like ants and termites, can also reduce the spread of disease (Fuxa, 1987).

A number of other host-related factors that influence the infectivity and virulence of entomopathogens include variation in host immune response and genetic resistance, as well as the host plant, host age, host life stage, and mutations in the host insect. For example, susceptibility of the Colorado potato beetle (*Leptinotarsa decemlineata*) to *B. bassiana* varied among different solanaceous hosts (Hare and Andreadis, 1983). Immature stages of the legume flower thrips (*Megalurothrips sjostedti*) and the German cockroach (*Blattella germanica*) were less susceptible than the adult stages to *M. anisopliae* (Ekesi and Maniania, 2000; Lopes and Alves, 2011). Molting, which removes fungal conidia penetrating the host cuticle, might have reduced the chances of infection in these cases. In contrast, developmental resistance is seen in some Lepidoptera, where mature larvae are less susceptible than young larvae to virus infection due to an increase in immunoresponsiveness with age. The reduced susceptibility of the Australian bollworm (*Helicoverpa punctigera*) to Heliothis zea nucleopolyhedrovirus (HzNPV) (Teakle et al., 1986) and of *L. dispar* to Lymantria dispar multinucleopolyhedrovirus (LdMNPV) (McNeil et al., 2010) in older larvae are two examples of developmental resistance. On the other hand, the diamondback moth (*Plutella xylostella*) developed resistance when Bt products were excessively used (Tabashnik et al., 1990).

13.2.6 Transmission

MCAs can be transmitted horizontally or vertically in host populations. In horizontal transmission, healthy insects are exposed to MCA propagules when they come in contact with infected individuals during mating (Toledo et al., 2007), grooming (Fuxa, 1987), foraging (Richards et al., 1999), cannibalism (Vasconcelos, 1996), or other interactions or behaviors (Sarfraz et al., 2011). Additionally, horizontal transmission may occur from host to environment to host. Environmental routes such as the growth of the fungal mycelia in the soil from infected cadavers are possible for some EPF (Gottwald and Tedders, 1984). Defecation and regurgitation from virus-infected insects (Vasconcelos,

1996) also contribute to MCA transmission. Vertical transmission from infected parent directly to the offspring occurs extensively in viruses as well as certain bacteria, fungi, and protists, but does not occur in EPNs (Andreadis, 1987; Tanada and Kaya, 1990). Vertical transmission can occur through the egg, such as on the egg surface (transovum) and sometimes within the egg via the ovary (transovarial, which can be seen as a special case of transovum).

13.2.7 Environmental Persistence

Persistence is an important issue but can be a double-edged sword. At a minimum, a pathogen needs to persist in the environment long enough to infect the target host (Shapiro-Ilan et al. 2012). The longer a pathogen can persist, the greater its efficacy, to the point that in some cases, greater environmental persistence may even compensate for lower virulence of some pathogens (Shapiro-Ilan et al. 2002). However, in some countries, regulatory authorities require inundatively applied MCAs such as EPF to decline and persist at lower, natural background levels in the soil so that they do not displace indigenous microorganisms, release toxic metabolites, or attack non-target organisms (Scheepmaker and Butt, 2010). Scheepmaker and Butt (2010) showed that natural background levels of indigenous EPF ranged from 740 to 830 colony-forming units (CFU)/g of soil and that inundatively released MCAs declined to natural background levels within 0.5–1.5 years for *B. bassiana*, 4 years for *Beauveria brongniartii*, and >10 years for *M. anisopliae*. It has also been reported that EPF can endophytically colonize and persist in plant tissues for several weeks (Batta, 2013; Dara et al., 2013). EPNs may only survive a matter of weeks or months, whereas Bt and certain baculoviruses may survive years or even decades in the soil (Van Cuyk et al., 2011; Shapiro-Ilan et al., 2012). Conversely, Fuller et al. (2012) estimated that persistence of the gypsy moth baculovirus on foliage was less than 3 days. However, when multiple generations were considered, it was shown that this virus may survive for longer periods even in the absence of complex persistence mechanisms, such as environmental reservoirs or covert (sublethal) infections. Further, Clark et al. (2005) reviewed the environmental persistence of Bt *Cry* toxins from transgenic plants and reported that *Cry1Ab* proteins had half-lives of 1.6 days within the soil but that low-level residues within various plant tissues lasted for months. There are many aspects in the environment that play a major role in entomopathogen persistence and thus determine microbial control efficacy. These include the intrinsic properties of the MCA (strain, propagule type, culture and storage conditions), as well as edaphic (soil texture and moisture, organic matter, pH), biotic (other soil organisms and microorganisms: synergists, antagonists, phoretic agents), climatic (temperature, moisture, UV), and cultural factors (tillage, crop type, fertilizers, pesticides) (Scheepmaker and Butt, 2010; Shapiro-Ilan et al., 2012). Formulation of MCAs can also influence persistence (see later in the chapter).

Conventional wisdom is that the long-term persistence of baculoviruses relies on their survival in the external environment in the form of occlusion bodies. These are proteinaceous matrices in which the virus particles are embedded, and which provide a degree of protection from UV irradiation. However, the cabbage moth, *Mamestra brassicae* has also been shown to harbor a persistent, nonlethal baculovirus infection (Mamestra brassicae nucleopolyhedrovirus, MbNPV) in laboratory culture, which may represent another putative persistence mechanism (Murillo et al., 2011). Such covert infections are also present and frequent in natural populations of the moth. The persistent infections were triggered into the lethal overt state by exposure to another baculovirus, and two closely related but different baculoviruses were subsequently identified as persistent infections within the populations sampled. These results have broad-ranging implications for our understanding of host–pathogen interactions in the field, the use of pathogens as biocontrol agents, and the evolution of virulence.

13.3   Methods to Improve Microbial Control

Based on the factors described in the previous section, the efficacy of microbial control can be enhanced by several approaches, including improvement of the organism, production practices, and formulation and application technologies.

13.3.1   Improving the Organism as a Microbial Control Agent

As the market for microbial pesticides continues to grow, bioprospecting – the process of discovering and commercializing biological organisms – efforts have increased. Since the virulence of the MCA and its environmental adaptability are both important, bioprospecting strains isolated from pest habitats can represent a good strategy. For example, a highly virulent and environmentally adapted *M. anisopliae* s.l. (strain F1-1045) isolated from the greyback canegrub (*Dermolepida albohirtum*) (Waterhouse) in Australian sugarcane was developed as BioCane for the control of scarab pests in sugarcane (Milner et al., 2002).

Directed strain selection or hybridization to improve certain traits can enhance the efficacy of MCAs. In directed strain selection, specific traits such as virulence, germination speed, formulation stability, and the ability to reproduce and persist under harsh environmental conditions (e.g., fungicide stress) are enhanced, as are tolerance to desiccation, thermal energy, and UV radiation (Berling et al., 2009; de Crecy et al., 2009; Shapiro-Ilan et al., 2011). A few examples include *M. anisopliae* isolates selected for thermostability at >35 °C to overcome behavioral fevers in locusts and grasshoppers (de Crecy, 2009), *B. bassiana* and *M. brunneum* with fungicide resistance (Shapiro-Ilan et al., 2011), and EPNs with improved host-seeking behavior (Gaugler et al., 1989), nematicide resistance (Glazer et al., 1997), and temperature tolerance (Ehlers et al., 2005). Baculovirus selection has been used to overcome codling moth resistance to a granulovirus (Berling et al., 2009). However, care should be taken to avoid inadvertent selection of unwanted traits, such as loss in storage capabilities while selecting for EPNs with enhanced host-finding abilities (Gaugler et al., 1990) or reduced reproductive ability in *B. bassiana* and *M. brunneum* with fungicide resistance (Shapiro-Ilan et al., 2011).

Hybridization is another way to improve MCAs. In Bt, conjugation allowed the transfer of compatible plasmids with toxin genes specific to different insect orders and generated improved strains with broader host ranges (Carlton and Gawron-Burke, 1993). Since genetic incompatibility is common in EPF, techniques such as protoplast fusion induced stable hybridization in *B. bassiana* for hypervirulence (Viaud et al., 1998) or expanded host ranges in species of *Lecanicillium* to improve biocontrol efficacy relative to parent isolates (Goettel et al., 2008). This approach (of protoplast fusion) has not yet been implemented commercially for EPF, although it has been commercially established for fungal control of plant pathogens (Sivan and Harman, 1990). Hybridization in EPNs has probably received the most attention, with improvements in heat tolerance in *H. bacteriophora* and targeted host-finding in *Steinernema carpocapsae* (Shapiro et al., 1997; Shapiro-Ilan et al., 2014). Meanwhile, a combination of trait selection (for desiccation tolerance) and hybridization (for improved storage stability) improved *S. feltiae* (Nimkingrat et al., 2013).

Recombinant DNA techniques to express genes-encoding toxins and other insecticidal proteins from other organisms can help improve the efficacy of MCAs (Inceoglu et al., 2006; Lu et al., 2008; St. Leger and Wang, 2010). Overexpression of scorpion neurotoxins and cuticle-degrading enzymes such as chitinases and proteases, which were inserted into the genomes of *M. anisopliae* and *B. bassiana*, improved fungal virulence and speed of kill (Fan et al., 2007; Lu et al., 2008; St. Leger and Wang, 2010). Genetically enhanced *H. bacteriophora* with a heat shock protein (HSP) from a free-living nematode species increased thermotolerance with no negative impact on field persistence when released in turfgrass field microplots (Gaugler et al., 1997). Similarly, the speed of kill of Autographa californica nucleopolyhedrovirus (AcNPV) was improved either by deleting the endogenous baculovirus gene *egt* (ecdysteroid UDP-glucosyltransferase) or by genetic modification to express insect hormones or insect-selective toxins from scorpions, spiders, sea anemones, and mites (Inceoglu et al., 2006; Harrison and Hoover, 2012). Toxicity of Bt was improved through modified *Cry1A* toxins (Mod toxins) (Soberón et al., 2007). While these examples demonstrate the potential of improving MCAs, such genetically modified entomopathogens may face significant registration challenges due to their potential environmental risks and consumer acceptability issues, and they are not used in agriculture today.

13.3.2 Improving Production Methods

The ease of commercial-scale production is important in the success of an MCA as a biopesticide, because such improvements reduce the cost of microbial control, allowing increased application rates and competitiveness with other integrated pest-management tactics. The global success of Bt is largely attributed to its economical means of commercial production through liquid batch fermentation using inexpensive ingredients (Federici, 1999; Couch and Jurat-Fuentes, 2014). In most cases, the most economical way of producing MCAs is through liquid-state fermentation *in vitro*; however, some pathogens are not amenable and require *in vivo* production, as is the case with all baculoviruses, the bacterium *P. popilliae*, and some fungi (Jackson et al. 2010; Jaronski, 2014; Shapiro-Ilan et al., 2014). Nutritional and environmental conditions during the mass-production of MCAs are also very important as they influence the form and efficacy of resulting propagules (Jackson, 1997; Shapiro-Ilan et al., 2014). Liquid media have been used for large-scale production of some hypocrealean Ascomycetes and EPNs (Jackson et al. 2010; Jaronski, 2014; Shapiro-Ilan et al., 2014). Specific liquid culture conditions led to the production of *M. anisopliae* microsclerotia, and formulation of these propagules improved insect control (Jaronski and Jackson, 2008; Jackson and Jaronski, 2009). *In vitro* production of baculoviruses in insect cell culture is a significant improvement in their production, due to inexpensive media that support cell growth, surfactants that reduce shear stresses during fermentation, and lack of disease problems that would affect an insect colony (Harrison and Hoover, 2012). However, this process needs optimization in developing cell lines that are suited for rapid growth in bioreactors, identifying viral isolates capable of producing large numbers of occlusion bodies, and limiting other issues such as mutations that reduce virulence.

13.3.3 Improving Formulation and Application Technologies

Microbial pesticides contain propagules or metabolites of organisms as active ingredients, in a state of either low or no metabolic activity, and changing technical aspects of the formulations can improve the stability, efficacy, and persistence of the organism and the safety and registration of the product (Burges, 1998; Ravensberg, 2011; Behle and Birthisel, 2014). Formulations vary depending on the mode of action of the MCA, the type of propagules used, and the habitat and ecology of the target pest (Jaronski, 2010; Ravensberg, 2011). Technologies that improve a biopesticide formulation include oxygen and carbon dioxide absorbers and water-absorbing gels (to improve shelf life), enzymes such as chitinases, proteases, and lipases (to increase virulence), UV blockers and hydrogels (to improve survival), oils and gums (to enhance secondary pick-up of inocula), and other additives such as growth-stimulating compounds, insect alarm pheromones, low doses of insecticides, attractants (sugars and baits), and repellents (garlic oil and aromatic compounds) (Ravensberg, 2011). Application of EPF such as *B. bassiana* and *M. anisopliae* in oils can improve control efficacy by increasing coverage on insect cuticle and conidial germination (Prior et al., 1988; Bateman et al., 1993). Oils are also preferred in ultra-low-volume applications at low humidity or in areas where water is scarce (Bateman et al., 1993; Douthwaite et al., 2001). Entrapment of Bti spores in calcium alginate microcapsules (Elcin, 1995) resulted in improved persistence in adverse environmental conditions. Similarly, encapsulation in the cells of the protist, *Tetrahymena pyriformis*, allowed targeted delivery to the water surface and resulted in faster time to death through increased toxicity against mosquito larvae (Manasherob et al., 1996).

Some commercially available products contain entomopathogenic organisms and/or their metabolites, and other compounds have been developed to provide improved biological control of insect pests. For example, the application and high efficacy of *S. carpocapsae* in a chitosan formulation (Biorend R, IDEBIO, Spain) against red palm weevil (*Rhynchophorus ferrugineus*) was demonstrated in a semi-commercial field trial (Llácer et al., 2009). Chitosan is a biodegradable product made by treating chitin from the shells of shrimp and other crustaceans with sodium hydroxide, which has the active ingredient N-acetyl-glucosamine, which can activate the defense mechanisms in plants (Hadwiger, 2013), increases lignification, and promotes the development of roots (Ait Barka et al., 2004). There are some other commercially available products based on bacteria and their metabolites. For example, Venerate (Marrone Bio Innovations, CA), which contains the heat-killed cells and fermented solids of the Gram-negative proteobacterium *Burkholderia rinojensis* (used to control foliar pests in orchards and row crops), controls pests, on contact or when ingested, through the enzymatic degradation of the insect exoskeleton, which leads to interference with molting. Grandevo PTO (Marrone Bio Innovation, CA) contains the bacterial metabolite of *Chromobacterium subtsugae* and is a broad-spectrum bioinsecticide used in turf and in ornamental and greenhouse markets. In New Zealand, a formulated, commercially available product called Beaublast, containing the spores and metabolites of *B. bassiana* strain K4B3, has been shown to have insecticidal knockdown activity toward diamondback moth caterpillars and aphids within 48 hours.

EPN applications can be enhanced significantly by optimizing application equipment or spraying aspects (e.g., nozzles, pumps, spray distribution) to provide enhanced pathogen survival and dispersion (Shapiro-Ilan et al., 2006). Delivery of EPNs inside infected host cadavers compared to the standard aqueous application can increase nematode dispersal, infectivity, survival, and efficacy (Shapiro and Glazer, 1996; Shapiro and Lewis, 1999; Shapiro-Ilan et al., 2003a; Kaplan et al., 2012). At least some of the advantages observed in cadaver-applied EPNs (such as increased infectivity and dispersal) have been explained by cues or signals that are given off from within the infected host, including nematode behavioral response (Shapiro and Lewis, 1999; Kaplan et al., 2012). Protection against UV radiation is one of the challenges in developing various MCA formulations. Multiple studies have demonstrated that plant extracts or oils, as well as microencapsulation of nucleopolyhedrovirus (NPV) occlusion bodies in lignin, confer resistance to UV inactivation (El Salamouny et al., 2009a,b; Shapiro et al., 2012), while application of gels (e.g., firegel) provides protection against UV and desiccation in EPN applications (Shapiro-Ilan et al., 2016). Additionally, formulation of viruses with optical brighteners (UV-absorbing fluorescent brightening compounds) not only conferred resistance to UV, but also increased the infectivity of the viruses 2.6–5731.0-fold (Shapiro and Robertson, 1992; Lasa et al., 2007).

Application technology for microbial pesticides can be improved in several areas (pathogen survival, dispersal, deposition rates, targeted applications, timing, equipment, mixing of attractants and other pesticides, autodissemination, employment of endophytes) to leverage their ecological success. Applying baculovirus products in smaller droplets promotes even coverage of foliage, but droplets tend to shrink rapidly due to evaporation, which necessitates the addition of compounds that prevent this (Ignoffo et al., 1976; Killick, 1990). Targeted applications of *B. bassiana* to the undersides of leaves, where pest whitefly larvae reside in cucurbits, significantly improved insect control (Wraight et al., 2000). A novel improvement in microbial control application came about with the discovery of the endophytic growth of *B. bassiana* in maize (Bing and Lewis, 1991), which launched a field of research that now looks at endophyte-mediated plant defense as a novel biological control mechanism in a wide range of crops (e.g., coffee, cotton, sorghum, etc.) (Vega et al., 2008; see Chapter 9). Autodissemination – an application strategy whereby an attractive trap (visual, chemical, or food) is artificially seeded with MCAs to be spread by insects – has greatly enhanced microbial control of fruit flies (Ekesi et al., 2007), diamondback moth (Furlong et al., 1995), and red palm weevils (El-Shafie et al., 2011); additionally, a commercial autodissemination device based on *B. bassiana* has recently been introduced and shown to be effective for mosquito control (Snetselaar et al., 2014).

13.3.4 Improving the Environment

Modifying the environment may be necessary in order to improve the efficacy and environmental persistence of MCAs. For example, application of MCAs in the evening, instead of in the morning, can avoid several hours of exposure to UV radiation and potentially warm temperatures. Modification of irrigation practices that promote pathogen efficacy and avoidance of pesticides and other agricultural chemicals that are incompatible with pathogens can help conserve MCAs. Irrigation prior to and within 24 hours after the application of *H. bacteriophora* enhanced control efficacy against *P. japonica* and the northern masked chafer (*Cyclocephala borealis*) in Kentucky bluegrass (Downing, 1994). Mulches and crop residues can be used to maintain moisture for EPN activity and infectivity (Sweeney et al., 1998; Shapiro et al., 1999). In Argentina, research centers use irrigation as an environmental manipulation strategy to promote natural infection of cicadas (*Proarna bergi*) by *Cordyceps* (= *Ophiocordyceps*) *sobolifera* (Alves et al., 2003).

Improving the texture of heavy clay soils could improve the efficacy of EPNs, as soil type and depth had a significant effect on the foraging efficacy of *S. riobrave* Poinar, Cabanillas, and Raulston in belowground applications (Kaspi et al., 2010). EPN and EPF efficacy can also be improved by adjusting the pH of soils (Fischer and Fuhrer, 1990).

Microbial control can be enhanced by using agrochemicals such as fertilizers and fungicides that are compatible with MCAs in tank mixes or in soil. Understanding the compatibility of agricultural chemicals with individual MCAs and avoiding those that are harmful, or applying them at an ideal time point, can also conserve MCAs. For example, the neonicotinoid insecticide imidacloprid has a synergistic interaction with certain EPF and EPNs but is antagonistic to *N. tanajoae* (Quintela and McCoy, 1998; Koppenhöfer et al., 2000; Dara and Hountondji, 2001). Similarly, several fungicides are compatible with *B. bassiana*, while Thiram and Captan negatively affected its infectivity (Dara et al., 2014). The efficacy of *B. bassiana* against the larvae of the red flour beetle (*Tribolium castaneum*) increased when it was used in combination with diatomaceous earth (Akbar et al., 2004), which is abrasive to cuticle and thus helps improve the infection process. Using compatible or synergistic chemicals and avoiding others is important for the success of MCAs. Host plants and potting media also have an impact. For example, *S. feltiae* was more efficacious against the fungus gnat (*Bradysia coprophila*) in impatiens than in poinsettia (Jagdale et al., 2004). Greenhouse potting mixes that contain combinations of peat, pine straw, vermiculite, sphagnum, and other ingredients can also influence nematode movement or survival, and thus substrates can be optimized to maximize efficacy (Jagdale et al., 2004).

13.4   Incorporating Microbial Control into Integrated Pest-Management Systems

13.4.1   Regulatory Issues

Recent concerns over human and environmental safety restricted the use of broad-spectrum chemical pesticides and increased opportunities for alternative control measures such as MCAs. However, regulatory issues can also be a challenge for the wider use of MCAs (Sundh and Goettel, 2013; Lacey et al., 2015). Even when the registration costs for MCAs are lower than those for chemical pesticides, the extent of regulatory requirements can still be a major barrier that prevents or limits the development of new products.

Regulations on the use of MCAs as pesticides are extensive in Europe and the United States, mainly because they are treated akin to chemical pesticides despite their known relative safety with respect to humans and the environment (European Commission, 2005; Sundh and Goettel, 2013; Lacey et al., 2015; FIFRA, 2016). However, EPNs are an exception and require minimal or no registration in most countries, as they are considered similar to arthropod predators and parasitoids (Ehlers, 2005; Grzywacz et al., 2014). Regulatory policies vary in less industrialized countries, and the lack of sufficient infrastructure in some countries hinders the ability to develop appropriate policies (Grzywacz et al., 2014). There is a need to streamline the regulation and registration processes in tune with the science that dictates the safety of MCAs. A step in that direction was incorporated into the European Union's Regulation of Biological Control Agents (REBECA) project, which was created to develop a more favorable regulatory system for biological control (Ehlers, 2007). Regulatory procedures will be enhanced through a greater understanding of entomopathogen ecology at community and ecosystem levels, as well as by advances in biogeography, invasion biology, and host–pathogen interactions (Chandler, 2008; Lacey et al., 2015).

13.4.2 Standalone vs. Integrated Approaches

MCAs can be used both as a standalone measure and as a tool in an integrated pest management (IPM) program, as they fit well with other tactics and collectively contribute to pest management. There are several examples where standalone inundative applications of MCAs provided good pest control. Bt applications successfully controlled the tomato leaf miner (*Tuta absoluta*) and protected tomato yields without the need for chemical pesticides in studies conducted in Spain (González-Cabrera et al., 2011). In Indonesia, the brown stink bug (*Riptortus liearis*), a major pest of soybeans, was effectively controlled by inundative application of *L. lecanii* (Prayogo, 2014). *Steinernema riobrave* and *Heterorhabditis indica* are highly virulent to the diaprepes root weevil (*Diaprepes abbreviatus*) in orchard crops, where larvae cause severe root damage (Duncan et al., 1996; Shapiro-Ilan et al., 2005). From the late 1990s to the 2000s, EPNs were the sole recommended control measure for *D. abbreviatus* that had become established in orchards, although their success depended on interactions with endemic EPNs, nematophagous fungi, and free-living nematodes (Shapiro-Ilan et al., 2005).

An example of an integrated approach includes using *S. riobrave* against the soil-dwelling stages of plum curculio (*Conotrachelus nenuphar*) in conjunction with trap-trees and chemical insecticides targeting adults (Leskey et al., 2008; Shapiro-Ilan et al, 2008a, 2013a). Another example of an integrated approach involved using *B. bassiana* with azadirachtin or lower label rates of chemical pesticides in an IPM program for the control of the western tarnished plant bug (*Lygus hesperus*) in California strawberries (Dara, 2015).

13.4.3 Case Studies

Several examples show the diversity of microbial control approaches adopted in various agroecosystems.

13.4.3.1 Orchard Crops

Orchard systems possess shading, temporal and structural stability, adequate soil moisture due to irrigation, and other aspects that are conducive to MCAs. A good example
of the use a combination of MCAs against different life stages of a pest can be seen in
the control of the pecan weevil (*Curculio caryae*), which is a key pest of pecans in the
southeastern United States and elsewhere (Shapiro-Ilan et al., 2007). Adult weevils
emerge from soil in the summer and feed on and oviposit into the nuts. Larvae develop
within the nut, drop to the soil, and form a soil cell, where they spend the next 2 or 3
years (90 and 10% of the population, respectively) before emerging as adults and recom-
mencing the cycle.

EPF, EPNs, and bacteria-based products can provide control of *C. caryae* (Lacey
and Shapiro-Ilan, 2008; Shapiro-Ilan et al., 2013b). EPNs, particularly *S. carpocapsae*,
are highly virulent to adult *C. caryae* and moderately virulent to larvae (Shapiro-Ilan
et al., 2003b; Shapiro-Ilan and Gardner, 2012). EPF, including *B. bassiana* and
Metarhizium spp., are also pathogenic to *C. caryae* adults and larvae (Shapiro-Ilan
et al., 2008b, 2003c). Applications of *B. bassiana* to the tree trunk, or to the ground in
conjunction with a cover crop intended to protect fungi from UV, resulted in ≥75%
adult *C. caryae* mortality (Shapiro-Ilan et al., 2008b). A product based on *C. subtsugae* applied as a canopy spray for *C. caryae* adults resulted in levels of control (e.g.,
>87%) that were equal to those produced by applications of the standard chemical
insecticides (i.e., carbaryl alternated with pyrethroids) (Shapiro-Ilan et al., 2013b,
unpublished data).

This approach targets different life stages of the pest, taking advantage of different
pathogen groups. Nematodes are applied to the understory in the spring, when temperatures are favorable, to penetrate *C. caryae*'s soil cell and kill the insect before it
emerges. Fungi are not likely to be effective in penetrating the soil cell due to the physical barrier and the antifungal properties associated with the cell (Shapiro-Ilan and
Mizell, 2015); thus, fungi are applied in late summer or early fall to create a barrier to
emerging adults or larvae entering soil. *Chromobacterium subtsugae* applied to the
orchard canopy targets the adults. This integrated approach has resulted in significant
reductions (e.g., >80%) of *C. caryae* infestations in organic pecan orchards (Shapiro-Ilan
et al., unpublished data).

13.4.3.2 Row Crops

Row crops are disturbed systems that vary considerably in their canopy structure and
microclimate; some of these systems are favorable environments for microbial control.
The velvetbean caterpillar (*Anticarsia gemmatalis*) is a major defoliator of soybean in
the Americas (Moscardi and Sosa-Gómez, 2007). Microbial control programs involving field production of Anticarsia gemmatalis multicapsid nucleopolyhedrovirus
(AgMNPV) have been established in countries like Brazil, where labor is relatively
inexpensive (Moscardi, 1999; Moscardi and Sosa-Gómez, 2007). Baculovirus-infected
caterpillars, which tend to climb to the top of the canopy and have obvious signs of
disease, were hand-collected and used for control of the target pest, and their efficacy
was found to be equivalent (e.g., 80%) to the use of chemical insecticides (Moscardi and
Sosa-Gómez, 2007). Frozen, infected larvae were initially distributed in the field before
a kaolin-based formulation was developed. Early pilot studies began in 1980, and
within a decade, the approach was used over 1 million hectares annually at $1.50 or less

per hectare (Moscardi, 1999). The use of AgMNPV against *A. gemmatalis* has been recognized as the largest virus-based biocontrol program in the world.

13.4.3.3 Forests

Similar to the orchard systems, forest environments provide shading and temporal and structural stability, yet management costs must be low in order for microbial control approaches to be viable. Natural epizootics of a fungus and a baculovirus, and inoculative applications of Bt and the baculovirus, contributed to the control of *L. dispar*, a major exotic pest of North American forests (Sharov et al., 2002; van Frankenhuyzen et al., 2007; Hajek et al., 2015).

LdMNPV, accidentally introduced into the United States in the early 1900s, is virulent to *L. dispar* (van Frankenhuyzen et al. 2007; Hajek et al., 2015). Positive results in field applications in the 1960s and 70s culminated in the USDA Forest Service registering an LdMNPV product named Gypchek in 1978 (van Frankenhuyzen et al. 2007). Subsequently, government-sponsored production and inoculative application of Gypchek was initiated by the USDA Forest Service and USDA APHIS, where up to 2000 ha were treated annually (van Frankenhuyzen et al., 2007). The use of the virus was also implemented on a commercial level in Canada and parts of Europe.

Entomophaga maimaiga is another accidentally introduced entomopathogen of *L. dispar*, first discovered in the United States causing natural epizootics in 1989 (Hajek et al., 1990). Since then, the fungus has spread both naturally and with some human intervention (i.e., releasing resting spores in some locations) (Hajek et al., 2015). Once established, *E. maimaiga* resting spores can persist in the environment for years (Hajek et al., 2015). Prior to the introduction of *E. maimaiga*, populations of *L. dispar* were regulated by epizootics of LdMNPV, but *E. maimaiga* is now the dominant pathogen (Hajek et al. 2015). Thus, both entomopathogens contribute to *L. dispar* population regulation, and their infection rates are significantly influenced by biotic and abiotic conditions.

13.4.3.4 Greenhouses

Greenhouses provide controlled environments that can be highly conducive to microbial control through the regulation of relative humidity (RH) and protection from UV radiation. Additionally, the high value and high plant density of greenhouse crops allow for inundative use of microbial agents that might be cost-prohibitive in other crop conditions.

Larvae of fungus gnats in the genus *Bradysia* are considered major pests of various greenhouse crops, such as ornamental plants. The use of EPN to control fungus gnats is a biocontrol success story (Tomalak et al., 2005; Georgis et al., 2006). Biotic and abiotic factors affecting efficacy have been extensively characterized and optimized, including EPN species, temperature, application rate, insect stage, host plant, and substrate. Although various EPN species have been tested for virulence against *Bradysia* spp., *S. feltiae* has been found to be most effective, providing levels of control similar to those from chemical insecticides (Harris et al., 1995; Georgis et al., 2006). Up to 100% control was achieved with a moderate application rate of *S. feltiae* at 2.5×10^5 IJ/m^2 (Jagdale et al., 2004), but only second- to fourth-instar larvae are susceptible, and not the pupae (Tomalak et al., 2005).

13.5 Conclusion

MCAs are sustainable alternatives to chemical pesticides in several crop–pest systems. Leveraging the success of microbial control against arthropod pests depends on the choice of MCA and the type of biological control approach used, whether it be classical, inundative, inoculative, or conservation/environmental manipulation. Since the biology, infection process, host and environmental requirements, and other parameters significantly vary among various MCAs, a thorough understanding of those factors is necessary to successfully choosing MCAs for a particular crop–pest system. Microbial control can be further enhanced by improving the MCA organism itself, mass-production methods, formulation and application approaches, or environmental modification, as well as by avoiding practices that could negatively affect its success. MCAs can be either standalone strategies or invaluable parts of an IPM strategy. By substituting one or more chemical pesticide applications with MCAs, environmental risks associated with chemical pesticides can be reduced without impacting pest-control efficacy. In many parts of the world, MCAs are seen as suitable options for organic production systems, but several successful examples discussed in this chapter demonstrate their suitability for conventional agriculture and their role in IPM.

References

Ait Barka, E., Eullaffroy, P., Clement, C., Vernet, G., 2004. Chitosan improves development, and protects *Vitis vinifera* L. against *Botrytis cinerea*. Plant Cell Rep 22, 608–614.

Akbar, W., Lord, J.C., Nechols, J.R., Howard, R.W., 2004. Diatomaceous earth increases the efficacy of *Beauveria bassiana* against *Tribolium castaneum* larvae and increases conidia attachment. J. Econ. Entomol. 97, 273–280.

Alves, S.B., Pereira, R.M., Lopes, R.B., Tamai M.A., 2003. Use of entomopathogenic fungi in Latin America, in: Upadhyay, R.K. (ed.), Advances in Microbial Control of Insect Pests. Kluwer Academic Plenum Publishers, New York, pp. 193–211.

Andreadis, T.G., 1987. Transmission, in: Fuxa, J.R., Tanada, Y. (eds.), Epizootiology of Insect Diseases. John Wiley & Sons, New York, pp. 159–176.

Bateman, R. P., Carey, M., Moore, D., Prior, C., 1993. The enhanced infectivity of *Metarhizium flavoviride* in oil formulations to desert locusts at low humidities. Ann. Appl. Biol. 122, 145–152.

Batta, Y.A., 2013. Efficacy of endophytic and applied *Metarhizium anisopliae* (Metch.) Sorokin (Ascomycota: Hypocreales) against larvae of *Plutella xylostella* L. (Yponomeutidae: Lepidoptera) infesting *Brassica napus* plants. Crop Prot. 44, 128–134.

Behle, R., Birthisel, T., 2014. Formulations of entomopathogens as bioinsecticides, in: Morales-Ramos, J.A., Guadalupe Rojas, M., Shapiro-Ilan, D.I. (eds.), Mass Production of Beneficial Organisms: Invertebrates and Entomopathogens. Elsevier, New York, pp., 483–518.

Braga, G.U.L., Flint, S.D., Miller, C. D., Anderson, A. J., Roberts, D.W., 2001. Both solar UVA and UVB radiation impair conidial culturability and delay germination in the entomopathogenic fungus *Metarhizium anisopliae*. Photochem. Photobiol. 74, 734–739.

Berling, M.C., Blachere-Lopez, C., Soubabere, O., Lery, X., Bonhomme, A., Sauphanor, B., Lopez-Ferber. M., 2009. *Cydia pomenella* granulovirus genotypes overcome virus resistance in the codling moth and improve virus efficacy by selection against resistant hosts. Appl. Environ. Microbiol. 75, 925–930.

Bing, L.A., Lewis, L.C., 1991. Suppression of *Ostrinia nubilalis* (Hubner) (Lepidoptera, Pyralidae) by endophytic *Beauveria bassiana* (Balsamo) Vuillemin. Environ. Entomol. 20, 1207–1211.

Burges, H.D., 1998. Formulation of mycoinsecticides, in: Burges, H.D. (ed.), Formulation of Microbial Biopesticides. Kluwer Academic Publishers, Dordrecht, pp. 131–185.

Carlton, B.C., Gawron-Burke, C., 1993. Genetic improvement of *Bacillus thuringiensis* for bioinsecticide development, in: Kim, L. (ed.), Advanced Engineered Pesticides. Marcel Dekker, New York, pp. 43–61.

Chandler, D., Davidson, G., Grant, W.P., Greaves, J., Tatchell, G.M., 2008. Microbial biopesticides for integrated crop management: an assessment of environmental and regulatory sustainability. Food Sci. Technol. 19, 275–283.

Clark, B.W., Phillips, T.A., Coats, J.R., 2005. Environmental fate and effects of *Bacillus thuringiensis* (Bt) proteins from transgenic crops: a review. J. Agric. Food Chem. 53, 4643–4653.

Couch, T.L., Jurat-Fuentes, J.L., 2014. Commercial production of entomopathogenic bacteria, in: Morales-Ramos, J.A., Guadalupe Rojas, M., Shapiro-Ilan, D.I. (eds.), Mass Production of Beneficial Organisms: Invertebrates and Entomopathogens, Elsevier, New York, pp. 415–436.

Dara, S.K., 2015. Integrating chemical and non-chemical solutions for managing lygus bug in California strawberries. CAPCA Adviser 18, 40–44.

Dara, S.K., Dara, S.R., Dara, S.S., 2013. Endophytic colonization and pest management potential of *Beauveria bassiana* in strawberries. J. Berry Res. 3, 203–211.

Dara, S.S.R., Dara, S.S., Sahoo, A., Bellam, H., Dara, S.K., 2014. Can entomopathogenic fungus, *Beauveria bassiana* be used for pest management when fungicides are used for disease management? UC ANR eNewsletter Strawberries and Vegetables. Available from: http://ucanr.edu/blogs/blogcore/postdetail.cfm?postnum=15671 (accessed May 8, 2017).

Dara, S.K., Hountondji, F.C.C., 2001. Effects of formulated imidacloprid on two mite pathogens, *Neozygites floridana* (Zygomycotina: Zygomycetes) and *Hirsutella thompsonii* (Deuteromycotina: Hyphomycetes). Insect Sci. Applic. 21, 133–138.

Dara, S.K. Semtner, P. J., 1996. Artificial introduction of *Pandora neoaphidis* (Zygomycotina: Entomophthorales) for the control of *Myzus nicotianae* (Homoptera: Aphididae) on flue-cured tobacco. In: 29th Annual meetings of the Society for Invertebrate Pathology, September 1–6, Córdoba, Spain, p. 18.

de Crecy, E., Jaronski, S., Lyons, B., Lyons, T.J., Keyhani, N.O., 2009. Directed evolution of a filamentous fungus for thermotolerance. BMC Biotechnol. 9, 1–11.

Douthwaite, B., Langewald, J., Harris, J., 2001. Development and Commercialization of the Green Muscle Biopesticide. Ibadan: International Institute of Tropical Agriculture.

Duncan, L.W., McCoy, C.W.,Terranova, A.C., 1996. Estimating sample size and persistence of entomogenous nematodes in sandy soils and their efficacy against the larvae of *Diaprepes abbreviatus* in Florida. Journal of Nematology 28, 56–67.

Downing, A. S., 1994. Effect of irrigation and spray volume on efficacy of entomopathogenic nematodes (Rhabditida: Heterorhabditidae) against white grubs (Coleoptera: Scarabaeidae). J. Econ. Entomol. 87, 643–646.

Ehlers, R.-U., 2005. Forum on safety and regulation, in: Grewal, P.S., Ehlers, R.-U., Shapiro-Ilan, D.I. (eds.), Nematodes as Biocontrol Agents. CABI, Wallingford, pp. 107–114.

Ehlers, R-U., 2007. REBECA, Regulation of Biological Control Agents. http://www.rebeca-net.de/?p=320. Ehlers, R.U., Oestergaard, J., Hollmer, S., Wingen, M., Strauch, O., 2005. Genetic selection for heat tolerance and low temperature activity of the entomopathogenic nematode-bacterium complex *Heterorhabditis bacteriophora-Photorhabdus lumnescens*. BioControl 50, 699–716.

Ekesi, S., Dimbi, S., Maniania, N.K., 2007. The role of entomopathogenic fungi in the integrated management of fruit flies (Diptera: Tephritidae) with emphasis on species occurring in Africa, in: Ekesi, S., Maniania, N.K. (eds.), Use of Entomopathogenic Fungi in Biological Pest Management. Research Signpost 37/661, 239–274.

Ekesi, S., Maniania, N.K., 2000. Susceptibility of *Megalurothrips sjostedti* developmental stages to *Metarhizium anisopliae* and the effects of infection on feeding, adult fecundity, egg fertility and longevity. Entomol. Exp. Appl. 94, 229–236.

Ekesi, S., Shah, P. A., Clark, S. J., Pell, J. K., 2005. Conservation biological control with the fungal pathogen *Pandora neoaphidis*: implications of aphid species, host plant and predator foraging. Agric. Forest Entomol. 7, 21–30.

Elcin, Y.M., 1995. Control of mosquito larvae by encapsulated pathogen *Bacillus thuringiensis* var. *israelensis*. J. Microencaps. 12, 515–523.

El Salamouny, S., Ranwala, D., Shapiro, M., Shepard, M.B., Farrar, B.R., 2009a. Tea, coffee, and cocoa as ultraviolet radiation protectants for the beet armyworm nucleopolyhedrovirus. J. Econ. Entomol. 102, 1767–1773.

El Salamouny, S., Shapiro, M.,Ling, K.S., Shepard, B.M., 2009b. Black tea and lignin as ultraviolet protectants for the beet armyworm nucleopolyhedrovirus. J. Entomol. Sci. 44, 1–9.

El-Shafie, H.A.F., Faleiro, J.R., Al-Abbad, A.H., Stoltman, L., Mafra-Neto, A., 2011. Baitfree attract and kill technology (Hook™ RPW) to suppress red palm weevil, *Rhynchophorus ferrugineus* (Coleoptera: Curculionidae) in date palm. Fl. Entomol. 94, 774–778.

European Commission, 2005. Council Directive 2005/25/EC of March 14, 2005 Amending Annex VI of Directive 91/414/EEC as Regards Plant Protection Products Containing Micro-Organisms. Criteria for Evaluation and Authorisation of Plant Protection Products Containing Micro-Organisms.

Fan, Y., Fang, W., Guo, S., Pei, X., Zhang, Y., Xiao, Y., Li, D., Jin, K., Bidochka, M. J., Pei, Y., 2007. Increased insect virulence in *Beauveria bassiana* strains overexpressing an engineered chitinase. Appl. Environ. Microbiol. 73, 295–302.

Federici, B.A., 1999. *Bacillus thuringiensis* in Biological Control, in: Fisher, T. (ed.), Handbook of Biological Control. Academic Press, New York, pp. 575–593.

FIFRA, 2016. USEPA. Available from: https://www.epa.gov/enforcement/federal-insecticide-fungicide-and-rodenticide-act-fifra-and-federal-facilities/ (accessed May 8, 2017).

Fischer, P., Fiihrer, E., 1990. Effect of soil acidity on the entomophilic nematode *Steinernema kraussei* Steiner. Biol. Fertil. Soils 9, 174–177.

Fuller, E., Elderd, B.D., Dwyer, G. 2012. Pathogen persistence in the environment and insect baculovirus interactions: disease-density thresholds, epidemic burnout and insect outbreaks. Am Nat. 179.

Furlong, M. J., Pell, J.K., Ong, P.C. Syed, A.R., 1995. Field and laboratory evaluation of a sex pheromone trap for the autodissemination of the fungal entomopathogen *Zoophthora radicans* (Entomophthorales) by the diamondback moth, *Plutella xylostella* (Lepidoptera: Yponomeutidae). Bull. Entomol. Res. 85, 331–337.

Fuxa, J. R., 1987. Ecological considerations for the use of entomopathogens in IPM. Annu. Rev. Entomol. 32, 225–251.

Gaugler, R., Campbell, J.F., McGuire, T.R., 1989. Selection for host finding in *Steinernema feltiae*. J. Invertebr. Pathol. 54, 363–372.

Gaugler, R., Campbell, J.F., McGuire, T.R., 1990. Fitness of a genetically improved entomopathogenic nematode. J. Invertebr. Pathol. 56, 106–116.

Gaugler, R., Wilson, M., Shearer, P., 1997. Field release and environmental fate of a transgenic entomopathogenic nematode. Biol. Control. 9, 75–80.

Georgis, R., Koppenhöfer, A.M., Lacey, L.A., Bélair, G., Duncan, L.W., Grewal, P.S., Samish, M., Tan, L., Torr P., van Tol, R.W.H.M., 2006. Successes and failures in the use of parasitic nematodes for pest control. Biol. Control 38, 103–123.

Glazer, I., Salame, L., Segal, D., 1997. Genetic enhancement of nematicide resistance in entomopathogenic nematodes. Biocontr. Sci. Tech. 7, 499–512.

Goettel, M.S., Koike, M., Kim, J.J., Aiuchi, D., Shinya, R., Brodeur, J., 2008. Potential of *Lecanicillium* spp. for management of insects, nematodes and plant diseases. J. Invertebr. Pathol. 98, 256–261.

González-Cabrera, J. Mollá, O., Montón, H., Urbaneja, A., 2011. Efficacy of *Bacillus thuringiensis* (Berliner) in controlling the tomato borer, *Tuta absoluta* (Meyrick) (Lepidoptera: Gelechiidae). BioControl 56, 71–80.

Gottwald, T.R., Tedders, W.L., 1984. Colonization, transmission, and longevity of *Beauveria bassiana* and *Metarhizium anisopliae* (Deuteromycotina: Hypomycetes) on pecan weevil larvae (Coleoptera: Curculionidae) in the soil. Environ. Entomol. 13, 557–560.

Grzywacz, D., Moore, D., Rabindra, R.J., 2014. Mass production of entomopathogens in less industrialized countries. In: Morales-Ramos, Juan A., Guadalupe Rojas, M., Shapiro-Ilan, David I. (eds.), Mass Production of Beneficial Organisms. Elsevier, Amsterdam, pp. 519–553.

Gullan, P.J., Cranston, P.S., 2014. The insects: an outline of entomology, 5th edn. John Wiley & Sons, Chichester.

Hadwiger, L.A., 2013. Multiple effects of chitosan on plant systems: solid science or hype. Plant Sci. 208, 42–9.

Hajek, A.E., Humber, R.A., Elkinton, J.S., May, B., Walsh, S.R.A., Silver, J.C., 1990. Allozyme and RFLP analyses confirm *Entomophaga maimaiga* responsible for 1989 epizootics in North American gypsy moth populations. Proc. Natl. Acad. Sci. USA 87, 6979–6982.

Hajek, A.E., Bauer, L., McManus, M.L., Wheeler, M.M., 1998. Distribution of resting spores of the *Lymantria dispar* pathogen *Entomophaga maimaiga* in soil and on bark. BioControl 43, 189–200.

Hajek, A.E., McManus, M.L, Junior, ID., 2005. Catalogue of introductions of pathogens and nematodes for classical biological control of insects and mites. FHTET, USDA Forest Service.

Hajek, A.E., Junior, I.D., McManus, M.M., 2007. Introduction of exotic pathogens and documentation of their establishment and impact, in: Lacey, L. A., Kaya, H. K. (eds.), Field Manual of Techniques in Invertebrate Pathology. Springer, Dordrecht, pp. 299–325.

Hajek, A.E., Tobin, P.C., Haynes, K.J., 2015. Replacement of a dominant viral pathogen by a fungal pathogen does not alter the collapse of a regional forest insect outbreak. Oecologia 177, 785–797.

Hare, J.D., Andreadis, T.G., 1983. Variation in the susceptibility of *Leptinotarsa decemlineata* (Coleoptera: Chrysomelidae) when reared on different host plants to the fungal pathogen, *Beauveria bassiana* in the field and laboratory. Environ. Entomol. 12, 1892–1897.

Harris, M.A., Oetting, R.D., Gardner, W.A. 1995. Use of entomopathogenic nematodes and new monitoring technique for control of fungus gnats, *Bradysia coprophila* (Diptera: Sciaridae), in floriculture. Biol. Control 5, 412–418.

Harrison, R., Hoover, K., 2012. Baculoviruses and other occluded insect viruses, in: Vega, F.E., Kaya, H.K. (eds.), Insect Pathology. Elsevier, Amsterdam, pp. 74–110.

Helyer, N., Cattlin, N.D., Brown, K.C., 2014. Biological Control in Plant Protection: A Color Handbook, 2nd edn. CRC Press, Boca Raton, FL.

Hemmati, F., Pell, J.K., McCartney, H.A., Clark, S.J., Deadman, M.L., 2001. Conidial discharge in the aphid pathogen *Erynia neoaphidis*. Mycol. Res. 105, 715–722.

Hountondji, F.C.C., Lomer, C.J., Hanna, R., Cherry, A.J., Dara, S.K., 2002. Field evaluation of Brazilian isolates of *Neozygites floridana* (Entomophthorales: Neozygitaceae) for the microbial control of cassava green mite in Benin, West Africa. Biocontr. Sci. Tech. 12, 361–370.

Humber, R.A., 20016. Entomophthoromycota: a new overview of some of the oldest terrestrial fungi, in: Li, D.W. (ed.), Biology of Microfungi. Springer, Switzerland, pp. 127–145.

Hummel, R.L., Walgenbach, J.F., Barbercheck, M.E., Kennedy, G.G., Hoyt, G.D., Arellano, C., 2002. Effects of production practices on soil-borne entomopathogens in Western North Carolina vegetable systems. Environ. Entomol. 31, 84–91.

Inceoglu, A.B., Kamita, S.G., Hammock, B.D., 2006. Genetically modified baculoviruses: a historical overview and future outlook. Adv. Virus Res. 68, 323–360.

Inglis, G.D., Goettel, M., Butt, T., Strasser, H., 2001. Use of hyphomycetous fungi for managing insect pests, in: Butt, T.M., Jackson, C.W., Magan, N. (eds.), Fungi as Biocontrol Agents: Progress, Problems and Potential. CABI, Wallingford, pp. 23–70.

Ignoffo, C.M., Hostetter, D.L., Smith, D.B., 1976. Gustatory stimulant, sunlight protectant, evaporation retardant: three characteristics of a microbial insecticidal adjuvant. J. Econ. Entomol. 69, 207–210.

Jackson, M.A., 1997. Optimizing nutritional conditions for the liquid culture production of effective fungal biological control agents. J Ind. Microbiol. Biotechnol. 19, 180–187.

Jackson, M.A., Jaronski, S.T., 2009. Production of microsclerotia of the fungal entomopathogen *Metarhizium anisopliae* and their potential for use as a biocontrol agent for soil-inhabiting insects. Mycol. Res. 113, 842–850.

Jackson, M.A., Dunlap, C.A., Jaronski, S.T., 2010. Ecological considerations in producing and formulating fungal entomopathogens for use in insect biocontrol. BioControl. 55, 129–145.

Jagdale, G.B., Casey, M.L., Grewal, P.S., Lindquist, R.K., 2004. Application rate and timing, potting medium and host plant effects on the efficacy of *Steinernema feltiae* against the fungus gnat, *Bradysia coprophila*, in floriculture. Biol. Control 29, 296–305.

Jaronski, S.T. 2010. Ecological factors in the inundative use of fungal entomopathogens. BioControl. 55, 159–185.

Jaronski, S.T., 2014. Mass production of entomopathogenic fungi: state of the art, in: Morales-Ramos, J.A., Guadalupe Rojas, M., Shapiro-Ilan, D.I. (eds.), Mass Production of Beneficial Organisms. Elsevier, Amsterdam, pp. 357–413.

Jaronski, S.T., Jackson, M.A., 2008. Efficacy of *Metarhizium anisopliae* microsclerotial granules. Biocontr. Sci. Tech. 18, 849–863.

Jones, K.A., Moawad, G., McKinley, D.J., Grzywacz, D., 1993. The effect of natural sunlight on *Spodoptera littoralis* nuclear polyhedrosis virus. Biocontr. Sci. Tech. 3, 189–197.

Kaplan, F., Alborn, H.T., von Reuss, S.H., Ajredini, R., Ali, J.G., Akyazi, F., Stelinski, L.L., Edison, A.S., Schroeder, F.C., Teal, P.E., 2012. Interspecific nematode signals regulate dispersal behavior. PLoS ONE, 7, e38735.

Kaspi, R., Ross, A., Hodson, A.K., Stevens, G.N., Kaya, H.K., Lewis, E.E., 2010. Foraging efficacy of the entomopathogenic nematode *Steinernema riobrave* in different soil types from California citrus groves. Appl. Soil. Ecol. 45, 243–253.

Killick, H.J., 1990. Influence of droplet size, solar ultraviolet light and protectants, and other factors on the efficacy of baculovirus sprays against *Panolis flammea* (Schiff.) (Lepidoptera: Noctuidae). Crop Prot. 9, 21–28.

Klein, M.G., Grewal, P.S., Jackson, T.A., Koppenhöfer, A.M., 2007. Lawn, turf and grassland pests, in: Lacey, L. A., Kaya, H. K. (eds.), Field Manual of Techniques in Invertebrate Pathology. Springer, Dordrecht, pp. 655–675.

Koppenhöfer, A.M., Grewal, P.S., Kaya, H.K., 2000. Synergism of imidacloprid and entomopathogenic nematodes against white grubs: the mechanism. Entomol. Exp. Appl. 94, 283–293.

Kung, S.-P., Gaugler, R., Kaya, H.K., 1991. Effects of soil temperature, moisture, and relative humidity on entomopathogenic nematode persistence. J. Invertebr. Pathol. 57, 242–249.

Lacey, L. A., Grzywacz, D., Shapiro-Ilan, D.I.,Frutos, R.,Brownbridge, M., Goettel, M.S., 2015. Insect pathogens as biological control agents: back to the future. J. Invertebr. Pathol. 132, 1–41.

Lacey, L. A. Shapiro-Ilan, D. I., 2008. Microbial control of insect pests in temperate orchard systems: potential for incorporation into IPM. Annu. Rev. Entomol. 53, 121–144.

Lasa, R., Ruiz-Portero, C., Alcázar, M.D., Belda, J.E., Williams, T., Caballero, P., 2007. Efficacy of optical brightener formulations of *Spodoptera exigua* multiple nucleopolyhedrovirus (SeMNPV) as a biological insecticide in greenhouses in southern Spain. Biol. Control. 40, 89–96.

Leskey, T.C., Pinero, J.C., Prokopy, R.J., 2008. Oder-baited trap trees: a novel management tool for plum curculio (Coleoptera: Curculionidae). J. Econ. Entomol. 101, 1302–1309.

Llácer, E., Martínez de Altube, M.M., Jacas, J.A. 2009. Evaluation of the efficacy of *Steinernema carpocapsae* in a chitosan formulation against the red palm weevil, *Rhynchophorus ferrugineus*, in *Phoenix canariensis*. BioControl 54, 559–565.

Lopes, R.B., Alves, S.B., 2011. Differential susceptibility of adults and nymphs of *Blattella germanica* (L.) (Blattodea: Blattellidae) to infection by *Metarhizium anisopliae* and assessment of delivery strategies. Neotrop. Entomol.40, 368–374.

Lu, D., Pava-Ripoll, M., Li, Z., Wang, C., 2008. Insecticidal evaluation of *Beauveria bassiana* engineered to express a scorpion neurotoxin and a cuticle degrading protease. Appl. Microbiol. Biotechnol. 81, 515–522.

Lyon, W.F., 1973. A plant-feeding mite *Mononychellus tanajoa* (bondar) (Acarina: Tetranychidae) new to the African continent threatens cassava (*Manihot esculenta* Crantz) in Uganda, East Africa. Pest. Artic. & News Summ. 19, 36–37.

Manasherob, R., Ben-Dov, E., Margalit, J., Zaritsky, A., Barak, Z., 1996. Raising activity of *Bacillus thuringiensis* var. *israelensis* against *Anopheles stephensi* larvae by encapsulation in *Tetrahymena pyriformis* (Hymenostomatida: Tetrahymenidae). J. Am. Mosq. Control Assoc. 12, 627–631.

McNeil, J., Cox-Foster, D., Slavicek, J., Hoover, K., 2010. Contributions of immune responses to developmental resistance in *Lymantria dispar* challenged with baculovirus. J. Insect Physiol. 56, 1167–1177.

Milner, R.J., Samson, P.R., Bullard, G.K., 2002. FI–1045: A profile a commercially useful isolate of *Metarhizium anisopliae* var. *anisopliae*. Biocontr. Sci. Tech. 12, 43–58.

Moscardi, F., 1999. Assessment of the application of baculoviruses for control of Lepidoptera. Annu. Rev. Entomol. 44, 257–289.

Moscardi, F., Sosa Gómez, D. R., 2007. Microbial control of insect pests of soybean. In: Lacey, L. A., Kaya, H. K. (eds.), Field Manual of Techniques in Invertebrate Pathology. Springer, Dordrecht, pp. 411–426.

Murillo, R., Hussey, M.S., Possee, R.D., 2011. Evidence for covert baculovirus infections in a *Spodopterra exigua* laboratory culture. J. Gen. Virol. 92, 1061–1070.

Nair, K.S.S. McEwen, F.L., 1973. *Strongwellsea castrans* (Phycomycetes: Entomophthoraceae), a fungal parasite of the adult cabbage maggot, *Hylemya brassicae* (Diptera: Anthomyiidae). J. Invertebr. Pathol. 22, 442–449.

Nimkingrat, P., Straucha, O., Ehlers, R.U., 2013. Hybridisation and genetic selection for improving desiccation tolerance of the entomopathogenic nematode *Steinernema feltiae*. Biocontr. Sci. Tech. 23, 348–36.

Parkman, J.P. Smart, Jr., G.C., 1996. Entomopathogenic nematodes, a case study: introduction of *Steinernema scapterisci* in Florida. Biocontr. Sci. Tech. 6, 413–419.

Pell, J.K., Eilenberg, J., Hajek, A.E., Steinkraus, D.C., 2001. Biology, ecology and pest management potential of Entomophthorales, in: Butt, T.M., Jackson, C.W., Magan, N. (eds.), Fungi as Biocontrol Agents: Progress, Problems and Potential. CABI, Wallingford, pp. 71–153.

Prayogo, Y.S., 2014. Integration of botanical pesticide and entomopathogenic fungi to control the brown stink bug *Riptortus linearis* F. (Hemiptera: Alydidae) in soybean. J. Hama Penyakit Tumbuhan Tropika 14, 41–50.

Prior, C., Jollands, P., le Patourel, G., 1988. Infectivity of oil and water formulations of *Beauveria bassiana* (Deuteromycotina: Hyphomycetes) to the cocoa weevil pest *Pantorhytes plutus* (Coleoptera: Curculionidae). J. Invertebr. Pathol. 52, 66–72.

Quintela, E. D. McCoy, C.W., 1998. Synergistic effect of imidacloprid and two entomopathogenic fungi on the behavior and survival of larvae of *Diaprepes abreviatus* (Coleoptera: Curculionidae) in soil. J. Econ. Entomol. 91, 110–122.

Ravensberg, W.J., 2011. A roadmap to the successful development and commercialization of microbial pest control products for control of arthropods, in: Ravensberg, W.J. (ed.), Progress in Biological Control. Springer, Dordrecht, pp. 59–117.

Richards, A., Cory, J., Speight, M., Williams, T., 1999. Foraging in a pathogen reservoir can lead to local host population extinction: a case study of a Lepidoptera virus interaction. Oecologia 118, 29–38.

Sarfraz, R.M., Cervantes, V., Myers, J.H., 2011. The effect of host plant species on performance and movement behaviour of the cabbage looper *Trichoplusia ni* and their potential influences on infection by *Autographa californica* multiple nucleopolyhedrovirus. Agric. For. Entomol. 13, 157–164.

Scheepmaker, J.W.A., Butt, T.M., 2010. Natural and released inoculum levels of entomopathogenic fungal biocontrol agents in soil in relation to risk assessment and in accordance with EU regulations. Biocontr. Sci. Tech. 20, 503–552.

Shapiro-Ilan, D.I., Glazer, I., 1996. Comparison of entomopathogenic nematode dispersal from infected hosts versus aqueous suspension. Environ. Entomol. 25, 1455–1461.

Shapiro-Ilan, D.I., Lewis, E.E., 1999. Comparison of entomopathogenic nematode infectivity from infected hosts versus aqueous suspension. Environ. Entomol. 28, 907–911.

Shapiro, D.I., Glazer, I., Segal, D., 1997. Genetic improvement of heat tolerance in *Heterorhabditis bacteriophora* through hybridization. Biol. Control. 8, 153–159.

Shapiro, D. I., Obrycki, J.J., Lewis, L.C., Jackson, J.J., 1999. Effects of crop residue on the persistence of *Steinernema carpocapsae*. J. Nematol. 31, 517–519.

Shapiro, M., Robertson, J.L., 1992. Enhancement of gypsy moth (Lepidoptera: Lymantriidae) baculovirus activity by optical brighteners. J. Econ. Entomol. 85, 1120–1124.

Shapiro, M., El Salamouny, S., Jackson, D.M., Shepard, B.M., 2012. Field evaluation of a kudzu/cottonseed oil formulation on the persistence of the beet Armyworm Nucleopolyhedrovirus. J. Entomol. Sci. 47, 1–11.

Shapiro-Ilan, D., Brown, I., 2013. Earthworms as phoretic hosts for *Steinernema carpocapsae* and *Beauveria bassiana*: implications for enhanced biological control. Biol. Control 66, 41–48.

Shapiro-Ilan, D.I., Gardner, W.A., 2012. Improved control of *Curculio caryae* (Coleoptera: Curculionidae) through multi-stage pre-emergence applications of *Steinernema carpocapsae*. J. Entomol. Sci. 47, 27–34.

Shapiro-Ilan, D.I., Mizell, R.F., 2015. An insect pupal cell with antimicrobial properties that suppress an entomopathogenic fungus. J. Invertebr. Pathol. 124, 114–116.

Shapiro-Ilan, D.I., Gouge, D.H., Koppenhöfer, A.M., 2002. Factors affecting commercial success: case studies in cotton, turf and citrus, In: Gaugler, R. (ed.), Entomopathogenic Nematology. CABI, Wallingford, pp. 333–356.

Shapiro-Ilan, D.I., Lewis, E.E., Tedders, W.L., Son, Y., 2003a. Superior efficacy observed in entomopathogenic nematodes applied in infected-host cadavers compared with application in aqueous suspension. J. Invertebr. Pathol. 83, 270–272.

Shapiro-Ilan, D.I., Stuart, R., McCoy, C.W., 2003b. Comparison of beneficial traits among strains of the entomopathogenic nematode, *Steinernema carpocapsae*, for control of *Curculio caryae* (Coleoptera: Curculionidae). Biol. Control. 28, 129–136.

Shapiro-Ilan, D.I., Gardner, W.A., Fuxa, J.R., Wood, B.W., Nguyen, K.B, Adams, B.J., Humber, R.A. Hall, M.J., 2003c. Survey of entomopathogenic nematodes and fungi endemic to pecan orchards of the Southeastern United States and their virulence to the pecan weevil (Coleoptera: Curculionidae). Environ. Entomol. 32, 187–195.

Shapiro-Ilan, D.I., Duncan, L.W., Lacey, L.A., Han, R., 2005. Orchard crops, in: Grewal, P., Ehlers, R.-U., Shapiro-Ilan, D.I. (eds.), Nematodes as Biological Control Agents. CABI, Wallingford, pp. 215–230.

Shapiro-Ilan, D.I., Gouge, G.H., Piggott, S.J., Patterson Fife, J., 2006. Application technology and environmental considerations for use of entomopathogenic nematodes in biological control. Biol. Control. 38, 124–133.

Shapiro-Ilan, D.I., Lacey, L.A., Siegel, J.P., 2007. Microbial control of insect pests of stone fruit and nut crops, in: Lacey, L.A., Kaya, H.K. (eds.), Field Manual of Techniques in Invertebrate Pathology. Springer, Dordrecht, pp. 547–565.

Shapiro-Ilan, D. I., Mizell, R.F., III, Cottrell, T.E., Horton, D.L., 2008a. Control of plum curculio, *Conotrachelus nenuphar* with entomopathogenic nematodes: effects of application timing, alternate host plant, and nematode strain. Biol. Control 44, 207–215.

Shapiro-Ilan, D.I., Gardner, W.A., Cottrell, T.E., Behle, R.W., Wood, B.W., 2008b. A comparison of application methods for suppressing the pecan weevil (Coleoptera: Curculionidae) with *Beauveria bassiana* under field conditions. Environ. Entomol. 37, 162–171.

Shapiro-Ilan, D.I., Reilly, C.C., Hotchkiss, M.W., 2011. Comparative impact of artificial selection for fungicide resistance on *Beauveria bassiana* and *Metarhizium brunneum*. Environ. Entomol. 40, 59–65.

Shapiro-Ilan, D.I., Bruck, D.J., Lacey, LA., 2012. Principles of epizootiology and microbial control, in: Vega, F.E., Kaya, H.K. (eds.), Insect Pathology, 2nd edn. Elsevier, Amsterdam, pp. 29–72.

Shapiro-Ilan, D.I., Wright, S.E., Tuttle, A.F., Cooley, D.R., Leskey, T.C., 2013a. Using entomopathogenic nematodes for biological control of plum curculio, *Conotrachelus nenuphar*: Effects of irrigation and species in apple orchards. Biol. Control 67, 123–129.

Shapiro-Ilan, D.I., Cottrell, T.E., Jackson, M.A., Wood, B.W., 2013b. Control of key pecan insect pests using biorational pesticides. J. Econ. Entomol. 106, 257–266.

Shapiro-Ilan, D.I., Han, R., Qiu, X., 2014. Production of entomopathogenic nematodes, in: Morales-Ramos, J.A., Guadalupe Rojas, M., Shapiro-Ilan, D.I. (eds.), Mass Production of Beneficial Organisms: Invertebrates and Entomopathogens. Elsevier, New York, pp. 321–356.

Shapiro-Ilan, D.I., Cottrell, T.E., Mizell, R.F. III, Horton, D.L., 2016. Efficacy of *Steinernema carpocapsae* plus fire gel applied as a single spray for control of the lesser peachtree borer, *Synanthedon pictipes*. Biol. Control 94, 33–36.

Sharov, A. A., Leonard, D., Liebhold, A.M., Roberts, E.A., Dickerson, W., 2002. "Slow the Spread": a national program to contain the gypsy moth. J. Forestry July/ August, 30–35.

Shyam Prasad, G., Jayakumar, V., Ranganath, H.R., Bhagwat, V.R., 2008. Bio-suppression of coconut rhinoceros beetle, *Oryctes rhinoceros* L. (Coleoptera: Scarabaeidae) by *Oryctes* baculovirus (Kerala isolate) in South Andaman, India. Crop Protec. 27, 959–964.

Sivan, A., Harman, G.E., 1990. Improved rhizosphere competence in a protoplast fusion progeny of *Trichoderma harzianum*. J. Gen. Microbiol. 137, 23–29.

Six, D.L., Mullens, B.A., 1996. Distance of conidial discharge of *Entomophthora muscae* and *Entomophthora schizophorae* (Zygomycotina: Entomophthorales). J. Invertebr. Pathol. 67, 253–258.

Snetselaar, J., Andriessen, R., Suer, R.A., Osinga, A.J., Knols, B.G.J., Farenhorst, M., 2014. Development and evaluation of a novel contamination device that targets multiple life-stages of *Aedes aegypti*. Parasites & Vectors 7, 200.

Soberón, M., Pardo-López, L., López, I., Gómez, I., Tabashnik, B.E., Bravo, A., 2007. Engineering modified Bt toxins to counter insect resistance. Science 318, 1640–1642.

Steinkraus, D.C., 2007. Management of aphid populations in cotton through conservation: delaying insecticide spraying has its benefits, in: Vincent, C., Goettel, M.S. & Lazarovits, G. (eds.), Biological Control: A Global Pespective. CABI, Wallingford, pp. 383–391.

Steinkraus, D.C., Hollingsworth, R.G., Slaymaker, P.H., 1995. Prevalence of *Neozygites fresenii* (Entomophthorales: Neozygitaceae) on cotton aphids (Homoptera: Aphididae) in Arkansas cotton. Environ. Entomol. 24, 465–474.

St. Leger, R.J., Wang, C., 2010. Genetic engineering of fungal biocontrol agents to achieve efficacy against insect pests. Appl. Microbiol. Biotechnol. 85, 901–907.

Sundh, I., Goettel, M.S., 2013. Regulating biocontrol agents: a historical perspective and a critical examination comparing microbial and macrobial agents. BioControl 58, 575–593.

Sweeney, J., Gesner, G., Bennett, R., Vrain, T., 1998. Effect of mulches on persistence of entomopathogenic nematodes (*Steinernema* spp.) and infection of *Strobilomyia neanthracina* (Diptera: Anthomyiidae) in field trials. J. Econ. Entomol. 91, 1320–1330.

Tabashnik, B.E., Cushin, N.L., Finson, N., Johnson, M.W., 1990. Field development of resistance to *Bacillus thuringiensis* in diamondback moth (Lepidoptera: Plutellidae). J. Econ. Entomol. 83, 1671–1676.

Tanada, Y., Kaya, H.K., 1990. Insect Pathology. Academic Press, San Diego, CA.

Teakle, R.E., Jensen, J.M., Giles, J.E., 1986. Age-related susceptibility of *Heliothis puctiger* to a commercial formulation of nuclear polyhedrosis virus. J. Invertebr. Pathol. 47, 82–92.

Toledo, J., Campos, S.E., Flores, S., Liedo, P., Barrera, J.F., Villaseñor, A., Montoya, P., 2007. Horizontal transmission of *Beauveria bassiana* in *Anastrepha ludens* (Diptera: Tephritidae) under laboratory and field cage conditions. J. Econ. Entomol. 100, 291–297.

Tomalak, M., Piggott, S., Jagdale, G.B., 2005. Glasshouse applications. In: Grewal, P.S., Ehlers, R.-U., Shapiro-Ilan, D. (eds.), Nematodes as Biocontrol Agents. CABI, Wallingford, pp. 147–166.

Van Cuyk, S., Deshpande, A., Hollander, A., Duval, N. Ticknor, L., Layshock, J. Gallegos-Graves, L., Omberg, K.M., 2011. Persistence of *Bacillus thuringiensis* subsp. *kurstaki* in urban environments following spraying. Appl. Environ. Microbiol. 77, 7954–7961.

van Frankenhuyzen, K., Reardon, R.C., and Dubois, N.R., 2007. Forest defoliators. In: Lacey, L.A., Kaya, H.K. (eds.), Field Manual of Techniques in Invertebrate Pathology. Springer, Dordrecht, pp. 481–504.

Vasconcelos, S.D., 1996. Alternative routes for the horizontal transmission of a nucleopolyhedrovirus. J. Invertebr. Pathol. 6, 269–274.

Vega, F.E., 2008. Insect pathology and fungal endophytes. J. Invertebr. Pathol. 98, 277–279.

Viaud, M., Couteaudier, Y., Riba, G., 1998. Molecular analysis of hypervirulent somatic hybrids of the entomopathogenic fungi *Beauveria bassiana* and *Beauveria sulfurescens*. Appl. Environ. Microbiol. 64, 88–93.

Wraight, S.P., Carruthers, R.I., Jaronski, S.T., Bradley, C.A., Garza, C.J., Galaini-Wraight, S., 2000. Evaluation of the entomopathogenic fungi *Beauveria bassiana* and *Paecilomyces fumosoroseus* for microbial control of the silverleaf whitefly, *Bemisia argentifolii*. Biol. Control. 17, 203–217.

Zúbrik, M., Hajek, A., Pilarska, D., Špilda, I., Georgiev, G., Hrašovec, G., et al., 2016. The potential for the fungal pathogen *Entomophaga maimaiga* to regulate gypsy moth *Lymantria dispar* (L.) (Lepidoptera; Erebidae) in Europe. J.Appl. Entomol. 140, 565–579.

14

Prevention and Management of Diseases in Terrestrial Invertebrates

Jørgen Eilenberg and Annette Bruun Jensen

Department of Plant and Environmental Sciences, University of Copenhagen, Frederiksberg, Denmark

14.1   Introduction

Prevention and management of diseases caused by pathogenic microorganisms in terrestrial invertebrates covers the scientific background for all human efforts to avoid such diseases by either preventing infection or curing or managing diagnosed diseases. Such prevention and management is based on a deep understanding of the risk of transmission of diseases in indoor and outdoor production systems. In production of terrestrial arthropods, there is a constant pressure from diseases, which may harm or, in serious cases, result in a complete breakdown of the production. In this chapter, we cover diseases in arthropods (insects and some mites) used in commercial production. We use the term "insects" in the chapter to cover the class Insecta and the class Arachnida. Likewise, the term "insect pathogen" may also include mite pathogens throughout the chapter.

Insects and mites involved in production can be grouped into the following major categories:

1) *Pollination and honey*: An ecosystem service, pertaining to honeybees and other bees. Honeybees, in addition, produce honey.
2) *Silk*: Use of silkworms and other lepidopterans to produce silk for human clothing.
3) *Biological control*: Use of insects and mites either directly as biocontrol agents or in the production of biocontrol agents.
4) *Food and feed*: Use of insects as human food or animal feed.

Even though the four categories all deal with insect or mite production, the biological backgrounds and production environments for different types of insects can differ significantly. Even the language used to describe production parameters may differ among categories. As an example, the way to measure the magnitude of production differs between the four. For pollination, we speak about "number of hives established," "hectares pollinated," or "tons of honey." For silk production, the measurement is typically "tons of silk," while for biological control the unit is mostly "number of individuals." Concerning insects used for food and feed, a much-used unit for large-scale production is "kilos of insects," and now, in industrial production, "tons of insects." Our focus is on

Ecology of Invertebrate Diseases, First Edition. Edited by Ann E. Hajek and David I. Shapiro-Ilan.
© 2018 John Wiley & Sons Ltd. Published 2018 by John Wiley & Sons Ltd.

examples where the inclusion of insects and mites in production has (or may have) importance for the transmission of insect diseases either within the production systems or between production systems and the environment. We are aware that additional commercially relevant categories can also be defined (e.g., insects produced for zoological gardens, insects produced for pharmaceutical testing), but the major issues relating to pathology for these categories will overall be similar to those in the four main insect production systems, so they will not be specifically dealt with in this chapter.

Only pathogens that are infective to insects and are harmful (lowering fitness or even resulting in mortality) to the production stock will be treated, limiting our chapter to dealing with obligatory or facultative arthropod pathogens (Steinhaus, 1963; Vega and Kaya, 2012). Among insect pathogen groups, we will focus on viruses, bacteria, fungi, and microsporidia, since there groups have caused the major known problems in production systems and can all be transmitted. Insect pathogenic nematodes do not seem to be important as pathogens in insect production systems in general so they will not be included. Also, we will not cover transmission of microorganisms where invertebrates act as passive vectors, even though both viruses and bacteria passively transmitted by insects may be important hazards for humans (Chakrabarti et al., 2008; Belluci et al., 2013). There is no strict separation between pathogens infective to insects and microorganisms that are just passively dispersed or transmitted by insects. For example, house flies (*Musca domestica*) can vector *Escherichia coli*, and studies have shown that this bacterium may proliferate and accumulate in the flies (Kobayashi et al., 1999). However, it still needs to be documented that *E. coli* actually can act as an insect pathogen against *M. domestica*. *M. domestica* can indeed act as a microbial reservoir and vector fungal genera, which may be toxic to and/or cause infection in humans, as shown by Phoku et al. (2014).

14.1.1 Types of Production Facilities

Terrestrial invertebrates are produced using one of the following major types of facilities, although some production systems may be a mixture of all three:

1) closed production facilities;
2) semi-open facilities;
3) outdoor production facilities.

Closed production facilities are illustrated in Figs. 14.1a–c. The production insects in Fig. 14.1a, lesser mealworms (*Alphitobius diaperinus*), are produced in plastic trays, which are stacked separately in racks. Each rack has wheels, which are the only physical contact between it and the floor. The wheels provide the ability to move racks between rooms. Food for the insects is added to the trays, and feces are removed at regular intervals during the production period. There is virtually nothing else in the room other than racks, and the room is completely closed, with very little contact with the environment outside. The production insects in Fig. 14.1b, mealworms (*Tenebrio molitor*), are produced in the same way, in closed production facilities. The photo shows the process whereby larvae are separated from remaining food and feces, since the end product must consist of only the larvae. The climate-controlled hall (Fig. 14.1c) used for the production of black soldier flies (BSFs; *Hermetia illucens*) also harbors racks with separate trays. In this case, plastic nets have been added to ensure there is as little contact as possible between adult BSFs and the environment. Closed production systems like these also include production in large, closed growth chambers with very little contact with the environment, and thus little risk of transmission of insect pathogens. In general, the closed production facilities, often with rather dense populations of insects or mites in the same instar, can significantly enhance the risk of horizontal transmission if pathogens are introduced. Vertical transmission is also a risk, and vertically transmitted pathogens may initially occur at low prevalence and therefore not be diagnosed before a serious epidemic outbreak occurs.

Semi-open facilities (Figs. 14.1d–g) are often seen when arthropods are produced in warm regions of the world, where outdoor temperatures are sufficient for insect production and there is therefore no need for additional heating or lighting. They are cheaper in construction and maintenance than closed facilities and are flexible with respect to scaling up and scaling down of production. Figs. 14.1d and e show facilities used for the production of crickets: house crickets (*Acheta domesticus*) and black crickets (*Gryllus bimaculatus*). The production buildings are separated into smaller compartments where insects are produced in cardboard trays and food and water are supplied on plates or in bowls. These small compartments can be covered by a nylon

Fig. 14.1 Different types of closed, semi-open, and outdoor facilities for large- or medium-scale production of insects. (a) Closed production facilities for lesser mealworm (*Alphitobius diaperinus*). Larvae are reared in plastic boxes on separate shelves. The Netherlands. (b) Automatic separation of *Tenebrio molitor* larvae from substrate. The Netherlands. (c) Closed production facilities for black soldier fly (BSF; *Hermetia illucens*). Flies are reared and eggs laid in a climate-controlled room. South Africa. (d,e) Semi-open production facilities for house cricket (*Acheta domesticus*) and black cricket (*Gryllus bimaculatus*). New cohorts are started in the different sections at different time intervals. Thailand. (f) Semi-open production facilities for silkworms (*Bombyx mori*). Pupae are kept on open shelves. South Africa. (g) *Bombyx mori* larvae on mulberry leaves on open shelves. Fresh leaves are added regularly. South Africa. (h) Alfalfa leafcutter bee (*Megachile rotundata*). Loose cocoons during overwintering in trays before release in spring. United States. (i) Outdoor nest for mason bees (*Osmia rufa*). Bee pupae can be artificially added in the cardboard cylinders, but naturally occurring *O. rufa* will also use the nest. Denmark. (j) Outdoor nest at a site where both naturally occurring *Osmia rufa* and leafcutter bee (*Megachile* sp., holes covered with green leaves) are nesting. Denmark. (k) Outdoor hives for honeybees (*Apis mellifera*) near tropical forest with high biological diversity. Tanzania. (l) Outdoor hives for honeybees in an agricultural landscape with medium biological diversity. Denmark. (m) Outdoor hives for honeybees in an almond plantation with low biological diversity. California, USA. *Source:* (a,b) Courtesy of Proti-Farm, The Netherlands; (c) Courtesy of Agri-Protein, South Africa; (d,e) Courtesy of Afton Halloran, University of Copenhagen, Denmark; (f,g,i,j) Courtesy of Jørgen Eilenberg; (h) Courtesy of Rosalind James; (k–m) Courtesy of Annette Bruun Jensen, Univ. Copenhagen, Denmark. (*See color plate section for the color representation of this figure.*)

Fig. 14.1 (*Continued*)

mesh (Fig. 14.1e) to minimize contact with the environment. Nonetheless, the production facilities are basically in physical contact with the environment, leading to possibilities for pathogen transmission from outside of the facility. Silkworms (*Bombyx mori*) can be produced in both closed and semi-open facilities; the latter is shown in Figs. 14.1f and g. Pupae and larvae are placed on open shelves, and only adults are kept under fairly closed conditions, for copulation and egg-laying. The facilities shown in Figs. 14.1f and g offer many possibilities for interactions with the environment.

Outdoor facilities are typically used for production of insect pollinators (e.g., honeybees (*Apis mellifera*) and solitary bees like alfalfa leafcutter bees (*Megachile rotundata*) and mason bees (*Osmia rufa*)). *O. rufa* outdoor nests, with tubes for egg-laying and larval development, are shown in Figs. 14i and j. The nests are fixed to a wall, a pole, or a similar vertical construction outdoors, where they are in close contact with the environment, with many options for transmission of diseases. Other bees than the species planned for production can gain access to the larval tubes, as seen in Fig. 14.1j, where nests of a species in the genus *Megachile* are present among the nests of *O. rufa*. The nests of the former are easily recognized because of the green leaves used as covers, which result in close contact between adult females when they are depositing eggs and, later, newly emerged adults when they leave the nests. Disease transmission between different solitary bee species can happen during these events. Mass production of *M. rotundata* is carried out using both outdoor facilities (nests with holes or tubes) during the cropping season and closed facilities for storage of pupae during overwintering and before distribution of the pupae in the field (Fig. 14.1h). Honeybees (*A. mellifera*), which are social insects, are also produced outdoors. Worldwide, there are a huge variety of hives for honeybees, as seen in Figs. 14.1k–m. As with solitary bees, the production unit, the beehive, is in close contact with the environment, resulting in many options for transmission of diseases, which can be strongly influenced by the location of the hive. Some environments are close to natural, unmanaged vegetation, and can harbor high biodiversity (Fig. 14.1k). Others can harbor greater biodiversity, even if the habitat is still managed (Fig. 14.1l). Finally, some hives are surrounded by monocultures that are poor in biodiversity (Fig. 14.1m). We hypothesize that the risk of transmission of new pathogens from the environment into hives is larger when the environment harbors higher biodiversity.

14.1.2 Transmission of Insect Diseases in Production Facilities

The major routes of infection (oral uptake and via the cuticle) can all happen during production of insects and mites. Vertical transmission of insect pathogens is the direct transmission from parent to offspring, while horizontal transmission in the broad sense is all other types of transmission (see Chapters 1 and 3). Both types of transmission can take place within production facilities. For pathogens with oral uptake (viruses, bacteria, and microsporidia), the major route of transmission will mostly be via the food. Newly produced, (semi)sterile artificial food does not, by itself, pose a major risk of pathogen transmission, but food that is transferred from one place in the production system to another may enhance transmission of pathogens already present in the facilities. Things are different with natural foods like leaves and wet substrates like slurries (remnants from other biological production) that are brought from external sources into the production facilities. For insect pathogens with oral uptake, the risk of transmission from the environment into production facilities via food and other substrates can be high. For pathogens infecting via the cuticle (like most fungi), the risk of their being transmitted from the environment and into a production facility, or of their being transmitted within the facility, still exists, and, due to this infection route, transmission via food substrates is very important. For insect pathogens that are also frequently transmitted vertically (like viruses, microsporidia, and some bacteria), the presence of all host instars in the same production facilities and, in some cases, within close proximity increases the risk of such vertical transmission.

Within production facilities, transmission of insect pathogens between insects in a tray, between insects in an incubator, or between bees in an outdoor hive can be very high when population densities are high, facilitating horizontal transmission. Besides the previously mentioned oral and cuticle transmission routes, a third route of infection plays a major role in production facilities with very dense insect populations, namely wounding (Steinhaus, 1963). When insects are kept at high densities, and if the produced species can exhibit cannibalistic behavior, wounding can be a major cause for transmission of both obligate and facultative pathogens (see Chapter 3 for definitions).

This is, for example, the case with mealworms like *T. molitor* and *Zophobas morio*. Insects may be physically wounded by handling or may be subjected to biting by their neighbors. The wounds will then be open and available for infection by facultative pathogens like *Pseudomonas* spp. (G. Vergara, pers. comm.), which normally would probably not be able to infect individuals that were otherwise healthy.

When garbage, like waste food, substrate, and dead insects, is brought from the production facilities into the environment, there is a risk of transmitting pathogens to the homologous hosts living outside the facilities or to other susceptible species. This type of transmission can be of environmental concern. Also, in return, environments close to the production facilities can act as reservoirs for an insect pathogen, which can later be transmitted back into the facilities to infect the production insect stock.

Each of the three categories of production facilities shown in Fig. 14.1. offers different possibilities for transmission of pathogens, success of pathogens in infection, and dispersal of pathogens within the facilities. For the closed production facilities, the risk of pathogens entering from outside is low, provided that contamination of food and substrates by potential pathogens is controlled. The risk of transmission between rearing trays is relatively low or medium, while the risk of transmission within one tray can be high. Here, infection via wounds can be significant. For the semi-open facilities, there is a continuous risk of insect pathogens being transmitted via infected invertebrates, via invertebrates acting as vectors of insect pathogens, via wind, or via food and substrate. However, insects produced under semi-open conditions are often kept at lower densities compared to insects in closed conditions, which can lower the risk for epidemic development. For pollinators produced outdoors, interaction with the environment is an integral part of production. Honeybees and solitary bees leave their nests and search for pollen for their offspring. Obviously, this interaction significantly increases the risk of insect pathogens being brought into the beehive system, and it also increases the risk of insect pathogens being moved from one production facility (e.g., hive/nest) to another. In hives and nests, the population density is high or very high, enhancing transmission within these facilities and thus the risk of epizootic development.

14.2 Major uses of Insects and Mites in the Production and Transmission of Insect Pathogens within Production Systems

Table 14.1 provides a list of examples of different terrestrial insects and mites used in production. The list is not exhaustive, but shows representatives from each of the four production categories: pollination and honey, silk, biological control, and food and feed. The species-rich insect orders Lepidoptera, Hymenoptera, Diptera, and Coleoptera are all represented. Also produced are species from the insect orders Orthoptera, Hemiptera, and Neuroptera, as well as mites in the class Arachnida. Characteristics of each type of production and associated insect and mite pathogens and their transmission are described in this section, using selected examples.

14.2.1 Pollination and Honey Production

Pollination is an essential process in terrestrial ecosystems, including natural ecosystems and ecosystems managed by humans. Pollination is the transfer by pollinators of

Table 14.1 Examples of important insect and mite species produced commercially.

Insects: name and order	Purpose of production
Acheta domesticus, house cricket (Orthoptera)	Human food and pet food
Gryllus bimaculatus, black cricket (Orthoptera)	Human food
Locusta migratoria, migratory locust (Orthoptera)	Pet food and human food
Schizocerca gregaria, desert locust (Orthoptera)	Pet food and human food
Orius spp., pirate bugs (Hemiptera)	Biological control
Trialeurodes vaporariorum, greenhouse whitefly (Hemiptera)	Production of parasitoids for biological control
Aphis gossypii, cotton aphid (Hemiptera)	Production of parasitoids for biological control
Acyrthosiphum pisum, pea aphid (Hemiptera)	Production of parasitoids for biological control
Chrysoperla carnea, green lacewing (Neuroptera)	Biological control
Spodoptera littoralis, African cotton leafworm (Lepidoptera)	Production of virus for biological control
Cydia pomonella, codling moth (Lepidoptera)	Production of virus for biological control
Bombyx mori, silkworm (Lepidoptera)	Silk production and human food
Samia cynthia, Eri silkmoth (Lepidoptera)	Silk production
Lymantria dispar, gypsy moth (Lepidoptera)	Production of virus for biological control
Alphitobius diaperinus, lesser mealworm (Coleoptera)	Pet food and human food
Tenebrio molitor, mealworm (Coleoptera)	Pet food and human food
Zoophobas atratus/morio, giant mealworm (Coleoptera)	Pet food and human food
Adalia bipunctata, two-spotted ladybird (Coleoptera)	Biological control
Musca domestica, house fly (Diptera)	Animal feed
Hermetia illucens, black soldier fly (BSF) (Diptera)	Animal feed
Aphidoletes aphidimyza, aphid midge (Diptera)	Biological control
Apis mellifera, honeybee (Hymenoptera)	Honey production, pollination, and human food
Osmia rufa, mason bee (Hymenoptera)	Pollination
Megachile rotundata, alfalfa leafcutter bee (Hymenoptera)	Pollination
Bombus terrestris, buff-tailed bumblebee (Hymenoptera)	Pollination
Trichogramma sp. (Hymenoptera)	Biological control
Encarsia formosa (Hymenoptera)	Biological control

(Continued)

Table 14.1 (Continued)

Mites: name and order	Purpose of production
Phytoseiulus persimilis (Mesostigmata)	Biological control
Hypoaspis miles (Mesostigmata)	Biological control
Tyrophagus putrescentiae, mold mite (Sarcoptiformes)	Production of predatory mites for biological control
Tetranychus urticae, two-spotted spider mite (Trombidiformes)	Production of predatory mites for biological control

pollen grains from the male anther of one flower to the stigma of another. The stigma is the top of the female reproductive organ, called a pistil. Growth of a pollen tube from the stigma and release of sperm in the base of the pistil enables fertilization and reproduction in seed plants. Pollination occurs if pollen is moved between flowers of the same species and if the flowers are compatible. The majority of pollinators are insects, including bees (honeybees, solitary bees, and bumblebees), wasps, ants, flies, butterflies, moths, and beetles. Worldwide, 75% of our major crop species mainly cultivated for human consumption depend directly on pollinators (Klein et al., 2007). The value of pollination provided by honeybees, bumblebees, and wild pollinators to the production of crops used directly for human food has been estimated to be USD 172 billion, accounting for about 9.5% of the total value of human food production (Gallai et al., 2009). Although economic benefits are not estimated for this, pollination is also beneficial for non-food agricultural production and for natural vegetation.

Insect pollination can be considered an ecosystem service provided by wild pollinators, in particular wild bees (Kremen et al., 2002; Winfree et al., 2007), but it is also a production practice or management tool where farmers purchase or rent honeybees, bumblebees, and a few solitary bee species to supplement the local pollinator fauna. The economic benefit of insect pollination for farmers has created a well-developed and highly organized market of pollination services, with rental honeybee colonies in the United States (Sumner and Boriss, 2006) and Europe (Carreck et al., 1997) and with sale of bumblebee colonies all over the world (Velthuis and van Doorn, 2006). This practice clearly shows that for certain crops and in certain regions, there are not enough wild pollinators to ensure adequate pollination. Since insect pollinators are not confined in closed indoor facilities, the widespread sale and movement of, in particular, honeybee and bumblebee colonies has created avenues for new and rapid distribution of the parasites and pathogens they harbor to conspecific colonies or even to natural fauna (Graystock et al., 2016).

Honeybees have been domesticated since ancient times; in North Africa, beekeeping in pottery vessels began about 9000 years ago (Roffet-Salque et al., 2015). Beekeeping is depicted in Egyptian art from 4500 years ago, and excavations of the 3000-year-old town of Tel Rehov in Israel exposed hundreds of preserved beehives, demonstrating advanced beekeeping, including trade and transport of honeybees (Bloch et al., 2010). Settlers in North and South America, Australia, and New Zealand brought beekeeping with them, and today beekeeping is practiced worldwide, utilizing mostly the European honeybee species *A. mellifera*, which has been introduced by humans to areas outside its natural distribution of Europe and Africa (Moritz et al., 2005).

Honeybees are social insects that live in huge societies, with a single reproductive female, the queen, 200 to 20 000 fertile male drones (managed vs. natural colonies), and 200 000–800 000 infertile female workers in each colony. Modern beekeeping has evolved around several important inventions from the 1800s, including the movable frame hive and comb foundation (Crane, 1990). This equipment allows beekeepers to manage their hives, including increasing colony size by adding frames and supers, avoiding swarming, enabling a nondestructive honey harvest, splitting colonies, engaging in advanced queen breeding, and producing queens. Honeybees are often kept for the multiple products that can be harvested from the colonies: honey, wax, pollen, propolis, Gelé Royale (royal jelly), and venom. In certain regions, however, honeybees are kept more for their ecosystem services. Almond production in California, USA, which encompasses approximately 80% of the world's almond production (California Almonds, 2016), is a striking example of the importance of honeybee pollination in agriculture (Fig. 14.1m). Every spring, over 60% of all US commercial honeybee colonies are transported and rented out for almond pollination; in 2013–14, this accounted for approximately 1.6 million colonies (Cavigli et al., 2016).

Due to the vital role honeybees play in pollination, major concerns were raised when, in fall 2006 and spring 2007, many US beekeepers found themselves with colonies without adult bees, and with abandoned food and brood, resulting in colony collapse (Cox-Foster et al., 2007). It was widely believed that these were symptoms of a new highly virulent pathogen, and the term "colony collapse disorder" (CCD) was coined to describe this disease phenomenon when it occurred at colony level with an unknown cause. Over time, researchers learned that CCD is more complex than it first appeared, and cannot be explained by a single pathogen. Apparently, it is caused by a combination of factors, including, for example, xenobiotic stress from pesticides and biotic stress from honeybee viruses (Cornman et al., 2012; Cherjanovski et al., 2014; Hou et al., 2014).

High honeybee colony density and movement of colonies create unforeseen transmission possibilities for honeybee pests and diseases, which may be part of the general decline of pollinating bees worldwide (Goulson et al., 2015). Due to the high value of *A. mellifera*, diagnostic tools for the major diseases are constantly being improved. For example, Genersch (2010) reviewed the progress made over 20 years in the diagnosis of American foulbrood (*Paenibacillus larvae*) and its subtypes. It was suggested that epidemiological studies in conjunction with approved subtyping protocols should be initiated in order to allow better recognition of outbreaks and better monitoring of disease transmission.

Scientists and extension services working with diseases of *A. mellifera* and their diagnosis and management need to implement diagnostic tools constantly. Some diseases, like the fungus *Ascosphaera apis*, can be diagnosed based on external symptoms (Fig. 14.2i), while others may remain as latent infections over a certain period without being noticed. As seen in Fig. 14.3., since 1960, new virus diseases of *A. mellifera* have been found over short intervals; sometimes, these virus diseases can be very detrimental, such as deformed wing virus (DWV; Fig. 14.2k). We suggest that the production systems of honeybees in themselves (Figs. 14.1k–m) provide the background for virus diseases to be rapidly dispersed. First, honeybees may pick up new diseases through interaction with the environment; in particular, pathogens have been detected on

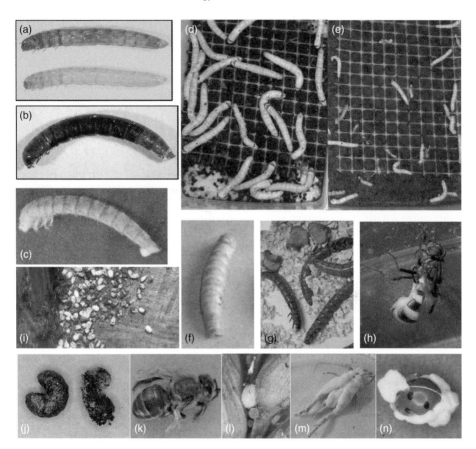

Fig. 14.2 Examples of external symptoms of diseases found among insects in production. (a) Mealworm (*Tenebrio molitor*). Upper: Uninfected larvae. Lower: Larva infected with nematode *Steinernema feltiae*. (b) *Tenebrio molitor* larva infected with unknown bacterium. (c) *Tenebrio molitor* larva infected with the fungus *Beauveria bassiana*. (d) Silkworm (*Bombyx mori*), fifth-instar larvae, uninfected. (e) *Bombyx mori*, fifth-instar larvae, infected with bacterium *Enterococcus mundtii*. (f) *Bombyx mori*, fifth-instar larvae, infected with virus Bombyx mori nucleopolyhedrovirus (BmNPV). (g) Giant mealworm (*Zophobas morio*) larvae infected with the bacterium *Pseudomonas* sp. (h) House fly (*Musca domestica*) adult, infected with the fungus *Entomophthora muscae*. (i) Honeybee (*Apis mellifera*). Specimens infected with fungus *Ascosphaera apis* outside hive. (j) Pupae of alfalfa leafcutter bees (*Megachile rotundata*) infected with *Ascosphaera aggregata* (left) and *Ascosphaera proliperda* (right). (k) *A. mellifera* adult infected with deformed wing virus (DWV). (l) Cereal aphids (*Sitobion avenae*). Central specimen infected with the fungus *Pandora neoaphidis*. (m) House cricket (*Acheta domesticus*), adult, infected with the fungus *Beauveria bassiana*. (n) Seven-spotted ladybird (*Coccinella septempunctata*) infected with the fungus *Beauveria bassiana*. *Source*: (a–f,h,k) Eilenberg et al. (2015). Reproduced with permission of Wageningen Academic Publishers; (g) Courtesy of Gabriela Maciel-Vergara; (i) Courtesy of Irfan Kandemir; (l) Courtesy of Jørgen Eilenberg; (m) Courtesy of Sunday Ekesi, ICIPE, Kenya; (n) Courtesy of Jørgen Eilenberg. (*See color plate section for the color representation of this figure.*)

flowers, where they will be contacted during nectar and pollen provisioning. Wild honeybees, solitary bees, and bumblebees may harbor insect pathogens that can actively or passively be transmitted to the production stocks in hives. Also, the intense moving around of honeybees by humans, such as in almond production and trading of bees

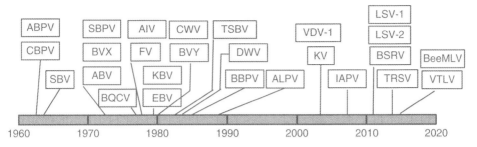

Fig. 14.3 Timeline of discovery of new virus diseases in honeybees (*Apis mellifera*). Virus names: acute bee paralysis virus, ABPV; chronic bee paralysis virus, CBPV; sacbrood virus, SBV; Arkansas bee virus, ABV; bee virus X, BVX; slow bee paralysis virus, SBPV; apis iridescent virus, AIV; filamentous virus, FV; black queen cell virus, BQCV; Kashmir bee virus, KBV; cloudy wing virus, CWV; bee virus Y, BVY; deformed wing virus, DWV; Berkeley bee virus, BBPV; aphid lethal paralysis virus, ALPV; Kakugo virus, KV; varroa destructor virus-1, VDV-1; Israeli acute paralysis virus, IAPV; Big Sioux River virus, BSRV; Lake Sinai virus 1 and 2, LSV-1 and LSV-2; bee macula-like virus, BeeMLV; tobacco ring spot virus, TRSV; varroa tymo-like virus, VTLV. *Source*: Eilenberg et al. (2015). Reproduced with permission of Wageningen Academic Publishers.

(honeybee queens, honeybee and bumble colonies, and solitary bees), may increase movement and transmission of bee pathogens. Cavigli et al. (2016) studied the prevalence of bee pathogens in honeybees in almond orchards. Among the most abundant were black queen virus (BQCV), Lake Sinai virus (LSV2), sacbrood virus (SBV), and the microsporidian *Nosema ceranae*. Parasitic Varroa mites can act as vectors assisting in disease spread, as exemplified by DWV (Bowen-Walker et al., 1999), which causes deformed wings in adult honeybees (Fig. 14.2k). A new virus complex, Macula-like virus (BeeMLV), was found in *A. mellifera* (de Miranda et al., 2015), and virus occurrence correlated with Varroa occurrence, suggesting that Varroa mites may be important for the spread of this virus.

Pathogen spillover from managed *A. mellifera* to other bees is also possible. A recent survey of pathogens in *A. mellifera* and in the Asian honeybee *A. cerana* in China and Vietnam indicated interspecific transmission of the Israeli acute paralysis virus (IAPV) and the bacterium European foulbrood (*Melissococcus plutonius*) from *A. mellifera* to sympatric *A. cerana* (Forsgren et al., 2015). Multiple RNA viruses associated with *A. mellifera* have been shown to be widespread in sympatric wild bumblebee populations in the United Kingdom, where virus prevalence in honeybees is a significant predictor of prevalence in bumblebees. Prevalence and individual virus loads in bumblebees were in certain cases higher than in honeybees, which brings into question the direction of disease spillover between managed and wild bees (McMahon et al., 2015).

Several bumblebee species from the genus *Bombus*, especially *B. terrestris*, are cultured commercially for pollination worldwide. Bumblebees are social insects like honeybees, although their colonies are smaller, containing 50–400 individuals. In the spring, new colonies are founded either above- or belowground by a mated queen that emerges from hibernation. Several decades of basic research in mating biology, storage of mated queens, circumvention or breaking of dormancy, and colony initiation paved the way for commercial production (Velthuis and Doorn, 2006). The commercial production of bumblebees started in 1987 (Velthuis and Doorn, 2006). Today, over

1 million bumblebee colonies are sold annually on a global scale, particularly for pollination of greenhouse crops. In 2006, the sale of commercial *Bombus* had an estimated value of USD62 million annually, and the value of the pollination performed by bumblebees has been estimated to be at least USD14 billion per year (Velthuis and Doorn, 2006).

The increasing use of commercial bumblebees within and beyond their natural ranges and the abundance of disease in commercial hives may allow pathogens to invade wild *Bombus* populations. The intestinal protozoan *Crithidia bombi* has been shown to spread horizontally when infected commercial bumblebees escape from greenhouses and share flowers with susceptible native bumblebees. Such spillovers can threaten wild pollinators, as modeled by Otterstatter and Thomson (2008). Graystock et al. (2013) found the presence of pathogens in 77% of *B. terrestris* colonies purchased from commercial producers. *Bombus terrestris* workers from uninfected colonies were fed feces or pollen from infected *B. terrestris* colonies, thereby documenting transmission of the microsporidia *Nosema bombi*, *N. apis*, and *N. ceranae* and the ascomycete fungus *Ascosphaera apis*; the three microsporidia are major honeybee pathogens. Also, survival of honeybees was significantly reduced when they were exposed to feces and pollen from diseased *B. terrestris*, and this demonstrates the risk of a dual direction in pathogen spillover. Recently, bumblebee queens of three species (*Bombus nevadensis*, *B. griseicollis*, and B. *vosnesenski*) were found to be infected by the bee specialist fungus *A. apis* when kept in captivity, where they were fed stored pollen from honeybee colonies (Maxfield-Taylor et al., 2015). *Ascosphaera apis* is a pathogen that causes a common honeybee disease, and Maxfield-Taylor et al. (2015) demonstrated that the potential risk of transmission from a production system to the environment is high, as is the possibility of transmission from one production system to another between honeybees and bumblebees.

Alfalfa leafcutting bees (*Megachile rotundata*) are solitary, cavity-nesting bees that are efficient pollinators of alfalfa, carrots, other vegetables, and some fruits. They are native to Europe but are managed on a large scale in the United States and Canada, where they are used for pollination in alfalfa seed production. Alfalfa seed growers release bees in their fields and provide the females with nesting boards, in which they construct cells in a linear series. Each cell is made of leaf pieces and contains one larva plus a food provision of pollen and nectar. The immature bees overwinter in the cells as prepupae after spinning a cocoon. In managed systems, these cells are stored during the winter and then released into fields in spring (Peterson et al., 1992). *Megachile rotundata* is susceptible to chalk brood infection caused by the fungus *Ascosphaera aggregata* (Fig 14.3j). In a study from the western United States (James and Pitts-Singer, 2005), 8% of *M. rotundata* larvae were infected with *A. aggregata*. However, this study documented a need to develop better broad-spectrum diagnostic tools, since more than 15% of the bees in the study were found to have died from unknown causes. Nest cells are commonly removed from the boards prior to overwintering and tumbled mechanically to remove excess debris – the "loose cell" system (Richards, 1984; Peterson et al., 1992). Most infected larvae die and sporulate before they form a cocoon (James, 2004), and cells containing chalkbrood cadavers easily break apart. Adults emerging from the loose cells are contaminated with *A. aggregata* spores, which will infect new larvae and increase disease spread (James and Pitts-Singer, 2005). Disinfection of the loose cells, followed by release of the cells in the field using clean nest material, did not prevent substantial disease transmission (James, 2011). James (2011) suggests that, in the field, pathogen transmission must occur from other, currently unknown sources. *Ascosphaera aggregata* has been found in the native leafcutting bees *M. relative* and *M. pugnata* (Goerzen et al., 1990, 1992), indicating that the disease may spread to and from such bee species when they fly into alfalfa leafcutting bee nest material.

14.2.2 Silk Production

Silk is produced worldwide. China is by far the biggest producer, but India is also a major source. Global production is measured by weight of silk and is currently several hundred thousand tons per year. *Bombyx mori* (silkworm or silk moth) is the most commonly used insect for the production of silk. Recent studies have documented that *B. mori* diverged from the Chinese wild silkworm *B. mandarina* (Li et al., 2010). The domestication of *B. mori* has, by use of gene-flow analysis, been calculated to have been initiated 7500 years ago and to have been terminated approximately 4000 years ago (Yang et al., 2014). Thus, *B. mori* has a long history of being domesticated and is mostly separated from the wild type, although some bilateral gene flow during the domestication period must have happened (Yang et al., 2014). Today, silkworm production can include more than one generation per year, according to the region of production and *B. mori* stock. Production facilities can be almost closed or semi-open (Figs. 14.1f and g).

Besides *B. mori*, other moths are used for silk production regionally or locally. Among the most prominent is Eri silkworm (*Samia ricini*), which is domesticated and produced in semi-open facilities and is an important silk-production insect in India (Zethner et al., 2012). Other important species in India are Tasar silk moth (*Antheraea mylitta*) and Muga silk moth (*Antheraea assamensis*). These moths have been in culture for several hundred years (Sarmah et al., 2010). Larvae are raised outdoors on plantation trees, with various levels of protection.

Knowledge about diseases and their transmission in *B. mori* is obviously much higher than that in the other moth species, due to the much longer history of domestication of *B. mori* and the much larger global production of its silk. *Bombyx mori* is subjected to a broad spectrum of diseases (James and Li, 2012; Eilenberg et al., 2015). For most important diseases, common names exist alongside the Latin ones for the causal insect pathogens, based on external symptoms that can be observed with light microscopy. Pebrine disease is caused by the microsporidian *Nosema bombycis*. Viral flacherie is caused by Bombyx mori infectious flacherie virus (BmIFV). There are clear symptoms in virus-infected larvae, which become swollen, as shown for Bombyx mori nucleopolyhedrovirus (BmNPV) (Fig. 14.2f). Sotto disease is caused by *Bacillus thuringiensis*. White muscardine is caused by the fungus *Beauveria bassiana*. Due to the high density of silkworms often found in production systems, facultative diseases (see Chapter 3 for discussion) can be important. Several bacteria can occur, like *Enterococcus mundtii*. There are symptoms discriminating healthy larvae (Fig. 14.2d) from infected larvae (Fig. 14.2e). In the case of infection, larvae shrink and do not thrive well. *Bacillus cereus* causes septicemia-like symptoms and results in high mortality rates of *B. mori* (Li et al., 2015). New virus diseases are also found in *B. mori* now and then, but much less commonly than in honeybees (*A. mellifera*) (Eilenberg et al., 2015).

Diagnostic methods are rapidly developing, since early detection of the microorganisms is essential for timely management of the diseases. Zhou et al. (2015) developed a rapid detection method for BmNPV using a set of four primers and a labeled probe. A loop-mediated isothermal amplification assay (LAMP) was successfully employed to detect the virus in larvae and feces, which pointed to the necessity of minimizing the risk of transmission via feces. Such a diagnosis of a single insect pathogen is very useful, if the particular microorganism is a common problem in a given production unit. Also, molecular methods can be useful for detecting the presence of a disease in areas where it occurs but is latent (= without clear symptoms; see Chapter 3). Vootla et al. (2013) were able to detect BmIFV in *B. mori* using only the molecular techniques real-time polymerase chain reaction (RT-PCR) and nested PCR. By screening *B. mori* from Karnataka, India, they proved, for the first time, the presence of the virus in that region.

It is more complicated to diagnose a disease if several insect pathogens could be involved in its outbreak. There can be valid general signs and symptoms to look for, allowing determination of the main causal insect pathogens (James and Li, 2012), yet it might be useful to have rapid molecular tools that allow simultaneous detection of several insect pathogens via one assay. Ravikumar et al. (2011) used a multiplex PCR, which proved effective for simultaneous detection of the microsporidians *N. bombycis* and two viruses – BmNPV and Bombyx mori densovirus (BmDNV) – in *B. mori*. The authors suggested that this method was also applicable to other silkworm species.

A relevant question to ask is how insect pathogens are transmitted within regions producing *B. mori*, as well as which forms of transmission might be occurring between silk production units and the environment. As seen in Figs. 14.1f and g, *B. mori* can be produced in semi-open facilities that allow pathogen interaction with the environment, including potential transmission. Other silkworm production units are closed facilities. The geographical range of silk production facilities and the diversity of these production systems with respect to the level of interaction with the surrounding environment and the distance to neighboring facilities make it difficult to generalize about the most important means of transmission.

Liang et al. (2013) studied BmNPV transmission among *B. mori* in China. They found three clades of the virus in different parts of Guangxi province. Interestingly, one clade was widely distributed within the province, while the two other covered separate regions. In Guangxi, there was local transmission of the virus between production facilities. Besides wind, rain, and predators, the authors suggested that humans also play a role in transmission. Silkworm production is organized in the region as follows: Larval production takes place at certain places specialized for this, and later, larvae are delivered to many small producers that produce pupae and harvest silk. This system allows for a rapid transmission of virus diseases, unless there is a strict control program at the facilities producing small larvae.

The microsporidian *N. bombycis* can be found in many silkworm production systems in China. By collecting samples of *N. bombycis* from different regions of China and Japan and analyzing the isolates via multicopy rDNA, Liu et al. (2013) found that isolates from Sichuan were very different from isolates from South East China. Interestingly, the South East China isolates proved to be similar to those from Japan.

For generalist insect pathogens, it is of major significance if they can be transmitted from one insect species to another. A major study compared isolates of the generalist fungus *B. bassiana* from silkworms and pine caterpillars (*Dendrolimus punctatus*)

(Chen et al., 2015). In *B. mori*, *B. bassiana* may cause high mortality. The lepidopteran species *D. punctatus* is a major pest in forestry in silk-producing areas, and biological control includes inundative spraying with *B. bassiana*. In the same study, sampling of infected specimens of the two insect species was performed in nine provinces in China. In total, 428 isolates of *B. bassiana* from *B. mori* and 327 isolates from *D. punctatus* were obtained and genetically compared. A 3D principal component analysis proved that these isolates fell into two distinct groups, related to the host insect. *Beauveria bassiana* isolates from *B. mori* were clearly different from *B. bassiana* isolates from *D. punctatus*. *B. bassiana* infections in *B. mori* were therefore not caused by heavy biological control programs using *B. bassiana* for control of *D. punctatus* in the same regions.

Less is known about diseases in other silkworm species. Muga silkworm (*Antheraea assamemsis*) is affected by a set of diseases: microsporidia, virus, bacteria, and fungi (Tikader et al., 2013). The Chinese oak silkworm (*Antheraea pernyi*) can be infected by *Nosema antheraeae* (Wang et al., 2006). This microsporidian is closely related to *N. bombycis*, which opens the possibility of its transmission between *A. pernyi* and *B. mori*. Eri silkworm is susceptible to Autographa californica multiply-enveloped nucleopolyhedrovirus (AcMNPV) from the lepidopteran species *Autographa californica* when inoculated intrahemocoelically (Hosamani et al., 2015). This host transmission can be seen as a potential threat to Eri silkworm production if Eri silkworms get into contact with virus originating from other lepidopteran hosts, as might occur in the neighborhood of semi-open production facilities. So far, however, there are no published reports of virus transmission from a naturally occurring lepidopteran species to *S. ricini* in a production facility.

14.2.3   Biological Control

Mass production of insects and mites for biological control of pests has been performed for decades, and much experience has been obtained in preventing disease outbreaks in production stocks of insects and mites. Van Lenteren (2012) summarized the status of species used in biological control and compiled an extensive list of arthropod predator and parasitoid species that were (and still are) produced commercially and sold for biological control. Several insect orders are produced for this purpose: Hymenoptera, Coleoptera, Thysanoptera, Diptera, Neuroptera, and Acari. Furthermore, Nematoda and Mollusca are included, with a few species in mass production. In total, van Lenteren (2012) listed 230 species of invertebrates. Among them are parasitoids like *Encarsia formosa* parasitizing whiteflies and *Trichogramma* spp. parasitizing lepidopteran eggs, which are produced worldwide. Predators like the mites *Phytoseiulus persimilis* and *Amblyseiulus swirskii*, both of which feed on pest mites, are also globally produced. A few examples of insects and mites produced for biological control are listed in Table 14.1.

Raising parasitoids and predators for biological control often requires rearing both the biological control agent itself and the insects that are needed as its food; these latter are frequently herbivorous. Both can be raised side by side as two trophic levels of production in one facility. However, at both trophic levels, there are potentially problems with insect diseases. Obviously, the occurrence of diseases and the risks of insect pathogen transmission will differ a lot between these diverse groups of organisms. It is well known that species of ladybirds can become infected with generalist insect pathogenic fungi (Fig. 14.2n). However, our knowledge about diseases differs a lot between species, and while much is known regarding commonly produced species, there are hardly any documented records of diseases for several of the less-studied insects or mites used for biological control. Much knowledge is merely anecdotal stories about problems in the production of certain insect and mite species, which have not been fully studied and published. In general, producers of biological control organisms are very careful to maintain healthy insects and mites in their production facilities, and they will typically have entomologists among their staff or will have close contact with such expertise (e.g., in academia). Still, new insect pathogens are often discovered and described in biological control production systems. Bjørnson et al. (2013) described the microsporidian *Nosema chrysoperlae* infecting the lacewing *Chrysoperla carnea* in commercial biological control rearing. The morphology, ultrastructure, and molecular characterization were described by the authors after finding this pathogen, although it is not yet known how widespread this microsporidian is or whether it is a serious threat to lacewing production.

Insects produced for biological control are usually raised on their natural host plants (if phytophagous), prey (if predatory), or insect hosts (for parasitoids and viruses); alternatively, they can be raised using artificial diets, where possible (Morales-Ramos et al., 2014). The level of automation that is possible differs depending on the insect species and the numbers of insects to be produced per time unit. Insects intended for biological control can be produced in closed facilities, semi-open facilities (called "greenhouse-based systems" in Morales-Ramos et al., 2014), or outdoor facilities (called "field cage systems" in Morales-Ramos et al., 2014). Examples of insects reared in semi-open production are *P. persimilis* and *E. formosa*, while an example of an insect produced outdoors is *Eretmocerus hayati* parasitizing *Bemisia tabaci* (Goolsby et al., 2014). One main point to consider with regard to minimizing the risk of diseases in facilities used for the production of predators and parasitoids is the quality of food and of the environment (Morales-Ramos et al., 2014; Shimoda et al., 2015). Obviously, semi-open and open facilities offer many opportunities for insect pathogens to enter into the production system from outside and vice versa.

Biological control agents are shipped after production in order to be released elsewhere, and handling during transportation and after storage is of great importance, since handling and storing provide a risk for further transmission if the insects being shipped are already infected (Bueno et al., 2014).

Mites are subjected to many insect pathogens (van der Geest et al., 2000; Bjørnson, 2008). Pathogens of mites include various bacteria (e.g., *Serratia marcescens*, *Acaricomes phytoseiuli*), fungi from the Entomophthorales, hypocrealean fungi (e.g., eight *Hirsutella* species), and microsporidia (e.g., nine *Nosema* species). A well-studied predatory mite species is *P. persimilis*. Overall poor health of *P. persimilis* in production facilities was noted by Bjørnson et al. (1997), who surveyed specimens from 14 commercial production facilities. They found evidence of several insect pathogen groups infesting the mites (viruses, bacteria, microsporidia). Schütte and Dicke (2008) listed verified and potential pathogens of this mite and other predatory mite species. Bacterial pathogens recorded in *P. persimilis* included *Wolbachia* species, *A. phytoseiuli*, and an underdetermined bacterial species. *A. phytoseiuli* seems potentially important as a pathogen, as it severely inhibits the production of healthy female *P. persimilis* (Schütte et al. 2006,

2008). Infected females shrink, cease oviposition, and finally die from infection. The authors tested horizontal transmission and concluded that it would not take place via food, but merely via feces and debris from infected specimens. Transmission seems to be hampered by sterilization of females. It is not known how widespread this bacterial disease is across production systems, but we would argue that it could be the reason for production problems at many places.

Microsporidia seem also to be important in *P. persimilis* production (Schütte and Dicke, 2008). The species *Microsporidium phytoseiuli* and at least two unidentified microsporidia are listed as pathogens of *P. persimilis*. Females infected with *M. phytoseiuli* produced fewer eggs and died sooner than uninfected females and may therefore be less effective for biological control (Bjørnson and Keddie, 1999). The microsporidia may remain in the production stock without being directly noticed due to the absence of clear external symptoms. Bjørnson and Keddie (2001) recommended that individual mites should be routinely examined for these microsporidia, as well as for other pathogens. In situations with low infection prevalence, such examinations can assist in detection, leading to avoidance of sudden disease outbreaks that decimate production systems.

There are examples of diseases in parasitoids, too. Schuld et al. (1999) reported an interesting example concerning the complexity of transmission and effect of an insect pathogen in mass rearing of *Trichogramma chilonis* in *Plutella xylostella*. A microsporidium, *Vairimorpha* sp., infected *P. xylostella* transovarially but had little impact on the fitness of the host. However, the parasitoids suffered when raised on infected *P. xylostella*.

Transmission of diseases from one insect species to another during production is a possibility. Steele and Bjørnson (2012) studied two microsporidian species, *Tubulinosema hippodamiae* and an undescribed species, infecting the ladybird species *Adalia bipunctata*. Via horizontal transmission, microsporidia infecting *A. bipunctata* could be transmitted to another ladybird beetle, *Hippodamia convergens*. Escapes of infected ladybirds from production facilities may enhance transmission of these diseases to naturally occurring ladybird species.

Mass production of obligate insect pathogens for biological control, like insect viruses, requires production of the host itself, to serve as substrate for the virus. All currently available commercial virus products intended for biological control are based on the production of viruses *in vivo* (in the host) in closed production facilities. This expands the width of insect species in commercial production to include lepidopteran and hymenopteran (sawflies) species. These insects are not themselves sold, but are the essential source of production for insect viruses. Therefore, the occurrence of contaminating diseases in these virus hosts (besides the viruses that are eventually intended for production), and their potential transmission within a production facility, is important. We are aware of examples (several biological control producers, pers. comm.) showing that insect virus production needs very careful management in order to avoid virus transmission into the production stock of insects. The same can be said regarding the *in vivo* production of nematodes for biological control (Shapiro-Ilan et al., 2014): care must be taken to avoid infection in the insect host used to produce the nematodes. Insects and mites are also sometimes used to produce insect parasitoids, insect and mite predators, and nematodes for biological control (Table 14.1.), although these insects are not the target pest, but are solely used as substrates for production. For example, aphids used as a substrate for such production can be subjected to infection by contaminating specialist insect pathogens like fungi from Entomophthorales, which may cause massive mortality in the production stock (Fig. 14.2l). Also, since the biocontrol industry has to produce a range of insect and mite species, there is a risk of transmission of generalist insect pathogens like the fungus *B. bassiana* (Fig. 14.2n) between the organisms used as substrates and the organisms being produced for use in biological control, across trophic levels.

14.2.4 Production of Insects for Food and Feed

Insects have for millennia been used as human food, and are still used as such in large parts of the world today (van Huis et al., 2013). Insects are also part of the natural diet of livestock species like pigs and poultry (Jozefiak et al., 2016). Recently, there has been a rapid increase in the production of insects as human food and animal feed. Some of the most important insect species for food and feed are listed in Table 14.1. In essence, there are three main reasons for using insects as food and feed:

1) *The environmental argument*: Some insect species emit fewer greenhouse gases and less ammonia than cattle or pigs and require significantly less land and water than cattle rearing. So, insects may "feed the world" in a more environmentally friendly way than mammalian livestock.
2) *The socioeconomic argument*: Insect harvesting/rearing can be a low-tech, low-capital investment option that offers entry even to the poorest (or less developed) sections of society. More high-tech production facilities may offer new jobs in more developed sections of society.
3) *The nutritional/gastronomic argument*: Insects can be healthy, nutritious alternatives to mainstream staples such as chicken, pork, beef, and even fish. Insects may also provide us with new tastes, textures, and flavors in our food.

A global perspective on insects as "mini-livestock" was presented by the UN Food and Agriculture Organization (FAO) in 2013 (van Huis et al., 2013). Since then, many scientific papers have been published further investigating global perspectives (e.g., van Zanten et al., 2015; Tabassum-Abassi et al., 2016). The global advantages of insects as mini-livestock can be summarized as lower greenhouse gas emissions, less land use, higher bioconversion rates, and lower risk of human disease spread (e.g., Oonincx and de Boer, 2012; van Zanten et al., 2015). Makkar et al. (2014) strongly argued that we need alternatives to soymeal and fish meal in order to avoid exhaustion of arable land and global fisheries. Many waste products exist that can have potential for insect production (Joensuu and Sinkko, 2015). Analyses of the sustainability of insect production for food and feed have just begun, and so far it is not really possible to compare sustainability in detail between different production systems and different insect species, due to a lack of data (Halloran et al., 2016).

The BSF (*Hermetia illucens*) has received attention worldwide as the most promising potential feed ingredient for animal husbandry and aquaculture, due to its high feed conversion efficiency (Oonincx et al., 2015). Furthermore, BSF larvae have the advantage that they can be reared on various organic wastes (Lalander et al., 2013). This species is efficient in treating compost and can degrade up to 70% of organic waste and convert almost 12% of biomass (Diener et al., 2011; Lalander et al., 2015). Mealworms

(*T. molitor*) and house crickets (*A. domesticus*) are also commonly used for the production of food and feed. These two species have been cultivated for decades as pet food globally. *Tenebrio molitor* can be raised on grain products, but it normally develops faster if also provided with additional proteins. *Acheta domesticus* is raised on a diet based on a mixture of leaves and other plant materials. The general positive profiles of proteins, vitamins, and micronutrient compounds in insects used for food and feed were outlined by van Huis et al. (2013). Comparable studies comparing nutritional profiles for different insect species and different instars of the same species are still lacking (Zielinska et al 2015).

The insect species most commonly used for food and feed production (Table 14.1) are well-known species that mostly have been cultivated for many years and have in some cases also been used as model insects for understanding infection, pathogen virulence, host immunity, and other factors of importance for understanding host–pathogen interactions. Some pathogens are well known, since they are generalists, have striking external symptoms, and are commonly found in insects in production. One example is the generalist fungus *B. bassiana*, which, for example, can infect *A. domesticus* (Fig. 14.2m). Specialist pathogens with striking external symptoms are also well known, like the fungus *Entomophthora muscae*, which infects house flies (Fig. 14.2h). Still, many insect pathogens have been little studied, and some of these may suddenly become important in production systems in the future. Welling et al. (1995) found natural infections of the fungus *Sorosporella* sp. in *Locusta migratoria* in Madagascar. An iridovirus from the crickets *Gryllus campestris* and *Acheta domesticus* was isolated by Jakob et al. (2002). Cito et al. (2014) isolated *Metarhizium pingshaense* from red palm weevils (*Rhychoptera ferrugineus*) and proved in bioassays that this fungus was pathogenic to the host. In a review, Mazza et al. (2014) listed microorganisms that they called "natural enemies" due to their antagonistic effect on palm weevil species. Their list is long and includes insect pathogenic species (obligate or facultative) like *Serratia marcescens*, *Bacillus thuringiensis*, *Pseudomonas aeriginosa*, *Lysinibacillus* (= *Bacillus*) *sphaericus*, *B. bassiana*, and *M. anisopliae*; several of the species have broad host ranges. Furthermore, the list also contains genera and species of microorganisms that may have toxic or infectious effects on weakened (e.g., wounded) hosts, such as *Aspergillus*, *Fusarium*, *Penicillium*, and *Bacillus laterosporus*.

Some of the insect species produced for food and feed are at the same time regarded as pest species in nature (e.g., *M. domestica*, *G. mellonella*, *T. molitor*, and *L. migratoria*), and biological control options include entomopathogens. *Tenebrio molitor* is susceptible to generalist insect pathogenic fungi, bacteria, and nematodes (Figs. 14.2a–c), which has resulted in their use as "insect baits" to isolate fungi from the genera *Metarhizium*, *Beauveria*, and *Isaria* from the environment (Zimmermann, 1986; Meyling et al., 2006; Medo and Cagán, 2011). Coleopteran larvae in production can be affected with facultative insect pathogenic bacteria. An example is seen in Figure 14.2g, where larvae of *Z. moria* have been killed by bacteria from the genus *Pseudomonas*.

Some of the insect species produced for food and feed have been subjected to studies concerning the diversity of microorganisms normally associated with them. The bacterial community in pulverized *T. molitor* included *Propionibacterium*, *Haemophilus*, *Staphylococcus*, and *Clostridium*, while the bacterial community in *L. migratoria* included *Weissella*, *Lactococcus*, and *Yersinia* (Stoops et al., 2016). *Tenebrio molitor* is actually a useful model insect for studying infections by opportunistic clinical pathogens infecting humans (e.g., *Candida albicans* and *Cryptococcus neo-formans*) (de Souza et al., 2015).

Locusta migratoria is susceptible to fungi like *Metarhizium acridium*, and this insect–fungus system is used as a model to understand immune responses. Mullen and Goldsworthy (2006) found that immune responses in nymphs were much stronger than in adults. Also as a model insect, the house fly *M. domestica* was used by Charles and Killian (2015) to understand the immune responses against different challenges, including infection by the bacterium *S. marcescens*. Comparative studies on immune responses may lead to novel insights into the risk of disease transmission from one arthropod class to another. Noonin et al. (2010) proved that for the bacterium *Aeromonas hydrohila*, virulence factors (lipopolysaccharides) were the same after infection of *T. molitor* and of the crayfish *Pacifastacus leniusculus*, indicating a similar immune response by these very different hosts.

Now, turning to the challenges in production systems, in general little is known about which disease problems are most urgent to solve and which transmission routes are most significant. As seen in Figs. 14.1a–c, insects raised for food and feed can be produced in closed facilities. In closed facilities with very high densities of, for example, mealworms, there is high potential for horizontal transmission of diseases within trays, where infective units like spores and virions may survive and come into contact with uninfected insects when contaminated trays are used for the next insect batch. Also, with very high densities of insects in a closed facility, problems such as cannibalism and physical injuries due to handling of larvae significantly increase the risk of disease transmission through wounds. Our research team at the University of Copenhagen has received samples of weakened or dead insects from companies producing insects for food and feed, and we assume in some cases that wounding is a major cause of pathogen transmission. This was, for example, our conclusion regarding disease outbreaks of facultative pathogenic bacteria in *Zophobas* beetles.

The constant and dense insect populations in production facilities open opportunities for interactions between different groups of pathogens or between one pathogen and several related or unrelated hosts. The bacterium *Pseudomonas fluorescens* did not have an impact on *G. mellonella* development, while *B. bassiana* had a negative impact (Meikle et al., 2013). However, when *G. mellonella* was subjected to both *P. fluorescens* and *B. bassiana*, the bacterial treatment had a positive effect on survival of *B. bassiana*-treated individuals. Such results suggest that in some cases the presence of microbial contaminants like *P. fluorescens* can assist in decreasing the effects of infection by fungi.

Densoviruses are known for their broad host ranges (Cotmore et al., 2014), so for this virus family there is a risk of transmission between different arthropod hosts like shrimp and insects. In phylogenetic studies by Roekring et al. (2002), densovirus from shrimp fell into the clades containing insect densoviruses, a clear indication of relatedness. The risk of transmission will, however, be heavily influenced not only by the virulence of a specific strain, but also by the production facilities. Closed production facilities for insects that are long distances from aquatic shrimp production facilities will prevent contact between the organisms in these different arthropod classes. On the other hand, close proximity between semi-open and open production facilities may increase the risk of a sudden transmission, which will be very harmful for the new host, whether a crustacean or an insect. The virus family Parvoviridae, subfamily Densovirinae, includes species infective to shrimp, like Spawner-isolated mortality virus (SMVmon), which is infective to black tiger shrimp (*Penaeus monodon*) (Fraser and Owens, 1996). Also in this family are viruses infective to insects, such as Galleria mellonellla densovirus (GmDNV) (Gross et al., 1990) and Bombyx mori densovirus (BmDNV) (James and Li, 2012). Penaeus merguiensis densovirus (PmergDNV) can infect *A. domesticus* and *T. molitor* when insects are challenged by injection (la Fauce and Owens, 2008). It can be found in infected insects of both species, but at higher levels in *A. domesticus*. Whether densoviruses can spill over from shrimp to insects, or vice versa, under natural conditions remains to be seen, but the likelihood could be increased if production facilities were near each other.

There is potential for transmission of generalist insect pathogens from production facilities to the environment. Ivie et al. (2002) did a field-based study of potential non-target effects from control of *L. migratoria* using entomopathogenic fungi, challenging more than 200 non-target coleopteran species with *B. bassiana* and *Metarhizium flavovriride*. Non-target effects differed, since one isolate of *M. flavoviride* had significant effects on several coleopterans, while another isolate of the same fungus and of *B. bassiana* had no effect. It has yet to be proven whether there will be any significant population-level effects in the field.

In 2014, we undertook a survey among companies producing insects for food and feed asking about their observations of insect diseases (if any) and what actions they took to prevent or control such diseases. We received more than 30 responses, with various levels of specific information. Companies were promised full confidentially, so the table of results (Table 14.2) does not include any company information. It appears that the disease pressures from insect pathogens differed significantly between the

Table 14.2 Selected results from a survey (2014–16) of insect diseases found by employees in production facilities for insects used as food and feed.

Production insect	Disease	Symptoms	Action
Acheta domesticus	Bacterium sp.	Increased mortality, red appearance	Cleaning of cages
Acheta domesticus	*Metarhizium* sp. and *Beauveria bassiana*	Some mortality in population	Quarantine, new breeding stock
Acheta domesticus	Cricket paralysis virus (CrPV)	Collapse of cricket population: the virus seems to spread globally	Switching to new breeding stock or even new cricket species
Gryllus bimaculatus	*Gryllus bimaculatus* iridovirus	Swollen abdomen, strikingly sluggish, mortality close to 100%	Occurs occasionally
Tenebrio molitor	*Beauveria bassiana*	Some mortality in population	Cleaning, removal of dead larvae, quarantine
Zophobas morio	*Pseudomonas* sp.	Increased mortality: recurrent problem	Removal of dead larvae
Musca domestica	*Entomophthora muscae*	Dead adult flies with spores, epidemic	Cleaning, removal of dead flies, quarantine
Hermetia illucens	Unknown (bacterium?)	Elongated, rounded mature larvae, moving slowly before dying	Quarantine

different insect species produced for food and feed. Apparently, BSFs were in general very resistant to diseases, since no companies reported significant problems with this species. House flies and mealworms suffered from diseases, but these could mostly be solved by relatively simple means like cleaning trays and other equipment with soap and water. For house crickets, however, the cricket paralysis virus (CrPV) is disastrous, and infection may result in complete collapse of production stock.

14.3 Status of Diagnostic Services

For honeybee production, diagnostic and extension services to detect and manage important diseases have existed for decades all over the world. Honeybee diseases may be so disastrous that actions for disease prevention and control are demanded by laws and regulation. American foulbrood, caused by the bacterium *P. larvae*, is a notifiable disease due to its highly virulent and devastating nature and is one of the bee diseases listed in the OIE (Office International des Epizooties – the World Organization for Animal Health) Terrestrial Animal Health Code (2011). Member countries and territories are obliged to report occurrences of American foulbrood outbreaks, and most countries have specific measures that are required upon detection of this pathogen. Measures most often include destruction and burning of infected colonies. For silkworms, diagnostic services are also well developed, and methods are disseminated for producers in all major silk-producing countries (like China; Liu, 2013) and regions (like Tamil Nadu in India; Tamil Nadu Agricultural University, 2016).

There are no general formalized public or private bodies or official institutions for handling insect disease problems within the biological control industry. Within the International Organization for Biological and Integrated Control (IOBC), West Palearctic Regional Section (WPRS), possibilities for biological control companies to ship diseased insects and receive advice on disease problems exist, including the service "Determination and Identification of Entomophagous Insects and Insect Pathogens" (IOBC, 2015). Insect pathologists often assist biological control producers by providing diagnoses and advice about treatments when serious diseases appear (often simply as "in-kind service").

The situation is similar for the expanding industry of insects as food and feed. There are no public or private bodies to give formalized advice on disease problems. We assume that the rapid expansion of this type of insect production will need a formalized diagnostic service. A European initiative (so far named "INSECTPATH") to set up a procedure for assisting this new industry with insect disease problems is currently being planned.

14.4 Ensuring Production of Healthy Insects

In this section, we present some general advice for preventing or managing disease transmission within a production facility and for limiting disease transmission between production facilities and the environment.

- Inspections of insects shipped between production facilities can help avoid transmission of pathogens. Of special concern are latent and vertically transmitted insect pathogens occurring mostly at low prevalence and without external symptoms on infected hosts, which can avoid detection before epidemic outbreaks.

- It is essential to maintain clean and well-managed production facilities. Surface areas, trays, and equipment should be kept clean.
- Rearing conditions (temperature, moisture, food, light, density, etc.) should be optimal for maintaining healthy insects of all instars, and therefore maintaining healthy production stock.
- Maintaining an optimal level of genetic variation within the production insect population is a matter of balancing different elements. On one hand, a high level of genetic homogeneity may help define the best production environment for the selected genotype of production stock. On the other, a high level of heterogeneity may assist in keeping the insect population diverse and therefore more robust to infections. The silkworm is an example of an insect maintained successfully in culture over thousands of years.
- The production stock should have as little contact with the external environment as possible, and ideally, production should occur in closed facilities. This is to avoid an influx of diseases via the air or infected insects. Obviously, this advice does not apply fully to semi-open production facilities (e.g., *A. domesticus* production in Thailand) or to outdoor production of pollinators like honeybees, bumblebees, and solitary bees, where interaction with the environment is an integral element in production and cannot be avoided.
- Food for insects and other substrate brought from outside should be carefully checked to avoid introduction of diseases. Generalist potential insect pathogens in particular may pose a problem in slurry and other wastes, and a check should include screening for such microorganisms, typically fungi or bacteria.
- The production stock should be separated into different subunits, which should have as little contact with one another as possible (e.g., keeping each subunit in a different room, chamber, or incubator). This lowers the risk of transmission within a production facility and makes effective cleaning in case of disease problems easier. It is essential to stop the possibility of vertical transmission of diseases (typically viruses and microsporidia) by checking all instars of the insects.
- Small subpopulations of the production stock should always be kept isolated from the main stock and maintained in small, separate boxes or similar. These boxes should not be handled at the same time as the production stock. Such reserves can also be used for maintenance of a certain genetic variability and can be very useful when restarting a culture. It may be a good idea to occasionally add insects from other production facilities or from the environment to the production stock in order to maintain genetic heterogeneity. This can work well provided that there is a careful check to ensure that no insect pathogens of any type are brought into the production facilities. Here, the inspection should not be restricted to a check for the known pathogens of the production insect: new and unknown pathogens may appear, as seen in Table 14.2.
- In daily work, it is essential to be observant. Are insects thriving? Is the production stock performing as expected? Workers should be trained to notice and report if some insects die or exhibit reduced fitness. If an insect disease is suspected as being the causal agent of such problems, the disease should be allowed to develop in some individuals (after their isolation from the stock) for diagnosis. Diseases that do not kill

insects and/or do not cause external symptoms, but which merely lower fitness (e.g., reduced fertility) can be very hard to detect.

- Except for individuals used for diagnosis, all diseased (or just "strange-looking") insects should be removed immediately and destroyed. In most cases, the entire chamber, incubator, or tray should be emptied, the insects removed and destroyed, and the equipment cleaned and disinfected.

- Symptoms of infected hosts should be noted, using a magnifying glass and, if possible, a dissecting microscope and a compound microscope for glass slides. The success of diagnosis at the production facility will depend on whether the daily managers are sufficiently aware of available methods for identifying pathogens. In any case, a number of suspected diseased insects should be maintained by different methods (e.g., dry, in alcohol, in a freezer) to ensure that a proper diagnosis can be conducted by a specialist later.

- Further steps in disease diagnosis are best conducted by specialized laboratories. These can include detailed microscopic observations of spores or other propagules in infected insects (Lacey, 2012). Other tools may be needed (detection using enzyme-linked immunosorbent assay (ELISA) or checking of genetic identity using polymerase chain reaction (PCR)) for precise determination of the etiological agent, to genus, species, or strain level. Insect pathologists may therefore be needed to properly determine the causal agent. New and previously unknown diseases may appear, in which case identification of pathogens will take longer.

- Of additional concern is whether insects are infected by generalist bacteria or fungi (potential pathogens), with infection occurring via wounds. Such generalist bacterial and fungal diseases might, besides harming the production insects, present a hazard to human health. This is of particular importance in the production of insects as food and feed.

- Cleaning of facilities will depend on the insect pathogens diagnosed. Soap and disinfecting agents are sometimes sufficient for cleaning equipment, incubators, trays, and rooms, if the pathogens are bacteria and fungi. Sterilization using ultraviolet (UV) light or stronger antibiotics (antiviral agents) may be needed.

- Quick and dramatic action may be necessary with abundant disease or a virulent pathogen. As an extreme example, the decision might be that the entire production stock should be destroyed in situations where the production has already collapsed or it seems to be on its way to collapse. A new production stock can be introduced after cleaning of facilities and re-evaluation of procedures for production.

- A joint analysis involving production managers and insect pathologists in order to evaluate, in retrospect, why a disease caused as much damage as it did can be very beneficial. Such analysis can include experimental work like bioassays and biological studies of host–pathogen interactions. Production and management system design may be reconsidered. In such cases, specific guidelines should be developed or updated for the production facility and general guidelines should be made available to interested producers.

- Staff involved in insect production should be educated in basic diagnostics of insect diseases. Some specialist staff members should learn how to recognize early symptoms of disease prevalence in a production population and how to take the necessary action to prevent further transmission.

- Infections among production insects can never be fully avoided. It is especially difficult to prevent and diagnose in due course, especially for (i) latent infections, (ii) diseases with few or no external symptoms, and (iii) new, previously unknown diseases.

14.5   Conclusion

Insects and mites are produced commercially for several reasons: pollination and honey production, silk production, biological control, and use in food and feed. Insects and mites in production may suffer from diseases that need proper action in order to prevent outbreaks and manage these diseases. The three main types of insect and mite production – closed, semi-open, and open facilities – are different with respect to risk of transmission of diseases but similar in the major reasons for disease spread, based on exposure to the environment and/or levels of health. A biological understanding of transmission routes of the different insect pathogens, as well as knowledge about disease symptoms, is essential in order to minimize risk. For honeybees and silkworms, knowledge about diseases, prevention, and treatment is vast, and the services allowing for detection and management are well organized. For other insect production systems, such services are gradually being developed.

Acknowledgments

We thank Margot Calis, Mark van der Zanden, Afton S. Halloran, David Drew, Rosalind R. James, E. Erin Morris, Irfan Kandemir, Silvia Cappellozza, Sunday Ekesi, Ruedi Ritter, Gabriela M. Vergara, and Henrik H. de Fine Licht for their photos, used for Figs. 14.1 and 14.2. The *Journal of Insects as Food and Feed* is thanked for permission to use Fig. 14.3 (in revised form) and Figs. 14.2a–f and k (from Eilenberg et al., 2015). Our work was in part supported by the project "Discerning Taste: Deliciousness as an Argument for Entomophagy," grant number 32726, funded by the Velux Foundation and the Committee for Development Research of the Danish International Development Agency (13-06KU – GREEiNSECT), the Danish Ministry of Foreign Affairs.

References

Belluci, S., Losasso, C., Maggioletti, M., Alonzi, C.C., Paoletti, M.G., Ricci, A., 2013. Edible insects in a food safety and nutritional perspective: a critical review. Comp. Rev. Food Sci. Food Safe. 12, 296–312.

Bjørnson, S., 2008. Natural enemies of mass-reared predatory mites (family Phytoseiidae) used for biological pest control. Exp. Appl. Acarol. 46, 299–306.

Bjørnson, S., Keddie, B.A., 1999. Effects of *Microsporidium phytoseiuli* (Microsporidia) on the performance of the predatory mites, *Phytoseiulus persimilis* (Acari: Phytoseiidae). Biol. Contr. 15, 153–161.

Bjørnson, S., Keddie, B.A., 2001. Disease prevalence and transmission of *Microsporidium phytoseeiuli* infecting the predatory mite, *Phytoseiulus persimilis* (Acari: Phytoseiidae). J. Invertebr. Pathol. 77, 114–119.

Bjørnson, S., Steiner, M.Y., Keddie, B.A., 1997. Birefringent crystals and abdominal discoloration in the predatory mite *Phytoseiulus persimilis* (Acari: Phytoseiidae). J. Invertebr. Pathol. 69, 85–91.

Bjørnson, S., Steele, T., Hu, Q., Ellis, B., Saito, T., 2013. Ultrastructure and molecular characterization of the microsporidium, *Nosema chrysoperlae* sp. nov, from the green lacewing, *Chrysoperla carnea* (Stephens) (Neuroptera: Chrysopidae) used for biological pest control. J. Invertebr. Pathol. 114, 53–60.

Bloch, G., Francoy, T.M., Wachtel, I., Panitz-Cohen, N., Fuchs, S., Mazar, A., 2010. Industrial apiculture in the Jordan valley during Biblical times with Anatolian honeybees. Proc. Natl. Acad. Sci. U.S.A. 107, 11240–11244.

Bowen-Walker, P.L., Martin, S.J. and Gunn, A., 1999. The transmission of deformed wing virus between Honeybees (*Apis mellifera* L.) by the ectoparasitic mite *Varroa jacobsoni* Oud. J. Invertebr. Pathol. 73, 101–106.

Bueno, V.H.P., Carvalho, L.M., van Lenteren, J.C.V., 2014. Performance of *Orius insidiosus* after storage, exposure to dispersal material, handling and shipment processes. Bull. Insect. 67, 175–183.

California Almonds, Almond Board of California (ABC), 2016. Almond Almanac, 2014. Available from: http://www.almonds.com/processors/resources/almond-almanac (accessed May 8, 2017).

Carreck, N.L., Williams, I.H., Little, D.J., 1997. The movement of honey bee colonies for crop pollination and honey production by beekeepers in Great Britain. Bee World 78, 67–77.

Cavigli, I., Daughenbaugh, K.F., Martin, M., Lerch, M., Banner, K., Garcia, E., et al., 2016. Pathogen prevalence and abundance in honey bee colonies involved in almond pollination. Apidologie 47, 251–266.

Chakrabarti, S., King, D.J., Cardona, C.J., Gerry, A.C., 2008. Persistence of exotic Newcastle Disease Virus (ENDV) in laboratory infected *Musca domestica* and *Fannia canicularis*. Avian Dis. 52, 375–379.

Charles, H.A., Killian, K.A., 2015. Response of the insect immune system to three different immune challenges. J. Ins. Physiol. 81, 97–109.

Chen. X., Huang, C., He, L., Zhang, S., Li, Z., 2015. Molecular tracing of white muscardine in the silkworm, *Bombyx mori* (Linn.) II. Silkworm white muscardine is not caused by artificial release or natural epizootic of *Beauveria bassiana* in China. J. Invertebr. Pathol. 125, 16–22.

Cherjanovski, N., Ophir, R., Schwager, M.S., Slebezki, Y., Grossman, S., Cox-Foster, D., 2014. Characterization of viral siRNA populations in honey bee colony. Virology 454–455, 176–183.

Cito, A., Mazza, G., Strangi, A., Benvenuti, C., Barzanti, G.P., Dreassi, E., et al., 2014. Characterization and comparison of *Metarhizium* strains isolated from *Rhynchoporus ferrugineus*. FEMS Microbiol. Lett. 355, 108–115.

Cornman, R.S., Tarpy, D.R., Chen, Y., Jeffreys, L., Lopez, D., Pettis, J.S., et al., 2012. Pathogen webs in collapsing honey bee colonies. PLoS ONE 7(8), e43562.

Cotmore, S.F., Agbandje-McKenna, M., Chiorini, J.A., Mukha, D.V., Pintel, D.J., Qiu, J., et al., 2014. The family Parvoviridae. Arch. Virol. 159, 1239–1247.

Cox-Foster, D.L., Conlan, S., Holmes, E.C., Palacios, G., Evans, J.D., Moran, N.A., et al., 2007. A metagenomic survey of microbes in honey bee colony collapse disorder. Science 318(5848), 283–287.

Crane, E., 1990. Bees and Beekeeping: Science, Practice and World Resources. Heinemann Newnes, Oxford.

de Miranda, J.R., Cornman, R.S., Evans, J.D., Semberg, E., Haddad, N., Neumann, P., Gauthier, L., 2015. Genome characterization, prevalence and distribution of a macula-like virus from *Apis mellifera* and *Varroa destructor*. Viruses 7, 3586–3602.

de Souza, P.C., Morey, A.T., Castanheira, G.M., Bocate, K.P., Panagio, L.A., Ito, F.A., et al., 2015. *Tenebrio molitor* (Coleoptera: Tenebrionidae) as an alternative host to study fungal infections. J. Microbiol. Meth. 118, 182–186.

Diener, S., Solano, N.M.S., Gutiérrez, F.R., Zurbrügg, C., Tockner, K., 2011. Biological treatment of municipal organic waste using black soldier fly larvae. Waste Biomass Valoriz. 2, 357–363.

Eilenberg, J., Vlak, J.M., Nielsen-LeRoux, C., Cappellozza, S., Jensen, A.B., 2015. Diseases in insects produced for food and feed. J. Ins. Food Feed 1, 87–102.

Forsgren, E., Wei, S., Guiling, D., Zhiguang, L., Tran, T., Tang, P.T., et al., 2015. Preliminary observations on possible pathogen spill-over from *Apis mellifera* to *Apis cerana*. Apidologie 46, 265–275.

Fraser, C.A., Owens. L., 1996. Spawner-isolated mortality virus from Australian *Penaeus monodon*. Dis. Aquatic Organ. 27, 141–148.

Gallai, N., Salles, J.M., Settele, J., Vaissière, B.E., 2009. Economic valuation of the vulnerability of world agriculture confronted with pollinator decline. Ecol. Econ. 68, 810–821.

Genersch, E., 2010. Honey bee pathology: current threats to honey bees and beekeeping. Appl. Microbiol. Biotech. 87, 87–97.

Goerzen, D.W., Erlandson, M.A., Bissett, J., 1990. Occurrence of chalkbrood caused by *Ascosphaera aggregata* Skou in a native leafcutting bee, *Megachile relativa* Cresson (Hymenoptera, Megachilidae), in Saskatchewan. Can. Entomol. 122, 1269–70

Goerzen, D.W., Dumouchel, L., Bissett, J., 1992. Occurrence of chalkbrood caused by *Ascosphaera aggregata* Skou in a native leafcutting bee, *Megachile pugnata* Say (Hymenoptera: Megachilidae), in Saskatchewan. Can. Entomol. 124, 557–558.

Goolsby, J.A., Ciomperlik, M.A., Simmons, G.S., Pickett, C.J., Gould, J.A., Hoelmer, K.A., 2014. Mass-rearing *Bemisia* parasitoids for support of classical and augmentative biological control programs, in: Morales-Ramos, J.A., Rojas, M.G., Shapiro-Ilan, D.I. (eds.), Mass Production of Beneficial Organisms. Academic Press, Amsterdam, pp. 145–162.

Goulson, D., Nicholls, E., Botías, C., Rotheray, E.I., 2015. Bee declines driven by combined stress from parasites, pesticides, and lack of flowers. Science 347, 1435–1445.

Graystock, P., Yates, K., Evison, S.E., Darvill, B., Goulson, D., Hughes, W.O., 2013. The Trojan hives: pollinator pathogens, imported and distributed in bumblebee colonies. J. Appl. Ecol. 50, 1207–1215.

Graystock, P., Blane, E.J., McFrederick, Q.S., Goulson, D., Hughes, W.O., 2016. Do managed bees drive parasite spread and emergence in wild bees? Internatl J. Parasitol.: Para. Wildlife 5, 64–75.

Gross, O., Tijssen, P., Weinberg, D., Tal, J., 1990. Expresssion of densonucleosis virus GmDNV in *Galleria mellonella* larvae: size analysis and in vitro translation of viral transcription products. J. Invertebr. Pathol. 56, 175–180.

Halloran, A., Roos, N., Eilenberg, J., Cerutti, A., Bruun, S., 2016. Life cycle assessment of edible insects for food protein: a review. Agron. Sustain. Devel. 36, 57.

Hosamani, M., Basagoudanavar, S.H., Sreenivasa, B.P., Inumaru, S., Ballal, C.R., Venkataramanan, R., 2015. Eri silkworm (*Samia ricini*), a non-mulberry host system for AcMNPV mediated expression of recombinant proteins. J. Biotechnol. 216, 76–81.

Hou, C., Rivkin, H., Slabezki, Y., Chejanovski, N., 2014. Dynamics of the presence of Israeli Acute Paralysis Virus in honey bee colonies with colony collapse disorder. Viruses 6, 2012–2027.

International Organization for Biological and Integrated Control (IOBC), 2015. Determination and identification of entomophagous insects and insect pathogens. Available from: https://www.iobc-wprs.org/expert_groups/c_identification.html (accessed May 8, 2017).

Ivie, M.A., Pollock, D.A., Gustafson, D.L., Rasolomandimby, J., Ivie, L.L., Swearingen, W.D., 2002. Field-based evaluation of biopesticides impacts on native biodiversity: Malagasy Coleoptera and anti-locust entomopathogenic fungi. J. Econ. Entomol. 95, 651–660.

Jakob, N.J., Kleespies, R.G., Tidona, C.A., Müller, K., Gelderblom, H.R., Darai, G., 2002. Comparative analysis of the genome and host range characteristics of two insect iriroviruses: Chilo iridescent virus and a cricket iridovirus isolate. J. Gen. Virol. 83, 463–470

James, R.R., 2004. Temperature and chalkbrood development in the alfalfa leafcutting bee. Apidologie 36, 15–23.

James, R.R., 2011. Chalkbrood transmission in the alfalfa leafcutting bee: the impact of disinfecting bee cocoons in loose cell management systems. Environ. Entomol. 40, 782–787.

James, R.R., Li, Z., 2012. From silkworms to bees: diseases of beneficial insects, in: Vega, F.E., Kaya, H.K. (eds.), Insect Pathology, 2nd edn. Elsevier, Amsterdam, pp. 425–459.

James, R.R., Pitts-Singer, T.L., 2005. *Ascosphaera aggregata* contamination on alfalfa leafcutting bees in a loose cell incubation system. J. Invertebr. Pathol. 89, 176–178.

Joensuu, K., Sinkko, T., 2015. Environmental sustainability and improvement options for agribiomass chains: straw and turnip rape. Biomass Bioenergy 83, 1–7.

Józefiak, D., Józefiak, A., Kierronczyk, B., Rawski, M., Swiatkiewitcz, S., Dlogusz, J., Engberg, R.M., 2016. Insects – a natural nutrient source for poultry – a review. Ann. Anim. Sci. 16, 297–313.

Klein, A.M., Vaissiere, B.E., Cane, J.H., Steffan-Dewenter, I., Cunningham, S.A., Kremen, C., Tscharntke, T., 2007. Importance of pollinators in changing landscapes for world crops. Proc. Roy. Soc. London B 274, 303–313.

Kobayashi, M., Saski, T., Saito, N., Tamura, K., Suzuki, K., Watanabe, H., Agui, N., 1999. Houseflies: not simple mechanical vectors of enterohemorrhagic *Escherichia coli* O157:h7. Am. J. Trop. Med. Hyg. 61, 625–629.

Kremen, C., Williams, N.M., Thorp, R.W., 2002. Crop pollination from native bees at risk from agricultural intensification. Proc. Natl. Acad. Sci. U.S.A. 99, 16812–16816.

la Fauce, K.S., Owens, L., 2008. The use of insects as a bioassay for *Penaeus merguiensis* densovirus (PmergDNV). J. Invertebr. Pathol. 98, 1–6.

Lalander, C.H., Diener, S., Magri, M.E., Zurbrügg, C., Lindström, A., Vinnerås, B., 2013. Faecal sludge management with the larvae of the black soldier fly (*Hermetia illucens*) – from a hygiene aspect. Sci. Total Environ. 458–460, 312–318.

Lalander, V.H., Fidjeland, J., Diener, S., Eriksson, S., Vinnerås, B., 2015. High waste-to-biomass conversion and efficient *Salmonella* spp. reduction using black soldier fly for waste recycling. Agron. Sustain. Dev. 35, 262–271.

Li, D., Guo, Y., Shao, H., Tellier, L., Wang, J., Xiang, Z., Xia, Q., 2010. Genetic diversity, molecular phylogeny and selection evidence of the silkworm mitochondria implicated by complete resequencing of 41 genomes. BMC Evol. Biol. 10, 81.

Li, G.N., Xia, X.J., Zhao, H.H., Sendegeya, P., Zhu, Y., 2015. Identification and Characterization of *Bacillus cereus* SW7−1 in *Bombyx mori* (Lepidoptera: Bombycidae). J. Insect Sci. 15: 136.

Liu, J.P., 2013. Current sericulture situation and the silkworm diseases control in China. Available from: http://www.bacsa-silk.org/user_pic/file/Dr_%20Jiping%20Liu%20pres_2.pdf (accessed May 8, 2017).

Liang, X., Lu., Z.-L., Wei, B.-X., Feng, J.-L., Qu, D., Luo, T.R., 2013. Phylogenetic analysis of *Bombyx mori* nucleopolyhedrosis polyhedrin and p10 genes in wild isolates from Guangxi Zhuang Autonomous Region, China. Virus Genes 46, 140–151.

Liu, H., Pan, G., Luo, B., Li, T., Yang, Q., Vossbrink, C.R., Debrunner-Vossbrink, B., et al., 2013. Intraspecific polymorphism of rDNA among five *Nosema bombycis* isolates from different regions in China. J. Invertebr. Pathol. 113, 63–69.

Makkar, H.P.S., Tran, G., Heuzé, V., Ankers, P., 2014. State-of-the-art on use of animals as animal feed. Anim. Feed Sci. Technol. 197, 1–33.

Maxfield-Taylor, S.A., Mujic, A.B., Rao, S., 2015. First detection of the larval chalkbrood disease pathogen *Ascosphaera apis* (Ascomycota: Eurotiomycetes: Ascosphaerales) in adult bumble bees. PloS ONE 10(4), e0124868.

Mazza, G., Francardi, V., Simoni, S., Benvenuti, C., Cervo, R., Feleiro, J.R., et al., 2014. An overview on the natural enemies of *Rhynchophorus* palm weevils, with focus on *R. ferrugineus*. Biol. Contr. 77, 83–92.

McMahon, D.P., Fürst, M.A., Caspar, J., Theodorou, P., Brown, M.J., Paxton, R.J., 2015. A sting in the spit: widespread cross-infection of multiple RNA viruses across wild and managed bees. J. Anim. Ecol. 84, 615–624.

Medo, J., Cagán, L., 2011. Factors affecting the occurrence of entomopathogenic fungi in soils of Slovakia as revealed using two methods. Biol. Contr. 56, 200–208.

Meikle, W.G., Bon, M., Cook, S.C., Gracia, C.G., Jaronski, S., 2013. Two strains of *Pseudomonas fluorescens* bacteria differentially affect survivorship of waxworm (*Galleria mellonella*) larvae exposed to an arthropod fungal pathogen, *Beauveria bassiana*. Biocont. Sci.Techn. 23, 220–233.

Meyling, N.V., Eilenberg, J., 2006. Occurrence and distribution of soil borne entomopathogenic fungi within a single organic ecosystem. Agric. Ecosyst. Environ. 113, 336–341.

Morales-Ramos, J.A., Rojas, M.G., Shapiro-Ilan, D.I. (eds.), 2014. Mass Production of Beneficial Organisms. Academic Press, Amsterdam.

Moritz, R.F., Härtel, S., Neumann, P., 2005. Global invasions of the western honeybee (*Apis mellifera*) and the consequences for biodiversity. Ecoscience 12, 289–301.

Mullen, L.M., Goldsworthy, G.J., 2006. Immune responses of locusts to challenge with the pathogenic fungus *Metarhizium* or high doses of laminarin. J. Insect Physiol. 52, 389–398.

Noonin, C., Jiravanichpaisal, P., Söderhäll, I., Merino, S., Tomás, J.M., Söderhäll, K., 2010. Melanization and pathogenicity in the insect, *Tenebrio molitor*, and the crustacean, *Pacifasciatus leniusculus*, by *Aeromonas hydrophila* AH-3. PLoS ONE 5(12), e15728.

Otterstatter, M.C., Thomson, J.D., 2008. Does pathogen spillover from commercially reared bumble bees threaten wild pollinators? PLoS ONE 3(7), e2771.

Oonincx, D.G.A.B., de Boer, I.J.M., 2012. Environmental impact of the production of mealworms as a protein source for humans – a life cycle assessment. PLoS ONE 7(12), e1145.

Oonincx, D.G.A.B., van Broekhoven, S., van Huis, A., van Loon, J.J.A., 2015. Feed conversion, survival and development, and composition of four insect species on diets composed by food by-products. PLoS ONE 10, e0144601.

Peterson, S.S., Baird, C.R., Bitner, R.M., 1992. Current status of the alfalfa leafcutting bee, *Megachile rotundata*, as a pollinator of alfalfa seed. Bee Sci. 2, 135–142.

Phoku, J.Z., Barnard, T.G., Potgieter, N., Dutton, M.F., 2014. Fungi in housefly (*Musca domestica* L.) as a disease risk indicator – a case study in South Africa. Acta Tropica 140, 158–165.

Ravikumar, G., Raje Urs, S., Vijaya Prakash, N.B., Rao, C.G.P., Vardhana, K.V. 2011. Development of a multiplex polymerase chain reaction for the simultaneous detection of microsporidians, nucleopolyhedrovirus, and densovirus infecting silkworms. J. Invertebr. Pathol. 107, 193–197

Richards, K.W., 1984. Alfalfa Leafcutter Bee Management in Western Canada. Agriculture Canada Publication 1495/E, Ottawa, Agriculture Canada.

Roekring, S., Nielsen, L., Owens, L., Pattanakitsakul, S., Malasit, P., Flegel, T.W., 2002. Comparison of paenaid shrimp and insect parvoviruses suggests that viral transfers may occur between two distantly related arthropod groups. Virus Res. 87, 79–87.

Roffet-Salque, M., Regert, M., Evershed, R.P., Outram, A.K., Cramp, L.J., Decavallas, O., et al., 2015. Widespread exploitation of the honeybee by early Neolithic farmers. Nature 534, 1–2.

Sarmah, M.C., Rahman, S.S.S., Barah, A., 2010. Traditional practices and terminologies in Muga and Eri culture. Indian J. Trad. Knowl. 9, 448–452.

Schuld, M., Madel, G., Schmuck, R., 1999. Impact of *Vairimorpha* sp. (Microsporidia: Burnellidae) on *Trichogramma chilonis* (Hymenoptera, Trichogrammalidae), a hymenopteran parasitoids of the cabbage moth, *Plutella xylostella* (Lepidoptera, Yponomeutidae). J. Invertebr. Pathol. 74, 120–126.

Schütte, C., Dicke, M., 2008. Verified and potential pathogens of predatory mites (Acari: Phytoseiidae). Exp. Appl. Acarol. 46, 307–328.

Schütte, C., Poitevin, O., Negash, T., Dicke, M., 2006. A novel disease affecting the predatory mite *Phytoseiulus persimilis* (Acari, Phytoseiidae): 2. Disease transmission by adult females. Exp. Appl. Acarol. 39, 85–103.

Schütte, C., Gols, R., Kleespies, R.G., Poitevin, O., Dicke, M., 2008. Novel bacterial pathogen *Acaricomes phytoseiuli* causes severe disease symptoms and histological changes in the predatory mite *Phytoseiulus persimilis* (Acari, Phytoseiidae). J. Invertebr. Pathol. 98, 127–135.

Shapiro-Ilan, D.I., Han, R., Qiu, X., 2014. Production of entomopathogenic nematodes, in: Morales-Ramos, J.A., Rojas, M.G., Shapiro-Ilan, D.I. (eds.), Mass Production of Beneficial Organisms. Academic Press, Amsterdam, pp. 321–355.

Shimoda, T., Kobori, Y., Yara, K., Hinomoto, N., 2015. A simple method of rearing insect natural enemies of spider mites. Biol. Contr. 80, 70–76.

Steele, T., Bjørnson, S., 2012. The effect of two microsporidian pathogens on the two-spotted lady beetle, *Adalia bipunctata* L. (Coleoptera, Coccinellidae). J. Invertebr. Pathol. 109, 223–228.

Steinhaus, E.A., 1963. Insect Pathology, an Advanced Treatise, Vol. 1. Academic Press, New York.

Stoops, J., Crauwels, S., Waud, M., Claes, J., Lievens, B., 2016. Microbial community assessment of mealworm larvae (*Tenebrio molitor*) and grasshoppers (*Locusta migratoria migratorioides*) sold for human consumption. Food Microbiol. 53, 122–127.

Sumner, D.A., Boriss, H., 2006. Bee-economics and the leap in pollination fees. Agric. Res. Econ. Update 9, 9–11.

Tabassum-Abbasi, Abbasi, T., Abbasi, S.A., 2016. Reducing the global environmental impact of livestock production; the mini-livestock option. J. Cleaner Prod. 112, 1754–1766.

Tamil Nadu Agricultural University (TNAU) AgritechPortal, 2016. Sericulture: Disease Management: Mulberry Silkworm. Available from: http://agritech.tnau.ac.in/sericulture/disese%20mgt_silkworm.html (accessed May 8, 2017).

Tikader, A., Kunjupillai, K., Saratchandra, B., 2013. Muga silkworm, *Antheraea assamensis* (Lepidoptera: Saturniidae) – an overview of distribution, biology and breeding. Eur. J. Entomol. 110, 293–300.

van der Geest, L.P., Elliot, S.L., Breeuwer, J.A., Beerling, E.A., 2000 Diseases of mites. Exp. Appl. Acarol. 24:497–560.

van Huis, A., van Itterbeck, J., Klunder, H., Mertens, E., Halloran, A., Muir, G., Vantomme, P., 2013. Edible Insects: Future Prospects for Food and Feed Security. FAO Forestry Paper 171. Food and Agricultural Organization of the United Nations, Rome.

van Lenteren, J.C., 2012. The state of commercial augmentative biological control: plenty of natural enemies, but a frustrating lack of uptake. BioControl 57, 1–20.

van Zanten, H., Mollenhorst, H., Klootwick, C.W., van Middelaar, C.E., de Boer, I.J.M., 2015. Global food supply: land use efficiency of livestock systems. J. Life Cycle Assess. 21, 747–758.

Vega, F.E., Kaya, H.K. (eds.), 2012. Insect Pathology, 2nd edn, Elsevier, Amsterdam.

Velthuis, H.H., van Doorn, A., 2006. A century of advances in bumblebee domestication and the economic and environmental aspects of its commercialization for pollination. Apidologie 37, 421–451.

Vootla, S.K., Lu, X.M., Kari, N., Gadwala, M., Lu, Q., 2013. Rapid detection of infectious flacherie virus of the silkworm, *Bombyx mori*, using RT-PCR and nested PCR. J. Ins. Sci. 13, 120.

Wang, L.L., Chen, K.P., Zhang, Z., Yao, Q., Gao, G.T., Zhao, Y., 2006. Phylogenetic analysis of *Nosema antheraeae* (Microsporidia) isolated from Chinese oak silkworm, Antheraea pernyi. J Eukaryot Microbiol. 53, 310–313.

Welling, M., Zelazny, B., Scherer, R., Zimmermann, G., 1995. First record of the entomopathogenic fungus *Sorosporella* sp. (Deuteromycotina: Hyphomycetes) in *Locusta migratoria* (Ortoptera: Acrididae) from Madagascar: symptoms of infection, morphology and infectivity. Biocontr. Sci. Technol. 5, 465–474.

Winfree, R., Williams, N.M., Dushoff, J., Kremen, C., 2007. Native bees provide insurance against ongoing honey bee losses. Ecol. Lett. 10, 1105–1113.

Yang, S.-Y., Han, M.-J., Kang, L.-F., Li, Z.-W., Shen, Y.-H., Zhang, Z., 2014. Demographic history and gene flow during silkworm domestication. BMC Evol. Biol. 14, 185.

Zethner, O., Koustrup, R., Barooah, D., 2012. Indian Ways of Silk. Bhabani Print & Publications, Guwahati.

Zhou, Y., Wu, J., Lin, F., Chen, N., Yuan, S., Ding, L., et al., 2015. Rapid detection of *Bombyx mori* nucleopolyhedrovirus (BmNPV) by loop-mediated isothermal amplification assay combined with a lateral flow dipstick method. Molec. Cell Probes 29, 389–395.

Zielinska, E., Baraniak, B., Karas, M., Rybczynska, K., Jacubczyk, J., 2015. Selected species of edible insects as a source of nutrient composition. Food Res. Internatl. 77, 460–466.

Zimmermann, G., 1986. The *Galleria* bait method for detection of entomopathogenic fungi in soil. J. Appl. Entomol. 102, 213–215.

15

Prevention and Management of Infectious Diseases in Aquatic Invertebrates

Jeffrey D. Shields

Department of Aquatic Health Sciences, Virginia Institute of Marine Science, The College of William & Mary, Gloucester Point, VA, USA

15.1 Scope

15.1.1 Myriad Pathogens Infect Aquatic Invertebrates

Numerous pathogens use aquatic invertebrates as hosts. By some accounts, more than half of the parasites on earth use or transit through aquatic ecosystems. The diversity of agents spans the viral, bacterial, fungal, protistan, and animal kingdoms, and in many cases the pathogens have received little attention. The diversity of host–parasite associations is extraordinary, and likely explains why we know so little about them. Some pathogens have a high degree of host specificity, whereas others are host generalists, infecting many host species within a taxon or even among taxa. Some associations are highly evolved and include significant modifications of host physiology and behavior, in contrast to simpler associations where the host is used as a phoretic substrate or where both host and symbiont live in a commensal relationship. The nature of these associations largely dictates effects on host physiology and ecology, and ultimately influences potential management efforts.

In general, parasites are natural enemies of their hosts that live at the expense of the host's metabolic resources (i.e., usually deriving energy from the host or imposing a metabolic cost to the host). Pathogens can include parasites, but are typically microorganisms (viruses, bacteria, and fungi) that produce disease under normal conditions of host resistance (Onstad et al., 2006). Microbial pathogens are sometimes referred to as microparasites, which reproduce rapidly within their hosts, usually by means of asexual division or replication; for modeling purposes the infected host is the unit of measure (Anderson and May, 1981). Macroparasites are typically metazoans (helminths, arthropods), which have longer generation times; for modeling purposes, their populations are related to their immigration and death, with the individual parasites being the typical unit of measure. Macroparasites often exhibit density-dependent pathology, with disease arising in heavy infections. An infectious disease is an abnormal imbalance or dysfunction in a host caused when a pathogen occurs at high densities and so disrupts the physiological function of the individual host. Endemic pathogens occur normally within an area, whereas exotic or introduced pathogens do not occur normally and thus

Ecology of Invertebrate Diseases, First Edition. Edited by Ann E. Hajek and David I. Shapiro-Ilan.
© 2018 John Wiley & Sons Ltd. Published 2018 by John Wiley & Sons Ltd.

have either invaded or been accidentally introduced. An epizootic is an outbreak of a pathogen or parasite in a host population that occurs above the normal background levels of the agent. Widespread epizootics are referred to as pandemics.

Disease ecology is fundamentally different between terrestrial and aquatic systems (Table 15.1). However, there is surprisingly little theoretical work exploring their differences with respect to parasites, symbionts, and pathogens. Aquatic systems are broadly defined to include freshwater, estuarine, and marine ecosystems. Marine ecosystems tend to be more diverse with respect to host phyla, have a large number of clonal, colonial animals, are often colonized by planktonic larvae, and tend to have large, open populations when compared to terrestrial and freshwater ecosystems (McCallum et al., 2004; Poulin et al., 2016). Marine ecosystems also tend to have a higher diversity of parasitic castrators than terrestrial systems, and the latter tend to have a higher diversity of pathogens that have vertical transmission, vectors, and parasitoids. This indicates that transmission has evolved very different modes, or routes, between systems, largely in response to adaptations to terrestrial existence (e.g., dessication, other environmental factors, dispersal). Freshwater ecosystems tend to have higher endemism and diversity compared to terrestrial ecosystems, but they have similar disturbances and environmental inputs as adjacent terrestrial habitats (Johnson and Paull, 2011). Freshwater ecosystems often serve as a nexus between aquatic and terrestrial organisms, and pathogens exploit this nexus in times of drought, during seasonal mating activities, or during seasonally high temperatures, when host populations require access to freshwater.

Landscape ecology is the study of the relationships between ecological processes, their ecosystems, and the environment. The principles of landscape ecology have been applied to the ecology of vector-borne diseases for many years. These principles are often studied as separate components: the ecosystem or habitat, the hosts (vectors, intermediate and definitive hosts), and the pathogen or parasite of interest. They will be compared separately in the sections that follow, along with anthropogenic effects. Although the principles of landscape ecology are well known in terrestrial systems, they have received far less attention as "seascape" ecology in marine systems (Boström et al., 2011). In terms of the environment, there are many homologous features between terrestrial and aquatic systems. For example, the physiography of terrain is a key component in most disease systems (e.g., scrub typhus, rodents, and tall grasslands; or malaria, mosquitoes, and swamps). The same applies to many aquatic systems, where the benthic physiography can constrain water masses, resulting in marked changes to water quality, residence time, stratification, and entrainment (*Hematodinium* sp., snow crabs, and fjords). With marine pathogens, confined habitats such as lagoons and fjords often have relatively "closed" host populations (i.e., with little immigration and emigration of juveniles and adults), restricted water exchange (i.e., narrow channels with shallow sills, deep confined areas, entrained water masses), stressful physical conditions (i.e., summer and winter thermal stress, high salinity, seasonal hypoxia, intense fishing pressure), and a pathogen that can amplify rapidly within a population (Kuris and Lafferty, 1992; Shields, 2012). However, unlike terrestrial systems, organisms in marine ecosystems rely on dissolved oxygen for survival. Physical features such as temperature, salinity, and depth can dramatically affect dissolved oxygen levels, which, when lowered due to high temperatures, can impose widespread physiological stress (hypoxia) to shallow-water fauna, making them more susceptible to disease and mortality (Shields, 2013).

Table 15.1 Components of landscape ecology associated with disease agents in relation to different ecosystems. The components are split broadly into environmental, anthropogenic, host, and pathogen themes, for comparison.

Components	Terrestrial	Freshwater	Marine
Environment	Landscape physiography	Benthic physiography	Benthic physiography
	Rainfall	Runoff	Residence time
	Humidity	Hardness/softness	Hypoxia
	Temperature	Temperature	Temperature
	Wind (speed, vector)	Mixing	Mixing
	Percent cover	Percent cover	Percent cover
	Insolation?	Physical size	Currents
			Depth
			Salinity
			Storms/hurricanes
Anthropogenic effects	Soil modification	Eutrophication	Eutrophication
	Salination	Acidification	Acidification
	Pesticides	Pesticides	Juvenilization of fished populations
	Contaminants	Erosion	Fishery stress: injuries, traps
	Introductions	Contaminants	Contaminants
		Introductions	Introductions
Host	Fewer phyla	Fewer phyla	Diverse phyla
	Airborne dispersal	Endemism, direct development	Planktonic larvae, high dispersal
	Rarely clonal	Clonal organisms	Clonal organisms
	Closed populations	Closed populations	Open populations
	Cuticle-covered invertebrates	Soft-bodied invertebrates with external absorption	Soft-bodied invertebrates with external absorption
	Internal absorption	Gills, mucus	Gills, mucus (enhanced surface area for transmission, exposed)
			Diverse feeding modes
Pathogen	Fewer phyla	Fewer phyla	Diverse phyla
	Vertical transmission	Multiple host cycles	Multiple host cycles
	Vectors	Active host-finding abilities?	Parasitic castrators
	Parasitoids	Flagellated, ciliated, and spore stages for dispersal	Flagellated, ciliated and spore stages for dispersal
	Spores resist dessication for dispersal	Low Reynolds numbers dominate	Reynolds numbers

Human-induced influences, or anthropogenic effects, can profoundly alter ecosystems (Table 15.1.). Negative anthropogenic effects in terrestrial systems include habitat fragmentation, soil modification, salination, contaminants, species introductions, and reductions in diversity through monoculture. In aquatic systems, runoff from agricultural areas and urban centers causes eutrophication, acidification, and the spread of contaminants. These are known stressors that reduce the susceptibility of hosts by compromising barriers to infection, particularly in soft-bodied invertebrates. In addition, fishing imposes a very different type of effect. Removal of adult hosts causes the "juvenilization" of host populations, and fishing stress from handling, culling, and bycatch release imposes additional stress. In heavily fished populations, these stressors can have profound, population-level impacts that manifest as emergent infectious diseases in compromised hosts (Shields, 2013). Anthropogenic effects, therefore, must be included in studies of disease ecology, because they are common disturbances and have become increasingly implicated in disease etiology.

Host factors are very important in disease ecology. Aquatic ecosystems have very high host diversity, particularly with respect to soft-bodied invertebrates such as corals, worms, mollusks, and echinoderms. These hosts exhibit a broad range of life-history adaptations that may facilitate transmission and infection by pathogens. For example, many marine invertebrates are detritus, suspension, or filter feeders, feeding on fecal pellets and organic wastes released by other animals. This feeding mode may make them particularly vulnerable to microbial infections or contaminants and facilitate the lifecycles of many parasites. Only recently have ecologists incorporated filter feeding and particle diffusion into models of marine diseases (Bidegain et al., 2016a,b). Many marine phyla also have adaptations that make them more susceptible to infection. These include soft, absorptive integuments, bodies covered in sticky mucus, and exposed, external gills. Soft integuments or exposed gills may make many phyla more susceptible to pathogens than the hard, protective cuticles required to resist dessication in many terrestrial invertebrates. Indeed, the hemolymph of many marine invertebrates is not sterile – a feature often overlooked in studies of their autoecology, physiology, and genomics. Host dispersal also differs between terrestrial and aquatic systems. Arthropods make an interesting comparison. Terrestrial arthropods often have fewer larval life-history stages, and adults have evolved wings for long-distance dispersal. Aquatic arthropods often have several larval life-history stages (nauplius and cypris, or zoeae and megalopa) for planktonic dispersal, whereas adults have adaptations for a benthic lifestyle. Pathogens have evolved different strategies for host exploitation depending on the dispersal method used by their hosts. For example, cannibalism is relatively common in arthropods, and several pathogens use it as a transmission route: the eating of younger conspecifics increases the probability of infection (Rudolf and Antonovics, 2007).

Parasites, symbionts, and pathogens have some of the most complex lifecycles of the animal world. These lifecycles can be extraordinarily difficult to unravel (e.g., Myxospora in freshwater oligochaetes and fishes; Haplosporidia in bivalves), exhibit remarkable adaptations to trophic transfer via paratenic hosts, and often exhibit marked morphological changes in response to environmental and seasonal cues. At least for eukaryotic parasites, there are differences in the modes of transmission and host exploitation among terrestrial and aquatic systems. For example, the transmission stage in many terrestrial pathogens is a cyst or spore that is highly resistant to dessication or that has a sticky surface to enhance dispersal. Many microbial pathogens in terrestrial systems also use vertical transmission and vectors for transmission. Water, however, is a more viscous medium than air in terms of dispersal and transmission; hence, physical aspects such as boundary layers and low Reynolds numbers (a metric for measuring movement or flow in viscous environments) dominate the probability of successful transmission in aquatic systems (Poulin et al., 2016). The transmission stage in aquatic pathogens is often motile, having flagellae or cilia, with adaptations for rapid and directed host finding, attachment, and penetration. When coupled with a filter-feeding host and anthropogenically altered host densities (i.e., aquaculture), motile spore stages can enhance transmission, resulting in high prevalence levels of pathogens in several marine invertebrates (Bidegain et al., 2016a,b).

Most work on diseases in aquatic invertebrates has focused on hosts in commercial production, such as aquaculture and managed fisheries. Several pathogens have caused serious damage to populations of shellfish hosts, presenting resource managers with seemingly intractable problems in mitigating the effects of serious epidemics in managed host populations or cultured stocks. Much effort is spent on rebuilding affected stocks by closing fisheries, restricting catches, or regulating culture production. Also, a great deal of work has been done to examine pathogens in aquatic invertebrates that serve as intermediate hosts or vectors for human pathogens. Invertebrate vectors present significant problems and require costly management programs to effect control of their populations. Note that controlling an invertebrate vector is based on eradication of its population, whereas managing a pathogen in a commercially or ecologically important invertebrate is based on conservation of the host; the two are diametrically opposed and represent very different approaches.

In this chapter, I present an overview of some select pathogens that have occurred in outbreaks in aquatic systems, focusing primarily on host species of commercial or ecological importance. I do not cover the aspects of control and mitigation of pathogens that use invertebrates as vectors of disease in humans. Aspects of the management of outbreaks, as well as their prevention, are discussed when possible, but one issue facing many systems is the paucity of suitable mitigation and control techniques available for appropriate responses. Recent work in this area has called for enhanced reporting mechanisms to improve responses, *in silico* modeling (when possible), improved strategies for dealing with potential spread of pathogens during outbreaks, and access to suitable funding sources for rapid responses to emerging disease issues (Fig. 15.1) (Plowright et al., 2008; Groner et al., 2016).

15.1.2 Overview of Disease Issues in Assessing Epidemics in Aquatic Invertebrates

Many populations of commercially and ecologically important marine invertebrates have been damaged by new or recurring outbreaks of pathogens (Table 15.2). Several of these outbreaks have been associated with climate change and commensurate anthropogenic stressors (Harvell et al., 2007; Webster, 2007; Shields, 2012; Burge et al., 2014). The impact of disease can be enormous, resulting in recurrent losses or nonrecovery of fisheries (Shields, 2012) and billions of dollars in commercial losses in aquaculture (Lightner et al., 2012a; Stentiford et al., 2012; Lafferty et al., 2015).

Disease epizootics present challenging issues when they occur in aquatic invertebrates. In many cases, the etiological agents are poorly known, rendering management

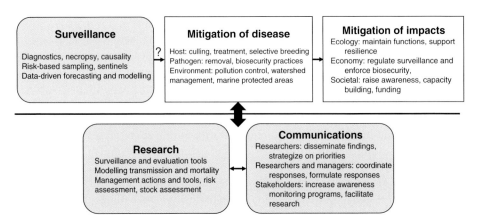

Fig. 15.1 Framework for managing disease outbreaks in aquatic systems. Surveillance, research, and communications (rounded boxes) are vital, continuing processes that enable detection, evaluation, and development of assessments. Mitigation actions (rectangular boxes) are responsive tasks catalyzed by informed, effective communication among stakeholders. *Source*: Groner et al. (2016). Reproduced with permission of Jeff Shields.

scenarios difficult (Shields, 2012; Groner et al., 2016). Establishing causality is important in understanding the nature of a disease, and in some cases there may be several possible scenarios that need to be evaluated in order to fully assess etiology. Causality may not be fully established, but rather inferred through an iterative process involving laboratory investigation, field studies, and *in silico* modeling (Plowright et al., 2008). In addition, a disease can have an overt presentation, or proximate causality, which is often associated with a pathogen or suite of opportunists, and an underlying cause, or ultimate causality, comprising strong environmental factors or other stressors that compromise host defences and ultimately allow the host to be colonized by pathogens (Shields, 2013). Identifying underlying environmental factors requires access to long-term data sets or field studies for assessment (e.g., Shields et al., 2007).

A common feature in most aquaculture systems and fisheries is that many pathogens remain at very low levels in their host populations until they are triggered to proliferate by external stimuli. Such stimuli include abiotic changes (e.g., increases in temperature and salinity) and biotic changes (e.g., fluctuations in host density or diet) (Shields, 2012). Baseline surveys for pathogens in important aquatic invertebrates are typically lacking, and this impedes our understanding of the impact diseases have on a host population during an outbreak. Baseline surveys are important because they can indicate the presence of a pathogen, its distribution and abundance, and possibly its potential for damage. Longer-term surveys can provide details as to whether an outbreak is an emerging phenomenon or a common component in the host population. Extensive survey data are often available for commercially important hosts, but in many cases they do not include early juvenile stages – stages that are often affected severely by pathogens (Shields et al., 2005, 2007; Behringer, 2012; Shields, 2012).

During an epizootic, initial disease assessments are important in establishing causality and severity, but they may underestimate heavy infections because sick animals do not enter traps, are more difficult to catch, and can die prior to assessment. Because causality can be difficult to assess, diagnostic tools may initially rely on exploratory

Table 15.2 Examples of marine invertebrate hosts that have experienced epizootic outbreaks, with the pathogens implicated, environmental correlates, scale of the epizootics, and regions in which epizootics occurred.

Host	Pathogen	Scale of outbreaks	Implicated stressors	Region	Key references
Sponges					
Several commercial species	?	Focal, regional	?	Bahamas, Cuba, Caribbean	Galtsoff et al. (1939)
Ircinia spp.	*Vibrio* spp.	Focal, regional	Temperature, seasonality	Mediterranean Sea	Maldonado et al. (2010); Stabili et al. (2012)
Ircinia variabilis, Sarcotragus spinosulus, Spongia officinalis	Cyanobacteria	Focal	Temperature	Adriatic Sea	Di Camillo et al. (2012)
Spongia spp., *Hippospongia* spp.	Bacteria	Focal, regional?	Temperature, shallow depths, fishing pressure	Mediterranean Sea	Gaino et al. (1992)
Commercial dictyoceratids, *Ircinia* spp., *Sarcotragus* spp., *Spongia* spp.	Bacteria	Focal, regional	?	Mediterranean Sea	Vacelet et al. (1994); Pronzato (1999)
Several species	Bacteria	Focal, regional	Abiotic factors	Caribbean	Wulf (2006)
Xestospongia muta, Geodia spp., *Ircinia* spp., *V. gigantea, Callyspongia plicifera*	Microbial consortium	Focal	?	Caribbean	Gammill and Fenner (2005); Cowart et al. (2006)
Corals (stony and soft)					
Acropora spp.	*Vibrio* spp.	Regional	Anthropogenic, storms	Caribbean	Gladfelter (1982); McClanahan and Muthiga (1998)
Acropora palmata	*Serratia marcescens*	Regional	Nutrient enhancement, human waste	Caribbean	Patterson et al. (2002)

(Continued)

Table 15.2 (Continued)

Host	Pathogen	Scale of outbreaks	Implicated stressors	Region	Key references
Gorgonia spp.	Aspergillus sydowii	Regional	Temperature, dust	Caribbean	Smith et al. (1996); Nagelkerken et al. (1997); Geiser et al. (1998); Alker et al. (2001)
Paramuricea clavata	Bacterial bleaching, Vibrio coralliilyticus	Regional	Temperature	Mediterranean Sea	Bally and Garrabou (2007)
Corals, octocorals, many species	Black band disease, microbial consortium	Regional	Anthropogenic	Caribbean	Rützler et al. (1983); Carlton and Richardson (1995); Voss and Richardson (2006)
Corals, many species	Aurantimonas coralicida	Focal, regional	Seasonal, hurricane	Caribbean	Richardson et al. (1998); Denner et al. (2003)
Gorgonians, several species	Opportunistic Vibrio spp.	Focal, regional	Temperature	Mediterranean	Cerrano et al. (2000)
Mollusks					
Crassostrea virginica	Perkinsus marinus	Regional, extensive	Temperature, salinity, anthropogenic introduction?	Mid-Atlantic USA	Andrews (1988); Burreson and Calvo Ragone (1996); Soniat (1996)
Crassostrea virginica	Haplosporidium nelsoni	Regional, extensive	Temperature, salinity, anthropogenic introduction	Mid-Atlantic USA	Andrews (1962); Carnegie and Burreson (2011)
Crassostrea virginica	Roseiovarius crassostreae – juvenile oyster disease	Focal	Temperature (>20°C)	Northeastern USA and Canada	Maloy et al. (2007); Ford and Borrero (2001)
Crassostrea virginica	Vibrio spp. (e.g., V. parahaemolyticus, V. vulnificus)	Focal	Temperature, salinity, anthropogenic input	Eastern USA	DePaola et al. (2000); Vezzuli et al. (2016)

Host	Pathogen/disease	Spread	Environmental factors	Region	References
Crassostrea gigas	Oyster herpes virus (OHV1)	Focal, regional	Temperature, anthropogenic spread	Western Europe, western USA	Renault et al. (1994); Burge et al. (2006)
Crassostrea gigas	*Vibrio splendidus* – summer mortality	Focal	Temperature, seasonality	Western Europe	Lacoste et al. (2001)
Ostrea edulis	*Bonamia ostreae*	Focal, regional	Temperature (warm autumn), anthropenic spread	Western Europe	van Banning (1991); Engelsma et al. (2010)
Saccostrea glomerata	*Bonamia roughleyi*	Focal	Low temperature, high salinity	Eastern Australia, New Zealand	Farley et al. (1988); Hine (1996)
Saccostrea glomerata	*Marteilia sydneyi*	Focal, regional	Unknown	Eastern Australia	Anderson et al. (1994)
Ostrea edulis	*Marteilia refringens*	Focal, regional?	Temperature, salinity	Western Europe	Balouet (1979;) Grizel and Heral (1991)
Mercenaria mercenaria	QPX (Quahog parasite unknown)	Focal, regional	Low temperature, seasonality, high salinity	Northeastern USA and Canada	Smolowitz et al. (1998); Lyons et al. (2007); Perrigault et al. (2010)
Pactinopecten yessoensis	*Perkinsus qugwadi*	Focal, regional?	Low temperature	Western Canada	Blackbourn et al. (1998); Bower et al. (1998)
Ruditapes philippinarum, *Ruditapes decussatus*	*Perkinsus olseni* (*P. atlanticus*)	Focal, regional	Temperature, salinity, substrate?	Western Europe, Korea	Villalba et al. (2005); Park and Choi (2001)
Ruditapes philippinarum, *R. decussatus*	*Vibrio tapetis* (brown ring disease)	Focal, regional	Cold temperature	Western Europe, Mediterranean	Paillard et al. (1994); Borrego et al. (1996)
Argopecten gibbus	*Marteilia* sp.	Focal	?	Florida, Northern Caribbean	Moyer et al. (1993)
Haliotis spp. – abalone	*Xenohaliotis haliotidis*	Focal, regional	Temperature, anthropogenic spread?	Eastern Pacific	Friedman et al. (1997, 2000); Braid et al. (2005)

(Continued)

Table 15.2 (Continued)

Host	Pathogen	Scale of outbreaks	Implicated stressors	Region	Key references
Haliotis rubra	*Perkinsus olseni*	Regional	Temperature	Australia	Goggin and Lester (1995)
Haliotis spp.	Abalone herpesvirus	Focal, regional	Anthropogenic spread	South East Asia	Chang et al. (2005); Hooper et al. (2007)
Haliotis diversicolor	Abalone-shriveling syndrome virus	Focal, regional	Anthropogenic spread	China	Zhuang et al. (2010)
Crustaceans					
Crayfish, many species	*Aphanomyces invadans*	Focal, regional, extensive	Anthropogenic spread	Western and Central Europe	Alderman and Polglase (1988); Holdich (2003)
Penaeid shrimp	White spot syndrome virus	Pandemic, extensive	Anthropogenic spread	China, Asia, the Americas	Zhan et al. (1998); Lightner (1999); Flegel (2006a)
Penaeid shrimp	Taura syndrome virus	Extensive, the Americas	Anthropogenic spread	The Americas	Lightner et al. (1995); Tu et al. (1999); Nielsen et al. (2005)
Penaeid shrimp	Infectious hypodermal and hematopoeitic necrosis virus	Extensive	Anthropogenic spread	Australasia, the Americas	Lightner (1996); Tank and Lightner (2002)
Penaeid shrimp	Yellowhead virus	Regional	Anthropogenic spread	Thailand, South East Asia	Boonyaratpalin et al. (1993)
Penaeid shrimp	Infectious myonecrosis virus	Regional	Anthropogenic spread	The Americas, Indonesia	Lightner (2003); Poulos et al. (2006); Senapin et al. (2007)
Penaeid shrimp	*Vibrio parahaemolyticus*	Extensive	Anthropogenic spread	China, Asia, Mexico	Tran et al. (2013); Joshi et al. (2014)
Macrobrachium rosenbergi	Macrobrachium nodavirus	Regional	Anthropogenic spread	South East Asia, Caribbean	Arcier et al. (1999); Bonami and Widada (2011)

Species	Pathogen/stressor	Extent	Stressor	Region	References
Chionoecetes spp.	*Hematodinium* sp.	Extensive	Temperature, physiographic features	Boreal distribution	Meyers et al. (1987); Shields et al. (2007); Mullowney et al. (2011)
Nephrops norvegicus	*Hematodinium* sp.	Regional	Temperature?, physiographic features	Western Europe	Field et al. (1992); Stentiford et al. (2001b)
Callinectes sapidus	*Hematodinium perezi*	Regional	Seasonality, temperature, salinity, physiographic features	Eastern USA	Messick and Shields (2000); Lee and Frischer (2004)
Necora (Liocarcinus) puber	*Hematodinium perezi*	Focal, regional	?	France	Wilhelm and Mialhe (1996)
Panulirus argus	Panulirus argus virus 1	Focal	Temperature?, anthropogenic spread	Florida Keys, Caribbean	Shields and Behringer (2004); Behringer et al. (2012)
Homarus americanus	*Aerococcus viridans* var. *homari*	Focal, regional	Temperature, anthropogenic spread	Northeastern North America	Stewart and Rabin (1970); Stewart et al. (1966); Stewart (1980)
Homarus americanus	Epizootic shell disease	Regional	Temperature, anthropogenic input	Eastern USA, southern New England	Castro and Angell (2000); Glenn and Pugh (2006); Shields (2013)
Paralithodes spp.	*Carcinonemertes* spp.	Regional	Anthropogenic activity	Southern Alaksa	Kuris et al. (1991); Kuris (1993)
Echinoderms					
Strongylocentrotus droebachensis	*Paramoeba invadans*	Regional	Temperature, storms	Northeast Canada	Miller and Colodey (1983); Schiebling et al. (1997)
Strongylocentrotus droebachensis	*Echinomermella matsi* – nematode	Focal	?	Norway	Christie et al. (1995); Skadsheim et al. (1995)
Diadema antillarum	?	Regional	Temperature	Caribbean	Lessios et al. (1984, 2005)
Heliaster kubiniji	?	Focal, high mort	Temperature	Gulf of California	Dungan et al. (1982)
Sea stars – 22 species	Densovirus	Regional, pandemic	Temperature	Northeastern Pacific	Hewson et al. (2014); Eisenlord et al. (2016)

?, unknown pathogen or stressor.

histological assessment, microbial culture work, or simple polymerase chain reaction (PCR) diagnostics with universal primers for bacterial and protistan parasites. These tools work best when samples are collected from live hosts, as there is less risk of interpretation of post-mortem changes or microbial contamination. For example, many invertebrates normally have bacteria in their hemolymph (see Shields et al., 2015 for examples in crabs). When the host dies, the bacteria rapidly expand in their host tissues.

Viral pathogens present additional challenges. For example, there are no continuous cell lines for crustaceans that can be used to establish viral etiology, and very few are available for mollusks. Primary cell cultures can be used to quantify viral loads (e.g., Li and Shields, 2007), but they have several limitations. However, viral infections in invertebrates often reach high levels, making it possible to purify viruses from host tissues for molecular characterization (Mari and Bonami, 1986; Deng et al., 2012). Not all viruses reach high intensity levels, and yet they still cause disease by attacking a vital organ or pathway (Flowers et al., 2015). Further, many viruses infect and cause minor pathologies to individual cells, but they are not pathogenic at the organismal level. Therefore, infection or exposure studies are needed to assess pathogenicity. For example, the simple presence of a virus in a metagenomic survey may not necessarily indicate that the agents are pathogenic or that they are even in the appropriate host. Positive assays or genomic "hits" can simply reflect viral adherence to external surfaces or benign passage through the digestive tract (Burreson, 2008). Additional studies are required to establish causality.

Advances in molecular biology have increased the number and types of diagnostic tools available for etiological investigations. Advanced molecular tools such as in-depth genomic analyses using high-throughput sequencing have been used to examine viral etiologies in two crabs, *Eriocheir sinensis* and *Scylla serrata*, in China (Chen et al., 2011; Deng et al., 2012; Guo et al., 2013), and have implicated a new densovirus in seastar wasting disease in the state of Washington, USA (Hewson et al., 2014). Such metagenomic tools allow for the rapid identification of certain viral agents, particularly RNA viruses, and when combined with transcriptomics have broad applications in disease research. When coupled with infection trials, they can serve to establish causality (e.g., Hewson et al., 2014).

In terms of disease pathogenesis in invertebrates, acute infections arise quickly and rapidly kill their host. In acute infections, pathogens with a predilection for a target organ or tissue rapidly overwhelm the target and cause it to fail, resulting in metabolic dysfunction and death. Survivors can recover, but are stunted or make easy prey. Chronic infections, by their very nature, arise more slowly, take longer to affect their host, and often cause distinctive and unusual gross pathologies. The pathogen may spread from a target tissue into secondary organs and tissues as it develops. Invertebrate hosts with chronic infections rarely survive once the infection has advanced beyond the ability of their defensive responses to effect recovery. The effects of chronic infections on host populations can be difficult to assess. Changes in long-term patterns of host abundance (e.g., larval settlement and juvenile recruitment indices) have been used to investigate disease impacts, particularly when environmental correlates are ruled out, but these are correlative patterns and must be interpreted with caution (Wahle et al., 2009). Survival studies can be used to gauge disease impacts under laboratory or mesocosm conditions (Shields and Squyars, 2000; Shields et al., 2005), and more sophisticated mark–recapture methods can be used to establish estimates of relative survival and mortality (Hoenig et al., 2017).

From the perspective of the resource manager, farmer, or culture facility, the most notable pathogens are those that directly affect the production or loss of harvestable animals. These are direct losses that can be quantified in terms of economic value. Well-managed fisheries and aquaculture programs often have long-term data on biomass, settlement, and recruitment that can be used to develop scenarios for initial assessment and mitigation of disease impacts. However, given the difficulty in establishing causality, implementation of mitigation and control techniques may seem intractable. Initial mitigation responses may appear rudimentary largely because there are no prescribed measures that fit all situations in aquatic environments and responses can be difficult to coordinate depending on the scale of an outbreak (e.g., withering disease in abalone, wasting disease in seastar) or the nature of the host (e.g., deep-water snow crabs); hence, methods are often developed piecemeal in response to specific host–pathogen systems. Nonetheless, there are actions that can be taken to mitigate impacts, and some of these have been quite successful in controlling outbreaks and facilitating expanded food production or resource conservation.

Due to the intensive nature of their aquaculture production, mollusks (specifically oysters), other bivalves, and crustaceans (primarily shrimp and crayfish) offer the best examples in terms of suitable approaches to the management and prevention of disease. Indeed, staggering economic losses worth several billion US dollars have been imposed on bivalve and shrimp industries by the presence of pathogens. Differences in production between oysters and shrimp dictate very different approaches to the management of diseases in each host. For example, oyster seed or spat is generally produced in hatcheries, sold as "spat on shell," and then seeded directly on to natural oyster beds or leased oyster grounds in estuarine or coastal waters for grow-out and harvest. Spat can also be collected on shells as cultch, which is then relayed on to viable oyster beds for grow-out. In these grow-out areas, oysters obtain pathogens directly from natural sources or reservoirs of infection. Similarly, shrimp postlarvae are now generally grown in hatcheries and sold as "postlarvae," which are then seeded into large grow-out ponds or impoundments for grow-out. They obtain pathogens from contaminated broodstock, infected live feeds, contaminated water sources, and several mechanical and biological vectors. In the remainder of the chapter, I give examples of the major disease issues facing these industries, before covering some of the mitigation and management methods used to control outbreaks of diseases in these disparate systems.

15.2   Oyster Diseases

At least 42 species of bivalves are in production worldwide (Food and Agriculture Organization, 2012a); this is therefore not an exhaustive treatment of the diseases in bivalves. Rather, I present an overview of the best-known examples in oyster production. The literature on diseases in mollusks, particularly oysters, is extensive. For a major review of the parasites and pathogens of bivalves, see Lauckner (1983). A fascinating review of the decline of oyster production due to disease is presented by Andrews (1988). Aspects of management of pathogens in oysters is reviewed in Andrews and Ray (1988), Ford and Haskin (1988), Krantz and Jordan (1996), and, more recently, Elston and Ford (2011).

Most disease problems in bivalve populations arise from the introduction of pathogens through movement of live mollusks, mainly via farming activities (Berthe, 2008; Elston and Ford, 2011; Pernet et al., 2012). Ironically, such transplants are undertaken to improve seed stock or enhance harvests. Disease problems may also arise via introduction in ballast water or other modes of dispersal, but there is less evidence for this (Berthe, 2008; Elston and Ford, 2011). Unfortunately, there are many examples of pathogens crossing host barriers in bivalve production and farming. Four pathogens will serve to highlight this issue. In France in the early 1980s, outbreaks of two protozoan parasites, *Marteilia refringens* and *Bonamia ostreae*, in European flat oyster (*Ostrea edulis*) inflicted losses estimated at US$31 million and a 20% decline in employment in the aquaculture sector, primarily through a 90% reduction in oyster production (Meuriot and Grizel, 1984; Grizel and Héral, 1991). Although *Marteilia refringens* was likely endemic in France and emerged with increasing aquaculture production (Balouet, 1979), *Bonamia ostreae* was introduced there through importation and translocation of infected *Crassostrea gigas* from California, USA (Cigarria and Elston, 1997; Grizel, 1997). Both parasites spread rapidly throughout Western Europe through introduction of infected stock. The oyster industry in Western Europe is now based largely on *C. gigas* because of the losses inflicted on *O. edulis* by these pathogens.

On the East Coast of the United States, two protozoan parasites, *Haplosporidium nelsoni* and *Perkinsus marinus*, decimated the industry for the American oyster (*Crassostrea virginica*) (Fig. 15.2). MSX, or "multinucleated sphere unknown," the common name for *H. nelsoni*, emerged in the late 1950s in wild oysters in Delaware Bay, causing mass mortality (95%) in commercial stocks. It then spread into Chesapeake Bay

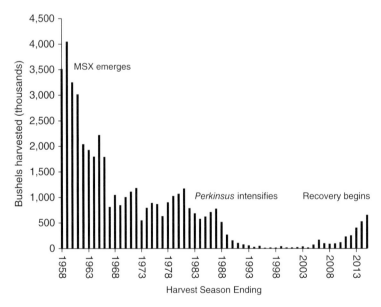

Fig. 15.2 Impact of *Haplosporidium nelsoni* (MSX) and *Perkinsus marinus* (dermo) on landings of the Virginia oyster (*Crassostrea virginica*) over time. Production is recovering due to improved management techniques and the use of triploid oysters with improved growth rates. *Source*: Data from Virginia Marine Resources Commission 2016, http://mrc.virginia.gov/SMAC/VA-Oyster-Harvests.pdf.

in the early 1960s, where it heavily damaged the industry and even now continues to impose significant losses on oyster production (Ford and Haskin, 1987; Andrews, 1988; Burreson and Ford, 2004; Carnegie and Burreson, 2011). The pathogen has a salinity range of 5–30 psu (although it rarely occurs below 10 psu), and is therefore capable of infecting *C. virginica* throughout its salinity range (Ford, 1985). In the late 1980s and early 1990s, "dermo," the common name for *Perkinsus* (= *Dermocystidium*) *marinus*, emerged as a significant pathogen in the oyster industry in Chesapeake Bay, spreading into Delaware Bay shortly thereafter (Burreson and Calvo, 1996). Unlike *H. nelsoni*, *P. marinus* can survive and thrive at low salinities (Chu et al., 1993). These two pathogens brought a lucrative major industry to near collapse (Andrews, 1988; Burreson and Calvo, 1996). As with *B. ostreae*, *H. nelsoni* was introduced to the East Coast through the translocation of infected *Crassostrea gigas*. However, the haplosporidian is not pathogenic in *C. gigas*, but it is highly pathogenic in its abnormal host *C. virginica*, in which it causes disease and major mortality (Burreson et al., 2000). In contrast, *P. marinus* is endemic in *C. virginica* from Chesapeake Bay (Andrews and Hewatt, 1957); its emergence was apparently fueled by environmental changes, including temporal periods of drought and increased summer temperatures (Burreson and Calvo, 1996; Ford, 1996). Its intensification in Chesapeake Bay led to widespread dispersal into previously nonendemic areas, including Delaware Bay, and increased mortalities in naïve host populations not previously exposed to the pathogen (Andrews, 1988; Burreson and Calvo, 1995). The combination of oyster translocations and changes in environmental conditions led to the rapid expansion of the range of *P. marinus* along the eastern seaboard, with highly variable and somewhat unpredictable impacts on oyster production (Burreson, 1991; Burreson and Ragone Calvo, 1996; Ford, 1996).

Managing persistent outbreaks of pathogens in oyster populations is complicated because the approach will differ depending on whether a wild fishery or cultured stock is being managed and on which pathogen is causing mortalities. Prior to the establishment of selective breeding programs, most strategies revolved around modifications to farming procedures in wild stocks (Andrews and Ray, 1988; Krantz and Jordan, 1996). Population models were developed to predict and establish best procedures for growth and farming activities, much of which was centered on reducing disease pressure through appropriate seasonal timing of the planting of seed oysters and the relaying of transplants into high-salinity endemic zones for faster growth or grow-out (Powell et al., 1998). The tradeoff involved lowering the legal market size in order to allow producers to raise and sell smaller oysters, because a shorter growing season with smaller individuals reduced mortality to tolerable levels (Andrews and Ray, 1988). For example, in some systems, oysters were planted in the late fall so that they could establish and grow in a relatively disease-free season, before being relayed to high-salinity areas where they could grow faster but would obtain disease throughout the summer season, and finally being harvested in the following fall. However, some of these mitigation efforts were in conflict with legislation protecting natural oyster stocks for conservation (see Villalba et al., 2004).

Oysters have been farmed for food production for several hundred years, but their domestication through the culture of broodstock and selective breeding for beneficial traits has only recently been established. With respect to disease resistance in *C. virginica*, selective breeding initially focused on the selection of oyster lines, or families, that were resistant to *H. nelsoni* (Haskin and Ford, 1979; Ford and Haskin, 1987). Resistance was not complete, but it was selective and inheritable, and it resulted in the production of several lines that exhibited partial resistance to the pathogen. Until the mid-1980s, populations of *H. nelsoni* and *P. marinus* did not overlap in distribution in their oyster hosts. Unfortunately, at about the time the *H. nelsoni*-resistant strains were being developed, *P. marinus* was introduced into Delaware Bay, and the *H. nelsoni*-tolerant strains were highly susceptible to the new pathogen (Burreson, 1991), which effectively stopped attempts at selective breeding for resistance to *H. nelsoni* for several years (Burreson and Ford, 2004). More recently, oysters in endemic zones where both parasites overlap have exhibited disease resistance to *H. nelsoni* (Carnegie and Burreson, 2011). In 1997, in Virginia, USA, the Aquaculture Genetics and Breeding Technology Center (ABC) was created in order to promote the domestication of *C. virginica* (Frank-Lawale et al., 2014). A priority in that initiative was to establish and selectively breed disease-resistant oysters, and in particular to select for resistance to *P. marinus.* The ABC has developed a triploid line that has good growth characteristics that allow oysters to reach a marketable size before dying to disease. Although the ABC is still establishing the characteristics of the different oyster families, several lines look promising in terms of survival and growth under intense disease pressure (Frank-Lawale et al., 2014).

Hatchery and culture systems are important to bivalve production because natural reproduction and production of bivalves have declined in many areas as a result of human impacts: urbanization, industrialization, and residential development (Elston and Ford, 2011). Hatcheries for marine bivalves typically rely on seawater drawn from estuaries within areas affected by anthropogenic activities, and it can be difficult to remove infectious agents that occur naturally within such waters. Moreover, hatchery and production systems change the environmental conditions that are normally encountered by the oyster in the field. These systems essentially produce larvae and juveniles at very high densities, creating opportunities for pathogens such as *Vibrio tubiashii*, other opportunistic bacteria (e.g., Paillard et al., 2004; Elston et al., 2008), oyster velar virus disease (Elston, 1979), and oyster herpes virus (e.g., Friedman et al., 2005; Burge et al., 2006, 2007) that might otherwise occur at very low levels in natural settings. Many of these agents are opportunists, essentially emerging as pathogens in crowded production systems. For example, *Vibrio tubiashii* occurs cryptically in cultures of algae used to feed oyster larvae and transfers to oysters in culture systems, where it causes mortalities (Elston et al., 2008). Therefore, management of affected hatcheries requires rigorous attention to destroying affected stock and disinfection of contaminated food sources, cultures, and equipment. Hatcheries and nurseries must focus on ensuring the production of healthy, uninfected animals for use as seed or transplant stock (Elston and Ford, 2011).

With the emergence of epizootics in many oysters, several programs have been initiated to produce specific pathogen-free (SPF) spat, with certification programs for export across regulatory (state, region, national) boundaries and enhanced communication networks for the provision of information on new and emerging issues. One of the first attempts at developing an oyster health-management program was initially designed for seed growers in the Pacific Northwest (Elston and Cheney, 2004), based on a loosely integrated approach involving surveillance, documentation, scheduled examination of stock diseases, adherence to operations protocols, characterization of the disease status of water sources and culture areas, maintenance of broodstock integrity, operations protocols regarding health management, and a response plan for outbreaks.

At the international scale, the World Organisation for Animal Health (formerly Office Internationale Epizooties, and still known as the OIE) now recognizes several pathogens of bivalves and offers specific advice on aspects of surveillance, identification, risk analysis, import/export procedures, contingency planning, sanitation, and other aspects of disease control (see Section 15.5).

A critical aspect of health management in bivalve production is disease surveillance. Surveillance can provide important baseline data on many pathogens, as well as critical information on specific agents. It relies on a funding base to be successful, and must be undertaken regularly to provide information on pathogen levels in oyster seed or stocks being moved over legal boundaries (Elston and Ford, 2011). Surveillance is usually focused on commercially cultured species and is not typically applied to closely related host species or vectors; this can present a problem, as alternate hosts are often reservoirs for pathogens. Long-term surveillance and data on health assessments are critical in evaluating the distribution of pathogens, potential changes in their prevalence and virulence in relation to environmental changes and management actions, and modeling efforts to better understand disease dynamics (Powell et al., 1998; Villalba et al., 2004). Other aspects of surveillance and reporting are covered in the Section 15.5.

15.3   Crustacean Diseases

Crustaceans provide the basis for an expanding market in the seafood trade, supplying commodities ranging from whole, live animals to highly processed, frozen, and canned products. Collectively, shrimp, crabs, lobsters, and crayfish support billion-dollar industries through intensive industrial-scale aquaculture and fisheries sectors. They also provide significant value at smaller, artisanal scales in terms of local economies and subsistence fishing, which can be difficult to quantify. In terms of scale, the production of fished and farmed crustaceans currently exceeds 10 million metric tons, with an ex-vessel value approaching $40 billion (Food and Agriculture Organization, 2009, 2014). Production of shrimp is split about equally between fisheries and aquaculture; that for crabs and lobsters is mostly in the fisheries sector. Fisheries are based on capture of existing wild stocks, whereas aquaculture is based on pond-, farm-, or cage-rearing of domesticated lines (livestock), typically from egg to adult. Marine shrimp ($12 billion) account for the largest component of fisheries production, followed by crabs ($3.7 billion), lobsters ($2.4 billion), and freshwater shrimp and crayfish ($1 billion) (Food and Agriculture Organization, 2012a). Marine shrimp also dominate the aquaculture sector for crustacean commodities, with production estimated at $12 billion in value at first sale, followed by cultured freshwater shrimp and crayfish ($4.7 billion) and freshwater crabs ($1 billion), with few others cultured at large commercial scales. The production of cultured shrimp has expanded rapidly, shifting from *Penaeus* (= *Litopenaeus*) *monodon* to *Penaeus* (= *Litopenaeus*) *vannamei*, which now makes up most of the commercial aquaculture market.

The intensive and rapid expansion of shrimp aquaculture through Asia and South America has led to a plethora of local, regional, and larger environmental, management, and disease issues. The scale of some of these issues has had profound implications for the shrimp industry and its markets. For example, environmental issues have included large-scale destruction of mangroves and wetlands to build multi-hectare commercial farms and problems with water management and water quality from waste disposal associated with farms (reviewed in Walker and Mohan, 2009). Management issues have included socioeconomic problems associated with low wages, export of high-paying jobs and products from local job sites, harvest of suboptimal animals in response to disease, and poor siting of farms. However, viral diseases emerged as a major impediment to the shrimp aquaculture industry in the 1990s and early 2000s. The use of wild broodstock, crowded culture conditions for postlarvae, and uncontrolled transportation of live shrimp across regional and international boundaries resulted in the unprecedented and widespread dispersal of several pathogens through Asia into North and South America. Pandemics of white spot syndrome virus (WSSV) caused an estimated US$6 billion in damage to Chinese production alone, with outbreaks reported in virtually every country producing shrimp. The cost of WSSV has been estimated at more than US$15 billion over 20 years (Flegel, 2006a, 2012; Lightner et al., 2012a). Moreover, shrimp imported from Asia into Mexico as broodstock inadvertently escaped and released infectious hypodermal and hematopoietic necrosis virus (IHHNV) into the Gulf of California, resulting in significant damage to a commercial shrimp fishery there (Morales-Covarrubias et al., 1999; Pantoja et al., 1999).

Pathogens and parasites are not new to crustacean hosts; what is new is the intensive nature of aquaculture and the demand for and transportation of postlarvae to initiate culture production. Rearing of animals under stressful conditions such as high densities and poor water quality, coupled with their propensity for cannibalism and the use of wild stocks, have presented opportunities for many pathogens to emerge under stressful culture conditions. Disease issues have been further exacerbated by poor monitoring of pathogens in large farm systems and weak biosecurity measures that allow dispersal of infected hosts into new culture systems and their escape into natural systems (Lightner 1996, 2011; Flegel 2012).

Crustaceans have a disparate flora and fauna of pathogens ranging from viruses to bacteria, protists, helminths, and other crustaceans. Viruses have caused the most damage to crustacean aquaculture, but bacteria and protists have also been implicated in epidemic outbreaks. Although there are fewer data available on viral infections in fisheries, viruses, bacteria, and protistan parasites have also caused losses to crustacean fisheries. The diversity of viral pathogens may appear high, but with few exceptions we really know very little about them. Over 50 viruses are known to infect shrimp, crabs, and crayfish, with 20 reported from shrimp (Lightner et al., 2012a; Flegel, 2012), 30 from crabs (Shields et al., 2015), and a few from crayfish (Edgerton et al., 2004; Longshaw, 2011). Surprisingly, only one virus has been found in natural infections in lobsters (Shields and Behringer, 2004). In this section, I provide insights into the mitigation and management of pathogens in crustacean aquaculture and fisheries. Disease issues in aquaculture are distinctly different than those in fisheries; hence, they will be discussed as separate examples.

15.3.1 Outbreaks in Shrimp Aquaculture

Eight penaeid species have been successfully bred in aquaculture (Bailey-Brock and Moss, 1992). For different reasons, two of these have dominated production: *Penaeus monodon* and *P. vannamei*. Culture of penaeid shrimp began in earnest in the late 1970s and early 1980s, peaking in the early 1990s, before several viral and bacterial pathogens emerged. Indeed, losses due to disease were estimated to be as high as 40% per year in the late 1980s and early 1990s, even during peak production (Lundin, 1996). The impact of viral infections, particularly WSSV, on shrimp aquaculture led to the development of extraordinary scientific advances in our understanding of the molecular genetics of shrimp and their viruses, as well as the development of sensitive diagnostic tools for the rapid identification of other shrimp pathogens. The shrimp aquaculture industry has largely recovered from the large epizootics of the 1990s and 2000s and is now in a new phase of growth, but several viruses remain serious pathogens and threaten shrimp production (Food and Agriculture Organization, 2012a,b; Lightner et al., 2012a). Because of the high food value of shrimp and their domestication for culture, their pathogens continue to receive much attention. For in-depth reviews of the developing problems with pathogens in the field, see Flegel (2006, 2009, 2012), Lightner (1999, 2011), Lightner et al. (2012), Lightner and Redman (2010), Walker and Mohan (2009), and Thitamadee et al. (2016). In this section, I briefly review the primary pathogens, before discussing aspects of their control and management in aquaculture.

Several viruses are serious pathogens of *P. vannamei* and *P. monodon* in culture in Asia, and a few are significant pathogens in the Americas. WSSV, the cause of white spot disease, is the most serious viral agent in shrimp (Lightner et al., 2012a; Flegel, 2012). In 1992, white spot disease emerged in China and Taiwan and caused the *P. monodon* industry to stagnate (Fig. 15.3) (Zhan et al., 1998). The syndrome, later shown to be caused by WSSV, soon became a pandemic in the shrimp industry. It infects all commercial species and spreads rapidly among hosts and between culture ponds,

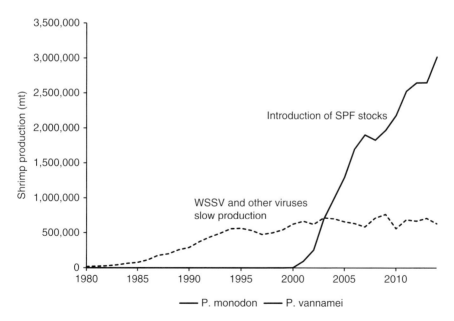

Fig. 15.3 Shrimp production in Asia over time. The expansion in the 1990s was largely halted by the emergence of white spot syndrome virus (WSSV) and several other viruses. The industry adopted the use of specific pathogen-free (SPF) stocks of *Penaeus vannamei* in the early 2000s and production expanded rapidly, effectively tripling in 10 years. *Source*: Data from Food and Agriculture Organization (2016).

causing heavy mortality (>80%) in most systems (Flegel, 2006a, 2012; Walker and Mohan, 2009). Surviving shrimp remain carriers for life. The virus is highly infectious and pathogenic, and because it can be transmitted easily via contaminated postlarvae and broodstock, it spread to most shrimp-producing countries in Asia within 2–3 years, and to the Americas within a decade (Lightner, 1999). Fortunately, it has not been reported from Australia (Walker and Mohan, 2009; Lightner, 2012).

WSSV is a host generalist and can infect other crustaceans, in some of which it can replicate, remain active, but not cause pathogenicity, allowing them to serve as reservoirs for dispersal to new systems. WSSV has also been reported from a variety of invertebrates, rotifers, polychaetes, insect larvae, algae, and bivalves, but it does not appear to be replicated in these hosts. Poor biosecurity practices – namely, the transportation of contaminated shrimp and shrimp products – can cause widespread dispersal. For example, WSSV reached the Americas from Asia through frozen shrimp products (Hasson et al., 2006) and was present in aquaculture there by 1999 (Lightner, 1996). In the Americas, there are reports of WSSV escaping from aquaculture production and causing pathology to native penaeids in commercial fisheries (Chapman et al., 2004; Dorf et al., 2005; Cavalli et al. 2010, 2011), but its pathogenicity and the scale of damage to local fisheries have not been well established.

WSSV is temperature-sensitive (Vidal et al., 2001; Granja et al., 2006): holding infected postlarvae at 32 °C for 7 days effectively clears them of the virus (Wongmaneeprateep et al., 2010). The nature of this sensitivity is based on aldehyde dehydrogenase and heat shock protein 70 (HSP-70), although the exact mechanism remains unclear (Lin et al., 2011). In any case, temperature treatment is not an effective biosecurity measure, because shrimp can reacquire the pathogen from contaminated rearing ponds or water sources (Corsin et al., 2002, 2003).

Specific management and control measures have been implemented against WSSV, and they are largely successful as long as biosecurity measures remain enforced (Walker and Mohan, 2009). The primary control measures remain rigorous surveillance using molecular diagnostics, preventive methods such as quarantine, and development of SPF stocks (Flegel et al., 2008; Cuellar Anjel et al., 2010; Flegel, 2010; Moss et al., 2012). This involves constant surveillance to screen seed stock for WSSV, disinfection of contaminated ponds, and use of SPF stocks in reseeding of decontaminated ponds. Disease resistance to WSSV appears to be a heritable trait, and some early success has been reported in establishing disease-resistant lines of *P. vannamei* (Cuellar Anjel et al., 2011).

Yellowhead virus (YHV) is also a significant pathogen of commercial penaeid shrimp. It was initially described as a pathogen from *P. monodon* in Thailand (Boonyaratpalin et al., 1993), but it also infects *P. vannamei*, and it has caused heavy and rapid mortality in both species of shrimp (Walker et al., 2001; Dhar et al., 2004; Sittidilokratna et al., 2008). It occurs throughout most of South East Asia. There are several strains of YHV, including a closely related species known as gill-associated virus (GAV), but only one, YHV-genotype 1, is highly virulent and occurs in outbreaks (Wijegoonawardane et al., 2008). YHV-genotype 1 also has two strains, 1a and 1b, both of which are highly pathogenic to shrimp (Senapin et al., 2010). Another pathogenic strain, YHV-type 7, has recently been described from Australia (Mohr et al., 2015). YHV genotype-1 infects penaeid and palaemonid shrimp, while its closely related genotypes infect only *P. monodon* (for review, see Walker and Mohan, 2009). Shrimp that survive infection may be carriers for life. In South East Asia, YHV outbreaks continue to occur even with rigorous biosecurity measures in place. A recent study has implicated mechanical vectoring or airborne transmission, possibly from birds moving between ponds (see Thitamadee et al., 2016).

Infectious myonecrosis virus (IMNV) is a significant pathogen of *P. vannamei* in the Americas and in Indonesia. It was first identified in infected *P. vannamei* from Brazil, where it caused widespread mortality (Lightner et al., 2003; Poulos et al., 2006). In 2006, it was introduced to Indonesia from illegally transported broodstock (Senapin et al., 2007). It has so far not spread from Indonesia to other shrimp-producing countries in Asia. Production losses to IMNV have been estimated at from $100 million to over $1 billion (Lightner et al., 2012a). Over time, there has been a decrease in the severity of IMNV in Indonesia, with some evidence of viral tolerance (Senapin et al., 2011). In initial outbreaks, mortality approached 80%, whereas later outbreaks had 30% mortality (Poulos et al., 2006). Most shrimp aquaculture facilities in Asian countries now quarantine and screen imported *P. vannamei* for IMNV and other viruses as a precaution against the release of viruses into new regions (Flegel, 2012). However, smuggling of postlarvae and broodstock continues to be a problem, so vigilance in screening imports is necessary (Thitamadee et al., 2016). A dsRNA "vaccine" targeting the capsid protein and RNA-dependent RNA polymerase genes is in development against IMNV (Bartholomay et al., 2012; Loy et al., 2012, 2013), but it has not progressed beyond the development stage (Thitamadee et al., 2016).

As the name implies, monodon slow growth syndrome (MSGS) causes severe growth retardation in *P. monodon*. The syndrome was conservatively estimated to cause a loss of over US$310 million in 2002 (Chayaburakul et al., 2004). The emergence of MSGS was a primary reason for the shift from *P. monodon* to SPF stocks of *P. vannamei* in Thailand (Flegel, 2012), but it is not clear if it affects *P. monodon* culture in other Asian countries (Rai et al., 2009). The etiology of the syndrome has not been fully resolved. Affected shrimp exhibit a wide range of conditions, making case definition for individuals difficult. The case definition is based on growth characteristics within a pond and not necessarily in individual shrimp (Flegel, 2012). Work on MSGS led to the characterization of Laem–Singh virus (LSNV) (Sritunyalucksana et al., 2006) and of a cofactor, an integrase-containing mobile element, that was found in *P. monodon* with the syndrome (Panphut et al., 2011). However, the etiology of LSNV and the integrase in MSGS remains unclear (Pratoomthai et al., 2008; Flegel, 2012). Use of dsRNA constructs against LSNV in shrimp feed improved growth in shrimp with LSNV and reduced viral loads in affected shrimp, implicating the virus in the syndrome (Saksmerprome et al., 2013). More work is needed to fully establish the etiology of MSGS and the role of LSNV and the integrase in promoting this disease.

IHHNV is endemic to *P. monodon* in the Australasian region. In this host, it causes little pathology (Flegel, 1997). It was first discovered in cultured *P. monodon* in Hawaii and was later shown to have originated from a culture of *P. monodon* imported into the Americas from the Philippines in the 1970s (Lightner, 1996; Tang and Lightner, 2002). Because it is benign in *P. monodon* and was not initially included in surveillance programs, over time the virus spread throughout the Americas with the importation of contaminated postlarvae and broodstock. In *P. stylirostris*, IHHNV causes severe pathology and mortality (Lightner et al., 1983; Lightner and Redman, 1991). The virus escaped into wild populations of *P. stylirostris* along the Pacific coast, where it caused significant damage to fisheries in Mexico (Morales-Covarrubias et al., 1999; Pantoja et al., 1999). It eventually became established in *P. vannamei* throughout most of the Americas (Lightner and Redman, 2010). Infected shrimp remain carriers for life, and there is evidence for vertical transmission. In *P. vannamei*, IHHNV causes a form of stunting known as runt deformity syndrome (RDS) (Bell and Lightner, 1984; Lightner, 1996). RDS was one of the primary reasons for the development of SPF stocks of *P. vannamei*, but these were also critical in controlling outbreaks of WSSV, TSV, and other pathogens, as they provided disease-free stock. The SPF stock is also resistant to TSV, IHHNV, and other pathogens (Lightner, 2005; Thitamadee et al., 2016).

Taura syndrome virus (TSV) emerged in penaeid shrimp from Ecuador in the early 1990s, where it was associated with severe mortality of *P. vannamei* (Lightner et al., 1995). It caused significant damage to shrimp production in the Americas before it was introduced into Asia in the mid-1990s (e.g., Tu et al., 1999; Nielsen et al., 2005). Shrimp that survive infection remain carriers, and mechanical transmission can occur via several routes. The SPF lines of *P. vannamei* are resistant to TSV (Lightner et al., 2009) and their use in Asia has resulted in the virus no longer being considered a disease threat there (Lightner et al., 2012a; Sookruksawong et al., 2013).

The most recent pathogen to emerge is acute hepatopancreatic necrosis disease (AHPND), which is caused by toxins released by strains of *Vibrio parahaemolyticus* (Tran et al., 2013). The bacterium responsible carries a unique plasmid, pVA1, which produces Pir-like toxins (Yang et al., 2014; Lai et al., 2015; Lee et al., 2015). The bacterium colonizes the foregut of the shrimp, releasing toxins that cause massive sloughing of the hepatopancreas, particularly the tubule epithelial cells. The disease is associated with a condition known as early mortality syndrome (EMS) but can be separated from other possible causes of EMS by stringent observations of pond conditions and histopathological aberrations (Lightner et al., 2012b). AHPND was first reported in shrimp from China in 2009, then rapidly spread through much of Asia (Tran et al., 2013; Joshi et al., 2014) and even into Mexico (Nunan et al., 2014; Soto-Rodriguez et al., 2015). Shrimp production in Thailand dropped by 66% due to mortality imposed by AHPND, as well as economic concerns over the continuing impact of the pathogen in shrimp culture (Thitamadee et al., 2016). Several highly sensitive diagnostic tools are now available for its detection (reviewed in Thitamadee et al., 2016). Because water treatment (ozonation, filtration, etc.) in most shrimp hatcheries removes bacterial and viral pathogens, the most likely route of entry of the bacterium into culture systems appears to be through live feeds, but a specific food item or source has not been clearly implicated (Thitamadee et al., 2016). Thus, control should emphasize screening postlarvae and feed sources to ensure the absence of AHPND. One possibility is to freeze or pasteurize feed sources (Thitamadee et al., 2016). Stocking tilapia into shared water sources with shrimp ponds may also improve control of the syndrome (Cadiz et al., 2016).

15.3.2 Disease Management in Shrimp Aquaculture

Over the years, the shrimp aquaculture industry has been plagued by several emerging pathogens despite expanded control efforts and improvements in management and production systems. Unfortunately, disease emergence and spread continues to be a problem due to lapses in biosecurity and poor industry practices (Stentiford et al., 2012). The need to place farms adjacent to natural waterways, with their resident fauna, places shrimp near potential pathogens. This was the case with WSSV, YHV, and TSV, all of which may have originated in nonpenaeid resident crustaceans or other reservoir hosts (Walker and Mohan, 2009). Shrimp diseases can largely be prevented by rigorous management of their culture conditions, including vigilance in biosecurity, water quality, stocking density, aeration, feed schedule, and quality of the broodstock and postlarvae (Flegel, 2012; Thitamadee et al., 2016). Flegel et al. (2008) listed several priorities for research to improve disease prevention in shrimp culture: (i) development of domesticated, genetically improved SPF stocks for all cultivated species; (ii) standardization and increased use of diagnostic tests in screening for pathogens; (iii) improvements in and application of rigorous biosecurity measures; (iv) better control over transboundary movements of crustaceans used in culture; (v) investigations into the efficacy of probiotics, immunostimulants, and "vaccines" in field trials; (vi) a more complete understanding of the host–pathogen interaction in shrimp and their primary pathogens; (vii) enhanced understanding of the epizootiology of shrimp pathogens; and (viii) improvements in molecular and biochemical engineering to control microbial dynamics in shrimp ponds and hatchery tanks.

15.3.2.1 Switching Species and Specific Pathogen-Free (SPF) Stocks

Because of continuing disease issues surrounding the culture of *P. monodon*, around the year 2000 most shrimp production in Asia switched to *P. vannamei*, due to its resistance to TSV and MSGS. Improvements in the domestication of *P. vannamei* led to the development of SPF broodstock that also reduced the impact of WSSV (Fig. 15.3.). This change in species, along with improved aquaculture management practices, was an effective intervention that more than tripled production in Thailand over a 5-year period (Flegel, 2012). Unlike *P. monodon*, SPF stocks of *P. vannamei* tolerate TSV and IHHNV under most conditions. However, the SPF shrimp are still susceptible to WSSV, YHV-type 1, and IMNV.

With the advent of SPF stocks of *P. vannamei*, the industry switched from producing postlarvae from wild broodstock to producing postlarvae from domesticated lines that had been bred for resistance to specific pathogens (Lightner et al., 2009; Moss and Moss, 2009). The use of SPF broodstock reduced disease outbreaks in culture ponds, took pressure off harvesting broodstock, removed the possibility of contaminating systems with pathogens from wild stock, and essentially saved the aquaculture industry from extensive losses, allowing it to continue to grow into a sustainable industry. This single development – SPF stocks – has been credited as the main contributor to the industry's recovery and subsequent expansion (Food and Agriculture Organization, 2006; Cock et al., 2009; Lightner et al., 2012a). The SPF stocks focused attention on the development of genetic improvements in penaeid shrimp, because they grew faster than wild *P. vannamei* and were genetically selected to be tolerant of TSV and IHHNV. Given the benefits of the SPF stocks, it is no surprise that a large segment of the shrimp culture industry has adopted their use and production.

The introduction of SPF stocks of *P. vannamei* was a major innovation, but it also caused complacency in biosecurity issues among shrimp farmers. Prior to the use of SPF shrimp, captured wild broodstock had to be quarantined and monitored for pathogens before it could be used in a hatchery, and postlarvae were usually given a health exam prior to use. With the introduction of SPF stocks, these practices mostly stopped (Thitamadee et al., 2016). At the same time, live feed was used widely in many hatcheries, particularly live polychaetes and clams because they resulted in higher yields in postlarvae. However, several pathogens can be vectored through filter-feeding invertebrates (as "sterile" mechanical vectors), and this use of live feed put the whole industry at high risk for the emergence of new pathogens such as AHPND (Thitamadee et al., 2016). Thus, the feed source must be considered in biosecurity plans given recent outbreaks of AHPND in Asia and its importation to Central America through the careless movement of live shrimp and live feeds (Flegel, 2006b).

Another example of complacency has been reported. Farmers have used SPF postlarvae to produce broodstock, but the broodstock is then bred with other animals to save money by circumventing the purchase of SPF stock. This has resulted in the loss of resistance genes in the progeny. Moreover, the progeny can carry other infections (*Enterocytozoon hepatopenaei* and AHPND) into other stocks (Thitamadee et al., 2016). The postlarvae from these animals are not reared in an SPF system, have been outbred with other animals, and therefore are not SPF animals. This increases the biosecurity risk by reducing resistance and possibly introducing other pathogens when these non-SPF postlarvae are used to stock new farms.

15.3.2.2 Surveillance

Importing exotic crustaceans for aquaculture has significant risks. The animals should be kept in quarantine situations and tested for known and unknown pathogens (Flegel, 2012). This should be done under virtually all circumstances, even when SPF stocks are used, because they can acquire pathogens during culture operations. Disease surveillance is a crucial component of aquaculture programs and fisheries surveys. In shrimp aquaculture, surveillance has several components. First, it is the stringent testing of postlarvae and broodstock, as well as their live feed, with a battery of histological and molecular tests to establish the presence or absence of pathogens. Second, it is the appropriate use of quarantine of transported stock to protect existing stock from pathogens. Third, it is the regular evaluation of cultured animals to ensure that they are disease-free prior to harvest. Fourth, it is the rigorous testing of animals that will be shipped across borders or regional boundaries. Surveillance is not cheap. It requires expensive resources, staff training in disease detection, and quarantine facilities. However, given the losses imposed upon the shrimp industry over the last 3 decades, it is a prudent component of any aquaculture program.

Molecular tools are in place for all of the major pathogens of penaeid shrimp used in aquaculture. Molecular diagnostics provide exquisite sensitivity and specificity to the pathogens of interest. However, not all PCR methods are alike, and in most cases they have not been standardized. A good example of this is the developing discussion around the diagnostic tools for AHPND (see Thitamadee et al., 2016). Moreover, because of the specificity of molecular diagnostics, not all pathogens are initially detectable with PCR-based tools. Emerging disease issues can be difficult to diagnose, identify, and assess. In some diseases, a case definition is required to establish clear criteria associated with a particular set of pathologies related to the emerging pathogen (Flegel, 2012; Lightner et al., 2012). For example, MSGS and IHHNV both require explicit case definitions incorporating aberrant histological conditions, mortalities, and specific conditions within ponds to establish disease association.

Unlike many vertebrate–pathogen systems, a feature of shrimp–virus systems is that the surviving hosts serve as carriers of persistent, low-level infections (Flegel 2006ab;

Walker and Mohan, 2009; Lightner and Redman, 2010). This means that shrimp should be certified free of serious pathogens when they are transported into new regions where the pathogens do not occur. With some agents, there are specific guidelines for what can and cannot be transported based on whether they are listed as a notifiable pathogen by the OIE. It also means that SPF stocks should be used whenever possible to curtail the acquisition of new pathogens from wild stocks.

15.3.2.3 Development of "Vaccines"

As with all invertebrates, shrimp lack an adaptive response with specific immune memory. However, shrimp do have defense pathways that respond to specific classes or groups of agents, such as Gram-negative bacteria, some viruses, and other pathogens (Hauton, 2012; Li and Xiang, 2013). Because they lack the typical vertebrate adaptive response, classical vaccines that stimulate immunity do not elicit long-term protection in shrimp. Nonetheless, specific foreign proteins prepared as vaccines have been shown to enhance protection and survival. A review of the burgeoning field of shrimp immunology is beyond the scope of this chapter; hence, only a few cases are mentioned here. Several capsid proteins of WSSV have been developed as candidate vaccines. One, VP28, can provide short-term protection over 4–7 weeks, with reduced mortality during this period (e.g., Witteveldt et al., 2004, 2006). Another approach has been the use of dsRNA constructs as interference RNA (RNAi). Injection of short RNAi homologues that code for viral mRNA can effectively block WSSV, TSV, and YHV (Robalino et al., 2005; Yodmuang et al., 2006; Xu et al., 2007). A dsRNA construct against LSNV showed efficacy in treating MSGS (Saksmerprome et al., 2013). Another RNAi-based construct was very effective in treating IMV in *P. vannamei* (Fiejo et al., 2015). Although none of these vaccines has progressed into commercial development, they may provide an avenue for specific treatments or prophylaxes (Thitamadee et al., 2016).

15.3.2.4 Ecological and Biological Control

Shrimp culture ponds are highly modified ecological habitats – the aquatic analog of the feed lot. They contain other invertebrates, have bird predators, and are subject to water quality and environmental issues. In some cases, controlling other organisms within ponds can alter or ameliorate the effects of disease. For example, removing two fish species from culture ponds reduced the prevalence of *Agmasoma penaei*, a microsporidian parasite that was causing mortality (Pasharawipas and Flegel, 1994). Not only did this remove the pathogen, but it also revealed a possible two-host lifecycle involving fish and crustaceans that is rarely observed in Microsporidia. Adding fish can also have benefits under the right circumstances. For example, the introduction of tilapia in adjacent connected pond systems greatly reduced the prevalence of AHPND in *P. vannamei* (Cadiz et al., 2016). The nature of this interaction is unknown, but its effect is remarkable. In another example, other crustaceans and rotifers were implicated as reservoirs for WSSV in shrimp ponds. Controlling these organisms in production systems can be difficult, and it can also remove other potential food sources for the shrimp. This highlights that culture ponds are altered communities and that some members of the communities can serve to help or harm production.

In considering ecological control, one must also examine such issues as physical changes (or physical control) to the culture system. Covering ponds with netting or mesh to reduce bird predation on shrimp has the added bonus of hindering the spread of viruses from birds through mechanical vectoring in fecal material (Lightner, 1999; Flegel et al. 2008; Flegel, 2010). Allowing ponds to "fallow" by drying them out, allowing sunlight to kill viruses, and liming to kill pathogens before restocking with SPF shrimp all allow for the exclusion of some pathogens (Walker and Mohan, 2009). Applying suitable filtration (ultraviolet (UV) or ozone) to influent and effluent may be necessary to keep pathogens from entering culture ponds and from escaping into wild shrimp and potential reservoirs. An appropriate biofilter can also remove excess nitrogenous wastes from the effluent.

Shrimp aquaculture has gone through significant changes over the last 50 years. It has metamorphosed into a massive multi-billion-dollar industry in Asia and the Americas, but only because of changes in production practices in response to impediments caused by several diseases. The emergence of serious pathogens highlights that countries importing stocks for culture should invest in appropriate centers for surveillance, screening, and health exams in order to further reduce the risk of repeated pathogen introductions through importation of potentially contaminated stock (Flegel, 2012; Thitamadee et al., 2016). For additional consideration of the effects of shrimp viruses on international trade and surveillance, see Section 15.5.

15.3.3 Crayfish and Krebspest

Crayfish plague, or krebspest, is caused by *Aphanomyces astaci*, an obligate parasitic oomycete of crayfish. The pathogen was introduced into Europe in the 1850s from the United States, and over several decades spread throughout most of southern Europe and into northern Europe (Alderman and Polglase, 1988). Native crayfish, notably the European noble crayfish (*Astacus astacus*), are susceptible to *A. astaci*, but American species are not susceptible, due, in part, to their thicker cuticles (Unestam, 1969, 1972).

In Scandinavian countries, crayfish plague has devastated fisheries for native crayfish (Edgerton et al., 2002; Aquiloni et al., 2011) and damaged populations of rare and threatened species (Holdich, 2003). Attempts to revive the crayfish industry resulted in a reintroduction of the pathogen from the United States into Europe in the 1960s and 1970s, when *Pacifastacus leniusculus* and *Procambarus clarkii* were imported to Sweden and Norway (Alderman, 1996; Holdich, 2003). This pathogen is considered highly invasive and has been reported throughout most of Europe (Edgerton et al., 2002; Holdich, 2003). Its effect on local species can be particularly damaging. For example, in the early 1900s, crayfish harvests in Scandinavian and Baltic countries totaled more than 2000 tons annually, but this declined to 200 tons following introduction of the pathogen (Skurdal et al., 1999).

In terms of management, sustained outbreaks of crayfish plague deplete local host populations very quickly. Epidemics eventually burn out due to extirpation of suitable hosts within a lake or stream system. Fortunately, the propagules of the pathogen only remain viable in host cadavers for a few weeks; therefore, management involves fallowing affected areas for a season, then reseeding them with new, uninfected stock (e.g., Taugbøl et al., 1993; Spink and Frayling, 2000). In complex river systems, the pathogen can exist at low levels among dispersed hosts, making fallowing less suitable for mitigation (Taugbøl et al., 1993). As with bivalve culture, reseeding efforts require rigorous adherence to rearing, quarantine, and disinfection protocols (e.g., Alderman et al., 1987; Dieguez-Uribeondo, 2006; Jussila et al., 2011), as well as surveillance and certification using sensitive and specific diagnostic tools (e.g., Oidtmann et al., 2004, 2006). Scrupulous attention to diagnostic sampling and certification must be implemented whenever live crayfish are to be transported across national and regional boundaries.

Crayfish farming has expanded rapidly in recent years, with production based on two species, the American red swamp crayfish (*Procambarus clarkii*) and the Australian redclaw crayfish (*Cherax quadricarinatus*), but several other species are also in production. Most farming occurs in the United States and China, though Australia and Mexico also have sizable production (Food and Agriculture Organization, 2012a). Indigenous species are also harvested, but extensive farming uses domesticated hosts such as the red swamp crayfish and the red-claw crayfish. Given the focus on introducing these highly productive, fast-growing domesticated crayfish, there will be an increased threat of introduction of crayfish plague into new regions; thus, appropriately controlled broodstock programs and certification systems should be established to limit the spread of this highly invasive pest.

There is an expansive literature on *A. astaci*, particularly in terms of host immunology, epidemics, control of infection, and the effects of the pathogen in different host species (Holdich, 1999). For focused reviews on crayfish plague, see Alderman and Polglase (1988), Alderman (1996), Holdich (2003), Holdich et al. (2009), and Diéguez-Uribeondo (2006). Other pathogens of crayfish have occurred in focal outbreaks. These have been reviewed by Edgerton et al. (2002, 2004) and Longshaw (2011), with recommended actions for certain host species and pathogens given by the Food and Agriculture Organization of the United Nations (FAO; www.fao.org).

15.3.4 Disease Emergence in Culture of the Chinese Mitten Crab

The Chinese mitten crab (*Eriocheir sinensis*) was successfully cultured in the early 1990s. Crab culture has since expanded in China, Taiwan, and other parts of eastern Asia, with production estimated at over 796 thousand metric tons, valued at over US$5.5 billion in 2014 (Food and Agriculture Organization, 2016). As production was expanding in the late 1990s, "tremor" disease emerged as a significant cause of mortality in the culture of the crab in China. The causative agent was later shown to be *Spiroplasma eriocheiris*, the only spiroplasma known from a crustacean (Wang and Gu, 2002; Wang et al., 2004; Wang, 2011). Infections have a short time course (7–14 days) and show a strong relationship between severity and increased temperature (Wang et al., 2004). The same spiroplasma also infects and causes disease in cultured crayfish (*Procambarus clarkia*) and shrimp (*Penaeus vannamei*) in China (Wang et al., 2005; Bi et al., 2008).

Several highly specific and sensitive diagnostic tests have been developed to study the pathogen, including sequence analyses, PCR primers and probes, and an enzyme-linked immunosorbent assay (ELISA) (Ding et al., 2007, 2013; Bi et al., 2008; Wang et al., 2009). Two potential virulence factors, a spiralin-like protein (SLP31) and an adhesion-like protein, have been identified (Meng et al., 2010ab). Oxytetracycline has been shown to have efficacy in the treatment of trembling disease (Liang et al., 2009), and the drug has well-established pharmacokinetics (Feng et al., 2011). Several antibiotics (quinolones, sulfonamides, and macrolides) have been found in the tissues of crabs from markets and urban locations in China; however, none was above regulatory levels for China (Liu et al., 2012; Wang et al., 2015). It is not clear if these were used in food production or were contaminants from sewage outfall.

The Chinese mitten crab was introduced into Europe in the late 1920s, and again in the late 1950s (Herborg et al., 2003), and into California, USA in the early 1990s (Cohen and Carleton, 1997). These introductions may have included the microsporidian *Hepatospora eriocheir*, because this parasite has been found in China and the United Kingdom (Stentiford et al., 2011). Given the relative ease of introductions of this crab via ballast water, and the fact that its pathogens are relatively unknown or have a poorly known host range (e.g., Sobecka et al. 2011; Ding et al., 2016), there should be concern about this crab possibly transporting new pathogens into new areas as it becomes more established in its expanded range.

15.4   Crustacean Fisheries

The economic aspects of diseases in fisheries are dwarfed by those occurring in aquaculture; nonetheless, several pathogens have plagued important crustacean fisheries, causing significant commercial losses at local and regional scales. As in aquaculture, pathogens attack larval, juvenile, and adult hosts with varying effects on each ontogenetic stage. The most notable pathogens are those that directly affect the catch, or harvest, of legal-sized animals; however, pathogens can impose significant losses to the unfished juvenile and female segments of the population – stages that are not frequently monitored in biomass surveys for fisheries (Behringer, 2012; Shields, 2012).

From the perspective of the resource manager, fisheries attempt to maximize F, the mortality due to fishing (harvest), while accounting for M, the natural mortality occurring due to other sources. Because of this, fisheries models treat all natural mortality as a fixed constant. During epidemics, natural mortality can be very high, significantly altering fishery assessments for M. Moreover, pathogens cause direct and indirect losses to fisheries, not just increases in M. Direct losses are mortalities inflicted on the "recruits" or the fished population by pathogens. Indirect losses are more common in some systems and have more insidious, cryptic effects than direct losses. They can limit larval supply through egg predation and castration (e.g., Brattey et al., 1985; Wickham, 1986; Blower and Roughgarden, 1989a,b), sterile matings (Shields and Wood, 1993), stunting (Hawkes et al., 1986), loss of recruits (Wahle et al., 2009), and predation risk (Butler et al., 2014). Indirect losses can be difficult to assess, because they require intensive surveys or mark–recapture studies on juveniles hosts (Shields, 2012; Hoenig et al., 2017). They are largely overlooked or ignored by the fishing industry, because its focus is on pre-recruits, or recruits to the fisheries; i.e., animals approaching or above the legal size. Moreover, in cold-water fisheries, the lag between early juvenile stages and recruitment into the fishery can take several years, masking the effect of disease-associated mortality, and increasing stochasticity in catch or production.

Several crustacean fisheries are affected by pathogens that cause stunting or other chronic, visually noticeable diseases. For example, rhizocephalan barnacles stunt their hosts and cause a cessation of molting, or anecdysis. Stunting, discoloration, or even epizootic shell disease (ESD) can reach high levels in some commercially fished populations, including among others blue crabs with the rhizocephalan barnacle *Loxothylacus texanus* (see Shields and Overstreet, 2007), lithodid crabs with the rhizocephalan *Briarosaccus callosus* (Hawkes et al., 1986), snow crabs with bitter crab disease (Shields et al., 2007), and lobsters with ESD (Castro and Somers, 2012). Stunted and visually diseased animals do not enter the fishery, are often culled back, and may accumulate in fishing grounds, which artificially inflates the prevalence of the pathogen (Hawkes et al., 1986; Meyers, 1990). Culled animals may also serve as foci for transmission to new hosts (Hawkes et al., 1986; Meyers, 1990). The accumulation of diseased animals in a fished population has led to the suggestion that fisheries should actively engage in their removal (Hawkes et al., 1986; Kuris and Lafferty, 1992; Shukalyuk et al., 2005); this would entail changing regulations to hold or kill undersized animals, and disposal issues may present a problem in terms of scale. One model found that removal of diseased abalone suffering from withering syndrome enhanced yield in the unaffected portion of the population (Ben-Horin et al., 2015). Such an approach might benefit other fisheries, particularly if transmission could be broken.

15.4.1   Snow Crabs and Bitter Crab Disease

Several important crab and lobster fisheries have been impacted by parasitic dinoflagellates in the genus *Hematodinium*. These parasites live in the hemocoel of their host and cause chronic disease leading to mortality. Infections in the American blue crab are usually fatal (Shields and Squyars, 2000). The taxon is split into two clades: a boreal clade that infects cold-water decapods and a temperate clade that infects temperate hosts (Small et al., 2012). Only two species have been described, but several others are undescribed and are referred to as *Hematodinium* or *Hematodinium*-like species. Species of *Hematodinium* are host generalists. At least 38 decapods are known hosts for the parasites (Stentiford and Shields, 2005; Small, 2012), and several amphipods are known hosts as well (Johnson, 1986; Pagenkopp et al., 2012).

Bitter crab disease, or bitter crab syndrome, is caused by an undescribed *Hematodinium*-like parasite that infects Tanner crabs (*Chionoecetes bairdi*) and snow crabs (*Chionoecetes opilio*). The syndrome is named after an unusual aspirin-like flavor that renders the infected crabs unpalatable. The parasite has a strong association with "closed" ecosystems such as the fjords of southeastern Alaska, USA in the Pacific (Meyers et al., 1987, 1990, 1996) and the fjord-like bays of northern Newfoundland, Canada in the Atlantic (Taylor and Khan, 1995; Dawe, 2002; Shields et al., 2005; Mullowney et al., 2011). This pattern in the ecology of the "seascape" has been reported in several other pathogens that affect commercial fisheries, including the rhizocephalan barnacles *Briarosaccus callosus* (Sloan et al., 1984), *Loxothylacus texanus* (Alvarez and Calderon, 1995), and *Sacculina granifera* (Shields and Wood, 1993) and the nemertean *Carcinonemertes regicides* (Kuris et al., 1991).

Many outbreaks of this parasite have been documented in Tanner and snow crab fisheries, and these have caused significant damage. In the fjords of southeastern Alaska, approximately one-third of the commercial fishery comprises infected crabs (Meyers et al., 1990), but offshore sites have much lower prevalence levels (Meyers et al., 1996). Outbreaks in the coastal bays of Newfoundland reach similar high levels (Shields et al., 2005, 2007), but prevalence levels in offshore sites are much lower (Mullowney et al., 2011). Outbreaks have been associated with several host factors and with increases in bottom temperature (Shields et al., 2007). An epidemic in 2003–2005 was associated with a 1 °C increase in temperature, and may have been related to an increase in molting activity in the crab hosts during the warming period. In the Tanner crab fishery of southeastern Alaska, an epidemic of *Hematodinium* was estimated to cost over US$250 000 (Meyers et al., 1987), but indirect losses to females and juveniles were not assessed.

Several other fisheries have suffered economic losses due to outbreaks of *Hematodinium*, including those for the American blue crab (*Callinectes sapidus*) (Messick and Shields, 2000; Lee and Frischer, 2004), the velvet crab (*Necora puber*) (Wilhelm and Mialhe, 1996), the European spider crab (*Maja squinado*; Majidae), the edible crab (*Cancer pagurus*) (Chualáin et al., 2009; Smith and Rowley, 2015; Smith et al., 2015), and the Norway lobster (*Nephrops norvegicus*) (see Stentiford and Shields, 2005), as well as several decapods in culture in China (Xu et al., 2007, 2010; Li et al., 2013b). Prevalence levels can be very high (70–100%) during outbreaks, but losses to fisheries can be difficult to estimate because dead crabs and lobsters are not landed; one must examine changes in host density over time (Stentiford et al., 2001a). Monetary losses due to *Hematodinium* in the Norway lobster fishery off Scotland have been estimated at £2–4 million annually (Field et al., 1992, 1998; Stentiford et al., 2001b). In the coastal bays of Virginia and Maryland, USA, annual losses may exceed $500 000 in non-epidemic years (Stentiford and Shields, 2005). Prevalence levels can be very high in seasonal outbreaks in the American blue crab (*Callinectes sapidus*) (Messick, 1994; Messick and Shields, 2000; Lee and Frischer, 2004).

From the fisher's perspective, crabs and lobsters with *Hematodinium* are unmarketable. Bitter crab disease ruins the flavor of its crab hosts, and a single infected crab can affect the flavor of an entire batch of cooked crabs. The flesh of infected Norway lobsters and blue crabs is pasty and unappetizing, and fishers can sometimes recognize heavily infected animals and remove them prior to their reaching markets. Infected snow crabs can be diagnosed macroscopically and are then culled at sea or disposed of in landfills (Pestal et al., 2003). In the Newfoundland fishery, resource managers have advised fishers to destroy infected crabs by disposal in landfills (D. Taylor, Department of Fisheries and Oceans, Canada, pers. comm., cited in Shields, 2012). However, during outbreaks, there can be too many crabs to efficiently destroy in this manner.

From the resource manager's perspective, the mortality imposed by a pathogen can far exceed fishery-based estimates of natural mortality, *M*, resulting in gross overestimation of harvestable biomass in an impacted fishery. This can lead to overfishing or collapse of a fishery if the epidemics are severe or recurring. In snow crabs, a mark–recapture method in Newfoundland was used to show that the relative survival of *Hematodinium*-infected crabs was less than 1% that of uninfected animals (Hoenig et al., 2017). The recommendation was that in years with disease outbreaks, enhanced monitoring should be undertaken to determine if quotas for the snow crab should be changed to reflect the increased mortality. In a modeling study of snow crab stocks in Alaskan waters, the presence of the parasite was shown to negatively impact recovery efforts in depleted stocks (Siddeek et al., 2010). More of these types of studies are needed to better gauge the effect of the pathogen. Such studies can provide resource managers with tools for disease assessments in stocks of these commercially important crabs (Hoenig et al., 2017).

15.4.2 American Lobster and Epizootic Shell Disease

The American lobster (*Homarus americanus*) supports one of the largest shellfish fisheries in the world. American landings are US$300–400 million annually (National

Marine Fisheries Service, 2016), and Canadian landings are CAN$650–900 million (Fisheries and Oceans Canada, 2016). Lobsters are fished, then typically held in holding facilities or seawater pounds prior to their sale. They can be held for several months.

Given the size of the fishery, it has had surprisingly few major disease issues (for review, see Shields, 2013; Shields et al., 2006). In the past, outbreaks of gaffkemia caused significant losses to the industry, but changes in handling and marketing practices have significantly reduced mortality to the disease. Gaffkemia is a bacterial disease caused by *Aerococcus viridans*, a common benthic (or soil) bacterium. It is highly pathogenic and can kill lobsters very quickly – within days of infection at temperatures >10 °C (Stewart et al., 1980). The bacterium requires a portal of entry, and this was readily provided when lobster claws were pegged by the industry prior to sale. Nowadays, rubber bands are used to restrain the claws, reducing injuries to the animals and thus limiting outbreaks of the pathogen. Background levels of the bacterium are very low in natural populations (Lavallée et al., 2001; Stebbing et al., 2012), but focal outbreaks occasionally occur in holding facilities and can be quite damaging (Menard and Myrand, 1987).

Recently, several diseases have affected fisheries and management of the American lobster stock off southern New England (waters off Rhode Island, Connecticut, and southern Massachusetts). In 1997, ESD emerged as a significant disease in lobsters from eastern Long Island Sound. It has fulminated over several years (see later). In 1999, an unusual protist, *Neoparamoeba pemaquidensis*, decimated the lobster population in western Long Island Sound (Mullen et al., 2004; Pearce and Balcom, 2005). The resulting mortality was associated with several environmental stressors that weakened the immune status of the lobsters, essentially making them susceptible to invasion by a facultatively parasitic benthic protist (Shields, 2013). In 2002, a physiological disorder known as calcinosis was described in lobsters from central Long Island Sound (Dove et al., 2004). It was associated with temperature stress and resulted in several smaller mortalities. Lobsters from Long Island Sound also have high prevalence levels (˜50%) of idiopathic blindness (Maniscalco and Shields, 2006; Magel et al., 2009; Shields et al., 2012). Although its etiology is unknown, this is likely an environmental disease, perhaps associated with hypoxic reactive metals (Shields, 2013). Several environmental stressors have contributed to the emergence of these diseases, including increased bottom temperatures, extensive eutrophication with concomitant hypoxia, storm-induced destratification, possible exposure to contaminants, and fishery-induced stressors.

ESD is a significant disease risk to lobsters. It emerged in eastern Long Island Sound and Buzzards Bay, MA in the late 1990s (Fig. 15.4.). It occurs primarily on lobsters off southern New England (Castro and Angell, 2000; Castro and Somers, 2012; Howell, 2012), and to a lesser extent northward into the southern Gulf of Maine (Glenn and Pugh, 2006). Off Rhode Island and eastern Long Island Sound, prevalence of ESD reaches 25–35%, and up to 65% in ovigerous females (Howell et al., 2005; Castro and Somers, 2012). Lobsters with the disease have been found in the northern Gulf of Maine, but at low levels. Prevalence off Maine has recently risen from 0.1 to 0.3%, and there is significant concern that it will continue to rise with increasing water temperatures (Maynard et al. 2016).

ESD is an environmental disease associated with a complex suite of factors, primarily increased temperature, widespread exposure to contaminants, and a dysbiotic bacterial flora on the lobster cuticle (Tlusty et al., 2007; Chistoserdov et al., 2012; Shields, 2013).

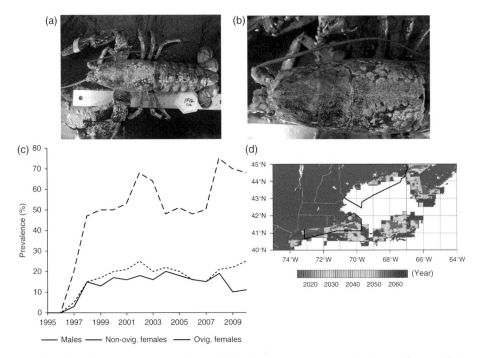

Fig. 15.4 Aspects of epizootic shell disease (ESD) of the American lobster: (a) heavy infection of the dorsal carapace of a lobster from southern Massachusetts; (b) close-up showing the necrotic appearance of the shell, the loss of the rostrum in the animal, and the dark melanotic response in the heavily damaged central portion of the lesion; (c) prevalence of ESD on sub-legal lobsters from inshore waters of Rhode Island – note the predilection for ovigerous (egg-bearing) females that cannot molt out of the disease; (d) forecast of when summer bottom temperatures will be over 12 °C in the Gulf of Maine and Long Island Sound, color-coded by year. *Source*: (a,b) Photos from author's collection; (c) Castro and Somers (2012). Reproduced with permission of National Shellfisheries Association; (d) Maynard, http://rstb.royalsocietypublishing.org/content/371/1689/20150208. Used under CC BY https://creativecommons.org/licenses/by/4.0/. (*See color plate section for the color representation of this figure.*)

Lobsters with moderate to severe cases of ESD are esthetically compromised; their carapace, claws, and abdomen are covered with melanotic lesions associated with bacterial necrosis of the cuticle (Smolowitz et al., 2005; Shields et al., 2012). High prevalence levels in lobsters from eastern Long Island Sound and Buzzards Bay suggests that widespread phenomena are involved in the disease, rather than point-source contaminants or local problems with water quality. However, one group of contaminants, the alkylphenols, have been found in lobsters and sediments from throughout the region (Biggers and Laufer, 2004; Laufer et al., 2005; Jacobs et al., 2012), and these may be associated with the disease because they interfere with the tanning of the cuticle (Laufer et al., 2012). Our current understanding of the etiology is that increased temperatures negatively affect host defensive responses, with contaminants potentially weakening the cuticle, making it more susceptible to colonization by a dysbiotic bacterial community brought about by anthropogenic effects to water quality (Shields, 2013).

ESD is associated with increasing ocean temperatures, especially with bottom temperatures above 16 °C (Glenn and Pugh, 2006); a laboratory-induced form of shell disease begins to emerge at temperatures over 10 °C (Tlusty et al., 2012). Because of the emergence of ESD and its effect on the southern New England lobster stock, there is concern that it will expand rapidly into the Gulf of Maine. Maynard et al. (2016) used a climate model to forecast bottom temperatures, surface temperatures, and the prevalence of ESD in relation to climate change. In this model, temperatures conducive to the spread of the disease will occur within the next 20 years in most of the Gulf of Maine fishery zones. The recent increases in ESD prevalence in Maine occur primarily in shallow bays, where waters are warming most quickly. Thus, long-term projections suggest that ESD will intensify into the region as temperatures increase.

Some lobsters can molt successfully out of ESD, and thus rid themselves of it, but they can also rapidly reacquire it (Castro and Angell, 2000). Ovigerous females are particularly vulnerable to the disease because they hold their eggs for 9–11 months and then do not molt until 9–12 months later; they apparently suffer increased mortality from the disease because they do not molt out of it (Hoenig et al., 2017). The direct impact of the disease is that the infected lobsters cannot be sold in the live trade because their lesions make them unmarketable; instead, they are either released or used in the less lucrative market for processed tail meat. More importantly, the indirect effect of ESD has been a major reduction in commercial landings (Castro and Somers, 2012; Howell, 2012), because of presumptive reductions in egg production (Wahle et al., 2009). The southern New England lobster stock has effectively experienced a recruitment failure and is now considered critically depleted, with increased temperature stress and ESD contributing to this decline in the fishery (Atlantic States Marine Fisheries Commission, 2010). One potential management tool was discussed as this fishery was collapsing: a moratorium was proposed on the fishing of female lobsters, but it was not imposed (Atlantic States Marine Fisheries Commission, 2010).

15.4.3 Spiny Lobsters and PaV1

In the late 1990s and early 2000s, declines in landings of the Caribbean spiny lobster (*Panulirus argus*) coincided with the finding of Panulirus argus virus 1 (PaV1), the first naturally occurring pathogenic virus from a lobster (Shields and Behringer, 2004). As with many crustacean viruses, PaV1 is mostly pathogenic to juvenile life-history stages; it occurs in adults but may not be pathogenic in them (Shields and Behringer, 2004; Huchin-Mian et al., 2009). It is transmitted via contact with diseased lobsters or from waterborne sources (Butler et al., 2008), but infections have been reported in adults and postlarvae, indicating the possibility of vertical transmission (Moss et al., 2011; Lozano-Álvarez et al., 2015). Mortality can be very high in early benthic juvenile stages, but in larger juveniles and adults the disease is subacute and can result in life-long infections (Shields and Behringer, 2004; Butler et al., 2008; Moss et al., 2011).

In the Florida Keys, prevalence has remained 6–10% over several years (Shields and Behringer, 2004; Behringer et al., 2011; Moss et al., 2011, 2012). Off Puerto Morelos and Chinchorro Bank, Mexico, it ranges from 2.5 to 10.9% (Lozano-Álvarez et al., 2008; Huchin-Mian et al., 2009). It occurs throughout much of the northern portions of the Caribbean Sea, including the Florida Keys, US Virgin Islands, Mexico, Cuba, and Belize (Huchin-Mian et al., 2008, 2009; Moss et al., 2012).

Healthy spiny lobsters can detect and avoid diseased animals (Behringer et al., 2006; Candia-Zulbarán et al., 2015). Spiny lobsters aggregate in dens during the day, for protection from predators. Avoidance behavior and social aggregation have implications for the transmission of the virus. Infected lobsters are often found alone, shunned by their conspecifics. Models of the behavior indicate that avoidance behavior alone can impede the spread of the virus, causing it to go extinct; hence, other avenues of transmission and dispersal – such as vertical (maternal) transmission or age-dependent infectivity – likely operate to maintain the virus in the lobster population (Dolan et al., 2014). The avoidance behavior, termed "behavioral immunity," has been credited with suppressing an epidemic during a stressful harmful algal bloom that lowered the effective density of dens in the affected area (Butler et al., 2015).

In the Florida Keys, fishers use live juvenile lobsters as a "bait" in a pot fishery for these socially gregarious animals. This fishing practice facilitates transmission of PaV1, and transmission rates are apparently high because of it (Behringer et al., 2012). The combination of lobsters confined in traps and juveniles held in live wells plus handling and temperature stress likely enhances transmission of the disease (Shields, 2012). Fishers are allowed to trap and transport large numbers of juvenile lobsters for use as social bait. This practice has caused some concern to the fishery, given that fishers are spreading the virus through their baiting activity. In Mexico, the lobster fishery employs casitas, or artificial dens, to enhance the number of protected habitats for lobster recruitment and aggregate lobsters for easier fishing, but no risk factors have been identified with the use of casitas in spreading the pathogen (Briones-Fourzán et al., 2012; Huchin-Mian et al., 2013).

PaV1 is a host specialist that only infects *P. argus*, and is unlikely to infect other crustaceans, including a close congener, *P. guttatus* (Behringer et al., 2008). The virus was found at high prevalence levels in frozen lobster tails from Belize bound for international markets (Huchin-Mian et al., 2009). Most spiny lobsters are marketed locally or shipped as frozen tails, so there is a low probability of introduction to new regions; nonetheless, live lobsters should be quarantined or screened for the virus if they are to be used as broodstock or to be released into new areas.

15.5   Agencies for Disease Management

Several international agencies provide policy and regulatory oversight to international trade in cultured fish and shellfish. Foremost among them is the OIE, which began in the 1920s to oversee the control of viral diseases in livestock, which were being increasingly transported across regional and international boundaries. In an attempt to prevent the spread of highly contagious or pathogenic organisms, the OIE established a framework to protect regions from the unwanted spread of livestock pathogens. The OIE regulations give protocols for appropriate surveillance and procedures for trade allowances, and provide a list of notifiable pathogens and their diseases that represent significant hazards to international commerce (World Organisation for Animal Health, 2016). Although the list focuses on diseases of vertebrate livestock, several pathogens of mollusks (primarily oysters and abalone) and crustaceans (primarily shrimp and crayfish) are included.

In Europe, European Union Council Directive 2006/88/EC, which facilitates protection of native stocks, applies to crustaceans, mollusks, and fish (Stentiford et al., 2010). In Asia, the OIE regulations are used, but there is a rapid communication network known as the Network of Aquaculture Centres in Asia-Pacific (NACA, www.enaca.org)

that also provides publications, pathogen alerts, and information on the availability of new diagnostic tests. The OIE, EU Directive 2006/88/EC, and NACA, as well as state and federal importation regulations in the United States, provide the regulatory basis for surveillance and control of unwanted pathogens.

Several shrimp and oyster pathogens highlight the problems posed by uncontrolled introductions of agents into new areas. With shrimp viruses, such introductions had unforeseen consequences in the form of several pandemics that caused billions of dollars of losses to the shrimp industry. With oysters, they hampered the development and establishment of aquaculture industries in Europe and the United States. Listed pathogens from mollusks include abalone herpesvirus, *Bonamia exitiosa*, *B. ostreae*, *Marteilia refingens*, *Perkinsus marinus*, *P. olseni*, and *Xenohaliotis californiensis*. Two of these are in abalone – abalone herpesvirus and *X. californiensis* – and the others are protists that use different oysters and clams as hosts. Listed pathogens from shrimp include viral and bacterial agents: AHPND caused by *Vibrio parahaemolyticus*, YHV, IHHNV, IMNV, necrotizing hepatopancreatitis (a rickettsial agent), TSV, WSSV, and white tail disease caused by Macrobrachium rosenbergii nodavirus (MrNV). There is also one very highly invasive pathogen from crayfish, the oomycete *Aphanomyces astaci*.

The Aquatic Animal Health Code (World Organisation for Animal Health, 2015) provides the framework for notifiable pathogens in aquatic animals (including fish and shellfish) and their products, as well as criteria for listing, surveillance, reporting, and other measures to prevent introductions of pathogens. To be listed by the OIE, pathogens undergo a vetting process, with specific criteria for inclusion. For example, a listed agent must have an appropriate case definition that clearly indicates its potential for crossing international boundaries. Although the OIE guidelines focus almost exclusively on commercially important hosts and pathogens that cause significant production losses to harvested populations, they also contain reference to pathogens that could damage endangered or ecologically important hosts. However, in practice, the listed pathogens pose significant risk of commercial losses because of their virulence, transmissibility, and difficulty of control in aquaculture species.

The Code also provides stringent guidelines for surveillance. This typically involves examining a set or subset of animals that are to be transported. The positive detection of a listed pathogen is notifiable to the authorities of the importing country, even in the absence of clinical signs of disease, and significant surveillance and reporting is required to monitor infected broodstock, seed, and infected animals that are to be used in commercial trade across regional zones or jurisdictional boundaries. Countries that are members of the OIE provide information on prevention measures, quarantine measures, and any restrictions on movement of aquatic animals and their products. Live commercial products that are imported into disease-free regions require diagnostic surveillance and health certifications guaranteeing the disease-free status of the animals in question. The Code also provides an extensive framework for surveillance in terms of methods and applications. It represents a guiding principle for health risk management, rational surveillance, prevention, and control. It also includes detailed steps for each listed pathogen (e.g., *Aphanomyces astaci*; Chapter 9.1 in World Organisation for Animal Health, 2015). Member countries have significant responsibility in reporting listed and emerging pathogens.

Although the Aquatic Animal Health Code provides an excellent framework for disease prevention and control, several weaknesses have been identified in its application to oyster pathogens, and these same weaknesses broadly apply to pathogens from other aquatic hosts. Carnegie et al. (2016) highlight them as three paradoxes: the paradox of what is a notifiable disease, the paradox of advanced diagnostics, and the paradox of addressing uncertainty in shellfish disease management. These paradoxes present significant challenges to surveillance, diagnostics, certification, importation, and risk management.

In terms of the paradox of notifiable disease, several organizations regulate international trade of livestock, and each maintains its own list of notifiable pathogens. These discrepancies can lead to confusion as to what risks certain pathogens have, and thus to complications in their transportation (Carnegie et al., 2016). For example, the EU list includes *Mikrocytos mackini* because of its known spread from California to Europe in imported *Crassostrea gigas*, but *M. mackini* was removed from the OIE list. *Perkinsus olseni*, a pathogen of clams, was removed from the EU Directive 2006/88/EC list but is included on the OIE list of notifiable pathogens. Oyster herpesvirus 1 (OsHV-1) appears to have a cosmopolitan distribution, but the highly virulent "μvar" genotype only occurs in Europe, Australia, and New Zealand (Segarra et al., 2010; Jenkins et al., 2013). Listing OsHV-1 would enable trade among endemic areas, but would risk spread of μvar to new areas. However, listing μvar alone would allow transfer of oysters with other genotypes of OsHV-1 (Carnegie et al., 2016). The European Union has listed μvar, but the OIE list does not include it. Such discrepancies between agencies could have consequences in terms of trade restrictions that might increase complacency about reporting emerging pathogens.

In terms of the paradox of advanced diagnostics, there are many excellent methods available for improved surveillance. However, several of the latest techniques have not had appropriate assessment or validation in terms of specificity, sensitivity, quality assurance, and quality control (Carnegie et al., 2016; Thitamadee et al., 2016). A "long history of use" is certainly appropriate for diagnostic tests, but a full assessment is necessary to rule out potential confounds (Carnegie et al., 2016). The Aquatic Animal Health Code (World Organisation for Animal Health, 2015) offers specific guidelines on assessment and validation, yet several assays have not had formal validation procedures.

The paradox of advanced diagnostics is also an indication of the increasing reliance on single-pathogen or specific-pathogen molecular techniques (Carnegie et al., 2016; Thitamadee et al. 2016). Most PCR methods can only detect specific pathogens; they cannot be applied to the exclusion of other tools, particularly if emerging agents are suspected. Histological assessment, molecular confirmation, and health certification are major requirements in working with notifiable pathogens. This highlights another major feature of the paradox: there is a scarcity of experts in invertebrate pathology with training in specific host taxa. Thus, training, funding, and retention of appropriate personnel working on disease surveillance and control should be a high priority in aquaculture production.

In terms of the paradox of shellfish disease management, Carnegie et al. (2016) succinctly state that management errors have two outcomes: increasing the spread of disease or unnecessarily limiting production. Risk-averse strategies based on the precautionary principle may be too strict, limiting trade and economic development, particularly when knowledge of the transmission or ecology of the host–pathogen interaction is lacking. On the other hand, risky strategies should be avoided because we have too many examples of the spread of pathogens due to complacency or ignorance in allowing trade between regions (e.g., Thitamadeed et al., 2016). This means that the stakeholders must be educated to prevent the illegal importation of potentially disease-bearing animals destined for culture and have appropriate surveillance in place for feed and feed products that are used in production.

15.6 Conclusion

Over the last several decades, disease outbreaks have been increasingly reported from several commercially important shellfish. Severe outbreaks have occurred in both aquaculture production and fisheries. In aquaculture, the breeding of SPF or disease-resistant stock appears critical to moving forward under intense disease situations such as those seen in WSSV in shrimp and MSX and dermo in oysters. Stringent adherence to biosecurity is essential in keeping pathogens out of vulnerable systems. In fisheries, changes in quotas and fishing out of diseased animals may provide some resiliency against pathogens, but more work is needed to model the effect of pathogens in fished stocks. Finally, we cannot forget the importance of baseline surveys of pathogens in important animal populations. It is important to know what pathogens are present in order to gauge their potential, and it is important to know whether the pathogens in an outbreak are part of an emergent event or a cyclical phenomenon. Pathogens often act in concert with environmental stressors, and the underlying stressors should be investigated as causal components of disease.

Acknowledgments

I thank Ryan Carnegie for help with Fig. 15.2. This research was funded in part by an EID Program Grant, NSF OCE #0723662, and a NOAA, Saltonstall-Kennedy Program grant, NA14NMF4270044. This is contribution # 3561 of the Virginia Institute of Marine Science, The College of William and Mary.

References

Alderman, D.J., 1996. Geographical spread of bacterial and fungal diseases of crustaceans. Revue Scientifique et Technique Office International des Epizooties 15, 603–632.

Alderman, D.J., Polglase, J.L., 1988. Pathogens, parasites and commensals, in: Holdich, D.M., Lowery, R.S. (eds.), Freshwater Crayfish Biology Management and Exploitation. Chapman & Hall, London, pp. 283–308.

Alderman, D.J., Polglase, J.L., Frayling, M., 1987. *Aphanomyces astaci* pathogenicity under laboratory and field conditions. J. Fish Dis. 10, 385–393.

Alker, A.P., Smith, G.W., Kim, K., 2001. Characterization of *Aspergillus sydowii* (Thom et Church), a fungal pathogen of Caribbean sea fan corals. Hydrobiologia 460, 105–111.

Alvarez, F., Calderón, J., 1996. Distribution of *Loxothylacus texanus* (Cirripedia, Rhizocephala) parasitizing crabs of the genus *Callinectes* in the southwestern coast Gulf of Mexico. Gulf Res. Rep. 9, 205–210.

Anderson, R.M., May, R.M., 1981. The population dynamics of microparasites and their invertebrate hosts. Phil. Trans. Roy. Soc. London, B, Biol. Sci. 291, 451–524.

Anderson, T.J., Wesche, S., Lester, R.J.G., 1994. Are outbreaks of *Marteilia sydneyi* in Sydney rock oysters, *Saccostrea commercialis*, triggered by a drop in environmental pH? Mar. Freshw. Res. 45, 1285–1287.

Andrews, J.D., 1962. Oyster mortality studies in Virginia IV. MSX in James River public seed beds. Proc. Natl. Shellfish. Assoc. 53, 65–84.

Andrews, J.D., 1988. Epizootiology of the disease caused by the oyster pathogen *Perkinsus marinus* and its effects on the oyster industry. Am. Fish. Soc. Spec. Publ. 18, 47–63.

Andrews, J.D., Hewatt, W.G., 1957. Oyster mortality studies in Virginia. II. The fungus disease caused by *Dermocystidium marinum* in oysters of Chesapeake Bay. Ecol. Monogr. 27, 2–25.

Andrews, J.D., Ray, S.M., 1988. Management strategies to control the disease caused by *Perkinsus marinus*. Am. Fish. Soc. Spec. Publ. 18, 257–264.

Aquiloni, L., Martín, M.P., Gherardi, F., Diéguez-Uribeondo, J., 2011. The North American crayfish *Procambarus clarkii* is the carrier of the oomycte *Aphanomyces astaci* in Italy. Biol. Invasions 13, 359–367.

Arcier, J.M., Herman, F., Lightner, D.V., Redman, R.M., Mari, J., Bonami, J.R., 1999. A viral disease associated with mortalities in hatchery-reared postlarvae of the giant freshwater prawn *Macrobrachium rosenbergii*. Dis. Aquat. Org. 38, 177–181.

Atlantic States Marine Fisheries Commission, 2010. Recruitment failure in the Southern New England lobster stock. Available from: http://www.asmfc.org/uploads/file/april2010_SNE_Recruitment_Failure_TCmemoB.pdf (accessed May 8, 2017).

Bailey-Brock, J.H., Moss, S.M., 1992. Penaeid taxonomy, biology and zoogeography, in: Fast, A.W., Lester, L.J. (eds.), Marine Shrimp Culture: Principles and Practices. Elsevier, Amsterdam, pp. 9–28.

Bally, M., Garrabou, J., 2007. Thermodependent bacterial pathogens and mass mortalities in temperate benthic communities: a new case of emerging disease linked to climate change. Global Change Biol. 13, 2078–2088.

Balouet, G., 1979. *Marteilia refringens* – considerations of the life cycle and development of Abers disease in *Ostrea edulis*. Mar. Fish. Rev. 41, 64–66.

Bartholomay, L.C., Loy, D.S., Loy, J.D., Harris, D.L., 2012. Nucleic-acid based antivirals: augmenting RNA interference to "vaccinate" *Litopenaeus vannamei*. J. Invert. Pathol. 110, 261–266.

Behringer, D.C., 2012. Diseases of wild and cultured juvenile crustaceans: insights from below the minimum landing size. J. Invert. Pathol. 110, 225–233.

Behringer, D.C., Butler, M.J., Shields, J.D., 2006. Ecology: avoidance of disease by social lobsters. Nature, 441, 421–421.

Behringer, D.C., Butler, M.J., Shields, J.D., 2008. Effect of PaV1 infection on Caribbean spiny lobster (*Panulirus argus*) movement, condition, and survival. Mar. Ecol. Prog. Ser. 359, 26–33.

Behringer, D.C., Butler, M.J., Shields, J.D., 2011. A review of the lethal spiny lobster virus PaV1 – ten years after its discovery. Proc. Gulf & Carib. Fish. Inst. 62, 370–376.

Behringer, D.C., Butler, M.J., Moss, J., Shields, J.D., 2012. PaV1 infection in the Florida spiny lobster fishery and its effects on trap function and disease transmission. Can. J. Fish. Aquat. Sci. 69, 136–144.

Bell, T.A., Lightner, D.V., 1984. IHHN virus: infectivity and pathogenicity studies in *Penaeus stylirostris* and *Penaeus vannamei*. Aquaculture 38, 185–194.

Ben-Horin, T., Lafferty, K.D., Bidegain, G., Lenihan, H.S., 2016. Fishing diseased abalone to promote yield and conservation. Phil. Trans. R. Soc., London, B, 371, 20150211.

Berthe, F.C., 2008. New approaches to effective mollusc health management, in: Diseases in Asian Aquaculture VI. Fish Health Section. Asian Fisheries Society, Manila, pp. 343–352.

Bi, K., Huang, H., Gu, W., Wang, J., Wang, W., 2008. Phylogenetic analysis of Spiroplasmas from three freshwater crustaceans (*Eriocheir sinensis*, *Procambarus clarkii* and *Penaeus vannamei*) in China. J. Invert. Pathol. 99, 57–65.

Bidegain, G., Powell, E.N., Klinck, J.M., Ben-Horin, T., Hofmann, E.E., 2016a. Microparasitic disease dynamics in benthic suspension feeders: Infective dose, non-focal hosts, and particle diffusion. Ecol. Model. 328, 44–61.

Bidegain, G., Powell, E.N., Klinck, J.M., Ben-Horin, T., Hofmann, E.E., 2016b. Marine infectious disease dynamics and outbreak thresholds: contact transmission, pandemic infection, and the potential role of filter feeders. Ecosphere 7(4), e01286.

Biggers, W.J., Laufer, H., 2004. Identification of juvenile hormone-active alkylphenols in the lobster *Homarus americanus* and in marine sediments. Biol. Bull. 206, 13–24.

Blackbourn, J., Bower, S.M., Meyer, G.R., 1998. *Perkinsus qugwadi* sp. nov. (incertae sedis), a pathogenic protozoan parasite of Japanese scallops, *Patinopecten yessoensis*, cultured in British Columbia, Canada. Can. J. Zool. 76, 942–953.

Blower, S., Roughgarden, J., 1989a. Population dynamics and parasitic castration, test of a model. Am. Nat. 134, 848–858.

Blower, S., Roughgarden, J., 1989b. Parasites detect host spatial pattern and density, a field experimental analysis. Oecologia 78, 138–141.

Bonami, J.R., Widada, J.S., 2011. Viral diseases of the giant fresh water prawn *Macrobrachium rosenbergii*: a review. J. Invertebr. Pathol. 106, 131–142.

Boonyaratpalin, S., Supamattaya, K., Kasornchandra, J., Direkbusaracom, S., Aekpanithanpong, U., Chantanachooklin, C., 1993. Non-occluded Baculo-like virus, the causative agent of Yellow Head Disease in the black tiger shrimp (*Penaeus monodon*). Fish Pathol. 28, 103–109.

Borrego, J.J., Castro, D., Luque, A., Paillard, C., Maes, P., Garcia, M.T., Ventosa, A., 1996. *Vibrio tapetis* sp. nov., the causative agent of the brown ring disease affecting cultured clams. Int. J. Syst. Evol. Microbiol. 46, 480–484.

Boström, C., Pittman, S.J., Simenstad, C., Kneib, R.T., 2011. Seascape ecology of coastal biogenic habitats: advances, gaps, and challenges. Mar. Ecol. Prog. Ser. 427, 191–217.

Bower, S.M., Blackbourn, J., Meyer, G.R., 1998. Distribution, prevalence, and pathogenicity of the protozoan *Perkinsus qugwadi* in Japanese scallops, *Patinopecten yessoensis*, cultured in British Columbia, Canada. Can. J. Zool. 76, 954–959.

Braid, B.A., Moore, J.D., Robbins, T.T., Hedrick, R.P., Tjeerdema, R.S., Friedman, C.S., 2005. Health and survival of red abalone, *Haliotis rufescens*, under varying temperature, food supply, and exposure to the agent of withering syndrome. J. Invertebr. Pathol. 89, 219–231.

Brattey, J., Campbell, A., Bagnall, A.E., Uhazy, L.S., 1985. Geographic distribution and seasonal occurrence of the nemertean *Pseudocarcinonemertes homari* on the American lobster, *Homarus americanus*. Can. J. Fish. Aquat. Sci. 42, 360–367.

Briones-Fourzán, P., Candia-Zulbarán, R.I., Negrete-Soto, F., Barradas-Ortiz, C., Huchin-Mian, J.P., Lozano-Álvarez, E., 2012. Influence of local habitat features on disease avoidance by Caribbean spiny lobsters in a casita-enhanced bay. Dis. Aquat. Org. 100, 135–148.

Burge, C.A., Griffin, F.J., Friedman, C.S., 2006. Mortality and herpesvirus infections of the Pacific oyster *Crassostrea gigas* in Tomales Bay, California, USA. Dis. Aquat. Org. 72, 31–43.

Burge, C.A., Judah, L.R., Conquest, L.L., Griffin, F.J., Cheney, D.P., Suhrbier, A., et al., 2007. Summer seed mortality of the Pacific oyster, *Crassostrea gigas* Thunberg grown in Tomales Bay, California, USA: the influence of oyster stock, planting time, pathogens, and environmental stressors. J. Shellfish Res. 26, 163–172.

Burge, C.A., Eakin, C.M., Friedman, C.S., Froelich, B., Hershberger, P.K., Hofmann, E.E., et al., 2014. Climate change influences on marine infectious diseases: implications for management and society. Ann. Rev. Mar. Sci. 6, 249–277.

Burreson, E.M., 1991. Effects of *Perkinsus marinus* infection in the eastern oyster, *Crassostrea virginica*: I. Susceptibility of native and MSX-resistant stocks. J. Shellfish Res. 10, 417–423.

Burreson, E.M., 2008. Misuse of PCR assay for diagnosis of mollusc protistan infections. Dis. Aquat. Org. 80, 81–83.

Burreson, E.M., Ford, S.E., 2004. A review of recent information on the Haplosporidia, with special reference to *Haplosporidium nelsoni* (MSX disease). Aquat. Living Resour. 17, 499–517.

Burreson, E.M., Ragone Calvo, L.M., 1996. Epizootiology of *Perkinsus marinus* disease of oysters in Chesapeake Bay, with emphasis on data since 1985. J Shellfish Res. 15, 17–34.

Burreson, E.M., Stokes, N.A., Friedman, C.S., 2000. Increased virulence in an introduced pathogen: *Haplosporidium nelsoni* (MSX) in the eastern oyster *Crassostrea virginica*. J. Aquat. Anim. Health 12, 1–8.

Butler, M.J., Behringer, D.C., Shields, J.D., 2008. Transmission of a pathogenic virus (PaV1) and its effects on the survival of juvenile Caribbean spiny lobster, *Panulirus argus*. Dis. Aquat. Org. 79, 173–182.

Butler, M.J., Tiggelaar, J.M., Shields, J.D., Butler V, M.J., 2014. Effects of the parasitic dinoflagellate *Hematodinium perezi* on blue crab (*Callinectes sapidus*) behavior and predation. J. Exptl. Mar. Biol. Ecol. 461, 381–388.

Butler, M.J., Behringer, D.C. Jr., Dolan, T.W., Moss, J., Shields, J.D., 2015. Behavioral immunity suppresses an epizootic in Caribbean spiny lobsters. PLoS ONE 10, e0126374.

Cáceres, C.E., Hall, S.R., Duffy, M.A., Tessier, A.J., Helmle, C., MacIntyre, S., 2006. Physical structure of lakes constrains epidemics in *Daphnia* populations. Ecology 87, 1438–1444.

Cadiz, R.E., Traifalgar, R.F.M., Sanares, R.C., Andrino-Felarca, K.G.S., Corre, V.L. Jr., 2016. Comparative efficacies of tilapia green water and biofloc technology (BFT) in suppressing population growth of green Vibrios and *Vibrio parahaemolyticus* in the intensive tank culture of *Penaeus vannamei*. AACL Bioflux 9, 195–203.

Candia-Zulbarán, R.I., Briones-Fourzán, P., Lozano-Álvarez, E., Barradas-Ortiz, C., Negrete-Soto, F., 2015. Caribbean spiny lobsters equally avoid dead and clinically PaV1-infected conspecifics. ICES J. Mar. Sci., fsu249.

Carlton, R.G., Richardson, L.L., 1995. Oxygen and sulfide dynamics in a horizontally migrating cyanobacterial mat: black band disease of corals. FEMS Microbiol. Ecol. 18, 155–162.

Carnegie, R.B., Burreson, E.M., 2011. Declining impact of an introduced pathogen: *Haplosporidium nelsoni* in the oyster *Crassostrea virginica* in Chesapeake Bay. Mar. Ecol. Prog. Ser, 432, 1–15.

Carnegie, R.B., Arzul, I., Bushek, D., 2016. Managing marine mollusc diseases in the context of regional and international commerce: policy issues and emerging concerns. Phil. Trans. R. Soc. London, B, 371, 20150215.

Castro, K.M., Angell, T.E., 2000. Prevalence and progression of shell disease in American lobster, *Homarus americanus*, from Rhode Island waters and the offshore canyons. J. Shellfish Res. 19, 691–700.

Castro, K.M., Somers, B.A., 2012. Observations of epizootic shell disease in American lobsters, *Homarus americanus*, in southern New England. J. Shellfish Res. 31, 423–430.

Cavalli, L.S., Nornberg, B.F.D.S., Netto, S.A., Poersch, L., Romano, L.A., Marins, L.F., Abreu, P.C., 2010. White spot syndrome virus in wild penaeid shrimp caught in coastal and offshore waters in the southern Atlantic Ocean. J. Fish Dis. 33, 533–536.

Cavalli, L.S., Romano, L.A., Marins, L.F., Abreu, P.C., 2011. First report of White Spot Syndrome Virus in farmed and wild penaeid shrimp from Lagoa dos Patos estuary, southern Brazil. Brazilian J. Microbiol 42, 1176–1179.

Cerrano, C., Bavestrello, G., Bianchi, C.N., Cattaneo-Vietti, R., Bava, S., Morganti, C., et al., 2000. A catastrophic mass-mortality episode of gorgonians and other organisms in the Ligurian Sea (North-western Mediterranean), summer, 1999. Ecol. Letters, 3, 284–293.

Chang, P.H., Kuo, S.T., Lai, S.H., Yang, H.S., Ting, Y.Y., Hsu, C.L., Chen, H.C., 2005. Herpes-like virus infection causing mortality of cultured abalone *Haliotis diversicolor supertexta* in Taiwan. Dis. Aquat. Org. 65, 23–27.

Chapman, R.W., Browdy, C.L., Savin, S., Prior, S., Wenner, E., 2004. Sampling and evaluation of white spot syndrome virus in commercially important Atlantic penaeid shrimp stocks. Dis. Aquat. Org. 59, 179–185.

Chayaburakul, K., Nash, G., Pratanpipat, P., Sriurairatana, S., Withyachumnarnkul, B., 2004. Multiple pathogens found in growth-retarded black tiger shrimp *Penaeus monodon* cultivated in Thailand. Dis. Aquat. Org. 60, 89–96.

Chen, J., Xiong, J., Yang, J., Mao, Z., Chen, X., 2011. Nucleotide sequences of four RNA segments of a reovirus isolated from the mud crab *Scylla serrata* provide evidence that this virus belongs to a new genus in the family Reoviridae. Arch Virol. 156, 523–528.

Chistoserdov, A.Y., Quinn, R.A., Gubbala, S.L., Smolowitz, R., 2012. Bacterial communities associated with lesions of shell disease in the American lobster, Homarus americanus Milne-Edwards. J. Shellfish Res. 31, 449–462.

Christie, H., Leinaas, H.P., Skadsheim, A., 1995. Local patterns in mortality of the green sea urchin, *Strongylocentrotus droebachiensis*, at the Norwegian coast, in: Skjoldal, H.R., Hopkins, C., Erikstad, K.E., Leinaas, H.P. (eds.), Ecology of Fjords and Coastal Waters. Elsevier, Amsterdam, pp. 573–584.

Chu, F.L.E., La Peyre, J.F., Burreson, C.S., 1993. *Perkinsus marinus* infection and potential defense-related activities in eastern oysters, *Crassostrea virginica*: salinity effects. J. Invert. Pathol. 62, 226–232.

Chualáin, C.N., Hayes, M., Allen, B., Robinson, M. 2009. *Hematodinium* sp. in Irish *Cancer pagurus* fisheries: infection intensity as a potential fisheries management tool. Dis. Aquat. Org. 83, 59–66.

Cigarría, J., Elston, R., 1997. Independent introduction of *Bonamia ostreae*, a parasite of *Ostrea edulis*, to Spain. Dis. Aquat. Org. 29, 157–158.

Cock, J., Gitterle, T., Salazar, M., Rye, M., 2009. Breeding for disease resistance of penaeid shrimps. Aquaculture 286, 1–11.

Cohen, A.N., Carlton, J.T., 1997. Transoceanic transport mechanisms: the introduction of the Chinese mitten crab *Eriocheir sinensis* to California. Pacific Sci. 51, 1–11.

Corsin, F., Phi, T.T., Phuoc, L.H., Tinh, N.T.N., Hao, N.V., Mohan, C.V., et al., 2002. Problems and solutions with the design and execution of an epidemiological study of white spot disease in black tiger shrimp (*Penaeus monodon*) in Vietnam. Prev. Vet. Med. 53, 117–132.

Corsin, F., Thakur, P.C., Padiyar, P.A., Madhusudhan, M., Turnbull, J., Mohan, C.V., et al., 2003. Relationship between white spot syndrome virus and indicators of quality in *Penaeus monodon* postlarvae in Karnataka, India. Dis. Aquat. Org. 54, 97–104.

Corsin, F., Turnbull, J.F., Mohan, C.V., Hao, N.V., Morgan, K.L., 2005. Pond-level risk factors for white spot disease outbreaks, in: Diseases in Asian Aquaculture V. Asian Fisheries Society, Manila, pp.75–92.

Cowart, J.D., Henkel, T.P., McMurray, S.E., Pawlik, J.R., 2006. Sponge orange band (SOB): a pathogenic-like condition of the giant barrel sponge *Xestospongia muta*. Coral Reefs 25, 513.

Cuéllar-Anjel, J., Corteel, M., Galli, L., Alday-Sanz, V., Hasson, K.W., 2010. Principal shrimp infectious diseases, diagnosis and management, in: Alday-Sanz, V. (ed.), The Shrimp Book. Nottingham University Press, Nottingham, pp. 517–621.

Cuéllar-Anjel, J., Chamorro, R., White-Noble, B., Schofield, P., Lightner, D.V., 2011. Testing Finds Resistance to WSSV in Shrimp from Panamanian Breeding Program. Global Aquaculture Advocate, Portsmouth, NH, pp. 65–66.

Dawe, E.G., 2002. Trends in prevalence of bitter crab disease caused by *Hematodinium* sp. in snow crab (*Chionoecetes opilio*) throughout the Newfoundland and Labrador continental shelf, in: Paul, A.J., Dawe, E.G., Elner, R., Jamieson, G.S., Kruse, G.H., Otto, R.S., et al. (eds.), Crabs in Cold Water Regions: Biology, Management, and Economics. University of Alaska Sea Grant, AK-SG-01-01, pp. 385–399.

Deng, X.X., Lü, L., Ou, Y.J., Su, H.J., Li, G., Guo, Z.X., et al., 2012. Sequence analysis of 12 genome segments of mud crab reovirus (MCRV). Virology 422, 185–194.

Denner, E.B., Smith, G.W., Busse, H.J., Schumann, P., Narzt, T., Polson, S.W., et al., 2003. *Aurantimonas coralicida* gen. nov., sp. nov, the causative agent of white plague type II on Caribbean scleractinian corals. Int. J. Syst. Evol. Microbiol. 53, 1115–1122.

DePaola, A., Kaysner, C.A., Bowers, J., Cook, D.W., 2000. Environmental investigations of *Vibrio parahaemolyticus* in oysters after outbreaks in Washington, Texas, and New York (1997 and 1998). Appl. Environ. Microbiol. 66, 4649–4654.

Dhar, A.K., Cowley, J.A., Hasson, K.W., Walker, P.J., 2004. Genomic organization, biology, and diagnosis of Taura syndrome virus and yellowhead virus of penaeid shrimp. Adv. Virus Res. 63, 353–423.

Di Camillo, C.G., Bartolucci, I., Cerrano, C., Bavestrello, G., 2013. Sponge disease in the Adriatic Sea. Mar Ecol 34, 62–71.

Diéguez-Uribeondo J., 2006. Pathogens, parasites and ectocommensals, in: Souty-Grosset, C., Holdich, D.M., Noël, P.Y., Reynolds, J.D., Haffner, P. (eds.), Atlas of Crayfish in Europe. National Museum of Natural History, Paris, pp. 133–149.

Ding, Z., Bi, K., Wu, T., Gu, W., Wang, W., Chen, J., 2007. A simple PCR method for the detection of pathogenic spiroplasmas in crustaceans and environmental samples. Aquaculture 265, 49–54.

Ding, Z., Yao, W., Du, J., Ren, Q., Li, W., Wu, T., et al., 2013. Histopathological characterization and in situ hybridization of a novel spiroplasma pathogen in the freshwater crayfish *Procambarus clarkii*. Aquaculture 380, 106–113.

Ding, Z., Meng, Q., Liu, H., Yuan, S., Zhang, F., Sun, M., et al., 2016. First case of hepatopancreatic necrosis disease in pond-reared Chinese mitten crab, *Eriocheir sinensis*, associated with microsporidian. J. Fish Dis. 39, 1043–1051.

Dolan, T.W., Butler, M.J., Shields, J.D., 2014. Host behavior alters spiny lobster-viral disease dynamics: a simulation study. Ecology 95, 2346–2361.

Dorf, B.A., Hons, C., Varner, P., 2005. A three-year survey of penaeid shrimp and callinectid crabs from Texas coastal waters for signs of disease caused by white spot syndrome virus or Taura syndrome virus. J. Aquat. Anim. Health, 17, 373–379.

Dove, A.D.M., LoBue, C., Bowser, P., Powell, M., 2004. Excretory calcinosis, a new fatal disease of wild American lobsters *Homarus americanus*. Dis. Aquat. Org. 58, 215–221.

Dungan, M.L., Miller, T.E., Thomson, D.A., 1982. Catastrophic decline of a top carnivore in the Gulf of California rocky intertidal zone. Science 216, 989–991.

Edgerton, B.F., Evans, L.H., Stephens, F.J., Overstreet, R.M., 2002. Synopsis of freshwater crayfish diseases and commensal organisms. Aquaculture 206, 57–135.

Edgerton, B.F., Henttonen, P., Jussila, J., Mannonen, A.R.I., Paasonen, P., Taugbøl, T., et al., 2004. Understanding the causes of disease in European freshwater crayfish. Conserv. Biol. 18, 1466–1474.

Elston, R., 1979. Viruslike particles associated with lesions in larval Pacific oysters (*Crassostrea gigas*). J. Invert. Pathol. 33, 71–74.

Elston, R.A., Cheney, D.P., 2004. Shellfish high health program – building success through health management and disease prevention. World Aquaculture 35, 48–52.

Elston, R.A., Ford, S.E. 2011. Shellfish diseases and health management, in: Shumway, S.E. (ed.), Shellfish Aquaculture and the Environment. John Wiley & Sons, Chichester, pp. 359–394.

Elston, R.A., Hasegawa, H., Humphrey, K.L., Polyak, I.K., Häse, C.C., 2008. Re-emergence of *Vibrio tubiashii* in bivalve shellfish aquaculture: severity, environmental drivers, geographic extent and management. Dis. Aquat. Org. 82, 119–134.

Engelsma, M.Y., Kerkhoff, S., Roozenburg, I., Haenen, O.L., Van Gool, A., Sistermans, W., et al., 2010. Epidemiology of *Bonamia ostreae* infecting European flat oysters *Ostrea edulis* from Lake Grevelingen, The Netherlands. Mar. Ecol. Prog. Ser. 409, 131–142.

European Union Council Directive 2006/88/EC, October 24, 2006, On animal health requirements for aquaculture animals and products thereof, and on the prevention and control of certain diseases in aquatic animals. Available from: http://eur-lex.europa.eu/LexUriServ/LexUriServ.do?uri=OJ:L:2006:328:0014:0056:en:PDF (accessed May 8, 2017).

Farley, C.A., Wolf, P.H., Elston, R.A., 1988. A long-term study of "microcell" disease in oysters with a description of a new genus, *Mikrocytos* (g. n.), and two new species, *Mikrocytos mackini* (sp. n.) and *Mikrocytos roughleyi* (sp. n.). Fish. Bull. US 86, 581–593.

Feijó, R.G., Maggioni, R., Martins, P.C., de Abreu, K.L., Oliveira-Neto, J.M., Guertler, C., et al., 2015. RNAi-based inhibition of infectious myonecrosis virus replication in Pacific white shrimp *Litopenaeus vannamei*. Dis. Aquat. Org. 114, 89–98.

Feng, Q., Wu, G.H., Liang, T.M., Ji, H.Y., Jiang, X.J., Gu, W., Wang, W., 2011. Pharmacokinetics of oxytetracycline in hemolymph from the Chinese mitten crab, *Eriocheir sinensis*. J. Vet. Pharmacol. Therapeutics 34, 51–57.

Field, R.H., Chapman, C.J., Taylor, A.C., Neil, D.M., Vickerman, K., 1992. Infection of the Norway lobster *Nephrops norvegicus* by a *Hematodinium*-like species of dinoflagellate on the west coast of Scotland. Dis. Aquat. Org. 13, 1–15.

Field, R.H., Hills, J.M., Atkinson, R.J.A., Magill, S., Shanks, A.M., 1998. Distribution and seasonal prevalence of *Hematodinium* sp. infection of the Norway lobster (*Nephrops norvegicus*) around the west coast of Scotland. ICES J. Mar. Sci. 55, 846–858.

Fisheries and Oceans Canada, 2016. Seafisheries landings. Available from: http://www.dfo-mpo.gc.ca/stats/commercial/sea-maritimes-eng.htm (accessed May 8, 2017).

Flegel, T.W., 1997. Major viral diseases of the black tiger prawn (*Penaeus monodon*) in Thailand. World J. Microbiol. Biotechnol. 13, 433–442.

Flegel, T.W., 2006a. Detection of major penaeid shrimp viruses in Asia, a historical perspective with emphasis on Thailand. Aquaculture 258, 1–33.

Flegel, T.W., 2006b. The special danger of viral pathogens in shrimp translocated for aquaculture. Science Asia, 32, 215–221.

Flegel, T.W., 2009. Review of disease transmission risks from prawn products exported for human consumption. Aquaculture 290, 179–189.

Flegel, T.W., 2010. Importance of host-viral interactions in the control of shrimp disease outbreaks, in: Alday-Sanz, V. (ed.), The Shrimp Book. Nottingham University Press, Nottingham, pp. 623–654.

Flegel, T.W., 2012. Historic emergence, impact and current status of shrimp pathogens in Asia. J. Invert. Pathol. 110, 166–173.

Flegel, T.W., Lightner, D.V., Lo, C.F., Owens, L., 2008. Shrimp disease control: past, present and future, in: Diseases in Asian Aquaculture VI. Fish Health Section. Asian Fisheries Society, Manila, pp. 355–378.

Flowers, E.M., Simmonds, K., Messick, G.A., Sullivan, L., Schott, E.J., 2015. PCR-based prevalence of a fatal reovirus of the blue crab, *Callinectes sapidus* (Rathbun) along the northern Atlantic coast of the USA. J. Fish Dis. 39, 705–714.

Food and Agriculture Organization, 2006. The state of world fisheries and aquaculture 2006. Available from: http://www.fao.org/docrep/009/A0699e/A0699e00.htm (accessed May 8, 2017).

Food and Agriculture Organization, 2009. The state of world fisheries and aquaculture 2008. Available from: http://www.fao.org/docrep/011/i0250e/i0250e00.htm (accessed May 8, 2017).

Food and Agriculture Organization, 2012a. FAO Yearbook. Fishery and Aquaculture Statistics. Available from: http://www.fao.org/3/a-i3740t/index.html (accessed May 8, 2017).

Food and Agriculture Organization, 2012b. The state of world fisheries and aquaculture 2012. Available from: http://www.fao.org/docrep/016/i2727e/i2727e.pdf (accessed May 8, 2017).

Food and Agriculture Organization, 2014. The state of world fisheries and aquaculture 2014. Available from: http://www.fao.org/3/a-i3720e.pdf (accessed May 8, 2017).

Food and Agriculture Organization, 2016. Fishery Statistical Collections. Global Aquaculture Production. [Several search terms: Asia, America, China, Japan, Taiwan, Korea, *Eriocheir sinensis, Penaeus monodon, P. vannamei*.] Available from: http://www.fao.org/fishery/statistics/global-aquaculture-production/en (accessed May 8, 2017).

Ford, S.E., 1996. Range extension by the oyster parasite *Perkinsus marinus* into the northeastern United States: response to climate change? J.Shellfish Res. 15, 45–56.

Ford, S.E., Borrero, F.J., 2001. Epizootiology and pathology of juvenile oyster disease in the eastern oyster, *Crassostrea virginica*. J. Invertebr. Pathol. 78, 141–154.

Ford, S.E., Haskin, H.H., 1987. Infection and mortality patterns in strains of oysters *Crassostrea virginica* selected for resistance to the parasite *Haplosporidium nelsoni* (MSX). J. Parasitol. 73, 368–376.

Ford, S.E., Haskin, H.H., 1988. Management strategies for MSX (*Haplosporidium nelsoni*) disease in eastern oysters. Am. Fish. Soc. Spec. Publ, 18, 249–256.

Ford, S.E., Kraeuter, J.N., Barber, R.D., Mathis, G., 2002. Aquaculture-associated factors in QPX disease of hard clams: density and seed source. Aquaculture 208, 23–38.

Frank-Lawale, A., Allen, S.K. Jr., Degremont, L., 2014. Breeding and domestication of Eastern oyster (*Crassostrea virginica*) lines for culture in the mid-Atlantic, USA: line development and mass selection for disease resistance. J. Shellfish Res. 33, 153–165.

Friedman, C.S., Thomson, M., Chun, C., Haaker, P.L., Hedrick, R.P., 1997. Withering syndrome of the black abalone, *Haliotis cracherodii* (Leach): water temperature, food availability, and parasites as possible causes. J. Shellfish Res. 16, 403–412.

Friedman, C.S., Andree, K.B., Beauchamp, K.A., Moore, J.D., Robbins, T.T., Shields, J.D., Hedrick, R.P., 2000. "Candidatus *Xenohaliotis californiensis*," a newly described pathogen of abalone, *Haliotis* spp., along the west coast of North America. Int. J. Syst. Evol. Microbiol. 50, 847–855.

Friedman, C.S., Estes, R.M., Stokes, N.A., Burge, C.A., Hargove, J.S., Barber, B.J., et al., 2005. Herpes virus in juvenile Pacific oysters *Crassostrea gigas* from Tomales Bay, California, coincides with summer mortality episodes. Dis Aquat Org. 63, 33–41.

Gaino, E., Pronzato, R., Corriero, G., Buffa, P., 1992. Mortality of commercial sponges: incidence in two Mediterranean areas. Ital. J. Zool. 59, 79–85.

Galtsoff, P.S., Brown, H.H., Smith, C.L., Smith, F.W., 1939. Sponge mortality in the Bahamas. Nature 143, 807–808.

Gammill, E.R., Fenner, D., 2005. Disease threatens Caribbean sponges: report and identification guide. Available from: http://www.reefbase.org/resource_center/publication/pub_24912.aspx (accessed May 8, 2017).

Geiser, D.M., Taylor, J.W., Ritchie, K.B., Smith G.W., 1998. Cause of sea fan death in the West Indies. Nature 394, 137–138.

Gladfelter, W.B., 1982. White-band disease in *Acropora palmate*: implications for the structure and growth of shallow reefs. Bull. Mar. Sci. 32, 639–643.

Glenn, R.P., Pugh, T.L., 2006. Epizootic shell disease in American lobster (*Homarus americanus*) in Massachusetts coastal waters: interactions of temperature, maturity, and intermolt duration. J. Crustacean Biol. 26, 639–645.

Goggin, C.L., Lester, R.J.G., 1995. *Perkinsus*, a protistan parasite of abalone in Australia: a review. Mar. Freshw. Res. 46, 639–646.

Granja, C.B., Vidal, O.M., Parra, G., Salazar, M., 2006. Hyperthermia reduces viral load of white spot syndrome virus in *Penaeus vannamei*. Dis. Aquat. Org. 68, 175–180.

Grizel, H., 1997. Les maladies des mollusques bivalves: Risqus et prevention. Rev. Sci. Tech. Off. Int. Epiz. 16, 161–171.

Grizel, H., Héral, M., 1991. Introduction into France of the Japanese oyster (*Crassostrea gigas*). ICES J. Mar. Sci. 47, 399–403.

Groner, M.L., Maynard, J., Breyta, R., Carnegie, R.B., Dobson, A., Friedman, C.S., et al., 2016. Managing marine disease emergencies in an era of rapid change. Frontiers in ecology and environment. Philos. Trans. R. Soc. Lond. B Biol. Sci. 371, 20150365.

Guo, Z.X., He, J.G., Xu, H.D., Weng, S.P., 2013. Pathogenicity and complete genome sequence analysis of the mud crab dicistrovirus-1. Virus Res. 171, 8–14.

Harvell, D., Jordán-Dahlgren, E., Merkel, S., Rosenberg, E., Raymundo, L., Smith, G.W., et al., 2007. Coral disease, environmental drivers, and the balance between coral and microbial associates. Oceanography 20, 172–195.

Haskin, H.H., S.E. Ford. 1979. Development of resistance to *Minchinia nelsoni* (MSX) mortality in laboratory-reared and native oyster stocks in Delaware Bay. Mar. Fish. Rev. 41, 54–63.

Hasson, K.W., Fan, Y., Reisinger, T., Venuti, J., Varner, P.W., 2006. White-spot syndrome virus (WSSV) introduction into the Gulf of Mexico and Texas freshwater systems through imported, frozen bait-shrimp. Dis. Aquat. Org. 71, 91–100.

Hauton, C., 2012. The scope of the crustacean immune system for disease control. J. Invert. Pathol. 110, 251–260.

Hawkes, C.R., Meyers, T.R., Shirley, T.C., Koeneman, T.M., 1986. Prevalence of the parasitic barnacle *Briarosaccus callosus* on king crabs of southeastern Alaska. Trans. Am. Fish. Soc. 115, 252–257.

Herborg, L.M., Rushton, S.P., Clare, A.S., Bentley, M.G., 2003. Spread of the Chinese mitten crab (*Eriocheir sinensis* H. Milne Edwards) in Continental Europe: analysis of a historical data set, in: Jones, M.B., Ingólfsson, A., Ólafsson, E., Helgason, G.V., Gunnarsson, K., Svavarsson, J. (eds.), Migrations and Dispersal of Marine Organisms. Springer, Amsterdam, pp. 21–28.

Hewson, I., Button, J.B., Gudenkauf, B.M., Miner, B., Newton, A.L., Gaydos, J.K., et al., 2014. Densovirus associated with sea-star wasting disease and mass mortality. Proc. Nat. Acad. Sci. 111, 17 278–17 283.

Hine, P.M., 1996. Southern hemisphere utbrea diseases and an overview of associated risk assessment problems. Rev. Sci. Tech. 15, 563–577.

Hoenig, J.M., Groner, M.L., Smith, M.W., Vogelbein, W.K., Taylor, D.M., Landers Jr., D.F., et al., 2017. Impact of disease on survival of three commercially fished species. Ecol. Appl. In press.

Holdich, D.M., 1999. The negative effects of established crayfish introductions. Crust. Issues 11, 31–48.

Holdich D.M., 2003. Crayfish in Europe – an overview of taxonomy, legislation, distribution, and crayfish plague outbreaks, in: Holdich D.M., Sibley P.J. (eds.), Management & Conservation of Crayfish, Proceedings of a Conference held in Nottingham on 7th November 2002. Environment Agency, Bristol, pp. 15–34.

Holdich, D.M., Reynolds, J.D., Souty-Grosset, C., Sibley, P.J., 2009. A review of the ever increasing threat to European crayfish from non-indigenous crayfish species. Knowl. Manag. Aquat. Ecosys. 11, 1–46.

Hooper, C., Hardy-Smith, P., Handlinger, J., 2007. Ganglioneuritis causing high mortalities in farmed Australian abalone (*Haliotis laevigata* and *Haliotis rubra*). Austral. Vet. J. 85, 188–193.

Howell, P., 2012. The status of the southern New England lobster stock. J. Shellfish Res. 31, 573–579.

Howell, P., Gianni, C., Benway, J., 2005. Status of shell disease in Long Island Sound, in: Tlusty, M.F., Halvorson, H.O., Smolowitz, R., Sharma, U. (eds.), Lobster Shell Disease

Workshop. Aquatic Forum Series 05-1. New England Aquarium, Boston, MA, pp. 106–114.

Huchin-Mian, J.P., Rodríguez-Canul, R., Arias-Bañuelos, E., Simá-Álvarez, R., Pérez-Vega, J.A., Briones-Fourzán, P., Lozano-Álvarez, E., 2008. Presence of Panulirus argus Virus 1 (PaV1) in juvenile spiny lobsters *Panulirus argus* from the Caribbean coast of Mexico. Dis. Aquat. Org. 79, 153–156.

Huchin-Mian, J.P., Briones-Fourzán, P., Simá-Álvarez, R., Cruz-Quintana, Y., Pérez-Vega, J.A., Lozano-Álvarez, E., et al., 2009. Detection of Panulirus argus Virus 1 (PaV1) in exported frozen tails of subadult-adult Caribbean spiny lobsters *Panulirus argus*. Dis. Aquat. Org. 86, 159–162.

Huchin-Mian, J.P., Rodríguez-Canul, R., Briones-Fourzán, P., Lozano-Álvarez, E., 2013. Panulirus argus virus 1 (PaV1) infection prevalence and risk factors in a Mexican lobster fishery employing casitas. Dis. Aquat. Org. 107, 87–97.

Jacobs, M., Laufer, H., Stuart, J., Chen, M., Pan, X. 2012. Endocrine disrupting alkylphenols are widespread in the blood of lobsters from southern New England and adjacent offshore areas. J. Shellfish Res. 31, 563–572.

Jenkins, C., Hick, P., Gabor, M., Spiers, Z., Fell, S.A., Gu, X., et al., 2013. Identification and characterisation of an ostreid herpesvirus-1 microvariant (OsHV-1 μ-var) in *Crassostrea gigas* (Pacific oysters) in Australia. Dis. Aquat. Org. 105, 109–126.

Johnson, P.T., 1986. Parasites of benthic amphipods: dinoflagellates (Duboscquodinida: Syndinidae). Fish. Bull. 84, 605–614.

Johnson, P.T., Paull, S.H., 2011. The ecology and emergence of diseases in fresh waters. Freshwater Biol. 56, 638–657.

Jordan, S.J., Coakley, J.M., 2004. Long-term projections of eastern oyster populations under various management scenarios. J. Shellfish Res. 23, 63–73.

Joshi, J., Srisala, J., Truong, V.H., Chen, I.T., Nuangsaeng, B., Suthienkul, O., et al., 2014. Variation in *Vibrio parahaemolyticus* isolates from a single Thai shrimp farm experiencing an outbreak of acute hepatopancreatic necrosis disease (AHPND). Aquaculture 428, 297–302.

Jussila, J., Makkonen, J., Kokko, H., 2011. Peracetic acid (PAA) treatment is an effective disinfectant against crayfish plague (*Aphanomyces astaci*) spores in aquaculture. Aquaculture 320, 37–42.

Krantz, G.E., Jordan, S.J., 1996. Management alternatives for protecting *Crassostrea virginica* fisheries in *Perkinsus marinus* enzootic and epizootic areas. J. Shellfish Res. 15, 167–176.

Kuris AM., 1993. Life cycles of nemerteans that are symbiotic egg predators of decapod Crustacea: adaptations to host life histories. Hydrobiologia 266, 1–14.

Kuris, A.M., Lafferty, K.D., 1992. Modelling crustacean fisheries: effects of parasites on management strategies. Can. J. Fish. Aquat. Sci. 49, 327–336.

Kuris, A.M., Blau, S.F., Paul, A.J., Shields, J.D., Wickham, D.E., 1991. Infestation by brood symbionts and their impact on egg mortality in the red king crab, *Paralithodes camtschatica*, in Alaska, Geographic and temporal variation. Can. J. Fish. Aquat. Sci. 48, 559–568.

Lacoste, A., Jalabert, F., Malham, S., Cueff, A., Gelebart, F., Cordevant, C., et al., 2001. A *Vibrio splendidus* strain is associated with summer mortality of juvenile oysters *Crassostrea gigas* in the Bay of Morlaix (North Brittany, France). Dis. Aquat. Org. 46, 139–145.

Lafferty, K.D., Harvell, C.D., Conrad, J.M., Friedman, C.S., Kent, M.L., Kuris, A.M., et al., 2015. Infectious diseases affect marine fisheries and aquaculture economics. Ann. Rev. Mar. Sci. 7, 471–496.

Lai, H.C., Ng, T.H., Ando, M., Lee, C.T., Chen, I.T., Chuang, J.C., Mavichak, R., et al., 2015. Pathogenesis of acute hepatopancreatic necrosis disease (AHPND) in shrimp. Fish Shellfish Immunol. 47, 1006–1014.

Lauckner, G., 1983. Diseases of Mollusca: Bivalvia, in: Kinne, O. (ed.), Diseases of Marine Animals, II. Biologische Anstalt Helgoland, Helgoland, pp. 477–961.

Laufer, H., Demir, N., Pan, X., 2005. Shell disease in the American lobster and its possible relation to alkylphenols, in: Tlusty, M.F., Halvorson, H.O., Smolowitz, R., Sharma, U. (eds.), Lobster Shell Disease Workshop. Aquatic Forum Series 05-1. New England Aquarium, Boston, MA, pp. 72–75.

Laufer, H., Chen, M., Johnson, M., Demir, N., Bobbitt, J.M., 2012. The effect of alkylphenols on lobster shell hardening. J. Shellfish Res. 31, 555–562.

Lavallée, J., Hammell, K.L., Spangler, E.S., Cawthorn, R.J., 2001. Estimated prevalence of *Aerococcus viridans* and *Anophryoides haemophila* in American lobsters *Homarus americanus* freshly captured in the waters of Prince Edward Island, Canada. Dis. Aquat. Org. 46, 231–236.

Lee, C.T., Chen, I.T., Yang, Y.T., Ko, T.P., Huang, Y.T., Huang, J.Y., et al., 2015. The opportunistic marine pathogen *Vibrio parahaemolyticus* becomes virulent by acquiring a plasmid that expresses a deadly toxin. Proc. Nat. Acad. Sci. U.S.A. 112, 10798–10803.

Lee, R.F., Frischer, M.E., 2004. Where have the blue crabs gone, evidence for a drought induced epidemic of a protozoan disease. Am. Sci. 92, 547–553.

Lessios, H.A., 2005. *Diadema antillarum* populations in Panama twenty years following mass mortality. Coral Reefs 24, 125–127.

Lessios, H.A., Robertson, D.R., Cubit, J.D., 1984. Spread of *Diadema* mass mortality through the Caribbean. Science 226, 335–337.

Li, C., Shields, J.D., 2007. Primary culture of hemocytes from the Caribbean spiny lobster, *Panulirus argus*, and their susceptibility to Panulirus argus Virus 1 (PaV1). J. Invert. Pathol. 94, 48–55.

Li, C., Song, S., Liu, Y., Chen, T., 2013. *Hematodinium* infections in cultured Chinese swimming crab, *Portunus trituberculatus*, in northern China. Aquaculture 396, 59–65.

Li, F., Xiang, J., 2013. Recent advances in researches on the innate immunity of shrimp in China. Develop. Comp. Immunol. 39, 11–26.

Li, W., Shi, Y., Gao, L., Liu, J., Cai, Y., 2012. Occurrence of antibiotics in water, sediments, aquatic plants, and animals from Baiyangdian Lake in North China. Chemosphere 89, 1307–1315.

Liang, T., Feng, Q., Wu, T., Gu, W., Wang, W., 2009. Use of oxytetracycline for the treatment of tremor disease in the Chinese mitten crab *Eriocheir sinensis*. Dis. Aquat. Org. 84, 243–250.

Lightner, D.V., 1996. Epizootiology, distribution and the impact on international trade of two penaeid shrimp viruses in the Americas. Rev. Sci. Tech. 15, 579–601.

Lightner, D.V., 1999. The penaeid shrimp viruses TSV, IHHNV, WSSV, and YHV: current status in the Americas, available diagnostic methods, and management strategies. J. Appl. Aquacult. 9, 27–52.

Lightner, D.V., 2003. The penaeid shrimp viral pandemics due to IHHNV, WSSV, TSV and YHV: history in the Americas and current status, in: Sakai, Y., McVey, J.P., Jang, D.,

McVey, E., Caesar, M. (eds.), Proceedings of the Thirty-second US Japan Symposium on Aquaculture. US–Japan Cooperative Program in Natural Resources. US Department of Commerce, NOAA, Silver Spring, MD, pp. 6–24.

Lightner, D.V., 2005. Biosecurity in shrimp farming: pathogen exclusion through use of SPF stock and routine surveillance. J. World Aquacult. Soc. 36, 229–248.

Lightner, D.V., 2011. Virus diseases of farmed shrimp in the Western Hemisphere (the Americas): a review. J. Invert. Pathol. 106, 110–130.

Lightner, D.V., Redman, R.M., 1998. Shrimp diseases and current diagnostic methods. Aquaculture 164, 201–220.

Lightner, D.V., Redman, R.M., 2010. The global status of significant infectious diseases of farmed shrimp. Asian Fish. Sci. 23, 383–426.

Lightner, D.V., Redman, R.M., Bell, T.A., 1983. Infectious hypodermal and hematopoietic necrosis, a newly recognized virus disease of penaeid shrimp. J. Invert. Pathol. 42, 62–70.

Lightner, D.V., Redman, R.M., Hasson, K.W., Pantoja, C.R., 1995. Taura syndrome in *Penaeus vannamei* (Crustacea: Decapoda): gross signs, histopathology and ultrastructure. Dis. Aquat. Org. 21, 53–53.

Lightner, D.V., Redman, R.M., Arce, S., Moss, S.M., 2009. Specific pathogen-free shrimp stocks in shrimp farming facilities as a novel method for disease control in crustaceans, in: Shumway, S.E., Rodrick, G.E. (eds.), Shellfish Safety and Quality. CSIRO Press, Clayton, pp.384–424.

Lightner, D.V., Redman, R.M., Pantoja, C.R., Tang, K.F.J., Noble, B.L., Schofield, P., et al., 2012a. Historic emergence, impact and current status of shrimp pathogens in the Americas. J. Invert. Pathol. 110, 174–183.

Lightner, D.V., Redman, R.M., Pantoja, C.R., Noble, B.L., Tran, L., 2012b. Early mortality syndrome affects shrimp in Asia. Global Aquaculture Advocate, 15, 40.

Lin, Y.R., Hung, H.C., Leu, J.H., Wang, H.C., Kou, G.H., Lo, C.F., 2011. The role of aldehyde dehydrogenase and hsp70 in suppression of white spot syndrome virus replication at high temperature. J. Virol. 85, 3517–3525.

Longshaw, M., 2011. Diseases of crayfish: a review. J. Invert. Pathol. 106, 54–70.

Loy, J.D., Mogler, M.A., Loy, D.S., Janke, B., Kamrud, K., Scura, E.D., et al., 2012. dsRNA provides sequence-dependent protection against infectious myonecrosis virus in Litopenaeus vannamei. J. Gen. Virol. 93, 880–888.

Loy, J.D., Loy, D.S., Mogler, M.A., Janke, B., Kamrud, K., Harris, D.L., Bartholomay, L.C., 2013. Sequence-optimized and targeted double-stranded RNA as a therapeutic antiviral treatment against infectious myonecrosis virus in *Litopenaeus vannamei*. Dis. Aquat. Org. 105, 57–64.

Lozano-Álvarez, E., Briones-Fourzán, P., Ramírez-Estévez, A., Placencia-Sánchez, D., Huchin-Mian, J.P., Rodríguez-Canul, R., 2008. Prevalence of Panulirus argus virus 1 (PaV1) and habitation patterns of healthy and diseased Caribbean spiny lobsters in shelter-limited habitats. Dis. Aquat. Org. 80, 95–104.

Lozano-Álvarez, E., Briones-Fourzán, P., Huchin-Mian, J.P., Segura-García, I., Ek-Huchim, J.P., Améndola-Pimenta, M., Rodríguez-Canul, R., 2015. Panulirus argus virus 1 detected in oceanic postlarvae of Caribbean spiny lobster: implications for disease dispersal. Dis. Aquat. Org. 117, 165–170.

Lundin, C.G., 1996 [dated October 1995]. Global attempts to address shrimp disease. Land, Water and Natural Habitats Division, Environment Department, The World Bank.

Available from: http://library.enaca.org/Shrimp/Publications/ShrimpDisease.pdf (accessed May 8, 2017).

Lyons, M.M., Smolowitz, R.M., Gomez-Chiarri, M., Ward, J.E., 2007. Epizootiology of quahog parasite unknown (QPX) disease in northern quahogs (= hard clams) *Mercenaria mercenaria*. J. Shellfish Res. 26, 371–381.

Magel, C.R., Shields, J.D., Brill, R.W., 2009. Idiopathic lesions and visual deficits in the American lobster (*Homarus americanus*) from Long Island Sound, NY. Biol. Bull. 217, 95–101.

Maldonado, M., Sánchez-Tocino, L., Navarro, C., 2010. Recurrent disease outbreaks in corneous demosponges of the genus *Ircinia*: epidemic incidence and defense mechanisms. Mar. Biol. 157, 1577–1590.

Maloy, A.P., Ford, S.E., Karney, R.C., Boettcher, K.J., 2007. *Roseovarius crassostreae*, the etiological agent of Juvenile Oyster Disease (now to be known as Roseovarius Oyster Disease) in *Crassostrea virginica*. Aquaculture 269, 71–83.

Maniscalco, A.M., Shields, J.D., 2006. Histopathology of idiopathic lesions in the eyes of *Homarus americanus* from Long Island Sound. J. Invert. Pathol. 91, 88–97.

Mari, J., Bonami, J.R., 1986. Les infections virales du crabe *Carcinus mediterraneus* Czerniavski, 1984, in: Vivares, C.P., Bonami, J. R., Jasper, E. (eds.), Pathology in Marine Aquaculture. European Aquaculture Society, Oostende, pp. 283–293.

Maynard, J., van Hooidonk, R., Harvell, C.D., Eakin, C.M., Liu, G., Willis, B.L., et al., 2016. Improving marine disease surveillance through sea temperature monitoring, outlooks and projections. Philos. Trans. R. Soc. Lond. B Biol. Sci.l 371, 20150208.

McCallum, H.I., Kuris, A., Harvell, C.D., Lafferty, K.D., Smith, G.W., Porter, J., 2004. Does terrestrial epidemiology apply to marine systems? Trends Ecol. Evol. 19, 585–591.

McClanahan, T.R., Muthiga, N.A., 1998. An ecological shift in a remote coral atoll of Belize over 25 years. Environm. Conserv. 25, 122–130.

Menard, J., Myrand, B., 1987. L'incidence d'*Aerococcus viridans* var. *homari* dans le stock naturel de homards (*Homarus americanus*) des Iles de la Madeleine (Quebec), suite a une epidemie de gaffkemie dans un elevage in situ. Can. J. Fish. Aquat. Sci. 44, 368–372.

Meng, Q., Gu, W., Bi, K., Ji, H., Wang, W., 2010a. Spiralin-like protein SLP31 from *Spiroplasma eriocheiris* as a potential antigen for immunodiagnostics of tremor disease in Chinese mitten crab *Eriocheir sinensis*. Folia Microbiologica 55, 245–250.

Meng, Q., Li, W., Liang, T., Jiang, X., Gu, W., Wang, W., 2010b. Identification of adhesin-like protein ALP41 from *Spiroplasma eriocheiris* and induction immune response of *Eriocheir sinensis*. Fish Shellfish Immunol. 29, 587–593.

Messick, G.A., 1994. *Hematodinium* perezi infections in adult and juvenile blue crabs *Callinectes sapidus* from coastal bays of Maryland and Virginia, USA. Dis. Aquat. Org. 19, 77–82.

Messick, G.A., Shields, J.D., 2000. Epizootiology of the parasitic dinoflagellate *Hematodinium* sp. in the American blue crab *Callinectes sapidus*. Dis. Aquat. Org. 43, 139–152.

Meuriot, E., Grizel, H., 1984. Note sur l'impact économique des maladies de l'huître plate en Bretagne. Rapports Techniques de l'Institut Scientifique et Technique des Pêches Maritimes 12, 1–20.

Meyers, T.R., 1990. Diseases caused by protistans, in: Kinne, O. (ed.), Diseases of Marine Animals, Vol. III, Diseases of Crustacea. Biologische Anstalt Helgoland, Helgoland, pp. 350–368.

Meyers, T.R., Koeneman, T.M., Botelho, C., Short, S., 1987. Bitter crab disease, a fatal dinoflagellate infection and marketing problem for Alaskan Tanner crabs *Chionoecetes bairdi*. Dis. Aquat. Org. 3, 195–216.

Meyers, T.R., Botelho, C., Koeneman, T.M., Short, S., Imamura, K., 1990. Distribution of bitter crab dinoflagellate syndrome in southeast Alaskan Tanner crabs *Chionoecetes bairdi*. Dis. Aquat. Org. 9, 37–43.

Meyers, T.R., Morado, J.F., Sparks, A.K., Bishop, G.H., Pearson, T., Urban, D., Jackson, D., 1996. Distribution of bitter crab syndrome in tanner crabs (*Chionoecetes bairdi, C. opilio*) from the Gulf of Alaska and the Bering Sea. Dis. Aquat. Org. 26, 221–227.

Miller, R.J., Colodey, A.G., 1983. Widespread mass mortalities of the green sea urchin in Nova Scotia, Canada. Mar. Biol. 73, 263–267.

Mohr, P.G., Moody, N.J., Hoad, J., Williams, L.M., Bowater, R.O., Cummins, D.M., et al., 2015. New yellow head virus genotype (YHV7) in giant tiger shrimp *Penaeus monodon* indigenous to northern Australia. Dis. Aquat. Org. 115, 263–268.

Morales-Covarrubias, M.S., Nunan, L.M., Lightner, D.V., Mota-Urbina, J.C., Garza-Aguirre, M.C., Chávez-Sánchez, M.C., 1999. Prevalence of Infectious Hypodermal and Hematopoietic Necrosis Virus (IHHNV) in wild adult blue shrimp *Penaeus stylirostris* from the northern Gulf of California, Mexico. J. Aquat. Anim. Health 11, 296–301.

Moss, J., Butler, M.J., Behringer, D.C., Shields, J.D., 2011. Genetic diversity of the Caribbean spiny lobster virus, Panulirus argus virus 1 (PaV 1), and the discovery of PaV 1 in lobster postlarvae. Aquat. Biol. 14, 223–232.

Moss, J., Behringer, D., Shields, J.D., Baeza, A., Aguilar-Perera, A., Bush, P.G., et al., 2013. Distribution, prevalence, and genetic analysis of Panulirus argus virus 1 (PaV1) from the Caribbean Sea. Dis. Aquat. Org. 104, 129–140.

Moss, S.M., Moss, D.R., 2009. Selective breeding of penaeid shrimp, in: Shumway, S.E., Rodrick, G.E. (eds.), Shellfish Safety and Quality. CRC Press, Boca Raton, FL, pp. 425–452.

Moss, S.M., Moss, D.R., Arce, S.M., Lightner, D.V., Lotz, J.M., 2012. The role of selective breeding and biosecurity in the prevention of disease in penaeid shrimp aquaculture. J. Invert. Pathol. 110, 247–250.

Moyer, M.A., Blake, N.J., Arnold, W.S., 1993. An ascetosporan disease causing mass mortality in the Atlantic calico scallop, *Argopecten gibbus* (Linnaeus, 1758). J. Shellfish Res. 12, 305–310.

Mullen, T.E., Russell, S., Tucker, M.T., Maratea, J.L., Koerting, C., Hinckley, L., et al., 2004. Paramoebiasis associated with mass mortality of American lobster *Homarus americanus* in Long Island Sound, USA. J. Aquat. Anim. Health 16, 29–38.

Mullowney, D.R., Dawe, E.G., Morado, J.F., Cawthorn, R.J., 2011. Sources of variability in prevalence and distribution of bitter crab disease in snow crab (*Chionoecetes opilio*) along the northeast coast of Newfoundland. ICES J. Mar. Sci. 68, 463–471.

Nagelkerken, I., Buchan, K., Smith, G.W., Bonair, K., Bush, P., Garzon-Ferreira, J., et al., 1997. Widespread disease in Caribbean sea fans: II. Patterns of infection and tissue loss. Mar. Ecol. Prog. Ser. 160, 255–263.

National Marine Fisheries Service, 2016. Commercial fisheries statistics. Available from: http://www.st.nmfs.noaa.gov/st1/commercial/index.html (accessed May 8, 2017).

Nielsen, L., Sang-oum, W., Cheevadhanarak, S., Flegel, T.W., 2005. Taura syndrome virus (TSV) in Thailand and its relationship to TSV in China and the Americas. Dis. Aquat. Org. 63, 101–106.

Nunan, L., Lightner, D., Pantoja, C., Gomez-Jimenez, S., 2014. Detection of acute hepatopancreatic necrosis disease (AHPND) in Mexico. Dis. Aquat. Org. 111, 81–86.

Oidtmann, B., Schaefers, N., Cerenius, L., Söderhäll, K., Hoffmann, R.W., 2004. Detection of genomic DNA of the crayfish plague fungus *Aphanomyces astaci* (Oomycete) in clinical samples by PCR. Vet. Microbiol. 100, 269–282.

Oidtmann, B., Geiger, S., Steinbauer, P., Culas, A., Hoffmann, R.W., 2006. Detection of *Aphanomyces astaci* in North American crayfish by polymerase chain reaction. Dis. Aquat. Org. 72, 53–64.

Onstad, D.W., Fuxa, J.R., Humber, R.A., Oestergaard, J., Shapiro Ilan, D.I., Gouli, V.V., Anderson, R.S., et al., 2006. An Abridged Glossary of Terms Used in Invertebrate Pathology, 3rd edn. Available from: http://www.sipweb.org/resources/glossary.html (accessed May 8, 2017).

Pagenkopp Lohan, K.M., Reece, K.S., Miller, T.L., Wheeler, K.N., Small, H.J., Shields, J.D., 2012. The role of alternate hosts in the ecology and life history of *Hematodinium* sp., a parasitic dinoflagellate of the blue crab (*Callinectes sapidus*). J. Parasitol. 98, 73–84.

Paillard, C., Maes, P., Oubella, R., 1994. Brown ring disease in clams. Ann. Rev. Fish Dis. 4, 219–240.

Paillard, C., Le Roux, F., Borrego, J.J., 2004. Bacterial disease in marine bivalves, a review of recent studies: trends and evolution. Aquat. Living Resour. 17, 477–498.

Panphut, W., Senapin, S., Sriurairatana, S., Withyachumnarnkul, B., Flegel, T.W., 2011. A novel integrase-containing element may interact with Laem-Singh virus (LSNV) to cause slow growth in giant tiger shrimp. BMC Vet. Res. 7, 18.

Pantoja, C.R., Lightner, D.V., Holtschmit, K.H., 1999. Prevalence and geographic distribution of Infectious Hypodermal and Hematopoietic Necrosis Virus (IHHNV) in wild blue shrimp *Penaeus stylirostris* from the Gulf of California, Mexico. J. Aquat. Anim. Health 11, 23–34.

Park, K.I., Choi, K.S., 2001. Spatial distribution of the protozoan parasite *Perkinsus* sp. found in the Manila clams, *Ruditapes philippinarum*, in Korea. Aquaculture 203, 9–22.

Pasharawipas, T., Flegel, T.W., 1994. A specific DNA probe to identify the intermediate host of a common microsporidian parasite of *Penaeus merguiensis* and *P. monodon*. Asian Fish. Sci, 7, 157–167.

Patterson, K.L., Porter, J.W., Ritchie, K.B., Polson, S.W., Mueller, E., Peters, E.C., et al., 2002. The etiology of white pox, a lethal disease of the Caribbean elkhorn coral, *Acropora utbrea*. Proc. Natl. Acad. Sci. U.S.A. 99, 8725–8730.

Pearce, J., Balcom, N., 2005. The 1999 Long Island Sound lobster mortality event: findings of the comprehensive research initiative. J. Shellfish Res. 24, 691–698.

Pernet, F., Barret, J., Le Gall, P., Corporeau, C., Dégremont, L., Lagarde, F., et al., 2012. Mass mortalities of Pacific oysters *Crassostrea gigas* reflect infectious diseases and vary with farming practices in the Mediterranean Thau lagoon, France. Aquacult. Environ. Interactions 2, 215–237.

Perrigault, M., Buggé, D.M., Allam, B., 2010. Effect of environmental factors on survival and growth of quahog parasite unknown (QPX) in vitro. J. Invertebr. Pathol. 104, 83–89.

Pestal, G.P., Taylor, D.M., Hoenig, J.M., Shields, J.D., Pickavance, R., 2003. Monitoring the presence of the lethal parasite *Hematodinium* sp. in snow crabs from Newfoundland. Dis. Aquat. Org. 53, 67–75.

Plowright, R.K., Sokolow, S.H., Gorman, M.E., Daszak, P., Foley, J.E., 2008. Causal inference in disease ecology: investigating ecological drivers of disease emergence. Frontiers Ecol. Environ. 6, 420–429.

Poulin, R., 2011. The many roads to parasitism: a tale of convergence. Adv. Parasitol. 74, 1–40.

Poulin, R., Blasco-Costa, I., Randhawa, H.S., 2016. Integrating parasitology and marine ecology: seven challenges towards greater synergy. J. Sea Res. 113, 3–10.

Poulos, B.T., Tang, K.F., Pantoja, C.R., Bonami, J.R., Lightner, D.V., 2006. Purification and characterization of infectious myonecrosis virus of penaeid shrimp. J. Gen. Virol. 87, 987–996.

Powell, E.N., Klinck, J.M., Hofmann, E.E., Ford, S., 1998. Varying the timing of oyster transplant: implications for management from simulation studies. Fish. Oceanogr. 6, 213–237.

Pratoomthai, B., Sakaew, W., Sriurairatana, S., Wongprasert, K., Withyachumnarnkul, B., 2008. Retinopathy in stunted black tiger shrimp *Penaeus monodon* and possible association with Laem-Singh virus (LSNV). Aquaculture 284, 53–58.

Pronzato, R., 1999. Sponge-fishing, disease and farming in the Mediterranean Sea. Aquat. Conserv: Mar. Freshw. Ecosys. 9, 485–493.

Ragone Calvo, L.M., Ford, S.E., Kraeuter, J.N., Leavitt, D.F., Smolowitz, R., Burreson, E.M., 2007. Influence of host genetic origin and geographic location on QPX disease in Northern quahogs (= hard clams), *Mercenaria mercenaria*. J. Shellfish Res. 26, 109–119.

Rai, P., Pradeep, B., Karunasagar, I., Karunasagar, I., 2009. Detection of viruses in *Penaeus monodon* from India showing signs of slow growth syndrome. Aquaculture 289, 231–235.

Renault, T., Le Deuff, R.M., Cochennec, N., Maffart, P., 1994. Herpesviruses associated with mortalities among Pacific oyster, *Crassostrea gigas*, in France – comparative study. Rev. Méd. Vétérin. 45, 735–742.

Richardson, L.L., Goldberg, W.M., Kuta, K.G., Aronson, R.B., Smith, G.W., Ritchie, K.B., et al., 1998. Florida's mystery coral-killer identified. Nature 392, 557–558.

Robalino, J., Bartlett, T., Shepard, E., Prior, S., Jaramillo, G., Scura, E., et al., 2005. Double-stranded RNA induces sequence-specific antiviral silencing in addition to nonspecific immunity in a marine shrimp: convergence of RNA interference and innate immunity in the invertebrate antiviral response? J. Virol. 79, 13561–13571.

Rudolf, V.H., Antonovics, J., 2007. Disease transmission by cannibalism: rare event or common occurrence? Proc. Biol. Sci. 274, 1205–1210.

Rutzler, K., Santavy, D.L., Antonius, A., 1983. The black band disease of Atlantic reef corals. Mar. Ecol. 4, 329–358.

Saksmerprome, V., Thammasorn, T., Jitrakorn, S., Wongtripop, S., Borwornpinyo, S., Withyachumnarnkul, B., 2013. Using double-stranded RNA for the control of Laem-Singh Virus (LSNV) in Thai *P. monodon*. J. Biotechnol. 164, 449–453.

Scheibling, R.E., Hennigar, A.W., 1997. Recurrent outbreaks of disease in sea urchins *Strongylocentrotus droebachiensis* in Nova Scotia: evidence for a link with large-scale meteorologic and oceanographic events. Mar. Ecol. Prog. Ser. 152, 155–165.

Segarra, A., Pépin, J.F., Arzul, I., Morga, B., Faury, N., Renault, T., 2010. Detection and description of a particular Ostreid herpesvirus 1 genotype associated with massive mortality outbreaks of Pacific oysters, *Crassostrea gigas*, in France in 2008. Virus Res. 153, 92–99.

Senapin, S., Phewsaiya, K., Briggs, M., Flegel, T.W., 2007. Outbreaks of infectious myonecrosis virus (IMNV) in Indonesia confirmed by genome sequencing and use of an alternative RT-PCR detection method. Aquaculture 266, 32–38.

Senapin, S., Thaowbut, Y., Gangnonngiw, W., Chuchird, N., Sriurairatana, S., Flegel, T.W., 2010. Impact of yellow head virus outbreaks in the whiteleg shrimp, *Penaeus vannamei* (Boone), in Thailand. J. Fish Dis. 33, 421–430.

Senapin, S., Phiwsaiya, K., Gangnonngiw, W., Flegel, T.W., 2011. False rumours of disease outbreaks caused by infectious myonecrosis virus (IMNV) in the whiteleg shrimp in Asia. J. Negat. Results Biomed. 10, 10.

Shields, J.D., 1994. The parasitic dinoflagellates of marine Crustacea. Ann. Rev. Fish Dis. 4, 241–271.

Shields, J.D., 2012. The impact of pathogens on exploited populations of decapod crustaceans. J. Invert. Pathol. 110, 211–224.

Shields, J.D., 2013. Complex etiologies of emerging diseases in lobsters (*Homarus americanus*) from Long Island Sound. Can. J. Fish. Aquat. Sci. 70, 1576–1587.

Shields, J.D., Behringer, D.C., 2004. A new pathogenic virus in the Caribbean spiny lobster *Panulirus argus* from the Florida Keys. Dis. Aquat. Org. 59, 109–118.

Shields, J.D., Overstreet, R.M., 2007. Parasites, symbionts, and diseases, in: Kennedy, V., Cronin, L.E. (eds.), The blue crab *Callinectes sapidus*. University of Maryland Sea Grant College, College Park, MD, pp. 299–417.

Shields, J.D., Squyars, C.M., 2000. Mortality and hematology of blue crabs, *Callinectes sapidus*, experimentally infected with the parasitic dinoflagellate *Hematodinium* perezi. Fish. Bull. NOAA 98, 139–152.

Shields, J.D., Wood, F.E.I., 1993. Impact of parasites on the reproduction and fecundity of the blue sand crab *Portunus pelagicus* from Moreton Bay, Australia. Mar. Ecol. Prog. Ser. 92, 159–170.

Shields, J.D., Taylor, D.M., Sutton, S.G., O'Keefe, P.G., Ings, D., Party, A., 2005. Epizootiology of bitter crab disease (*Hematodinium* sp.) in snow crabs, *Chionoecetes opilio* from Newfoundland, Canada. Dis. Aquat. Org. 64, 253–264.

Shields, J.D., Stephens, F.J., Jones, J.B., 2006. Pathogens, parasites and other symbionts, in: Phillips, B.F. (ed.), Lobsters: Biology, Management, Aquaculture and Fisheries. Blackwell Scientific, Oxford, pp. 146–204.

Shields, J.D., Taylor, D.M., O'Keefe, P.G., Colbourne, E., Hynick, E., 2007. Epidemiological determinants in outbreaks of bitter crab disease (*Hematodinium* sp.) in snow crabs, *Chionoecetes opilio* from Newfoundland, Canada. Dis. Aquat. Org. 77, 61–72.

Shields, J.S., Wheeler, K.N., Moss, J., 2012. Histological assessment of lobsters in the "100 Lobster" Project. J. Shellfish Res. 31, 439–447.

Shields, J.D., Williams, J.D., Boyko, C.B., 2015. Parasites and pathogens of Brachyura, in: Castro, P., Davie, P.J.F., Guinot, D, Schram, F.R., von Vaupel Klein, J.C. (eds.), Vol. 9B, The Crustacea. Treatise on Zoology/Traite de Zoologie. Brill, Leiden, pp. 639–774.

Shukalyuk, A.I., Isaeva, V.V., Pushchin, I.I., Dolganov, S.M., 2005. Effects of the *Briarosaccus callosus* infestation on the commercial golden king crab *Lithodes aequispina*. J. Parasitol. 91, 1502–1504.

Siddeek, M.S.M., Zheng, J., Morado, J.F., Kruse, G.H., Bechtol, W.R., 2010. Effect of bitter crab disease on rebuilding in Alaska Tanner crab stocks. ICES J. Marine Science 67, 2027–2032.

Sittidilokratna, N., Dangtip, S., Cowley, J.A., Walker, P.J., 2008. RNA transcription analysis and completion of the genome sequence of yellow head nidovirus. Virus Res. 136, 157–165.

Skadsheim, A., Christie, H., Leinaas, H.P., 1995. Population reductions of *Strongylocentrotus droebachiensis* (Echinodermata) in Norway and the distribution of its endoparasite *Echinomermella matsi* (Nematoda). Mar. Ecol. Prog. Ser. 119, 199–209.

Skurdal, J, Taugbøl, T., Burba, A., Edsman, L., Soderback, B., Styrrishave, B., Tuusti, J., Westman, K., 1999. Crayfish introductions in the Nordic and Baltic countries, in: Gherardi, F., Holdich, D.M. (eds.), Crayfish in Europe as Alien Species. How to Make the Best of a Bad Situation? A.A. Balkema, Rotterdam, pp. 193–219.

Sloan, N.A., 1984. Incidence and effects of parasitism by the rhizocephalan barnacle, *Briarosaccus callosus* Boschma, in the golden king crab, *Lithodes aequispina* Benedict, from deep fjords in northern British Columbia, Canada. J. Exptl. Mar. Biol. Ecol. 84, 111–131.

Small, H.J., 2012. Advances in our understanding of the global diversity and distribution of *Hematodinium* spp. – significant pathogens of commercially exploited crustaceans. J. Invert. Pathol. 110, 234–246.

Small, H.J., Shields, J.D., Reece, K.S., Bateman, K., Stentiford, G.D., 2012. Morphological and molecular characterization of *Hematodinium perezi* (Dinophyceae: Syndiniales), a dinoflagellate parasite of the harbour crab, *Liocarcinus depurator*. J. Eukaryotic Microbiol. 59, 54–66.

Smith, A.L., Hirschle, L., Vogan, C.L., Rowley, A.F., 2015. Parasitization of juvenile edible crabs (*Cancer pagurus*) by the dinoflagellate, *Hematodinium* sp.: pathobiology, seasonality and its potential effects on commercial fisheries. Parasitology 142, 428–438.

Smith, A.L., Rowley, A.F., 2015. Effects of experimental infection of juvenile edible crabs *Cancer pagurus* with the parasitic dinoflagellate *Hematodinium* sp. J. Shellfish Res. 34, 511–519.

Smith, G.W., Ives, L.D., Nagelkerken, I.A., Richie, K.B., 1996. Caribbean sea-fan mortalities. Nature 383, 487.

Smolowitz, R., Leavitt, D., Perkins, F., 1998. Observations of a protistan disease similar to QPX in *Mercenaria mercenaria* (hard clams) from the coast of Massachusetts. J. Invert. Pathol. 71, 9–25.

Smolowitz, R., Chistoserdov, A.Y., Hsu, A., 2005. A description of the pathology of epizootic shell disease in the American lobster, *Homarus americanus*, H. Milne Edwards 1837. J. Shellfish Res. 24, 749–756.

Sobecka, E., Hajek, G., Skorupiński, Ł., 2011. Four pathogens found associated with *Eriocheir sinensis* H. Milne-Edwards, 1853 (Crustacea: Brachyura: Grapsidae) from Lake Dąbie (Poland). Oceanological and Hydrobiological Studies 40, 96–99.

Soniat, T.M., 1996. Epizootiology of *Perkinsus marinus* disease of eastern oysters in the Gulf of Mexico. J. Shellfish Res. 15, 35–43.

Sookruksawong, S., Sun, F., Liu, Z., Tassanakajon, A., 2013. RNA-Seq analysis reveals genes associated with resistance to Taura syndrome virus (TSV) in the Pacific white shrimp *Litopenaeus vannamei*. Develop. Comp. Immunol. 41, 523–533.

Soto-Rodriguez, S.A., Gomez-Gil, B., Lozano-Olvera, R., Betancourt-Lozano, M., Morales-Covarrubias, M.S., 2015. Field and experimental evidence of *Vibrio parahaemolyticus* as the causative agent of acute hepatopancreatic necrosis disease of cultured shrimp

(*Litopenaeus vannamei*) in Northwestern Mexico. Appl. Environ. Microbiol. 81, 1689–1699.

Spink, J., Frayling, M., 2000. An assessment of post plague reintroduced native white clawed crayfish, *Austropotamobius pallipes*, on the Sherston Avon and Tetbury Avon, Wiltshire. Freshwater Forum 14, 59–69.

Sritunyalucksana, K., Apisawetakan, S., Boon-nat, A., Withyachumnarnkul, B., Flegel, T.W., 2006. A new RNA virus found in black tiger shrimp *Penaeus monodon* from Thailand. Virus Res., 118, 31–38.

Stabili, L., Cardone, F., Alifano, P., Tredici, S.M., Piraino, S., Corriero, G., Gaino, E., 2012. Epidemic mortality of the sponge *Ircinia variabilis* (Schmidt, 1862) associated to proliferation of a *Vibrio* bacterium. Microbial. Ecol. 64, 802–813.

Stebbing, P.D., Pond, M.J., Peeler, E., Small, H.J., Greenwood, S.J., Verner-Jeffreys, D., 2012. Limited prevalence of gaffkaemia (*Aerococcus viridans* var. *homari*) isolated from wild-caught European lobsters *Homarus gammarus* in England and Wales. Dis. Aquat. Org. 100, 159–167.

Stentiford, G.D., Shields, J.D., 2005. A review of the parasitic dinoflagellates *Hematodinium* species and *Hematodinium*-like infections in marine crustaceans. Dis. Aquat. Org. 66, 47–70.

Stentiford, G.D., Neil, D.M., Atkinson, R.J.A., 2001a. Alteration of burrow-related behaviour of the Norway lobster, *Nephrops norvegicus* during infection by the parasitic dinoflagellate *Hematodinium*. Mar. Freshw. Behav. Physiol. 34, 139–156.

Stentiford, G.D., Neil, D.M., Atkinson, R.J.A., 2001b. The relationship of *Hematodinium* infection prevalence in a Scottish *Nephrops norvegicus* population to seasonality, moulting and sex. ICES J. Mar. Sci. 58, 814–823.

Stentiford, G.D., Oidtmann, B., Scott, A., Peeler, E.J., 2010. Crustacean diseases in European legislation: implications for importing and exporting nations. Aquaculture 306, 27–34.

Stentiford, G.D., Neil, D.M., Peeler, E.J., Shields, J.D., Small, H.J., Flegel, T.W., et al., 2012. Disease will limit future food supply from the global crustacean fishery and aquaculture sectors. J. Invert. Pathol. 110, 141–157.

Stewart, J.E., 1980. Diseases, in: Cobb, J.S., Phillips, B.F. (eds.), The Biology and Management of Lobsters. Academic Press, New York, pp. 301–342.

Stewart, J.E., Rabin, H., 1970. Gaffkemia, a bacterial disease of lobsters (Genus *Homarus*), in: Snieszko, S.F. (ed.), A symposium on Diseases of Fishes and Shellfishes. Special Publication No. 5. American Fisheries Society, Washington, DC, pp. 431–439.

Stewart, J.E., Cornick, J.W., Spears, D.I., McLeese, D.W., 1966. Incidence of *Gaffkya homari* in natural lobster (*Homarus americanus*) populations of the Atlantic region of Canada. J. Fish. Res. Bd. Can. 23, 1325–1330.

Tang, K.F., Lightner, D.V., 2002. Low sequence variation among isolates of infectious hypodermal and hematopoietic necrosis virus (IHHNV) originating from Hawaii and the Americas. Dis. Aquat. Org. 49, 93–97.

Taugbøl T., Skurdal J., Håstein, T., 1993. Crayfish plague and management strategies in Norway. Biol. Cons. 63, 75–82.

Taylor, D.M., Khan, R.A., 1995. Observations on the occurrence of *Hematodinium* sp. (Dinoflagellata, Syndinidae), the causative agent of Bitter Crab Disease in the Newfoundland snow crab (*Chionoecetes opilio*). J. Invert. Pathol. 65, 283–288.

Thitamadee, S., Prachumwat, A., Srisala, J., Jaroenlak, P., Salachan, P.V., Sritunyalucksana, K., et al., 2016. Review of current disease threats for cultivated penaeid shrimp in Asia. Aquaculture 452, 69–87.

Tlusty, M.F., Metzler, A., 2012. Relationship between temperature and shell disease in laboratory populations of juvenile American lobsters (*Homarus americanus*). J. Shellfish Res., 31, 533–541.

Tlusty, M.F., Smolowitz, R.M., Halvorson, H.O., DeVito, S.E., 2007. Host susceptibility hypothesis for shell disease in American lobsters. J. Aquat. Anim. Health 19, 215–225.

Tran, L., Nunan, L., Redman, R.M., Mohney, L.L., Pantoja, C.R., Fitzsimmons, K., Lightner, D.V., 2013. Determination of the infectious nature of the agent of acute hepatopancreatic necrosis syndrome affecting penaeid shrimp. Dis. Aquat. Org. 105, 45–55.

Tu, C., Huang, H.T., Chuang, S.H., Hsu, J.P., Kuo, S.T., Li, N.J., et al., 1999. Taura syndrome in Pacific white shrimp *Penaeus vannamei* cultured in Taiwan. Dis. Aquat. Org. 38, 159–161.

Unestam, T., 1969. Studies on the crayfish plague fungus *Aphanomyces astaci*. II. Factors affecting zoospores and zoospore production. Physiol. Plantae 19, 1110–1119.

Unestam T., 1972. On the host range and origin of the plague fungus. Rep. Inst. Freshwat. Res. Drottingholm 52, 192–198.

Vacelet, J., Vacelet, E., Gaino, E., Gallissian, M.F., 1994. Bacterial attack of spongin skeleton during the 1986–1990 Mediterranean sponge disease, in: VanSoest, R.W.M., van Kempen, T.M.G., Braekman, J.-C. (eds.), Sponges in Time and Space. Balkema, Rotterdam, pp. 355–362.

Van Banning, P., 1991. Observations on bonamiasis in the stock of the European flat oyster, *Ostrea edulis*, in the Netherlands, with special reference to the recent developments in Lake Grevelingen. Aquaculture 93, 205–211.

Vezzulli, L., Previati, M., Pruzzo, C., Marchese, A., Bourne, D.G., Cerrano, C., 2010. *Vibrio* infections triggering mass mortality events in a warming Mediterranean Sea. Environm. Microbiol. 12, 2007–2019.

Vezzulli, L., Grande, C., Reid, P.C., Hélaouët, P., Edwards, M., Höfle, M.G., et al., 2016. Climate influence on *Vibrio* and associated human diseases during the past half-century in the coastal North Atlantic. Proc. Natl. Acad. Sci. U.S.A. 113, E5062–E5071.

Vidal, O.M., Granja, C.B., Aranguren, F., Brock, J.A., Salazar, M., 2001. A profound effect of hyperthermia on survival of *Litopenaeus vannamei* juveniles infected with white spot syndrome virus. J. World Aquacult. Soc. 32, 364–372.

Villalba, A., Reece, K.S., Ordás, M.C., Casas, S.M., Figueras, A., 2004. Perkinsosis in molluscs: a review. Aquat. Living Resour., 17, 411–432.

Villalba, A., Casas, S.M., López, C., Carballal, M.J., 2005. Study of perkinsosis in the carpet shell clam *Tapes decussatus* in Galicia (NW Spain). II. Temporal pattern of disease dynamics and association with clam mortality. Dis. Aquat. Org. 65, 257–267.

Virginia Marine Resources Commission, 2016. Virginia's oyster ground production. Available from: http://mrc.virginia.gov/SMAC/VA-Oyster-Harvests.pdf (accessed May 8, 2017).

Voss, J.D., Richardson, L.L., 2006. Nutrient enrichment enhances black band disease progression in corals. Coral Reefs 25, 569–576.

Wahle, R.A., Gibson, M., Fogarty, M.J., 2009. Distinguishing disease impacts from larval supply effects in a lobster fishery collapse. Mar. Ecol. Prog. Ser. 376, 185–192.

Walker, P., Cowley, J., Spann, K., Hodgson, R.R., Hall, M.M., Withyachumnarnkul, B.B., 2001. Yellow head complex viruses: transmission cycles and topographical distribution in the Asia-Pacific region, in: Browdy, C.L., Jory, D.E. (eds.), The New Wave: Proceedings of a Special Session on Sustainable Shrimp Farming, 2001. World Aquaculture Society, Baton Rouge, LA, pp. 227–237.

Walker, P.J., Mohan, C.V., 2009. Viral disease emergence in shrimp aquaculture origins, impact and the effectiveness of health management strategies. Rev. Aquacult. 1, 125–154.

Wang, H., Wang, B., Zhou, Y., Zhao, Q., Wang, N., Fu, C., et al., 2015. [Detection and assessment of antibiotic and sex hormone residues in *Eriocheir sinensis* sold in markets in Shanghai.] Zhonghua Liu Xing Bing Xue Za Zhi 36, 445–449.

Wang, J., Huang, H., Feng, Q., Liang, T., Bi, K., Gu, W., et al., 2009. Enzyme-linked immunosorbent assay for the detection of pathogenic spiroplasma in commercially exploited crustaceans from China. Aquaculture 292, 166–171.

Wang, W., 2011. Bacterial diseases of crabs: a review. J. Invert. Pathol. 106, 18–26.

Wang, W., Chen, J., 2007. Ultrastructural study on a novel microsporidian, *Endoreticulatus eriocheir* sp. nov. (Microsporidia, Encephalitozoonidae), parasite of Chinese mitten crab, *Eriocheir sinensis* (Crustacea, Decapoda). J. Invert. Pathol. 94, 77–83.

Wang, W., Gu, W., 2002. Rickettsia-like organism associated with tremor disease and mortality of the Chinese mitten crab, *Eriocheir sinensis*. Dis. Aquat. Org. 48, 149–153.

Wang, W., Wen, B., Gasparich, G.E., Zhu, N., Rong, L., Chen, J., 2004. A spiroplasma associated with tremor disease in the Chinese mitten crab (*Eriocheir sinensis*). Microbiol. 150, 3035–3040.

Wang, W., Gu, W., Ding, Z., Ren, Y., Chen, J., Hou, Y., 2005. A novel Spiroplasma pathogen causing systemic infection in the crayfish *Procambarus clarkii* (Crustacea: Decapod), in China. FEMS Microbiol. Lett. 249, 131–137.

Webster, N.S., 2007. Sponge disease: a global threat? Environm. Microbiol, 9, 1363–1375.

Wickham, D.E., 1986. Epizootic infestations by nemertean brood parasites on commercially important crustaceans. Can. J. Fish. Aquat. Sci. 43, 2295–2302.

Wijegoonawardane, P.K., Cowley, J.A., Phan, T., Hodgson, R.A., Nielsen, L., Kiatpathomchai, W., Walker, P.J., 2008. Genetic diversity in the yellow head nidovirus complex. Virology 380, 213–225.

Wilhelm, G., Mialhe, E., 1996. Dinoflagellate infection associated with the decline of *Necora puber* crab populations in France. Dis. Aquat. Org. 26, 213–219.

Witteveldt, J., Cifuentes, C.C., Vlak, J.M., van Hulten, M.C., 2004. Protection of *Penaeus monodon* against white spot syndrome virus by oral vaccination. J. Virol. 78, 2057–2061.

Witteveldt, J., Vlak, J.M., van Hulten, M.C., 2006. Increased tolerance of *Litopenaeus vannamei* to white spot syndrome virus (WSSV) infection after oral application of the viral envelope protein VP28. Dis. Aquat. Org. 70, 167–170.

Wongmaneeprateep, S., Chuchird, N., Baoprasertkul, P., Prompamorn, P., Thongkao, K., Limsuwan, C., 2010. Effect of high water temperature on the elimination of white spot syndrome virus in juveniles of *Litopenaeus vannamei*. Kasetsart Univ. Fish. Res. Bull. 34, 14–26.

World Organisation for Animal Health, 2015. Aquatic Animal Health Code (2016). Available from: http://www.oie.int/en/international-standard-setting/aquatic-code/access-online/ (accessed May 8, 2017).

World Organisation for Animal Health, 2016. OIE-listed diseases, infections and infestations in force in 2016. Available from: http://www.oie.int/animal-health-in-the-world/oie-listed-diseases-2016/ (accessed May 8, 2017).

Wulff, J.L., 2006. Rapid diversity and abundance decline in a Caribbean coral reef sponge community. Biol. Conserv. 127, 167–176.

Xu, W., Sheng, X., Xu, H., Shi, H., Li, P., 2007. Dinoflagellates *Hematodinium* sp. parasitizing the mud crab *Scylla serrata*. Periodical of Ocean University of China, 37, 9166920.

Xu, W., Xie, J., Shi, H., Li, C., 2010. *Hematodinium* infections in cultured ridgetail white prawns, *Exopalaemon carinicauda*, in eastern China. Aquaculture 300, 25–31.

Yang, Y.T., Chen, I.T., Lee, C.T., Chen, C.Y., Lin, S.S., Hor, L.I., et al., 2014. Draft genome sequences of four strains of *Vibrio parahaemolyticus*, three of which cause early mortality syndrome/acute hepatopancreatic necrosis disease in shrimp in China and Thailand. Genome Announc 2(5), e00816-14.

Yodmuang, S., Tirasophon, W., Roshorm, Y., Chinnirunvong, W., Panyim, S., 2006. YHV-protease dsRNA inhibits YHV replication in *Penaeus monodon* and prevents mortality. Biochem. Biophys. Res. Comm. 341, 351–356.

Zhan, W.B., Wang, Y.H., Fryer, J.L., Yu, K.K., Fukuda, H., Meng, Q.X., 1998. White spot syndrome virus infection of cultured shrimp in China. J. Aquat. Anim. Health, 10, 405–410.

Zhuang, J., Cai, G., Lin, Q., Wu, Z., Xie, L., 2010. A bacteriophage-related chimeric marine virus infecting abalone. PLoS ONE 5(11), e13850.

16

Ecology of Emerging Infectious Diseases of Invertebrates

Colleen A. Burge, Amanda Shore-Maggio and Natalie D. Rivlin

Institute of Marine and Environmental Technology, University of Maryland Baltimore County, Baltimore, MD, USA

16.1   Introduction

Infectious diseases are important drivers of population dynamics within both natural and human-based systems, and large-scale disease outbreaks can reshape ecological systems with direct and indirect economic consequences. Although disease is a natural component of ecosystem health, over the last 30 years epidemics of infectious disease have become more frequent (Wilcox & Gubler, 2005). Increased attention has been given to emerging or re-emerging diseases in human and wildlife populations, and a framework has been laid out for defining emerging infectious diseases (EIDs) (Daszak et al., 2000, 2001). EIDs in humans and wildlife are defined as those which have recently rapidly increased in incidence or geographic range, moved into new host populations, and been discovered, as well as those that are associated with a newly evolved pathogen(s) (Daszak et al., 2000). Examples of EIDs in humans are: severe acute respiratory syndrome (SARS), H1N1 influenza, Nipah virus, West Nile virus (WNV), and methicillin-resistant *Staphylococcus aureus* (MERS). Key contemporary vertebrate wildlife examples include: chytridomycosis of amphibians (Mutschmann, 2015) and white-nose syndrome of bats (Frick et al., 2016). In invertebrates, crayfish plague (*Aphanomyces astaci*) (Dieguez-Uribeondo et al., 2015) and varoosis (*Varroa destructor*) of honeybees (Nazzi and Le Conte, 2016) are examples of past (crayfish plague) and contemporary (varoosis) EIDs.

Pathogen transmission from a reservoir host to a novel host commonly underpins infectious disease emergence, which is also known as "host jumping" or "spillover". In vertebrate EIDs, host jumping frequently results in zoonosis (infectious disease of animals transmitted to humans) caused by spillover from domestic or wildlife populations to humans (60.3% of EIDs of humans are zoonotic, with 71.8% of zoonoses originating from wildlife as opposed to domestic animals) (Jones et al., 2008). Three recent EIDs of humans are zoonotic (with examples of spread from both wildlife and domestic animals): SARS (from bats), H1N1 (from pigs and birds), and Nipah (from pigs), all caused by viruses. Interestingly, H1N1 was the result of recombination among human, swine, and avian influenza viruses. The recombination of segmented genomes of RNA viruses, such as influenza, facilitates the increased chance of host jumping (Pulliam, 2008).

Ecology of Invertebrate Diseases, First Edition. Edited by Ann E. Hajek and David I. Shapiro-Ilan.
© 2018 John Wiley & Sons Ltd. Published 2018 by John Wiley & Sons Ltd.

Movement of hosts because of human commerce can be a key factor in "spillover" between wild and domestic populations (and vice versa). Two invertebrate examples include the introduction and spread of *Aphanomyces astaci* into susceptible European crayfish populations and the global spread of the *Varroa destructor* mite within *Apis mellifera* (see Section 16.3.4).

Emerging diseases can progress in a succession of steps defined by the host–pathogen relationship (Parrish et al., 2008; Morens and Fauci, 2013). First, pathogens must spread to a new ecosystem or adapt to a new host. This first step may lead to a "dead end" if the new environment or host is not conducive to transmission. Second, successful transmission and infection of a pathogen in a new host or within a new ecosystem must lead to measureable disease. Strength of disease is defined by the microorganism's pathogenicity (the ability of a pathogen to cause disease in its host) and virulence (the quantifiable disease-producing power of a microorganism that allows it to invade and injure host tissues). In extreme cases (i.e., highly pathogenic and virulent pathogens), diseases build to an epidemic (widespread occurrence within a population) or pandemic (occurrence across a country or the world). Third, over time, certain pathogens may become "endemic" (omnipresence within the host population), existing at low prevalence, and eventually potentially nonpathogenic or even commensal (but see Ewald et al., 1995 for an opposing view regarding evolution toward lower virulence). For example, previously explosive human diseases such as smallpox, measles, and syphilis had become endemic in pre-modern European societies before the development of vaccines against these agents (Diamond, 1999). However, introductions of these diseases decimated Native American societies after contact with Spanish conquistadors and colonial English settlers, starting the EID cycle afresh in the Americas (Ramenofsky, 1993; Diamond, 1999). The evolutionary timescale in which an emerging pathogen moves through these steps is likely dependent on several key characteristics, including but not limited to pathogen type (e.g., mutation rates of bacteria vs. viruses) and host–pathogen interactions (e.g., domestic versus wildlife zoonoses in humans) (Ramenofsky, 2003).

EIDs can be newly emerging or re-emerging. Emerging diseases are those that are affecting hosts for the first time, such as HIV/AIDS (Morens et al., 2004). Re-emerging diseases are those that have historically infected a host species but are now appearing in new locations or in drug-resistant forms, or that are reappearing after apparent control or elimination (Morens and Fauci, 2013). Examples of re-emerging EIDs of humans include the mosquito-vectored WNV, which re-emerged through geographic spread and "host jumping," and both community and hospital-spread MERS (Morens and Fauci, 2013). In invertebrates, we often lack baseline data or a confirmed etiology to determine whether a disease is newly emerging or re-emerging.

To date, most attention on EIDs has focused on humans and charismatic vertebrates (e.g., game animals, amphibians, and bats), and not so much on invertebrates (Daszak et al., 2001). For the context of invertebrate EIDs, we have adapted terminologies used in the literature that primarily describe human EIDs, but in this chapter we will discuss disease dynamics within an ecological (as opposed to epidemiological) context.

There are many challenges for the control or elimination of EIDs in both vertebrate and invertebrate wildlife, due to the complexity of ecological systems and a lack of treatment options. For some domesticated animals, including certain invertebrates (see Chapters 14), treatment options do exist and may be key to stopping the spread of EIDs.

However, for marine diseases (see Chapter 15), treatment is often either not practical or unavailable. For some infectious diseases (of both terrestrial and marine vertebrates and invertebrates), the World Organisation for Animal Health (OIE; acronym based on previous name: Office International des Epizooties) requires reporting of a specific list of 118 terrestrial and aquatic animal diseases (World Organisation for Animal Health, 2016c). For example, EIDs such as infections of honeybees with *V. destructor* or of abalone with herpesviruses are listed, reportable diseases, but other diseases, such as black band disease of corals (see Section 16.3.5.1) are not. To date, the OIE has focused on diseases of economically important organisms. In addition, the OIE provides diagnostic manuals, treatment options (where they exist), and reference laboratories for listed diseases and other specific pathogens of interest (e.g., "Infection with Ostreid herpesvirus 1 microvariants" in bivalve mollusks or "Nosemosis of honeybees").

EIDs of invertebrates have to date been poorly defined as such, due in part to a lack of baseline data and to difficult etiologies leading to a lack of specific and sensitive diagnostic tools for regular surveillance (Burge et al., 2016). Despite the difficulty of defining EIDs in invertebrates, Table 16.1 contains several important examples. The sheer number and diversity of invertebrate species and the lack of population data on invertebrate host taxa means that changes in host populations may only be noted when an organism is absent. In addition, studies of the natural diversity of parasites within invertebrate species are lacking, meaning that in many if not most cases, the first description of a parasite occurs during an epizootic. Where samples are available, it is possible to conduct retrospective analyses to determine whether a pathogen was present prior to a disease outbreak. Within invertebrate taxa, infectious diseases of economically important invertebrates (e.g., shrimp, oysters, pollinators, etc.) are better characterized, although difficulties in original disease diagnoses may occur (see Section 16.3.4). In addition, evolution of pathogens may hinder diagnoses or our understanding of a specific diagnosis, especially when either a symbiont or a pathogen becomes more virulent (see Section 16.3.1.1 or 16.3.2). Recent work has highlighted the necessity of understanding infectious diseases affecting other (non-commercially important) invertebrate taxa, especially as they relate to losses of biodiversity (Altizer et al., 2003). Reef-building corals represent an example where biodiversity is at risk and there is an inherent difficulty in defining causation for diseases that appear to be emerging, are thought to be infectious, and for which polymicrobial disease may play a key role (see Section 16.3.5.1). Recent attention has been focused on climate change and other anthropogenic influences on infectious diseases, including those infecting invertebrate hosts (Harvell et al., 1999, 2002; Altizer et al., 2013; Burge et al., 2014). In fact, diseases infecting sessile marine invertebrates can be more easily correlated with climate change than their vertebrate host counterparts, such as finfish or marine mammals, in part due to the sedentary nature of most of their adult lifeforms (Burge et al., 2014). Complementary and multidisciplinary approaches to disease diagnosis are necessary, applying both classic (i.e., microscopy, gross descriptions, and field surveys) and modern (i.e., high-throughput sequencing, digital photo analysis, and citizen science) approaches to the diagnosis of emerging disease (Burge et al., 2016).

In this chapter, we consider EIDs in terrestrial and marine invertebrates, focusing on factors contributing to disease emergence and disease spread in invertebrates, such as ecological context, environmental change, and human-related risks. In several case studies, we will focus on the following questions: What is/was the ecology of disease

Table 16.1 Examples of invertebrate emerging infectious diseases (EIDs) and factors contributing to their emergence.

EIDs	Host organism[a]	Etiology	Dates active	Region	Main contributing factor(s) and mechanism	References
Terrestrial						
A. domesticus-associated densovirus	*Acheta domesticus* (Arthropoda, Insecta) (D)	Densovirus	1970s–present	USA, Canada, Europe	Direct human activities; movement of infected individuals and high density of culture conditions	Szelei et al. (2011); Weissman et al. (2012)
Varroa mite infestations	*Apis mellifera* (Arthropoda; Insecta) (D/W)	*Varroa destructor* (arthropod)	1960s–present	Global	Direct human activities; movement of infected individuals	Reviewed in Rosenkranz et al. (2010); Goulson and Hughes (2015)
Honeybee wing deformities	*Apis mellifera* and *Bombus* spp. (Arthropoda; Insecta) (D/W)	Deformed wing virus (DWV)	1980s–present	Global	Direct human activities; movement of infected bees and mites; infection of *V. destructor* mites	Reviewed in de Miranda and Genersch (2010)
Type C nosemosis	*Apis mellifera* and *Bombus* spp. (Arthropoda; Insecta) (D/W)	*Nosema ceranae* (fungus)	1990s–present	Asia, North America, Europe	Direct human activities; movement of infected individuals, "spillover" into wild populations	Reviewed in Higes et al. (2010); Goulson and Hughes (2015)
					Climate change; changes in distribution of parasite	Goulson and Hughes (2015)
Beetle parasitism	*Apis mellifera* (Arthropoda; Insecta) (D)	*Aethina tumida* (arthropod)	1990s–present	North America, Australia	Direct human activities; movement of infected individuals	Reviewed in Cuthbertson et al. (2013)
Japanese gypsy fungus	*Lymantria dispar* (Arthropoda; Insecta) (W)	*Entomophaga maimaiga* (fungus)	1980s–present	Northeast USA	Direct human activities; movement of infected individuals	Andreadis and Weseloh (1990); Hajek (1999)

Aquatic

Disease	Host	Pathogen	Time	Location	Drivers	References
Ostreid herpesvirus 1 and variants	*Crassostrea gigas* and others (Mollusca; Bivalvia) (D/W)	Herpesvirus	1993–present	Global	Direct human activities; movement of infected individuals Pathogen evolution; emergence of new variants within populations	Reviewed in Pernet et al. (2016), Arzul et al. (2017) Segarra et al. (2010)
Abalone viral ganglioneuritis	*Haliotis spp.* (Molluska; Gastropoda) (D/W)	*Haliotis* herpesvirus	2005–present	Australia, Taiwan	Domestication; unknown	Chang et al. (2005); Hooper et al. (2007); Arzul et al. (2017)
Sea star wasting disease	Several species of sea stars (Echinodermata; Asteroidae) (D/W)	Densovirus	2014–present	West coast of North America	Unknown; pathogen evolution?; temperature (increased and high) magnifies disease	Eisenlord et al. (2016)
Black band disease of corals	Several species of coral and sea fans (Cnidaria; Scleractinia and Octocorallia) (W)	Polymicrobial infection (bacteria)	1970s–present	Global	Climate change; thermal stress Human activities; eutrophication	Reviewed in Sato et al. (2016) Voss and Richardson (2006); Kuta and Richardson (2002)
Acropora serratiosis	*Acropora palmata* (Cnidaria; Scleractinia) (W)	*Serratia marcescens* (bacterium)	1996–present	Caribbean Sea	Human activities; sewage pollution	Patterson et al. (2002); Sutherland et al. (2010)
Acute *Montipora* white syndrome	*Montipora capitata* (Cnidaria; Scleractinia) (W)	Unknown, possible bacterial infection	2010–present	Kaneohe Bay, Oahu, HI	Unknown	Ushijima et al. (2014); Aeby et al. (2016)
Sea-fan aspergillosis	*Gorgonia ventalina* and *G. flabellum* (Cnidaria; Octocorallia) (W)	*Aspergillus sydowii* and other fungal species	1995–present	Caribbean Sea	Human activities?; terrestrial runoff?	Reviewed by Burge et al. (2013)

(Continued)

Table 16.1 (Continued)

EIDs	Host organism[a]	Etiology	Dates active	Region	Main contributing factor(s) and mechanism	References
Acute hepatopancreatic necrosis disease of shrimp	*Penaeus spp.* (Arthropoda; Decopoda) (D/W)	*Vibrio parahaemolyticus* (bacterium)	2009–present	Global	Pathogen evolution; acquisition of new genetic element for toxin	Tran et al. (2013); Nunan et al. (2014); Lee et al. (2015)
Crayfish plague	*Astacus spp.* and *Austropotamobius spp.* (Arthropoda; Decopoda) (D/W)	*Aphanomyces astaci* (fungus)	Mid–1800s	Europe	Human activities; introduction of non-native species (in which infection is nonlethal and enzootic)	Reviewed in Edgerton et al. (2002)
White spot disease of shrimp	Several shrimp species (Arthropoda; Decopoda) (D/W)	White spot syndrome virus (WSSV) (*Whispovirus*)	1990s–present	Global	Human activities; movement of infected individuals and high density of culture conditions	Reviewed in Sanchez-Martinez et al. (2007)
Aplysina red band syndrome	*Aplysina* spp. (Porifera) (W)	Unknown, possible cyanobacterial infection	2004–present	Caribbean Sea	Unknown	Olson et al. (2006)
Sponge orange band disease	*Xestospongia muta* (Porifera) (W)	Unknown, possible cyanobacterial infection	2005–present	Caribbean Sea	Unknown	Cowart et al. (2006); Angermeier et al. (2011)
Sponge necrosis syndrome	*Callyspongia (Euplacella) aff biru* (Porifera) (W)	Co-infection with *Rhabdocline* sp. (fungus) and Rhodobacteraceae (bacterium)	2011–present	Maldives	Unknown	Sweet et al. (2015)

a) Populations affected.
D, domesticated; W, wild.

emergence as it relates/related to anthropogenic influences? Are there aspects of domestication (or method of culture) that played a role in disease emergence? What are some potential management responses and practices for EIDs?

16.2   Host–Pathogen Relationships and Anthropogenic Change

16.2.1   Ecological Context of Invertebrate Host–Pathogen Relationships

Stable host–pathogen–environment relationships (classically depicted with the disease triad represented by a Venn diagram) underlie disease ecology, where a disruption of this balance can lead to disease and in some cases to emerging disease. Changes in the host–pathogen relationship are possible through natural ecological and evolutionary processes, and recently through anthropogenic change (see Section 16.2.2). Host–pathogen interactions are commonly described as a coevolutionary arms race in which each partner is adapting to maximize its own reproductive output and fitness (Roy et al., 2009). We will consider factors for each component of the host–pathogen interaction that can play a role in the development of disease. For the host, such factors include (but are not limited to) genetics, previous exposure, nutrition, social behavior, life history, and density. For the pathogen, they include virulence factors, dose, strain, specificity, and evolution. The environment can play a role in creating an optimal or suboptimal state for the host or the pathogen, where physical, biological, or chemical characteristics can play a role in disease outcomes. Without prior knowledge of a specific host–pathogen relationship, the ecological contexts facilitating disease emergence and spread are unknown.

Infectious disease emergence and spread can be impacted by interconnected host–pathogen disease ecology, including (but not limited to) mode of pathogen transmission (horizontal or vertical), population density and diversity (host or pathogen), host range (or pathogen specificity), and the ability to live/spread in the environment (Roy et al., 2009). Horizontal transmission (from host to host, including vectors) may be more density-dependent than vertical transmission, favoring dense host populations; in low host-density populations, pathogens are unable to effectively spread and emerge. Pathogens often exist at low levels that can be described as "covert" infections and may go undiscovered until an "overt" infection occurs, usually after an alteration in the host–pathogen relationship. Additionally, pathogens may be a ubiquitous part of the environment as constituent members of communities in soils or sediments, water (fresh, brackish, or sea), or the air. For example, *Vibrio harveyi* is pathogenic to a variety of invertebrates but is commonly found in the surrounding seawater environment of their hosts (reviewed in Austin and Zhang, 2006). Host range, including both the primary infected hosts and reservoir hosts, may also play a role in the emergence of diseases. Both the host and the pathogen exist within a range of environmental conditions (i.e., specific temperatures, salinities, pH, etc). An optimal condition for the host is more likely to be conducive to a covert infection and an optimal condition for a pathogen is more likely to be conducive to an overt one. Invertebrates are often thermoconformers (although some insects may behaviorally be thermoregulators), and therefore both invertebrate hosts and their pathogens will metabolically track ambient temperatures. Examples of thermal optima across pathogen type (e.g., viruses, fungi, bacteria, and protists) are available for both terrestrial and marine invertebrates (Roy et al., 2009; Altizer et al., 2013; Burge et al., 2014). Similarly, many marine invertebrates are osmoconformers (meaning they maintain their internal salinity to be equal to the surrounding seawater), and optimas exist for both marine invertebrates and their pathogens (Burge et al., 2014; Fuhrmann et al., 2016). Therefore, the disease outcome relies on the environmental optima of both the host and the pathogen and is an interplay between host immunity (or fitness) and pathogen virulence. When diseases do emerge, we describe the "ecology of the disease" through observations relating to the host–pathogen relationship, such as environmental conditions, population or community structure (either host or pathogen), organismal biology (including gross disease signs), tissue and cellular architecture (including potential pathogens), and genes (either host or pathogen) (Burge et al., 2016).

16.2.2 Anthropogenic Change and Disease Emergence

The human-coupled environment that invertebrates and pathogens occupy can lead to disequilibrium of the host–pathogen relationship and emergence of disease. In Fig. 16.1, we've considered factors that may be involved in disease emergence and overlaid them on the classic disease triad; factors that are in bold may be particularly impactful for invertebrate EIDs. For many EIDs, documented changes in host–pathogen relationships are affected by direct (e.g., movement of hosts) and indirect (e.g., climate change via CO_2 emissions) human activities. Although changes to host–pathogen relationships are interconnected, in this section we will consider the factors of each component of the disease triad (outlined in Fig. 16.1) that may be important for disease emergence in human-coupled systems.

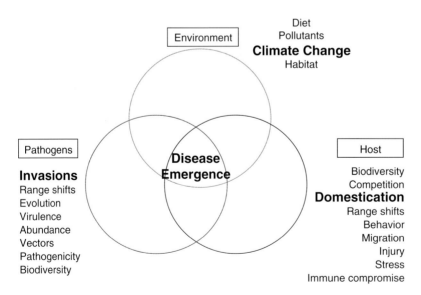

Fig. 16.1 Disease emergence in invertebrates. Emergence is linked to key and pathogen traits and changes in the environment in which the host and pathogen reside. Terms in bold may play a relatively larger role in disease emergence.

16.2.2.1   Host Factors

Movement of organisms, often coupled with increased densities, is frequently cited as a major cause of disease emergence in domesticated invertebrates that are used directly for food (e.g., shrimp or oysters) or services (e.g., pollinators) (see Table 16.1) and in the natural populations of invertebrates they interact with. Although domestication has played a large role in disease emergence, invasions of hosts may also occur through range shifts associated with climate change or habitat destruction. For diseases with horizontal transmission, shared resources such as food or substrates may facilitate emergence from domestic to wild populations. For example, in bivalve mollusks, disease transmission can be facilitated by filter-feeding (Ben-Horin et al., 2015) that occurs when infected individuals (often domesticated) share water resources with naïve populations (domesticated or wild). Similarly, disease transmission between pollinator populations (domestic to wild, or vice versa) can be facilitated through landing on the same flower for pollination or gathering of nectar (Goulson and Hughes, 2015). Clearly, basic host ecology can play a large role in disease transmission among invertebrates in both terrestrial and aquatic systems. Human activities that result in a compromised host population (immune-compromised, injured, or otherwise stressed) may also facilitate disease emergence, especially for ubiquitous, opportunistic pathogens (reviewed by Burge et al., 2013).

16.2.2.2   Pathogen Factors

Like the movement of hosts, "microbial traffic" (Morse, 2004) – or the movement of infected hosts, vectors/intermediate hosts, or contaminated equipment – is also a common route for disease emergence in invertebrates (see Table 16.1). Urban development can also lead to contamination of pathogens into ecosystems, especially via waterways. For example, sewage contamination is the source of a coral pathogen, *Serratia marcescens*, to coral reef ecosystems in the Florida Keys, USA (see Section 16.3.5.2). Increasing temperature conditions due to anthropogenic climate change can affect the distribution of ubiquitous pathogens (see Section 16.2.2.3). Antibiotic use to prevent disease outbreaks in aquaculture, especially shrimp production, has been practiced extensively since the 1990s (Moriarty et al., 1999; Holmstrom et al., 2003); however, this has led to antibiotic resistance (Tendencia et al., 2001; Le et al., 2005; Defoirdt et al., 2007). In particular, antibiotic-resistant *Vibrio* spp. have caused mass-mortality events in shrimp hatcheries (Karunasagar et al., 1994; Jayasree et al., 2006).

Newly evolved pathogens are less reported in invertebrate EIDs (as opposed to human EIDs), but mutation or gene-transfer events allow invertebrate pathogens to evade immune systems of hosts, increase host range, or acquire new virulence factors. As in the development of antibiotic resistance, human activities create environments that promote mutation or gene transfer among potential pathogens (Alonso et al., 2001; Gillings, 2016). Eutrophication of pond water and crowding of hosts in aquaculture facilities increase bacterial abundances, which in turn increases the probability of horizontal gene transfer (see Section 16.3.2).

16.2.2.3   Environment Factors

The homeostasis of an invertebrate and its pathogens is directly impacted by the surrounding environment, and changes such as those ascribed to climate change (i.e., changes in temperature, pH, salinity, humidity, and precipitation patterns, coupled with increased exposure to storms or cyclones) can play a role or even drive disease emergence (Altizer et al., 2013; Burge et al., 2014). We've established that invertebrates may be thermo- and/or osmoconformers, meaning they will be directly impacted by stress (leading to more susceptible host populations) caused by changes in temperature and/or salinity (for marine invertebrates). Stressors linked to temperature or salinity changes are easier to monitor (in some cases) and easier to predict for invertebrates than for vertebrates. Increasing temperature conditions due to anthropogenic climate change can affect habitat range (temporally or spatially) or increase the virulence of pathogens – bacterial pathogens, in particular. For example, warm winter conditions allowed the spread of the oyster pathogen *Perkinsus marinus* further north along the eastern coast of the United States (Ford, 1996; Cook et al., 1998), while increased sea-surface temperatures enhance the virulence of the coral pathogen *Vibrio shiloi*, so that coral bleaching at lower temperatures switches to into coral mortality at high temperatures (Kushmaro et al., 1998). The roles of other stressors associated with climate change (i.e., precipitation, humidity, and climate variability) are more difficult to predict (Altizer et al., 2013). In the coastal environment, increased exposure to terrestrial runoff, which can introduce potential pathogens and pollutants, may play a role in disease emergence. It is hypothesized, for example, that the emergence of *Aspergillosis* (caused by opportunistic fungi) in immune-compromised sea fans in the Florida Keys and the Caribbean is linked to terrestrial runoff (reviewed by Burge et al., 2013).

Organism health (and populations losses) and/or disease emergence may be impacted by anthropogenic stressors such as changes in climate (i.e., climate change), pollutants, habitat loss, and diet; these factors may be difficult to decouple, and in reality they must act together as "multi-stressors." The effect of a combination of multiple stressors (global or local) on disease dynamics is difficult to investigate, and therefore is understudied. However, changes in disease dynamics (or host–pathogen relationships) are unlikely to occur under the regime of a single stressor. For example, many coastal coral reef ecosystems deal with pollutants associated with terrestrial runoff (local stressor) and are increasingly experiencing thermal stress (global stressor), both of which factors are individually correlated to disease emergence in corals (Harvell et al., 2007; Sutherland et al., 2011; Miller et al., 2014). Although EIDs are recognized as playing an important role in losses of wild and domesticated pollinator populations, "multi-stressors" (a combination of bee stressors such as pollutants (including pesticides), habitat loss, food quality, shipping fever, competition, climate change, pathogens, etc.) are likely playing a role in declining bee health (reviewed in Goulson et al., 2015). A better understanding of how multiple stressors lead to disease emergence will be required in order to predict future disease outbreaks and establish preventive measures or treatment options.

16.3   Case Studies of Invertebrate Disease Emergence

The increasing number of EIDs in invertebrates warrants thorough investigation of the mechanisms behind disease emergence and the ecological consequences of major EID outbreaks. The majority of the information we have on invertebrate EIDs has been gathered from outbreaks that significantly disrupted commercial ventures. Some of the best examples of known EIDs come from aquaculture or marine systems.

Over the past 50 years, the aquaculture industry has grown faster than any other food-producing sector, and it is estimated that by 2050, aquatic production will need to reach almost double its current level in order to supply global demand (Bostock et al., 2010; Stentiford et al., 2012). Ocean fisheries are limited in the amount of supply they are able to output over a given time, and have reached capacity. Aquaculture, particularly of marine species, is a potential solution to fill the deficit and to provide the next food revolution (Duarte et al., 2009; Stentiford et al., 2012). Marine invertebrates play an important role in the growing aquaculture industry, especially species such as oysters (i.e., *Crassostrea gigas*) and shrimp (i.e., *Penaeus monodon* and *Litopenaeus vannamei*), and these species are affected by emerging diseases associated with human-related activities such as movement of infected animals and culture-related practices (i.e., factors related to domestication, such as density and water quality).

Invertebrate EIDs are a major problem not just for marine systems, but also for terrestrial ecosystems. EIDs of terrestrial species that affect agriculture are the best understood. The importance of bees for the production of many crops (especially fruits and vegetables) has led to the domestication and management of a few species (e.g. Apis millefera and Bombus species). Although crop pollination by wild bee populations does still occur, several species of honey bee bumblebee are produced commercially in factories and exported on a global scale (Goulson and Hughes, 2015). Therefore, emergent diseases in wild and domesticated insect pollinators not only have the potential to cause significant economic loss but also threaten food security. Like diseases in aquaculture species, emergent diseases in insect pollinators are heavily linked to human activities and practices relating to domestication.

Emergent diseases in wildlife species of invertebrates are the least understood of all EIDs, due in part to a lack of monitoring efforts in these populations. Coral diseases may be an exception (although many difficulties are found in establishing etiology), with a large number of different diseases described for scleractinian and gorgonian corals (Sutherland et al., 2004). Even though EIDs in wildlife populations are the least well characterized, they have the greatest potential to significantly alter the ecosystems in which they occur, as compared to EIDs in domesticated species.

The case studies described in this section are but a few examples of major EIDs of commercial and wildlife species of invertebrates. In each case, we will outline the basic ecology of the disease as it affects the host, the etiology of the disease (if known), and the human activities or environmental covariates associated with its emergence and/or transmission (if known). In addition, we will describe any attempted or suggested preventative measures. With these case studies, we hope to highlight the knowledge gaps in the study of invertebrate EIDs and to compare host–pathogen interactions within different EID systems.

16.3.1 Molluscan Herpesvirus Infections of Bivalves

Molluscan herpesviruses belong to Herpesvirus Class III, in the family Malacoherpesviridae (Davison et al., 2009), which is composed of two mollusk viruses: Ostreid herpesvirus 1 (OsHV-1) and Haliotid herpesvirus 1 (HHV-1) (ICTV, 2014). OsHV-1 and its variants are emergent pathogens of bivalves globally, and HHV-1 is an emergent pathogen of abalone (*Haliotis* sp.) in Australia and Tawian (World Organisation for Animal Health, 2016a).

16.3.1.1 OsHV-1 Infections of Pacific Oysters and Other Bivalves

The Pacific oyster (*Crassostrea gigas*) is the primary species impacted by mass mortalities caused by OsHV-1 and its variants, which can infect larval, juvenile, and adult stages of the host. It was first introduced as a food product from Asia to grow-out areas globally and is now one of the most important species of bivalve cultivated in aquaculture (Hedgecock et al., 2005). In 2012, global production of *C. gigas* totaled 4.7 million tons (97% of global oyster production), with a value of US$4 billion (Food and Agriculture Organization, 2014). Despite the global movement and scale of Pacific oyster culture, the most notable global mortality threats to the species include the general "summer mortality syndrome" ("multi-stressor" mass-mortality events that have not been attributed to a single factor, although pathogens play a role in some cases) of adults and the recent emergence of OsHV-1. The increased incidence of OsHV-1 and its geographic spread indicate it is an EID.

OsHV-1 is a double-stranded DNA virus (207 kb genome; ~116 nm) that was purified, described, and sequenced from infected oyster larvae collected in France in 1995 (Le Deuff and Renault, 1999; Davison et al., 2005). Prior to its sequencing and the development of molecular diagnostic tools, electron microscopy was used to describe herpes-like viral infections in a variety of oyster species globally, including adult *Crassostrea virginica* in 1970 (the year of first herpes-like virus description in oysters) (Farley et al., 1972), *C. gigas* larvae in New Zealand (Hine et al., 1992) and France (Nicolas et al., 1992) in 1991, and adult flat oysters (*Ostrea angasi*) in Australia in 1994 (Hine and Thorne, 1997). The relationship of these herpes-like viruses to the characterized OsHV-1 is unknown. In France and the United States, OsHV-1-associated mortalities have occurred in seed Pacific oysters during summer months since 1993 (Renault et al., 1994a,b; Burge et al., 2006, 2007; Garcia et al., 2011; Renault et al., 2014). Infections are short in duration, with 3–15 days until host death (dependent on the OsHV-1 strain, host genetics, and environmental factors), and can reach up to 100% mortality. The host range of OsHV-1 is not limited to Pacific oysters, and mortalities have at times been associated with larval losses of scallops and clams in hatcheries in France (Arzul et al., 2001a; Renault et al., 2001). Laboratory studies have confirmed that both interspecies transmission (Arzul et al., 2001b,c) and transmission between life stages are possible (Burge and Friedman, 2012). The effect of OsHV-1 on native, nondomesticated bivalve species is relatively unknown.

OsHV-1 sequence data indicate that multiple strains of OsHV-1 occur, in both Pacific oysters (Renault et al., 2012) and other bivalve species (Ren et al., 2013; Xia et al., 2015). Since 2008, an economically devastating increase in *C. gigas* mortality in France has been associated with the apparent emergence of a new viral genotype, OsHV-1 μvar (Segarra et al., 2010). Microvariants of OsHV-1 are defined by the OIE as those viral strains having sequence variation in a microsatellite locus upstream of open reading frame (ORF) 4 and within ORF 4 and ORF 42/43, as compared to the OsHV-1 reference sequence (World Organisation for Animal Health, 2016b). A specific variant showing all the mutations reported by Segarra et al. (2010) is considered "OsHV-1 μvar" (World Organisation for Animal Health, 2016b), while similar variants are referred to as OsHV-1 μvars but have not been named as specific variants as of yet (as a note, "OsHV-1 μvar" will be listed in quotes within this section to avoid confusion between singular and plural uses of μvar and μvars, as other strain names have not been defined). By this definition, the OIE has recognized microvariants (μvars) infecting Pacific oysters in

Europe, New Zealand, Australia, and Asia (World Organisation for Animal Health, 2016b). Multiple studies have indicated variation between strain types globally, although Asia may be the point of origin, as *C. gigas* was exported from the continent globally. However, more data are necessary in order to understand the initial emergence of OsHV-1 and the spread of OsHV-1 μvars. OsHV-1 μvars are continuing to spread in Europe, and similar variants are causing losses in Australia, New Zealand, and Asia (Pernet et al., 2016). In 2016, an OsHV-1 μvar had devastating impacts on the Pacific oyster aquaculture industry in Australia (DPI, 2016). Although it is assumed that OsHV-1 μvars have increased virulence compared to OsHV-1, no study has been conducted to confirm this, or to compare the virulence of OsHV-1 μvars either within a relatively "local" geographic area (e.g., multiple OsHV-1 μvars detected in France) or between geographic locations.

Full viral genome sequences are currently only published for the reference strain (Davison et al., 2005) and for variants that cause mass mortalities of scallops (*Chlamys farreri*) (acute viral necrosis virus; Ren et al., 2013) and blood clams (*Scaphara broughtonii*) (OsHV-1 SB; Xia et al., 2015) in China. A genome of OsHV-1 μvar or other μvars infecting oysters is not currently available. In France, the reference strain of OsHV-1 has rarely been detected since 2008, with "OsHV-1 μvar" the primary OsHV-1 variant infecting *C. gigas* since 2008 (Segarra et al., 2010), although other OsHV-1 μvars exist in the country (Martenot et al., 2011, 2012).

Detection of OsHV-1 and mortalities caused by the virus in France, New Zealand, and the United Studies in the early 1990s indicated an initial emergence in this time period. The source of these original outbreaks is unknown, and samples for DNA analysis of early outbreaks are limited. Each summer, OsHV-1-associated mortalities move from south to north with warming water temperatures in France (Garcia et al., 2011). Interestingly, in North America, despite extensive Pacific oyster culture on the west coast of the United States, OsHV-1 has only been detected in Tomales Bay and nearby Drakes Bay in California (Friedman et al., 2005; Burge et al., 2011) and in Mexico (Vasquez-Juarez et al., 2006). In Tomales Bay, mortalities occur yearly, although the extent and magnitude of OsHV-1-related mortalities vary (Burge et al., 2006, 2007; C.A. Burge and C.S. Friedman, unpublished data). OsHV-1 was detected at low prevalence in multiple oyster species (*C. ariakensis*, *C. sikamea*, *C. gigas*, and *C. hongkongensis*) in Asia (Japan, South Korea, and China) from samples collected between 1999 and 2005, although reports of mortality were not made at that time (Moss et al., 2007).

Since the initial emergence of OsHV-1 μvar in France, OsHV-1 μvars and mortalities have been reported to spread geographically in Europe (first dates of detection), including to Spain (2008) (Roque et al., 2012), Ireland (2008) (Morrissey et al., 2015), the United Kingdom (2008) (Peeler et al., 2012), Italy (2010) (Dundon et al., 2011), the Netherlands (2011) (Gittenberger et al., 2016), Portugal (2011) (Batista et al., 2015), Sweden (2014) (Mortensen et al., 2016), and Norway (2014) (Mortensen et al., 2016). Although specific regulatory frameworks exist in Europe to control movements of live animals and trade (reviewed in Pernet et al., 2016), shipments of Pacific oysters from France into several of these countries coincide with the detection of an OsHV-1 microvariant genotype, with detection occurring following shipments to Spain (Roque et al., 2012), Ireland (Morrissey et al., 2015), the United Kingdom (Peeler et al., 2012), and Italy (Dundon et al., 2011). In contrast, in other locations, no evidence has been presented for the introduction of OsHV-1 μvar or μvars following live animal shipments, including of wild *Crassostrea angulata* in Portugal and feral *Crassostrea gigas* in the Netherlands/ Scandinavia. Interestingly, OsHV-1 µvars were detected in both Norway and Sweden prior to OsHV-1-associated mortalities in 2014, but cooler water temperatures may have limited viral infection and spread (Mortensen et al., 2016).

OsHV-1 µvars causing *C. gigas* mortalities were first detected in 2010 in New Zealand (Bingham et al., 2013; Keeling et al., 2014) and Australia (Jenkins et al., 2013). In New Zealand, OsHV-1 has only been detected to date in the North Island, with spread most likely occurring through animal transfers between farms (Bingham et al., 2013). Although OsHV-1 µvars have spread in Australia, both within New South Wales and, in 2016, to Tasmania (~850 km from New South Wales), neither the initial emergence nor the spread has been linked to animal transfers. Although OsHV-1 µvars have been described in Asia (China, and Japan), routine monitoring of OsHV-1 has not been conducted, and reports are limited (Bai et al., 2015; Shimahara et al., 2012). In South Korea, a recent report indicated OsHV-1 was involved in mortalities in an oyster hatchery (Hwang et al., 2013).

OsHV-1 mortalities occur during the spring to summer season, as is typical of the nonspecific summer mortality syndrome. Temperature is frequently positively associated with OsHV-1, although the synergism between temperature and OsHV-1 variant may vary based on the variant and the system ecology. In France, epizootics occur when seawater temperature is between 16 and 24 °C, although onset may occur below 16 °C (reviewed in Pernet et al., 2016). In California, OsHV-1 and mass mortalities were only detected after peak seasonal water temperatures reached ~18–20 °C, with maxima of 24–25 °C (Burge et al., 2006). In Australia and New Zealand, OsHV-1 mortalities have been detected at between 23 and 26 °C (reviewed in Pernet et al, 2016). The temperature necessary for disease onset is currently unknown. Temperature may facilitate replication of latent OsHV-1 infection in oysters that have survived mortality events (Burge et al, 2006).

Control measures such as culling or vaccination are either not practical (culling) or not possible (vaccination), and therefore are not being used to control OsHV-1 (reviewed by Pernet et al., 2016). Interestingly, early data suggest that the viral mimic Poly I:C protects oysters in subsequent OsHV-1 µvar exposures (Green and Montagnani, 2013). Poly I:C production and treatment are not practical or cost-effective. However, a recent study suggests treatment of oyster broodstock with Poly I:C enhances survival of larvae exposed to an OsHV-1 µvar in Australia (Green et al., 2016).

Legislation limiting the movement of oysters has also been used as a control measure. In 2010, the European Union passed legislation to limit the movement of OsHV-1 µvar by setting containment zones surrounding areas of OsHV-1 µvar detection (detection of µvar defined by a 12 bp deletion in ORF 4 of the genome as compared to OsHV-1) (European Union, 2010). However, as already shown, animal movements are clearly common both within and between European countries, leading to the spread of OsHV-1 µvars. In Australia, quarantine and prevention of movement of infected stocks was implemented in New South Wales (Paul-Pont et al., 2013). The spread of OsHV-1 µvars without (known) animal movements in both Europe and Australia suggests the role of other disease vectors in spread, such as larval recruitment (Europe) and shipping (Australia), although these are only hypotheses. Breeding for resistance is an applied solution to limiting OsHV-1 mortalities in areas where OsHV-1 has become established (reviewed by Degremont et al., 2015). Experimental studies in France have shown success (in limiting both virus infection and mortalities), and this approach would certainly be a path toward preparation for future outbreaks due to the current spread of OsHV-1. Additionally, several commercial projects on OsHV-1 resistance in *C. gigas* exist in France, along with an experimental program in California and industry-based programs in Australia and New Zealand, but no reports on their success have been released to date (Degremont et al., 2015).

16.3.2 Acute Hepatopancreatic Necrosis Disease of Shrimp

Of the farmed species in the global aquaculture industry, marine shrimp form the most significant proportion, ranking in the top 10 species by production quantity and being number one for production effort (Stentiford et al., 2012). Two species of penaeid shrimp, Pacific white shrimp (*Litopenaeus vannamei*) and black tiger shrimp (*Penaeus monodon*), dominate production (Food and Agriculture Organization, 2014), and are susceptible to multiple viral and bacterial pathogens (Flegel, 2012). Chapter 15 provides an excellent review of viral and bacterial shrimp pathogens, so we focus here on an emerging pathogen causing shrimp disease: acute hepatopancreatic necrosis disease (AHPND).

In 2014, a global shrimp survey by the Global Aquaculture Alliance revealed a rough overall loss to disease of approximately 22% in a single year (Anderson et al., 2014), likely linked to the emergence of a new bacterial disease affecting penaeid shrimp culture. It is estimated that approximately 60% of disease losses in shrimp aquaculture have been caused by viral pathogens, while 20% have been caused by bacterial pathogens (Flegel, 2012), although this value predates the recent outbreak in species *P. monodon* and *L. vannamei* caused by *Vibrio parahaemolyticus*. Bacterial diseases of the *Vibrio* genus have become a major source of concern for the cultured shrimp industry because of their close association with low survival rates in hatcheries and grow-out ponds (Saulnier et al., 2000).

AHPND initially emerged in China in 2009, where it was reported to be causing significant production losses in shrimp farms (Tran et al., 2013). By 2010, the range of affected farms in China had expanded, and by 2011, the disease was detected in Vietnam and Malaysia. The onset of AHPND was confirmed in Thailand in 2012 and then in the Philippines in 2013. Most recently, AHPND has decimated *L. vannamei* culture in the northern states of Mexico, including Nayarit, Sinaloa, and Sonora (Nunan et al., 2014). The spread of ADHPND is a major concern to shrimp producers, as mortality can approach up to 100% in farmed species of both *L. vannamei* and *P. monodon* (Han et al., 2015).

AHPND has already led to tremendous losses in penaeid shrimp production in both Asia and the Western Hemisphere (Tran et al., 2013). Typically appearing within 20–30 days of stocking ponds with postlarvae, AHPND causes progressive dysfunction of the hepatopancreas (HP). Clinical signs include anorexia, lethargy, discoloration of the HP, and a soft, generally darker shell and mottling of the carapace (Lightner et al., 2012).

Progressive dysfunction of the HP results from necrotic lesions, which reflect degeneration and dysfunction of the tubule epithelial cells. In the terminal stages of the disease, a severe secondary infection caused by *V. parahaemolyticus* bacteria occurs in the sloughed masses of epithelial cells in the HP tubule lumens (Lightner et al., 2012). Although AHPND induces massive sloughing of HP cells into HP tubules, there is no sign of any bacteria in the lesions, which strongly suggests a toxin etiology common to *Vibrio* infections in a variety of species. Microbiological analyses also indicate a low presence of bacteria in the hemolymph and a high load of *Vibrio* in the stomach (Soto-Rodriguez et al., 2015).

Emergence of AHPND has been directly linked to the evolution of a pathogen. *Vibro parahaemolyticus* is ubiquitous in the marine environment, with both pathogenic and benign strains existing. In the case of AHPND, the opportunistic *V. parahaemolyticus* became highly virulent by acquiring a unique AHPND-associated plasmid, which encodes a binary toxin that induces cell death (Lee et al., 2015). *Vibrio* species are well known for engaging in horizontal gene transfer (via conjugation) and acquisition of exogenous DNA (via natural competence or phage transduction) (reviewed in Sun et al., 2013). For example, the emergence of *Vibrio cholerae* strain O139 was the result of horizontal gene transfer between highly pathogenic (O1) and nonpathogenic (non-O1) strains of *V. cholerae*, which changed the antigenic properties of the O139 strain (Bik et al., 1995). In addition, some fish-pathogenic *Vibrio anguillarum* strains have acquired plasmids for iron sequestration that contribute to virulence (Crosa, 1980). Interestingly, the AHPND-associated toxin in *V. parahaemolyticus* is a homolog of an insecticidal toxin produced by bacteria (PirAB; produced by *Photorhabdus asymbiotica*); before the emergence of AHPND, this toxin (or any homologs) had never been reported in a marine organism (Lee et al., 2015). Utilization of a homolog of a terrestrial-based toxin by a marine bacterium raises concerns about the exchange of pathogens/virulence factors between biomes and how human activities may be creating selective pressures for pathogen evolution via horizontal gene transfer (Gillings, 2016).

The emergence of AHPND may be linked to multiple environmental and handling stressors (Pragnell et al., 2016). *Vibrio* infections are often thought to be opportunistic, and the stress of harvest and restocking, combined with water-quality stressors, are factors that likely depress the shrimp immune system and result in increased vulnerability to *Vibrio* infection (Prangnell et al., 2016). Transmission of AHPND occurs orally through the digestive tract, and the high densities of shrimp in production ponds encourage cannibalism of sick individuals (Soto-Rodriguez et al., 2015). In addition, *Vibrio* densities are generally high in shrimp production ponds (up to 10^5 cells per mL pond water) (Soto-Rodriguez et al., 2015; Prangnell et al., 2016). These two aspects of production (high host density and high potential pathogen density) may increase the likelihood of pathogen evolution via horizontal gene transfer and thereby increase the likelihood of transmission once an infected individual is present (Gillings, 2016).

Several mechanisms for preventing AHDNP are feasible. Transfer of broodstock, larvae, and commodity products associated with shrimp aquaculture has led to the global distribution of shrimp pathogens, including *V. parahaemolyticus* (Tran et al., 2013). Screening of these products for shrimp pathogens and/or greater regulation in the transfer of these shrimp products may reduce the spread of *V. parahaemolyticus*. Production activities that decrease host density or improve water quality will help reduce the likelihood of outbreaks by increasing host health and resistance to infection. In particular, preventing bacterial blooms may be a better option for limiting disease outbreaks caused by *V. parahaemolyticus*. Early practices to prevent *Vibrio* infections relied on antibiotic addition to feed or pond water as a means of reducing bacterial loads. This method is no longer recommended due to concerns about food safety and because it has resulted in antibiotic resistance among pathogens (both of shrimp and human) (Park et al., 1994). More recently, probiotic bacteria have been supplemented to limit *Vibrio* growth in pond waters or within shrimp intestinal tracts (Kumar et al., 2016).

16.3.3 Emerging Densoviruses of Arthropods and Echinoderms

In both terrestrial and marine environments, the impact of Densovirus (DNV)-related diseases has been tremendous. DNV is a genus in a subfamily of Parvoviridae with over 30 members that are known to infect a variety of arthropods. Terrestrial arthropods affected by DNV include Lepidoptera (moths and butterflies) (Kouassi et al., 2007), Diptera (true flies) (Afanasiev et al., 1991), Orthoptera (crickets, locusts) (Liu et al., 2011a), and Dictyoptera (termites and cockroaches) (Mukha and Schal, 2003). DNVs affect marine arthropods, including commercially important decopod crustaceans such as penaeid shrimp (Lightner and Redman, 1998) and freshwater crayfish (La Fauce and Owens, 2007; Bochow et al., 2015). Recently, DNVs have also been associated with echinoderms, including Echinoidea (sea urchins) (Gudenkauf et al., 2014; Hewson et al., 2014), Ophiuroidea (brittle stars) (Hewson et al., 2014), and Asteroidea (sea stars) (Hewson et al., 2014).

DNVs are small, non-enveloped viruses with a genome of nonsegmented single-stranded DNA. These viruses are highly pathogenic in their hosts and have been associated with many epizootics described since the 1960s, especially in insects (Tijssen and Bergoin, 1995; Fédière, 2000; Bergoin and Tijssen, 2010). The host range of insect-associated DNVs varies considerably (Fédière 2000): some have restricted ranges, while others can infect insect species in different families (Lebedeva et al., 1973; Suto, 1979). Transmission can likewise vary, including routes such as fecal–oral and intraspecific predation (van Munster et al., 2003; Mutuel et al., 2010). Recent epizootics in crickets and sea stars are considered emergent and are economically or ecologically impactful. This section will outline the information currently available about the distribution, ecology, and impact of these emergent Densovirus diseases.

16.3.3.1 Acheta Domesticus Densovirus

An emergent insect DNV disease is Acheta domesticus densovirus (AdDNV). *Acheta domesticus*, the European house cricket, was introduced to the United States from Europe in the 18th century, and today commercial cricket farms are a multimillion-dollar industry that supplies live animals for a variety of uses, such as pet feed, fishing bait, and research. AdDNV was first isolated from commercial production facilities in Europe (Switzerland) (Meynadier et al., 1977), and past epizootics have been mostly confined to Europe, except for a small one in the southern United States in 1988 (Styer and Hamm, 1991). Recently, epizootics have decimated the US and Canadian production of *A. domesticus* (Weissman et al., 2012). Mass mortalities during an AdDNV epizootic in 2009 were reported from several regions of North America, including Quebec and Alberta in Canada and Washington, California, and Florida in the United States (Szelei et al., 2011).

It is unclear what triggered the 2009 AdDNV epizootic in North America, but phylogenetic studies have shown that the original European AdDNV strain (isolated in 1977) and American-derived AdDNV isolates from the 2009 epizootic have 99% sequence identity (Szelei et al., 2011). In addition, all of the American-derived isolates have 100% sequence identity to one another (Liu et al., 2011b; Szelei et al., 2011). These data suggest that disease emergence in North America is not a result of the evolution of a more "virulent" strain of AdDNV. Instead, human activities, such as importation and trade of diseased stock between suppliers, are likely the original sources of AdDNV in North America, as well as the subsequent sources for different regions within North America. Epizootics continue, with reports of mass mortalities in 2011 that have bankrupted companies (Weissman et al., 2012). Because of the high rate of production in rearing facilities and the high rate of mortality (close to 90% in 2 days) (Szelei et al., 2011), facilities that can normally ship millions of crickets per week have often had to cull entire harvests. Current strategies to reduce the risk of AdDNV include decreasing the density of individuals in rearing cages and switching production to other cricket varieties not susceptible to AdDNV, such as *Gyllodes* spp. (Szelei et al., 2011; Weissman et al., 2012; Eilenberg et al., 2015).

16.3.3.2 Sea Star-Associated Densovirus

Sea star wasting disease (SSWD) is a good example of an emerging disease of invertebrates that has recently received attention in both the popular and the scientific communities. SSWD presents initially as lesions in the dermis that progress to cause detachment of limbs, deflation (loss of turgor), and eventual mortality (Dungan et al., 1982; Bates et al., 2009; Eisenlord et al., 2016). SSWD has impacted local populations of sea star species since the 1970s (Dungan et al., 1982; Eckert, 1999; Bates et al., 2009), but a major epizootic of SSWD occurred along the west coast of North America from 2013 through 2015, with peak prevalence during the summer of 2014 (Menge et al., 2016). The 2014 epizootic was startling in both the number of species affected and the geographic range of affected populations. Over 20 species of asteroids were affected (Eisenlord et al., 2016), compared to one or a few species (<10) per affected region in previous events (Dungan et al., 1982; Eckert et al., 1999; Bates et al., 2009). Previous reports of SSWD came from British Columbia (Bates et al., 2009) and California (Dungan et al., 1982; Eckert et al., 1999), and reported cases were isolated to these specific regions. In contrast, over the course of the 2014 epizootic, reports of infected sea stars ranged from the Baja peninsula of Mexico to southern Alaska (Eisenlord et al., 2016; Menge et al., 2016). Unlike the previous reports, which never confirmed disease causation, the massive mortalities seen in 2014 were linked to DNV infection, subsequently named sea star-associated densovirus (SSaDV) (Hewson et al., 2014). An epizootic of this magnitude (both geographically and in terms of severity) had never before been described in marine ecosystems for a noncommercial species (Eisenlord et al., 2016). Sea stars are keystone species in the rocky intertidal and subtidal ecosystem of the west coast of North America (Paine, 1966; Menge et al., 1994), and the ecological impact of this massive mortality event is currently under investigation.

A key factor linked to SSWD risk is temperature. Previous reports of SSWD, as well as the recent epizootic, occurred during warmer temperature regimes (Eckert et al., 1999; Bates et al., 2009; Staehli et al., 2009; Eisenlord et al., 2016). Summer sea-surface temperatures in 2014 were the warmest ever recorded for the northwest coast of North America (Eisenlord et al., 2016). Laboratory experiments showed that risk of mortality and rate of disease progression were higher with increased temperatures (at 12 vs. 19 °C) (Eisenlord et al., 2016), and that lower temperatures (8 vs. 12 °C) resulted in slower progression rates (Kohl et al., 2016). Transmission and mechanisms of infection of SSaDV are largely unknown. Laboratory experiments have shown that SSaDV is waterborne, and it is not thought to be vector-borne, although other echinoderms may be reservoirs of the virus (Hewson et al., 2014). The high abundance of individuals affected simultaneously across many regions suggests a general increase in disease susceptibility across sea star populations, perhaps linked to the increased temperatures observed during disease outbreaks. Interestingly, the effect of temperature in other Densovirus infection systems has either not been studied or has produced conflicting results. For example, one study found that the replication of infectious hypodermal and hematopoietic necrosis virus (IHHNV, Densoviridae) in shrimp was reduced in warm conditions (24 vs. 32 °C), suggesting that warm temperatures will not lead to outbreaks of IHHNV (Montgomery-Brock et al., 2007). However, in another study reporting the same temperature range, IHHNV prevalence was found to be best correlated (positively) to temperature (Chai et al., 2016). In light of this, the link between increased temperature and emergence of SSWD and how it might influence host–pathogen dynamics remains to be further investigated.

The 2014 SSWD epizootic has shifted the population structure for many sea star species and is predicted to have major impacts on ecosystem function in the intertidal shores along the west coast of North America (Eisenlord et al., 2016; Menge et al., 2016). Many questions remain about this epizootic, especially regarding why there was such an increased host range and whether this epizootic was triggered by changes in virulence of SSaDV. No management strategies aimed at reducing spread or mortality were tested during the course of the epizootic; however, research organizations and management agencies utilized citizen science reporting to great effect in monitoring the spatial distribution of the disease. The level of public involvement, media attention, and concern that was displayed is heartening and has rallied political support and funding for the investigation and mitigation of future SSWD epizootics. In response to the 2014 epizootic, a bill was introduced to the US House of Representatives, the Marine Disease Recovery Act of 2015, that, if passed, will allow for rapid responses to future marine disease outbreaks.

16.3.4 Emerging Pathogens of Pollinators

Bees and other insect pollinators are both ecologically and economically important, and increased attention has been brought recently to perceived declines of domestic and wild bee populations. Available evidence supports the perception of losses of wild bee populations and of changes in domesticated populations (losses in Europe and North America, but increases in Asia, leading to a net gain) (Goulson et al., 2015). However, the best-characterized EIDs of pollinators are those of domesticated bees, especially in Europe and North America, and EIDs and other pathogens linked with bee health are of growing concern. Colony losses associated with colony collapse disorder (CCD) (see Section 16.3.4.3) in the United States or with overwintering mortality in the North Hemisphere can be related to EIDs in some cases. Globally, insect pollinators benefit the production of 75% of crops to the tune of an estimated US$215 billion (Gallai et al., 2009). Due to the necessity of pollinators in crop production, movement and release of bees have occurred for thousands of years, allowing for the invasion of non-native bee species and their parasites (Goulson and Hughes, 2015). Historic movements occurred without knowledge of natural parasite distributions in terms of geographic distribution and host specificity, making it difficult to determine when a non-native bee parasite arrived and spread in a particular population. The most widespread pollinator is the honeybee (*Apis mellifera*) (believed to be native to Africa, western Asia, and South East Asia), which has been deliberately introduced to every continent except Antarctica (Michener, 1979). Bumblebees (*Bombus* spp.) are also important pollinators, and are produced and exported on a global scale (Goulson and Hughes, 2015). Although most bee parasites were likely spread historically, contemporary evidence suggests EIDs in multiple honeybee and bumblebee species have occurred recently via multiple spillover events facilitated by the domestication and movement of pollinators globally (see review by Goulson and Hughes, 2015). Many of these bee parasites (e.g., deformed wing virus (DWV) and *Nosema ceranae*) appear to have broad host ranges, which likely facilitates emergence in new species.

16.3.4.1 *Varroa destructor* and Deformed Wing Virus

The best-known example of an EID in bees is spillover of the mite *Varroa destructor* (formerly *V. jacobsoni*) from *Apis cerana* to *A. mellifera*, beginning in approximately the 1960s (likely when *A. mellifera* colonies were transported to Eastern Russia or East Asia). As of 2016, *V. destructor* has a nearly cosmopolitan distribution (all major land masses with substantial honeybee populations, except Australia and Newfoundland) (McMahon et al., 2016). Four separate species of *Varroa* mites infect bee species, although *V. destructor* is the only mite of significant economic importance (Rosenkranz et al., 2010). *V. destructor* infestation can lead to colony death within 2–3 years in temperate regions if mite control is not practiced by bee-keepers (see Rosenkranz et al., 2010, table 2, for common and potential treatment options). The host specificity of *V. destructor* is fortunately quite narrow, and to date this mite is unable to survive on bees outside the genus *Apis*.

V. destructor also acts as a vector of several RNA viruses, most notably DWV (McMahon et al., 2015), but also viruses belonging to the acute bee paralysis virus complex, Kashmir bee virus, Israeli acute bee paralysis virus, black queen cell virus, sacbrood virus, and the slow bee paralysis virus (Mondet et al., 2014). The combined effect of *V. destructor* plus DWV and other diseases is a major contributor to recent honeybee colony losses in North America and Europe (Rosenkranz et al., 2010).

V. destructor's association as a vector of DWV has led to increased viral virulence and prevalence of DWV, which otherwise persists at low levels in the hive (reviewed by de Miranda and Genersch, 2010). Infection with DWV leads to development of "deformed wings" (within adults), hence the name "deformed wing virus." When feeding, *V. destructor* mites inject DWV directly into the host hemolymph (de Miranda and Genersch, 2010; Ryabov et al., 2014), thus bypassing natural barriers to horizontal and some types of vertical transmission, such as the exoskeleton and membranes of the digestive tract. In addition to association with *V. destructor*, DWV can also be transmitted between bees by horizontal (fecal–cannibal–oral) and vertical transmission, and can infect all life stages (reviewed by de Miranda and Genersch, 2010). DWV also appears to amplify in *V. destructor*, either directly through replication (Gisder et al., 2009; Ryabov et al., 2014) or via accumulation of virus particles in the gut, or both (Erban et al., 2015). DWV is globally distributed with honeybees, and, based on phylogentically inferred migration of DWV and the known global spread of *V. destructor* (Wilfert et al. 2016), may be a re-emerging pathogen, spread from a common ancestor, *A. mellifera* (as compared to *V. destructor* or *A. cerana*, but not other hosts) (Wilfert et al., 2016). Recent invasions of *V. destructor* into Hawaii and New Zealand led to an increase of DWV prevalence concomitant with higher viral load (Martin et al., 2012; Mondet et al., 2014), and, interestingly, to the dominance of a single DWV strain in the Hawaiian introduction (Martin et al., 2012). These invasions indicate that the presence of *V. destructor* increases the spread of DWV (Wilfert et al., 2016). DWV is also present in honeybee populations where *V. destructor* is not present, and its spread may be mediated by other human-mediated spread (Wilfert et al., 2016).

Interestingly, multiple strains of DWV (DWV-A, B, and C) exist (Martin et al., 2012, 2016; Mordecai et al., 2016), and DWV is considered – like many other RNA viruses – to exist as a quasi-species with high genetic heterogeneity (Mordecai et al., 2016). It is unclear whether the quasi-species nature of DWV allows it to continually evade host immunity, enhance virulence, or move into new host species. The global phylogeography of DWV strains is not well understood, and strain is not always indicated in analyses. McMahon et al. (2016) recently demonstrated increased virulence of DWV-B (as compared to DWV-A) and the potential for increased viral evolution in an *in vivo* model. The knowledge of DWV genotypes and diversity may be critical in further understanding the spread of DWV (McMahon et al., 2016).

In addition to the re-emergence and increased importance of DWV as a pathogen of domesticated *A. mellifera*, it is likely also an emerging pathogen of wild pollinators, especially bumblebees (which can also be infected in commercial production). DWV has also been detected in solitary bees, although the infectivity or pathology is unknown in this case (reviewed by Goulson and Hughes, 2015).

16.3.4.2 Spillover and Spread of *Nosema ceranae*

Like the spillover of *V. destructor*, the microsporidian fungus *Nosema ceranae* is believed to have jumped from *A. cerana* to *A. mellifera* and native bumblebee and honeybee hosts (Graystock et al., 2013; Goulson and Hughes, 2015; Goulson et al., 2015). Within bee hosts, several *Nosema* spp. are believed to be natural parasites; for example, *N. apis* is the natural parasite of *A. mellifera*, and *N. ceranae* of *A. cerana*. In acute cases of *Nosema* infection, often called "nosemosis," dysentery signs or brown fecal marks are seen on the comb and the front of the hive, with sick and dead bees in the vicinity, but often these outward signs are not present even when the disease is sufficient to cause bee mortalities and losses in honey production and pollination efficiency (World Organisation for Animal Health, 2013). "Nosemosis" in *A. mellifera* hives was originally attributed to *N. apis*, although molecular tools have since determined *N. ceranae* to be widespread among tested populations in the Americas, Europe, Asia, and Oceania (Huang et al., 2007; Klee et al., 2007; Chen et al., 2008), with *N. ceranae* infections named nosemosis type C and *N. apis* infections nosemosis type A (Higes et al., 2010). In fact, the clinical dysentery signs on hive structures may be a sign of *N. apis* infection rather than *N. ceranae*, which appears to cause colony mortalities without clinical signs (Higes et al., 2010), and leads to mortalities year-round rather than die-offs in the winter (Klee et al., 2007). *Nosema ceranae* is primarily a parasite of adult bees, and transmission is by the fecal–oral route. *Nosema ceranae* spores are present on bees, pollen, and hive materials and can remain viable for at least a year (Fenoy et al., 2009; Higes et al., 2010). Bees ingest these spores, which germinate in the midgut and infect epithelial cells, where they replicate. Bees subsequently excrete spores in the feces, which infect additional hosts.

In the last 20 years, *N. ceranae* has become prevalent and widespread in Europe and the Americas (reviewed by Goulson et al., 2015). In addition, it has been detected in several Asian *Apis* species, in wild bumblebees in Europe, China, and South America, and in solitary bees in Europe (see Goulson et al., 2015, figure 2). Although the impact of *N. ceranae* on wild populations is unknown, in the laboratory, Graystock et al. (2013) found that *N. ceranae* is more virulent to buff-tailed bumblebees (*Bombus terrestris audax*) than to honeybees. Human-mediated contact between *A. mellifora* and wild bee species likely mediated the spread of *N. ceranae* through infected colonies, equipment, and pollinators.

16.3.4.3 Multi-stressors, Bee Mortalities and Control Measures

CCD, which is a specific case example of colony mortality, cannot be tied to any one cause (even EIDs such as *V. destructor* and associated viruses or *N. ceranae*) (vanEngelsdorp et al., 2009). Other causes of bee colony losses include habitat loss, pesticides, poor diet, shipping fever, competition, and climate change (reviewed in Goulson et al., 2015). CCD is considered an interaction of these stressors. For example, poor diet (caused by mass monoculture production of crops) may lead to immune-compromised bees, which can reduce the bees' ability to cope with toxins and pathogens (Goulson et al., 2015). The interactions of pesticides and pathogens can also lead to compromised immunity, and pesticide exposure alone (i.e., neonicotinoids and the insecticide fipronil) can lead to increased infection and mortality in bees exposed to either DWV or *N. ceranae* (Goulson and Hughes, 2015; Goulson et al., 2015).

Control measures for mitigating the effects of the future emergence of diseases of pollinators are important, especially given the past spread of bee pathogens. Goulson and Hughes (2015) provide a thorough discussion of potential control measures. In order to directly mitigate disease, potential control measures include the use of parasite-free bees, feeding with sterile food, practice of good hygiene (i.e., sterilization of hive materials and clothing), use of native bee species, reduction of interactions between domestic and wild populations (i.e., limiting escapes), use of pathogen control measures (e.g., treatment for *V. destructor*), and limiting of stressors such as shipping. In addition, disease screening should occur prior to the shipping of bees and upon their arrival (since shipping stress may lead to disease expression), and periodically over the production cycle. In addition, natural bee populations should be monitored for bee parasites both prior to and after development of an apiary.

16.3.5 Emergent Coral Diseases

Coral reefs are among the most productive and biologically diverse ecosystems on earth (Connell, 1978). Corals are the keystone species that protect coastal land, provide beach sand, and drive industries such as tourism and fisheries. In addition, corals are valuable indicator organisms for evaluating the impact of environmental change (Shinn 1966; Johannes, 1975; Rogers, 1990). However, they are declining at an alarming rate. Coral cover on Caribbean reefs has decreased by an average of 80%, and Indo-Pacific reefs have lost an estimated 50% cover in the last 30 years (Gardner et al., 2003; Hughes et al., 2003; Bruno and Selig, 2007). Coral diseases are considered a main contributor to these reported declines (Harvell et al., 2002).

Coral diseases are becoming more widespread, and reports of novel coral diseases have increased worldwide (reviewed in Sutherland et al., 2004). Commonly, biotic diseases in corals present as progressive tissue loss, which exposes the coral skeleton, and these diseases are usually described as "white band diseases" or "white syndromes." Of the numerous coral diseases characterized, only a few etiologies have been elucidated (Sutherland et al., 2004). Pathogens of corals include bacteria, cyanobacteria, fungi, protists, and multicellular parasites, although most of the coral pathogens identified to date are culturable bacteria (reviewed in Rosenberg et al., 2007). Some bacterial coral pathogens include the coliform *Serratia marcescens*, which infects *Acropora palmata* (Patterson et al., 2002), *Vibrio shiloi*, which attacks the zooxanthellae of *Oculina patago-nica* (Kushmaro et al., 1997), and *Vibrio coralliilyticus*, which causes tissue loss in a variety of coral species (Ben-Haim et al., 2003; Sussman et al., 2008; Ushijima et al., 2014). Increasingly, research on coral diseases has aimed at better predicting disease outbreaks and identifying mechanisms of disease resistance. However, there are many aspects of coral diseases that are poorly understood, such as the relationship between human activities and their incidence/emergence (see Table 16.1). These links will need to be better understood if we are to preserve our remaining coral reef ecosystems. Although many coral diseases have been described, two of the most important, and the two best investigated, are black band disease (BBD) and acroporid serratiosis (APS). These examples highlight how coral diseases can have vastly different ecologies and different interactions with anthropogenic activities.

16.3.5.1 Black Band Disease

BBD was the first tissue-loss disease ever described in coral, reported in 1973 off the coast of Brazil (Antonius, 1973). Although this disease has been known since the 1970s, it can still be considered an emerging disease because novel outbreaks are being reported at new geographic locations. For example, it was not confirmed in the Hawaiian archipelago until 2012 (Aeby et al., 2015). BBD is present in many regions of the world, usually at low prevalence (Edmunds, 1991; Bruckner et al., 1997; Dinsdale, 2002; Weil et al., 2002; Page and Willis, 2006). It is considered impactful because it can persist in an ecosystem for years, contributing to its long-term decline (Bruckner et al., 1997; Page and Willis, 2006; Montano et al., 2015). BBD is not host-specific, meaning it affects a wide variety of coral species in different families with disparate life-history strategies (Richardson, 2004; Sutherland et al., 2004). The etiology of BBD is complex, being caused by a consortium of microbial pathogens that includes a filamentous cyanobacterium that creates the characteristic black band, sulfide-oxidizing bacteria, sulfate-reducing bacteria, and numerous heterotrophic bacteria (Cooney et al., 2002; Frias-Lopez et al., 2003; Barneah et al., 2007; Arotsker et al., 2009; Rasoulouniriana et al., 2009; Aeby et al., 2015). In addition to the complex makeup of the BBD disease lesion, the interactions among the BBD microbial consortia are complex, with the cyanobacteria and sulfide-oxidizing bacteria displaying vertical migration within the microbial mat in a diel cycle, which results in dynamic vertical gradients of toxic concentrations of oxygen and sulfide. Changes in BBD prevalence display seasonality, with most active infections occurring during late summer and fall, and with prevalence related to summer seawater temperatures (Rutzler et al., 1983; Edmunds, 1991; Kuta and Richardson, 2002). This correlation between seawater temperature and BBD prevalence shows that the disease will likely be enhanced by global anthropogenic-induced climate change (Hansen et al., 2006; Sato et al., 2016).

16.3.5.2 Acroporid Serratosis

A coral disease of concern, displaying disparate disease ecology, is APS, a form of white pox disease (WPX) (Patterson et al., 2002). The diagnostic criterion for APS is a WPX lesion for which the bacterial pathogen has been positively identified. This criterion is important because not all apparent white pox lesions have resulted in detectable *S. marcescens* (Sutherland et al., 2010). WPX is a progressive tissue-loss disease that was first described in 2002 and has caused in excess of 70% loss of coral cover in the Florida Keys, USA (Patterson et al., 2002). Currently, APS has only been positively diagnosed within the Caribbean basin, with the highest number of confirmed cases in the Florida Keys. APS is highly host-specific, only infecting the coral *Acropora palmata*, and its etiology is limited to particular strains of one pathogen, the enteric bacterium *S. marcescens* (Sutherland et al., 2011). Although the rate of progression of the APS lesions is faster with higher seawater temperature (Patterson et al., 2002), WPX prevalence was historically not higher during the warmer summer months (Santavy et al., 2001), although recent surveys have shows peak WPX prevalence during the summer (Sutherland et al., 2016). The source of *S. marcescens* has been found to be leakage of untreated human sewage into coastal waters (Sutherland et al., 2010, 2011). Therefore, a local human impact – specifically, poor wastewater management – is a major driver of APS. Furthermore, advanced wastewater treatment has been found to reduce *S. marcescens* to undetectable levels, which represents an effective strategy for reducing coral exposure to this pathogen and aiding the recovery of *A. palmata*, which is now listed as an endangered species (Hogarth, 2006).

The different etiologies and ecologies of coral diseases, highlighted by the examples of BBD and APS, emphasize the need to investigate coral diseases with a multifaceted approach. More importantly, the differing human impacts related to BBD versus APS shows that many tools and strategies (on both local and global scales) will be needed to not only better understand why these diseases have emerged in coral reef ecosystems but also mitigate future disease outbreaks.

16.3.5.3 Problems Facing Coral Disease Investigations

Even though BBD and APS are well understood, the majority of coral diseases are not. Many challenges face coral disease epizootiologists. The limited visual responses to infection make in-field observations of lesions hard to link with specific etiologies. This is highlighted in the example of lesions described as "white pox" not being considered APS unless there is positive identification of *S. marcescens* in the lab. Another example includes tissue loss in *Montipora capitata*, found in the Hawaiian Islands, which can be associated with multiple potential etiologies, including bacterial infections, ciliates, and an interspecific coral chimerism (Work et al., 2011, 2012; Ushijima et al., 2012, 2014). The limited variations of gross lesions in corals and the lack of histological investigation into the cellular response of reported lesions have also resulted in much confusion around the naming system for coral diseases (Work and Aeby, 2006; Pollock et al., 2011). For example, the phrase "white plague" is used to describe different types of lesions and is categorized based on either host species or rate of progression (Sutherland et al., 2004), with the latter being prone to variability due only to differential subjective assessment in the field. The lack of a systematic classification of coral lesion types can result in cases of misidentified diseases, name changes for the same disease, and predation scars being mistaken for disease. Many coral diseases do not have established etiologies, although they are assumed to be caused by infectious agents, making further investigation about transmission of pathogens impossible. For those coral diseases with established etiologies, another impediment to investigation is the lack of robust and efficient diagnostic measures (Pollock et al., 2011; Sutherland et al., 2016). Addressing these challenges will help reef managers discern the threats that enhance the severity of diseases and perhaps look into preventative measures (Bruckner, 2002).

Few preventative measures commonly used in terrestrial ecosystems (e.g., culling, vaccination, quarantine) are possible or practical for use in marine ecosystems or with invertebrates; however, a few methods have been conducted to limit the spread of coral disease. Individual treatment of coral colonies with active lesions has been attempted with varying degrees of success. Attempted treatments have included lesion occlusion of BBD, where the cyanobacterial mat is removed from the coral surface and the exposed lesion front is covered with a marine epoxy (Hudson, 2000; Aeby et al., 2015), and lesion removal, where tissue-loss lesions or growth anomalies are mechanically removed from the affected colony (Dalton et al., 2010; Williams, 2013). However, the authors warn that these treatments do not address the underlying sources of pathogen transmission or the environmental cofactors that facilitate disease emergence. Furthermore, they stress that the treatments will only be effective at local scales (Dalton et al., 2010; Aeby et al., 2015). Additional management actions that have been discussed include boosting coral immunity, performing phage therapy, and administering probiotics (summarized in Beeden et al., 2012).

A preventative strategy that has the potential to address concerns has been implemented in the Florida Keys. In response to human fecal contamination of nearshore coastal and reef environments, the State of Florida passed legislation in 1999 to upgrade all wastewater facilities to advanced treatment by 2010. Advanced wastewater treatment successfully removes *S. marcescens*, the causative agent of APS, from human sewage effluent (Sutherland et al., 2010; Joyner et al., 2015). In addition, confirmation of *S. marcescens* in WPX sessions has declined in recent years (Sutherland et al., 2015). Although many reports throughout the Caribbean still show low population sizes of *A. palmata*, some studies have reported recoveries in the US Virgin Islands (Muller et al., 2014).

16.4   Conclusion

Human-coupled environments have provided opportunities for the emergence of infectious diseases in the most diverse and abundant group of organisms, the invertebrates. With growing human populations and climate change, the implications of increased anthropogenic effects on both wild and domesticated invertebrate populations are clear. Management of infectious diseases and multi-stressors affecting invertebrates is necessary to continue production of both terrestrial and marine crops. In addition to EIDs affecting vertebrate animals, invertebrates can vector human or animal pathogens, which should be a consideration in disease management. An understanding of natural pathogens in wild populations is increasingly necessary to protecting the biodiversity of both invertebrates (e.g., coral reef ecosystems, natural pollinators, sea stars) and their associated communities of organisms, at both the microbial and the organismal level. Disease diagnoses should take advantage of modern and classical techniques to discover pathogens and develop robust diagnostic methods. Public awareness of EIDs in invertebrates may further our ability to rapidly diagnose and stop the spread of invertebrate disease or to facilitate collection of the epidemiological data necessary to understand transmission and spread and to develop models and forecasting tools. In addition, public awareness may lead to legislative changes that can additionally impact human health. Disease can be impacted by pollutants in the environment, but we still know very little about the interaction of pathogens and pollutants. The magnitude of invertebrate mortality seen in recent EIDs is staggering. Unfortunately, changes in ecosystem structure and function have already been reported as a consequence of some of these invertebrate EIDs. The long-term implications of these ecosystem changes will need to be investigated but are likely to negatively impact biodiversity and human use of natural resources. Given the high impact of invertebrate EIDs, we need to be ready to detect and respond to the next invertebrate EID.

Acknowledgments

This research was financed in part by funding to C. Burge, including a grant from the National Institute of Standards and Technology (0010509) and start-up funds provided by the University of Maryland Baltimore County and the University of Maryland Baltimore.

References

Aeby, G.S., Callahan, S., Cox, E.F., Runyon, C., Smith, A., Stanton, F.G., et al., 2016. Emerging coral diseases in Kane'ohe Bay, O'ahu, Hawai'i (USA): two major disease outbreaks of acute *Montipora* white syndrome. Dis. Aquat. Org. 119, 189–198.

Aeby, G.S., Work, T.M., Runyon, C.M., Shore-Maggio, A., Ushijima, B., Videau, P., et al., 2015. First record of black band disease in the Hawaiian archipelago: response, outbreak status, virulence, and a method of treatment. PLoS ONE 10, e0120853.

Afanasiev, B.N., Galyov, E.E., Buchatsky, L.P., Kozlov, Y.V., 1991. Nucleotide sequence and genornic organization of *Aedes* densonucleosis virus. Virology 185, 323–336.

Alonso, A., Sanchez, P., Martinez, J.L., 2001. Environmental selection of antibiotic resistance genes. Environ. Microbiol. 3, 1–9.

Altizer, S., Harvell, D., Friedle, E., 2003. Rapid evolutionary dynamics and disease threats to biodiversity. Trends Ecol. Evol. 18, 589–596.

Altizer, S., Ostfeld, R.S., Johnson, P.T., Kutz, S., Harvell, C.D., 2013. Climate change and infectious diseases: from evidence to a predictive framework. Science 341, 514–519.

Anderson, J.L., Valderrama, D., Jory, D., 2014. Global shrimp production review. Global Aquac. Advocate 17, 10–11.

Andreadis, T.G., Weseloh, R.M., 1990. Discovery of *Entomophaga maimaiga* in North American gypsy moth, *Lymantria dispar*. Proc. Natl. Acad. Sci. U.S.A. 87, 2461–2465.

Angermeier, H., Kamke, J., Abdelmohsen, U.R., Krohne, G., Pawlik, J.R., Lindquist, N.L., Hentschel, U., 2011. The pathology of sponge orange band disease affecting the Caribbean barrel sponge *Xestospongia muta*. FEMS Microbiol. Ecol. 75, 218–230.

Antonius, A., 1973. New observations on coral destruction in reefs. Assoc. Isl. Mar. Lab. Caribb. 10, 3.

Arotsker, L., Siboni, N., Ben-Dov, E., Kramarsky-Winter, E., Loya, Y., Kushmaro, A., 2009. *Vibrio* sp. as a potentially important member of the Black Band Disease (BBD) consortium in *Favia* sp. corals. FEMS Microbiol. Ecol. 70, 515–524.

Arzul, I., Corbeil, S., Morga, B., Renault, T., 2017. Viruses infecting marine molluscs. J. Invertebr. Pathol. https://doi.org/10.1016/j.jip.2017.01.009.

Arzul, I., Nicolas, J.L., Davison, A.J., Renault, T., 2001a. French scallops: a new host for ostreid herpesvirus-1. Virology 290, 342–349.

Arzul, I., Renault, T., Lipart, C., 2001b. Experimental herpes-like viral infections in marine bivalves: demonstration of interspecies transmission. Dis. Aquat.Org. 46, 1–6.

Arzul, I., Renault, T., Lipart, C., Davison, A.J., 2001c. Evidence for interspecies transmission of oyster herpesvirus in marine bivalves. J. Gen. Virol. 82, 865–870.

Austin, B., Zhang, X.H., 2006. *Vibrio harveyi*: a significant pathogen of marine vertebrates and invertebrates. Lett. Appl. Microbiol. 43, 119–124.

Bai, C., Wang, C., Xia, J., Sun, H., Zhang, S., Huang, J., 2015. Emerging and endemic types of Ostreid herpesvirus 1 were detected in bivalves in China. J. Invertebr. Pathol. 124, 98–106.

Barneah, O., Ben-Dov, E., Kramarsky-Winter, E., Kushmaro, A., 2007. Characterization of black band disease in Red Sea stony corals. Environ. Microbiol. 9, 1995–2006.

Bates, A.E., Hilton, B.J., Harley, C.D., 2009. Effects of temperature, season and locality on wasting disease in the keystone predatory sea star *Pisaster ochraceus*. Dis. Aquat. Org. 86, 245–251.

Batista, F.M., López-Sanmartín, M., Grade, A., Morgado, I., Valente, M., Navas, J.I., et al., 2015. Sequence variation in ostreid herpesvirus 1 microvar isolates detected in dying and asymptomatic *Crassostrea angulata* adults in the Iberian Peninsula: insights into viral origin and spread. Aquaculture 435, 43–51.

Beeden, R., Maynard, J.A., Marshall, P.A., Heron, S.F., Willis, B.L., 2012. A framework for responding to coral disease outbreaks that facilitates adaptive management. Environ. Manag. 49, 1–13.

Ben-Haim, Y., Thompson, F.L., Thompson, C.C., Cnockaert, M.C., Hoste, B., Swings, J., Rosenberg, E., 2003. *Vibrio coralliilyticus* sp. nov., a temperature-dependent pathogen of the coral *Pocillopora damicornis*. Int. J. Syst. Evol. Microbiol. 53, 309–315.

Ben-Horin, T., Bidegain, G., Huey, L., Narvaez, D.A., Bushek, D., 2015. Parasite transmission through suspension feeding. J. Invertebr. Pathol. 131, 155–176.

Bergoin, M., Tijssen, P., 2010. Densoviruses: a highly diverse group of arthropod parvoviruses, in: Asgari, S., Johnson, K.N. (eds.), Insect Virology. Caster Academic Press, Norwich, pp. 59–72.

Bik, E., Bunschoten, A., Gouw, R., Mooi, F., 1995. Genesis of the novel epidemic *Vibrio cholerae* O139 strain: evidence for horizontal transfer of genes involved in polysaccharide synthesis. EMBO J. 14, 209–216.

Bingham, P., Brangenberg, N., Williams, R., van Andel, M., 2013. Investigation into the first diagnosis of ostreid herpesvirus type 1 in Pacific oysters. Surveill. 40, 20–24.

Bochow, S., Condon, K., Elliman, J., Owens, L., 2015. First complete genome of an Ambidensovirus; *Cherax quadricarinatus* densovirus, from freshwater crayfish *Cherax quadricarinatus*. Mar. Genom. 24, 305–312.

Bostock, J., McAndrew, B., Richards, R., Jauncey, K., Telfer, T., Lorenzen, K., Little, D., et al., 2010. Aquaculture: global status and trends. Philos. Trans. R. Soc. Lond. B Biol. Sci. 365, 2897–2912.

Bruckner, A.W., 2002. Priorities for effective management of coral diseases. NOAA Fisheries. Available from: http://www.bio-nica.info/biblioteca/bruckner-managementcoraldiseases.pdf (accessed May 8, 2017).

Bruckner, A.W., Bruckner, R.J., Williams, E.H. Jr., 1997. Spread of black-band disease epizootic through the coral reef system in St. Ann's Bay, Jamaica. Bull. Mar. Sci. 61, 919–928.

Bruno, J.F., Selig, E.R., 2007. Regional decline of coral cover in the Indo-Pacific: timing, extent, and subregional comparisons. PLoS ONE 2, e711.

Burge, C.A., Friedman, C.S., 2012. Quantifying Ostreid herpesvirus (OsHV-1) genome copies and expression during transmission. Microb. Ecol. 63, 596–604.

Burge, C.A., Friedman, C.S., Getchell, R., House, M., Lafferty, K.D., Mydlarz, L.D., et al., 2016. Complementary approaches to diagnosing marine diseases: a union of the modern and the classic. Philos. Trans. R. Soc. Lond. B Biol. Sci. 371, 20150207.

Burge, C.A., Griffin, F.J., Friedman, C.S., 2006. Mortality and herpesvirus infections of the Pacific oyster *Crassostrea gigas* in Tomales Bay, California, USA. Dis. Aquat. Org. 72, 31–43.

Burge, C.A., Judah, L.R., Conquest, L.L., Griffin, F.J., Cheney, D.P., Suhrbier, A., et al., 2007. Summer seed mortality of the Pacific oyster, *Crassostrea gigas* Thunberg grown in Tomales Bay, California, USA: the influence of oyster stock, planting time, pathogens, and environmental stressors. J. Shellfish Res. 26, 163–172.

Burge, C.A., Kim, C.J., Lyles, J.M., Harvell, C.D., 2013. Special issue oceans and human health: the ecology of marine opportunists. Microb. Ecol. 65, 869–879.

Burge, C.A., Mark Eakin, C., Friedman, C.S., Froelich, B., Hershberger, P.K., Hofmann, E.E., et al., 2014. Climate change influences on marine infectious diseases: implications for management and society. Ann. Rev. Mar. Sci. 6, 249–277.

Burge, C.A., Strenge, R.E., Friedman, C.S., 2011. Detection of the oyster herpesvirus in commercial bivalves in northern California, USA: conventional and quantitative PCR. Dis. Aquat. Org. 94, 107–116.

Chai, C., Liu, Y., Xia, X., Wang, H., Pan, Y., Yan, S., Wang, Y., 2016. Prevalence and genomic analysis of infectious hypodermal and hematopoietic necrosis virus (IHHNV) in *Litopenaeus vannamei* shrimp farmed in Shanghai, China. Arch. Virol 161, 3189–3201.

Chang, P.H., Kuo, S.T., Lai, S.H., Yang, H.S., Ting, Y.Y., Hsu, C.L., Chen, H.C., 2005. Herpes-like virus infection causing mortality of cultured abalone *Haliotis diversicolor supertexta* in Taiwan. Dis. Aquat. Org. 65, 23–27.

Chen, Y., Evans, J.D., Smith, I.B., Pettis, J.S., 2008. *Nosema ceranae* is a long-present and wide-spread microsporidian infection of the European honey bee (*Apis mellifera*) in the United States. J. Invertebr. Pathol. 97, 186–188.

Chen, Y.-Y., Chen, J.-C., Tseng, K.-C., Lin, Y.-C., Huang, C.-L., 2015. Activation of immunity, immune response, antioxidant ability, and resistance against *Vibrio alginolyticus* in white shrimp *Litopenaeus vannamei* decrease under long-term culture at low pH. Fish Shellfish Immunol. 46, 192–199.

Connell, J.H., 1978. Diversity in rainforests and coral reefs. Science 199, 1302–1310.

Cook, T., Folli, M., Klinck, J., Ford, S., Miller, J., 1998. The relationship between increasing sea-surface temperature and the northward spread of *Perkinsus marinus* (Dermo) disease epizootics in oysters. Estuar. Coast. Shelf Sci. 46, 587–597.

Cooney, R.P., Pantos, O., Le Tissier, M.D.A., Barer, M.R., O'Donnell, A.G., Blythell, J.C., 2002. Characterization of the bacterial consortium associated with black band disease in coral using molecular microbiological techniques. Environ. Microbiol. 4, 401–413.

Cowart, J.D., Henkel, T.P., McMurray, S.E., Pawlik, J.R., 2006. Sponge orange band (SOB): a pathogenic-like condition of the giant barrel sponge, *Xestospongia muta*. Coral Reefs 25, 513–513.

Crosa, J.H., 1980. A plasmid associated with virulence in the marine fish pathogen *Vibrio anguillarum* specifies an iron-sequestering system. Nature 284, 566–568.

Cuthbertson, A.G., Wakefield, M.E., Powell, M.E., Marris, G., Anderson, H., Budge, G.E., et al., 2013. The small hive beetle *Aethina tumida*: a review of its biology and control measures. Curr. Zool. 59, 644–653.

Dalton, S.J., Godwin, S., Smith, S.D.A., Pereg, L., 2010. Australian subtropical white syndrome: a transmissible, temperature-dependent coral disease. Mar. Freshwater Res. 61, 342–350.

Daszak, P., Cunningham, A.A., Hyatt, A.D., 2000. Wildlife ecology – emerging infectious diseases of wildlife – threats to biodiversity and human health. Science 287, 443–449.

Daszak, P., Cunningham, A.A., Hyatt, A.D., 2001. Anthropogenic environmental change and the emergence of infectious disease in wildlife. Acta Trop. 78, 103–116.

Davison, A.J., Eberle, R., Ehlers, B., Hayward, G.S., McGeoch, D.J., Minson, A.C., et al., 2009. The order Herpesvirales. Arch. Virol. 154, 171–177.

Davison, A.J., Trus, B.L., Cheng, N.Q., Steven, A.C., Watson, M.S., Cunningham, C., et al., 2005. A novel class of herpesvirus with bivalve hosts. J. Gen. Virol. 86, 41–53.

de Miranda, J.R., Genersch, E., 2010. Deformed wing virus. J. Invertebr. Pathol. 103(Suppl. 1), S48–S61.

Defoirdt, T., Boon, N., Sorgeloos, P., Verstraete, W., Bossier, P., 2007. Alternatives to antibiotics to control bacterial infections: luminescent vibriosis in aquaculture as an example. Trends Biotech. 25, 472–479.

Degremont, L., Garcia, C., Allen, S.K., 2015. Genetic improvement for disease resistance in oysters: a review. J. Invertebr. Pathol. 131, 226–241.

Diamond, J., 1999. Guns, Germs, and Steel: The Fates of Human Societies. W.W. Norton and Co., New York.

Diéguez-Uribeondo, J., 2015. The biology of crayfish plague pathogen *Aphanomyces astaci*: Current answers to most frequent questions, in: Kawai, T., Faulkes, Z., Scholts, Z. (eds.), Freshwater Crayfish: A Global Overview. CRC Press, Boca Raton, FL, pp. 182–204.

Dinsdale, E., 2002. Abundance of black-band disease on corals from one location on the Great Barrier Reef: a comparison with abundance in the Caribbean region. Proceedings of the 9th International Coral Reef Symposium 2, 1239–1243.

Department of Primary Industries, 2016. NSW Oyster Industry – Update. Available from: http://www.dpi.nsw.gov.au/__data/assets/pdf_file/0012/595596/nsw-oyster-industry-update-poms-26-feb-2016.pdf/ (accessed May 8, 2017).

Duarte, C.M., Holmer, M., Olsen, Y., Soto, D., Marbà, N., Guiu, J., et al., 2009. Will the oceans help feed humanity? Bioscience 59, 967–976.

Dundon, W.G., Arzul, I., Omnes, E., Robert, M., Magnabosco, C., Zambon, M., et al., 2011. Detection of Type 1 Ostreid Herpes variant (OsHV-1 mu var) with no associated mortality in French-origin Pacific cupped oyster *Crassostrea gigas* farmed in Italy. Aquaculture 314, 49–52.

Dungan, M.L., Miller, T.E., Thomson, D.A., 1982. Catastrophic decline of a top carnivore in the Gulf of California rocky intertidal zone. Science 216, 989–991.

Eckert, G., Engle, J.M., Kushner, D., 1999. Sea star disease and population declines at the Channel Islands. Proceedings of the 5th California Islands Symposium 5, 390–393.

Edgerton, B.F., Evans, L.H., Stephens, S.J., Overstreet, R.M., 2002. Synopsis of freshwater crayfish disease and commensal organisms. Aquaculture 206, 57–135.

Edmunds, P.J., 1991. Extent and effect of black band disease on a Caribbean reef. Coral Reefs 10, 161–165.

Eilenberg, J., Vlak, J.M., Nielsen-LeRoux, C., Cappellozza, S., Jensen, A.B., 2015. Diseases in insects produced for food and feed. J. Insects Food Feed 1, 87–102.

Eisenlord, M.E., Groner, M.L., Yoshioka, R.M., Elliott, J., Maynard, J., Fradkin, S., et al., 2016. Ochre star mortality during the 2014 wasting disease epizootic: role of population size structure and temperature. Philos. Trans. R. Soc. Lond. B Biol. Sci. 371, 20150212.

Erban, T., Harant, K., Hubalek, M., Vitamvas, P., Kamler, M., Poltronieri, P., et al., 2015. In-depth proteomic analysis of *Varroa destructor*: Detection of DWV-complex, ABPV, VdMLV and honeybee proteins in the mite. Sci. Rep. 5, 13907.

European Union, 2010. Commission Regulation EU No. 175/2010: Implementing Council Directive 2006/88/EC as regards measures to control increased mortality in oysters of the species *Crassostrea gigas* in connection with the detection of Ostreid herpesvirus 1 μvar (OsHV-1 μvar). Off. J. Eur. Union. 52, 1–12.

Ewald, P.W., 1995. The evolution of virulence: a unifying link between parasitology and ecology. J. Parasitol. 81, 659–669.

Farley, C.A., Banfield, W.G., Kasnic, G., Foster, W.S., 1972. Oyster herpes-type virus. Science 178, 759–760.

Fédière, G., 2000. Epidemiology and pathology of Densovirinae, in: Faisst, S., Rommelaere, J. (eds.), Parvoviruses: From Molecular Biology to Pathology and Therapeutic Uses. Karger, Basel, pp. 1–11.

Fenoy, S., Rueda, C., Higes, M., Martin-Hernandez, R., del Aguila, C., 2009. High-level resistance of *Nosema ceranae*, a parasite of the honeybee, to temperature and desiccation. Appl. Environ. Microbiol. 75, 6886–6889.

Flegel, T.W., 2012. Historic emergence, impact and current status of shrimp pathogens in Asia. J. Invert. Pathol. 110, 166–173.

Food and Agriculture Organization of the United Nations, 2014. Fisheries statistics extracted with FishStatJ (Copyright 2013). Aquaculture production quantities 1950–2012; Aquaculture production values 1984–2012; Capture production 1960–2012; Fisheries commodities production and trade 1976–2011. Available from: http://www.fao.org/fishery/statistics/software/fishstatj/en (accessed May 8, 2017).

Ford, S.E., 1996. Range extension by the oyster parasite *Perkinsus marinus* into the northeastern United States: response to climate change? Oceanogr. Lit. Rev. 12, 1265.

Frias-Lopez, J., Bonheyo, G.T., Jin, Q., Fouke, B.W., 2003. Cyanobacteria associated with coral black band disease in Caribbean and Indo-Pacific reefs. Appl. Environ. Microbiol. 69, 2409–2413.

Frick, W.F., Puechmaille, S.J., Willis, C.K.R., 2016. White-nose syndrome in bats, in: Voigt, C.C., Kingston, T. (eds.), Bats in the Anthropocene: Conservation of Bats in a Changing World. Springer, Cham, pp. 245–262.

Friedman, C.S., Estes, R.M., Stokes, N.A., Burge, C.A., Hargove, J.S., Barber, B.J., et al., 2005 Herpes virus in juvenile Pacific oysters *Crassostrea gigas* from Tomales Bay, California, coincides with summer mortality episodes. Dis. Aquat. Org. 63, 33–41.

Fuhrmann, M., Petton, B., Quillien, V., Faury, N., Morga, B., Pernet, F., 2016. Salinity influences disease-induced mortality of the oyster *Crassostrea gigas* and infectivity of the ostreid herpesvirus 1 (OsHV-1). Aquac. Environ. Interact. 8, 543–552.

Gallai, N., Salles, J.-M., Settele, J., Vaissière, B.E., 2009. Economic valuation of the vulnerability of world agriculture confronted with pollinator decline. Ecol. Econ. 68, 810–821.

Garcia, C., Thebault, A., Degremont, L., Arzul, I., Miossec, L., Robert, M., et al., 2011. Ostreid herpesvirus 1 detection and relationship with *Crassostrea gigas* spat mortality in France between 1998 and 2006. Vet. Res. 42, 73–84.

Gardner, T.A., Cote, I.M., Gill, J.A., Grant, A., Watkinson, A.R., 2003. Long-term region-wide declines in Caribbean corals. Science 301, 958–960.

Gillings, M.R., 2016. Lateral gene transfer, bacterial genome evolution, and the Anthropocene. Ann. NY Acad. Sci. 1389, 20–36.

Gisder, S., Aumeier, P., Genersch, E., 2009. Deformed wing virus: replication and viral load in mites (*Varroa destructor*). J. Gen. Virol. 90, 463–467.

Gittenberger, A., Voorbergen-Laarman, M., Engelsma, M., 2016. Ostreid herpesvirus OsHV μVar in Pacific oysters *Crassostrea gigas* (Thunberg 1793) of the Wadden Sea, a UNESCO world heritage site. J. Fish Dis. 39, 105–109.

Goulson, D., Hughes, W.O.H., 2015. Mitigating the anthropogenic spread of bee parasites to protect wild pollinators. Biol. Conserv. 191, 10–19.

Goulson, D., Nicholls, E., Botias, C., Rotheray, E.L., 2015. Bee declines driven by combined stress from parasites, pesticides, and lack of flowers. Science 347, 1255957.

Graystock, P., Yates, K., Darvill, B., Goulson, D., Hughes, W.O., 2013. Emerging dangers: deadly effects of an emergent parasite in a new pollinator host. J. Invertebr. Pathol. 114, 114–119.

Green, T.J., Helbig, K., Speck, P., Raftos, D.A., 2016. Primed for success: oyster parents treated with poly(I:C) produce offspring with enhanced protection against Ostreid herpesvirus type I infection. Mol. Immunol. 78, 113–120.

Green, T.J., Montagnani, C., 2013. Poly I:C induces a protective antiviral immune response in the Pacific oyster (*Crassostrea gigas*) against subsequent challenge with Ostreid herpesvirus (OsHV-1 μvar). Fish Shellfish Immunol. 35, 382–388.

Gudenkauf, B.M., Eaglesham, J.B., Aragundi, W.M., Hewson, I., 2014. Discovery of urchin-associated densoviruses (family Parvoviridae) in coastal waters of the Big Island, Hawaii. J. Gen. Virol. 95, 652–658.

Hajek, A.E., 1999. Pathology and epizootiology of the Lepidoptera-specific mycopathogen *Entomophaga maimaiga*. Microbiol. Molecul. Biol. Rev. 63, 814–835.

Han, J.E., Tang, K.F., Tran, L.H., Lightner, D.V., 2015. *Photorhabdus* insect-related (Pir) toxin-like genes in a plasmid of *Vibrio parahaemolyticus*, the causative agent of acute hepatopancreatic necrosis disease (AHPND) of shrimp. Dis. Aquat. Org. 113, 33–40.

Hansen, J.D., Sato, M., Ruedy, R., Lo, K., Lea, D.W., Medina-Elizade, M., 2006. Global temperature change. Proc. Natl. Acad. Sci. U.S.A. 103, 14 288–14 293.

Harvell, C.D., Kim, K., Burkholder, J.M., Colwell, R.R., Epstein, P.R., Grimes, D.J., et al., 1999. Emerging marine diseases – climate links and anthropogenic factors. Science 285, 1505–1510.

Harvell, D., Jordán-Dahlgren, E., Merkel, S., Rosenberg, E., Raymundo, L., Smith, G., et al., 2007. Coral disease, environmental drivers, and the balance between coral and microbial associates. Oceanography 20, 172–195.

Harvell, C.D., Mitchell, C.E., Ward, J.R., Altizer, S., Dobson, A.P., Ostfeld, R.S., Samuel, M.D., 2002. Ecology – climate warming and disease risks for terrestrial and marine biota. Science 296, 2158–2162.

Hedgecock, D., Gaffney, P.M., Goulletquer, P., Guo, X., Reece, K., Warr, G.W., 2005. The case for sequencing the Pacific oyster genome. J. Shellfish Res. 24, 429–441.

Hewson, I., Button, J.B., Gudenkauf, B.M., Miner, B., Newton, A.L., Gaydos, J.K., et al., 2014. Densovirus associated with sea-star wasting disease and mass mortality. Proc. Natl. Acad. Sci. U.S.A. 111, 17 278–17 283.

Higes, M., Martín-Hernández, R., Meana, A., 2010. *Nosema ceranae* in Europe: an emergent type C nosemosis. Apidologie 41, 375–392.

Hine, P.M., Thorne, T., 1997. Replication of herpes like viruses in haemocytes of adult flat oysters *Ostrea angasi*: an ultrastructural study. Dis. Aquat. Org. 29, 189–196.

Hine, P.M., Wesney, B., Hay, B.E., 1992. Herpesviruses associated with mortalities among hatchery-reared larval Pacific oysters *Crassostrea-gigas*. Dis. Aquat. Org. 12, 135–142.

Hogarth, W.T., 2006. Endangered and threatened species: final listing determinations for elkhorn coral and staghorn coral. Fed. Regist. 71, 26 852–26 861.

Holmstrom, K., Graslund, S., Wahlstrom, A., Poungshompoo, S., Bengtsson, B.-E., Kautsky, N., 2003. Antibiotic use in shrimp farming and implications for environmental impacts and human health. Int. J. Food Sci. Tech. 38, 255–266.

Hooper, C., Hardy-Smith, P., Handlinger, J., 2007. Ganglioneuritis causing high mortalities in farmed Australian abalone (*Haliotis laevigata* and *Haliotis rubra*). Aust. Vet. J. 85, 188–193.

Huang, W.F., Jiang, J.H., Chen, Y.W., Wang, C.H., 2007. A *Nosema ceranae* isolate from the honeybee *Apis mellifera*. Apidologie 38, 30–37.

Hudson, J.H., 2000. First aid for massive corals infected with black band disease, *Phormidium corallyticum*: an underwater aspirator and post-treatment sealant to curtail reinfection. Proceedings of the American Academy of Underwater Science, 20th Symposium, pp. 10–11.

Hughes, T.P., Baird, A.H., Bellwood, D.R., Card, M., Connolly, S.R., Folke, C., et al., 2003. Climate change, human impacts, and the resilience of coral reefs. Science 301, 929–933.

Hwang, J.Y., Park, J.J., Yu, H.J., Hur, Y.B., Arzul, I., Couraleau, Y., Park, M.A., 2013. Ostreid herpesvirus 1 infection in farmed Pacific oyster larvae *Crassostrea gigas* (Thunberg) in Korea. J. Fish. Dis. 36, 969–972.

ICTV (International Committee on Taxonomy of Viruses), 2014. Virus Taxonomy, 2014 Release. Available from: http://www.ictvonline.org/virusTaxonomy.asp?msl_id=29 (accessed May 8, 2017).

Jayasree, L., Janakiram, P., Madhavi, R., 2006. Characterization of *Vibrio* spp. associated with diseased shrimp from culture ponds of Andhra Pradesh (India). J. World Aquac. Soc. 37, 523–532.

Jenkins, C., Hick, P., Gabor, M., Spiers, Z., Fell, S.A., Gu, X., et al., 2013. Identification and characterisation of an ostreid herpesvirus–1 microvariant (OsHV- μvar) in *Crassostrea gigas* (Pacific oysters) in Australia. Dis. Aquat. Org. 105, 109–126.

Johannes, R.E., 1975. Pollution and degradation of coral reef communities, in: Wood, E.J.F., Johannes, R.E. (eds.), Elsevier Oceanography Series. Elsevier, Amsterdam, pp. 13–51.

Jones, K.E., Patel, N.G., Levy, M.A., Storeygard, A., Balk, D., Gittleman, J.L., Daszak, P., 2008. Global trends in emerging infectious diseases. Nature 451, U990–U994.

Joyner, J.L., Sutherland, K.P., Kemp, D.W., Berry, B., Griffin, A., Porter, J.W., et al., 2015. Systematic analysis of white pox disease in *Acropora palmata* of the Florida Keys and role of *Serratia marcescens*. Appl. Environ. Microbiol. 81, 4451–4457.

Karunasagar, I., Pai, R., Malathi, G., Karunasagar, I., 1994. Mass mortality of *Penaeus monodon* larvae due to antibiotic-resistant *Vibrio harveyi* infection. Aquaculture 128, 203–209.

Keeling, S.E., Brosnahan, C.L., Williams, R., Gias, E., Hannah, M., Bueno, R., et al., 2014. New Zealand juvenile oyster mortality associated with ostreid herpesvirus 1 – an opportunistic longitudinal study. Dis. Aquat. Org. 109, 231–239.

Klee, J., Besana, A.M., Genersch, E., Gisder, S., Nanetti, A., Tam, D.Q., et al., 2007. Widespread dispersal of the microsporidian *Nosema ceranae*, an emergent pathogen of the western honey bee, *Apis mellifera*. J. Invertebr. Pathol., 96, 1–10.

Kohl, W.T., McClure, T.I., Miner, B.G., 2016. Decreased temperature facilitates short-term sea star wasting disease survival in the keystone intertidal sea star *Pisaster ochraceus*. PLoS ONE 11, e0153670.

Kouassi, N., Peng, J.-X., Li, Y., Cavallaro, C., Veyrunes, J.-C., Bergoin, M., 2007. Pathogenicity of *Diatraea saccharalis* densovirus to host insets and characterization of its viral genome. Virol. Sin. 22, 53–60.

Kumar, V., Roy, S., Meena, D.K., Sarkar, U.K., 2016. Application of probiotics in shrimp aquaculture: importance, mechanisms of action, and methods of administration. Rev. Fish. Sci. Aquac. 24, 342–368.

Kushmaro, A., Rosenberg, E., Fine, M., Loya, Y., 1997. Bleaching of the coral *Oculina patagonica* by *Vibrio* AK-1. Mar. Ecol. Prog. Ser. 147, 159–165.

Kushmaro, A., Rosenberg, E., Fine, M., Haim, Y.B., Loya, Y., 1998. Effect of temperature on bleaching of the coral *Oculina patagonica* by *Vibrio* AK-1. Mar. Ecol. Prog. Ser. 171, 131–137.

Kuta, K.G., Richardson, L.L., 2002. Ecological aspects of black band disease of corals relationships between disease incidence and environmental factors. Coral Reefs 21, 393–398.

La Fauce, K., Owens, L., 2007. Investigation into the pathogenicity of *Penaeus merguiensis* densovirus (PmergDNV) to juvenile *Cherax quadricarinatus*. Aquaculture 271, 31–38.

Le, T.X., Munekage, Y., Kato, S., 2005. Antibiotic resistance in bacteria from shrimp farming in mangrove areas. Sci. Total Environ. 349, 95–105.

Le Deuff, R.M., Renault, T., 1999. Purification and partial genome characterization of a herpes-like virus infecting the Japanese oyster, *Crassostrea gigas*. J. Gen. Virol. 80, 1317–1322.

Lebedeva, O., Kuznetsova, M., Zelenko, A., Gudz-Gordan, A., 1973. Investigation of a virus disease of the densonucleosis type in a laboratory culture of *Aedes aegypti*. Acta Virol. (Engl. Ed.). 17, 253–256.

Lee, C.T., Chen, I.T., Yang, Y.T., Ko, T.P., Huang, Y.T., Huang, J.Y., et al., 2015. The opportunistic marine pathogen *Vibrio parahaemolyticus* becomes virulent by acquiring a plasmid that expresses a deadly toxin. Proc. Natl. Acad. Sci. U.S.A. 112, 10798–10803.

Lightner, D.V., Redman, R., Pantoja, C., Noble, B., Tran, L., 2012. Early mortality syndrome affects shrimp in Asia. Glob. Aquac. Advocate 15, 40.

Lightner, D.V., Redman, R.M., 1998. Shrimp diseases and current diagnostic methods. Aquaculture 164, 201–220.

Liu, K., Li, Y., Jousset, F.X., Zadori, Z., Szelei, J., Yu, Q., et al., 2011. The *Acheta domesticus* densovirus, isolated from the European house cricket, has evolved an expression strategy unique among parvoviruses. J. Virol. 85, 10 069–10 078.

Liu, S., Vijayendran, D., Bonning, B.C., 2011. Next generation sequencing technologies for insect virus discovery. Viruses 3, 1849–1869.

Martin, S.J., Highfield, A.C., Brettell, L., Villalobos, E.M., Budge, G.E., Powell, M., et al., 2012. Global honey bee viral landscape altered by a parasitic mite. Science 336, 1304–1306.

Martenot, C., Fourour, S., Oden, E., Jouaux, A., Travaille, E., Malas, J.P., Houssin, M., 2012. Detection of the OsHV-1 μVar in the Pacific oyster Crassostrea gigas before 2008 in France and description of two new microvariants of the Ostreid Herpesvirus 1 (OsHV-1). Aquaculture 338, 293–296.

Martenot, C., Oden, E., Travaille, E., Malas, J.P., Houssin, M., 2011. Detection of different variants of Ostreid Herpesvirus 1 in the Pacific oyster, *Crassostrea gigas* between 2008 and 2010. Virus Res. 160, 25–31.

McMahon, D.P., Furst, M.A., Caspar, J., Theodorou, P., Brown, M.J., Paxton, R.J., 2015. A sting in the spit: widespread cross-infection of multiple RNA viruses across wild and managed bees. J. Anim. Ecol. 84, 615–624.

McMahon, D.P., Natsopoulou, M.E., Doublet, V., Furst, M., Weging, S., Brown, M.J., et al., 2016. Elevated virulence of an emerging viral genotype as a driver of honeybee loss. Proc. Biol. Sci. 283, 20160811.

Menge, B.A., Berlow, E.L., Blanchette, C.A., Navarrete, S.A., Yamada, S.B., 1994. The keystone species concept: variation in interaction strength in a rocky intertidal habitat. Ecol. Monogr. 64, 249–286.

Menge, B.A., Cerny-Chipman, E.B., Johnson, A., Sullivan, J., Gravem, S., Chan, F., 2016. Sea star wasting disease in the keystone predator *Pisaster ochraceus* in Oregon: insights into differential population impacts, recovery, predation rate, and temperature effects from long-term research. PLoS ONE 11, e0153994.

Meynadier, G., Galichet, P., Veyrunes, J., Amargier, A., 1977. Mise en évidence d'une densonucléose chez *Diatraea saccharalis* [Lep.: Pyralidae]. Entomophaga 22, 115–120.

Michener, C.D., 1979. Biogeography of the bees. Ann. Mo. Bot. Gard. 66, 277–347.

Miller, A.W., Richardson, L.L., 2015. Emerging coral diseases: a temperature-driven process? Mar. Ecol. 36, 278–291.

Mondet, F., de Miranda, J.R., Kretzschmar, A., Le Conte, Y., Mercer, A.R., 2014. On the front line: quantitative virus dynamics in honeybee (*Apis mellifera* L.) colonies along a new expansion front of the parasite *Varroa destructor*. PLoS Pathog. 10, e1004323.

Montano, S., Strona, G., Seveso, D., Maggioni, D., Galli, P., 2015. Slow progression of black band disease in *Goniopora* cf. *columna* colonies may promote its persistence in a coral community. Mar. Biodivers. 45, 857–860.

Montgomery-Brock, D., Tacon, A.G., Poulos, B., Lightner, D., 2007. Reduced replication of infectious hypodermal and hematopoietic necrosis virus (IHHNV) in *Litopenaeus vannamei* held in warm water. Aquaculture 265, 41–48.

Morens, D.M., Fauci, A.S., 2013. Emerging infectious diseases: threats to human health and global stability. PloS Pathog. 9, e1003467.

Morens, D.M., Folkers, G.K., Fauci, A.S., 2004. The challenge of emerging and re-emerging infectious diseases. Nature 430, 242–249.

Moriarty, D. J., 1999. Disease control in shrimp aquaculture with probiotic bacteria, in: Bell, C.R., Brylinsky, M., Johnson-Green, P. (eds.), Proceedings of the 8th International Symposium on Microbial Ecology, Atlantic Canada Society for Microbial Ecology, Halifax, Canada, pp. 237–243.

Mordecai, G.J., Wilfert, L., Martin, S.J., Jones, I.M., Schroeder, D.C., 2016. Diversity in a honey bee pathogen: first report of a third master variant of the deformed wing virus quasispecies. ISME J. 10, 1264–1273.

Morrissey, T., McCleary, S., Collins, E., Henshilwood, K., Cheslett, D., 2015. An investigation of ostreid herpes virus microvariants found in *Crassostrea gigas* oyster producing bays in Ireland. Aquaculture 442, 86–92.

Morse, S.S., 2004. Factors and determinants of disease emergence. Rev. Sci. Tech. Off. Int. Epiz. 23, 443–452.

Mortensen, S., Strand, A., Bodvin, T., Alfjorden, A., Skar, C.K., Jelmert, A., et al., 2016. Summer mortalities and detection of ostreid herpesvirus microvariant in Pacific oyster *Crassostrea gigas* in Sweden and Norway. Dis. Aquat. Org. 117, 171–176.

Moss, J.A., Burreson, E.M., Cordes, J.F., Dungan, C.F., Brown, G.D., Wang, A., et al., 2007. Pathogens in *Crassostrea ariakensis* and other Asian oyster species: implications for non-native oyster introduction to Chesapeake Bay. Dis. Aquat. Org. 77, 207–223.

Mukha, D., Schal, K., 2003. A densovirus of German cockroach *Blattella germanica*: detection, nucleotide sequence, and genome organization. Mol. Biol. 37, 513–523.

Muller, E.M., Rogers, C.S., Van Woesik, R., 2014. Early signs of recovery of *Acropora palmata* in St. John, US Virgin Islands. Mar. Biol. 161, 359–365.

Mutschmann, F., 2015. Chytridiomycosis in amphibians. J. Exot. Pet Med. 24, 276–282.

Mutuel, D., Ravallec, M., Chabi, B., Multeau, C., Salmon, J.-M., Fournier, P., Ogliastro, M., 2010. Pathogenesis of *Junonia coenia* densovirus in *Spodoptera frugiperda*: a route of infection that leads to hypoxia. Virology 403, 137–144.

Nazzi, F., Le Conte, Y., 2016. Ecology of *Varroa destructor*, the major ectoparasite of the western honey bee, *Apis mellifera*. Annu. Rev. Entomol. 61, 417–432.

Nicolas, J.L., Comps, M., Cochennec-Laureau, N., 1992. Herpes-like virus infecting Pacific oyster larvae, *Crassostrea gigas*. Bull. Eur. Assoc. Fish Pathol. 191, 11–13.

Nunan, L., Lightner, D., Pantoja, C., Gomez-Jimenez, S., 2014. Detection of acute hepatopancreatic necrosis disease (AHPND) in Mexico. Dis. Aquat. Org. 111, 81–86.

Olson, J.B., Gochfeld, D.J., Slattery, M., 2006. *Aplysina* red band syndrome: a new threat to Caribbean sponges. Dis. Aquat. Org. 71, 163–168.

Page, C.A., Willis, B., 2006. Distribution, host range and large-scale spatial variability in black band disease prevalence on the Great Barrier Reef, Australia. Dis. Aquat. Org. 69, 45–51.

Paine, R.T., 1966. Food web complexity and species diversity. Am. Nat., 65–75.

Park, E.D., Lightner, D.V., Park, D.L., 1994. Antimicrobials in shrimp aquaculture in the United States: regulatory status and safety concerns, in: Ware, G.W. (ed.), Reviews of Environmental Contamination and Toxicology. Springer, New York, pp. 1–20.

Parrish, C.R., Holmes, E.C., Morens, D.M., Park, E.C., Burke, D.S., Calisher, C.H., et al., 2008. Cross-species virus transmission and the emergence of new epidemic diseases. Microbiol. Mol. Biol. Rev. 72, 457–470.

Patterson, K.L., Porter, J.W., Ritchie, K.E., Polson, S.W., Mueller, E., Peters, E.C., et al., 2002. The etiology of white pox, a lethal disease of the Caribbean elkhorn coral, *Acropora palmata*. Proc. Natl. Acad. Sci. U.S.A. 99, 8725–8730.

Paul-Pont, I., Dhand, N.K., Whittington, R.J., 2013. Influence of husbandry practices on OsHV-1 associated mortality of Pacific oysters *Crassostrea gigas*. Aquaculture 412–413, 202–214.

Peeler, E.J., Reese, R.A., Cheslett, D.L., Geoghegan, F., Power, A., Thrush, M.A., 2012. Investigation of mortality in Pacific oysters associated with Ostreid herpesvirus-1 μVar in the Republic of Ireland in 2009. Prev. Vet. Med. 105, 136–143.

Pernet, F., Lupo, C., Bacher, C., Whittington, R.J., 2016. Infectious diseases in oyster aquaculture require a new integrated approach. Philos. Trans. R. Soc. Lond. B Biol. Sci. 371, 20150213.

Pollock, F.J., Morris, P.J., Willis, B.L., Bourne, D.G., 2011. The urgent need for robust coral disease diagnostics. PloS Pathog. 7, e1002183.

Prangnell, D.I., Castro, L.F., Ali, A.S., Browdy, C.L., Zimba, P.V., Laramore, S.E., Samocha, T.M., 2016. Some limiting factors in superintensive production of juvenile Pacific white shrimp, *Litopenaeus vannamei*, in no-water-exchange, biofloc-dominated systems. J. World Aquac. Soc. 47, 396–413.

Pulliam, J.R.C., 2008. Viral host jumps: moving toward a predictive framework. EcoHealth 5, 80.

Ramenofsky, A., 1993. Diseases of the Americas, 1492–1700, in: Kiple, K.F. (eds.), The Cambridge World History of Human Disease. Cambridge University Press, Cambridge, pp. 317–327.

Ramenofsky, A., 2003. Native American disease history: past, present and future directions. World Archaeol. 35, 241–257.

Rasoulouniriana, D., Siboni, N., Ben-Dov, E., Kramarsky-Winter, E., Loya, Y., Kushmaro, A., 2009. *Pseudoscillatoria coralii* gen. nov., sp. nov., a cyanobacterium associated with coral black band disease (BBD). Dis. Aquat. Org. 87, 91–96.

Ren, W.C., Chen, H.X., Renault, T., Cai, Y.Y., Bai, C.M., Wang, C.M., Huang, J., 2013. Complete genome sequence of acute viral necrosis virus associated with massive mortality outbreaks in the Chinese scallop, *Chlamys farreri*. Virol. J. 12, 110.

Renault, T., Cochennec, N., Le Deuff, R.-M., Chollet, B., 1994. Herpes-like virus infecting Japanese oyster (*Crassostrea gigas*) spat. Bull. Eur. Assoc. Fish Pathol. 14, 64–66.

Renault, T., Ledeuff, R.M., Cochennec, N., Maffart, P., 1994. Herpesviruses associated with mortalities among Pacific Oyster, *Crassostrea-gigas*, in France – comparative study. Rev. Med. Vet. (Toulouse, Fr.). 145, 735–742.

Renault, T., Lipart, C., Arzul, I., 2001. A herpes-like virus infecting *Crassostrea gigas* and *Ruditapes philippinarum* larvae in France. J. Fish Dis. 24, 369–376.

Renault, T., Moreau, P., Faury, N., Pepin, J.F., Segarra, A., Webb, S., 2012. Analysis of clinical Ostreid Herpesvirus 1 (Malacoherpesviridae) specimens by sequencing amplified fragments from three virus genome areas. J. Virol. 86, 5942–5947.

Renault, T., Tchaleu, G., Faury, N., Moreau, P., Segarra, A., Barbosa-Solomieu, V., Lapague, S., 2014. Genotyping of a microsatellite locus to differentiate clinical Ostreid herpesvirus 1. Vet. Res. 45, 1–8.

Richardson, L.L. 2004. Black band disease, in: Rosenberg, E., Loya, I. (eds.), Coral Health and Disease. Springer, Berlin, pp. 325–336.

Rogers, C.S., 1990. Responses of coral reefs and reef organisms to sedimentation. Mar. Ecol. Prog. Ser. 62, 185–202.

Roque, A., Carrasco, N., Andree, K.B., Lacuesta, B., Elandaloussi, L., Gairin, I., et al., 2012. First report of OsHV-1 microvar in Pacific oyster (*Crassostrea gigas*) cultured in Spain. Aquaculture 324, 303–306.

Rosenberg, E., Koren, O., Reshef, L., Efrony, R., Zilber-Rosenberg, I., 2007. The role of microorganisms in coral health, disease and evolution. Nat. Rev. Microbiol. 5, 355–362.

Rosenkranz, P., Aumeier, P., Ziegelmann, B., 2010. Biology and control of *Varroa destructor*. J. Invertebr. Pathol. 103(Suppl. 1), S96–S119.

Roy, H.E., Hails, R.S., Hesketh, H., Roy, D.B., Pell, J.K., 2009. Beyond biological control: non-pest insects and their pathogens in a changing world. Insect Conserv. Divers. 2, 65–72.

Rutzler, K., Santavy, D.L., Antonius, A., 1983. The black band disease of Atlantic reef corals. Mar. Ecol. 4, 329–358.

Ryabov, E.V., Wood, G.R., Fannon, J.M., Moore, J.D., Bull, J.C., Chandler, D., et al., 2014. A virulent strain of deformed wing virus (DWV) of honeybees (*Apis mellifera*) prevails after *Varroa destructor*-mediated, or in vitro, transmission. PLoS Pathog. 10, e1004230.

Sánchez-Martínez, J.G., Aguirre-Guzmán, G., Mejía-Ruíz, H., 2007. White Spot Syndrome Virus in cultured shrimp: a review. Aquac. Res. 38, 1339–1354.

Saulnier, D., Haffner, P., Goarant, C., Levy, P., Ansquer, D., 2000. Experimental infection models for shrimp vibriosis studies: a review. Aquaculture 191, 133–144.

Santavy, D.L., Mueller, E., Peters, E.C., MacLaughlin, L., Porter, J.W., Patterson, K.L., Campbell, J., 2001. Quantitative assessment of coral diseases in the Florida Keys: strategy and methodology. Hydrobiologia 460, 39–52.

Sato, Y., Civiello, M., Bell, S.C., Willis, B.L., Bourne, D.G., 2016. Integrated approach to understanding the onset and pathogenesis of black band disease in corals. J. Appl. Environ. Microbiol. 18, 752–765.

Segarra, A., Pepin, J.F., Arzul, I., Morga, B., Faury, N., Renault, T., 2010. Detection and description of a particular Ostreid herpesvirus 1 genotype associated with massive mortality outbreaks of Pacific oysters, *Crassostrea gigas*, in France in 2008. Virus Res. 153, 92–99.

Shimahara, Y., Kurita, J., Nishioka, T., Yuasa, K., Kawana, M., Kamaishi, T., Oseko, N., 2012. Surveillance of type 1 Ostreid Herpesvirus (OsHV-1) variants in Japan. Fish Pathol. 47, 129–136.

Shinn, E.A., 1966. Coral growth-rate, an environmental indicator. J. Paleontol. 40, 233–240.

Soto-Rodriguez, S.A., Gomez-Gil, B., Lozano-Olvera, R., Betancourt-Lozano, M., Morales-Covarrubias, M.S., 2015. Field and experimental evidence of *Vibrio parahaemolyticus* as the causative agent of acute hepatopancreatic necrosis disease of cultured shrimp (*Litopenaeus vannamei*) in Northwestern Mexico. J. Appl. Environ. Microbiol. 81, 1689–1699.

Staehli, A., Schaerer, R., Hoelzle, K., Ribi, G., 2009. Temperature induced disease in the starfish *Astropecten jonstoni*. Mar. Biodivers. Rec. 2.

Stentiford, G., Neil, D., Peeler, E., Shields, J., Small, H., Flegel, T., et al., 2012. Disease will limit future food supply from the global crustacean fishery and aquaculture sectors. J. Invertebr. Pathol. 110, 141–157.

Styer, E., Hamm, J., 1991. Report of a densovirus in a commercial cricket operation in the southeastern United States. J. Invert. Pathol. 58, 283–285.

Sun, Y., Bernardy, E.E., Hammer, B.K., Miyashiro, T., 2013. Competence and natural transformation in vibrios. Mol. Microbiol. 89, 583–595.

Sussman, M., Willis, B.L., Victor, S., Bourne, D.G., 2008. Coral pathogens identified for White Syndrome (WS) epizootics in the Indo-Pacific. PLoS ONE 3, e2393.

Sutherland, K.P., Berry, B., Park, A., Kemp, D.W., Kemp, K.M., Lipp, E.K., Porter, J.W., 2016. Shifting white pox aetiologies affecting *Acropora palmata* in the Florida Keys, 1994–2014. Philos. Trans. R. Soc. Lond. B Biol. Sci. 5, 371.

Sutherland, K.P., Lipp, E.K., Porter, J.W., 2015. Acroporid Serratiosis, in: Woodley, C.M., Downs, C.A., Bruckner, A.W., Porter, J.W., Galloway, S.B. (eds.), Diseases of Coral. Wiley-Blackwell, Hoboken, NJ, pp. 221–230.

Sutherland, K.P., Porter, J.W., Torres, C., 2004. Disease and immunity in Caribbean and Indo-Pacific zooxanthellate corals. Mar. Ecol. Prog. Ser. 266, 273–302.

Sutherland, K.P., Porter, J.W., Turner, J.W., Thomas, B.J., Looney, E.E., Luna, T.P., et al., 2010. Human sewage identified as likely source of white pox disease of the threatened Caribbean elkhorn coral, *Acropora palmata*. J. Appl. Environ. Microbiol. 12, 1122–1131.

Sutherland, K.P., Shaban, S., Joyner, J.L., Porter, J.W., Lipp, E.K., 2011. Human pathogen shown to cause disease in the threatened elkhorn coral *Acropora palmata*. PLoS ONE 6, e23468.

Suto, C., 1979. Characterization of a virus newly isolated from the smoky-brown cockroach, *Periplaneta fuliginosa* (Serville). Nagoya J. Med. Sci. 42, 13–25.

Sweet, M., Bulling, M., Cerrano, C., 2015. A novel sponge disease caused by a consortium of micro-organisms. Coral Reefs 34, 871–883.

Szelei, J., Woodring, J., Goettel, M.S., Duke, G., Jousset, F.X., Liu, K.Y., et al., 2011. Susceptibility of North-American and European crickets to *Acheta domesticus* densovirus (AdDNV) and associated epizootics. J. Invertebr. Pathol. 106, 394–399.

Tendencia, E.A., de la Peña, L.D., 2001. Antibiotic resistance of bacteria from shrimp ponds. Aquaculture 195, 193–204.

Tijssen, P., Bergoin, M., 1995. Densonucleosis viruses constitute an increasingly diversified subfamily among the parvoviruses. Semin. Vir. 6, 347–355.

Tran, L., Nunan, L., Redman, R.M., Mohney, L.L., Pantoja, C.R., Fitzsimmons, K., Lightner, D.V., 2013. Determination of the infectious nature of the agent of acute hepatopancreatic necrosis syndrome affecting penaeid shrimp. Dis. Aquat. Org. 105, 45–55.

Ushijima, B., Smith, A., Aeby, G.S., Callahan, S.M., 2012. *Vibrio owensii* induces the tissue loss disease *Montipora* white syndrome in the Hawaiian reef coral *Montipora capitata*. PLoS ONE 7, e46717.

Ushijima, B., Videau, P., Burger, A.H., Shore-Maggio, A., Runyon, C.M., Sudek, M., et al., 2014. *Vibrio coralliilyticus* Strain OCN008 is an etiological agent of acute *Montipora* White Syndrome. J. Appl. Environ. Microbiol. 80, 2102–2109.

van Munster, M., Dullemans, A., Verbeek, M., van den Heuvel, J., Reinbold, C., Brault, V., et al., 2003. Characterization of a new densovirus infecting the green peach aphid *Myzus persicae*. J. Invert. Pathol. 84, 6–14.

vanEngelsdorp, D., Evans, J.D., Saegerman, C., Mullin, C., Haubruge, E., Nguyen, B.K., et al., 2009. Colony collapse disorder: a descriptive study. PLoS ONE 4, e6481.

Vasquez-Juarez, R., Hernandez-Lopez, J., Neftali-Guitierrez, J., Coronado-Molinda, D., Mazon-Saustegui, J. M., 2006. Incidence of herpes-like virus in Pacific oyster *Crassostrea gigas* from farms in northwestern Mexico, in: Palacios, C., Lora, C., Ibarra, A.M., Maeda-Martinez, A.N., Racotta, I. (eds.), Recent Advances in Reproduction, Nutrition, and Genetics of Mollusks. Proceedings of the International Workshop on Nutrition of Molluscs, La Paz, Mexico, November 6–9, 2006.

Voss, J.D., Richardson, L.L., 2006. Nutrient enrichment enhances black band disease progression in corals. Coral Reefs, 25, 569–576.

Weil, E., Urreiztieta, I., Garzón-Ferreira, J., 2002. Geographic variability in the incidence of coral and octocoral diseases in the wider Caribbean. Proceedings of the 9th International Coral Reef Symposium 2, 1231–1237.

Weissman, D.B., Gray, D.A., Pham, H.T., Tijssen, P., 2012. Billions and billions sold: Pet-feeder crickets (Orthoptera: Gryllidae), commercial cricket farms, an epizootic densovirus, and government regulations make for a potential disaster. Zootaxa 3504, 67–88.

Wilcox, B.A., Gubler, D.J., 2005. Disease ecology and the global emergence of zoonotic pathogens. Environ. Health Prev. Med. 10, 263–272.

Wilfert, L., Long, G., Leggett, H.C., Schmid-Hempel, P., Butlin, R., Martin, S.J.M., Boots, M., 2016. Deformed wing virus is a recent global epidemic in honeybees driven by *Varroa* mites. Science 351, 594–597.

Williams, G.J., 2013. Contrasting recovery following removal of growth anomalies in the corals *Acropora* and *Montipora*. Dis. Aquat. Org. 106, 181–185.

Work, T.M., Aeby, G.S., 2006. Systematically describing gross lesions in corals. Dis. Aquat. Org. 70, 155–160.

Work, T.M., Forsman, Z.H., Szabo, Z., Lewis, T.D., Aeby, G.S., Toonen, R.J., 2011. Inter-specific coral chimerism: genetically distinct multicellular structures associated with tissue loss in *Montipora capitata*. PLoS ONE 6, e22869.

Work, T.M., Russell, R., Aeby, G.S., 2012. Tissue loss (white syndrome) in the coral *Montipora capitata* is a dynamic disease with multiple host responses and potential causes. Proc. Biol. Sci. 279, 4334–4341.

World Organisation for Animal Health, 2013. Nosemosis of honey bees, in: Manual of Diagnostic Tests and Vaccines for Terrestrial Animals. Available from: http://www.oie.int/fileadmin/Home/eng/Health_standards/tahm/2008/pdf/2.02.04_NOSEMOSIS_FINAL.pdf (accessed May 8, 2017).

World Organisation for Animal Health, 2016a. Infection with abalone herpesvirus, in: Manual of Diagnostic Tests for Aquatic Animals. Available from: http://www.oie.int/fileadmin/Home/eng/Health_standards/aahm/current/chapitre_abalone_herpesvirus.pdf (accessed May 8, 2017).

World Organisation for Animal Health, 2016b. Infection with Ostreid herpesvirus 1 microvariants, in: Manual of Diagnostic Tests for Aquatic Animals. Available from: http://www.oie.int/fileadmin/Home/eng/Health_standards/aahm/current/chapitre_ostreid_herpesvirus_1.pdf (accessed May 8, 2017).

World Organisation for Animal Health, 2016c. OIE-listed diseases, infections and infestations in force in 2016. Available from: http://www.oie.int/animal-health-in-the-world/oie-listed-diseases-2016 (accessed May 8, 2017).

Xia, J., Bai, C., Wang, C., Song, X., Huang, J., 2015. Complete genome sequence of Ostreid herpesvirus-1 associated with mortalities of *Scapharca broughtonii* broodstocks. Virol. J. 12, 110.

17

Conclusions and Future Directions

David Shapiro-Ilan[1] and Ann E. Hajek[2]

[1] *USDA-ARS, SEA, SE Fruit and Tree Nut Research Unit, Byron, GA, USA*
[2] *Department of Entomology, Cornell University, Ithaca, NY, USA*

In this book, we have covered diverse aspects of invertebrate pathogen ecology. In this final chapter, we offer conclusions on the increasing need to expand studies on the ecology of invertebrate pathogens, including the role of pathogen ecology in microbial control, the invasion biology of pathogens, and the role of pathogens in the conservation of beneficial species. Based on the preceding chapters and our own analyses, we summarize the current status of the field and provide some suggestions for future research.

17.1   The Increasing Urgency of the Study of Invertebrate Pathogen Ecology

17.1.1   Food Security and the Role of Microbial Control

The world population could surpass 12 billion by the end of the century (Gerland et al., 2014), and the capacity of agricultural activities to feed all of these people is being stretched. Indeed, the United Nations considers food security a primary challenge to the well being of humanity (Ban, 2012). Several forces have greatly exacerbated agricultural food security issues, including pressures on water availability (Fader et al., 2013; Ercin and Hoekstra, 2014), shortfalls in production land (Ray et al., 2013), and unpredictable environmental variability due to climate change (Wheeler and von Braun, 2013). Therefore, there is an urgent need to focus on the development of sustainable climate-smart agricultural systems (Wheeler and von Braun, 2013).

   The use of biorational approaches to pest control in agriculture, such as the application of invertebrate pathogens in microbial biological control, can contribute significantly to meeting the mounting food security challenges. Unlike chemical pesticides, microbial agents are climate-smart, because they are not innately dependent on fossil fuels for their manufacture. Furthermore, chemical pesticides are inferior to microbial agents because they may cause secondary pest outbreaks, promote the development of insecticide resistance, destroy natural enemies, or create hazards for humans and the environment (DeBach, 1974; Hajek, 2004). Worldwide, there are approximately

26 million human poisonings and 220 000 deaths annually due to chemical pesticides (Richter, 2002). While resistance to pathogens being used for microbial control has been identified in a few instances, development of resistance to these pathogens appears to be uncommon, and the microbial control industry has developed solutions for this eventuality (Berling et al., 2009; Lundgren and Jurat-Fuentes, 2012).

The role of ecological studies is critical to enhancing microbial control technology in order to meet world food challenges. Microbial control is based on the generation of epizootics in targeted pest populations, and the factors that affect epizootiology are rooted in ecology. Therefore, knowledge of the ecological processes that impact the survival, dispersal, infectivity, and virulence of invertebrate pathogens is required to facilitate improved microbial control efficacy.

There are a number of steps that can be taken to advance microbial control, and the ecology behind the discipline, to meet food security challenges. First, we must consider the primary barriers to wider adoption of microbial control techniques. Barriers include cost competitiveness with competing tactics (e.g., chemical pesticides), persistence in the environment, and (in some cases) consistent levels of field efficacy that meet grower/producer demands. These hindrances can be overcome using a variety of approaches, including direct improvement of the microbial agent (via discovery, genetic selection, hybridization, or manipulation), enhancement of production, formulation, and application technologies, and manipulation of the environment to increase its suitability for the microbial agents and their transmission to hosts (Chapter 13) (Shapiro-Ilan et al., 2012). Improvements in one area are likely to also impact others (e.g., improvements in production technology can reduce costs but also enhance product quality, and hence efficacy; enhancements in the virulence of an organism can increase field efficacy but may also reduce the rates of applications that are required, and therefore costs). Likely, based on recent developments and directions in research, we can expect the greatest benefits to microbial control to be from the discovery of new superior invertebrate pathogen strains and species, and advances in formulation technology. Additionally, leveraging synergies among different invertebrate pathogens or with other biotic or abiotic agents, and maximizing conservation biocontrol, will also bring great advances. The role of transgenic organisms is also likely to play a major part in expanding the role of microbial control in pest-management systems, but the impact will vary depending on regulatory issues and public perception within specific countries or regions. In all these endeavors, understanding invertebrate pathogen ecology will provide the basis for maximizing efficiency and efficacy. Ecological knowledge is essential to determining which biotic or abiotic factors must be considered, sought, or improved upon to achieve wider utility in microbial control agents. Thus, substantial expansions of ecological studies in relation to microbial control are needed. The characterization of individual parameters that influence microbial control ecology must continue to be elucidated, but more in-depth field studies looking at system-wide interactions are also required.

17.1.2 Conservation of Beneficial Organisms

In contrast to the utilization of invertebrate pathogens to augment disease in pestiferous organisms, ecological studies are also required to help protect various beneficial invertebrates from contracting disease (see Chapters 14–16). For example, pollinators, particularly honeybees (*Apis mellifera*) and bumblebee species (*Bombus* spp.), are experiencing severe declines (e.g., colony collapse disorder in honeybees), and invertebrate pathogens have been identified as significant causal factors (Ellis et al., 2010; Goulson et al., 2015; Cameron et al., 2016). However, the dynamics of pathogens as they interact with other stress factors (e.g., parasites, pesticides, abiotic influences, and climate change) in causing pollinator decline will require a substantial amount of additional research if we are to identify the major drivers among groups of pathogens and parasites and different abiotic conditions.

Invertebrate pathogens also affect organisms that are used directly as human food. For example, aquatic organisms such as oysters and other mollusks and crustaceans, including shrimp and crayfish, are plagued with diseases; these diseases have increased in their distribution and severity due to various factors, including increased human transportation of organisms (intentional or not) and climate change (Chapter 15). Challenges include developing resistant populations, maintaining disease-free stock, and elucidating the ecology of transmission modes in order to better understand the spread of disease and produce expanded and improved mechanisms of detection. Based on the difficulty of studying aquatic populations (e.g., due to accessibility and costs), ecological studies on aquatic organisms will require some greater level of effort and creativity relative to the study of their terrestrial counterparts.

The use of insects as food for humans (and as animal feed) is also expanding; this entails challenges in terms of invertebrate disease control (Chapter 14) (Shockley and Dossey, 2014). Insects, as food for humans, are highly nutritious, economically efficient to produce, and climate-smart relative to various other high-protein food sources, such as vertebrates (Shockley and Dossey, 2014). Thus, entomophagy promises to be an integral part of meeting food security challenges in the future. The impact of invertebrate pathogen ecology on insect production systems is a burgeoning field that will continue to grow as the industry expands.

17.2   The Future for Invasive and Native Invertebrate Pathogens

Our increasing human population and increasing global trade and travel have led to unprecedented movements of non-native organisms of all types around the world. Those with an ecological impact – generally, a negative impact – are often referred to as "invasive" (Lockwood et al., 2013). Uncontrolled movements of diverse species often include microorganisms, and the case could be made that perhaps microbes are being introduced to new places more frequently than we are aware, as they are, by definition, very difficult to detect. However, movement around the globe is often thought to have begun on a large scale in the Age of Exploration (in the 15th century), so microorganisms have certainly been moved around the globe for a long time. In fact, due to such movements, many microorganisms are cryptogenic, meaning that we do not know whether they are endemic or alien to an area.

Invasive invertebrate pathogens that have been detected are generally those causing a significant impact on populations of managed invertebrates. One primary recent example would be the honeybee pathogens that have been moved around the world with devastating impacts, especially when they are combined with Varroa mites and pesticides (see Chapters 14 and 16). Chapters 15 and 16 report pathogens of mollusks and crustaceans that have been unwittingly moved to new areas, sometimes evolving into new variants (see Chapter 16). In fact, the specter of emerging infectious diseases (Chapter 16), either with causative pathogens that are newly introduced invasives or via evolution of new pathogen variants, will require concentrated efforts to prevent problems that could negatively impact human populations and ecosystems.

Movement of pathogens has also resulted in some spillover, often on to hosts that are native to areas of introduction. For example, pathogens from managed honeybees have spilled over to native bumblebee populations, and this has been suggested as one of the reasons behind the global decline in *Bombus* spp. (Fürst et al., 2014).

We know much less about introduced invertebrate pathogens that become invasive and impact non-managed invertebrate populations. An example would be a fungal pathogen, *Entomophaga maimaiga*, of the forest-defoliating gypsy moth (*Lymantria dispar*) (Chapter 9); this pest is an introduced invasive impacting urban and natural forests. We are uncertain how or when it was introduced to North America (Nielsen et al., 2005), although the location and timing of the introduction have been postulated based on distributional and climatic data (Weseloh, 1998). In this case, the pathogen is not having a negative impact, but rather is providing control for this invasive pest. However, it has also been documented that this fungus has taken over the dominant role among natural enemies of gypsy moth in the northeastern United States, a role previously held by a viral pathogen that had been introduced accidentally around 1906 (Hajek et al., 2015).

We can only expect that under the increasing global trade that has been predicted (Fouré et al., 2012), more invertebrate pathogens will be moved around the world, perhaps as hitchhikers alongside managed species. However, improved methods for detection of pathogens that are known to be problematic in systems where regular programs for detection are instituted could help curb problems. The extent that invertebrate pathogens could be introduced and become invasive is completely unknown, but knowledge of the identity and normal prevalence of naturally occurring pathogens in native invertebrate populations will be necessary in order to know when an epizootic is being caused by an invader or by a native species.

In this time when biodiversity is being threatened on many levels, we must remember that many species of invertebrates have not yet been described, and probably even more species of pathogens attacking those invertebrates are undescribed. Our lack of knowledge about pathogen diversity has become evident as we have investigated the causes of the recent declines in honeybees. Two of the general arguments for the importance of retaining biodiversity are the involvement of species in ecosystem services (in this case, both in providing control when it is needed and in attacking beneficial species) and the potential of future human uses for unknown species. While unknown species of invertebrate pathogens could therefore be either beneficial or harmful to humans, scientists should continue to learn more about undescribed species. Modern molecular methods also make these kinds of studies more possible than ever before.

17.3 New Directions and Novel Tools for Studying Invertebrate Ecology

17.3.1 Molecular Tools

When considering new tools that have accelerated our ability to conduct meaningful ecological studies on invertebrate pathogens, the advancements made in molecular methods are clearly at the forefront. As mentioned in Chapter 1, invertebrates (particularly terrestrial invertebrates) are used as models for ecological studies due to their ease of study (e.g., based on their abundance and ease of collection), and for this reason, the study of pathogens in association with their invertebrate hosts has also flourished. However, the study of invertebrate pathogen ecology in the absence of hosts has lagged behind due to the challenging nature of investigating microbial agents in cryptic habitats (an issue that relates to microbial ecology as a whole). Yet, the recent advent of various molecular techniques has facilitated in-depth studies of invertebrate pathogens in their natural terrestrial or aquatic habitats at levels that were not possible previously (Chapter 2).

Indeed, advancements in molecular methodology have led to enhanced capabilities in the detection, identification, and quantification of invertebrate pathogens, as well as insights into host–pathogen relationships. Recent progress has stemmed from technologies such as next-generation sequencing, quantitative real-time polymerase chain reaction (qPCR), and transcriptomics (Chapter 2). All of these tools can be used at various levels within the ecosystem, from studies of individuals to multi-species communities. For example, among other things, next-generation sequencing allows for the study of community-level biodiversity using metagenomics, while qPCR is employed for the species-specific identification and quantification of target pathogen species, either in the host or in the environment, and single nucleotide polymorphisms (SNPs) can be used to detect variability within species with high levels of resolution.

Certainly, novel molecular tools will continue to expand our ability to explore invertebrate pathogen ecology. At the same time, traditional (or "classical") methods should not be ignored as they will continue to be useful and necessary. For example, survey techniques employing traditional soil baiting (using susceptible sentinel hosts) and bioassay provide useful information on pathogen activity/virulence, whereas molecular surveys may only provide information on pathogen presence and quantity (albeit more extensively than baiting). Baiting is also likely more useful for discovering new pathogens. A balanced use of traditional and molecular techniques that complement one another will allow for the greatest advances in invertebrate pathogen ecology.

17.3.2 Chemical Ecology and Signaling

Substantial progress has been made in learning how invertebrate pathogens communicate with one another and with other organisms in the environment. In particular, advances have been made in elucidating the chemical ecology and signaling of entomopathogenic nematodes (EPNs) (Chapter 11). For example, EPN dispersal from the infected host has been shown to be induced by nematode pheromones called ascarosides (Kaplan et al., 2012), and it is believed that (like other nematodes) ascarosides are also responsible for regulating various other behaviors in EPNs, such as group infection and foraging (Shapiro-Ilan et al., 2014).

EPN signaling also traverses multiple trophic levels (e.g., through "call-for-help signals"). Insect feeding on roots can cause the release of volatile compounds that attract EPNs to the site; the EPNs then infect the damaging insects and thereby protect the plant (Hiltpold et al., 2010). Insects attacking aboveground portions of a plant were found to elicit a belowground response and so recruit EPNs (Filgueiras et al., 2016), the presumptive adaptive value of this being that the EPNs infect the belowground stages of the host, which otherwise would cause additional damage to the plant. Furthermore, EPNs can "learn" which compounds are important in localized habitats; that is, prior exposure to volatile compounds primes the nematodes in a compound-specific manner, increasing preference for the compounds they were exposed to and decreasing attraction to other compounds (Willett et al., 2016).

Signaling has also been reported in other pathogen groups. For example, in *Bacillus thuringiensis*, quorum sensing (a biological process used to coordinate gene expression in response to population density) is required for completion of the infectious cycle (Slamti et al., 2014). Quorum sensing can also be important in fungal pathogens; for example, quorum sensing in *Metarhizium rileyi* regulates the hyphal bodies-to-mycelia transition stage and is apparently influenced by nutritional factors (*in vivo* vs. *in vitro*) (Boucias et al., 2016).

17.3.3 Exploring Other Novel Biotic Associations

Investigating multiple or complex interactions involving invertebrate pathogens within communities, beyond just the host–pathogen relationship, provides great insights into overall ecological processes. One example in relation to invertebrate pathogen ecology is chemical signaling across trophic levels. Other biotic interactions that have also recently come into the spotlight include endophytic relationships, synergisms, non-target effects, and antagonists.

Endophytic relationships between entomopathogenic fungi and plants have been recognized for some time (e.g., with *Beauveria bassiana* in corn; Bing and Lewis, 1991), but the topic has received increased attention in recent years and now extends to other fungi, such as *Metarhizium* spp. (Chapters 3 and 9) (Vega et al., 2012; Behie et al., 2015). According to Scopus (www.scopus.com), approximately two-thirds of all scientific papers on *Beauveria* or *Metarhizium* as endophytes have been published in the past 4 years. Despite the increase in the number of associations reported, there is a dearth of information on the role that endophytic lifestyles play in the overall biology and ecology of the fungi, and on the impact of these relationships on invertebrate populations and other organisms within the community. Furthermore, although established endophytes have been shown to cause mortality in targeted invertebrate pests, the approach has yet to be used in practice (i.e., in applied microbial control programs). Clearly, endophytic relationships involving entomopathogenic fungi are highly intriguing, and there is much to explore in terms of mechanisms of establishment, impact on fitness among all organisms involved, and potential to manipulate the relationships for human benefit.

Another area of expanded interest is the interrelationship among entomopathogens in dual infections. From an ecological standpoint, dual infection and competition are of great general interest in terms of intraguild competition and cooperation. From the standpoint of leveraging these relationships to enhance microbial control efficacy, a great amount of attention has focused on seeking synergistic interactions among entomopathogens (Chapter 13). Yet, despite these efforts, very little progress has been made in the commercialization of combined entomopathogenic organisms. Additional research on pathogen–pathogen interactions is needed, as is an exploration of the mechanisms that might be used to manipulate these relationships to provide improved biological control. Moreover, the impact of multi-pathogen infections (rather than just two agents) also promises to be fruitful in terms of augmentative or conservation microbial control (e.g., Jabbour et al., 2011) observed increased host mortality in relation to increased entomopathogen species richness). In addition, increasing our understanding of interactions between pathogens and other parasites (e.g., Hajek and van Nouhuys, 2016) can lead toward improved prediction of the impacts of natural enemy communities, both on improved pest control and on the protection of managed host species.

The intentional (or accidental) introduction or augmentation of invertebrate pathogens in a system can result in non-target effects (i.e., unintended negative impacts on other species in the community). Thus, it is the responsibility of researchers to assess non-target risks when considering the application or enhancement of invertebrate pathogens in microbial control. For example, based on numerous studies, the application of EPNs has been considered safe to non-target invertebrates (Ehlers, 2005), but outcomes can vary in different systems and it would not be wise to consider EPNs safe for all introductions simply because no net negative effects have been reported thus far; new assessments need to be made, particularly when novel associations are expected or untested ecosystems are involved. Interestingly, in a recent assessment of risk to non-targets in a clear-fell forest ecosystem, inundative application of two exotic EPN species (*Steinernema carpocapsae* and *Heterorhabditis downesi*) was found to present a lower risk than the application of native species and strains (of *Steinernema feltiae*); the result was attributed primarily to shorter persistence of the introduced species in their new environment (Harvey et al., 2016).

Non-target effects are not always detrimental; they can be beneficial as well. For example, EPNs (or their symbiotic bacteria) can contribute to the suppression of harmful plant parasitic nematodes (Lacey et al., 2015). Furthermore, symbiotic bacteria (or their byproducts) from EPNs, and certain entomopathogenic fungi, have been shown to suppress a variety of plant pathogenic fungi (Chapter 6) (Ownley et al., 2010; Lacey et al., 2015). Continued exploration of these relationships can expand microbial control efforts, and wider investigations of these novel interactions and their impact in field ecology will be instigated in the future.

The role of antagonists and their ability to regulate invertebrate pathogen populations has received relatively little attention to date. For example, EPNs have various invertebrate predators (such as certain collembolans and mites), as well as protozoan parasites and nematophagous fungi (Kaya, 2002), and bacteria may be susceptible to phages (Krieg, 1971; Kaya, 2002). Insect pathogenic fungi are attacked by mycoparasites (e.g., Posada et al., 2004; Castrillo and Hajek, 2015), but the impact of mycoparasites both directly on fungal populations and indirectly on host populations has yet to be investigated. Recent studies described complex dynamics among soil types, nematophagous fungi, and endemic EPN species in Florida citrus orchards (Campos-Herrera et al., 2016). Additional research along these lines, characterizing interactions with antagonists and other biotic or abiotic factors, is needed.

17.3.4 Interdisciplinary Studies

Research on the ecology of invertebrate pathogens benefits from interdisciplinary studies. The capacity to bring together different perspectives and different abilities and tools enhances the breadth of the studies conducted. The number of interdisciplinary studies in invertebrate pathology has been increasing: the number of studies that involved a chemist (i.e., a scientist identified with a chemistry department or unit) increased approximately twofold in 2006–2016, for example, relative to the previous 10-year period (www.scopus.com). Studies involving chemistry have facilitated the elucidation of various issues related to invertebrate pathogen ecology, such as chemical volatiles and signaling, and modes of action or virulence (e.g., identification of toxins). Beyond chemistry, various other disciplines have also contributed substantially to our understanding of invertebrate pathogen ecology, such as various subdisciplines within biology (genomics, microbiology, etc.), mathematics (for modeling, see Chapter 12), and even physics. For example, incorporation of physics in ecological research resulted in kinetic studies on the relationship between relative humidity and *Bacillus thuringiensis* spore size (Westphal et al., 2003), and in the discovery that EPNs respond directionally to electromagnetic signals (Ilan et al., 2013).

The study of invertebrate pathogen ecology encompasses many fundamental aspects of broader ecological theory, such as host–pathogen relationships, transmission, epizootiology, symbiosis and competition, multitrophic interactions, food webs, and community-level dynamics. Thus, the study of invertebrate pathogens can be used to develop more comprehensive model systems in ecology. Continued advances in the ecology of invertebrate pathogens, individually or as model systems, will be fostered by using new tools (including those described in this chapter) and new methods and by taking creative approaches to answering new questions, through collaborations with interdisciplinary programs.

References

Ban, K., 2012. A message from the UN Secretary General for the opening session of the 39th Session of the Committee on World Food Security, Rome, 15–20 October 2012. Available from http://www.un-foodsecurity.org/node/1356 (accessed May 8, 2017).

Berling, M., Blachere-Lopez, C., Soubabere, O., Lery, X., Bonhomme, A., Sauphanor, B., Lopez-Ferber, M., 2009. *Cydia pomonella* granulovirus genotypes overcome virus resistance in the codling moth and improve virus efficiency by selection against resistant hosts. Appl. Environ. Microbiol. 75, 925–930.

Bing, L.A., Lewis, L.C., 1991. Suppression of *Ostrinia nubilalis* (Hubner) (Lepidoptera, Pyralidae) by endophytic *Beauveria bassiana* (Balsamo) Vuillemin. Environ. Entomol. 20, 1207–1211.

Behie, S.W., Jones, S.J., Bidochka, M.J., 2015. Plant tissue localization of the endophytic insect pathogenic fungi *Metarhizium* and *Beauveria*. Fung. Ecol. 13, 112–119.

Boucias, D., Liu, S., Meagher, R., Baniszewski, J., 2016. Fungal dimorphism in the entomopathogenic fungus *Metarhizium rileyi*: detection of an in vivo quorum-sensing system. J. Invertebr. Pathol. 136, 100–108.

Cameron, S.A., Lim, H.C., Lozier, J.D., Duennes, M.A., Thorp, R., 2016. Test of the invasive pathogen hypothesis of bumble bee decline in North America. Proc. Natl. Acad. Sci. 113, 4386–4391.

Campos–Herrera, R., El–Borai, F.E., Rodriguez Martin, J.A., Duncan, L.W., 2016. Entomopathogenic nematode food web assemblages in Florida natural areas. Soil Biol. Biochem. 93, 105–114.

Castrillo, L.A., Hajek, A.E., 2015. Detection of presumptive mycoparasites associated with *Entomophaga maimaiga* resting spores in forest soils. J. Invertebr. Pathol. 124, 87–89.

DeBach, P., 1974. Biological Control by Natural Enemies. Cambridge University Press, London.

Ehlers, R-U., 2005. Forum on safety and regulation, in: Grewal, P.S., Ehlers, R.-U., Shapiro-Ilan, D.I. (eds.), Nematodes as Biological Control Agents. CABI Publishing, Wallingford, pp. 107–114.

Ellis, J.D., Evans, J.D., Pettis, J., 2010. Colony losses, managed colony population decline, and Colony Collapse Disorder in the United States. J. Apic. Res. 49, 134–136.

Ercin, A.E., Hoekstra, A.Y., 2014. Water footprint scenarios for 2050: a global analysis. Environ. International 64, 71–82.

Fader, M., Gerten, D., Krause, M., Lucht, W., Cramer, W., 2013. Spatial decoupling of agricultural production and consumption: quantifying dependences of countries on food imports due to domestic land and water constraints. Environ. Res. Lett. 8, 014046.

Filgueiras, C.C., Willett, D.S., Pereira, R.V., Moino Junior, A. Pareja, M., Duncan, L.W., 2016. Eliciting maize defense pathways aboveground attracts belowground biocontrol agents. Sci. Reports 6, 36484.

Fouré, J., Bénassy-Quéré, A., Fontagné, L., 2012. The Great Shift: Macroeconomic Projections for the World Economy at the 2050 Horizon. Centre d'Etudes Prospectives et D'Informations Internationales, Paris.

Fürst, M.A., McMahon, D.P., Osborne, J.L., Paxton, R.J., Brown, M.J.F., 2013. Disease associations between honeybees and bumblebees as a threat to wild pollinators. Nature 506, 364–366.

Gerland, P., Raftery, A.E., Ševčíková, H., Li, N., Gu, D., 2014. World population stabilization unlikely this century. Science 346, 234–237.

Goulson, D., Nicholls, E., Botías, C., Rotheray, E.L., 2015. Bee declines driven by combined stress from parasites, pesticides, and lack of flowers. Science 347, 1255957.

Hajek, A.E., 2004. Natural Enemies. Cambridge University Press, Cambridge.

Hajek, A.E., van Nouhuys, S., 2016. Interactions among fatal diseases and parasitoids driven by density of a shared host. Proc. R. Soc. B 283, 20160154.

Hajek, A.E., Tobin, P.C., Haynes, K.J., 2015. Replacement of a dominant viral pathogen by a fungal pathogen does not alter the synchronous collapse of a forest insect outbreak. Oecologia 177, 785–797.

Harvey, C.D., Williams, C.D., Dillon, A.B., Griffin, C.T., 2016. Inundative pest control: how risky is it? A case study using entomopathogenic nematodes in a forest ecosystem. Forest Ecol. Manage. 380, 242–251.

Hiltpold, I., Toepfer, S., Kuhlmann, U., Turlings, T.C.J., 2010. How maize root volatiles influence the efficacy of entomopathogenic nematodes against the western corn rootworm? Chemoecology 20, 155–162.

Ilan, T., Kim-Shapiro, D.B., Bock, C., Shapiro-Ilan, D.I., 2013. The impact of magnetic fields, electric fields and current on the directional movement of *Steinernema carpocapsae*. Int. J. Parasitol. 43, 781–784.

Jabbour, R., Crowder, D.W., Aultman, E.A., Snyder, W.E., 2011. Entomopathogen biodiversity increases host mortality. Biol. Contr. 59, 277–283.

Kaplan, F., Alborn, H.T., von Reuss, S.H., Ajredini, R., Ali, J.G., Akyazi, F., et al., 2012. Interspecific nematode signals regulate dispersal behavior. PLoS ONE, 7, e38735.

Kaya, H.K., 2002. Natural enemies and other antagonists, in: Gaugler, R. (ed.), Entomopathogenic Nematology. CABI Publishing, Wallingford, pp. 189–204.

Krieg, A., 1971. Interactions between pathogens, in: Burges, H.D., Hussey, N.W. (eds.), Microbial Control of Insects and Mites. Academic Press, London, pp. 459–468.

Lacey, L.A., Grzywacz, D., Shapiro-Ilan, D.I., Frutos, R., Brownbridge, M., Goettel, M.S., 2015. Insect pathogens as biological control agents: back to the future. J. Invertebr. Pathol. 132, 1–41.

Lockwood, J.L., Hoopes, M.F., Marchetti, M.P., 2013. Invasion Ecology, 2nd edn. Wiley-Blackwell, Oxford.

Lundgren, J.G., Jurat-Fuentes, J.L., 2012. Physiology and ecology of host defense against microbial invaders, in: Vega, F.E., Kaya, H.K. (eds.), Insect Pathology, 2nd edn. Academic Press, Amsterdam, pp. 461–480.

Nielsen, C., Milgroom, M.C., Hajek, A.E., 2005. Genetic diversity in the gypsy moth fungal pathogen *Entomophaga maimaiga* from founder populations in North America and source populations in Asia. Mycol. Res. 109, 941–950.

Ownley, B.H., Gwinn, K.D., Vega, F.E., 2010. Endophytic fungal pathogens with activity against plant pathogens: ecology and evolution. BioControl 55, 113–128.

Posada, F., Vega, F.E., Rehner, S.A., Blackwell, M., Weber, D., Suh, S.O., Humber, R.A., 2004. *Syspastospora parasitica*, a mycoparasite of the fungus *Beauveria bassiana* attacking the Colorado potato beetle *Leptinotarsa decemlineata*: a tritrophic association. J. Insect Sci. 4, 24.

Ray, D.K., Mueller, N.D., West, P.C., Foley, J.A., 2013. Yield trends are insufficient to double global crop production by 2050. PLoS ONE, 8, e66428.

Richter, E.D., 2002. Acute human pesticide poisonings, in: Pimentel, D. (ed.), Encyclopedia of Pest Management. Marcel Dekker, New York, pp. 3–5.

Shapiro-Ilan, D.I., Bruck, D.J., Lacey, LA., 2012. Principles of epizootiology and microbial control, in: Vega, F.E., Kaya, H.K. (eds.), Insect Pathology, 2nd edn. Elsevier, Amsterdam, pp. 29–72.

Shapiro-Ilan, D.I., Lewis, E.E., Schliekelman, P., 2014. Aggregative group behavior in insect parasitic nematode dispersal. Int. J. Parasitol. 44, 49–54.

Shockley, M., Dossey, A.T., 2014. Insects for human consumption, in: Morales, J.A., Rojas, G.M., Shapiro-Ilan, D.I. (eds.), Mass Production of Beneficial Organisms: Invertebrates and Entomopathogens. Academic Press, Amsterdam, pp. 617–652.

Slamti, L., Perchat, S., Huillet, E., Lereclus, D., 2014. Quorum sensing in *Bacillus thuringiensis* is required for completion of a full infectious cycle in the insect. Toxins 6, 2239-2255.

Vega, F.E., Meyling, N.V., Luangsa-Ard, J.J., Blackwell, M., 2012. Fungal entomopathogens, in: Vega, F.E., Kaya, H.K. (eds.), Insect Pathology, 2nd edn. Academic Press, Amsterdam, pp. 171–220.

Weseloh, R.M., 1998. Possibility for recent origin of the gypsy moth (Lepidoptera: Lymantriidae) fungal pathogen *Entomophaga maimaiga* (Zygomycetes: Entomophthorales) in North America. Environ. Entomol. 27, 171–177.

Westphal, A.J., Price, P.B., Leighton, T.J., Wheeler, K.E., 2003. Kinetics of size changes of individual *Bacillus thuringiensis* spores in response to changes in relative humidity. Proc. Natl. Acad. Sci. 100, 3461–3466.

Wheeler, T., von Braun, J., 2013. Climate change impacts on global food security. Science 341, 508–513.

Willett, D.S., Alborn, H.T., Duncan, L.W., Stelinski, L.L., 2016. Social networks of educated nematodes. Sci. Reports 5, 14388.

Index

Page numbers in *italics* refer to illustrations; those in **bold** refer to tables

Ecology of Invertebrate Diseases, First Edition. Edited by Ann E. Hajek and David I. Shapiro-Ilan
© 2018 John Wiley & Sons Ltd. Published 2018 by John Wiley & Sons Ltd.